Fish & Fisheries Series

Volume 43

Series Editor

Kai Lorenzen, University of Florida, Gainesville, FL, USA

The Fish & Fisheries series aims to publish key books on all aspects of fish and fisheries science, conservation and management: biology and ecology of fishes and invertebrates of fisheries interest, fisheries assessment and management, aquaculture, and the social and economic dimensions of aquatic resource use and conservation. For over thirty years, the Fish & Fisheries series has been at the forefront of providing in-depth syntheses of knowledge in this vibrant and fast-moving field. The series is directed at researchers in the fish and fisheries sciences, senior undergraduate and postgraduate students, and at professionals in fisheries, aquaculture and marine and freshwater conservation.

* * *

Prospective authors and/or editors are encouraged to consult the Publishing Editor for more details. Comments or suggestions for future volumes are welcome:

Éva Lőrinczi
Publishing Editor
eva.loerinczi@springer.com

Vladica Simić • Snežana Simić • Vladimir Pešić
Editors

Ecological Sustainability of Fish Resources of Inland Waters of the Western Balkans

Freshwater Fish Stocks, Sustainable Use and Conservation

Springer

Editors
Vladica Simić
Faculty of Science, Department of Biology and Ecology
University of Kragujevac
Kragujevac, Serbia

Snežana Simić
Faculty of Science, Department of Biology and Ecology
University of Kragujevac
Kragujevac, Serbia

Vladimir Pešić
Faculty of Science, Department of Biology
University of Montenegro
Podgorica, Montenegro

ISSN 2542-890X ISSN 2542-8926 (electronic)
Fish & Fisheries Series
ISBN 978-3-031-36925-4 ISBN 978-3-031-36926-1 (eBook)
https://doi.org/10.1007/978-3-031-36926-1

This Springer imprint is published by the registered company Springer Nature Switzerland AG
The registered company address is: Gewerbestrasse 11, 6330 Cham, Switzerland

Preface

This monograph is the result of three decades of efforts of the Center for Fisheries and Biodiversity Conservation of inland waters-Aquarium (CERIKOB) (Organizational Unit within the Institute of Biology and Ecology, Faculty of Science, University of Kragujevac) for the conservation of biodiversity, fish stocks, and ecosystems of the inland waters of Serbia and the Balkan Peninsula. The efforts were mainly aimed at assessing the water quality status based on biological indicators (algae, macroinvertebrates, fish), conservation of aquatic biodiversity (especially crayfish and fish) under in situ and ex situ conditions, and improvement of ecological management methods, and strategies for sustainable use of fish stocks.

The content of the monograph is the result of the Center's cooperation with related scientific research institutions from the region, which has contributed to a broader understanding of the problems related to the conservation and sustainable use of biological resources of inland waters in the Western Balkans. Particularly significant is the cooperation with the Department of Environmental Science, Institute "Jožef Stefan" (Slovenia), Department of Biology, Faculty of Science, University of Zagreb (Croatia), Biotechnical faculty, University of Bihać and Faculty of Natural Sciences and Mathematics, University of Banja Luka (Bosnia and Herzegovina), Institute of Marine Biology and the Faculty of Natural Sciences and Mathematics, University of Montenegro (Montenegro), and Hydrobiological Institute Ohrid (North Macedonia). In the wider region, cooperations with the Department of Aquaculture Szent István University, Gödöllő (Hungary), and the Department of Zoology, National Museum, Prague (Czech Republic) are particularly valuable. In Serbia, the Center has achieved its goals through many joint projects, mainly in cooperation with the Institute for Biological Research "Siniša Stanković," University of Belgrade, Faculty of Biology, University of Belgrade, and the Department of Biology and Ecology, Faculty of Sciences and Mathematics, University of Niš, Department of Biology and Ecology, Faculty of Sciences, University of Novi Sad.

For the content of this monograph, in addition to the cooperation with scientific institutions, the most important is the cooperation with the industry and the managers of fishing areas, realized mainly through the development of Fishing Areas Management Programs for the management and sustainable use of fish stocks and the monitoring of fish stocks. Therefore, in this monograph, the experience of cooperation with managers of fishing areas, from public state companies such as SE "Srbijašume" to private companies and managers of fish stocks in protected areas, is presented on a scientific basis.

The Center's research and the results presented in this monograph were obtained by leading or participating in national and international projects. Among the national projects, the project "Ex situ conservation of inland water biodiversity" stands out, especially since the active database of biodiversity in aquatic ecosystems in Serbia was designed as a result of this project (BAES ex situ; http://baes.pmf.kg.ac.rs). Additionally, ex situ and in situ conservation programs were carried out for *Astacus astacus*, *Umbra krameri*, *Tinca tinca*, and *Hucho hucho.* International projects significant for this monograph include bilateral projects with Croatia on hydrobiological research of the Sava River and the Danube and Morava floodplains, joint international research on the Danube (IAD 2 and IAD 3), European Union FP6 project from the "ALARM" call, EC FP7 Project "GLOBAQUA" comparative overview of multiple stressors on the sustainability of ecological services of six major river basins (Adige (Italy), Anglian (Great Britain), Ebro (Spain), Evrotas (Greece), Sava (Slovenia, Croatia, Serbia, Bosnia and Herzegovina, and Montenegro) and Souss Massa (Morocco)), and "NATURA 2000" for Bulgaria.

The monograph is divided into six series, within which 19 chapters are arranged, corresponding to the main theme of each series. The series' topics refer to current problems of inland aquatic ecosystems and biological resources, with special reference to fish and fish resources and the ecological services they provide. Data on fish and fish resources of inland waters from Slovenia to Albania, their management, threats, conservation measures, and sustainable use are presented. A special series deals with species important for fisheries, such as sturgeons, trouts, and crayfish. The effects of various stressors on fish populations, including the ecological condition of water bodies, eutrophication, toxic pollution, and the effects of non-native and invasive species, are covered in the Stressors series. A special series is devoted to the conservation of aquatic ecosystems, with several chapters on topics that include aquaculture and its contribution to the conservation of wild fish populations and fish stocks, as well as conservation programs for endangered fish species under in situ and ex situ conditions. This series highlights the importance of maintaining huchen *Hucho hucho* populations in the waters of the Western Balkans, habitats that are critical for the global conservation of this endemic species for the Danube Basin. The series on the methodology for assessing the status of fish stocks and fisheries waters includes a chapter presenting the results of the model ES-HIPPO as a specific model based on the relationship between biota (ecological specialization of fish species, population structure, and fish community structure) and predominantly abiota with a range of HIPPO factors (habitat alteration, invasive species, pollution, human population growth, overfishing, and climate factors) to assess the ecological sustainability and conservation priority of fish species, populations, communities, and habitats down to the river basin level. The structure of the model allows its application not only at the local but also at the global scale. The final series is devoted to the social aspects of freshwater fisheries, providing a comparative overview of the entire area from country to country. In this series, there is also a chapter that presents the general conclusions drawn from the entire content of the monograph.

The monograph aims to highlight the importance of the preserved freshwater ecosystems of the Western Balkans for the ecological sustainability of fish stocks, which are important not only on a local but also on a global scale. The monograph is also an alarm call for urgent action to protect inland water ecosystems, fish, and other biological resources in the era of expansion and dominance of anthropogenic influences and climate change.

This monograph can be useful and engaging to a wide range of users, from scientists, researchers, students of biology and ecology (and their affiliates) to managers of fisheries areas and fish resources. Environmental activists and organizations can use the contents of this monograph to organize actions to protect inland waters and biodiversity.

We would like to thank Éva Loerinczi, Ph.D., for her positive response and attitude toward the proposed topic of this monograph, her advice, and her support in further writing. Additionally, we are grateful to Prof. Lorezen"s for very useful suggestions and proposals with the aim that the content of this monograph has not only regional but also international significance.

The monograph received its final form thanks to valuable reviews and suggestions from the reviewers, namely:

Prof. Dr. **Ákos Horváth**, Department of Aquaculture, Szent István University, Páter Károly U.1, 2100 Gödöllő, Hungary,
Prof. Dr. **Doru Bănăduc**, Applied Ecology Research Center, Lucian Blaga University of Sibiu, Sibiu, Romania,
Dr. **Radek Šanda**, Department of Zoology, National Museum, Václavské náměstí 68, 115 79 Prague, Czech Republic,
Prof. Dr. **Gordana Subakov Simić**, Faculty of Biology, University of Belgrade, Studentski trg 16, 11000 Belgrade, Serbia,
Prof. Dr. **Milica Stojković Piperac**, University of Niš, Faculty of Sciences and Mathematics, Department of Biology and Ecology, Niš, Serbia
Prof. Dr. **Đurađ Milošević**, University of Niš, Faculty of Sciences and Mathematics, Department of Biology and Ecology, Niš, Serbia
Dr. **Aleksandra Milošković**, Department of Science, Institute for Information Technologies Kragujevac, University of Kragujevac, Kragujevac, Serbia
Dr. **Milena Radenković**, Department of Biology and Ecology, Faculty of Science, University of Kragujevac, Kragujevac, Serbia.

Kragujevac, Serbia
Podgorica, Montenegro

Vladica Simić
Snežana Simić
Vladimir Pešić

Acknowledgments

The book entitled "Ecological Sustainability of Fish Resources of Inland Waters of the Western Balkans in the Anthropocene—Freshwater Fish Stocks, Sustainable Use and/or Conservation" is the result of the work of a total of 57 authors and co-authors from 35 institutions and ten countries. In addition to the scientific and professional results, the authors and co-authors have also contributed their significant field and research experiences to the chapters they have written, making a valuable contribution to the book's final content. The editors would like to thank all colleagues for their dedication toward the idea and realization of this "project" and their final contribution to the content of this book.

In addition to the editors, authors, and co-authors, many people have contributed in various ways to the realization of this idea. Not only during the preparation of this publication but also during all the years of research on the Western Balkans aquatic ecosystems, and especially on the ichthyofauna of this area, colleagues and collaborators from numerous institutions and organizations in Serbia and beyond were involved.

The editors would like to thank in particular Users of Fishing Areas in the Republic of Serbia, their managers, and professional collaborators: Petar Nećak, Aleksandra Komarnički Ćirlić, Alan Koljukaj, Ivana Stanković (SE "Srbijašume"), Milan Milošević ("Balkan Eco Team" Ltd.), Ninoslav Kostić, Žika Milenković, Živojin Stojilković, Nenad Đorđević (Association "Veternica—Vlasina" Ltd.), Nikola Kumanović ("Južna Morava 2" Ltd.), Predrag Ičelić ("Cassini fishing" Ltd.), Radosav Novčić, Suzana Komatović, Milosav Filipović (NP "Kopaonik"), Ivan Lazić, Olivera Aleksić ("Plus Sport" Ltd.), Miodrag Petrović (Anglers association "Timočka Krajina" Ltd.), Novica Stanković (LOEQ "Vlasina"), Igor Aleksić (LOEQ "Dolina Pčinje", Orthodox Diocese of Vranje), and Igor Petrović (NP "Stara Planina").

We also thank Dr. Milan Lugić and Miljko Glavinić from the Reprocenter "Braduljica" for the data provided and the very successful cooperation in joint projects in fisheries.

The content of this book could not be so complete without the continuous collaboration, research, and joint projects with former PhD students of the Institute of Biology, Faculty of Sciences, University of Kragujevac, and now very successful scientists: Dr. Momir Paunović, Principal Research Fellow, Institute for Biological Research "Siniša Stanković" (IBISS) University of Belgrade, Dr. Milica Stojković-Piperac and Dr. Đurađ Milošević, Associate Professors from the Department of Biology and Ecology, Faculty of Sciences and Mathematics, University of Niš.

In the past years, we had much help during field and laboratory research from a large number of technical staff working at the Center for Fisheries and Biodiversity Conservation of inland waters-Aquarium (CERIKOB) at the Department of Biology and Ecology, Faculty of Sciences from the inception until today. Special thanks to Marko Vlajković and Vitomir Anđelković.

For their previous cooperation and especially for their significant participation in shaping the final content of this book, we would like to thank: Research Associates Dr. Milena Radenković, Dr. Nataša Kojadinović, and Dr. Simona Đuretanović; Assistant Professors Dr. Ana Petrović and Dr. Nevena Đorđević; Teaching Assistant Tijana Veličković, and Research Assistants Marija Jakovljević, Marijana Nikolić, and Predrag Simović.

Special thanks go to the reviewers of this book for their promptness, valuable suggestions, and comments, which contributed significantly to the quality of the text.

The corresponding editor of this book, Dr. Vladica Simić, Full Professor, would like to thank his mentors who are no longer with us, Dr. Vera Mitrović-Tutundžić (Faculty of Agriculture, University of Belgrade), who introduced me to the world of hydrobiology and Dr. Draga Janković (IBISS), who introduced me to the fisheries science.

Vladica Simić
Snežana Simić
Vladimir Pešić

Introduction

Geographically, the Western Balkans occupy a large part of the Balkan Peninsula and, in the broadest sense, includes all the states to which the main topic of this monograph refers, from Slovenia to Albania (Fig. 1). However, Slovenia and Croatia are excluded from this area due to geopolitical terms. Without elaborating on these socio-political differences, the Western Balkans is ecologically characterized by a great regional heterogeneity and heterogeneity of climate (alpine, temperate-continental, sub-Mediterranean, Mediterranean), relief (Alps, Dinarsko-Šarsko Pindski, Balkan, Carpathian, and Rhodope mountain ranges; Pannonian, Metohija, and Vardar valley), and hydrography. The middle course of the Danube dominates the area with its sub-basins, the most important being the Sava Basin (the largest) and the Great Morava basin (which geographically occupies the area of the Central Balkans). The Mediterranean-Adriatic area includes mainly isolated river basins draining water from the karstic Dinaric mountains into the Adriatic Sea. The Mediterranean-Aegean-Ionian area includes basins draining water from the Šarska-Pinda and Rhodope mountain massifs into the Aegean Sea and a very small part into the Ionian Sea. Apart from the heterogeneity of aquatic habitats, the Western Balkans is considered the area with the best preserved aquatic ecosystems compared to other parts of the European continent.

From the perspective of species and ecosystem biodiversity, the Western Balkans region is one of the world's most important "hot-spot" areas for biodiversity conservation. The rivers and lakes of the Adriatic and Aegean basins are the centers of endemic fish species. It is believed that at least 20% of the endemic fish species of the European continent occur in this area. The high endemism of fish is the result of various paleoecological and paleogeographical events concentrated in the rivers of the Dinaric Massif (Croatia, Bosnia and Herzegovina) and the large Balkan lakes Skadar, Ohrid, Prespa, and Dojran. This area has endemic fish of great conservation importance and no fishing value as well as very attractive species for fishing (e.g., Adriatic trout *Salmo obtusirostris*, belvica *Salmo ohridanus*, Ohrid trout *Salmo letnica*, and species of the genus *Chondrostoma*), which have been exploited since

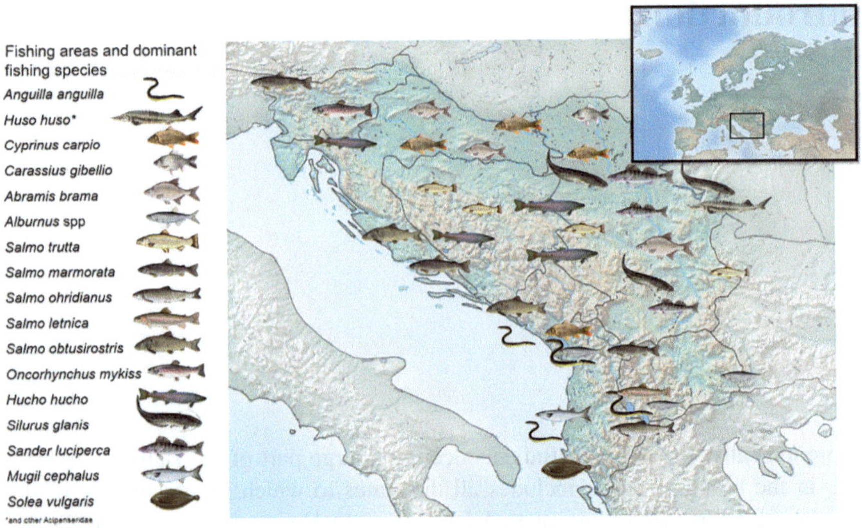

Fig. 1 Fishing areas and dominant fishing species of the Western Balkans (Figure design by Andrija Dumanović)

the beginning of human presence in this region. The Danube part of the Western Balkans does not have a large number of endemic fish species, but it is an important area of the European continent for the fisheries and fish resources, especially the major rivers Danube, Sava, and Tisza.

The great diversity of fishing waters and fish resources of the Western Balkans has also influenced the development of fisheries and fishing as a centuries-old human activity. Throughout the region, one can see a great variety of preserved traditional and modern fishing techniques and tools adapted to the dominant fish resources in certain fishing waters.

The presence of commercial and sport/recreational fishing is represented differently. In the Danube basin, commercial fishing is declining compared to recreational fishing, while in the Adriatic and Aegean basins, it is the opposite.

In the past, fisheries in the Danube basin focused on migratory (anadromous) sturgeons for consumption and caviar production and other high-value species such as sterlet sturgeon *Acipenser ruthenus*, wels catfish *Silurus glanis*, pikeperch *Sander lucioperca*, northern pike *Esox lucius*, carp *Cyprinus carpio,* as well as salmonids: huchen *Hucho hucho*, grayling *Thymallus thymallus*, and brown trout *Salmo trutta*. Today, depending on the area, the selection is limited to carp (most of which come from aquaculture), wels catfish, pikeperch, the increasingly frequent bream *Abramis brama*, and non-native species such as Prussian carp *Carassius gibelio*, gras carp *Ctenopharyngodon idella*, and bighead carp *Hypophthalmichthys nobilis*. In the Adriatic and Aegean, native salmonids are still exploited commercially, but to a much lesser extent due to lower stocks, which is also true for some cyprinids such as carp and bleak *Alburnus alburnus*. European eel *Anguilla anguilla* is also still fished despite significantly reduced stocks in the area.

In the first part of the book (Chaps. 1–7) present different models of fisheries management, with either a centralized or a heterogeneous model predominating. There is a constant lack of reliable data on catches and fishing pressure, especially for recreational fisheries. The lack of reliable fishing pressure data is a chronic global problem in freshwater fisheries. In this book are presented different models of catch data collection and processing by commercial and recreational fishermen in the Western Balkans.

Threats and impacts on fish stocks are highlighted in this book. Attention is drawn to fragmentation, habitat destruction, eutrophication, and hybridization of salmonid species, which is pronounced in the Danube River Basin. Impacts such as overfishing, poaching, and the presence of non-native invasive species are more pronounced in the fishing waters of the Adriatic and Aegean Seas, especially in the southern parts (Montenegro, North Macedonia, and Albania).

Contents

Contributors

Avdul Adrović Faculty of Science, University of Tuzla, Tuzla, Bosnia and Herzegovina

Subha Avdić Biotechnical Faculty, University of Bihać, Bihać, Bosnia and Herzegovina

Damjana Drobac Backović Department of Biology and Ecology, Faculty of Sciences, University of Novi Sad, Novi Sad, Serbia

Azra Bakrač Biotechnical Faculty, University of Bihać, Bihać, Bosnia and Herzegovina

Neven Bočić Department of Geography, Faculty of Science, University of Zagreb, Zagreb, Croatia

Miljanović Branko Department of Biology and Ecology, Faculty of Sciences, University of Novi Sad, Novi Sad, Serbia

Mimoza Çobani Ministry of Agriculture, Rural Development & Water Administration Albania, Tirana, Albania

Gorčin Cvijanović Institute for Multidisciplinary Research, University of Belgrade, Belgrade, Serbia

Radoslav Dekić Faculty of Natural Sciences and Mathematics, University of Banja Luka, Banja Luka, Bosnia and Herzegovina

Dhimitër Dhora University of Shkodra "Luigj Gurakuqi", Shkodër, Albania

Vesna Đikanović Institute for Biological Research "Siniša Stanković", University of Belgrade, Belgrade, Serbia

Nevena Đorđević Department of Biology and Ecology, Faculty of Science, University of Kragujevac, Kragujevac, Serbia

Samir Đug Faculty of Science, University of Sarajevo, Sarajevo, Bosnia and Herzegovina

Simona Đuretanović Department of Biology and Ecology, Faculty of Science, University of Kragujevac, Kragujevac, Serbia

Laszlo Galambos Institute for Nature Conservation of Vojvodina Province, Novi Sad, Serbia

Dragojla Golub Faculty of Natural Sciences and Mathematics, University of Banja Luka, Banja Luka, Bosnia and Herzegovina

Linda Grapci-Kotori Linda Grapci-Kotori University of Prishtina "H. Prishtina", Prishtina, Kosovo

Marija Jakovljević Department of Biology and Ecology, Faculty of Science, University of Kragujevac, Kragujevac, Serbia

Goran Jakšić AQUATIKA – KARLOVAC FRESHWATER AQUARIUM Public Institute, Karlovac, Croatia

Vera Kanlić Government Republic of Srpska, Ministry of Agriculture, Forestry and Water Management, Banja Luka, Bosnia and Herzegovina

Nataša Kojadinović Institute of Biology and Ecology, Faculty of Science, University of Kragujevac, Kragujevac, Serbia

Ivana Maguire Department of Biology, Faculty of Science, University of Zagreb, Zagreb, Croatia

Saša Marić Institute of Zoology, Faculty of Biology, University of Belgrade, Belgrade, Serbia

Zoran Marinović Department of Aquaculture, Hungarian University of Agriculture and Life Sciences, Gödöllő, Hungary

Zoran Z. Marković Faculty of Agriculture, University of Belgrade, Belgrade, Serbia

Dragana Milošević Faculty of Science and Mathematics, Department of Biology, University of Montenegro, Podgorica, Montenegro

Đurađ Milošević Department of Biology and Ecology, Faculty of Sciences and Mathematics, University of Niš, Niš, Serbia

Aleksandra Milošković Institute for Information Technologies Kragujevac, University of Kragujevac, Kragujevac, Serbia

Danilo Mrdak Faculty of Science and Mathematics, Department of Biology, University of Montenegro, Podgorica, Montenegro

Polona Pengal REVIVO, Institute for Ichthyological and Ecological Research PE, Ljubljana, Slovenia

Vladimir Pešić Department of Biology, Faculty of Sciences, University of Montenegro, Podgorica, Montenegro

Ana Petrović Department of Biology and Ecology, Faculty of Science, University of Kragujevac, Kragujevac, Serbia

Milica Stojković Piperac Department of Biology and Ecology, Faculty of Sciences and Mathematics, University of Niš, Niš, Serbia

Marina Piria Faculty of Agriculture, Department of Fisheries, Apiculture, Wildlife Management and Special Zoology, University of Zagreb, Zagreb, Croatia
Faculty of Biology and Environmental Protection, Department of Ecology and Vertebrate Zoology, University of Lodz, Poland

Kaja Pliberšek Fisheries Research Institute of Slovenia, Ljubljana, Slovenia

Vesna D. Poleksić Faculty of Agriculture, University of Belgrade, Belgrade, Serbia

Milena Radenković Department of Biology and Ecology, Faculty of Science, University of Kragujevac, Kragujevac, Serbia

Milka Rajković Environment Protection Agency of Montenegro, Podgorica, Montenegro

Radek Šanda Department of Zoology, National Museum, Prague, Czech Republic

Johannes Schöffmann St. Veit/Glan, Austria

Nenad Sekulić Institute for Nature Conservation of Serbia, Belgrade, Serbia

Spase Shumka Department of Biotechnology and Food, Faculty of Biotechnology and Food, Agricultural University of Tirana, Tirana, Albania

Snežana Simić Department of Biology and Ecology, Faculty of Science, University of Kragujevac, Kragujevac, Serbia

Vladica Simić Department of Biology and Ecology, Faculty of Science, University of Kragujevac, Kragujevac, Serbia

Predrag Simović Department of Biology and Ecology, Faculty of Science, University of Kragujevac, Kragujevac, Serbia

Rifat Škrijelj Faculty of Natural and Mathematical Science, University of Sarajevo, Sarajevo, Bosnia and Herzegovina

Marija Smederevac-Lalić Institute for Multidisciplinary Research, University of Belgrade, Belgrade, Serbia

Ivan Špelić Faculty of Agriculture, Department of Fisheries, Apiculture, Wildlife Management and Special Zoology, University of Zagreb, Zagreb, Croatia

Milica Stojković-Piperac Department of Biology and Ecology, Faculty of Sciences and Mathematics, University of Niš, Niš, Serbia

Belma Kalamujić Stroil University of Sarajevo-Institute for Genetic Engineering and Biotechnology, Sarajevo, Bosnia and Herzegovina

Josip Suić Ministry of Agriculture, Directorate of Fisheries, Zagreb, Croatia

Marina Talevska Department of Hydrobotany, PSI Hydrobiological Institute, Ohrid, Republic of North Macedonia

Trajče Talevski Department of Cyprinid Fish, PSI Hydrobiological Institute, Ohrid, Republic of North Macedonia

Orhideja Tasevska Department of Zooplankton, PSI Hydrobiological Institute, Ohrid, Republic of North Macedonia

Tone Tavčar Fisheries Research Institute of Slovenia, Ljubljana, Slovenia

Nada Tokodi Department of Biology and Ecology, Faculty of Sciences, University of Novi Sad, Novi Sad, Serbia

Blagoja Trajčeski Department of Cyprinid Fish, PSI Hydrobiological Institute, Ohrid, Republic of North Macedonia

Tijana Veličković Department of Biology and Ecology, Faculty of Science, University of Kragujevac, Kragujevac, Serbia

Katarina Zorić Department for Hydroecology and Water Protection, Institute for Biological Research "Siniša Stanković", National Institute of the Republic of Serbia, University of Belgrade, Belgrade, Serbia

Part I
Fish Resources and Fishing Areas of the Western Balkans

Summary

This part of the book contains chapters that present data on the state of the fish stock in the countries of the Western Balkans, starting from Slovenia to Albania. A historical overview of freshwater fisheries, changes that have occurred over time, as well as factors that negatively affect the fish stock, are presented. The chapters show the organization and management of fish resources. From this Part, the differences and effectiveness of applied management in the protection and sustainable use of the fish stock can be seen.

Fish Resources of Inland Waters and Fisheries in Slovenia: Management, Sustainability, and Conservation

Kaja Pliberšek and Tone Tavčar

Abstract Slovenia [20,230 km^2] encompasses five geographical regions with 26,989 km of watercourses (Natek 1998; Kolbezen 1998), inhabiting 95 freshwater fish species included in 14 families (Povž et al., 2015). There are 49 game species; 31 native and 21 permanently protected (RIBKAT, 2022). The most alluring native game species are *Hucho hucho*, *Salmo marmoratus*, *Salmo trutta*, *Thymallus* sp., *Esox lucius,* and *Silurus glanis.*

Slovenian fisheries management of inland waters is under State control (ZSRib). Fisheries Research Institute of Slovenia manages watercourses ($N = 9$) of special importance, while Angling clubs ($N = 64$) manage 67 fishing sub-regions divided into fishing districts. Favorable fish conservation status is sustained with fisheries management regulations, including population reinforcements based on applicability and research. The introduction of non-indigenous species is prohibited, except where stocking is controlled and limited to important game fish species (PUR, 2015). In the last eight years, the release of non-indigenous species has been reduced by 20% (LIFE16 NAT/SI/000644). Wild fish populations are threatened mostly by habitat destruction, piscivorous animals increase, hybridization, and climate change. Therefore, Management plans are in preparation, which focus toward endangered fish population reinforcements with native genes and efficient reduction of threats.

Keywords Fisheries economic impact · Freshwater fisheries management · Fish resources · Game fish conservation · Sustainable fisheries

---

K. Pliberšek (✉) · T. Tavčar
Fisheries Research Institute of Slovenia, Ljubljana, Slovenia
e-mail: kaja.plibersek@zzrs.si

V. Simić et al. (eds.), *Ecological Sustainability of Fish Resources of Inland Waters of the Western Balkans*, Fish & Fisheries Series 43,
https://doi.org/10.1007/978-3-031-36926-1_1

1 Slovenian Inland Fisheries: Geographical Features and Fish Resources

Slovenian territory encompasses 20,230 km^2 and it extends across five regions: Alpine (Fig. 1), Alpine foothill, Pannonian, Dinaric–Karst (Fig. 2), and Mediterranean (Natek 1998; Kolbezen 1998). Predominant bedrock is carbonate and silicate in the Pohorje Massif (ARSO 2022). This exceptional regional diversity on a small geographical area offers the occurrence of various surface water types, such as mountain torrents, springs, streams, rivers, wetlands, and the sea (Kolbezen 1998). The combined length of flowing surface waters is 26,989 km resulting in a 1.33 km/km^2 average of the river network density. Largest river drainages are Mura (1376 km^2), Drava (3253 km^2), Sava (11,744 km^2), and Soča with Vipava (2316 km^2), which belong to the Black and the Adriatic Sea basins (Fig. 3). The majority (81%) of watercourses in Slovenia belong to the Black sea basin (16,373 km^2) and drain into the Danube River. The Adriatic basin (3857 km^2) only covers 19% of Slovenia including the main river basin Soča and other tributaries, which flow directly into the sea (Kolbezen 1998; Čehić 2007). Geological and climatic conditions govern all four main types of river regimes: Nival, Nival-pluvial, Pluvio-nival, and Pluvial. Large rivers can have combined regimes, passing from one regime to another. Watercourses in Slovenia are generally characterized with

Fig. 1 Alpine region, Soča River. Photo by Krištof Istinič

Fig. 2 Dinaric–Karst region, Unica River. Photo by Andreja Ravnak in scope of LIFE for LASCA project

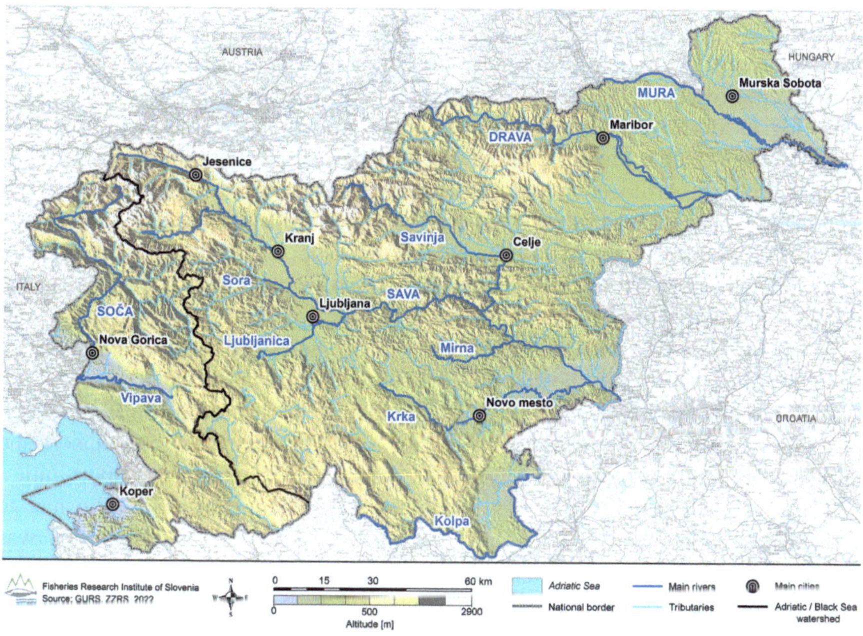

Fig. 3 Main rivers and basins of Slovenia

high fluctuations between low, medium, and high water discharges. Exception is the Karst region, where fluctuations are more subtle due to the region's indicative capability for water retention.

Geographical, hydrological, and morphological processes influence stream channel networks, riparian vegetation, and floodplains creating diverse habitats for fish (Cowx and Welcomme 1998). The diversity of Slovenian watercourses provides favorable conditions for the survival of salmonid and cyprinid fish species creating an important natural resource. Longitudinal patterns of rivers and streams cover all four European freshwater fish habitat zones; trout zone, grayling zone, barbell zone, and bream zone. Consequently, it does not come as a surprise that Slovenia inhabits 95 freshwater fish species included in 14 families (Povž et al. 2015). There are 49 game fish species, 31 native and 21 permanently protected (RIBKAT 2022). The most captivating native freshwater game species are Danube salmon (*Hucho hucho)*, marble trout (*Salmo marmoratus)*, brown trout (*Salmo trutta)*, grayling (*Thymallus* sp.*)*, Northern pike (*Esox lucius*), and wels catfish (*Silurus glanis)*. Management and exploitation of these resources have been traditionally present in Slovenia for well over a 140 years.

2 Fisheries Management in Slovenia: Organization, Implementation, and Applicability

2.1 Fisheries History in Slovenia

Fishing has accompanied humans since the beginning (Širnik 2012). Very early on, man has created various tools in order to improve fishing success, also evidenced by archeologists on Slovenian territory. The most numerous finds are from the Ljubljana Marsh, which is the place of the first settlers (pile dwellers) of the modern day capital Ljubljana. Artifacts, including bone hooks and harpoons prove that man has been fishing since the end of the Neolithic. Mastered skills such as net weaving and dugout canoe construction start to appear in the Copper Age, followed by metal hooks and harpoons. All of these evolutionary improvements kick-start man into the next era of fisheries resource exploitation.

In the Middle Ages, fishing rights belonged to the crown and were commonly passed to monasteries and the church, for who occupational servants, who did not have rights of their own, performed the actual fishing (Širnik 2012). Eventually, aristocracy started to expand fishing rights to lower nobilities, merchants, and markets. Common people eventually gained fishing rights and market opportunities. Control over fishing activities was conducted by "fishing masters," who had to follow instructions later devised into "fishing policies." The oldest policy was published in 1503 for the Gorizia diocese. The purpose of these policies was to protect spawning grounds and juvenile habitats, as well as to control practices that may be harmful to fish. Prohibited was the use of tightly woven traps, knight slings

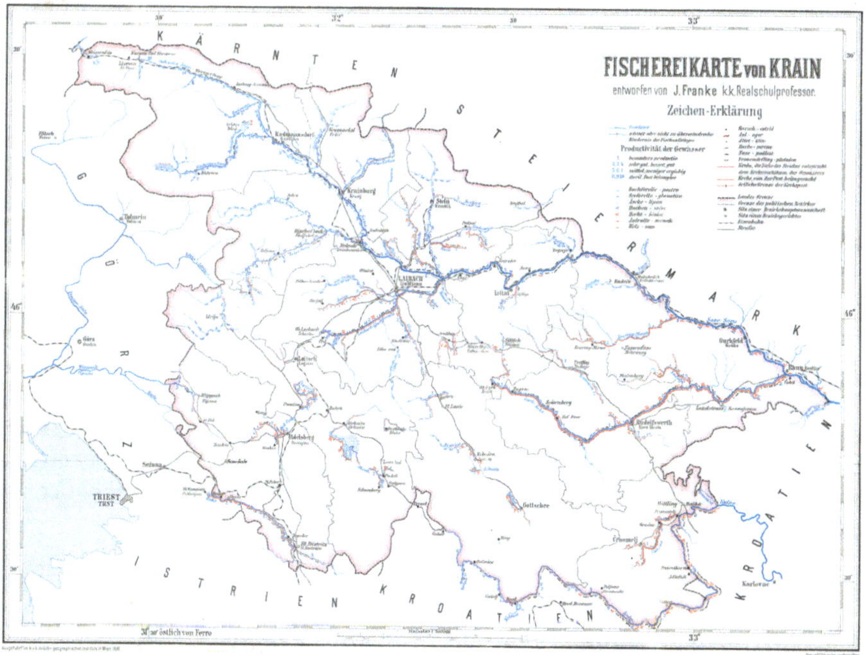

Fig. 4 A species distribution map of central Slovenia from 1892. The map is showing game species occurrence and production potential of watercourses. The map was issued as an appendix to an article by Franke J. (1892). Die Gewaesser in Krain und ihre nutsbare Fauna

and use of a light source at knight. Permitted was the use of nets with a determined design and mesh size. Floating and combined nets were not allowed, unless there was a need to translocate "noble species" (ex. huchen, grayling). Fish harvest size limits and defined seasons came into effect already in 1506. For instance, fishing for huchen and grayling was not allowed in March, and trout closed season was 14 days before and after October 13th. Punishment for violating these rules could also be imprisonment.

Organized fishery in Slovenia dates back for more than 140 years, with records starting in 1880 (Toš 2021; Fishing Association of Slovenia 2022). In the time of Austro–Hungarian empire, Slovenian anglers organized the first independent angling club with Professor Dr. Ivan Franke as the scientific authority (Fig. 4) (Munda 1927; Fishing Association of Slovenia 2022; Širnik 2012). They began with artificial breeding of brown trout, marble trout and later non-native brook trout (*Salvelinus fontinalis*) by importing roe from abroad. These were first recorded fish farming practices in Slovenia. Mills and small streams in vicinity of larger rivers served as hatcheries. Hatchlings and fingerlings were stocked into main rivers. In these times, the non-native rainbow trout (*Oncorhynchus mykiss*) was also introduced, which still successfully reproduces in the wild today. In 1888, Cesar Franc Jožet signed the "Fisheries law for the Duchy of Carniola," which came into full effect in 1890. Based on the law, all regions had to divide their flowing surface waters into fishing districts. The law determined a "fishing district" as a continuous stretch of a waterbody with suitable life conditions and of large enough size for healthy development of fish. Boundaries of districts were determined with natural barriers, bridges, tributary estuaries, and other permanently visible markers. Tributaries, oxbows, backwaters, and side channels could all belong to fishing districts. Fisheries activities stopped during both world wars (1914–1918, 1941–1945). Most anglers were mobilized; fish resources became an important food source for soldiers and civilians. Watercourses were depleted of fish, even with explosives. After wars, in 1946, the "Law for temporary Fisheries regulations" ignited the progress of fisheries management in Slovenia. Nine Angling cooperative societies were established, which managed with leased watercourses in their region. In 1953, societies were replaced with angling clubs, which were all united under the Fishing Association of Slovenia, which still exists today. In 1954, the leaseholds were finally eliminated and fisheries management became the responsibility of the civilian society. Even in those times, anglers emphasized, warned, and cared about consequences of pollution and other ecological problems.

2.2 *Fisheries Management Organization Today*

Slovenian fisheries management of inland waters is under State control, governed by the "Act of freshwater fisheries" declared in 2006 (ZSRib). The act determines tasks and responsibilities of fisheries managers, organizations, and anglers, as responsible caretakers of Slovenian waters with emphasis toward fisheries conservation for next generations. With this, anglers have become an indispensable part of environmental protection in Slovenia.

Slovenian waters are divided into 12 fisheries regions (Fig. 5) and 67 territories (Regulation on determining boundaries of fisheries regions and territories in the Republic of Slovenia). Each territory is then divided into fisheries districts. Based on the type of management, fisheries districts can be conservational, angling, without active management and affected. Conservational districts are subsequently divided into nursery districts for sustainable fish rearing, reserves for establishment or conservation of indigenous species, and reserves for broodstock conservation of indigenous species. Sport fishing is exclusively permitted in angling districts. Districts without active management are left to natural process without fisheries management interventions. Affected districts are watercourses or stretches of waters, where fish survival requirements are considerably diminished due to habitat degradation or pollution. Types of fisheries districts and their boundaries are determined with legally binding fisheries management plans ($N = 67$), which are prepared every 6 years by the Fisheries Research Institute of Slovenia (FRIS), and approved by the Ministry of Agriculture, Forestry and Food. Fisheries management is conducted by Angling clubs as concessionaires ($N = 64$), which are locally based non-governmental organizations comprised of 12,000 registered volunteer anglers.

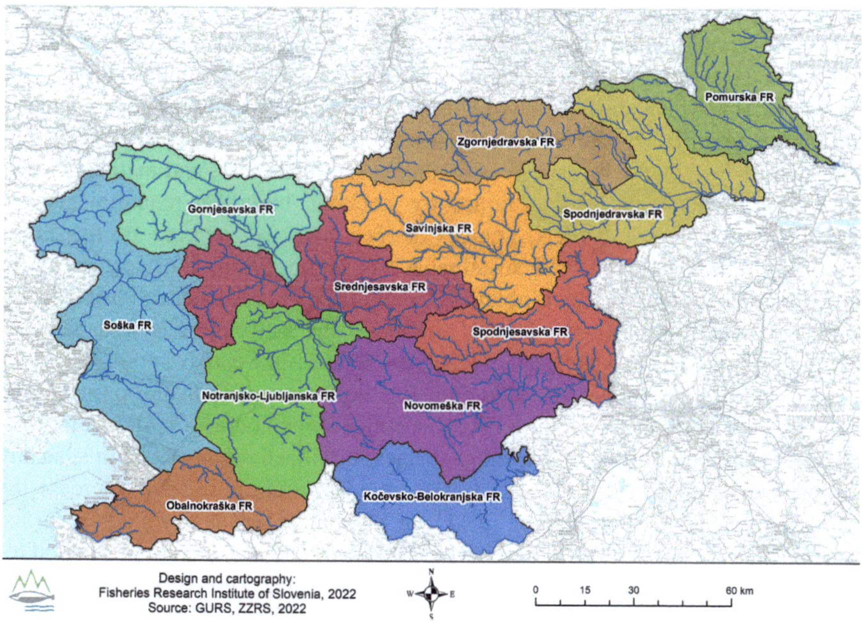

Fig. 5 Slovenia divided into 12 fishing regions

These clubs are united under The Slovenian Fisheries Association as their umbrella organization. Fisheries districts, which are excluded from clubs management, are "waters of special importance" ($N = 9$) managed by FRIS and "privately owned artificial waterbodies." Commercial exploitation of fisheries on inland freshwater bodies does not exist in Slovenia, apart from small recreational fishponds.

2.3 *Implementation and Applicability of Fisheries Management*

The long-term trend of fisheries management in Slovenia aims toward the reduction of non-indigenous species introduction and preservation of a good environmental status of indigenous species and their habitats (PUR 2015).

Slovenia is partitioned into 649 fisheries districts; 278 on lacustrine and 371 on riverine waterbodies (RIBKAT 2022). Based on the Fisheries Cadaster of Slovenia (RIBKAT), angling pressure in 2021 amounted to 122,674 realized angling days, 82% resident and 18% tourist angler days. The majority of angling pressure (66%) was attributed to cyprinid fishery. Nevertheless, salmonid fishery is more recognized and sought-after. In fisheries districts, which inhabit only indigenous species (salmonid and/or cyprinid), "catch and release" angling approach is promoted. These districts are becoming more popular with 33% of all realized angling days in year 2021. Conservational status of native game fish species is sustained and regulated with various limiting angling techniques, minimal size/bag/number limits, and closed seasons, which are all legally binding (ZSRib). Freshwater fisheries managers can introduce stricter conservational measures, but cannot reduce the limitations obligated by law. Game fish species population reinforcements and maintenance of favorable fish abundances is sustained by stocking, based on requirements and capabilities of fisheries districts. Sources of specimens for stocking are fish farms or nursery fisheries districts for sustainable fish rearing. Source of specimens for reproduction of native species is obtained locally from nature, hatchery broodstocks, and reserve districts for preservation of native species. Entry of indigenous genes is highly recommended and lately often practiced, whereas the entry of non-indigenous species/genes is becoming less favorable for most managers. In the past decade, mangers have been implementing the "Cocooning" method as a sustainable management technique for gravel spawning species. Fertilized roe is placed in specially designed boxes and buried into the watercourse substrate in order to increase the recruitment success (Holtzer et al. 2011). In Slovenia, there is not much recorded data on the success of this method. However, FRIS used the "cocooning" method for rearing brown trout and marble trout in the wild. Results were very successful for brown trout with an 80% recruitment rate in the Karst region watercourses (Petkovšek 2020). In the Alpine region, the method was unsuccessful for marble trout, due to torrential characteristics of streams and rivers, which exhibit annual flash floods. Cocooning boxes are frequently swept downstream and buried with inadequate substrate, which creates difficulties for monitoring and recruitment success. More commonly and traditionally used stocking practices for native species are wild release of alevins and YOY juvenile specimens (RIBKAT 2022) in order to insure a better adaptation of stocked fish in the wild. Stocking of different species in various adult sizes is also implemented in selected fisheries districts, based on angling pressure, animal predation, or anthropogenic effects.

Communication and awareness amongst public bodies and non-governmental organizations is of great importance in fisheries management. FRIS representatives in the scope of the LIFE for LASCA project (LIFE16 NAT/SI/000644) conducted two surveys in years 2017–2022 with all 64 angling clubs in Slovenia. The purpose of surveys was to establish communication bonds, to expose local management issues, gain feedback regarding Natura 2000 network, and raise awareness regarding possible negative implications of non-native species introduction. All the angling clubs, as active managers of fisheries districts, agreed that the biggest problems regarding fisheries management in Slovenia are population increase of piscivorous animals and extensive non-sustainable hydrological interventions (Pustovrh 2019; LIFE for Lasca 2021). The surveys also concluded that Natura 2000 sites are very well accepted amongst anglers. Managers state that protected habitats under Natura 2000 serve as an advantage for control over hydrological interventions and possible illegal activities.

3 Game Fish; Management and Future Prospective: Conservation and Sustainable Approaches

3.1 Indigenous Game Fish; Management and Future Prospective: Conservation and Sustainable Approaches

Slovenia is actively engaged in management of indigenous freshwater game fish. The objective is to preserve and establish a favorable fish conservation status in angling, as well as in non-angling fisheries districts. Angling clubs are officially obligated to annually report their activities. Fisheries management data such as stocking, spawning, angling pressure and harvest are collected in the public Fisheries cadaster of Slovenia, named RIBKAT. Additionally, FRIS as a public body regularly monitors fish communities across the entire country with an average of 800 sample points per year (BIOS 2022). Fisheries surveys data such as national monitoring for the ecological status of waters and Natura 2000 monitoring is collected in the FRIS Biological database, named BIOS. Results are at disposal to all managers and can be acquired for a more efficient, integrated, and sustainable management of fish resources. For certain more endangered game fish species, such as huchen, marble trout, brown trout, and grayling, national and European projects are being implemented. In the future, expanded research will also be focused toward Northern pike and native carp.

3.2 Huchen (Danube salmon) Hucho hucho *(Linnaeus, 1758); Salmoniformes, Salmonidae*

Huchen is an endemic species of the Danube river basin and the largest salmonid in Slovenia. Angling for this species has already been described in the year 1689 by Janez Vajkard Valvasor in the "The Glory of the Duchy of Carniola" (Omerzu 1999). It is one of the most endangered salmonids in Europe, included in the Annex II. and V. of the Habitats Directive, IUCN Red List, and Slovenian Red List (2002). It is categorized as an Endangered species with an Unfavorable conservation status (Reporting under the Article 17 of the Habitats directive 2022).

Distribution range of huchen in Slovenia is substantially smaller in comparison to years prior to 1900 (Holčík 1990; Franke 1892; Omerzu 1999; Bajec 2016). The reduction in its distribution range is the greatest in the Mura and Drava river basins, where huchen almost disappeared. Between years 1996 and 2021, electrofishing surveys conducted by FRIS recorded only 5 specimens in segregated starches of Drava River, and 12 specimens limited to an 11 km stretch of Mura River (BIOS 2022). Angling clubs in the region report 23 specimens (TL= > 80 cm) harvested in Mura River and 9 in Drava River since 1980 until 2021 (RIBKAT 2022). The most preserved huchen population in Slovenia is in the Sava River basin (Zabric et al. 2008; Bajec 2016; Podgornik 2018). However, the species has disappeared from

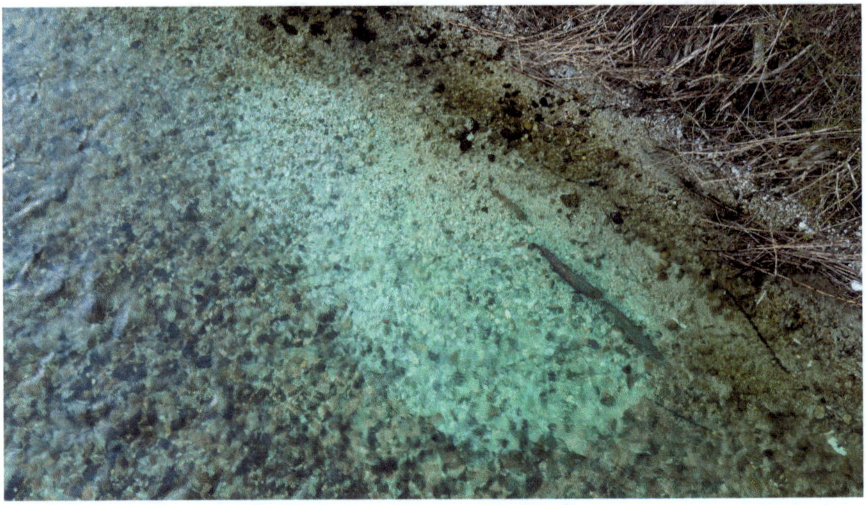

Fig. 6 Huchen spawning, Sava River basin. Photo: FRIS archive

certain Sava River tributaries and numbers have been reduced in rivers Savinja, Mirna, Krka, and Kolpa. In the stretch of the Sava River from the town Zidani Most to the Croatian border, ichthyological monitoring surveys in years 2006–2013 show that after the construction of the hydroelectric (HE) dam chain, the species does not live in this part of the river anymore (Bajec 2016). Despite this, the largest wild and most vital population of huchen in Slovenia remains in the Sava River basin, in central Slovenia between the town of Medvode and Kresnice (cca. 1800 km^2). Recent genetic analysis indicates a self-sustaining population in the region (Snoj et al. 2022). Here the species reproduces successfully and is constantly present (juvenile and adults) in field surveys and angling reports (Zabric et al. 2008; Bajec 2016; Podgornik 2018; RIBKAT 2022). In this approximate geographical surface area of 1800 km^2, FRIS in close collaboration with local angling clubs conducted the huchen spawning grounds survey in years 2013–2017 (BIOS 2022). An annual average of more than 50 active spawning grounds (reds) were recorded (Fig. 6), which remained at the same locations every year, with the same specimens often returning to the same spawning grounds (Fig. 7). Angling data for the same area reveals that 1517 specimens (TL => 80 cm) were harvested in years 1980–2021 (RIBKAT 2022). Unfortunately, there is another chain of five HE dams planned in this part of the Sava River in the near future, which if constructed will be devastating for the huchen population (Zabric et al. 2008; Bajec 2016; Podgornik 2018). The fact is, that the HE dam construction changes the water flow, to such an extent that severely damages the species habitat, including spawning and feeding grounds. These changes in water regimes highly endanger the survival of huchen and the assemblages of fish species, which it feeds upon. Protecting the huchen and its habitat with a "top predator down approach" also preserves the species lower on the trophic level. Historically, the two main threats to the species were excessive

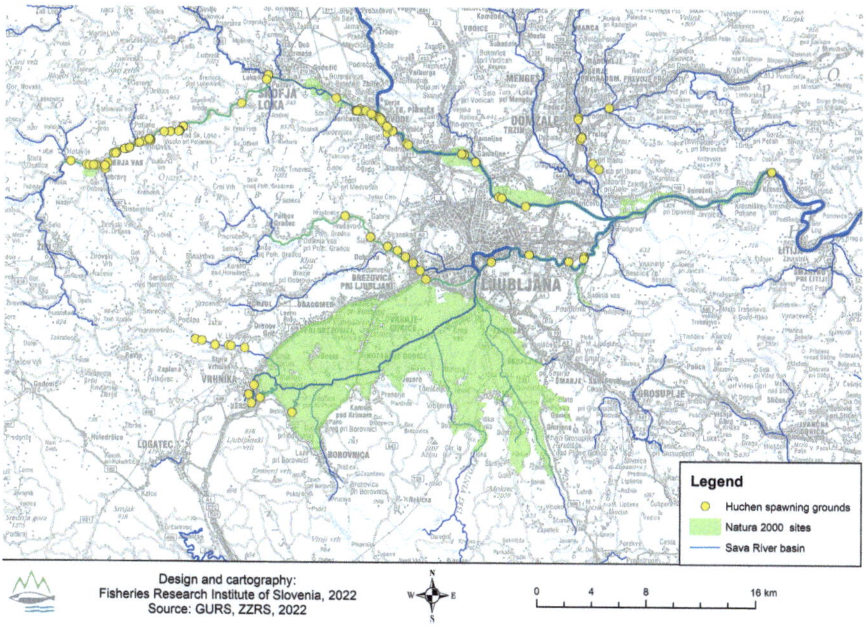

Fig. 7 Recorded Huchen spawning grounds (yellow) dots in middle Sava River basin

unsuitable fishing and pollution, which are no longer considered as severe as HE dam construction in Slovenia. Huchen angling is legally and strictly regulated (Angling regime regulations), and water quality has improved in last 20 years. Freyhof and Kottelat (2008) have estimated that, the most of the huchen populations are fragmented throughout its entire range and dependent on hatchery reared stocks. Natural reproductive success in rivers is very low. Therefore, the wild population of huchen in Sava River and its tributaries in Slovenia can be considered as a natural treasure, which is extremely important to protect. The most important among huchen conservation measures are those that deal with habitat preservation. Without a sufficient habitat size and suitability, which provides life requirements in all life stages, any conservation measure is ineffective. In Slovenia, there are 11 Natura 2000 sites declared, which are crucial for huchen conservation status (Regulation on Special Protection Areas –Natura 2000 sites). The last site (Sava–Medvode–Kresnice) was declared in the year 2013, which connected the Natura 2000 sites of Sava River tributary basins of Sora River and River Ljubljanica. This link protects the most suitable huchen habitat with the most vital part of the species population in Slovenia. Angling regime regulations are also a significant measure in huchen conservation. The lower harvest size limit is TL = 70 cm, which provides at least one spawning season before a specimen is harvested. However, in practice, most managers (angling clubs) increased the length limit to TL = 80 or 90 cm, encouraging greater spawning success. Closed season is from the 15th February to the 30th September, which provides protection during spawning and development of early life stages. More anglers are also practicing the catch and release approach of angling. Proportion of the catch and release approach compared to the harvest angling is annually increasing. In 2021, proportion of catch and release angling days for huchen amounted to 47% (RIBKAT 2022). In the time frame from 2012–2017, within the region of the most vital population, there were only 57 adult huchen (TL => 90 cm) harvested by anglers.

For the purpose of preserving wild populations, huchen in Slovenia is being bred in captivity for over 140 years (Lah 1998; Omerzu 2016; Širnik 2012; Omerzu 1999). Eggs and sperm are obtained (milting) locally in nature and usually fertilized on the spot. Reared specimens are stocked back into the watercourse for wild population reinforcements. Into the Sava River basin in the region of the most vital huchen population, there was an annual average of 5052 huchen of various size classes stocked in the last ten years (RIBKAT 2022). 60% of the stock included specimens TL < 5 cm. Stocking yearlings or younger specimens gives the fish the opportunity and time for adaptation in the wild until reaching the adult phase. In the current huchen management, more effective awareness activities for general public are still missing. Especially in terms of exposing the importance of the existing wild population and its conservation. Consequently, FRIS strives to unite public bodies and local communities including angling clubs, nature protection organizations, genetic laboratories, and other relevant scientific institutions. The goal is to redeem the huchen reputation of being the "King of waters" in Slovenia, as it was rooted in the past amongst the general public.

3.3 Marble trout Salmo marmoratus*; Salmoniformes, Salmonidae*

Marble trout (Fig. 8) is one of the most attractive game fish in Slovenia (Povž et al. 1996). It is an endemic species of the Adriatic basin and the second largest salmonid after huchen. It is included in the Annex II. of the Habitats Directive, IUCN Red list, and the Slovenian Red List (2002). At the end of the twentieth century, the species has almost completely disappeared from the Adriatic basin (Povž et al. 1996). The reason was hybridization with brown trout, which was introduced from the Danube River basin. In the year 1996, it was reported that few genetically pure populations still exist in the headwaters of the Soča River basin, which encompasses 7 km of watercourses. At that time, Slovenia issued an action plan for the conservation of marble trout (Povž et al. 1996). Local fish hatcheries established marble trout breeding, based on the species genetics, for the purpose of reinforcing wild populations. Introduction of the non-native brown trout was officially banned (Article 17. and 18. ZON). Angling regulations were revised (Berrebi et al. 2022). Within the hybridization zone, the marble trout size limit was gradually raised from 40 cm prior to 1996, to 50 cm in 1996–2004. Currently, the marble trout size limit is 60 cm, while brown trout and hybrids size limits remain the same at 26 cm. Daily bag

Fig. 8 Marble trout, Soča River basin. Photo by Krištof Istinič

limit for marble trout is only one specimen. Based on the recent genetic research, Berrebi and colleagues (2022) concluded that after a long struggle with hybridization, the success is significant. In the last 20 years, the percentage of the native marble trout strain in the wild has constantly been increasing for approximately 2% per year. This is considered to be a good result. Today is the time to update the marble trout action plan, with the aid of gained experiences and new approaches.

3.4 *Brown trout* Salmo trutta; *Salmoniformes, Salmonidae*

Brown trout (Fig. 9) is a widespread species in Slovenia (BIOS 2022). It was introduced into the Adriatic basin, while it is native in Danube River basin (BIOS 2022; Povž et al. 2015). The species inhabits cold water stretches of streams and rivers; it can also occur in lacustrine waterbodies (Freyhof and Kottelat 2007). Brown trout abundance in Alpine and Alpine foothill regions is on average 1800 specimens or 80 kg per hectare of a watercourse (Honsig-Erlenburg and Podgornik 2013). The species occurs up to 1281 meters above sea level. On the other hand, the species is rare in South-Eastern part of Slovenia (BIOS 2022), where only the upper and middle flow of the Mura River represents a suitable habitat for the species. Populations are threatened by habitat destruction and hybridization with domesticated Atlantic strains of brown trout (Povž et al. 2015; Honsig-Erlenburg and Podgornik 2013; Bravničar et al. 2017). It is listed in the Slovenian Red List (2002) as an "Endangered species". Angling for brown trout is very popular in Slovenia. The catch and release angling approach is common. However, an average of 5653 specimens per year is recorded in the angling harvest (RIBKAT 2022).

Fig. 9 Male brown trout in the wild. Sava River basin. Photo by Krištof Istinič

Brown trout is protected by the size limit (TL => 25 cm) and with the closed season from September 30th until the April 1st (Angling regime regulations). The official daily bag limit is three specimens. Although, most managers have changed the size limit up to 60 cm and lowered the daily bag limit down to 1 specimens. For stocking purposes, the species is bred in 19 fish farms (RIBKAT 2022). Annually, an average of 1,162,034 specimens are released into the wild, mainly intended for wild population reinforcements. 84% of released specimens are less than 9 cm in length. Breeding native wild strains of Brown trout in captivity compared to domesticated Atlantic strains is inefficient; growth rate and survival of offspring is low. Consequentially, breeding of domesticated Atlantic lines has persisted in the past and has almost completely supplanted breeding of native wild strains (Zabric et al. 2022). On the other hand, it turned out, the long-term survival (after one year) of domesticated Atlantic strains in the wild is on average three times lower than the survival of native wild strains (Weiss and Schmutz 2011; Wills 2006). Breeding in captivity causes negative selection on some adaptive alleles (Bravničar et al. 2017). For this reason, sustainable fish farming today is turning toward breeding wild strains, as much as possible (Zabric et al. 2022). However, in Slovenia, genetically pure populations are rare and confined to a few isolated streams in headwaters (Bravničar et al. 2017). Genetically pure brown trout were confirmed only in nine headwater watercourse sections, mostly in the Northern Drava River basin (Carinthia). All pure populations and also those with a high Danubian genetic share (100% mtDNA in >90% nDNA) are few in numbers and have low genetic diversity. Therefore, they have a low evolutionary potential, meaning these populations are not sufficient for repopulation of all brown trout hybrid zones in Slovenia. With the aim to improve the regional conservation management of these valuable genetic resources, currently a research project is in operation (Zabric et al. 2022). The final goal of the project is to determine areas representing the basic management units for brown trout. Within each unit, the breeding of native brown trout will be promoted, using wild breeders or internal broodstock fertilized by males from the wild. Selection of parents will be based on phenotype, and where feasible parents will be genotyped. Local breeding in "nursery streams for sustainable fish rearing" will be maintained. In the last twenty years, these nursery streams have produced around 200,000 specimens annually (RIBKAT 2022).

3.5 *Grayling* Thymallus *sp.; Salmoniformes, Salmonidae*

Grayling (Fig. 10) is for some people considered as the most captivating game fish. In Slovenia, the species is the most abundant in the Sava River basin; it also occurs in Soča, Drava, and Mura River basins (BIOS 2022). In general, grayling is fairly vulnerable (Jesenšek and Šumer 2004; Povž et al. 2015). It is endangered by river pollutions, dam constructions, and hydrological regulations, as well as climate change. Slovenian grayling is severely threatened by piscivorous animals and hybridization with Danube River genes in watercourses in the Adriatic Sea basin (IUCN Red list; Jesenšek and Šumer 2004). In the last decade, grayling populations have been seriously affected throughout Europe (Jesenšek and Šumer 2004). In Slovenia, the species conservation status is also unstable. Hatchery-raised grayling is stocked for strengthening wild populations (Ribkat 2022). The species is being bred in captivity for over 140 years (Lah 1998; Omerzu 2016; Širnik 2012; Omerzu 1999). Roe and sperm are obtained locally in the wild and fertilized on the spot. In the last decade, for the purpose of wild population reinforcements, there were 2,500,000 juvenile specimens (TL = 5 cm–15 cm) released into watercourses (RIBKAT 2022).

Two species of grayling live in Slovenia. Danube River basin is inhabited by the "European" grayling (*Thymallus thymallus*), while the Adriatic grayling (*Thymallus aeliani*) is native to the Adriatic basin (Cuvier and Valenciennes 1848). *Thymallus aeliani* was officially acknowledged as a species in the year 2020 (Duchi et al. 2020; Bravničar et al. 2020). The conservation status of the Adriatic grayling is critically endangered across the entire species range, which encompasses Northern Italy and West Slovenia (Duchi et al. 2020). Grayling in Slovenia in the Soča River basin is also in decline. In the lower flows of the river, in the region of HE chain of dams ($N = 7$) populations are very small, fragmented, and constricted to short stretches below dams. In the upper Soča River, the species distribution range is declining, based on FRIS ichthyologic surveys (BIOS 2022). In Soča River above the estuary of the tributary Koritnica (marginal area of its range), 40 specimens/ha were reported in the year 1985, 26 specimens/ha in the year 1999, and only 15 specimens/ha in year 2014 were recorded. In the year 2022, no specimens were recorded in the

Fig. 10 Grayling in the Sava River basin. Photo by Krištof Istinič

electrofishing survey at the same location. However, anglers do report sporadic catches of grayling in the same area (RIBKAT 2022).

The greatest pressure and a risk factor to the Adriatic grayling in its distribution range within Soča River basin is the hybridization with the European grayling from the Danube River basin (Bravničar 2021). In order to preserve the Adriatic grayling in Slovenia, in 2003, angling club Tolmin and the Italian partner Ente Tutela Pesca del Friuli Venezia Giulia in the scope of the "Program Phare CBC, SMF" carried out the project "Adriatic Grayling" (Jesenšek and Šumer 2004). The main goal of the project was to establish a grayling broodstock, in order to start breeding the species in captivity. The goal was reached successfully. This broodstock, with specimens possessing the majority of Adriatic genes, for years served as the source of specimens for wild population reinforcements in Slovenia and Italy. Unfortunately, due to a disease outbreak in 2016, the entire broodstock died. The source of specimens for wild pupation reinforcements collapsed. Natural populations are very weak and need strengthening in order to ensure their existence. Broodstocks should be reestablished, and out precaution separated in different locations to minimize the risk of losing the entire broodstock. Studies show that this would be quite challenging for Slovenia. In the Soča River basin, recent population research shows that the degree of hybridization with non-native genes in nature does not allow the possibility of isolating specimens with the majority of the Adriatic genes, even with selection in captivity (Bajić et al. 2018; Duchi et al. 2020; Bravničar et al. 2020; Bravničar 2021). More intimidating is that in the entire species range, only two small sub-populations of genetically pure Adriatic grayling are still left, both in Italy. One population is in the Sesia River, and the other smaller one in the headwaters of the Po River, in lower Pellice River (Duchi et al. 2020). Each sites does not measure more than 4 km in length. These populations could become the source for the new establishment of Slovenian broodstocks, which could then be used for reintroducing native genes back into the wild. Nonetheless, the success is not guaranteed. Because of local genetic/phenotypic adaptations, there is a possibility that genetically pure specimens from Italy would have a high mortality rate, when introduced into the new habitat in the Soča River basin. Currently, Slovenia is searching for possibilities to tackle this demanding task of acquiring genetically pure Adriatic grayling from Italy and gradually adapt specimens for the reintroduction. This is possible to achieve, first via hatcheries followed by a controlled natural environment, and later with a generation of specimens, which could be released into the wild. Until then, the only way to preserve grayling in the Soča River basin is to manage with the existing hybridized populations.

3.6 *Northern pike* Esox lucius*; Esociformes, Esocidae*

In Slovenia, the Northern pike (Fig. 11) is a native species in the Danube River basin and in the Adriatic basin (BIOS 2022). In Danube River basin it inhabits all larger rivers basins; Sava, Drava, and Mura. In the Adriatic basin, it inhabits Reka and Vipava River basins. Northern pike lives in slow-flowing or lacustrine waters among aquatic vegetation (Povž et al. 2015; Casselman and Lewis 1996). During spawning and fry development, it inhabits overgrown shallow waters and flood plains. Wild populations are mainly threatened by watercourse regulations and water pollution (Povž et al. 2015). Angling harvest reports show an average of 2864 specimens per year (RIBKAT 2022). Northern pike stock in Slovenia is regulated by prescribed size limit (TL $\geq$ 50 cm), closed fishing season is from February 1st until the end of April (Angling regime regulations). Even more, most of fisheries managers have risen the size limit for pike to TL $\geq$ 60 cm. For stocking purposes, Northern pike are being bred in 12 aquaculture facilities (RIBKAT 2022). Approximately 6800 specimens per year are stocked, 90% are more than 20 cm in length.

Genetic and morphological research of *Esox* species has shown that there are at least three different species of pike living in Europe; *E. lucius*, *E. aquitanicus*, and *E. cisalpinus* also named *E. flaviae* (Denys et al. 2014; Lucentini et al. 2011; Bianco and Delmastro 2011). According to pike photos taken during FRIS surveys and angling harvest, at least two different pike species are present in Slovenia; *E. lucius*—Northern pike and *E. cisalpinus*—Southern pike. They differ in body pattern, in the number of scales in lateral line, and in the number of submandibular pores. Northern pike have rounded dots on the sides, 48–105 scales in lateral line, and 5–5 submandibular pores (Denys et al. 2014; Bianco and Delmastro 2011). Southern pike have banded or reticulated pigmentation pattern on the sides, 92–107 scales in lateral line, and 4–4 submandibular pores. A reliable confirmation of the occurrence of different pike species in Slovenia, their distribution and possible hybridization still needs to be investigated.

Fig. 11 *Esox* sp. in Sava River basin. Photo by Krištof Istinič

3.7 *Common nase* **Chondrostoma nasus**; *Cypriniformes, Cyprinidae*

Common nase is a widespread species in Slovenia. It was introduced into the Adriatic basin, while it is native in Danube River basin (BIOS 2022; Povž et al. 2015). It inhabits middle starches of rivers with a gravel bottom (Povž et al. 2015; Freyhof and Kottelat 2007). Populations are threatened by watercourse regulations and by other interventions that interrupt species migration paths to spawning and feeding grounds (Povž et al. 2015). Fishing for common nase and sister species, such as cactus roach, barbels, and chubs, is a long tradition in this part of the world. Generally, the species are harvested for family food purposes. On average, more than 10,000 common nase specimens are harvested annually (RIBKAT 2022). Consequently, conservation actions are in place to conserve a favorable status of the species. In Slovenia, common nase is protected by the size limit (TL $\geq$ 30 cm) and by the closed season from February 1st until the end of May (Angling regime regulations). Additionally, for wild population reinforcements, common nase is bred in 8 fish farms (RIBKAT 2022). In practice, the eggs arrive from wild breeders. Annually, we release more than 100,000 specimens in the wild. 78% of introduced specimens are smaller than 9 cm in length to give specimens an opportunity to adapt to the local environment.

Common nase was introduced into the Adriatic basin (Soča River basin) in the 1960s to amplify the angling experience (Povž et al. 2015). According to Povž (1983), the introduced common nase expanded to such an extent that it outcompeted the native Natura 2000 species, the South European nase or lasca (*Protochondrostoma genei*) and caused its gradual disappearance. This was well documented in the Vipava River basin. Furthermore, FRIS revealed that extensive hydrological works in the 1980s also contributed to the disappearance of lasca, due to habitat degradation (LIFE for LASCA 2019). This is an executive example of how human actions, such as river interventions and alien species introduction, can cause a negative impact on nature. Today, after 40 years of natural processes, some water-course sections in the Vipava River valley have become overgrown with riparian vegetation and have turned into a semi-sustainable habitat. Common nase populations in the Vipava River valley are declining due to natural processes and population reduction actions (LIFE for LASCA 2022).

3.8 *Wels catfish* Silurus glanis*; Siluriformes, Siluridae*

In Slovenia, wild populations of Wels catfish (Fig. 12) naturally occur in lower stretches of the Sava, Drava, and Mura River basins (BIOS 2022; Povž et al. 2015). It was introduced in the Adriatic basin and many artificial ponds and reservoirs, where it successfully reproduces and thrives (Povž et al. 2015; BIOS 2022). Wels catfish is mainly threatened by river pollution, and to a lesser extent by watercourse regulations and angling (Povž et al. 2015). Angling harvest amounts to an average of 896 specimens per year, with specimens commonly reaching over a 100 cm in length (RIBKAT 2022). Catch and release fishing is becoming more popular, especially for larger specimens. It is protected by the size limit (TL $\geq$ 60 cm) and the closed fishing season from May 1st to end of June (Angling regime regulations).

3.9 *Sander* Sander lucioperca*; Perciformes, Percidae*

In Slovenia, wild populations of sander naturally occur in the Sava, Drava, and Mura River basins (BIOS 2022; Povž et al. 2015). The species has been introduced to many artificial waterbodies, such as commercial ponds and reservoirs. Sander is mainly threatened by river pollution, and to a lesser extent by watercourse regulations (Povž et al. 2015). Average angling harvest is 4106 specimens per year (RIBKAT 2022), which is compensated yearly by 5440 stocked specimens, mostly (86%) of larger sizes (TL >15 cm). Sander is bred in 51 aquaculture facilities, 13 have the license for stocking. The size limit for sander is TL $\geq$ 50 cm and the closed season is from March 1st to the end of May (Angling regime regulations).

Fig. 12 Wels catfish, lake Bled. Sava River basin. Photo by Krištof Istinič

4 Non-indigenous Game Fish; Management and Future Prospective

Law officially prohibits introduction of non-indigenous species in Slovenia (ZON). However, it exclusively permits stocking of specific non-native fish species under constant control of the competent Ministry (Article 17. and 18. ZON). This is bound to economically significant game fish species, such as farm-raised carp (*Cyprinus carpio*) and rainbow trout (*Oncorhynchus mykiss*) (PUR 2015). The two species represent more than 95.5% of non-native species entry into nature (RIBKAT 2022). Annual average of non-native species entry into Slovenian watercourses amounts to 145 tons or 200,000 specimens. The trend of entry is in decline, it has decreased by 20% in the last 8 years (LIFE for LASCA 2021).

Originally carp (Fig. 13) is a native species in Slovenia; it inhabits the Danube River basin (Povž et al. 2015). Today, these wild populations are critically endangered, mostly due to habitat destruction and due to the release of domesticated strains of carp, which are considered as non-native in Slovenia. Domesticated strains of carp were introduced into nature and many artificial waterbodies, such as commercial ponds and reservoirs, where it is still sustained by stocking (RIBKAT 2022). The occurrence of "wild-carp" in open waterbodies in Slovenia is very rare. Despite numerous field-monitoring sample points ($N > 8000$ sampling sites), performed by FRIS in the last decade, only 76 specimens of carp were recorded in open waters. Therefore, specimens for stocking are commonly imported from abroad. In angling fisheries districts, which inhabit carp, the fundamental management measure is to raise awareness regarding origin of specimens for stocking. Managers are encouraged to constantly search for the most appropriate native forms of carp from the Danube River basin in order to minimize the entry of specimens with an inappropriate origin.

Fig. 13 Carp, lake Bled. Sava River basin. Photo by Krištof Istinič

Rainbow trout (Fig. 14) has been present in Slovenia for 140 years (Munda 1927; Fishing Association of Slovenia 2022; Širnik 2012). Today, it is a widespread species that successfully reproduces in the wild (BIOS 2022). 40% of angling clubs in Slovenia, representing most managers, who manage with the species, consider rainbow trout as a naturalized game fish species without severe negative management implications (Pustovrh 2019; LIFE for LASCA 2021). However, officially the species is considered as non-native and all managements activities should be focused toward the species decline in nature (PUR 2015). On the bases of these guidelines, Slovenia has gradually implemented a temporary management measure in the last decade. Stocking sterile specimens of rainbow trout into "put and take" fisheries districts is gradually replacing the entry of fertile specimens. This measure has been reducing the entry of fertile rainbow trout specimens into the wild by 65% in the last 8 years.

Stocking of non-indigenous species into fisheries districts is temporally and spatially tied to angling districts, where species are reduced annually by harvesting (RIBKAT 2022). Average annual angler harvest of non-native species in Slovenia is approximately 77 tons and 70,000 specimens. Based on the last 8 years of the risk assessment study, the harvest increased by 10% (LIFE for LASCA 2021). Greater harvest value is for rainbow trout (15%), because the minimum size limit of the species has been abandoned (Angling regime regulations).

In the year 2011, FRIS performed a study on stocked sterile rainbow trout in the Alpine Radovna River (Ivanc 2011). It was discovered that stocked sterile rainbow trout diminishes within the season and does not affect the abundance of native brown trout. Sterile rainbow trout (TL > 20 cm) is released into angling districts for the

Fig. 14 Rainbow trout in Sava River basin. Photo by Krištof Istinič

purpose of “put and take fishery.” These specimens are not well adapted to wild conditions (ex. food seeking, hiding, predator avoidance), usually stay in one place and do not disperse along the river efficiently. Therefore, they are more susceptible to anglers and predation. In the Radovna River, there were 197 rainbow trout recorded in the angler harvest in the year 2011, of which 183 (93%) were from the same year’s stock. There were 2225 specimens of marked sterile rainbow trout released into the river, which means that the angler harvest amounted to 9% of the entire stock of the year. Despite low angler harvest compared to stocking numbers, the rainbow trout abundance in the river did not increase, but even decreased significantly (20%). During electrofishing surveys at the end of the season, no marked specimens were found. Simultaneously it was discovered that the native brown trout abundances increased on same sampling locations. Population reinforcement stocking with native brown trout in the same year was 460 specimens, which were not harvested due to conservation regulations for the river. Apart from angler harvest, the additional decrease in rainbow trout abundance in the river was attributed to natural mortality and processes, such as predation from piscivorous animals (cormorants, herons, and otters), as well as poaching. Reliable assessment of these threats is difficult to quantify and remains to be defined in the future.

5 Economic Impact of Slovenian Inland Fisheries

In Slovenian inland open waters, only non-commercial fishing permits exploitation of fisheries resources. Commercial use of the resource is conducted in commercial fishponds, which are privately owned. On inland open waters recreational sport fishing is carried out, where individuals can purchase fishing licenses for one day, three days, one week, and one year. In year 2021, the estimated revenue from fishing licenses amounted to EUR 1.5 million, which represents more than 50% of all fisheries managers revenue (RIBKAT 2022). Additional EUR 1.5 million of inland fisheries income comes from other sources, such as farm-raised fish commerce. Most of these funds (~70%) are intended for stocking, river keeping, fisheries association memberships, training courses, regular operating costs, and concessions to the state. In 2021, roughly EUR 300,000 came into the state treasury from concessions.

About EUR 7,000,000 of public budget funds are allocated annually in Slovenia for the purposes of the entire fishery (inland and marine) including fish farming; in 2020 EUR 6,878,184 (MKGP 2020). Regarding inland waters, these funds are intended toward the sustainable management of freshwater fisheries resources, especially for conservation and protection of native fish populations, as well as for sustaining recreational sport fishing (MKGP 2022). Sport fishing also has an important economic impact in connection with tourism activities, such as culinary experiences, shops, accommodations, and fish farming (Mavrič et al. 2021). A complete quantification of the recreational sport fishing economic impact is difficult to assess, since activities are scattered and do not have a basis in a combined organizational unit. Locations where sport fishing is connected to tourism are commonly along popular larger rivers (Soča, Kolpa, and Sava) and lakes (Bled and Bohinj), as well in close vicinity of fish farms and commercial fish ponds (Mavrič et al. 2021).

Fisheries managers, anglers, and fish farmers through European funds and interested stakeholders seek additional financial support for sustainable management of native game fish populations, and for connection of tourism industry with sport fishing. European funds are preferentially drawn from European Maritime and Fisheries Fund (EMFF), European Agricultural Fund for Rural Development (EAFRD), European Regional Development Fund, Cohesion Fund, Specific Target Research Projects, INTERREG—European territorial cooperation and LIFE mechanism. For example, FRIS as a public body in year 2021 received EUR 765,186 of budget funds, EUR 287,119 from fishing licenses sales, and EUR 386,522 from other marketing projects and fish farming (ZZRS 2021). At the beginning of the year, FRIS employed 44 people. Within the same year, the public body received EUR 1,777,982 of special purpose assets; of this EUR 131,587 was from the EMFF, EUR 1,209,841 from the Cohesion Fund, EUR 24,555 from the European Regional Development Fund, EUR 23,581 from Specific Target Research Projects, and EUR 388,418 from the LIFE mechanism. European funds are most commonly gained by submitting proposals to the national and European tender calls. Preparation of successful individual project proposals can be time-consuming and can take a few months up to a few years. These projects are non-profitable and in many cases require the investment of "own contribution" funds. EU projects are of a great importance for FRIS in terms of development, recognizability, improving bonds with local communities, connecting angling and scientific communities. One of the most important contributions of EU funds is to enable the possibility to focus research toward species-specific management of native game fish, which commonly surpasses regular budget funds. An example of one of the most successful Slovenian projects regarding the conservation of native game fish species is The Marble trout Project, by which Slovenia managed to sustain the species and change the species population conservation trend from decreasing to improving (Pajk and Podgornik 2014; Berrebi et al. 2022). In the ranks of most successful projects in terms of connecting recreational sport fishing and tourism are certainly The Marble trout Festival in Kobarid and Fly-fishing festival in Bohinj (Mavrič et al. 2021). These events exhibit fly-fishing, offer culinary fish specialties, workshops on different fishing techniques, and various entertainment and musical programs. An added economic value is also contributed by connecting fish farming, commercial recreational sport fishing—ponds with culinary opportunities. Some restaurants offer guests to catch their own fish, which are then prepared in various recipes. Examples of these good practices are commercial sport fishing—ponds located near cities Kamnik, Naklo, and Ptuj.

To conclude, Slovenia for its small geographical range covers many different approaches to promote and implement sustainable exploitation of inland fisheries, which will hopefully be beneficial for generations to come.

Acknowledgments The authors especially thank Primož Zupančič for motivation, and for creating the communication bond with prof. Vladica Simić from the University of Kragujevac. This paper and the information involved, it would not have been possible without the exceptional support and collaboration of all 64 Slovenian angling clubs. The authors are grateful to all the clubs for their devotion and voluntary work in fisheries management through the years. The authors sincerely thank our coworkers at the Fisheries Institute of Slovenia for all the help and opportunity to compose the paper, especially to Krištof Istinič for providing the photographs. The authors also applaud all the scientist mentioned in the paper for providing data through their devoted work.

References

Angling regime regulations/Pravilnik o ribolovnem režimu v ribolovnih vodah. (n.d.) Official Gazette of Republic of Slovenia, No. 99/07 and 75/10.

ARSO (2022) Agencija Republike Slovenije za okolje, Atlas okolja, Pedološka karta Slovenije. http://gis.arso.gov.si/atlasokolja/profile.aspx?id=Atlas_Okolja_AXL@Arso, 22.6.2022.

Bajec SS (2016) Izdelava strokovnih podlag za ohranjanje habitata in populacije sulca na območju srednje Save. Zaključno poročilo o rezultatih opravljenega raziskovalnega dela na projektu v Ciljnega raziskovalnega programa (CRP) »Zagotovimo.si hrano za jutri« 2011–2020. Univerza v Ljubljani, Biotehniška fakulteta, Oddelek za zootehniko. Ljubljana. 176 p.

Bajić A, Jojić V, Snoj A, Miljanović B, Askeyev O, Askeyev I, Marić S (2018) Comparative body shape variation of the European grayling *Thymallus thymallus* (Actinopterygii, Salmonidae) from wild populations and hatcheries. Zoologischer Anzeiger 272:73–80

Berrebi P, Jesenšek D, Laporte M, Crivelli AJ (2022) Restoring marble trout genes in the Soča River (Slovenia). Two decades of monitoring the introgression. Conserv Genet. 15 pages

Bianco PG, Delmastro GB (2011) Recenti novità tassonomiche riguardanti i pesci d'acqua dolce autoctoni in Italia e descrizione di una nuova specie di luccio. In: de Filippo G (ed) Researches on wildlife conservation, vol 2. (suppl.). IGF Publ., USA, pp 1–14

BIOS (2022) Biološka zbirka podatkov Zavoda za ribištvo Slovenije. Ljubljana, Zavod za ribištvo Slovenije. Available on Marčeta B. and Pliberšek ur. BiosWeb. [online] Ljubljana, Zavod za ribištvo Slovenije, 2014, [apdated on 29.09.2022], [Accessed on 29 Sept 2022], Dostopno na spletnem naslovu: www.biosweb.org, ISSN 2350-4757.

Bravničar J (2021) Populacijska genomika in varstvo jadranskega lipana Thymallus aeliani (Valenciennes, 1848) v reki Soči. Doktorska disertacija. Univerza v Ljubljani, Biotehniška fakulteta. Ljubljana. 151 p.

Bravničar J, Palandačić A, Bajec SS, Snoj A (2020) Neotype designation for *Thymallus aeliani* Valenciennes, 1848 from a museum topotype specimen and its affiliation with Adriatic grayling on the bases of mitochondrial DNA. ZooKeys 999:165–178

Bravničar J, Sušnik BS, Snoj A (2017) Varstvena genetika avtohtone potočne postrvi v Sloveniji. Zaključno poročilo o rezultatih opravljenega raziskovalnega dela na projektu v okviru Ciljnega raziskovalnega programa (CRP) »Zagotavljamo.si hrano za jutri« 2011–2020. Ljubljana, 56 p.

Casselman JM, Lewis CA (1996) Habitat requirements of northern pike (*Esox lucius*). Can J Fish Aquat Sci 53(Suppl. 1):161–174

Cowx IG, Welcomme RL (1998) Rehabilitation of rivers for fish. A study undertaken by the European Inland Fisheries Advisory Commission of FAO. 97 p.

Cuvier G, Valenciennes A (1848) Histoire naturelle des poissons, vol 21. Chez F.G. Levrault, Paris. Chapt. 7, pp 447–448

Čehić S (2007) Pogled na vode v Sloveniji. Statistični urad Republike Slovenije, Posebne publikacije, Okolje, Ljubljana, pp 9–27. 13–61

Denys GPJ, Dettai A, Persat H, Hautecoeur M, Keith P (2014) Morphological and molecular evidence of three species of pikes *Esox* spp. (Actinopterygii, Esocidae) in France, including the description of a new species. C R Biologies 337, 521–534.

Duchi A, Puzzi C, Volta P (2020) *Thymallus aeliani*. The IUCN Red List of Threatened Species 2020: e.T146280943A146280948. https://doi.org/10.2305/IUCN.UK.2020-2.RLTS.T146280943A146280948.en. Accessed 11 Aug 2022

Fishing Association of Slovenia/Ribiška zveza Slovenije (2022) Web page of Fisheries society of Slovenia. Available on ribiska-zveza.si/o-nas/zgodovina. Accessed 22 June 2022.

Franke J (1892) Die Gewaesser in Krain und ihre nutsbare Fauna. Buchdruckerei von Ig. V. Kleinmayer and Fed. Bamberg. Verlag der k. k. Staats-Oberrealschule, Laibach. 55 p.

Freyhof J, Kottelat M (2007) Handbook of European freshwater fishes. Publications Kottelat, Cornol and Freyhof, Berlin. 646 p

Freyhof J, Kottelat M (2008) *Hucho hucho*. The IUCN Red List of Threatened Species 2008: e.T10264A3186143. https://doi.org/10.2305/IUCN.UK.2008.RLTS.T10264A3186143.en. Accessed 29 Sept 2022

Holčík J (1990) Conservation of the huchen, *Hucho hucho* (L.), (Salmonidae) with special reference to Slovakian rivers. J Fish Biol 37:113–121

Honsig-Erlenburg W, Podgornik S (2013) The KARAFISH - Project. INTERREG IV. A Project Slovenia – Austria. High altitude distribution of brown trout and occurrence of bullhead in the mountain range Karawanke. AKL and ZZRS. 703 pp.

Holtzer G, Unfer G, Hinterhofer M (2011) "Cooconing" - eine alternative Methode zur fischereilichen Bewirtschaftung. Oesterreichs Fischerei 64:16–27

Ivanc M (2011) Določanje optimalne količine dopolnilno vložene šarenke v ribolovnem revirju Radovna (ZZRS). Fisheries Research Institute os Slovenia, Internal report, Spodnje Gameljne. 6 p.

Jesenšek D, Šumer S (2004) Jadranski lipan (*Thymallus thymallus*, Linnaeus, 1758) v porečju Soče v Sloveniji. Varstveni načrt. Ribiška družina Tolmin in Ente Tutela Pesca del Friuli Venezia Giulia. Tolmin. 32 p.
Kolbezen M (1998) Rečna mreža. Geološke značilnosti in gostota. Knjiga: Površinski vodotoki in vodna bilanca Slovenije, vol 5. Ministrstvo za okolje in prostor, Ljubljana, p 29
Lah A (1998) Sladkovodno ribištvo Slovenije v stodesetih letih razvoja 1888–1998. Ribiška zveza Slovenije, Ljubljana. 88 p.
LIFE for LASCA (2019) LIFE SAVING LASCA Urgent measure to conserve nearly extinct species Protochondrostoma genei. LIFE16 NAT/SI/000644. Obvestilo o ponovni naselitvi Natura 2000 vrste primorska podust (Protochondrostoma genei) v Močilnik, Jevšček, Ozlenšček in Konjščak ter poziv k sodelovanju v smeri omejitve večjih posegov v omenjene vodotoke v vsaj sedem letnem obdobju. Report, Spodnje Gameljne. 4 p. Available on www. lifeforlasca.eu.
LIFE for LASCA (2021) LIFE SAVING LASCA Urgent measure to conserve nearly extinct species Protochondrostoma genei. LIFE16 NAT/SI/000644. Vnos tujerodnih ribolovnih vrst v slovenske vode. Primerjava obdobij 2013–2016 in 2017–2020. Fisheries Research Institute of Slovenia, Report, Spodnje Gameljne. 13 p. Available on: www.zzrs.si/blog/vnos-tujerodnih-ribolovnih-vrst-v-slovenske-vode/.
LIFE for LASCA (2022) Web page of the Project. LIFE SAVING LASCA Urgent measure to conserve nearly extinct species Protochondrostoma genei. LIFE16 NAT/SI/000644. www. lifeforlasca.eu, viewed on 25 Sept 2022
Lucentini L, Puletti ME, Riccolini C, Gigliarelli L, Fontaneto G, Lanfanoli L, Blio F, Natali M, Panara F (2011) Molecular and Phenotypic Evidence of a new species of Genus *Esox* (Esocidae, Esociformes, Actinopterygii): The Southern Pike, *Esox flaviae*. PLoS One 6(12):e25218. https://doi.org/10.1371/journal.pone.0025218
Mavrič A, Bojnec Š, Ramšak A (2021) Izzivi razvoja ribištva v Sloveniji. Založba Univerze na Primorskem, Koper. 131 pages
MKGP (2020) 2330 - Ministrstvo za kmetijstvo, gozdarstvo in prehrano. Obrazložitev zaključnega računa proračuna za leto 2020. 244 pages
MKGP (2022) https://www.gov.si/podrocja/kmetijstvo-gozdarstvo-in-prehrana/ribistvo/; viewed on 26 Oct 2022.
Munda A (1927) Profesor Ivan Franke. Lovec. 398–401.
Omerzu S (1999) Sulec, slovenski sulčarji in sulčja lov do II. Svetovne vojne. Racoon, Ljubljana. 148 p.
Natek K (1998) O regionalizaciji Slovenije. Geografski vestnik. 70/139 - 150.
Omerzu S (2016) Sulec in sulčelov v Sloveniji od osvoboditve do osamosvojitve 1945–1991. Racoon, Ljubljana. 357 p
Pajk N, Podgornik S (2014) Monitoring populacij izbranih ciljnih vrst rib. Soška postrv (*Salmo marmoratus*). Poročilo Zavoda za ribištvo Slovenije, Ljubljana – Šmartno. 56 pages
Petkovšek M (2020) Valjenje iker potočne postrvi 2019–2020 na Malenščici. Fisheries Research Institute of Slovenia, Internal Report, Planina. 14 p.
Podgornik S (2018) Monitoring populacij izbranih ciljnih vrst rib. Sulec (*Hucho hucho*). Report of Fisheries Research Institute of Slovenia, Ljubljana-Šmartno. 87 p.
Povž M (1983) Spremembe v arealu dveh vrst podusti (Chondrostoma, Pisces) v Sloveniji. Biološki vestnik 31(1):45–52
Povž M, Gregori A, Gregori M (2015) Sladkovodne ribe in piškurji v Sloveniji. Zavod Umbra, Ljubljana. 293 p.
Povž M, Jesenšek D, Berrebi P, Crivelli AJ (1996) Soška postr, Salmo trutta marmoratus, Cuvier 1817, v porečju Soče. Tour du Valat. 65 p.
PUR (2015) Program upravljanja rib v celinskih vodah Republike Slovenije za obdobje do leta 2021. Republika Slovenije, Vlada Republike Slovenije, Ljubljana. 138 p.
Pustovrh K (2019) Ribiči in Natura 2000. Ribič. 5:128–129

Regulation on determining boundaries of fisheries regions and territories in the Republic of Slovenia/Uredba o določitvi meja ribiških območij in ribiških okolišev v Republiki Sloveniji. (n.d.) Official Gazette of Republic of Slovenia, No. 52/07.
Regulation on Special Protection Areas (Natura 2000 sites)/Uredba o posebnih varstvenih območjih (območjih Natura 2000). Official Gazette of Republic of Slovenia, No. 49/04, 110/04, 59/07, 43/08, 8/12, 33/13, 35/13 – popr., 39/13 – odl. US, 3/14, 21/16 in 47/18.
Reporting under the Article 17 of the Habitats directive (2022) Period 2007–2012. Species: Hucho. European Environment Agency. European Topic Centre on Biological Diversity. 7 p. Available online on www.eionet.europa.eu/etcs/etc-be/activities/reporting/article-17/outcomes-2007-2012.
RIBKAT (2022) Slovenian Fishing Cadaster database. Available online on webapl.mkgp.gov.si/apex/f?p=104:LOGIN_DESKTOP:254531670507. Accessed on 29 Sept 2022.
Širnik R.E. (2012) Slovensko sladkovodno ribištvo in ribiči skozi čas. Ribiška zveza Slovenije, Ljubljana. 223 p.
Slovenian Red List (2002) Pravilnik o uvrstitvi ogroženih rastlinskih in živalskih vrst v rdeči seznam. Official Gazette of Republic of Slovenia, No. 82/2002.
Snoj A, Bravničar J, Zabric D, Bajec SS (2022) Conservation genetics study of huchen in Slovenia recommends river system-based management and indicates self-sustainability of the middle Sava population. Aquatic Conserv Mar Freshw Ecosyst:1–13. https://doi.org/10.1002/aqc.3834
Toš M (2021) Ribolov in ribištvo v 19. Stoletju. Ribič. 1–2, 4–6.
Weiss S, Schmutz S (2011) Performance of Hatchery-Reared Brown Trout and Their Effects on Wild Fish in Two Small Austrian Streams. Trans Am Fish Soc 128:302–316
Wills TC (2006) Comparative abundance, survival, and growth of one wild and two domestic Brown Trout strains stocked in Michigan rivers. N Am J Fish Manag 26:535–544
Zabric D, Bajec SS, Bravničar J, Ramšak L, Leskošek T, Puklavec D (2022) Izdelava strategije upravljanja s potočno postrvjo v Sloveniji. Zaključno poročilo o rezultatih opravljenega raziskovalnega dela na projektu v Ciljnega raziskovalnega programa (CRP) »Zagotovimo.si hrano za jutri« 2011–2020. Univerza v Ljubljani, Biotehniška fakulteta, Oddelek za zootehniko in Zavod za ribištvo Slovenije, Ljubljana. Draft. 33 p.
Zabric D, Bertok M, Jenič A (2008) Stanje in varstvo sulca (Hucho hucho) v Sloveniji. Zavod za ribištvo Slovenije, Spodnje Gameljne. 62 p.
ZON: Zakon o ohranjanju narave. Official Gazette of Republic of Slovenia, No. 96/04 – uradno prečiščeno besedilo, 61/06 – ZDru-1, 8/10 – ZSKZ-B, 46/14, 21/18 – ZNOrg, 31/18, 82/20 in 3/22 – ZDeb.
ZSRib: Zakon o sladkovodnem ribištvu. Official Gazette of Republic of Slovenia, No. 61/06.
ZZRS (2021) Poslovno poročilo Zavoda za ribištvo Slovenije za leto 2021. Zavod za ribištvo Slovenije, Spodnje Gameljne, 114 pages

Inland Fisheries in Croatia: Historical Aspects, Fish Resources, Management, and Conservation

Marina Piria, Josip Suić, and Ivan Špelić

Abstract Freshwater aquatic ecosystems are the most vulnerable ecosystems due to pressures caused by human activities, while also having high importance for retaining biodiversity and social and economic significance. Inland commercial and recreational fishing in Croatia has had a long history and remains so today. The main purpose of this chapter is to give an overview of the Croatian inland fishery and to identify its future prospects. The commercial fishery in Croatia has persisted in only two large rivers in continental Croatia and faces a declining trend and strong impacts from ecosystem changes. On the contrary, the value of the recreational fishery is growing and today holds substantial economic and social importance. The changes in the catch of commercially important fish species were identified through history mainly due to habitat degradation (e.g., dam construction, pollution, or channelisation), overexploitation, introduction of non-native species, and extinction of several commercially important species (e.g., sturgeons). Today fishing relies mostly on native common bream *Abramis brama*, common carp *Cyprinus carpio,* northern pike *Esox lucius,* European catfish *Silurus glanis*, pike-perch *Sander lucioperca*, and large non-native fish species, such as gibel carp *Carassius gibelio* and grass carp *Ctenopharyngodon idella.* In order to preserve healthy fish stock,

Supplementary Information The online version contains supplementary material available at https://doi.org/10.1007/978-3-031-36926-1_2.

M. Piria (✉)
University of Zagreb Faculty of Agriculture, Department of Fisheries, Apiculture, Wildlife Management and Special Zoology, Zagreb, Croatia

Faculty of Biology and Environmental Protection, Department of Ecology and Vertebrate Zoology, University of Lodz, Lodz, Poland
e-mail: mpiria@agr.hr

J. Suić
Ministry of Agriculture, Directorate of Fisheries, Zagreb, Croatia

I. Špelić
University of Zagreb Faculty of Agriculture, Department of Fisheries, Apiculture, Wildlife Management and Special Zoology, Zagreb, Croatia

V. Simić et al. (eds.), *Ecological Sustainability of Fish Resources of Inland Waters of the Western Balkans*, Fish & Fisheries Series 43,
https://doi.org/10.1007/978-3-031-36926-1_2

digital data collection systems were introduced in the inland water fishery in both the commercial and recreational sectors. Fish stocking for recreational fishery purposes is regulated by several acts of legislation concerning environmental protection and fishery management. In order to continue fishing activities, both anglers and commercial fishers have to gain an understanding of environmental issues and non-native and protected fish resources. Whereas the prospect of recreational fishing appears considerable, the prospect of commercial exploitation has a dismal future and its sustainability is questionable.

Keywords Commercial fisheries · Recreational fisheries · Fish stock · Threat · Sustainability

1 Introduction

Freshwaters are the most vulnerable ecosystems due to multiple threats and ecological alterations caused by human activities (Craig et al. 2017; Dudgeon 2019; Piria et al. 2019a), particularly overexploitation, water pollution, flow modification, destruction or degradation of habitat, and invasion by exotic species (Dudgeon et al. 2006). Many efforts have to be made to stop degradation processes, preserve resources (Pham et al. 2019), and manage ecosystem functioning while still maintaining human livelihoods (Dudgeon and Smith 2006). However, freshwater ecosystems are important for socioeconomic development (Davis et al. 2015), including the recreational and commercial fishery (Lynch et al. 2016), as these ecosystems can not only provide recreational services and eco-tourism but also contribute to biodiversity conservation (Vehanen et al. 2020) through improved fishery management (McIntyre et al. 2016).

Freshwater fishing activities on the rivers and lakes of Europe through history have shown that freshwater fish exploited from local waters had great importance for livelihoods (Hoffmann 2005), particularly for inhabitants settled near freshwaters (Svanberg and Locker 2020). However, the opportunities for fishery development in European freshwaters, particularly for commercial fisheries, were fairly limited through history, and increased harvesting of the natural resources often led to their overexploitation (Sahrhage and Lundbeck 2012). Mainly a result of overfishing, but also other causes such as water pollution and protection of threatened species, the commercial inland fishery in Europe has declined and disappeared in many European countries (Arlinghaus et al. 2002; Cowx 2015; Vehanen et al. 2020). Nevertheless, it remains significant in Balkan countries, such as on Skadar Lake that is shared between Montenegro and Albania, in the Danube River in Bosnia and Herzegovina, Bulgaria, Croatia, Romania, and Serbia (Milošević and Talevski 2015; Sala 2016; Treer and Kubatov 2017). However, as elsewhere in Europe, a declining trend of the commercial inland fishery has been documented in most countries that perform fishing activities on the Danube River (Smederevac-Lalić et al. 2011;

Vehanen et al. 2020) and a decline in important commercial fishes has also been observed on Skadar Lake (Marić 2018). A commercial inland fishery in Croatia is also present on the Sava River, where a decline trend in catches of economically important species has also been documented (Vehanen et al. 2020), though the actual state and future prospects remain unknown.

The recreational fishery on inland waters developed much later and was insignificant in Europe until recent times. Its popularity, however, is on the rise (Lyach and Čech 2018), though under varying management practices (Vehanen et al. 2020). The inland recreational fishery is well developed in Croatia and has the capacity to increase its potential (Matulić et al. 2010; Treer and Kubatov 2017). However, with growing public awareness concerning nature conservation, the need has arisen to protect and better manage fish stocks. This has led to changes in management practices, improvements in fishery data collection, and development of the legislation for fish management and species protection.

This review gives an overview of the historical background and current state of the inland commercial and recreational fisheries in Croatia and identifies their future prospects.

2 Historical Overview of Inland Fisheries in Croatia

2.1 *Legislation Through History*

The oldest written document related to inland waters fishing activity in Croatia is the Deed of Gift to Vukovar County (*Darovnica Vukovarskoj županiji)* of Herceg Koloman (brother of King Bela IV) of 1231 on the right to fish without limitations on the Danube and Vuka Rivers. The official profession of fishers was mentioned for the first time in 1460 in the book *Čudesa Ivana Kapistrana* (The Wonders of Ivan Kapistran) found in the town of Ilok (Bojčić 1982). However, records on fishing were very rare until the eighteenth century, when the fishing profession became recognised and fishing guilds were organised. At the end of the eighteenth century, the commercial and recreational fisheries were recognised as separate activities.

News of fishing activities and related legislation became more accessible in the nineteenth century with appearance of first the Croatian publication specialising in fishing and hunting '*Lovačko-ribarski viestnik'* (Fishing and hunting newsletter), issued monthly and first published in 1892 with the support of Hunting and Fishing Society. In the late nineteenth century, the need arose to protect freshwater fish stocks and the problem of overfishing was addressed (Drnić 1896). Hence, in 1894, the first season closures were put into effect by the government for 13 fish species, and minimum market sizes were set for 33 freshwater fish species (Table 1), together with a ban on selling fish under minimal sizes, ban of fishing in flooded areas, and prohibition of the use of explosives (Kesterčanek 1894). Fish stocking occurred regularly with common carp *Cyprinus carpio*, pike-perch *Sander lucioperca*, huchen *Hucho hucho*, and brown trout *Salmo trutta*, together with experimental stocking of

Table 1 Closed seasons and minimum sizes for freshwater fish listed in Regulation on the protection of fish in freshwater fisheries adopted in 1894 (Kesterčanek 1894)

Fish species	Season closure (from–to)	Minimum permitted length (cm)
European perch *Perca fluviatilis*	16 March–16 June	20
Pike-perch *Sander lucioperca*	1 April–31 May; 1 October–31 October	35
Volga pike-perch *Sander volgensis*	1 April–31 May; 1 October–31 October	25
Common carp *Cyprinus carpio*	1 April–31 May; 1 October–31 October	25
Zingel *Zingel zingel*	1 April–31 May; 1 October–31 October	20
Barbel *Barbus barbus*	1 April–31 May; 1 October–31 October	30
Tench *Tinca tinca*	16 May–16 July	20
Nase *Chondrostoma nasus*	16 March–30 April	20
Danube bream *Abramis brama*	16 May–30 June	25
Brown trout *Salmo trutta*	1 October–31 December	20 (stream); 30 (lake)
Huchen *Hucho hucho*	1 March–30 May	40
Grayling *Thymallus thymallus*	1 March–30 May	20
Sterlet *Acipenser ruthenus*	1 May–30 June	35
Ruffe *Gymnocephalus cernua*		10
Striped ruffe *Gymnocephalus schraester*		10
Streber *Zingel streber*		10
Burbot *Lota lota*		25
European catfish *Silurus glanis*		50
Crucian carp *Carassius carassius*		20
Danube barbel *Barbus balcanicus*		15
Rudd *Scardinius erytropthalmus*		20
Ide *Leuciscus idus*		20
Roach *Rutilus rutilus*		15
Cactus roach *Rutilus virgo*		20
Chub *Squalius cephalus*		20
Dace *Leuciscus leuciscus*		15
Nase *Vimba vimba*		20
Zope *Ballerus ballerus*		20
Danube bream *Ballerus sapa*		20
White bream *Blicca bjoerkna*		20
Asp *Leuciscus aspius*		35
Sichel *Pelecus cultratus*		25
Northern pike *Esox lucius*		35

North American catfish *Ameiurus* sp. and translocation of *C. carpio* (Kesterčanek 1907). The first Freshwater Fisheries Act was adopted in the Kingdoms of Croatia and Slavonia in 1906 and it was valid until 1937 when it was replaced by the Freshwater Fisheries Act of the Kingdom of Yugoslavia. Initially there was a strong resistance in accepting this law, and some regions avoided to abide by the new legal regulations (Kesterčanek 1907).

After World War II, in 1945, all waterbodies and fish became the state property of SFR Yugoslavia. Several changes were made to the act in the following decades, defining freshwater fisheries and fishing zones. The first modern Croatia Fisheries Act was adopted in 2001, long after the proclamation of Croatian independence in 1991 (Zobundžija 2005). Today, this act has been improved and aligned with EU Directives, and in 2019 Croatia adopted the new Freshwater Fisheries Act, which together with its associated Ordinances regulates all fishing activities (Official Gazette 2019a).

2.2 *Overview of Commercial Fishing*

Knowledge of freshwater fisheries in Croatia dates back to Roman times when fish (mostly *C. carpio*) was kept in ponds called *piscinae* (Balon 2004). The Romans did not value or practice inland fisheries, presuming such a practice as sacrilegious, and stocking was banned. Nevertheless, the indigenous people practiced fishing to secure their livelihoods (Ćurčić 1910). Archaeological remains and evidence from the fifteenth to sixteenth century, such as bronze hooks, drawings (church mural painted with fish strung on a hook), carved *C. carpio* on tombstone and crafted ornaments of fish scales, indicated the existence of the fishing in inland waters of the Black Sea Basin in the Balkan region (Ćurčić 1910), especially on the Danube, Drava, and Sava Rivers (Bojčić 1982). It was assumed that commercial fishery associations existed before the eighteenth century (likely from the sixteenth) though their renewal officially started between 1740 and 1750 in the form of fishing guilds (*ribarski ceh*). At the end of the nineteenth century (1870), the fishing guild was replaced by artisan cooperatives (*obrtne zadruge*) when rules on fishing trades were adopted (Bojčić 1982). This was followed by the Fish Protection Act of 1894 (Kesterčanek 1894) and the Freshwater Fishery Act of 1906, when fishing cooperatives were registered in the cadastre (Kesterčanek 1907).

The records describe that in the eighteenth and nineteenth centuries, the commercial fishery was a profitable occupation due to the great abundance of fishes. The most valuable fish from the Danube, Drava, Sava, and Una Rivers was beluga *Huso huso*. Also, other valuable fishes were northern pike *Esox lucius*, *C. carpio*, sterlet *Acipenser ruthenus*, European catfish *Silurus glanis*, *Salmo trutta,* and European eel *Anguilla anguilla* (Bojčić 1982). At the end of the nineteenth century, with the development of aquaculture, commercial fishery began to lose its importance given its poor organisation. There was no knowledge or statistical catch data collected and the amount of the commercial catch was not known. At the beginning of the twentieth century, the total annual catch was estimated to be between 2500 and 3500 tonnes on the Danube, Sava, and Drava Rivers (Bojčić 1982).

In the second half of the twentieth century, changes in water quality and degradation of large rivers in general affected the commercial fishery sector in Croatia (Piria et al. 2019a). Fishers were faced with mass fish kills, and consequently their product was of questionable quality. The abundance of valuable fish species for commercial fishing was considerably reduced (e.g., *A. ruthenus, Hucho hucho*) or extinct (*Huso huso,* Danube sturgeon *Acipenser gueldenstaedtii,* fringebarbel sturgeon *Acipenser nudiventris,* Atlantic sturgeon *Acipenser sturio*) (Mrakovčić et al. 2006). In this period, the most important commercial fish were Danube bream *Abramis brama*, *C. carpio*, and *S. glanis* and the grass carp *Ctenopharyngodon idella* as new species introduced in the 1960s (Treer and Kubatov 2017; Vehanen et al. 2020). Furthermore, more recently the total catch of non-native species almost prevails that of native fish (Vehanen et al. 2020).

2.3 Overview of Recreational Fishing

The formal organisation of recreational-fishing activities was established in 1881 under the name *Društvo za gajenje lova i ribolova* (Hunting and Fishing Society, hereinafter: Society) in Zagreb, and angling was described as an activity that should be encouraged (Drnić 1897). During this time, Croatia was a part of the Austria-Hungary Empire and the Society was the third one registered in mainland Europe (Feldhofer 2021). The first fishing association in Croatia was established in 1894, under the name of *Zagrebačko ribarsko društvo* (Zagreb Fishing Society) (Mateš 2004). By the Act that came into force 1906, all fishing associations were registered in the cadastre as holders of the fishing right (Kesterčanek 1907).

Several local fishing associations were established in Croatia during the early twentieth century and there was a need to gather them under one larger organisation. Hence, the *Zagrebački ribolovni klub* (Zagreb fishing club) summoned in 1935 a Founding Assembly of the *Savez ribarskih društava Savske banovine* (Savska Banovina Association of Fishing Societies) in Zagreb, which had 1100 members organised in 14 smaller local angling associations (Feldhofer 2021). Savska Banovina County was a part of the former Kingdom of Yugoslavia which included the central and western parts of modern-day Croatia. This association still exists today as *Hrvatski športsko-ribolovni savez* (*HŠRS*, Croatian Sports Fishing Association). During 1990, *HŠRS* reached its peak, with over 70,000 members organised in 257 local associations. This number of members was greatly reduced during the Croatian War of Independence (1991–1995) but has been steadily increasing since. In 2020, *HŠRS* had around 40,000 members organised in 468 local fishing associations (Feldhofer 2021).

Since the beginning of organised angling, it seems that *Salmo trutta* was the most valued angling species with successful attempts of farming and stocking (Flögel 1896). Recreational fishing relied on native fish until the introduction of rainbow

Table 2 Year of the first introduction and translocation of most important freshwater recreational-fishing species in Croatia (*translocated from Black Sea Basin to Adriatic Basin within Croatia)

Common name	Scientific name	Time of introduction/translocation
Grass carp	*Ctenopharyngodon idella*	1963
Bighead carp	*Hypophthalmichthys molitrix*	1963
Silver carp	*Hypophthalmichthys nobilis*	1963
Gibel carp	*Carassius gibelio*	1970s
Rainbow trout	*Oncorhynchus mykiss*	1883
Largemouth bass	*Micropterus salmoides*	1930s
Common carp*	*Cyprinus carpio*	1800s
European pike*	*Esox lucius*	1750s
Brown trout*	*Salmo trutta/labrax*	1890s
Pike-perch *	*Sander lucioperca*	1970s
European catfish*	*Silurus glanis*	1948
Tench*	*Tinca tinca*	1750s
Grayling*	*Thymallus thymallus*	1870s

trout *Oncorhynchus mykiss* in the late nineteenth century, and this species became the first introduced species that anglers recognised as a valuable angling species (Horvat 2005). Since the early twentieth century, 30 non-native fish species have been introduced for fishing and other purposes (Piria et al. 2018). Also, native fish species were translocated within Croatia to the areas where they did not occur naturally, mostly from the Black Sea Basin to the Adriatic Basin (Table 2) (Pofuk et al. 2017; Mihinjač et al. 2019). Through the twentieth century, sturgeon species that were regularly caught (e.g., *H. huso, A. gueldenstaedtii, A. nudiventris, A. sturio*) became regionally extinct (Mrakovčić et al. 2006) and their self-sustaining populations extirpated due to river fragmentation, mainly because of the Iron Gates I and II hydropower plants constructed on the Danube River in 1985 (Piria et al. 2019a). Also, the *H. hucho* population was significantly reduced because of the industrialisation, dam construction, and habitat degradation (Mrakovčić et al. 2006). These human-mediated actions resulted with a change in game fish species, and anglers needed to adapt to it together with the acceptance of new fishing equipment and techniques. In recent times, in the Danube and Sava River Basins, the most popular freshwater recreational-fishing techniques are aimed towards *C. carpio, E. lucius, S. lucioperca*, and *S. glanis* (Matulić et al. 2010; Treer 2022).

Croatian freshwater anglers are successful competitors, winning multiple medals in international competitions, with the most success as the national carp fishing team and national casting team. Croatia has hosted numerous international fishing competitions, as early as 1997 until the present day (Feldhofer 2021).

2.4 Development of Fishing Gears

Some fishing techniques and gears used in the Pannonian region of Croatia were strictly prohibited by the end of the nineteenth century. Those techniques were related to collecting *C. carpio* and weatherfish *Misgurnus fossilis* by hand during spawning in the flood period, using a primitive noose trap (*zamka*) made from horse hair for catching northern pike *Esox lucius* (Fig. 1) and using a steer (Fig. 2) for catching *C. carpio*, huchen *Hucho hucho*, *S. glanis*, and European eel *Anguilla anguilla*. However, the possibility of using such methods indicated that the area (Sava River catchment) was very rich in different fish species in the eighteenth and nineteenth centuries, though this consequently may have contributed to a depletion of the wild fish stock and the first attempts to setting up the fishery regulation (Kesterčanek 1894). Furthermore, throughout history, specific fishing gears such as hook and line '*klapac*' or clonk '*bučalo*' (Figs. 3 and 4) were developed in the Danube River Basin near the large rivers and its tributary, the Sava River (Ćurčić 1910), and some of those gears remain in use today (Official Gazette 2022b).

Angling was done with the *klapac* (Rollangel) (Fig. 3), a piece of ash wood with a piece of string tied to it with a hook at the end for catching *E. lucius* and *S. glanis*. Similar versions of this gear were used in Hungary and throughout Europe. A more efficient technique was using the *kanaf* (Fig. 4) or *strug* where several hooks were used moored on a longline string. Those gears were used for *S. glanis* fishing, using *M. fossilis*, Hirudinea, or different worms as bait, and for *C. carpio* fishing using corn as bait (Ćurčić 1910).

A clonk, specific gear carved from gourd or hollowed wood, and known as *bučalo*, *bućkalo* or *bučkaljka* was used to catch *S. glanis* (Fig. 5). This gear seems to be known in Europe from the eighth century BC when it was mentioned by Homer in *The Iliad* (Ćurčić 1910) and it still is today used legally in Croatian inland waters (Official Gazette 2022b).

Fishing nets were also developed through this time. A primitive ring net type called *Račilo* (Fig. 6; originally made for crayfishing) was used after the crayfish plague in 1880 (Maguire et al. 2016) to catch European perch *Perca fluviatilis*, tench

Fig. 1 Noose made of horsehair used to fish for northern pike *Esox lucius* (Ćurčić 1910)

Fig. 2 Nineteenth century harpoon used in the Sava River Basin (Ćurčić 1910)

Fig. 3 'Klapac' type of hook and line fishing gear (Ćurčić 1910)

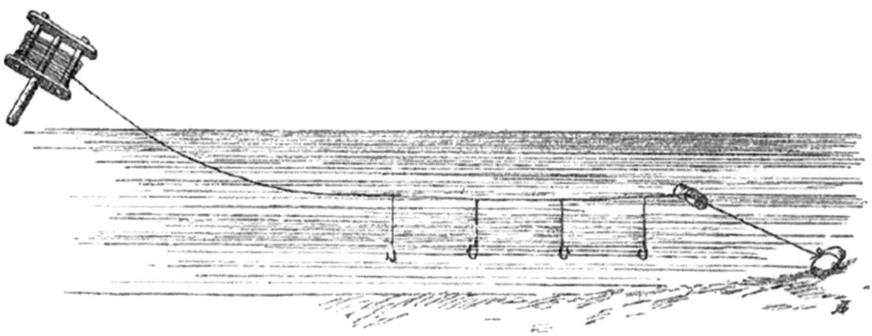

Fig. 4 'Kanaf' type of longline gear used in the Sava River around the town of Bosanska Gradiška (Ćurčić 1910)

Tinca tinca, crucian carp *Carassius carassius* hidden under the water lilies near the shore. Specific handheld lift net gears for catching *C. carpio* from a boat were developed and called a *čelinac* (Fig. 7) and *vlak virnjaš* (Fig. 8). A red rope was used to attract fish and by lifting the gear the fish remains in the net (Ćurčić 1910).

For pike and small pelagic fish, the handheld lift net *šajtar* was used from a boat (Figs. 9 and 10) and the trawl net *geca* was used for fishing during the winter to catch all bottom species (Fig. 11).

Fig. 5 'Bučkaljka' handmade clonk for summoning European catfish (photo by Mijo Vojković)

Fig. 6 'Račilo' type of ring net (Ćurčić 1910)

Fig. 7 'Čelinac' type of handheld lift net (Ćurčić 1910)

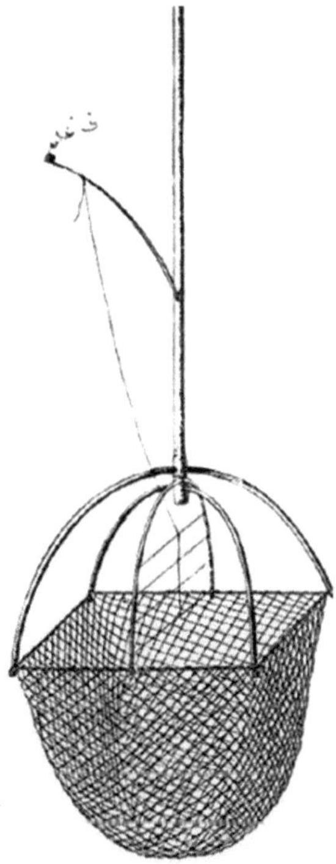

Fig. 8 ‘Vlak virnjaš’ type of handheld lift net (Ćurčić 1910)

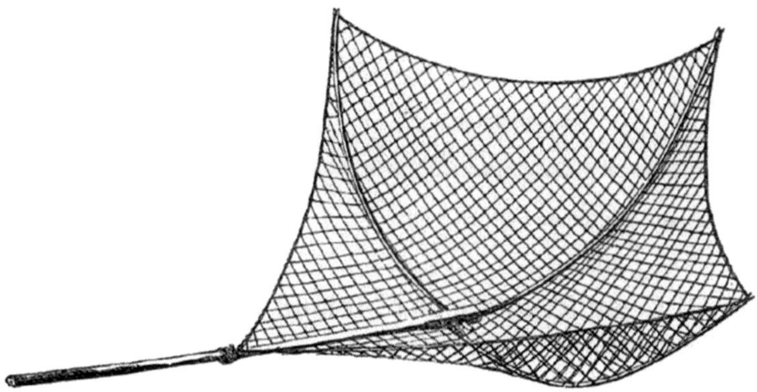

Fig. 9 Šajtar type of handheld lift net (Ćurčić 1910)

The *šajtar*, *geca*, and *bučkalo* are part of the prescribed fishing gears in Croatia that are permitted for use in the commercial fishery pursuant to the recent Freshwater Fisheries Act and its associated Ordinance (Official Gazette 2019a, 2022b)

Fig. 10 'Šajtar' handheld lift net in use (photo by Mijo Vojković)

Fig. 11 'Geca' trawl net (photo by Mijo Vojković)

3 Current Legal Regulations of Inland Commercial and Recreational Fishery

3.1 Legislation in Force

In the Croatian legislation, sport and recreational fishing describe different activities regarding marine fishing, though these terms are interchangeable when used in freshwater fishing. Since sport fishing mostly relates to fishing competitions, the term recreational fishing has a broader meaning (Potts et al. 2020) and in this paper is used as general term for both activities.

Performing recreational and commercial fishing in inland waters (freshwater fisheries) of Croatia is regulated by the Freshwater Fisheries Act (Official Gazette

2019a) and by-laws (ordinances): (i) Ordinance on the fishing and fish protection exam in freshwater fisheries (Official Gazette 2020a); (ii) Ordinance on the manner of drafting and implementing the management plan, revision, and supplement of the management plan and the programme for monitoring fish stocks in freshwater fisheries (Official Gazette 2020b); (iii) Ordinance on fishing licences in freshwater fisheries (Official Gazette 2020c); (iv) Ordinance on recreational (sport) fishing in freshwater fisheries (Official Gazette 2021); (v) Ordinance on boundaries and size of fishing zones for freshwater fisheries and restrictions and allowance of fishing (Official Gazette 2022a); and (vi) Ordinance on commercial freshwater fisheries (Official Gazette 2022b).

With a lack of a common EU freshwater fisheries policy, each European country may adopt its fisheries legislation in line with its not only geomorphological, biological, ecological, ichthyological but also socioeconomic circumstances and conditions. The new Freshwater Fisheries Act is aligned with Council Regulation (EC) No 1224/2009 of 20 November 2009 of a control system for ensuring compliance with the rules of the common fisheries policy and Regulation (EU) No 1379/2013 of the European Parliament and of the Council of 11 December 2013 on the common organisation of the markets in fishery and aquaculture products to secure freshwater fish and fish products market from commercial fisheries, and the traceability of fish and fish products (EC 2009; EU 2013).

Market regulation is of national importance to achieve effective control and surveillance over releasing fish and fish products on the market, furthermore, to decrease the black and grey market and to enable sufficient supply of freshwater fish and fish products to satisfy market demands, and most importantly, ensuring traceability for consumers and food/health issues. The Freshwater Fisheries Act also gives priority to the responsible and sustainable usage of natural freshwater fish stock resources, ecologically balanced, socioeconomically justified through measures for protection, conservation, and renewal of resources and freshwater ecosystems, which will contribute to the viability and development of local rural communities via fisheries tourism and the supply of healthy fish on the market along assurance of its traceability.

3.2 *Legal Background on Commercial Fisheries*

Commercial inland fishing is permitted only on the large lowland rivers, the Danube River, and lower parts of the Sava River. It is limited on the fishing grounds of the Danube River from the Croatian state border with the Republic Hungary downstream to the state border with Republic of Serbia and on the Sava River from Croatian state border with Republic Bosnia and Herzegovina downstream to the state border with the Republic of Serbia. The number of licences is limited and issued by the counties (local authorities) which have public authority according to the fisheries legislation (Official Gazette 2019a, 2022b). This is a form of recent decentralisation, since licences for commercial fishing were previously issued inclusively by the Ministry of Agriculture.

In total, 49 licences are allowed for seven fishing zones on the Sava River, and 50 licences for three fishing zones on the Danube River for commercial fishers or artisanal fisheries (Official Gazette 2019a, 2022b). Every licensee is required to have a licence for each selected zone; hence, the actual number of commercial fishers is smaller than the number of fishing zones. The current status is 11 licences for the Sava River and 25 for the Danube River (September 2022), after revision of all the licences. Some licensees have licences for several fishing zones (Josip Suić, pers. data).

Commercial inland fisheries in Croatia are permitted using the prescribed conventional fishing tools in all fishing zones and by the prescribed traditional fishing tools at only one fishing zone at the Sava River. The increased local interest to perform commercial fishing with traditional fishing gears was recognised and was recently included in the Ordinance for regulation of commercial fishery (Official Gazette 2022b). Traditional commercial fishing is recognised as important for the socioeconomic development of rural areas and the intention is for its use in tourism and educational purposes, and, also, for the maintaining and conservation of old gear-making crafts (Rastija 2018).

Commercial fishers or fisheries crafts are allowed to perform commercial fisheries if they: (i) hold the corresponding licence; (ii) are properly educated for this activity (possess the certificate of the commercial fisher exam); and (iii) possess the fishing tools and gear as listed on the licence that correspond to the purpose, type, and quantity of tools and gears (and properly marked) as per the provisions of the Freshwater Fisheries Act and its by-laws (Official Gazette 2022b).

Commercial fishers pay an annual fee to the county for the use of the quota on a particular fishing ground, and these funds are used to improve the state of local freshwater fisheries. Commercial fishers are also obliged to collect and submit catch data to the Ministry for certain fish species included in the quota system (Table 3) (Official Gazette 2022b).

3.3 Legal Background on Recreational Fisheries

In Croatia, there are no privately-owned waters (except some small artificial lakes on private grounds). State-owned running and standing waters are available for using in recreational fishing under the concession of the holder of the fishing rights. The fishing rights must be obtained to use and manage surface waters. These rights can be obtained through concession by the Ministry of Agriculture for a maximum 20-year period, by public tender (Vehanen et al. 2020)

Concessions are free of charge, i.e., no fee is paid to the state in advance to own a fishing right, but 40% of every fishing licence payment goes to the state budget for scientific monitoring of fish stocks. A further 40% is earmarked for the Croatian Anglers Association (HŠRS) for financing public affairs (angler education,

Table 3 List of species to be reported, and the catch quotas for commercially important fish species in the Sava and the Danube fishing zones according to the Freshwater fisheries Act from 2019 (*species reported until 2019)

Fish species	Catch quota (kg)	
	Sava	Danube
Common carp *Cyprinus carpio**	1500	9000
European catfish *Silurus glanis**	2800	22,000
Pike-perch *Sander lucioperca* *	1700	8000
Northern pike *Esox lucius**	1300	6000
Sterlet *Acipenser ruthenus**	100	500
Asp *Leuciscus aspius**	35,000	104,000
Danube bream *Abramis brama**		
White bream *Blicca bjoerkna*		
Ide *Leuciscus idus**		
Barbel *Barbus barbus*		
Common nase *Chondrostoma nasus*		
Cactus roach *Rutilus virgo*		
European eel *Anguilla anguilla*	–	–
Non-native species (Asian carps: *Ctenopharyngodon idella, Hypophthalmichthys molitrix, Hypophthalmichthys nobilis,* gibel carp *Carassius gibelio* and other)	No limit	No limit

improvement of recreational and sports fisheries, etc.), and the rest (20%) remains with the fishing right holder for financing of management plans, education, and informatisation. This is a small, symbolic fixed payment regulated by subordinate legislation, and there is a variable payment fee which covers all the real costs of enforcing the fishing right (e.g., restocking or bailiff service). The fishing rights holders are required to obtain a management plan, drafted by an authorised institution and confirmed by the Institute for Environment and Nature under the Ministry of Economy and Sustainable Development and Ministry of Agriculture (Official Gazette 2020b).

To conduct recreational fishing in a certain fishing area or fishing zone, anglers purchase licences from the fishing right holders in the fishing area or fishing zone for which the licensee has obtained a fishing right, as well as in the fishing area or fishing zone of another fishing licensee if there is a reciprocal fishing right agreement. There are 120 fishing right holders in Croatia, although there are approximately 450 active recreational-fishing associations, which are members of higher organisational units (county recreational-fishing federations or communities). Fishing licences are issued by the Ministry of Agriculture through the fishing right holders, who sell them with a validity period for one day, three days, and seven days or for current calendar year (annual licence), and an annual licence for Croatian Disabled War Veterans (HRVI). The daily licence is valid from 00.00 to 24.00 hours, the three-day licence from 00.00 hours of the first date to 24.00 hours of the last date and the seven-day licence from 00.00 hours of the first date to 24.00 hours of the last date. The annual permit is

Table 4 Fish species for which data collection is required (weight and numbers caught) by the holders of fishing rights for submission to the Ministry of Agriculture (*added in 2018)

Common name	Scientific name
Common carp	*Cyprinus carpio*
Grass carp	*Ctenophariyngodon idella*
Bighead carp	*Hypophthalmichthys nobilis*
Silver carp	*Hypophthalmichthys molitrix*
European catfish	*Silurus glanis*
Pike-perch	*Sander lucioperca*
Northern pike	*Esox lucius*
Tench	*Tinca tinca*
Common bream	*Abramis brama*
Ide	*Leuciscus idus*
Asp	*Leuciscus aspius*
Gibel carp	*Carassius gibelio*
Brown trout	*Salmo trutta /S. labrax/ S. farioides*
Rainbow trout	*Oncorhynchus mykiss*
Grayling	*Thymallus thymallus*
Huchen	*Hucho hucho*
Barbel*	*Barbus barbus*
Danube barbel*	*Barbus balcanicus*
Italian barbel*	*Barbus plebejus*
Sichel*	*Pelecus cultratus*
Cactus roach*	*Rutilus virgo*
Twaite shad*	*Alosa fallax*

valid from 00.00 on 1 January to 24.00 on 31 December of the year for which it was purchased (Official Gazette 2021).

The fishing right holder sells the fishing licence through an authorised application. The holder of the fishing right issues an annual permit to a Croatian or foreign citizen upon presentation of a certificate of passing the fishing exam. In addition to passing the fishing exam, the condition for purchasing an annual licence is membership in HŠRS (Croatian Sports Fishing Association), as well as membership of one of the holders of fishing right. The condition for purchasing a fishing licence for HRVI, in addition to the above, is the presentation of a certificate of status (Official Gazette 2021).

The holder of fishing right issues a daily, three-day, or seven-day licence to a Croatian or foreign citizen without presenting a certificate of passing the fishing exam and without the obligation to be a member of HŠRS or with the fishing right holder. In all types of licences, the condition for the start of recreational fishing is payment of all costs of enforcing the fishing right, which is determined by each holder of the fishing right pursuant to the provision of Article 44, paragraph 3 of the Freshwater Fisheries Act (Official Gazette 2019a). Recreational fishers are also obliged to collect and submit to the Ministry the catch data for certain fish species that are prescribed in the Ordinance (Official Gazette 2005) (Table 4).

Recreational fishing may be carried out with a maximum of three fishing rods and/or accessories with one hook each, and exceptionally an artificial lure which may have more than one hook, and a maximum of three. Recreational fishing of salmonids, *S. trutta*, *H. hucho*, grayling *Thymallus thymallus* may be carried out with only one fishing rod and one artificial fly or lure, and it is forbidden to fish in habitats of these species using natural live or dead bait (Official Gazette 2021).

When fishing for *S. glanis* from a vessel, one fishing rod, fishing rod or fishing line with one hook may be used. Fishing gear and equipment used for fishing must be under the constant and direct supervision of the fisher. An artificial lure may have up to three hooks (single-hook, double-hook, or triple-hook) with or without a counter hook. Artificial bait—the fly may have one hook with or without a counter hook. The size of the lure when fishing for *H. hucho* must not be less than 10 cm. It is forbidden to use a non-native species of fish (live or dead) as bait (Official Gazette 2021). In case of taking and retaining the catch from the fishing water, the angler is required to enter and report it through the mRibič application, which can be downloaded free from GooglePlay or AppleStore (Ministry of Agriculture 2022).

3.4 *Digitalisation of the Data Collection System*

The requirement to collect data on commercial and recreational catches started in 2004 when the Ministry of Agriculture prescribed fish logbooks for recreational and commercial fishers (Vehanen et al. 2020). However, with the adoption of the new Freshwater Fisheries Act in 2019, catch data are collected electronically (Official Gazette 2019a).

Data collection digitalisation is expected to improve and ease data delivery and communication and should also assure transparency of the purposeful use of funds and inhibit financial malversations. By introducing flexible financing, the holders of the fishing right will easily fulfil the financial obligations according to each management plan (restocking, bailiffs and other). The financial value of the licence is divided into two: one smaller, fixed part (for the licence itself) covering the financing of fisheries monitoring and the public affairs of Croatian Anglers Association, including the 5-year accumulation of financial funds for financing the management plan revision, and the larger, variable part created according to real holders of the fishing right needs covering restocking, bailiff service, etc. (Vehanen et al. 2020).

The interactive electronic database was created and the application for purchasing of electronic licences, cashless transactions, and allocation and distribution of financial funds was designed. The two main electronic applications were launched, one for collection of anglers' catch data (mRibič), and the other for collection of commercial fishers catch data (mAlas), both in mobile and web forms (Ministry of Agriculture 2022). These applications are of great importance because the holders of the fishing right receive instant data in real time (for planning restocking, bailiff activities, etc). In commercial fisheries, the Ministry receives catch data in order to monitor catch quotas and to use data for various reports. In addition, commercial fishers collect data about fishing effort (fishing tools, gear, motor and vessel usage, duration of fishing activities, landing sites and landing announcements, etc.). Also, those applications are intended to be of importance for the future Ordinance on fish trade, in terms of traceability and food safety. Finally, these applications may contribute to recording the prevalence and distribution of non-native fish species that may occur in both commercial and recreational fishing.

3.5 Fish Monitoring and Stock Management Plans

The programme for monitoring freshwater fisheries status has been conducted by the Ministry of Agriculture, Fisheries Directorate since 2003. The results obtained through the monitoring programme are used to implement the management plans for the holders of the fishing rights, which represents the scientific basis for the management of fish stocks. Fishing rights holders could improve their management plans and improve restocking, fishing regime, and quotas. Also, the results are used by the Ministry to create useful and rational subordinate legislation at the national level. The monitoring programme is financed from anglers fishing licences fees—40 per cent of every fishing licence fee (Official Gazette 2020b; Vehanen et al. 2020).

The legal standards listed in the Freshwater Fisheries Act (Official Gazette 2019a) concerning fish stock management plans (Official Gazette 2020b) are aligned with the nature protection (Official Gazette 2013a) and water management legislation (Official Gazette 2019c), in the process of bringing the acceptability evaluation for the ecological network and alignment with water management plans.

The fishing right holder is obliged to obtain a management plan no later than one year after signing the contract with the Ministry. Management plans are prepared by official scientific institutions. These studies contain all the relevant elements for conducting the fishing right (estimation on fish stocks of the fishing zone, maximum allowed catch quotas, restocking, bailiff service, conservation measures for spawning/migration grounds, etc.), and studies are verified by a special Ministry commission. Each fishing right holder may set up its own internal provisions (maximum allowable daily catch in kilograms or pieces, night fishing, etc.), which are stipulated in the valid management plan. To purchase permits as well as for any additional information, fishers must contact the holders of fishing right. Every six years, the management plan must be revised. Even before that period, yearly amendments of some elements are possible, in case of urgent need for protection of fish stocks (Official Gazette 2019a, 2020b).

The main goal during the preparation of Freshwater Fisheries Act (Official Gazette 2019a) was to transform the holders of the fishing right from 'licence sellers' into real managers responsible for fish stocks, but also capable of improving recreational fisheries and overall tourism facilities. The fishing right holders also received a new possibility of introducing certain closed waterbodies into 'pay lakes' or 'fee fishing lakes', to expand the recreational-fishing facilities and to enable recreational fishers to buy all the surplus catch they want. This opens a new aspect on some voluntary fisheries locations and widens the facilities from strict conservational 'catch and release' to 'pay lakes' (Vehanen et al. 2020).

There is an important novelty regulating the issue of the area validation of fishing licences, which are now strictly connected to the fishing zones assigned to each of 120 holders of the fishing right. The possibility that the minister of agriculture could regulate fisheries licences for a particular fishing water, several nearby fishing zones, or even big fishing areas was introduced. This measure could improve the management of those connected zones and ease the purchasing and control of the licences. Also, fishing rights holder may conclude reciprocity agreements amongst themselves.

One novelty deals with all the new fishing waters that appear after construction works, digging/exploitation of the gravel, clay, or any hydro-management works, accumulations, channels, etc. If a fishing rights holder wants to manage this new waterbody that appeared in the administrative territory for which he holds the fishing right, there is no need to conduct a public tender to obtain those fishing rights, only a simple administrative procedure is required. Afterwards a management plan is prepared for this waterbody. The main goal is to put those waters into service, enrich such waters in the ecological, biological, and ichthyological sense, and put them into the function of fisheries, to improve fisheries tourism and the development of local rural communities.

4 Inland Water Fish Resources

4.1 Commercial Fishery Catch and Data Analysis

The fish caught in commercial inland fisheries, and fish from aquaculture, is the only freshwater fish of Croatian origin legally permitted to be placed on the market. Unlike some other Croatian economic activities, commercial inland fisheries have low economic value and an insignificant impact on the economy (Mitchell et al. 2010). Competition for space and minor conflicts with recreational fisheries seem to further diminish the potential for maintaining and developing a viable commercial inland fishery in the future, although there is a demand for fresh fish from inland waters (Vehanen et al. 2020).

The Croatian legislation defines the catch quotas for each commercially important species of fish (Table 3) (Official Gazette 2022b), as well as their minimum catch sizes and season closures during the spawning period (Official Gazette 2005; see Chap. 5). Commercial fishers are obliged to report catches of asp *Leuciscus aspius, A. brama,* white bream *Blicca bjoerkna,* ide *Leuciscus idus,* barbel *Barbus barbus,* common nase *Chondrostoma nasus, A. anguilla,* and other native fish species that inhabit Croatian inland waters because of the obligation prescribed by European Commission (Vehanen et al. 2020). Also, the catch of non native fish species is not limited, but should also be reported because of the demand for information on their distribution and incidence.

The data of fish catch, using fish catch logbooks, and given licences in commercial inland fishery have been collected since 2004. The nine fish species were obliged to report before the new regulation entered into force (Vehanen et al. 2020; Official Gazette 2022b), but only data from 2004–2020 are available and comparable (Table 5). The average annual freshwater commercial fish catch is 10,000 kg from the Sava River and 50,000 kg from the Danube River. The most important fish species in the both the Sava and Danube Rivers are *A. brama*, *C. carpio*, *S. glanis*, other native and non-native fish species (Table 5 and Supplementary materials S1 and S2). The number of given licences in the Sava River varied between 6 and 18, and in the Danube River between 19 and 33. Although the number of licences increased in the Sava River, the total overall catch in the last few years has a declining trend, which was also noted in catch data for the Danube River (Fig. 12).

One reason for the declining trend of fish catch in the Sava River could be dam construction in 2018 in its upstream region (Piria et al. 2019a) with the consequence of several changes in water level daily (I. Špelić, pers. communication) that possibly had an impact on recruiting trends of important commercial fishes (i.e., *A. brama, C. carpio*). The second reason could be that some commercial fishers report their catch only partly or not at all. This is visible in the large oscillation in the catch data for some species in certain years, for example the variability of the catch biomass of *C. carpio* on the Sava River (Table 3), and was confirmed by recent monitoring data analysis (Piria et al. 2019b). Despite this declining trend of fish catch in the commercial inland fishery, the number of fishing zones and fishing licences issued has been considerably increased with new regulation that entered into force in 2022 (from the maximum 20 licences according to the old regulation, to 50). However, the total fish catch quota was not increased and remains at about 42 t for the Sava and 150 t for the Danube (see Official Gazette 2022b; Vehanen et al. 2020).

Abramis brama and *C. carpio* are the most important species for commercial inland fishers in this region but the catch of both species has a decreasing trend. Tench *Tinca tinca* was not recorded or was barely recorded in the commercial catch, thus this species was excluded from obligatory reporting commercial fishery catch data by the new Ordinance from 2022 (Official Gazette 2022b). *Tinca tinca* is a species characteristic for backwaters where commercial inland fisheries cannot be performed. An increased catch of *A. ruthenus* was recorded in 2018. Although *A. ruthenus* is the last known sturgeon species in the Black Sea Basin, it is very rare and it is not clear if the remaining population is the result of stocking, or refugees from aquaculture or natural reproduction; still the fishing demand in Croatia is considerable. Recent research suggests that the population of *A. ruthenus* is ageing in the Black Sea Basin due to a lower number of specimens caught each year and the increasing average weight of individuals (Kubala et al. 2021). Attempts to impose a fishing moratorium on sturgeons including *A. ruthenus* in Croatia based on the EU Strategy for the Danube Region (Bloesch et al. 2006) resulted in a huge resistance of commercial inland fishers, thus a compromise solution was to temporarily lower the catch quota.

The considerably high overall kept catch of non-native species is recorded, nearly 20% in the overall catch (Table 5, Supplementary materials S1 and S2). Such a result

Table 5 Amount (kg) and share of each species in the overall kept catch in commercial fishing in Croatia (2004–2020) in the Danube and Sava fishing zones

River	Year	Common carp *Cyprinus carpio*	European catfish *Silurus glanis*	Pike-perch *Sander lucioperca*	*Esox lucius*	Danube bream *Abramis brama*	Sterlet *Acipenser ruthenus*	Tench *Tinca tinca*	Ide *Leuciscus idus*	Asp *Leuciscus aspius*	Other native	Non-native
Sava	2004	24	472.5	344	168	3522.5	4	0	213	358	1564	26
Danube		1430	3612	1095.5	481.5	11,307	95	24.3	528.5	370.5	9149	10,726.5
Sava	2005	106.1	333.6	182	303.1	2877.1	0	3.7	441	230	1868	252.7
Danube		2480.9	2859.2	1283.55	1124.7	12,623.5	38	8.2	2077.5	627.2	11,979	9551.2
Sava	2006	53	646.95	380.6	377.37	2107.86	15.55	0.5	655.5	142.5	970.1	292.5
Danube		1461.5	2654.65	1299.5	2239	5320.5	191.5	47	497	218.5	5219.5	8403.5
Sava	2007	41.2	560.6	214.9	212.4	2296.5	0	0	118	211.5	2097.5	360.5
Danube		3676	3078.5	1452	1831.3	7281.9	270.9	14	252	217.3	6728.7	8193.4
Sava	2008	79	491.7	178.5	125	2223.9	0	0	136.9	15	2282.35	6272.35
Danube		2544.8	3263	1357	995.8	8468.5	121		693.5	628.8	5920.5	16,910
Sava	2009	27	211.4	125	56.5	1725	5	0	52	45	1579	190
Danube		4092.6	5272.4	1839.1	1445.6	9729	153	0	1064	927	7374	9244.5
Sava	2010	349.2	459.75	220.6	168	4312.4	0	0	295	170	2214	30
Danube		3118.5	4353.3	2388.5	2167.3	9101.5	61.5	1	1094.2	1483.5	7949.5	8206
Sava	2011	314.5	398	219.7	72.5	877	14	0	248	41.5	1174	473
Danube		3093.5	4353.3	2388.5	2167.3	9101.5	61.5	1	1094.2	1483.5	7949.5	8206
Sava	2012	998	654.3	196.3	128.5	4212.5	1	0	387	44	2486	754
Danube		3184.9	3320.6	884.25	783.5	5187.7	3.4	0	430	501	6189.7	4587.4
Sava	2013	472	432	369.5	290	3294	48	0	210	317	4573	529
Danube		2163.88	6783.45	1822.25	2532.68	10,298	279.9	0	584.7	520.9	10,512.4	6330.1
Sava	2014	632	209	310.2	672.3	2796.5	11	2	330	467	4540.5	1047
Danube		5314.9	7257.45	2124.4	2670.3	8141.05	95.07	0	416.7	370	7337.7	7598.5

(continued)

Table 5 (continued)

River	Year	Common carp *Cyprinus carpio*	European catfish *Silurus glanis*	Pike-perch *Sander lucioperca*	*Esox lucius*	Danube bream *Abramis brama*	Sterlet *Acipenser ruthenus*	Tench *Tinca tinca*	Ide *Leuciscus idus*	Asp *Leuciscus aspius*	Other native	Non-native
Sava	2015	275.5	424	454	623.27	2138.5	168.5	0	359.5	281.5	3351	593
Danube		4515.96	9806.14	4161.3	2249.42	9677.3	273.1	0	1197.8	706.15	6914.3	10,637.6
Sava	2016	412	436.5	204	241	1639.5	110.5	1	527.5	285.5	4298.5	236
Danube		8338.65	13,135.92	4343.52	2084.63	9657	26.16	0	686.7	512.8	8018.82	9032.2
Sava	2017	199.6	237.9	119	374	1153	4	0	303	127	2716	175
Danube		9579.62	10,936.98	4962.22	2036.73	13,533.5	41.1	0	472.6	717	7015.25	8986.3
Sava	2018	554	389	187	206	1033	13	0	122	199	1444.5	189.2
Danube		8053.6	10,340.03	3656.13	1455.7	9081	547	0	59	313	5684.17	5399.8
Sava	2019	112	236.9	132	414.4	1008.5	17.8	0	304.5	269	1787	160
Danube		5429	11,566.1	3680.3	1354.7	7707.1	275.7	0	105	308.8	3849.1	7946.5
Sava	2020	48	121.8	98.5	48.2	536.5	16.5	0	123.5	316	916.5	34
Danube		6217.5	9348.8	3121.4	1520.5	8332	189.4	0	284	168	6172.5	9245.5
Total Sava		4697.1	6715.9	3935.8	4480.54	37,754.26	428.85	7.2	4826.4	3519.5	39,861.95	11,614.25
Total Danube		74,705.81	111,941.82	41,859.42	29,140.66	154,548.1	2724.23	95.5	11,537.4	10,073.95	123,963.64	149,205
Total kept catch in kg		79,402.91	118,657.72	45,795.22	33,621.2	192,302.3	3153.08	102.7	16,363.8	13,593.45	163,825.59	160,819.25
% of overall catch		9.59	14.34	5.53	4.06	23.24	0.38	0.01	1.98	1.64	19.79	19.43

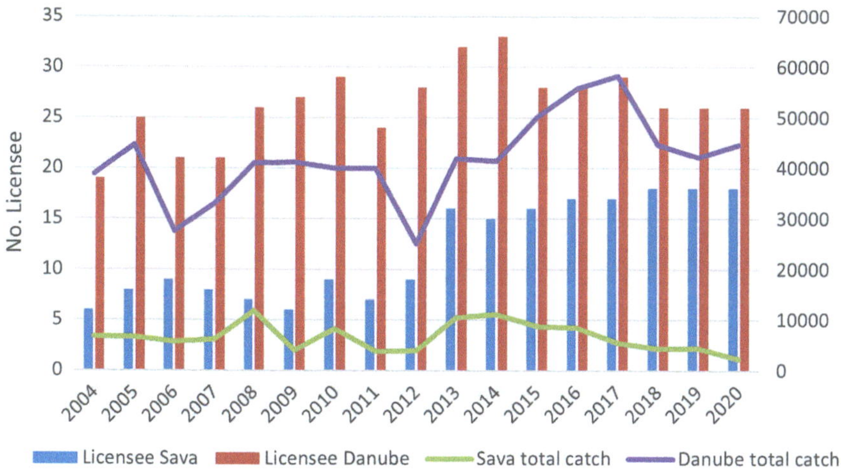

Fig. 12 Fluctuation of number of licensed commercial fishers *vs.* total annual kept catch in the Sava and Danube fishing zones for the period 2004–2020

implies that the inland waters are inhabited by a greater biomass of non-native species that can grow to larger body sizes. Although the non-native species are not recorded by collected data, these species are mostly represented by *C. idella* and *C. gibelio* (Smederevac-Lalić et al. 2011) together with the occasional catch of non-native fish species that are refugees from nearby aquaculture facilities (Safner et al. 2013; Piria et al. 2022).

4.2 Recreational Fishery Catch and Data Analysis

Systematic data on the number of recreational anglers and catch of most important fish species have been collected by the Ministry of Agriculture, Fisheries Directorate, since 2004 and includes only fish kept by the angler (Vehanen et al. 2020). The catch data include the most attractive recreational-fishing species and several endangered species for which catch rates need to be reported to the European Commission. The list includes 16 species for which data are available from 2004–2020 and an additional six species included from 2018–2020 (Table 4).

The catch data were collected by logbooks until 2020 and electronically just for 2021, and since the e-platform is still in the test phase, the collected data cannot be compared with previous years. Over 17 years of reporting, the average number of registered anglers was around 38,000, just over half the historic maximum in 1990. The minimal number of registered anglers was recorded in 2006 and maximal in 2020 (Figs. 13 and 14). The average annual catch by anglers fluctuates around 570,000 kilograms over these years. However, there is a visible decrease, especially from 2016 onward in the kept catch, despite the slight increase in the number of

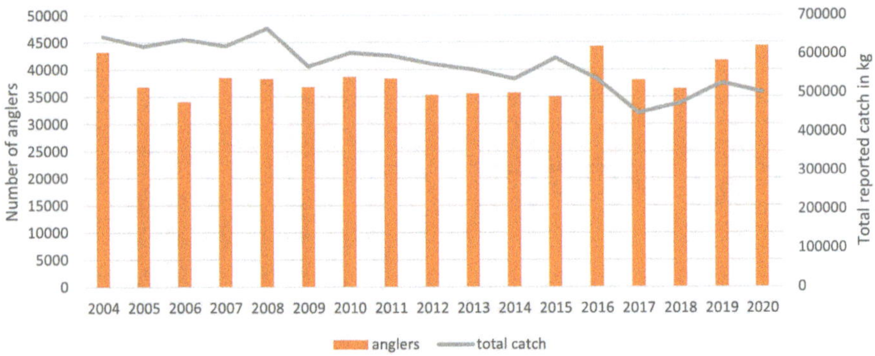

Fig. 13 Fluctuation of the number of anglers *vs.* total annual kept catch in the period 2004–2020 at the national level

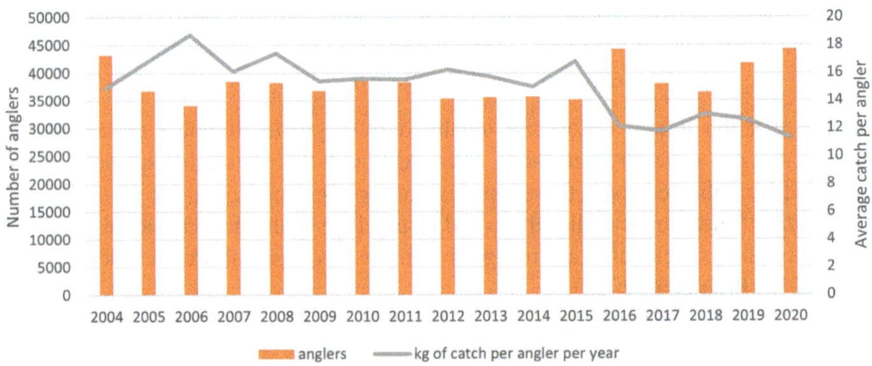

Fig. 14 Fluctuation of the number of anglers *vs.* average annual kept catch per angler in the period 2004–2020 for the Danube and Sava Rivers

registered anglers (Fig. 13). As expected, the average catch per angler is high when the number of anglers is low and vice versa, but the number is also overall lower from 2016 onward (Fig. 14).

The reason for this decrease in kept catch is not clear and may be a combination of several factors including lower wild stock recruiting in previous years, tighter restrictions of fishing right holders, or an increased number of catch and release style anglers. Restrictions on the number of permitted to be kept fish is the decision of every holder of fishing right and they are set on the base of expert opinion and production of the management plan by official scientific institutions (Vehanen et al. 2020). There is no central database for this type of data and there are several institutions and hundreds of fishing right holders involved. Catch and release-

Table 6 Amount (kg) and share of each species in the overall kept catch in freshwater recreational fishing in Croatia (2004–2020)

Species	Kept catch in kg	% of overall catch
Common carp *Cyprinus carpio*	2,914,989.90	30.05
Gibel carp *Carassius gibelio*	1,689,310.94	17.41
Grass carp *Ctenophariyngodon idella*	933,699.37	9.62
Northern pike *Esox lucius*	805,165.38	8.30
Common bream *Abramis brama*	790,375.43	8.15
European catfish *Silurus glanis*	600,194.20	6.19
Pike-perch *Sander lucioperca*	322,156.84	3.32
Asp *Leuciscus aspius*	166,137.63	1.71
Ide *Leuciscus idus*	146,966.23	1.51
Tench *Tinca tinca*	62,370.82	0.64
Brown trout *Salmo trutta*	52,271.14	0.54
Rainbow trout *Oncorhynchus mykiss*	49,703.12	0.51
Bighead carp *Hypophthalmichthys nobilis*	35,730.33	0.37
Silver carp *Hypophthalmichthys molitrix*	15,353.01	0.16
Grayling *Thymallus thymallus*	5474.34	0.06
Huchen *Hucho hucho*	540.00	0.01
Other species	1,110,628.30	11.45
Overall	**9,701,066.97**	**100.00**

based angling is gaining popularity worldwide (Arlinghaus et al. 2007) including Croatia, with an increased number of maximum size limit lakes or even exclusively catch and release fishing venues. Such an approach allows fishing right holders to develop very attractive fishing destinations with trophy fish, mainly *C. carpio*, *E. lucius*, *S. lucioperca*, and *S. glanis* (Jelkić et al. 2018), diversifying Croatian freshwater fishing tourism opportunities. Designated fishing forums, social network groups, and angler meetings imply that this approach is gaining popularity amongst new generations of anglers in Croatia.

Cyprinus carpio is the most popular recreational-fishing species in Croatia, making up almost a third of the overall kept catch in the period of 17 years, followed by *C. gibelio* and *C. idella* (Table 6, Supplementary material S3). Because of the high demand and limited natural recruitment, *C. carpio* is also the most stocked species. It is farmed in extensive aquaculture systems and sold to fishing right holders for yearly restocking (Eurofish 2015). There is a high number of specialist carp anglers and many carp fishing competitions across Croatia.

Carassius gibelio and *C. idella* are non-native species with a substantial share in the overall catch. *Carassius gibelio* is considered invasive and it is not legal to introduce it or release it after catch, but it still has a very wide distribution in Croatian waters (Tomljanović et al. 2012). Asian carps (*C. idella*, *H. molitrix*, and *H. nobilis*) are usually introduced for biocontrol purposes of macrophytes and plankton, and *C. idella* is a desirable angling species because of its fast growth and large size (Mihinjač et al. 2019). Stocking with Asian carps is only allowed in closed water systems upon approval of the Ministry of Economy and Sustainable Development, by the Act on the Prevention and Management of the Introduction and Spread of Alien and Invasive Alien Species (Official Gazette 2018, 2019b), since there is evidence of successful spawning in the wild in some European rivers (Mihinjač et al. 2019).

Other important angling species, such as *E. lucius* and *S. lucioperca,* are farmed in polyculture with *C. carpio,* but are not as often stocked because of their limited presence on the market and consequently higher price, thus their catch rates are not as high as for *C. carpio*. Both species are very desirable as a trophy (big specimens) and for human consumption.

Abramis brama has a significant share in overall catch, despite not being stocked, because of its abundant populations in the largest Croatian rivers: the Sava, Drava, and Danube (Treer et al. 2003; Jelić et al. 2012; Simonović et al. 2015).

Silurus glanis often raises negative connotations for most anglers because it is a large top predator (Vejřík et al. 2017). This species is not usually stocked and its selectively removal from lakes is permitted by the Regulation on recreational fishing in freshwater fisheries if their abundance becomes threat to stability of the ecosystem (Official Gazette 2021; Getto 2013). Nevertheless, there are specialist anglers that fish exclusively for *S. glanis* and for them it remains a desirable catch.

Other species are represented by lower catch rates, but it is noteworthy that, although not reaching high catch numbers, salmonids are presenting promising fishing tourism opportunities as fly fishing is gaining popularity in some regions such as Lika and Gorski Kotar (Vehanen et al. 2020).

4.3 Sustainability of the Most Important Commercial and Recreational Fish Species

Cyprinus carpio, *E. lucius*, *A. brama*, *S. glanis*, and *S. lucioperca* are recognised as the most important native freshwater fish species used for recreational and commercial fishing in Croatia, due to their high catch rates over the years. However, data collected are not absolutely reliable because the species may be misidentified by fishers and a considerable amount of the catch was not reported (catch and release fishing, poaching).

Since the overall catch is not correlated with the number of anglers or the number of commercial licences between the years, trends in fish population fluctuations across the analysed 17-year period are mostly informative. Commercial catch data are based solely on the fish caught in Croatian part of the Danube River and lower Sava River in Croatia so it can be used as a supplementary data for these areas, since the catch is on a much lower scale when compared to recreational fisheries.

Cyprinus carpio has, by far, the greatest importance in recreational fishing and it seems that *C. carpio* catch remains stable for years after initial decrease in 2005 (Fig. 15). An obvious increase in the *C. carpio* catch of commercial fishers from

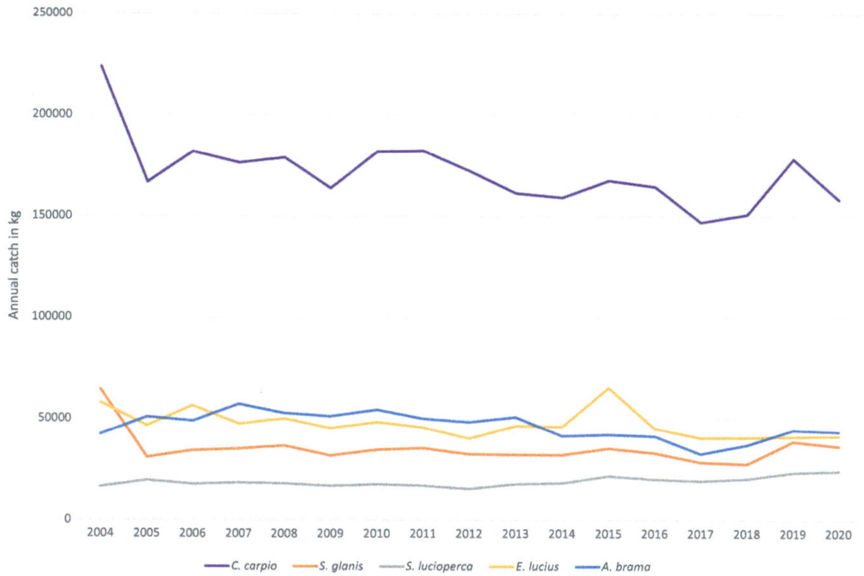

Fig. 15 Total annual kept catch of the most important species in recreational fishing in the period 2004–2020

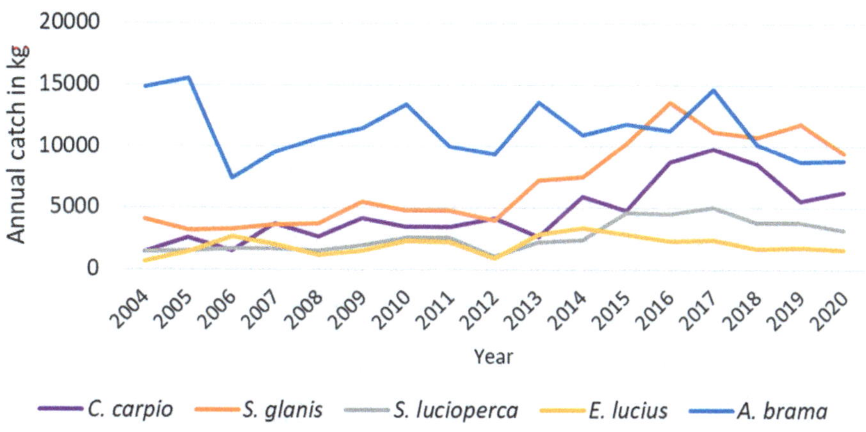

Fig. 16 Total annual catch of the most important species in commercial fishing in the period from 2004–2020 for the Sava and Danube Rivers

2013 to 2017 was noted, which could be a result of its successful reproduction in the Sava and Danube Rivers during higher-than-usual water levels in 2010 and 2014, replenishing the natural stocks of this rivers (Figs. 16 and 17). The Sava River water level, at the most important fish spawning area, Lonjsko Polje, was used as a proxy for yearly mean water levels (Fig. 17). This increase in the catch is not noted in recreational-fishing catch probably because the majority of *C. carpio* fishing is

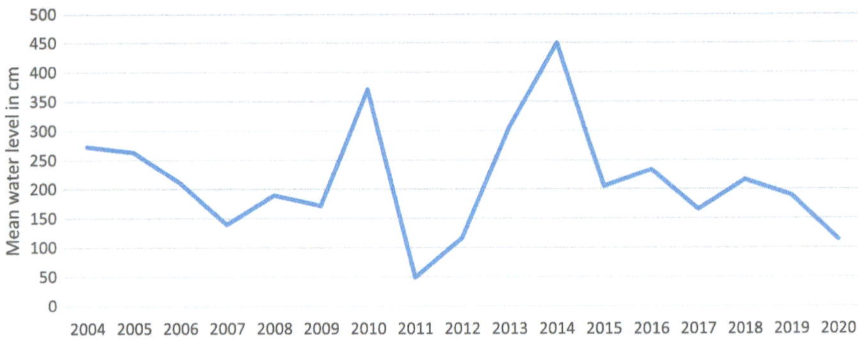

Fig. 17 Fluctuation of mean water levels of the Sava River at the Lonjsko Polje wetland in the period 2004–2020

practiced on stocked lakes and is not depending on natural recruitment. The lakes are usually stocked with substantial amount of farmed *C. carpio* enabling continuous high numbers of kept catch every year. In the last couple of decades, with catch and release of big specimens (trophy fish) practiced on several lakes, Croatia has become one of the most attractive carp fishing countries, with specimens reaching sizes over 40 kg consequently attracting fishing tourists from all around Europe (Slavinić 2021). Motivated by this practice with *C. carpio*, there are similar attempts to develop similar fishing venues with other attractive fish species such as *E. lucius*, *S. lucioperca*, or *S. glanis*. The wild form of *C. carpio* is endangered in Croatia and critically endangered in Europe because of hybridisation with farmed forms, reduction spawning habitat, and competition with non-native species, especially by *C. gibelio* (Mrakovčić et al. 2006).

Esox lucius is the most common predatory species in the kept catch of Croatian anglers. The annual catch usually is between 40 and 50 tons with two peaks: one smaller in 2006 and a substantial one in 2015 (Fig. 15). This higher catch rate may be related to the higher precipitation and the higher mean water level in 2004 and 2014 (Fig. 17) which possibly extended recruitment possibilities. High water levels benefit spawning of phytophile fish species that are abundant in Croatian waters (Mihaljević et al. 2011) and consequently produce high amount of available prey which may sustain abundant population of *E. lucius*, causing its high catch rate in the following year. The commercial catch of *E. lucius* in the Sava and Danube Rivers is the lowest amongst the important predatory species, likely because *E. lucius* is usually not well adapted to the environmental conditions in large rivers (deep, often turbid water, sparse vegetation) when compared to the other two apex predators, *S. lucioperca* and *S. glanis* (Greenberg et al. 1995; Ranaaker et al. 2014; Cucherousset et al. 2018). Also, the aquaculture production and stocking of *E. lucius* in Croatia is low and most of the catch relies on the natural recruitment in the inland waters (Habeković and Pažur 1998). Hence, it is important to protect natural spawning grounds in its native area to sustain or improve population of *E. lucius* in Croatia. On the other hand, *E. lucius* was translocated to non-native areas in Croatia where it is well adapted, it

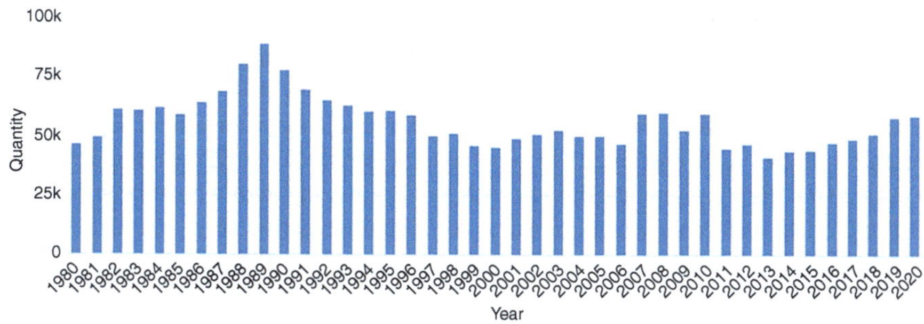

Fig. 18 Global fluctuation of the annual *A. brama* catch in the period 1980–2020 (taken from FAO 2012)

has a negative impact on native and endemic fauna and is still not managed properly (Pofuk et al. 2017).

Abramis brama is the most abundant species caught by commercial inland fishers in Croatia (Fig. 16), but it also has high catch rates in recreational fishing. There is no farming of *A. brama* in Croatia, hence all reported catch originates from wild populations (Treer et al. 2003). However, the annual recreational and commercial catch rates of *A. brama* fluctuate, with a visible decrease in recent years (Figs. 15 and 16). Similar fluctuations in capture are also visible on the global scale (Fig. 18) (FAO 2022a, 2022b). Nevertheless, the species is listed as least concern by IUCN (Kottelat and Freyhof 2007) and there is no indication of a possible decline in the future.

As one of the largest freshwater fish species in the world, *S. glanis* is exceptionally attractive for specialised anglers (Cucherousset et al. 2018). On the contrary, some anglers perceive it to be a pest species even in its native range. This species thrives in both lotic and lentic warm water conditions, it is an adaptive generalist (Copp et al. 2009) and there is no need to restock it in most of habitats. However, a substantial decrease in the annual catch of *S. glanis* by anglers was observed from 2004 to 2005, from over 60 tonnes in 2004 to 30 tonnes in 2005, followed by stable catch rates between 30 and 40 tonnes (Fig. 15). The initial drop in presented data is not clear, and there is no data about *S. glanis* recruitment or annual biomass production for comparison. On the contrary, there is no such decrease in the commercial catch of *S. glanis* on the Sava and Danube Rivers. The commercial catch of *S. glanis* has considerably increased since 2012: from around 5 t before 2012 to over 10 t in 2016, with a slight decrease in recent years (Fig. 16). Since there is virtually no restocking of *S. glanis* in Croatia, this data showed that the natural recruitment of this species is probably sufficient to sustain both recreational and commercial catch. *Silurus glanis* has been translocated to the Adriatic Basin of Croatia where it has already established populations and as top predator may have considerable impact on vulnerable Mediterranean freshwater ecosystems (Pofuk et al. 2017).

Sander lucioperca is one of the most popular freshwater food fish because of its tasty, low fat meat (Tönißen et al. 2022). Because of the high demand, it reaches high prices on the market (FAO 2022a, 2022b). It was historically under high fishing pressure in continental Croatia as the most desirable angling species. *Sander lucioperca* is farmed in polyculture with *C. carpio* and it is used as species for stocking, mainly lakes. Artificial nests with fertilised eggs are often deployed in desired habitats as a cost-effective replacement for stocking (Zakęś and Demska-Zakęś 2010). Recreational catch rates of *S. lucioperca* in Croatia have been increasing in the last five years when compared to the earlier period (Fig. 15). One reason for this increase may be related to the appearance of the Ponto-Caspian gobies in Croatian rivers (Delić et al. 2014; Piria et al. 2011) which becomes considerably abundant (Jakovlić et al. 2015; Piria et al. 2016). It is noted that invasive Ponto-Caspian gobies can become the main prey for *S. lucioperca* (Oesterwind et al. 2017) and they are observed as abundant in *S. lucioperca* stomachs in the Sava River (I. Špelić, pers. comm.). This new source of food may possibly sustain a larger populations of *S. lucioperca* in Croatian rivers. This observation can be supported by the data of the commercial fishery catch of *S. lucioperca* that have increased since 2010 (Fig. 16). The species has also been translocated to the Adriatic basin and it needs to be managed properly to reduce its possible impact to native fauna due its predatory behaviour (Piria et al. 2018).

5 Conservation Measures

Conservation of biodiversity in Croatia, including freshwater fish, is governed by the Nature Protection Act (Official Gazette 2013a, 2013b), and freshwater fish species are additionally protected by Freshwater Fisheries Act (Official Gazette 2019a) and its by-laws and regulations, which includes minimum sizes and closed season for species used commercially, listed in the Regulation on the protection of fish in freshwater fisheries (Official Gazette 2005) (Table 7). A substantial number of freshwater fish species (67) is under strict protection by the law in Croatia (Table 8) (Official Gazette 2013b, 2016). The reason for their strict protection is that they are endemic, rare, and either listed on the IUCN national/international lists or protected by international agreements. Harvesting of those species is not allowed but in case that they are unintentionally caught, must be immediately released back into the water unharmed. *Anguilla anguilla* is the only strictly protected species that has commercial value and is regularly harvested in Croatia. The species is protected only at two localities in the Adriatic Basin of Croatia: Krka National Park and Vransko Lake Nature Park, but not in other habitats such as the Neretva River Delta. The species is considered non-native in the Black Sea Basin, and there are no restrictions for its harvest.

Recreational fishing is generally allowed in protected areas, but not in national parks. The only exception is Risnjak National Park where limited catch and release angling is permitted, but only until the expiry of current concession (https://www.np-risnjak.hr/sportski-ribolov/).

Table 7 Season closures and minimum sizes (cm) for freshwater fish in Croatia listed in the Regulation on the protection of fish in freshwater fisheries, in effect since 2005

Fish species	Season closures (from–to)	Waterbody	Minimal catch size (cm)
Brown trout *Salmo trutta m. fario L.*	1 October–31 March	Kupa River Čabranka River Žumberak Streams and north of the Sava River Other fishing waters	30
	1 October–28/29 February	Gacka River	50
Brown trout *Salmo trutta m. lacustris L.*	1 October–28/29 February	All fishing waters	40
Adriatic trout *Salmo obtusirostris*	1 October–31 May	Jadro, Žrnovnica, Vrlika, and Krka Rivers	30
Huchen *Hucho hucho*	16 February–30 September	All fishing waters	80
Marble trout *Salmo marmoratus*	16 February–30 September	All fishing waters	70
Grayling *Thymallus thymallus*	16 October 15 May	Kupa and Kupica Rivers	32
	1 January–15 May	Other fishing waters	30
Northern pike *Esox lucius*	1 February–31 March	All fishing waters except in habitats with the presence of trout species, huchen, and grayling	40
Pike-perch *Sander lucioperca*	31 March–31 May	All fishing waters	40
Common carp *Cyprinus carpio*	1 April–31 May	Only for wild form	40
European catfish *Silurus glanis*	16 April–15 June	All fishing waters, except Danube River	60
Sterlet *Acipenser ruthenus*	1 March–31 May	All fishing waters	40
Asp *Leuciscus aspius*	1 April–31 May	Dunavci: Poretak, Ostrovski,Vučedolski, Opatovački i Renovski	40
Ide *Leuciscus idus*	1 May–31 May	Dunavci: Poretak, Ostrovski,Vučedolski, Opatovački i Renovski	25
Rainbow trout *Oncorhynchus mykiss*	–	–	30
Tench *Tinca tinca*	-	–	20
Barbel *Barbus barbus*	–	–	28

Table 8 Fish species strictly protected by the Nature Protection Act

Species	Threat Category	International agreement	Endemic species	Remarks
Eudontomyzon vladykovi (= *Eudontomyzon danfordi)*	precautionary principle			
Lampetra zanandreai (= *Lethenteron zanandreai)*	EN	BE2		
Petromyzon marinus	precautionary principle			
Acipenser gueldenstaedtii	RE			
Acipenser naccarii	CR (EN)	BE2, DS4		
Acipenser nudiventris	RE			
Acipenser stellatus	RE			
Acipenser sturio	RE	BE2		
Huso huso	RE	BE2		
Anguilla anguilla	CR (European Union)			Populations from Vransko Lake and Krka River (upstream of Skradinski Buk)
Alosa fallax	EN			Only from Natura 2000 HR5000031 and the Neretva Delta
Alosa immaculata (= *Alosa pontica)*	DD, precautionary principle			
Alburnus arborella (= *Alburnus albidus)*	VU			
Alburnus neretvae (= *Alburnus albidus)*	precautionary principle			
Alburnus sarmaticus (= *Chalcalburnus chalcoides)*	VU			
Aulopyge huegelii	EN			
Barbus plebejus	EN			Krupa and Zrmanja Rivers within Velebit Nature Park
Carassius carassius	VU			Only in Lonjsko Polje and Kopački Rit Nature Parks and the Mura Drava Regional Park

(continued)

Table 8 (continued)

Species	Threat Category	International agreement	Endemic species	Remarks
Chondrostoma phoxinus	CR			
Delminichtys adspersus (= *Phoxinellus adspersus*)	VU			
Delminichtys ghetaldii (= *Phoxinellus ghetaldii*)	EN			
Delminichtys jadovensis (= *Phoxinellus jadovensis*)	CR		Yes	
Delminichtys krbavensis (= *Phoxinellus krbavensis*)	CR		Yes	
Leucaspius delineatus	VU			
Phoxinellus alepidotus	DD, precautionary principle			
Phoxinellus dalmaticus	VU		Yes	
Romanogobio benacensis	precautionary principle			
Romanogobio kesslerii (= *Gobio kesslerii*)	precautionary principle			
Romanogobio uranoscopus (= *Gobio uranoscopus*)	precautionary principle			
Romanogobio vladykovi (= *Gobio albipinnatus*)	DD, precautionary principle			
Rutilus aula	NT			
Squalius microlepis	CR			
Squalius zrmanjae	VU		Yes	Krupa and Zrmanja Rivers within Velebit Nature Park
Telestes ukliva	CR		Yes	

(continued)

Table 8 (continued)

Species	Threat Category	International agreement	Endemic species	Remarks
Telestes croaticus (= Phoxinellus croaticus)	EN		Yes	
Telestes fontinalis (= Phoxinellus fontinalis)	EN		Yes	
Telestes karsticus			Yes	
Telestes metohiensis (= Phoxinellus metohiensis)	RE			
Telestes polylepis	CR		Yes	
Telestes souffia (= Leuciscus souffia)	VU			
Telestes turskyi	CR		Yes	
Cobitis bilineata (= Cobitis taenia)	precautionary principle			
Cobitis dalmatina (= Cobitis taenia)	VU		Yes	
Cobitis elongata	VU			
Cobitis illyrica (= Cobitis taenia)	VU		Yes	
Cobitis jadovensis (= Cobitis taenia)			Yes	
Cobitis narentana (= Cobitis taenia)	VU			
Misgurnus fossilis	VU			
Sabanejewia balcanica (= Sabanejewia aurata)	VU			
Aphanius fasciatus	EN	BE2		
Salaria fluviatilis	VU			
Gymnocephalus baloni	VU			
Gymnocephalus schraetzer	CR			
Zingel streber	VU			
Zingel zingel	VU			
Knipowitschia croatica	CR		Yes	
Knipowitschia mrakovcici	EN		Yes	

(continued)

Table 8 (continued)

Species	Threat Category	International agreement	Endemic species	Remarks
Knipowitschia panizzae	precautionary principle			
Knipowitschia radovici	DD		Yes	
Padogobius bonelli	EN			
Pomatoschistus canestrinii	EN	BE2		
Salmo dentex	CR			
Salmo farioides	EN			
Salmo marmoratus	CR			
Salmo visovacensis	EN		Yes	
Salmothymus obtusirostris (= *Salmo obtusirostris*)	CR			Krka, Jadro Vrljika Rivers and Neretva Delta
Cottus gobio	VU			Zrmanja River

There are two freshwater Ichthyological Special Reserves in Croatia: the upper part of the Jadro River and source of the Vrljika River in the Adriatic Sea Basin (Ćaleta et al. 2015). Both reserves were established to protect the endemic Adriatic trout *Salmo obtusirostris* and all fishing is banned. The Lonjsko Polje and Kopački Rit Nature Parks have great significance for the Black Sea Basin as important fish spawning sites. Both areas are floodplains nearly completely in their natural state, flooding yearly, and replenishing the Sava, Drava, and Danube Rivers with fish stocks (Mrakovčić et al. 2006).

Croatia is a part of European Natura 2000 network that protects certain habitat types and/or species in specially designated sites. Freshwater habitats are also protected under the Water Framework Directive and implemented in the Water Act (Official Gazette 2019c). Hence, freshwater fish species that are not strictly protected in Croatia are protected by Natura 2000 Directive and are listed in Annex II and V of the Council Directive 92/43/EEC (Supplementary material S4), (EC 1992). Those species usually have minimal or non-commercial value, though some are important angling species, such as *H. hucho* or *L. aspius*. The sites which any of listed species inhabits must be managed in accordance with the ecological needs of the species and maintained or restored at a favourable conservation status. That means that holders of fishing rights are required to implement measures that will contribute to species conservation, such as imposing precautionary measures on stocking or limitations and quotas on harvesting.

6 Conclusion

One century ago, the commercial fishery catch was measured in the thousands of tonnes but nowadays it has fallen to less than a thousand tonnes. The disappearance of historically important fish species, such as most sturgeons, has placed catch pressure on the remaining native species, such as *A. brama, C. carpio,* or *S. glanis* and even on the endangered *A. ruthenus*. The sustainability of harvesting natural populations of other native commercial important fish species from Croatian inland waters is also questionable and, in the future, may rely only on non-native fish that are considerably present in the catch of the commercial and recreational fishery.

Electronic data collection is currently in the experimental phase and data collected during 2021 did not yield the expected results. The entered data showed that users probably inadvertently entered several extra zeros for biomass, and the resulting data are unreliable and incomparable with data previously collected by the logbook. Currently amendments are under consideration with a return to logbooks for the commercial fishery and the collection of e-data on catch only for recreational fishery. In addition to this temporary problem, the system of e-data catch collection in Croatia has certain deficiencies that need to be addressed in the future. The data collected from the recreational fishery should be collected separately for running and standing waters. Currently, the data collected in recreational fishery is sorted only by fishing right holders, regardless of whether the association manages only standing water or both standing and running waters. Thus, data collected from the fishing right holders do not give a reliable overview of the fish catch of the waterbodies that they have under concession. However, the data collected from the inland commercial fishery likely do not rely on the actual situation on the ground due to a lack of accurate catch reports, but despite this, shows a general trend.

Finally, it seems that commercial fishery in inland waters of Croatia is declining and will likely remain in the form of traditional, ethnological, and cultural heritage. On the contrary, recreational fishing is showing prosperity, particularly in the catch and release technique and trophy fishing development.

Acknowledgements We would like to thank Mijo Vojković for providing photos of traditional fishing gears used in the Sava River. Special thanks to Linda Zanella for language editing and constructive comments. The writing of this chapter was supported by the EIFAAC/FAO.

References

Arlinghaus R, Mehner T, Cowx IG (2002) Reconciling traditional inland fisheries management and sustainability in industrialized countries, with emphasis on Europe. Fish Fish 3(4):261–316. https://doi.org/10.1046/j.1467-2979.2002.00102.x

Arlinghaus R, Cooke SJ, Lyman J, Policansky D, Schwab A, Suski C, Sutton SG, Thorstad EB (2007) Understanding the complexity of catch-and-release in recreational fishing: An integrative synthesis of global knowledge from historical, ethical, social, and biological perspectives. Rev Fish Sci 15(1–2):75–167. https://doi.org/10.1080/10641260601149432

Balon EK (2004) About the oldest domesticates among fishes. J Fish Biol 65:1–27. https://doi.org/10.1111/j.1095-8649.2004.00563.x

Bloesch J, Jones T, Reinartz R, Striebel B (2006) An action plan for the conservation of sturgeons (Acipenseridae) in the Danube River Basin. Österreichische Wasser-und Abfallwirtschaft 58(5): 81–88

Bojčić C (1982) Povijesni razvoj slatkovodnog ribarstva [Historical development of freshwater fisheries] pp. In: Bojčić C, Lj D, Vuković T, Jovanović Kršljanin B, Apostolski K, Ržaničanin B, Turk M, Volk S, Drecun Đ, Habeković D, Hristić Đ, Fijan N, Pažur K, Bunjevac I, Marošević Đ (eds) Slatkovodno ribarstvo. Poslovna zajednica slatkovodnog ribarstva Jugoslavije, Zagreb, pp 1, 605 pp–45

Ćaleta M, Buj I, Mrakovčić M, Mustafić P, Zanella D, Marčić Z, Duplić A, Mihinjač T, Katavić I (2015) Endemic Fishes of Croatia. Croatian Environment Agency, Zagreb, 116 pp

Copp GH, Britton JR, Cucherousset J, Garcıa-Berthou E, Kirk R, Peeler E, Stakenas S (2009) Voracious invader or benign feline? A review of the environmental biology of European catfish *Silurus glanis* in its native and introduced ranges. Fish Fish 10:252–282. https://doi.org/10.1111/j.1467-2979.2008.00321.x

Cowx IG (2015) Characterisation of inland fisheries in Europe. Fish Manag Ecol 22:78–87. https://doi.org/10.1111/fme.12105

Craig LS, Olden JD, Arthington AH, Entrekin S, Hawkins CP, Kelly JJ, Kennedy TA, Maitland B, Rosi EJ, Roy AH, Strayer DL, Tank JL, West AO, Wooten MS (2017) Meeting the challenge of interacting threats in freshwater ecosystems: A call to scientists and managers. Elem Sci Anth 5: 72. https://doi.org/10.1525/elementa.256

Cucherousset J, Horky P, Slavík O, Ovidio M, Arlinghaus R, Boulêtreau S, Britton R, García-Berthou E, Santoul F (2018) Ecology, behaviour and management of the European catfish. Rev Fish Biol Fish 28(1):177–190. https://doi.org/10.1007/s11160-017-9507-9

Ćurčić V (1910) Narodno ribarstvo u Bosni i Hercegovini. Glasnik zemaljskog muzeja u Bosni i Hercegovini, Sarajevo, 109p

Davis J, O'Grady AP, Dale A, Arthington AH, Gell PA, Driver PD, Bond N, Casanova M, Finlayson M, Watts RJ, Capon SJ, Nagelkerken I, Tingley R, Fry B, Page TJ, Specht A (2015) When trends intersect: The challenge of protecting freshwater ecosystems under multiple land use and hydrological intensification scenarios. Sci Total Environ 534:65–78. https://doi.org/10.1016/j.scitotenv.2015.03.127

Delić A, Šanda R, Bučar M, Mihoci I, Vilenica M, Vukić J, Lelo S, Kučinić M (2014) New data on distribution of the monkey goby, *Neogobius fluviatilis* (Pallas, 1814) in Bosnia and Herzegovina and Croatia with notes on ecology and associated fish fauna. Natura Croatica 23(2):297–304

Drnić M (1896) Ribarstvo i ribogojstvo u kraljevinah Hrvatskoj i Slavoniji. Lovačko-ribarski viestnik 5(1):6–9

Drnić M (1897) Šport udičarenja. Lovačko-ribarski viestnik 6(10):110–112

Dudgeon D (2019) Multiple threats imperil freshwater biodiversity in the Anthropocene. Curr Biol 29(19):R960–R967. https://doi.org/10.1016/j.cub.2019.08.002

Dudgeon D, Arthington A, Gessner M, Kawabata Z, Knowler D, Lévêque C, Naiman RJ, Prieur-Richard A-H, Soto D, Stiassny MLJ, Sullivan C (2006) Freshwater biodiversity: Importance, threats, status and conservation challenges. Biol Rev 81(2):163–182. https://doi.org/10.1017/S1464793105006950

Dudgeon D, Smith REW (2006) Exotic species, fisheries and conservation of freshwater biodiversity in tropical Asia: the case of the Sepik River, Papua New Guinea. Aquatic Conserv: Mar Freshw Ecosyst 16:203–215. https://doi.org/10.1002/aqc.713

Eurofish (2015) Analysis of the Croatian freshwater aquaculture sector focusing on carp and trout species. Ministry of Agriculture of Croatia. Available at: https://ribarstvo.mps.hr/UserDocsImages/akvakultura/Analysis%20of%20the%20Croatian%20freshwater%20aquaculture%20sector%20focusing%20on%20carp%20and%20trout%20species.pdf [visited on 04.07.2022]

EC (2009) Council Regulation (EC) No 1224/2009 of 20 November 2009 establishing a Community control system for ensuring compliance with the rules of the common fisheries policy, amending Regulations (EC) No 847/96, (EC) No 2371/2002, (EC) No 811/2004, (EC) No 768/2005, (EC) No 2115/2005, (EC) No 2166/2005, (EC) No 388/2006, (EC) No 509/2007, (EC) No 676/2007, (EC) No 1098/2007, (EC) No 1300/2008, (EC) No 1342/2008 and repealing Regulations (EEC) No 2847/93, (EC) No 1627/94 and (EC) No 1966/2006. Official Journal of the European Union, L 343/1. https://eur-lex.europa.eu/legal-content/EN/ALL/?uri=celex%3A32009R1224

EU (2013) Regulation (EU) No 1379/2013 of the European Parliament and of the Council of 11 December 2013 on the common organisation of the markets in fishery and aquaculture products, amending Council Regulations (EC) No 1184/2006 and (EC) No 1224/2009 and repealing Council Regulation (EC) No 104/2000. Official Journal of the European Union, L 354/1. https://eur-lex.europa.eu/legal-content/EN/TXT/?uri=celex%3A32013R1379

EC (1992) Council Directive 92/43/EEC of 21 May 1992 on the conservation of natural habitats and of wild fauna and flora. Official Journal of the European Communities L 206/7. https://eur-lex.europa.eu/legal-content/EN/TXT/?uri=celex%3A31992L0043

FAO (2022a) *Abramis brama* Linnaeus, 1758. Fisheries and Aquaculture Division [online]. Rome. Available at: https://www.fao.org/fishery/en/aqspecies/2153/en [accessed 27.07.2022]

FAO (2022b) *Sander lucioperca*. Cultured Aquatic Species Information Programme. Text by Zakęś, Z. Fisheries and Aquaculture Division [online]. Rome. Updated 2012-03-16 [accessed 29.7.2022]

Feldhofer T (2021) Hrvatski športsko ribolovni savez od 1935. do 2020. godine. Hrvatski športsko ribolovni savez, Zagreb

Flögel R (1896) Pastrve po potocih zagrebačke okolice. Lovačko-ribarski viestnik, year V, issue 5, 59 p

Getto I (2013) Puškašem pliva više od sto tona somova! Glas Slavonije. http://www.glas-slavonije.hr/185809/4/Puskasem-pliva-vise-od-sto-tona-somova [accessed 25.7.2022]

Greenberg LA, Paszkowski CA, Tonn WM (1995) Effects of prey species composition and habitat structure on foraging by two functionally distinct piscivores. Oikos 74(3):522–532

Habeković D, Pažur K (1998) Pike (*Esox lucius* L.), its characteristics and importance. Ribarstvo: znanstveno-stručni časopis za ribarstvo 56(2):55–60

Hoffmann RC (2005) A brief history of aquatic resource use in medieval Europe. Helgol Mar Res 59:22–30. https://doi.org/10.1007/s10152-004-0203-5

Horvat M (2005) Ribe i njihov život. In: Pažur K (ed) Priručnik za športski ribolov. Zagreb, Hrvatski športsko ribolovni savez

Jakovlić I, Piria M, Šprem N, Tomljanović T, Matulić D, Treer T (2015) Distribution, abundance and condition of invasive Ponto-Caspian gobies *Ponticola kessleri* (Gunther, 1861), *Neogobius fluviatilis* (Pallas, 1814), and *Neogobius melanostomus* (Pallas, 1814) in the River Sava basin, Croatia. J Appl Ichthyol 31(5):888–894. https://doi.org/10.1111/jai.12803

Jelić M, Jelić D, Žutinić P, Ćaleta M (2012) Fish fauna of the lower reaches of the river Drava and surrounding marshland habitats near Donji Miholjac (eastern Croatia). Croat J Fish 70(4): 153–167

Jelkić D, Lužaić R, Opačak A, Ozimec S, Tucak K (2018) Trophy structure of common carp, pike, wels catfish and pike-perch in Drava – Danube fishing zone. In: Rozman V., Antunović Z. (eds.) Zbornik radova 53. hrvatski i 13. međunarodni simpozij agronoma

Kesterčanek FŽ (1894) U zaštitu ribo- i rakolova. Lovačko-ribarski viestnik 3(5):79–80

Kesterčanek FŽ (1907) Ribo- I rakolov. Lovačko-ribarski viestnik 16(5):59

Kottelat M, Freyhof J (2007) Handbook of European Freshwater Fishes. Publications Kottelat, Cornol and Freyhof, Berlin, 646 pp

Kubala M, Farský M, Krajč T, Pekárik L (2021) Bayesian modelling suggests that the sterlet (*Acipenser ruthenus*, Linnaeus 1758) population is ageing in the middle Danube River. Aquatic Conserv: Mar Freshw Ecosyst 31:469–479. https://doi.org/10.1002/aqc.3515

Lyach R, Čech M (2018) A new trend in Central European recreational fishing: More fishing visits but lower yield and catch. Fish Res 201:131–137. https://doi.org/10.1016/j.fishres.2018.01.020

Lynch AJ, Cooke SJ, Deines AM, Bower SD, Bunnell DB, Cowx IG, Nguyen VM, Nohner J, Phouthavong K, Riley B, Rogers MW, Taylor WW, Woelmer W, Youn S-J, Beard TD Jr (2016) The social, economic, and environmental importance of inland fish and fisheries. Environ Rev 24(2):115–121. https://doi.org/10.1139/er-2015-006

Marić D (2018) The Ichthyofauna of Lake Skadar/Shkodra: Diversity, Economic Significance, Condition, and Conservation Status. In: Pešić V, Karaman G, Kostianoy A (eds) The Skadar/Shkodra Lake Environment, The Handbook of Environmental Chemistry, vol 80. Springer, Cham. https://doi.org/10.1007/698_2018_238

Maguire I, Jelić M, Klobučar G, Delpy M, Delaunay C, Grandjean F (2016) Prevalence of the pathogen *Aphanomyces astaci* in freshwater crayfish populations in Croatia. Dis Aquat Org 118(1):45–53. https://doi.org/10.3354/dao02955

Mateš A (2004) Zaneseni ribar, knjiga prva. Pastrve. J & B, Zagreb

Matulić D, Šprem N, Piria M, Tomljanović T, Treer T, Safner R, Aničić I (2010) Analysis of recreational fisheries in the Croatian areas of the Sava and Danube Rivers. Agric Conspec Sci 75(4):183–190

McIntyre PB, Liermann CAR, Revenga C (2016) Linking freshwater fishery management to global food security and biodiversity conservation. Proc Natl Acad Sci U S A 113:12880. https://doi.org/10.1073/pnas.1521540113

Mihaljević Z, Kerovec K, Mrakovčić M, Primc-Habdija B, Plenković-Moraj A, Alegro A (2011) Testiranje bioloških metoda ocjene ekološkog stanja (Okvirna direktiva o vodama, 2000/60/EC) u reprezentativnim slivovima panonske i dinaridske ekoregije. Elaborat, Biološki odsjek, Prirodoslovno - matematički fakultet, Sveučilište u Zagrebu, Zagreb, 350 str

Mihinjač T, Sučić I, Špelić I, Vucić M, Ješovnik A (2019) Strane vrste slatkovodnih riba u Hrvatskoj. Ministarstvo okoliša i energetike, Udruga Hyla, Zagreb

Milošević D, Talevski T (2015) Conservation status of native species in natural lakes of Drim system (Prespa, Ohrid and Skadar Lake) and dangers of commercial fishing. Bulg J Agric Sci 1: 61–67

Ministry of Agriculture (2022) Freshwater sports fisheries - angling in Croatia. https://ribarstvo.mps.hr/default.aspx?id=2802

Mrakovčić M, Brigić A, Buj I, Ćaleta M, Mustafić P, Zanella D (2006) Red Book of Freshwater Fish of Croatia. Ministry of Culture, State Institute for Nature Protection, Zagreb, Croatia

Mitchell M, Vanberg J, Sipponen M (2010) Commercial inland fishing in member countries of the European Inland Fisheries Advisory Commission (EIFAC). FAO, Rome, Italy. https://epub.sub.uni-hamburg.de/epub/volltexte/2013/16696/pdf/an222e.pdf

Official Gazette (2005) Ordinance on the protection of fish in freshwater fisheries (Naredba o zaštiti riba u slatkovodnom ribarstvu) Narodne Novine 2005 no. 82 (1584). https://narodne-novine.nn.hr/clanci/sluzbeni/2005_07_82_1584.html

Official Gazette (2013a) Nature Protection Act (Zakon o zaštiti prirode). Narodne Novine 2013 no. 80 (1658). https://narodne-novine.nn.hr/clanci/sluzbeni/2013_06_80_1658.html

Official Gazette (2013b) Ordinance on strictly protected species (Pravilnik o strogo zaštićenim vrstama). Narodne Novine 2013 no. 144 (3086). https://narodne-novine.nn.hr/clanci/sluzbeni/2013_12_144_3086.html

Official Gazette (2016) Ordinance on amendments to the Ordinance on strictly protected species (Pravilnik o izmjenama i dopunama Pravilnika o strogo zaštićenim vrstama). Narodne Novine 2016 no. 73 (1745). https://narodne-novine.nn.hr/clanci/sluzbeni/2016_08_73_1745.html

Official Gazette (2018) Act on the Prevention and Management of the Introduction and Spread of Alien and Invasive Alien Species (Zakon o sprječavanju unošenja i širenja stranih te invazivnih stranih vrsta i upravljanju njima). Narodne Novine 2018 no. 15 (310). https://narodne-novine.nn.hr/clanci/sluzbeni/2018_02_15_310.html

Official Gazette (2019b) Act on the Amendments to the Act on the Prevention and Management of the Introduction and Spread of Alien and Invasive Alien Species (Zakon o izmjenama i dopuni zakona o sprječavanju unošenja i širenja stranih te invazivnih stranih vrsta i upravljanju njima). Narodne Novine 2019 no. 14 (275). https://narodne-novine.nn.hr/clanci/sluzbeni/2019_02_14_275.html

Official Gazette (2019a) Freshwater Fisheries Act (Zakon o slatkovodnom ribarstvu). Narodne Novine 2019 no. 63 (1232). https://narodne-novine.nn.hr/clanci/sluzbeni/2019_06_63_1232.html

Official Gazette (2019c) Water Act (Zakon o zaštiti voda). Narodne Novine 2019 no. 66 (1285). https://narodne-novine.nn.hr/clanci/sluzbeni/2019_07_66_1285.html

Official Gazette (2020a) Ordinance on the fishing and fish protection exam in freshwater fisheries (Pravilnik o ribičkom i ribočuvarskom ispitu u slatkovodnom ribarstvu). Narodne Novine 2020 no 6 (86). https://narodne-novine.nn.hr/clanci/sluzbeni/2020_01_6_86.html

Official Gazette (2020b) Ordinance on the method of drafting and implementing the management plan, revision and supplement of the management plan and the programme for monitoring fish stocks in freshwater fisheries (Pravilnik o načinu izrade i provođenja plana upravljanja, revizije i dodatka plana upravljanja te programa praćenja stanja ribljeg fonda u slatkovodnom ribarstvu). Narodne Novine 2020 no. 79 (1492). https://narodne-novine.nn.hr/clanci/sluzbeni/2020_07_79_1492.html

Official Gazette (2020c) Ordinance on fishing licenses in freshwater fisheries (Pravilnik o ribičkim dozvolama u slatkovodnom ribarstvu). Narodne Novine 2020 no. 139 (2686). https://narodne-novine.nn.hr/clanci/sluzbeni/2020_12_139_2686.html

Official Gazette (2021) Ordinance on recreational (sport) fishing in freshwater fisheries (Pravilnik o sportskom ribolovu u slatkovodnom ribarstvu). Narodne Novine 2021 no. 81 (1507) https://narodne-novine.nn.hr/clanci/sluzbeni/2021_07_81_1507.html

Official Gazette (2022a) Ordinance on borders and size of fishing zones for freshwater fisheries and restrictions and allowance of fishing (Pravilnik o granicama i površini ribolovnih područja i ribolovnih zona za slatkovodni ribolov te o dopuštenosti i ograničenjima ribolova). Narodne Novine no. 14 (138). https://narodne-novine.nn.hr/clanci/sluzbeni/2022_02_14_138.html

Official Gazette (2022b) Ordinance on commercial freshwater fisheries (Pravilnik o gospodarskom ribolovu u slatkovodnom ribarstvu). Narodne Novine 2022 no. 21 (254). https://narodne-novine.nn.hr/clanci/sluzbeni/2022_02_21_254.html

Oesterwind D, Bock C, Förster A, Gabel M, Henseler C, Kotterba P, Menge M, Myts D, Winkler HM (2017) Predator and prey: the role of the round goby *Neogobius melanostomus* in the western Baltic. Mar Biol Res 13(2):188–197. https://doi.org/10.1080/17451000.2016.1241412

Pham HV, Torresan S, Critto A, Marcomini A (2019) Alteration of freshwater ecosystem services under global change – A review focusing on the Po River basin (Italy) and the Red River basin (Vietnam). Sci Total Environ 652:1347–1365. https://doi.org/10.1016/j.scitotenv.2018.10.303

Piria M, Simonović P, Zanella D, Ćaleta M, Šprem N, Paunović M, Tomljanović T, Gavrilović A, Pecina M, Špelić I, Matulić D, Rezić A, Aničić I, Safner R, Treer T (2019a) Long-term analysis of fish assemblage structure in the middle section of the Sava River – The impact of pollution, flood protection and dam construction. Sci Total Environ 651:143–153. https://doi.org/10.1016/j.scitotenv.2018.09.149

Piria M, Špelić I, Treer T, Matulić D, Tomljanović T, Gavrilović A, Radočaj T (2019b) Završno izvješće o provođenju programa praćenja stanja u slatkovodnom ribarstvu u 2019. godini, Grupa B) Ribolovno područje Sava. Naručitelj: Ministarstvo poljoprivrede, Izvršitelj: Sveučilište u Zagrebu Agronomski fakultet, Zagreb, 100 p

Piria M, Špelić I, Gavrilović A (2018) Distribution and impact of translocated predatory fish species in the Dalmatia ecoregion. In: Jelaska SD (ed) 3rd Croatian Symposium on Invasive Species, Book of Abstracts. Croatian Ecological Society, Zagreb, p 64

Piria M, Jakšić G, Jakovlić I, Treer T (2016) Dietary habits of invasive Ponto-Caspian gobies in the Croatian part of the Danube River basin and their potential impact on benthic fish communities. Sci Total Environ 540:386–395. https://doi.org/10.1016/j.scitotenv.2015.05.125

Piria M, Šprem N, Jakovlić I, Tomljanović T, Matulić D, Treer T, Aničić I, Safner R (2011) First record of round goby, *Neogobius melanostomus* (Pallas, 1814) in the Sava River, Croatia. Aquat Invasion Records 8(Supplement 1):153–157

Piria M, Jelkić D, Gavrilović A, Horváth Á, Kovác B, Enikő Balogh R, Špelić I, Radočaj T, Vilizzi L, Ozimec S, Opačak A (2022) Finding of hybrid African catfish 'Clariobranchus' in the Danube River. J Vertebr Biol 71:22008. https://doi.org/10.25225/jvb.22008

Potts WM, Downey-Breedt N, Obregon P, Hyder K, Bealey R, Sauer WHH (2020) What constitutes effective governance of recreational fisheries?—A global review. Fish Fish 21:91–103. https://doi.org/10.1111/faf.12417

Pofuk M, Zanella D, Piria M (2017) An overview of the translocated native and non-native fish species in Croatia: pathways, impacts and management. Manag Biol Invasions 8(3):425–435. https://doi.org/10.3391/mbi.2017.8.3.16

Ranaaker L, Persson J, Jönsson M, Nilsson PA, Brönmark C (2014) Piscivore-prey fish interactions: mechanisms behind diurnal patterns in prey selectivity in brown and clear water. PLoS One 9(11):e102002. https://doi.org/10.1371/journal.pone.0102002

Rastija K (2018) Žele spasiti tradicijski ribolov u Parku prirode Lonjsko polje [The intention to save traditional fishing in Lonjsko Polje Nature Park]. Agroklub https://www.agroklub.com/ribarstvo/zele-spasiti-tradicijski-ribolov-u-parku-prirode-lonjsko-polje/45272/

Safner R, Treer T, Aničić I, Piria M, Šprem N, Matulić D, Tomljanović T (2013) First record of palmeto bass (*Morone saxatilis* X *M. chrysops*) in the Croatian part of the River Danube. Arch Biol Sci 65(1):197–200

Sala A (2016) Review of the EU small-scale driftnet fisheries. Mar Policy 74:236–244. https://doi.org/10.1016/j.marpol.2016.10.001

Sahrhage D, Lundbeck J (2012) A history of fishing. Springer Science & Business Media, Springer-Verlag, Berlin/Heidelberg, 348 pp. https://doi.org/10.1007/978-3-642-77411-9

Simonović P, Povž M, Piria M, Treer T, Adrović A, Škrijelj R, Nikolić V, Simić V (2015) Ichthyofauna of the River Sava System. In: Milačić R, Ščančar J, Paunović M (eds) The Sava River, The Handbook of Environmental Chemistry. Springer-Verlag, Berlin/Heidelberg, pp 361–400

Slavinić S (2021) Ribolovni svjetski rekordi: uhvaćen najveći šaran na svijetu. Hrvatski glasnik. https://hrvatski-glasnik.com/2021/04/13/ribolovni-svjetski-rekordi-uhvacen-najveci-saran-na-svijetu/ Accessed 11 July 2022

Smederevac-Lalić M, Višnjić-Jeftić Ž, Pucar M, Mićković B, Skorić S, Nikčević M, Hegediš A (2011) Fishing circumstances on the Danube in Serbia. Water Res Manag 1(4):45–49

Svanberg I, Locker A (2020) Ethnoichthyology of freshwater fish in Europe: a review of vanishing traditional fisheries and their cultural significance in changing landscapes from the later medieval period with a focus on northern Europe. J Ethnobiol Ethnomed 16:68. https://doi.org/10.1186/s13002-020-00410-3

Tomljanović T, Fabijanić N, Treer T, Matulić D, Piria M, Šprem N, Aničić I, Safner R (2012) Utjecaj invazivne vrste riba babuške (*Carassius gibelio*) na sastav ihtiocenoze u Republici Hrvatskoj. Proceedings, 47. Croatian and 7. International Symposium on Agriculture, Opatija 13–17 February 2012, p. 634-636

Tönißen K, Pfuhl R, Franz GP, Dannenberger D, Bochert R, Grunow B (2022) Impact of spawning season on fillet quality of wild pikeperch (*Sander lucioperca*). Eur Food Res Technol 248(5): 1277–1285. https://doi.org/10.1007/s00217-022-03963-7

Treer T, Kubatov I (2017) The co-existence of recreational and artisanal fisheries in the central parts of the Danube and Sava rivers. Croat J Fish 75(3):116–127. https://doi.org/10.1515/cjf-2017-0012

Treer T, Opačak A, Aničić I, Safner R, Piria M, Odak T (2003) Growth of bream, *Abramis brama*, in the Croatian section of the Danube. Czech J Anim Sci 48(6):251–256

Treer T (2022) Trends in total anglers' catches at the Croatian section of the Sava River Basin. Croat J Fish 80:97–101. https://doi.org/10.2478/cjf-2022-0010

Vehanen T, Piria M, Kubečka J, Skov C, Kelly F, Pokki H, Eskelinen P, Rahikainen M, Keskinen T, Artell J, Romakkaniemi A, Suić J, Adámek Z, Heimlich R, Chalupa P, Ženíšková H, Lyach R, Berg S, Birnie-Gauvin K, Jepsen N, Koed A, Pedersen MI, Rasmussen G, Gargan P, Roche W, Arlinghaus R (2020) Data collection systems and methodologies for the inland fisheries of Europe. FAO Fisheries and Aquaculture Technical Paper No. 649. Budapest, FAO. https://doi.org/10.4060/ca7993en

Vejřík L, Vejříková I, Blabolil P, Eloranta AP, Kočvara L, Peterka J, Sajdlová Z, Chung SHT, Šmejkal M, Kiljunen M, Čech M (2017) European catfish (*Silurus glanis*) as a freshwater apex predator drives ecosystem via its diet adaptability. Sci Rep 7(1):15970. https://doi.org/10.1038/s41598-017-16169-9

Zakęś Z, Demska-Zakęś K (2010) Controlled reproduction of pikeperch *Sander lucioperca* (L.): A review. Fish Aquatic Life 17(4):153–170. https://doi.org/10.2478/v10086-009-0014-z

Zobundžija V (2005) Zakonska regulative prava na ribolov. In: Pažur K (ed) Priručnik za športski ribolov. Zagreb, Hrvatski športsko ribolovni savez

Fish Resources and Fisheries in Bosnia and Herzegovina: Overview, Sustainability, and Conservation

Radoslav Dekić, Dragojla Golub, Belma Kalamujić Stroil, Azra Bakrač, Subha Avdić, Vera Kanlić, Rifat Škrijelj, and Avdul Adrović

Abstract Bosnia and Herzegovina (B&H) has a dense and branched freshwater hydrographic network that consists of two drainage basins, the Adriatic and the Black Sea. B&H fish diversity has been relatively well researched. The fauna of freshwater fish of B&H is characterized by significant richness and diversity of species. It includes many endemic and relict taxa, resulting from spatial-ecological, geomorphological, hydrological, and climate diversity. Firstly, it recorded 108 freshwater fish species of B&H, but the most recent inventory registered 118 species. Today, fish are considered the most endangered group of vertebrates worldwide. As the most important causes of degradation of their habitats and reduction in the population density are the introduction of non-native invasive species, the decline of water quality, regulation of watercourses and land reclamation, construction of dams and creation of hydro-accumulations, and climate change. B&H is no exception in this regard, and the trend of deteriorating environmental conditions is current

R. Dekić (✉) · D. Golub
Faculty of Natural Sciences and Mathematics, University of Banja Luka, Banja Luka, Bosnia and Herzegovina
e-mail: radoslav.dekic@pmf.unibl.org

B. K. Stroil
Institute for Genetic Engineering and Biotechnology, University of Sarajevo, Sarajevo, Bosnia and Herzegovina

A. Bakrač · S. Avdić
Biotechnical Faculty, University of Bihać, Bihać, Bosnia and Herzegovina

V. Kanlić
Ministry of Agriculture, Forestry and Water Management, Government of the Republic of Srpska, Banja Luka, Bosnia and Herzegovina

R. Škrijelj
Faculty of Natural Sciences and Mathematics, University of Sarajevo, Sarajevo, Bosnia and Herzegovina

A. Adrović
Faculty of Natural Sciences and Mathematics, University of Tuzla, Tuzla, Bosnia and Herzegovina

V. Simić et al. (eds.), *Ecological Sustainability of Fish Resources of Inland Waters of the Western Balkans*, Fish & Fisheries Series 43,
https://doi.org/10.1007/978-3-031-36926-1_3

in both inland and coastal waters. The fisheries sector in B&H includes commercial and sport fishing in inland waters and sport fishing in coastal waters, with a very long tradition of fish farming. In the aquaculture sector, the most important fish species of salmonid ponds are brown trout and rainbow trout. Carp, grass carp, silver carp, and bighead carp are dominant species in cyprinid ponds. Salmonid aquaculture production occurs in the Federation of B&H and in the Republic of Srpska equally. Almost the total production of cyprinid aquaculture is realized in the Republic of Srpska. Given the favorable natural conditions and regular market demands, aquaculture, as a branch of agriculture, has the potential for more progressive development. Activities must focus on creating a more favorable business environment for both domestic and foreign investments in this area.

Keywords Ichthyofauna · Fish stock · Endemic species · Aquaculture · Endangerment and protection

1 Historical Overview of Ichthyological Studies in Bosnia and Herzegovina

The freshwater fish fauna of Bosnia and Herzegovina (B&H) is relatively well studied. The first ichthyological research in B&H was conducted in the second half of the nineteenth century by foreign researchers such as Heckel, Kner, and Steindachner (Heckel 1843; Heckel and Kner 1858; Steindachner 1895). The first written data on freshwater fish in Bosnia and Herzegovina date back to 1882 by Steindachner, and surveys gained momentum in 1928 with the works of Taler (Taler 1951a, b, 1954). The second period of ichthyofauna research in B&H began in the first half of the twentieth century with the prominent work of Karaman (1923, 1928). It should be noted that the Fisheries Institute of the People's Republic of Bosnia and Herzegovina was founded in 1953 to promote the development of freshwater fisheries in B&H. After integrating with the Biological Institute, the Department of Ichthyology and Fisheries of the Biological Institute, University of Sarajevo, was founded in 1964. The department, among other things, was responsible for promoting ichthyology and especially for developing salmonid and cyprinid fisheries in B&H. Studies surveyed morphological, taxonomic, and ecological specificities of fish from B&H waters, and special attention was paid to endemic fish species. Tonko Šoljan and Tihomir Vuković made an exceptional contribution to the development of ichthyology in the country and globally. Vuković, one of the most important ichthyologists of B&H and a well-known and respected name in European and world ichthyology, published around three hundred scientific and professional works in biosystematics, ecology, and phylogeny, mostly of freshwater fish. A special place in his work is occupied by studies on the endemic ichthyofauna of B&H. Between his books and monographs stand out "Fishes of Bosnia and Herzegovina" (Vuković 1968) and "Freshwater fishes of Yugoslavia," which he wrote together with Božina Ivanović (Vuković and Ivanović 1971). The tradition of ichthyological research was continued by the research teams of the Center for Ichthyology and Fisheries of the Faculty of Science, Faculty of Agriculture and Food, University of Sarajevo, Faculty of Science, University of Tuzla, and Faculty of Science and Faculty of Agriculture, University of Banja Luka. With the first decade of the twenty-first century, molecular genetic research on the B&H fish began, first with the works of researchers from the Biotechnical Faculty, University of Ljubljana, and later from the Institute for Genetic Engineering and Biotechnology, University of Sarajevo. While the earliest molecular genetic research was mainly on salmonid species of the Adriatic basin in Bosnia and Herzegovina, studies in the second decade of the twenty-first century focused primarily on endemic cyprinid species of the Adriatic basin (Adrović 2012; Bakrač-Bećiraj et al. 2005; Dekić et al. 2017; Kalamujić Stroil et al. 2020; Kosorić 1969; 1989; Kosorić et al. 1983; Mikavica et al. 2006; Škrijelj 2002; Tutman et al. 2009; Vuković et al. 2008, etc.).

2 Overview of the Freshwater Ichthyofauna of Bosnia and Herzegovina

The B&H fish fauna is biogeographically and ecologically differentiated into marine and freshwater. The division of the hydrological network into the Adriatic and Danube basins further differentiated freshwater fauna and derived specificities in terms of the distribution and number of species and the presence of endemic forms (Fig. 1, (Table 1)). The wealth of species, especially endemics, places B&H in the group of ichthyologically richest European countries. Geographical location, complex geological history, climatic features, and isolation of river systems caused the development of numerous and diverse freshwater fish fauna of Bosnia and Herzegovina (Glamuzina et al. 2010; Dekić et al. 2011).

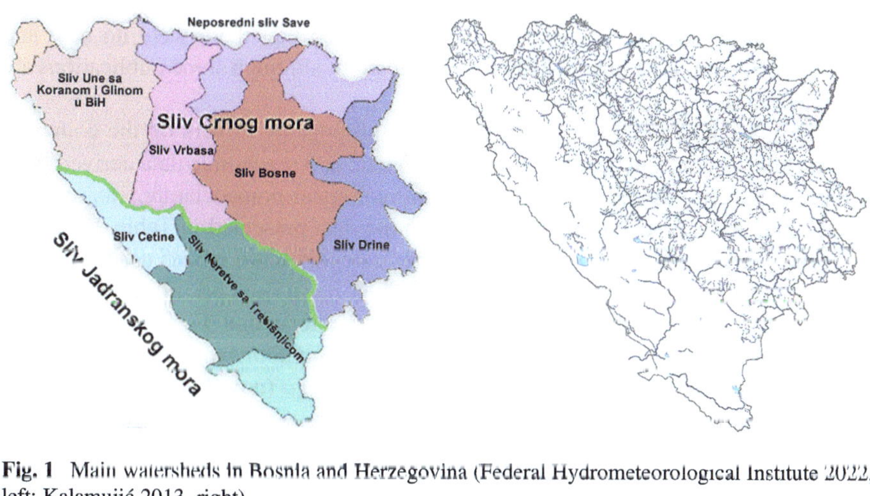

Fig. 1 Main watersheds in Bosnia and Herzegovina (Federal Hydrometeorological Institute 2022, left; Kalamujić 2013, right)

Table 1 Key hydrographic characteristics of Bosnia and Herzegovina (Federal Hydrometeorological Institute 2022)

Catchment areas of B&H				
Basin	Surface (km^2)	Length of watercourses longer than 10 km	Number of inhabitants (1991)	Average flow (m^3/s)
Black Sea Basin	38,719	7947.7	4,012,266	722
River Sava immediate basin	5506	1693.2	635,353	63
River Una basin with rivers Korana and Glina in B&H	9130	1480.7	620,373	240
River Vrbas basin	6386	1096.3	514,038	132
River Bosna basin	10,457	2321.9	1,820,080	163
River Drina basin in B&H	7240	1355.6	422,422	124
Adriatic Sea basin	12,410	1063.8	515,360	433
River Neretva basin in B&H River Trebišnjica basin	10,110	886.8	436,271	402
River Cetina basin in B&H	2300	177.0	79,089	31
Total	51,129	9011.5	4,527,626	1155

Analyzing data on the qualitative composition of the freshwater ichthyofauna of B&H some 50 years ago, it was established that the country's freshwaters were inhabited by four species of agnates and 104 species and subspecies of bony fish from 26 families (Vuković 1968).

According to data from 2009, the freshwater ichthyofauna of B&H includes about 118 species (and subspecies) of agnates and bony fishes (Sofradžija 2009). Taxa from 27 families are present; the largest number belongs to Cyprinidae, Salmonidae, Percidae, and Acipenseridae (Table 2). The highest species diversity was found within the family Cyprinidae (26 genera and 52 species) and Salmonidae (nine species).

The aforementioned work systematically provides an overview of the freshwater ichthyofauna of B&H. However, it should be noted that there were no new and revised publications of this type after its publication, although some publications are in preparation.

With new field ichthyological research, anthropogenic drivers on the country's aquatic ecosystems and beyond, new phylogenetic and taxonomic data derived from molecular genetic research, revisions in ichthyological nomenclature, etc., considerable changes can be expected compared to the data presented in Table 2.

It should be emphasized that the subspecies category has not been used for a long time by most modern ichthyologists, considering that it does not fit into the evolutionary concept of understanding the species (Kottelat and Freyhof 2007). For example, *Alosa fallax nilotica* (now *Alosa fallax*), *Salmothymus obtusirostris oxyrhynchus* (now *Salmo obtusirostris*), *Rutilus pigus virgo* (now *Rutilus virgo*) are no longer recognized in the ichthyological nomenclature. Such applies to many other

Table 2 Representation of agnates and bony fish in the freshwaters of B&H by families (according to Sofradžija 2009)

Family	Species (subspecies) number	
	Vuković 1963	Sofradžija 2009
Petromyzontidae	4	5
Acipenseridae	7	7
Clupeidae	1	2
Salmonidae	9	9
Thymallidae	1	1
Esocidae	1	1
Umbridae	1	1
Cyprinidae	43	52
Cobitidae	5	7
Siluridae	1	1
Ictaluridae	1	1
Anguillidae	1	1
Gadidae	1	1
Gasterosteidae	1	2
Sygnathidae	1	1
Poecilidae	1	1
Cyprinodontidae	/	1
Mugilidae	6	5
Atherinidae	2	1
Moronidae	1	1
Centrarchidae	1	1
Percidae	7	7
Sparidae	1	1
Blenniidae	1	1
Gobiidae	4	5
Pleuronectidae	1	1
Cottidae	1	1
27	**104**	**118**

taxa found in the literature about the freshwater ichthyofauna of B&H, even some more recent.

The distinction between lacustrine and fluviatic forms of brown trout (*Salmo trutta* m. *lacustris* and *Salmo trutta* m. *fario*), which today are described as one species, *Salmo trutta*, is considered incorrect. On the other hand, the Adriatic brook trout, *Salmo farioides* (Karaman 1928), which inhabits the upper part of the Neretva River, has been classified as *Salmo trutta* until now.

The number of species from the Acipenseridae family is also incorrect. It is considered that as many as six of its seven recorded species have become regionally extinct due to the impact of hydropower dams in preventing the upstream migration of these fish. Only the Sterlet sturgeon (*Acipenser ruthenus*, Fig. 2) is certain to inhabit the freshwaters of B&H (Ćaleta et al. 2019). The presence of *Acipenser ruthenus* has been confirmed in lower part of river Vrbas (AA 1975, 1985; Radević

Fig. 2 *Acipenser ruthenus* from river Sava near Brčko (Photo: Škrijelj R. 2006)

2000), in river Sava, near Kozarska Dubica, Bosanski Brod and Brčko (Project reports 2004), and in river Drina (Simonović et al. 2014).

The diversity of species from the Cobitidae family has also undergone major changes. The species *Cobitis taenia* is mentioned in many publications on the freshwater ecosystems of Bosnia and Herzegovina. However, this species does not inhabit the country's rivers; instead, *Cobitis elongata* and *Cobitis elongatoides* are encountered. Particularly confusing are the data for the species *Sabanejewia balcanica*, which some Bosnian ichthyologists believe occurs in several subspecies and belongs to a completely different genus, *Cobitis aurata balcanica* and *Cobitis aurata bosniaca* (Vuković and Ivanović 1971; Sofradžija 2009; Golub et al. 2016; Tutman et al. 2017).

Regarding the representatives of the agnates, Table 1 shows that four (Vuković 1963) or five species (Sofradžija 2009) were considered present in the freshwaters of B&H. Recent research indicates that three species inhabit the country's freshwaters, *Eudontomyzon vladykovi* in the Danube basin and *Lampetra soljani* and *Petromyzon marinus* in the Adriatic basin (Tutman et al. 2020a)

Now present in most of the freshwaters of B&H, the introduced invasive species of American catfish was wrongly classified as *Ameiurus nebulosus* (i.e., I*ctalurus nebulosus*). Only a few years ago, it was detected as *Ameiurus melas* (Halilović and Adrović 2015). Also, non-native trout perch (*Micropterus salmoides*) is not mentioned in previously published lists and publications on the freshwater ichthyofauna of B&H. However, data on its presence can be found in other sources (Lelo and Hamzić 2020). There are more such examples, especially when it comes to introduced species of fish whose arrival in our waters is somewhat more recent.

According to the modern taxonomic division, the once-monotypic family Thymallidae is classified as Salmonidae. Taxonomic changes also affected representatives of the genera *Telestes, Delminichthys* (Dekić et al. 2014; Francuski et al. 2019), *Chalcalburnus*, and others. Therefore, it can be concluded that a comprehensive revision of the composition of the freshwater ichthyofauna of B&H is necessary in terms of the number of species and families, as well as serious field and molecular research that would help solve many doubts and mistakes.

According to Kottelat and Freyhof (2007), the freshwater ichthyofauna of B&H includes about 20% of the European freshwater ichthyofauna.

3 Endemic Species of Freshwater Fish in Bosnia and Herzegovina

The ichthyofauna of B&H is characterized by a large number of endemic species, many of which are distributed only in certain localities. Most endemic species have a limited distribution area, which is the main reason for their endangerment (Glamuzina et al. 2010; Dekić et al. 2013b; Dekić et al. 2014). Endemic fish species are particularly sensitive and usually have little ability to adapt to changes in external environmental factors. They are mostly distributed in former glacial refugia and are closely related to the environment they inhabit. Looking at the basins, the largest part of the endemic ichthyofauna is related to the karst areas of the Adriatic basin (e.g., *Salmo marmoratus, Salmo obtusirostris, Telestes dabar, Telestes metohiensis, Squalius svallize, Aulopyge huegelli* (Fig. 3), *Delminichthys ghetaldii* (Fig. 4), *Cobitis narentana*).

Data on the number of endemic fish species in B&H are partially different in various publications. Glamuzina et al. (2010) state that 40 endemic species are represented in the ichthyofauna of B&H. It is considered that more than half of the freshwater fish species of the Adriatic basin live in the Neretva River and its tributaries (Glamuzina et al. 2010). Vuković and Sofradžija (1986) state the existence of 12 endemic species (two salmonids and ten cyprinids). However, Glamuzina et al. (2010) believe this number is much higher if the term endemism is viewed in a broader sense. According to that study, there are about 35 endemic species in this area: two endemics of the Mediterranean, nine endemics of the wider

Fig. 3 Endemic species *Aulopyge huegelli* (left) from Šatorsko Lake (right) (Photo: Dekić R. 2015)

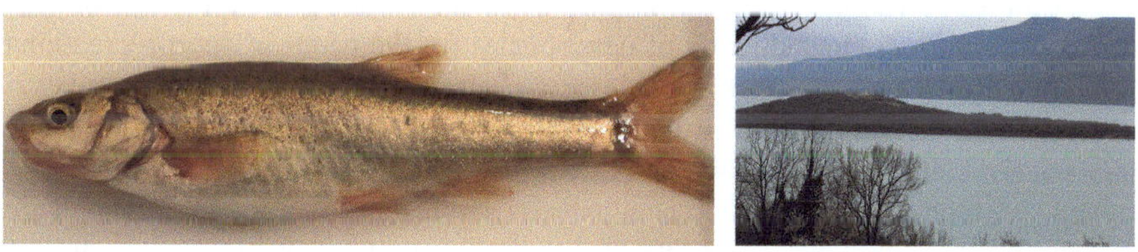

Fig. 4 *Delminichthys ghetaldii* (left) from Fatnica (right) (Photo: Dekić R. 2015)

Fig. 5 *Hucho hucho* (left) from Vrbas River, near Gornji Vakuf (right) (Photo: Golub D. 2017)

area of the Adriatic basin, 20 regional endemics of the narrower area of the Adriatic basin between Croatia and B&H, one endemic of B&H, as well as three potential endemics of B&H, which require additional research in order to determine their status with certainty.

Some of the endemic fish species of B&H have fishing importance, primarily in sport fishing. The most famous such species is the Danube salmon (*Hucho hucho,* Fig. 5), endemic of the Danube basin, a very desirable catch for sport anglers and unfortunately, a frequent target of illegal fishing. This species is protected by Regulation on strictly protected and protected wild species (Official Gazette of the Republic of Srpska 65/20) in category of strictly protected species. In order to protect this species, in 2020, the SRD of Banjaluka made a decision to ban fishing for Danube salmon for a period of three years.

4 Non-native Freshwater Fish Species in Bosnia and Herzegovina

Non-native fish species in wild water can affect the autochthonous ichthyofauna and lead to various changes. When non-native individuals reach open watercourses, they may enter reproductive and/or competitive relationships with indigenous populations. This is especially problematic if it concerns individuals of a foreign gene pool (hybridization) or those with a better ecological capacity to successfully conquer territory (competition). Non-native species can enter wild waters by accidental transfer (usually at the fry stage), deliberate stocking, or escapes from fish farms. Although the introduction of new species for the needs of sport-recreational fishing in B&H has long been considered a positive contribution to the number of species in an ecosystem, given that such introductions were natural, uncontrolled and scientifically unfounded, the end effect was utterly the opposite. Such species, mostly cyprinids, adapted well to new habitats and began suppressing or completely eradicating the autochthonous ichthyofauna. An example of this is the introduction of carp (*Cyprinus carpio*), Prussian carp (*Carassius gibelio*), and perch (*Sander lucioperca*) into the Neretva basin, which today have numerous populations in the Neretva reservoirs and dominate the native salmonid ichthyofauna (Škrijelj 2002).

Origin-wise, 105 recorded freshwater species are autochthonous, while 13 are allochthonous (Sofradžija 2009). Some non-native species have an invasive character. It should be noted that this analysis does not include natural hybrids and species whose presence has not been confirmed in the waters of B&H. If species that probably inhabit freshwaters of B&H were also taken into account, the introduced fish species are: bighead carp (*Hypophthalmichthys nobilis*), silver carp (*Hypophthalmichthys molitrix*), grass carp (*Ctenopharyngodon idella*), black bullhead (*Ameiurus melas*), brown bullhead (*Ameiurus nebulosus*), pumpkinseed (*Lepomis gibbosus*), stone moroko (*Pseudorasbora parva*), Amur sleeper (*Perccottus glenii*), largemouth bass (*Micropterus salmoides*), Prussian carp (*Carassius gibelio*), rainbow trout (*Oncorhynchus mykiss*), arctic charr (*Salvelinus alpinus*), brook trout (*Salvelinus fontinalis*), as well as all species of the genus *Neogobius*. These data primarily refer to the Black Sea Basin Rivers, and the Adriatic basin's total number would be even higher.

Hybridization of natural autochthonous populations can occur with individuals of non-native species (interspecies hybridization) or individuals of foreign evolutionary lineages of the same species (intraspecies hybridization). In both cases, individuals of foreign species/lineages reach the habitats of native species by accidental transfer (usually at the fry stage), deliberate stocking, or escapes from fish farms. An example of such a harmful effect on autochthonous ichthyodiversity is the introduction of the Atlantic lineage of brown trout (*Salmo trutta* L.), the most common breeding lineage of this species due to the characteristic of intensive growth in aquaculture conditions. The expansion of this lineage in both river basins, the Danube (Durmić-Pašić 2008; Simonović et al. 2017; Škraba et al. 2017) and the Adriatic (Durmić-Pašić 2008; Kalamujić 2013) resulted from the low ecological level of awareness among breeders, fishers, and the public (Piria et al. 2018, 2019), and the use of currently available, but genetically untested, juveniles for stocking. However, a realistic conclusion about the intensity of introgression of this lineage is difficult due to the small number of available studies.

5 Endangerment and Protection of Freshwater Ichthyofauna of Bosnia and Herzegovina

5.1 National Framework

Due to increasing human pressure on aquatic habitats, fish are nowadays one of the most endangered groups of vertebrates (Müller et al. 2018). Given that they represent the most diverse and numerous groups of vertebrates in general, they are a very important part of the total biodiversity on Earth. The importance of fish in natural aquatic ecosystems is not only reflected in the taxonomic diversity of species. As intermediate and often final links in food chains, fish are closely related to other members of aquatic life communities and have various roles in the aquatic ecosystem. Consequently, changes in fish communities' qualitative and quantitative composition inevitably affect other ecosystem components (Mrakovčić et al. 2006). It is known that freshwater fish are of great importance as a source of food (protein) for the human population. Fishing and aquaculture are key economic sectors for food security and poverty reduction at the global level. In many countries, sport fishing is an important component of freshwater fisheries. The importance of fish is manifested through economic importance in commercial fishing and aquaculture, sport fishing importance (hobby, tourism), source of genetic variability, aquaristics, etc. Precisely because of this, there is a clear need to assess the degree of endangerment of certain fish species and their populations and to establish models and carry out activities for their protection (Craig 2015).

When it comes to Bosnia and Herzegovina,[1] the area of nature protection is under the jurisdiction of the entities and Brčko District. Part of the competence in the field

[1] Bosnia and Herzegovina is defined as a decentralized state, which consists of two entities (the Federation of Bosnia and Herzegovina and the Republic of Srpska), and from March 8, 2000, also the Brčko District of Bosnia and Herzegovina (https://www.ustavnisud. ba/public/down/USTAV_BOSNE_I_HERCEGOVINE_bos.pdf). The Federation of Bosnia and Herzegovina consists of federal units (cantons), of which there are 10 each of which has its government and constitution (ustav-federacije-Bosne i Hercegovine.pdf (skupstinabd.ba). According to the data of the Federal Bureau of Statistics of the Federation of Bosnia and Herzegovina (http://fzs.ba/index.php/2022/08/04/karta-bosne-i-hercegovine/), there are 79 local self-government units (municipalities/cities) in the Federation of Bosnia and Herzegovina. On the link of the National Assembly of the Republic of Srpska, we find information on the territorial organization of this entity, which includes 64 local government units, six cities, and 58 municipalities (https://www.narodnaskupstinars.net/?q=la/akti/usvojeni-zakoni/zakon -on-amendments-and-amendments-to-the-law-on-the-territorial-organization-of-the-republic). Brčko District of Bosnia and Herzegovina is a separate administrative unit in Bosnia and Herzegovina that does not belong to any entity but to the state of Bosnia and Herzegovina. In this territorially organized Bosnia and Herzegovina, the management of water resources includes responsibilities at the state level, at the entity level, and at the Brčko District level, and in the Federation of Bosnia and Herzegovina at the cantonal level and at the local administration level, that is, at the municipal level. In the Republic of Srpska, in addition to the republican level, there is also a municipal level of water management. Observed according to different levels of management, the main institutions that participate in water resource management processes are: the Parliamentary Assembly of Bosnia and Herzegovina as the highest legislative body, which makes decisions on the ratification of documents important for the water sector. The Presidency of Bosnia and Herzegovina is responsible for conducting foreign policy and representing the state in international and European institutions and organizations. The Council of Ministers of Bosnia and Herzegovina, through the work and jurisdiction of the Ministry of Foreign Affairs, the Ministry of Foreign Trade and Economic Relations, the Ministry of Communications and Transport and the Directorate for European Integration, coordinates and implements activities related to international agreements and contracts in the water sector. The Ministry of Foreign Affairs is responsible for the implementation of the established policy of Bosnia and Herzegovina and for the development of international relations in accordance with the positions and guidelines of the Presidency of Bosnia and Herzegovina. The Ministry of Foreign Trade and Economic Relations is responsible for performing tasks in the field of defining policy, principles, coordinating activities and harmonizing the plans of entity authorities and institutions on the

of nature protection is regulated at the state level. It refers to the implementation of B&H's international obligations (monitoring and coordinating preparations for the conclusion and implementation of international agreements, coordination for projects in cooperation with international organizations, programs, and funds, coordination of cooperation with institutional structures in B&H, entities, and Brčko District). State-level responsibilities mainly belong to the Ministry of Foreign Trade and Economic Relations of Bosnia and Herzegovina (https://nasljedje.org/propisi-2/).

Republic of Srpska

Article 21 of the Water Law indicates that the Republic of Srpska manages the waters in the Republic of Srpska in the manner defined by that law, as well as fulfills the obligations that Bosnia and Herzegovina has as an international legal entity. The key institutions of the Republic of Srpska that participate in the processes that are essential for the management of water resources are:

- *The National Assembly of the Republic of Srpska, whose jurisdiction includes the adoption of the law on water, in the strategy of water management, the spatial plan and the budget.*
- *The President of the Republic of Srpska has the authority to promulgate laws and other general acts adopted by the National Assembly.*
- *The Government of the Republic of Srpska is responsible for policy implementation and law enforcement, proposing and giving recommendations in the field of legislation, proposing a water management strategy, as well as adopting regional river basin plans.*
- *The Ministry of Agriculture, Forestry and Water Management is responsible for administrative, professional, and other tasks established by law, which relate to the jurisdicton of the Republic of Srpska in the field of water management.*
- *The Ministry of Spatial Planning, Construction and Ecology is responsible for administrative, professional, and other tasks established by law that relate to the jurisdiction of the Republic of Srpska in the field of protection, preservation and improvement of the environment.*
- *The Ministry of Health and Social Protection, which is responsible for administrative, professional, and other tasks established by law relating to the jurisdiction of the Republic of Srpska regarding the healthiness of water for human use.*
- *The Republic's Administration for Inspection Affairs is in charge of conducting inspection supervision of the use of water resources.*
- *The Water Agency of the Sava Regional River Basin and the Water Agency of the Trebišnjica Regional River Basin, which are responsible for the implementation*

---

international level in the field of environmental protection, development, and use of natural resources. The Ministry of Communications and Transport is in charge of preparing international treaties, agreements, and other acts regulating navigation on the Sava River. The Directorate for European Integration is responsible for coordinating the process of integration of Bosnia and Herzegovina into the European Union. The Food Safety Agency is responsible for ensuring the quality of water intended for human consumption.

of water management tasks, which are placed under their jurisdiction by the Law on Waters of the Republic of Srpska and the regulations adopted on the basis of that law.

- *The Fund for Environmental Protection of the Republic of Srpska, which is responsible for collecting and distributing financial resources for environmental protection in the territory of the Republic of Srpska.*
- *City/municipal legislative bodies that are responsible for ensuring public water supply, for the collection and treatment of wastewater in the city/municipality area.*

In the Republic of Srpska (RS), many legal acts deal with nature protection, including the protection and endangerment of fish, where the Law on Nature Protection (Official Gazette of the Republic of Srpska 20/14) stands out, and especially the Regulation on the Red List of protected species of flora and fauna of the Republic of Srpska (Official Gazette of the Republic of Srpska 124/12) and Regulation on strictly protected and protected wild species (Official Gazette of the Republic of Srpska 65/20). Within the Regulation on the Red List of Protected Species of Flora and Fauna of the Republic of Srpska, there is a list with two species of jawless fish and 46 species of fish with a bony skeleton without estimated categories of threat. The text of this Regulation states that the endangered species referred to in this Regulation are protected under the Law on Nature Protection provisions. Based on the Regulation on strictly protected and protected wild species, the protection of strictly protected wild species (except for species that are protected following the regulations in the field of fishing) is implemented by prohibiting the use, destruction, and undertaking of all activities that may endanger these species and their habitats as well as by undertaking measures and activities on population management. The protection of protected wild species, except for species protected under fishing regulations, is carried out by restricting use, prohibiting destruction, and undertaking other activities that cause damage to species and their habitats, as well as by undertaking measures and activities to manage populations. Within this Regulation, the list of strictly protected species includes two species of jawless fish from the family Petromyzontidae and 29 species of fish with a bony skeleton, where 13 species are from the Cyprinidae, five from the Acipenseridae family, four from the Percidae family, three from the Salmonidae family, and one each from the families Anguillidae, Clupeidae, Umbridae, and Cobitidae. The list of protected species includes 20 species of bony fish, ten of which are from the Cyprinidae family, three from the Cobitidae family, and one each from the Acipenseridae, Salmonidae, Thymallidae, Lotidae, Gasterosteidae, Percidae, and Gobiidae families.

Federation of Bosnia and Herzegovina

Article 21 of the Water Law of the Federation of Bosnia and Herzegovina points to the fact that water management is the responsibility of Bosnia and Herzegovina, the Federation of Bosnia and Herzegovina, cantons, cities, and municipalities. The key institutions that participate in the processes essential for the management of water resources in the Federation of Bosnia and Herzegovina are:

- *The Parliament of the Federation of Bosnia and Herzegovina, in addition to other incompetencies, is also responsible for passing the Water Law and the Water Management Strategy.*
- *The President of the Federation of Bosnia and Herzegovina is responsible for signing decisions and ratifying international agreements after their adoption.*
- *The Government of the Federation of Bosnia and Herzegovina is responsible for implementing the policies and laws of the federal government, for proposing and giving recommendations in the area of legislation and for preparing budget proposals for the Parliament of the Federation, as well as for the adoption of water management plans for water areas and their forwarding to the Council of Ministers of Bosnia and Herzegovina.*
- *The Ministry of Agriculture, Water Management and Forestry, in addition to other responsibilities, is in charge of administrative, professional, and other tasks established by law that relate to the competences of the Federation in the field of water management.*
- *The Ministry of Tourism and Environment also has certain jurisdiction in administrative, professional, and other tasks established by law, which relate to the competences of the Federation in the field of protection, preservation, and improvement of the environment.*
- *The Ministry of Health is responsible for administrative, professional, and other tasks established by law that relate to the competences of the Federation related to the healthiness of water for human consumption.*
- *The Directorate for Inspection Affairs is responsible for conducting inspection supervision of the use of water resources.*
- *The Agency for the Water Area of the Sava River and the Agency for the Water Area of the Adriatic Sea, which are responsible for the implementation of water management tasks that, according to the Water Act of the Federation of Bosnia and Herzegovina and the regulations adopted on the basis of that law, are placed in their jurisdiction, including the preparation of management plans and program of measures for water areas under their jurisdiction.*
- *The Environmental Protection Fund of the Federation of Bosnia and Herzegovina is responsible for collecting and distributing financial resources for environmental protection on the territory of the Federation of Bosnia and Herzegovina.*
- *The cantonal legislative and executive bodies are responsible for water management, established by the Water Act and cantonal regulations, ensuring water supply and collection and purification of waste water in the canton.*
- *City/municipal legislative and executive bodies are responsible for water management tasks established by the Water Act and city/municipal regulations, ensuring public water supply as well as collecting and purifying waste water in the city/municipality area.*

In the Federation of Bosnia and Herzegovina (FB&H), the issue of nature protection is regulated by the Law on Nature Protection (Official Gazette of the Federation of Bosnia and Herzegovina 66/13) as well as numerous bylaws and regulations. One such act is the Red List of Fauna of the FB&H (Official Gazette of the Federation of Bosnia and Herzegovina 7/14), containing 115 freshwater and marine fish taxa. Regarding exclusively freshwater fish species, for the FB&H, seven species have been established in the critically endangered category, 16 species in the endangered category, seven in the vulnerable category, and seven species in the data deficient category. Based on the data taken from the portal http://e-prirodafbih.ba/en/protectedspecies/, it was established that 22 species of freshwater fish are protected in FB&H. As strictly protected species, seven species of bony fish (four from the Cyprinidae family, two from the Salmonidae family, and one from the Acipenseridae family) are listed, while two jawless fish species (both from the Petromyzontidae family) and 13 fish species with a bony skeleton are listed as protected species (eight from the family Cyprinidae, two from the family Acipenseridae, and one each from the families Salmonidae and Umbridae).

It is evident that a greater number of fish species are protected in the Republic of Srpska, in both protection regimes. Thus, only seven species of fish are strictly protected in the FB&H, and 31 in the RS, i.e. more than four times compared to the FB&H. A somewhat smaller, but still noticeable difference is also found in the number of protected species, as 15 species are protected in the FB&H, and 20 in the RS (Table 3). What has also been noticed is the uneven nomenclature, so there are several cases where a species is protected under one name in FB&H and under another in RS (e.g., *Pungitius platygaster* and *Gasterosteus aculeatus)*. Differences in the names of fish species are present in both scientific and domestic names.

According to Simonović et al. (2014) mudminnow (*Umbra krameri*) and Danube salmon (*Hucho hucho*) (Fig. 6) are considered the two most threatened fish species of the river Sava catchment, where various types of riverbed modifications, especially the damming, were seen the most prominent threatening factors for fish diversity.

For the preservation and sustainable use of fish species, the Law on Fisheries (Official Gazette of the Republic of Srpska 72/12), as well as certain bylaws, and above all, the Rulebook on fishing methods, fishing conditions, tools, equipment, and means of fishing are of great importance (Official Gazette of RS 62/20). A positive shift in the measures related to protecting certain fish species is noticeable. A temporary ban on hunting or ban on hunting individuals below a certain body length was established for 26 fish species. Hunting for around 20 species is permanently prohibited (all species of the genus *Eudontomyzon*, *Acipenser gueldenstaedtii*, *Acipenser nudiventris*, *Acipenser ruthenus*, *Acipenser stellatus*, *Acipenser sturio*, *Huso huso*, *Alosa immaculata*, *Anguilla anguilla*, *Carassius carassius*, *Cobitis elongata*, *Gymnocephalus baloni*, *Gymnocephalus schraetster*, *Leucaspius delineatus*, *Sabanejewia balcanica*, *Salmo marmoratus*, *Telestes souffia*, *Tinca tinca, Zingel streber*, and *Zingel zingel*). Also, it is stated that hunting non-native species of fish is unlimited.

The Law on Freshwater Fisheries (Official Gazette of the Federation of Bosnia and Herzegovina 64/04 and 27/10) regulates the protection and use of FB&H fish stocks. A new Law on Freshwater Fisheries of the Federation of Bosnia and Herzegovina from 2021 is currently in the form of a preliminary draft (Law on

Table 3 Overview of strictly protected and protected fish species in B&H (SP— strictly protected, P—protected)

Taxon	Protected in RS	Protected in FB&H
Lampetra fluviatilis	/	P
Lampetra planeri	/	P
Eudontomyzon danfordi	SP	/
Eudontomyzon mariae	SP	/
Acipenser naccarii	/	P
Acipenser sturio	SP	SP
Acipenser gueldenstaedtii	SP	/
Acipenser nudiventris	SP	/
Acipenser stellatus	SP	/
Acipenser ruthenus	P	/
Huso huso	SP	P
Anguilla anguilla	SP	/
Alosa immaculata	SP	/
Salmo marmoratus	SP	SP
Salmo obtusirostris	SP	SP
Salmo trutta	P	/
Hucho hucho	SP	P
Thymallus thymallus	P	/
Umbra krameri	SP	P
Lota lota	P	/
Pungitius platygaster	P	/
Gasterosteus aculeatus	/	P
Leucos basak	/	P
Phoxinellus alepidotus	/	P
Delminichthys ghetaldii	SP	P
Delminichthys adspersus	/	P
Leucaspius delineatus	SP	/
Leuciscus aspius	P	/
Leuciscus idus	P	/
Telestes metohiensis	SP	SP
Telestes turskyi	/	SP
Telestes dabar	SP	/
Telestes souffia	SP	/
Squalius svallize	SP	/
Squalius microlepis	/	SP
Pelecus cultratus	P	/
Chondrostoma knerii	/	P
Chondrostoma phoxinus	/	SP
Alburnoides bipunctulus	P	/
Ballerus sapa	P	/
Barbus balcanicus	P	/
Rhodeus amarus	P	/

(continued)

Table 3 (continued)

Taxon	Protected in RS	Protected in FB&H
Rutilus virgo	P	/
Vimba vimba	P	/
Gobio obtusirostris	P	/
Romanogobio uranoscopus	SP	P
Romanogobio kesslerii	SP	P
Romanogobio vladykovi	SP	/
Aulopyge huegelii	/	P
Alburnus sarmaticus	SP	/
Carassius carassius	SP	/
Cyprinus carpio	SP	/
Tinca tinca	SP	/
Gymnocephalus baloni	SP	/
Gymnocephalus schraetser	SP	/
Zingel streber	SP	/
Zingel zingel	SP	/
Sander volgensis	P	/
Cobitis narentana	SP	/
Cobitis elongata	P	/
Misgurnus fossilis	P	/
Sabanejewia balcanica	P	/
Proterorhinus semilunaris	P	/

Fig. 6 *Umbra krameri* (left) from Matura River, near Srbac (Sava river basin) (right) (Photo: Golub D. 2018)

Freshwater Fisheries of the Federation of Bosnia and Herzegovina 2021). Within the bylaws, the Rulebook on the way, tools, and means used for fishing (Official Gazette of the Federation of Bosnia and Herzegovina 63/05) as well as the Rulebook on amendments to the Rulebook on the way, tools, and means used for fishing (Official Gazette of the Federation of Bosnia and Herzegovina 31/09). These regulations set a fishing ban in a certain period of the year for 11 fish species, while a limitation in terms of body size is prescribed for 27 species. No fish species is protected by a permanent ban on hunting.

The Law on Freshwater Fisheries of the Brčko District of Bosnia and Herzegovina (Official Gazette of the Brčko District of Bosnia and Herzegovina 35/05 and 19/07) regulates the area of sustainable use and protection of the ichthyofauna of this part of B&H. The Ordinance on Fish Protection (Official Gazette of the Brčko District of BiH 40/07) prescribes the hunting season for five fish species in a certain part of the year and limits the fishing body length for 16 fish species. Tables 4 and 5 provide a comparative overview of measures for protecting fish species in both B&H entities and Brčko District.

The data analysis found that the measures for protecting the fish stock through hunting are relatively inconsistent, i.e. for the same species, a longer or shorter hunting period is prescribed. However, this often involves the same water bodies located in both entities and Brčko District (e.g., hunting grounds for *Silurus glanis*, for the Sava River). Another example of non-compliance relates to saplings (*Hucho hucho*), where in RS the hunting period is from February 1 until August 31, and in FB&H from January 1 until May 31. A similar situation was observed regarding the body length values of fish below which they may not be caught. One example is the minimum body length for *Hucho hucho*, which is 100 cm in the RS and 70 cm in the FB&H. It can also be noted that for introduced fish species in the Rulebook from RS, there are no minimum body lengths and no hunting period, i.e. the maximum possible catch is encouraged.

The largest number of species protected by the hunting period is in the RS, while the largest number of species for which the minimum body length is prescribed is in the FB&H. A certain number of species under a permanent ban on fishing in the RS regulations are allowed to be hunted in FB&H and Brčko District (e.g., *Acipenser ruthenus*, *Tinca tinca*, *Salmo marmoratus*, and others).

5.2 *International Framework*

International law, which builds on the legal protection of each country, is an essential mechanism in protecting freshwater fish species. B&H has acceded to numerous international conventions and agreements regulating the preservation and improvement of ichthyofauna, both through the protection of habitats and the protection of species. In this sense, the Convention on Biological Diversity stands out as a critical regime for protecting all aspects of biological diversity. Until now, B&H take part in preparation of several reports and strategic documents to the CBD (First National Report, Second National Report, Third National Report, Fourth National Report, Fifth National Report, Sixth National Report, and Strategy and Action Plan for the Protection of Biological Diversity of Bosnia and Herzegovina (2015–2020) (http://bih-chm-cbd.ba/?page_id=21).

As significant international agreements and documents, the following stand out: Council Directive 92/43/EEC on the conservation of natural habitats and of wild fauna and flora (by which special areas for habitat conservation are classified, i.e. based on this directive, an ecological network of areas—Natura 2000) is formed,

Table 4 Overview of fishing ban in the regulations of the entities of B&H and Brčko District

Period of a fishing ban		
Rulebook on fishing methods, fishing conditions, tools, equipment, and means for fishing (Official Gazette of the RS, 62/20)	Rulebook on methods, tools, and means used for fishing (Official Gazette of the Federation of Bosnia and Herzegovina, 63/05)	Rulebook on Fish Protection (Official Gazette of the Brčko District of Bosnia and Herzegovina, 40/07)
Cyprinus carpio 01/04/–31/05	*Cyprinus carpio* 01/04/–31/05	*Cyprinus carpio* 01/04/–31/05
Esox lucius 01/02/–31/03	*Esox lucius* 01/02/–31/03	*Esox lucius* 01/02/–31/03
Silurus glanis 01/05/–15/06	*Silurus glanis* 01/04/–15/06	*Silurus glanis* 16/04/–15/06
–	*Acipenser ruthenus* 01/03/–31/05	*Acipenser ruthenus* 01/03/–31/05
Sander lucioperca 01/03/–30/04	*Stizostedion lucioperca* 01/03/–31/05	*Stizostedion lucioperca* 01/03/–31/05
Hucho hucho 01/02/–31/08	*Hucho hucho* 01/01/–31/05	–
Salmo obtusirostris 01/10/–01/03	*Salmothymus obtusirostris oxyrhynchus* 01/10/–30/04	–
Salmo trutta 01/10/–01/03	*Salmo trutta m. fario* 01/10/–28/02	–
	Salmo trutta m. lacustris 01/10/–28/02	–
Thymallus thymallus 01/03/–31/05	*Thymallus thymallus* 01/01/–30/04	
–	*Salmo marmoratus* 01/10/–31/03	–
Abramis brama od 15/04/–31/05	–	–
Barbus balcanicus od 15/04/–31/05	–	–
Barbus barbus od 15/04/–31/05	–	–
Chalchalburnus chalcoides 15/04/–15/06	–	–
Chondrostoma nasus 15/04/–31/05	–	–
Leuciscus idus od 15/04/–31/05	–	–
Lota lota 01/12/–01/03	–	–
Rutilus pigus od 15/04/–31/05	–	–
Sander volgensis 01/03/–30/04	–	–
Squalius cephalus 15/04/–31/05	–	–
Squalius svallize 01/04/–31/05	–	

Table 5 Overview of the minimum values of the fish body length in the regulations of the entities of B&H and Brčko District. The gray part contains species that are listed only in the individual rulebooks

The permissible fishing body length of fish		
Rulebook on fishing methods, fishing conditions, tools, equipment, and means for fishing (Official Gazette of the RS, 62/20)	Rulebook on methods, tools, and means used for fishing (Official Gazette of the Federation of Bosnia and Herzegovina, 63/05)	Rulebook on Fish Protection (Official Gazette of the Brčko District of Bosnia and Herzegovina, 40/07)
Abramis brama 20 cm	*Abramis brama* 20 cm	*Abramis brama* 20 cm
Barbus barbus 25 cm	*Barbus barbus* 35 cm	*Barbus barbus* 20 cm
Chondrostoma nasus 20 cm	*Chondrostoma nasus* 25 cm	*Chondrostoma nasus* 20 cm
Cyprinus carpio 30 cm	*Cyprinus carpio* 35 cm	*Cyprinus carpio* 30 cm
Esox lucius 40 cm	*Esox lucius* 40 cm	*Esox lucius* 40 cm
Lota lota 25 cm	*Lota lota* 20 cm	*Lota lota* 20 cm
Rutilus pigus 20 cm	*Rutilus pigus virgo* 15 cm	*Rutilus pigus virgo* 20 cm
Silurus glanis 60 cm	*Silurus glanis* 60 cm	*Silurus glanis* 60 cm
Sander lucioperca 40 cm	*Stizostedion lucioperca* 40 cm	*Stizostedion lucioperca* 40 cm
–	*Aspius aspius* 20 cm	*Aspius aspius* 35 cm
–	*Acipenser ruthenus* 40 cm	*Acipenser ruthenus* 50 cm
–	*Leuciscus cephalus* 20 cm	*Leuciscus cephalus* 20 cm
–	*Tinca tinca* 20 cm	*Tinca tinca* 20 cm
Chalchalburnus chalcoides 20 cm	*Chalcalburnus chalcoides* 30 cm	–
Hucho hucho 100 cm	*Hucho hucho* 70 cm	–
Leuciscus idus 20 cm	*Leuciscus idus* 20 cm	–
Salmo obtusirostris 30 cm	*Salmothymus obtusirostris oxyrhynchus* 30 cm	–
Salmo trutta 25 cm	*Salmo trutta m. fario* 25 cm	–
	S. trutta m. lacustris 40 cm	–
Thymallus thymallus 30 cm	*Thymallus thymallus* 30 cm	–
*Vimba vimba*15 cm	*Vimba vimba* 20 cm	–
Barbus balcanicus 15 cm	*Oncorhynchus mykiss* 25 cm	*Abramis blicca* 20 cm
Gobio spp. 10 cm	*Salmo marmoratus* 60 cm	*Ctenopharyngodon idella* 30 cm
Leuciscus aspius 30 cm	*Salvelinus alpinus* 25 cm	*Hypophthalmichthys molitrix* 30 cm
Sander volgensis 25 cm	*Salvelinus fontinalis* 25 cm	–
Squalius albus 20 cm	*Zingel streber* 15 cm	–
Squalius cephalus 20 cm	*Zingel zingel* 25 cm	–
Squalius svallize 20 cm	–	–
Pelecus cultratus 18 cm	–	–
Perca fluviatilis 10 cm	–	–
Telestes turskyi 20 cm	–	–

Convention on the conservation of European wildlife and natural habitats or the Bern Convention (based on which the ecological network of the area—the Emerald network is formed), the Convention on the Conservation of Migratory Species of Wild Animals or the Bonn Convention, as well as the Washington Convention on International Trade in Endangered Species of Wild Fauna and Flora (CITES). Within the project "Support for the implementation of the Birds Directive and the Habitats Directive in BiH 2012–2015" a list of potential Natura 2000 areas with codes, areas, species, and habitats was created. Guidelines for the preparation of Management Plans for Natura 2000 areas, as well as indicative Management Plans for three potential Natura 2000 areas in different biogeographical regions, were drawn up. By Amendments and Supplements to the Spatial Plan of the Republic of Srpska until 2025, 63 potential areas of the Natura areas are identified. It will be an area with surface of 294,952.38 hectares (11.96% of the territory of the Republic of Srpska). On the other hand, Bern Convention, Bonn Convention, CITES and Habitats Directive, in addition to some other parameters, served as a basis for the selection of species that are categorized as strictly protected and protected (Regulation on strictly protected and protected wild species (Official Gazette of the Republic of Srpska 65/20)).

The Convention on Wetlands or the Ramsar Convention stands out as important. Wetland habitats are vital as centers of great ichthyological diversity, production, and feeding grounds for juveniles. Three Bosnia and Herzegovina wetland areas are also on the Ramsar list: Bardača (Ramsar site no. 1658, designated in 2007), Hutovo Blato (Ramsar site no. 1105, designated in 2001), and Livanjsko polje (Ramsar site no. 1786, designated in 2008). Also important is the EU Water Framework Directive, which represents the current European regulatory framework for water protection, and where fish fauna is used, among other things, as a monitoring indicator and an element for assessing the ecological state of waters (EU 2000). It should be emphasized that the national legislation is harmonized to a greater or lesser extent with the mentioned international documents (https://nasljedje.org/propisi-2/).

A certain number of activities in terms of preserving the freshwater ichthyofauna of B&H are carried out within the framework of the International Commission for the Protection of the Danube River and the International Sava River Basin Commission. The cooperation between the ICPDR and the International Sava River Basin Commission has helped Sava countries tackle issues of crucial importance for transboundary development in the Sava River Basin (https://www.icpdr.org/main/danube-basin/bosnia-herzegovina). Besides this, B&H participated (or still participating) in several international projects that contributed to a more adequate management of water resources, primarily the Sava river basin. Some of them are Drina River Basin Water Resources Management, (WBIF), West Balkans Drina River Basin Management (GEF-SCCF), Integrating Climate Change and Reducing the Risk of Flooding in the Vrbas River Basin (UNDP) (http://www.mvteo.gov.ba/Content/Read/vodni- resources-links).

In the context of the area of significance and importance of ichthyofauna conservation, the Food and Agriculture Organization (FAO), the UN agency for food and agriculture, should be mentioned. As a member of FAO, B&H took part in some activities concerning fish resources. The result of those activities was the document Analysis of the Fisheries and Aquaculture Sector in B&H (FAO 2015). On the other hand, the FAO report, The State of the World's Aquatic Genetic Resources for Food and Agriculture (FAO 2019), represents the first global assessment of the state of aquatic genetic resources for food and agriculture and focuses on farmed aquatic organisms and their wild relatives. B&H has not yet participated in preparing this report (http://www.fao.org/3/ca5256en/CA5256EN.pdf).

6 Models of Conservation of Fish Resources

Protection of ichthyofauna implies in-situ and ex-situ conservation that can be achieved at different levels, from protecting aquatic habitats to preserving cell cultures.

In-situ conservation is "the preservation of original ecosystems and natural habitats, as well as the maintenance and recovery of viable populations of species in their natural environment and, in the case of native or cultivated species, in the environment in which their characteristic traits have evolved" (CBD 1992). In-situ conservation is a highly desirable strategy because it maintains the population of aquatic organisms in a natural habitat or culture system that allows them to develop most adequately.

Protected areas that include aquatic ecosystems are one of the most widespread and adequate methods used for the in-situ conservation of fish species. This type of protection implies that these areas have multiple uses, i.e. that it is possible to engage in, for example, sport fishing as long as fish resources are sustainably used and protected. In B&H, 37 areas are protected on a total area of 105,602.18 hectares, representing less than 3% of the territory of B&H. A total of 32 areas are protected in the RS and 12 in the FB&H. No area has been declared protected in Brčko District. In addition, three areas in B&H are protected by the international Ramsar Convention on Wetlands (https://nasljedje.org/propisi-2/; https://cin.ba/baza-zasticenih-podrucja/). Well-managed aquatic protected areas represent in-situ gene banks for fish species, which is of great importance for farmed fish species and overall biological diversity. However, their role is often not sufficiently recognized. The example of in-situ protection of the fish resources through protected areas is River Una (Fig. 7), which is in a significant part of its course in a certain protection regime. Namely, the National Park "Una" was established in the FB&H, while the Una Nature Park was established in the RS. The area of the Una National Park is located in the western part of B&H and extends along the valley of the upper course of the Una River as well as around the canyon of the Unac River (the right tributary of the Una) (it also includes the slopes of the mountains Plješevica, Grmeč, and Osječenica). In accordance with the legal obligations of the National Park "Una," a Decision on fishing that takes place within the National Park is made on an annual basis. Decision on the requirements and methods of sports and recreational fishing

Fig. 7 Una River, near Novi Grad (left) and *Barbus barbus* (right) (Photo: Dekić R. 2022)

activities on watercourses in the National Park "Una" for the year 2022 defines the permitted methods of fishing in the established areas:

- fly fishing on the "catch and release" principle,
- fishing using a water ball (Wasserkugle) with a maximum of 3 (three) artificial baits-dry flies (buzz),
- spinning fishing with blinkers and other artificial baits.

This Decision defines six basic areas and one special area, and in 2022, seven permanent fishing prohibition zones were established in the area of the National Park "Una." In two areas, fishing is allowed exclusively on the "catch and release" principle. In three areas, where sport-recreational fishing is carried out exclusively by the method of using water balls, a maximum of two specimens of fish are allowed to be caught per day, in accordance with the permitted measures. Combined fishing is allowed for one area, where in a certain period it is allowed by the method of fly fishing, and in another by using a water ball (Official Gazette of the Federation of Bosnia and Herzegovina 44/08).

The Nature Park "Una" covers four municipalities in the Republic of Srpska: Krupa on Una, Novi Grad, Kostajnica, and Kozarska Dubica. The length of the Una River included in the protected area is 95 km (Official Gazette of the Republic of Srpska 79/19). According to the Protection Study, 39 species of fish have been found in the Una River, and traditional use of the area by the local population and sport fishing with the use of fishing tools and equipment for sport or recreation is allowed. In addition to the above, the damming of the Una River and the regulation of the riverbed and the strengthening of the river banks, the removal of natural aquatic vegetation and coastal vegetation, and the direct discharge of waste sewage water and the construction of permeable septic tanks for households, the entry of heavy machinery into the water course (extraction of gravel and other construction activities) and destruction of spawning grounds of endangered fish species are prohibited. All these mechanisms indirectly contribute to the preservation of the habitat quality of the fish species that inhabit the Una River (Nature Park Una Protection Study, 2019).

On-farm in-situ concept of conservation of fish species usually refers to aquaculture (farms located in their natural environment). However, producers engaged in fish farming prioritize the improvement of the characteristics of farmed fish species, not preserving the characteristics of wild species from which the farmed forms originate. As an economic branch that includes various forms of controlled fish farming, aquaculture is carried out in natural, salt, and freshwaters (seas, rivers, lakes, etc.) and artificial water systems. Aquaculture is used in about 140 countries, and about 200 different species are cultivated. Fish are most farmed, followed by turtles, clams, other molluscs, aquatic plants, frogs, etc. Aquaculture is still not sufficiently developed, especially concerning livestock and crop farming, but it is considered to improve in the future (FAO 2020b).

The main species grown in aquaculture in Bosnia and Herzegovina are salmonids, mainly rainbow trout (*Oncorhynchus mykiss*), grown in tanks and cages (FAO 2015).

When the previous preventive forms of protection of endangered species fail, i.e. when the species abundance in its natural habitat falls to the critical limit of extinction, some of the methods of ex-situ protection are applied. Ex-situ *conservation* means "conserving components of biological diversity outside their natural habitats" (CBD 1992). Ex-situ conservation of AqGR implies a series of procedures and methods for conservation, cultivation, and reproduction outside their natural habitats. Ex-situ habitats are zoos, private collections of living organisms (aquariums, etc.), gene banks, and specialized laboratories where living material collected in natural habitats can be maintained and reproduced on special nutrient media under strictly controlled conditions for a long time. Three approaches are available for preserving ichthyofauna in ex-situ model: in vivo (individuals or populations are kept in research institutions and aquariums), in vitro as cryopreserved spermatozoa (the most important technique available), and less often as embryos or tissues containing DNA. The goal of ex-situ protection methods is the return or reintroduction of endangered species to their natural habitats from which they have disappeared or are in great danger of disappearing.

Regarding the ex-situ protection of fish species, the situation in B&H is not at an enviable level. The first and only public aquarium in B&H is opened in 2002 at the Faculty of Science and Mathematics in Banja Luka. Currently, there are about 20 local fish species in the aquarium, among others (Fig. 8).

The function of the aquarium from the aspect of biodiversity protection in ex-situ conditions is of particular importance because, in such conditions, it is possible to maintain rare and endangered species and enable their reproduction and rearing of young individuals. This way, a contribution can be made to the survival and protection of rare and endangered species with the possibility of reintroduction and increasing numbers in natural populations (Dekić et al. 2013b). The educational role of the aquarium is also undeniable, given that a large number of visitors, primarily school-age children, visit it (Simić et al. 1999).

Gene banks containing samples of fish species do not exist in B&H.

Fig. 8 Public aquarium at the Faculty of Science and Mathematics, University of Banja Luka (Photo: Golub D 2017)

7 The Sector of Aquaculture and Freshwater Fisheries in Bosnia and Herzegovina

In Bosnia and Herzegovina, the beginning of aquaculture dates to the end of the 19th and the beginning of the twentieth century, with the establishment of the first trout and carp fish farms. The establishment of a fish farm called "Vrelo Bosne" near Ilidža in 1894 marked the beginning of modern fishing in B&H, and therefore it can be concluded that B&H has a long tradition of freshwater fish breeding. In B&H, hatcheries and fish farms are numerous and of different capacities, areas, and annual production. According to the type of fish, fish farms and ponds are divided into salmonid, cyprinid, and marine. The capacities of the farming facilities (fishponds and cages) fluctuated in the last decade (Fig. 9). In 2021, trout fishponds covered a total of 93,389 m^2, carp fishponds 2043.2 ha, while the capacity of the cage production was 89,550.0 m^3 (Agency for Statistics of Bosnia and Herzegovina 2022).

The gender structure of employees in the aquaculture sector in B&H is predominantly male (Fig. 10). The number of people employed in the sector mostly declined over the last decade (Table 6, Agency for Statistics of Bosnia and Herzegovina 2022).

The lack of effective regulation of fish farm licensing has been a long-lasting problem. In 2015, there were about 40 trout farms officially registered, but the field research by Hamzić et al. (FAO 2015) has shown that at least 97 such farms existed, located in mountainous areas of B&H. Of these farms, 80 had tank production, and 17 had cage production. In addition to rainbow trout, there are also 28 large pond farming farms that produce mainly carp and predatory fish (perch, European catfish,

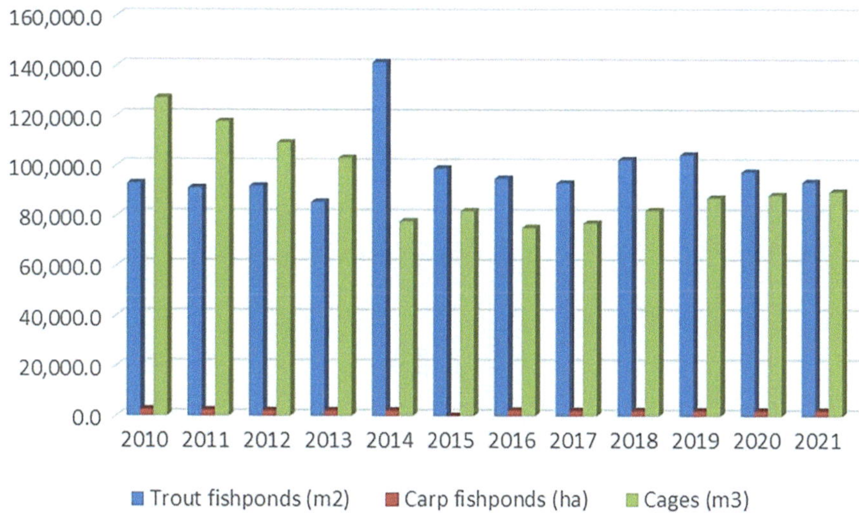

Fig. 9 Structure of facilities in exploitation in Bosnia and Herzegovina

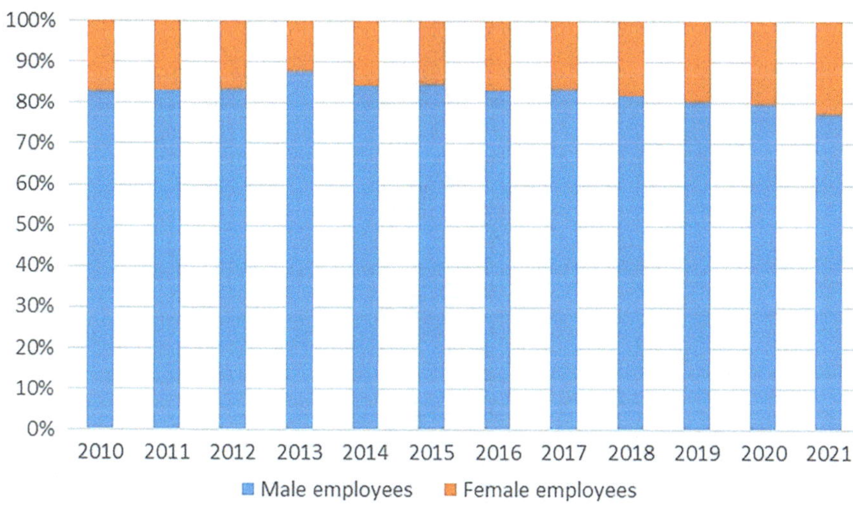

Fig. 10 The percentage of male and female workers in the aquaculture sector in B&H (Agency for Statistics of Bosnia and Herzegovina 2022)

pike) (FAO 2015). The latest data show that there are 46 business entities for the production of edible fish registered in B&H, of which 11 have significant facilities for the production of fry. In addition to registered companies, there are also small family fish farms, of which about 100 have an average production of 500–1500 kg of fish. Of all the production entities, two are marine, while the rest are inland freshwater farms (located on rivers and reservoirs).

Table 6 The number of male and female workers in the aquaculture sector in Bosnia and Herzegovina in the period 2010–2021

Employees	2010	2011	2012	2013	2014	2015	2016	2017	2018	2019	2020	2021
Males	378	326	340	303	295	256	313	330	309	311	234	257
Females	79	66	67	41	55	46	64	66	68	76	59	75

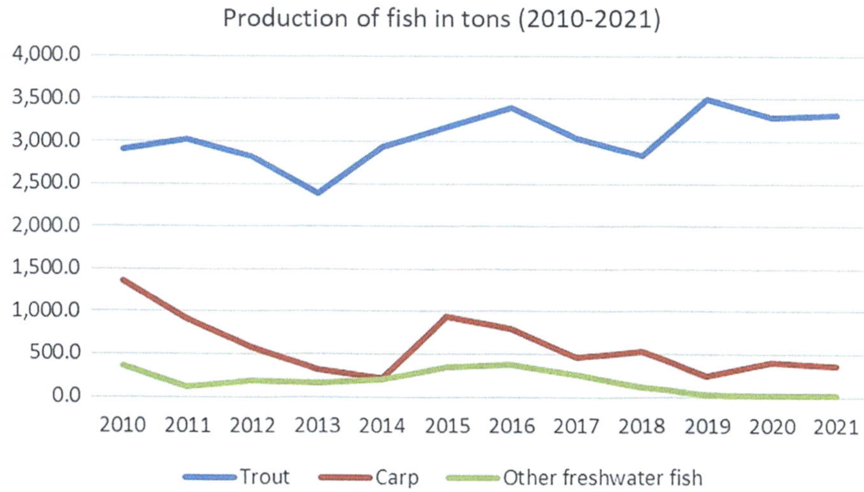

Fig. 11 Production of fish for consumption in B&H expressed in tons

According to the data Agency for Statistics of B&H (2022), the production of fish for consumption (trout) has grown considerably since 2010 (2900.9 t) in 2021 (3305.9 t) in B&H (Fig. 11). On the other hand, a decline in carp production was recorded in B&H, from 1355.8 t in 2010 to 356.2 t in 2022. A decline in the production of other freshwater fish was also recorded (355.0 t in 2010 compared to 20.4 t in 2021). According to FAO statistical data, B&H, i.e. FB&H (RS and BD have no registered production), is among the ten largest producers of edible brook trout (8th place) and edible brook trout (6th place) in the world.

In 2017, the average consumption of fish meat in Europe was 21.5 kg per capita, while B&H is unfortunately in last place with an annual consumption of 5.8 kg per capita. The largest amount of freshwater aquaculture products is intended for the domestic market, while a part is exported to the European market (CEFTE, EFTE countries, etc.). (Agency for Statistics of Bosnia and Herzegovina 2020). Fish and fish products are one of the strategic products of animal origin that B&H can export to the demanding EU market (Fig. 12). For a successful export, it is necessary to fulfill a number of conditions that include the application of international and European Union standards related to the area of food safety, both in legal regulations and in procedures, and the fulfillment of all necessary veterinary-hygienic and sanitary conditions (such as the HACCP standard for this branch industry), but also other required standards by the manufacturer/exporter.

In the freshwater aquaculture sector in B&H, the most important fish species are salmonid (rainbow trout, brown trout, and brook trout) and cyprinid (carp, silver carp and grass carp).

Of the cold-water fish species, the area of Bosnia and Herzegovina is characterized by the dominant commercial cultivation of rainbow trout (*Oncorhynchus mykiss*) due to its pronounced growth characteristics and favorable conditions for

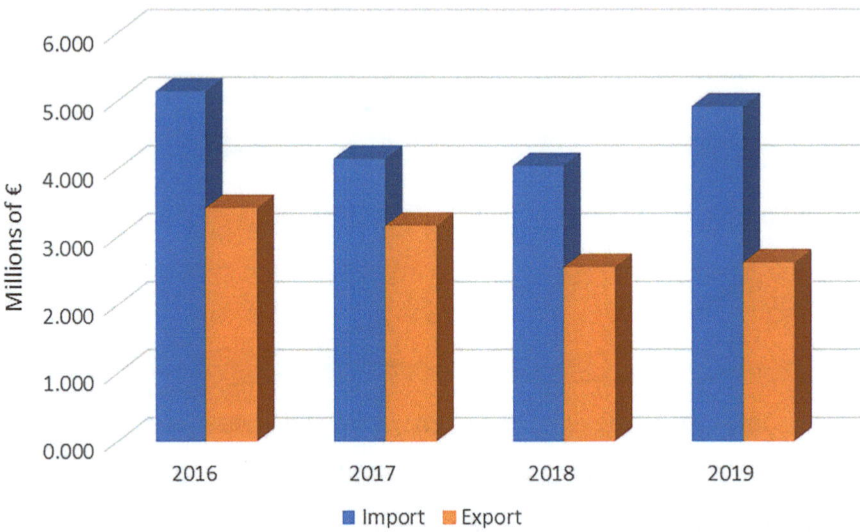

Fig. 12 Bosnia and Herzegovina's export and import of fish and fish products in the period 2016–2019 (adapted from the Agency for Statistics of Bosnia and Herzegovina 2020)

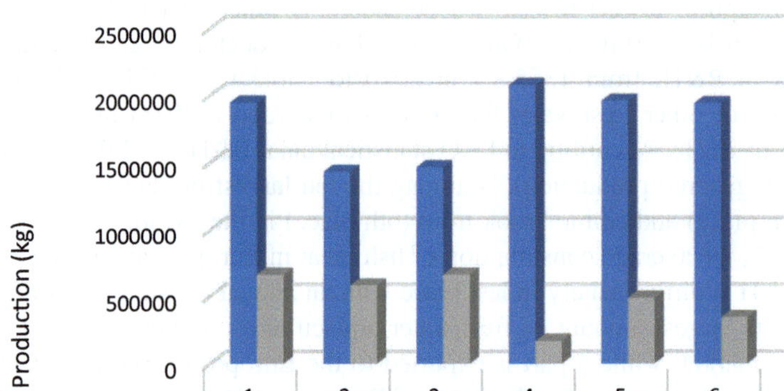

	1	2	3	4	5	6
Brown trout (kg)	1945777	1431942	1465187	2080405	1959223	1941675
Carp (kg)	663055	581719	664187	166638	488843	346294

Fig. 13 Salmonid and cyprinid fish production in Republic of Srpska from 2016 to 2021

its cultivation, which are primarily reflected in the quality and quantity of available water (Savić et al. 2013). When it comes to warm-water fish species, carp (*Cyprinus carpio*) is dominantly farmed in B&H freshwaters, with a smaller share of silver carp (*Hypophthalmichthys molitrix*), bighead carp (*Hypophthalmichthys nobilis*), grass carp (*Ctenopharyngodon idella*), and wels catfish (*Silurus glanis*) (FAO 2015).

In the RS, aquaculture includes exclusively cyprinid and salmonid freshwater ponds (Fig. 13). The dominant species grown in cyprinid ponds is carp (*Cyprinus carpio*). As accompanying species (in polyculture), there are wels catfish (*Silurus*

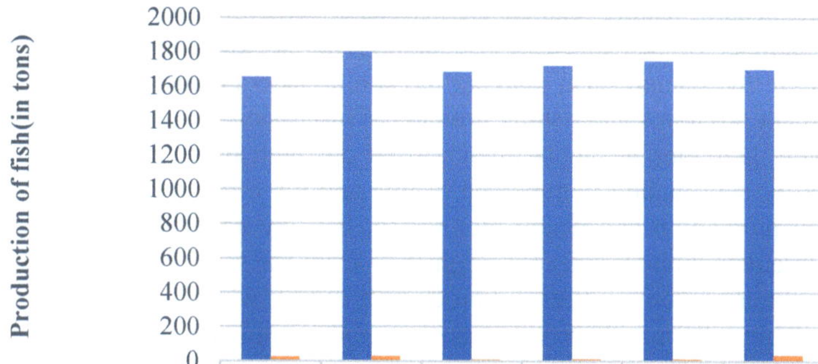

	2016	2017	2018	2019	2020	2021
■ Trout	1657	1803	1687	1722	1750	1701
■ other fish species	25	30	12	13	13.1	37

Fig. 14 Fish production in the FB&H from 2016 to 2021

glanis), pike (*Esox lucius*), pikeperch (*Stizostedion lucioperca*), grass carp (*Ctenopharyngodon idella*), bighead carp (*Hypophthalmichthys nobilis*), and silver carp (*Hypophthalmichthys molitrix*). In salmonid ponds, the dominant species is rainbow trout (*Oncorhynchus mykiss*), although, in recent times, ponds with brown trout (*Salmo trutta*) are becoming more common. There are several fishponds where brook trout is raised (e.g., EKO ZONA Ugarski brzaci d.o.o., Kneževo on river Ugar, Boračko jezero pond on Boračko jezero, LAKS EXPORT-IMPORT pond, d.o.o. Mostar on Buna River). It is certainly worth mentioning the hatchery of brown trout and Danube salmon Klašnik (Vrbas River), which functions as part of the Banja Luka sports-fishing society.

Only salmonid aquaculture is represented in the FB&H, and primarily rainbow trout is cultivated (Fig. 14).

The fisheries sector in B&H includes commercial and sport fishing in freshwaters and sport fishing in seawaters. About 50 species of the total freshwater ichthyofauna are significant in sport fishing. At the same time, commercial fishing is based on a smaller number of species, with the following being economically significant: carp, catfish, perch, pike, bream, roach, silver carp, Danubian roach, etc.

Commercial fishing in B&H is carried out in the Sava River (Republic of Srpska and Brčko District). In contrast, commercial fishing practically does not exist in the Federation of B&H territory. It is regulated by legal acts, and permits are issued by competent institutions. The number of registered commercial fishers has changes over the years. Most are in Gradiška, Bosanski Brod, Bijeljina (RS), and Brčko District (three registered commercial fishers). According to the latest FAO report (FAO 2020a) on the state of fisheries in the Mediterranean and Black Seas, B&H reported no operational fishing fleets.

Sport/recreational fishing is more prominent in the territory of B&H. The fishing waters are managed by Sports Fishing Associations, which manage a particular area following strategic documents (fishing fundamentals) and work plans. The rulebooks on sport fishing regulate the periods of the fishing ban, the allowed baits, and the number of fish that can be caught. The stocking process is carried out according to the management plans of the fishing associations, as well as fisheries development programs and fishing grounds as professional bases.

Stocking activities are most often carried out with those fish species that are attractive for sport and/or commercial fishing and for which there is organized and controlled production and health monitoring. It is recommended that stocking be done with autochthonous lineages originating from the same catchment area. Juveniles are obtained from registered reproductive centers when such stocking material is unavailable (absence of proven autochthonous populations, too low population density, etc.).

Sports-fishing associations function within the Association of Sports Anglers of B&H founded in 1906. The Association of Sports Anglers of B&H was re-established in 2005 under the Law on Associations and Foundations of Bosnia and Herzegovina and the Law on Sports of Bosnia and Herzegovina. The association includes the associations from the FB&H, RS, and Brčko District. According to the available data, the FB&H Sports Fishing Federation has 71 member associations, RS 54 members, and BD one member.

The preconditions for licensing (issuing fishing rights) for commercial and sport/recreational fishing are uniformly and clearly defined in the relevant fisheries laws in both entities and Brčko District. The laws also prescribe how the licensed water (fishing water) should be managed during the authorized period of 10, 15, and 20 years in the FB&H, BD, and RS, respectively.

The number of sports fishers varies annually, and accordingly, the number of licenses issued. Thus, in 2019, 9507 annual and 8220 daily, weekly, and monthly licenses were sold in the RS, while in 2020, there were 7507 seasonal licenses and 9277 others. In the FB&H, during 2019, 11,802 annual licenses (of which 695 were cadets) and 14,449 daily and weekend licenses were issued. For 2020, 15,324 annual licenses were issued (of which 582 were cadets) and 15,324 day and weekend licenses.

There is practically no monitoring and supervision in the fishery subsector (FAO 2015). Although all three major associations maintain websites, they are not regularly updated, so there is no clear, publicly available list of sport fishing territories in B&H. A reliable annual statistical evaluation of the catch is missing. FAO (2015) estimated that the annual catch, including non-reported catches, in FB&H is between 5–10 kg per person (61.5–123 MT per year) and up to 10 kg per person in RS.

The organizational problems in the fisheries sectors are multifold. There is no ministry for agriculture and rural development at the state level; therefore, the sector's representation and coordination are very weak. Consequently, the sector, and sport fishing in particular, is not seen as a priority and no country-wide strategy on fisheries has been developed, keeping the sector from national and international support programs.

Reflecting the complex structure of the country, the entities' laws on fisheries are not cross-harmonized, and sometimes they clash also with other laws such as concessions and labor. Some of the articles in the fishery laws are not well elaborated and are thus open for individual interpretations. Related secondary legislation in the form of implementation and enforcement guidelines and regulations is also missing (FAO 2015). Furthermore, local regulations and governments do not always obey higher-level decisions, hampering the proper law implementation. Although some paid positions exist, most fishers' associations' work is done voluntarily due to the lack of monetary funds. Insufficient funding primarily affects working conditions and equipment available to the fish guards. Fish guards are usually understaffed and unauthorized for immediate, on-the-spot actions. One of the significant threats to the fisheries sector is the lack of water cadasters and the unpredictable allocation of waters that prevents sport fishers' associations from making long-term investments (FAO 2015).

8 Quality and Ecological Status of Water: Importance for the Sustainability of Fish Resources

Ecological status, regardless of the type of water body, is an expression that defines ecosystem integrity (quality) based on biological, hydromorphological, and physico-chemical elements of water quality (Radulović and Teodorović 2011). The term "quality water" refers to water whose present components are optimal and enable the unhindered growth of plants and animals (Umerfaruq and Solanki 2015). Aquatic ecosystems are continuously in a dynamic state of change in terms of geological and geochemical characteristics. This dynamic balance is further disturbed by anthropogenic activities, which cause pollution of aquatic ecosystems and ultimately lead to fish death. The productivity of aquatic organisms depends on the physical and chemical characteristics of water and the sufficient amount of nutrients necessary for the growth of living organisms (Agbaire and Obi 2009; Verma et al. 2012). The main limiting factor for the productivity of fish resources in aquatic ecosystems is water quality (Niyoyitungiye et al. 2019). The effect of water quality on fish varies with their size, age, and condition. Changes in physical (temperature) and chemical water parameters (dissolved oxygen concentration, chemical consumption of oxygen, nitrates, and phosphates) provide important information about water quality and its impact on biodiversity. Also, water quality significantly influences the toxicity of certain substances dissolved in water. All living organisms have a certain degree of tolerance concerning the water quality parameters in which they can function normally.

Monitoring of surface waters carried out in the area of the FB&H and the RS shows that a significant number of watercourses have water of good quality and that there are also polluted watercourses whose water quality is impaired. Water quality research during 2020 shows that most of the questioned watercourses have satisfactory quality. An analysis of physical-chemical and sanitary-microbiological parameters of surface water quality was conducted in the territory of the RS at a total of 39 localities, while an analysis of ichthyofauna was carried out at 42 localities,

Observed in total based on values of physical-chemical and microbiological parameters it can be stated that most of the investigated watercourses have water of satisfactory quality and that the largest number of monitored parameters indicate water of the first or second quality class. In some primaries, the presence of deviations, i.e. values that point to the third or fourth class are noticeable. Based on the analysis of sanitary-microbiological parameters, ascertained are 4 classes of quality of water, which the largest number of parameters indicated to the second class. The research of ichthyofauna for the purposes of biological water monitoring in 2020 was carried out at a total of 42 locations, with a profile on each watercourse. The overall observed state of ichthyofauna in the mentioned localities is at a satisfactory level, with the absolute dominance of indigenous fish species. The results of the saprobic index for the studied watercourses indicate that most of these rivers have water of good quality, i.e. water of I or II quality class, i.e. oligosaprobic and β mesosaprobic waters. All investigated localities are within these creditworthiness categories, with a different saprobic index. The saprobic index values ranged from 0.4 to 2.71, with 5 localities belonging to the category of xenosaprobic waters, 10 investigated localities belonging to the category of oligosaprobic waters, 26 localities belonging to the category of β mesosaprobic waters, and one locality belonging to the category of α-mesosaprobic waters (Dekić et al. 2021).

The results of surface water quality monitoring in the water area of the Sava River in the FB&H indicate that the worst water quality was found in the sub-basin of the Bosna River. The main reasons cited are urban wastewater directly discharged into waterways without prior treatment and untreated industrial wastewater. The impact of pollution is particularly pronounced in the Tuzla industrial basin, affecting the quality of the Spreča River and making it the most polluted watercourse in the Sava River drainage in FB&H (https://www.voda.ba/laboratorija-za-vode-avp-sava-podsliv-bosne-ima-najlosiji-kvalitet-voda). Regarding RS, more than 70% of the water in the Trebišnjica basin is of excellent quality, while about 30% of the water in the Sava basin is of poor quality. Lower water quality is primarily associated with untreated municipal wastewater, and it is stated that industrial pollution does not have a major impact on water quality, primarily due to the reduced capacity of industry (http://www.voders.org/upravjanje-vodama/projekti/ surface water monitoring).

9 Key Threats to Ichthyofauna and Fisheries Resources in Bosnia and Herzegovina

Fish are sensitive to habitat degradation to varying degrees. The biggest negative impact on them is the introduction of foreign species, regulation of watercourses, construction of dams, pollution, and other habitat disturbance (Mrakovčić et al. 2006). The water quality is very important for fish existence. Fish live in a certain range of basic parameters (temperature, oxygen, pH, ammonia, water hardness, etc.) within which they can function. Any exceeding these parameters' minimum or maximum limits leads to a change in their status, which manifests at different levels (from biochemical, individual to population and ecosystem).

One of the most negative anthropogenic impacts is pollution. Industrial facilities, dumping toxins, sewage, and urban and agricultural runoff disturb environmental quality and make change in biodiversity (Dekić et al. 2020).

Research has shown that sturgeon and salmonid fish species are the most endangered in the freshwater ecosystems of B&H today (Mrakovčić et al. 2006; Sofradžija 2009).

Mini hydropower plants (MHPs) represent a special problem. The MHP construction process begins with the damming of watercourses and the formation of water-capturing buildings. Derivation pipelines with an average length of 1–3 km, but up to 5 km, are then installed, transporting water to the machine building with turbines, from where it is returned to the watercourse. During periods of low water, almost all the water is often directed to the derivation pipelines, after which the bed remains dry for several kilometers, which leads to the complete destruction of the living world in the watercourse. There are 69 MHPs in RS and 52 in FB&H. Even more devastating are the plans to build 352 more MHPs at the level of the entire B&H, whose most harmful effect will be on migratory fish species, like Danube salmon (*Hucho hucho*) (https://voda.ekoakcija.org/bs/brojMHEpoEntitetu; Freyhof et al. 2015)).

Changes in the physico-chemical parameters of water, especially the Neretva basin in B&H, were also caused by the formation of storage lakes with the associated hydroelectric power plants, which changed the liquid ecosystem into a stagnant one. With the creation of reservoirs, many microhabitats and spawning grounds disappear, corridors important for holobiont migratory species such as brown trout are cut, and the rheophilic (trout) fish community changes to a limnophilic one in general. Research by Pojskić (2005) showed that dams at hydropower plants represent a physical barrier between the soft-mouth trout populations of the upper and middle reaches of the Neretva River, which results in a clear differentiation of these populations. Durmić-Pašić (2008) also confirmed these results.

The most significant drivers of ichthyofauna are excessive and uncontrolled fishing, illegal fishing (poaching), the use of illegal fishing equipment, inadequate stocking, etc. (Mrakovčić et al. 2006).

Records on the number and types of violations related to fishing are kept by sports-fishing associations, based on official notes submitted to them by fish guards. For example, in 2002, 24 violations were found in the fishing area managed by the Banja Luka Sports Fishing Association. The types of violations were as follows: fishing with an underwater fly with a weight, fishing using natural bait, fishing below the permissible size, an excessive number of fish in the catch, fishing without a license, fishing a species of fish that is prohibited to be caught and fishing from a bridge.

Climate changes (e.g., drought or a significant drop in the water level in the summer months due to the absence of precipitation and long-term high temperatures) should also be considered. They can significantly affect fish reproduction, nutrition, and consequently abundance, which is especially important for endemics with a narrow ecological valence (i.e., show high sensitivity to any changes in the habitat, biotic, and abiotic factors). The effects of climate change on fish are reflected in different ways. Because of these changes, there may be an increase in water temperature and a change in gas concentration. Temperature represents one of the most important environmental factors that affect all living beings, and whose effect is particularly pronounced in poikilothermic organisms. It affects several physiological processes in the body, changing the speed of their running (Dekić et al. 2013a). Freshwater fish are ectothermic organisms that cannot regulate their body temperature by physiological means (Moyle and Cech 2004) and whose body temperature is identical or close to the temperature of the environment in their specific location. Accordingly, the speed of physiological reactions is highly dependent on body temperature, as are all aspects of fish physiology including growth, reproduction, and activity (Ficke et al. 2005). Fish can regulate body temperature only indirectly, choosing thermally heterogeneous microhabitats (Brett 1971). For this reason, changes in the temperature of freshwater ecosystems will cause changes in their growth, reproduction, metabolism, food consumption, and generally maintained homeostasis. The impact of the changes depends on the fish species, but it will also affect stenothermic and eurythermic organisms (Ficke et al. 2007). Factors that cause stress in fish are naturally associated with changes in the physical, chemical, and biological factors of the water environment, which affects the increased susceptibility of fish to diseases (Kubilay and Ulukoy 2002).

Some rivers and reservoirs are already "polluted" by alien fish species (biological contamination), jeopardizing native species' survival by competing with them for food and microhabitats or predating on their eggs and fry. Highly reproducing and adaptive invasive species are especially dangerous, not only for fish but also for other aquatic native organisms. By introducing species for cultivation purposes, other species can be co-introduced that quickly acclimatize to new environmental conditions, become invasive, and cause great damage to aquatic ecosystems. Such an example is the Prussian carp (*Carassius gibelio*). This non-native species spreads quickly in B&H thanks to great reproductive adaptive capabilities and high tolerance to habitat changes. In B&H's territory, there are many lakes and reservoirs of different origins and purposes, which represent important habitats for different species of non-native fish, where they have settled and found optimal conditions. Some of the most represented allochthonous fish species in the freshwaters of B&H, which undeniably have an invasive character and negatively affect indigenous fish species, are *Carassius gibelio, Lepomis gibbosus, Pseudorasbora parva, Oncorhynchus mykiss,* and species from genus *Neogobius*. In addition to the mentioned species, it is expected that some other invasive species registered in the neighboring countries, primarily the Chinese sleeper (*Perccottus glenii*), recorded in Croatia, will appear in the waters of B&H (Adrović and Skenderović 2007; Vuković et al. 2008; Glamuzina et al. 2017; Tutman et al. 2020b; Dekić et al. 2022).

10 Conclusions

According to the latest most comprehensive data, the freshwater ichthyofauna of B&H counts 118 fish species and includes about 20% of the European freshwater ichthyofauna. The species abundance, especially endemic, places B&H among the richest European countries. The largest number of endemic species of ichthyofauna is related to the karst areas of the Adriatic basin. Fish species distribution and morphology are relatively well studied, and the missing data mainly refer to population, genetic, and molecular research. Reflecting the complex structure of the country, the entities' laws, which, among others, include the area of fish protection, are not cross-harmonized, and sometimes they clash also with other laws. For example, 22 species of fish are protected and strictly protected in the FB&H, and 51 in the RS. The freshwater fisheries sector in B&H includes commercial and sport fishing. About 50 species of the total freshwater ichthyofauna are significant in sport fishing. Commercial fishing in B&H is carried out only in the area of the Sava River (Republic of Srpska and Brčko District). Sport fishing is much more represented in B&H. The main species grown in aquaculture in B&H are salmonids, mainly rainbow trout, grown in tanks and cages. The most significant pressures on ichthyofauna diversity are pollution, different types of habitat destruction (regulation of watercourses, construction of dams), alien fish species, etc. Monitoring of freshwater quality carried out in the B&H shows that a significant number of watercourses have water of good quality and the worst water quality is noticed in the sub-basin of the Bosna River.

References

Adrović A (2012) Ribe Modraca, Ihtiološka monografija. B&H: NAM Tuzla, Tuzla

Adrović A, Skenderović I (2007) Allochthonous Ichthyofauna of certain water bodies in North-Eastern Bosnia. Acta Agric Serb XII 23:37–46

Agbaire PO, Obi CG (2009) Seasonal variation of some physico-chemical properties of River Ethiope water in Abraka, Nigeria. J Appl Sci Environ Manag 13:55–57

Agency for Statistics of Bosnia and Herzegovina (2020) Trade of B&H with foreign countries in 2019, Thematic Bulletin TB06, ISSN 1840-104x

Agency for Statistics of Bosnia and Herzegovina (2022) Annual report on aquaculture (in Bosnian and English language), 11, XI, 20.05.2022

Anonymus (1975) Ribarska osnova za slivno područje srednjeg i donjeg toka Vrbasa (područje SO Mrkonjić Grad, Skender Vakuf, Banja Luka, Kotor varoš, Čelinac i Srbac i OOUR "Incel" Banja Luka i OOUR PD: "Motajica" – Srbac). Biološki Institut Univerziteta Sarajevo

Anonymus (1985) Ribarska osnova za ribolovno područje „Vrbas II" (Banja Luka, Kotor varoš, Čelinac, Lakraši i Srbac). Biološki Institut Univerziteta Sarajevo

Bakrač-Bećiraj A, Škrijelj R, Trožić-Borovac S (2005) Sastav i struktura ribljih populacija u ocjeni kvaliteta vode rijeke Klokot. Veterinaria 54(1–2):67–97

Brett JR (1971) Energetic responses of salmon to temperature. A study of some thermal relations in the physiology and freshwater ecology of sockeye salmon (*Oncorhynchus nerka*). Am Zool 11: 99–113

Ćaleta M, Marčić Z, Buj I et al (2019) A review of extant Croatian freshwater fish and lampreys - annotated list and distribution. Croat J Fish 77:137–234. https://doi.org/10.2478/cjf-2019-0016
CBD (1992) Convention on biological diversity. Retrieved from: https://www.cbd.int/doc/legal/cbd-en.pdf
Craig FJ (ed) (2015) Freshwater fisheries ecology. Wiley, Chichester. https://doi.org/10.1002/9781118394380
Dekić R, Ivanc A, Lolić S et al. (2011) The recent state of distribution of endemic fish species in Eastern Herzegovina. V International conference "Aquaculture and fishery", Belgrade-Zemun, June 1–3. Conference proceedings, pp. 195–199
Dekić R, Ivanc A, Mandić M et al (2013a) The effect of temperature change on erythrocyte profile of Carassius gibelio. Croat J Fish 71:59–66
Dekić R, Trbojević I, Čolić V (2013b) Vodič Javni akvarijum, Prirodno-matematički fakultet
Dekić R, Ivanc A, Erić Ž et al (2014) Hematological characteristics of *Delminichthys ghetaldii* (Steindachner 1882) inhabiting the karst region of Eastern Herzegovina. Arch Biol Sci 66(4): 1423–1430. https://doi.org/10.2298/ABS1404423D
Dekić R, Friščić J, Manojlović M et al. (2017) Ichthyofaunal diversity oh the Drinjača catchment area. VII International Syposium of ecologists of Montenegro – The proceedings, Sutomore, Montenegro, pp. 77–81
Dekić R, Ivanc A, Ćetković D, Lolić S (2020) Anthropogenic impact and environmental quality of different tributaries of the River Vrbas (Bosnia and Herzegovina). In: Doru B, Angela C-B, Franco P, Cianfaglione K, Akeroyd J (eds) Human impact on Danube watershed biodiversity in the XXI century, Hardcover. Springer International Publishing, Cham, pp 169–213. https://doi.org/10.1007/978-3-030-37242-2
Dekić R, Trbić G, Gnjato R et al. (2021) Ispitovanje kvaliteta voda vodotoka u Republici Srpskoj u 2020. godini – ihtiofauna. Prirodno – matematički fakultet, Izvještaj 1–293
Dekić R, Golub D, Manojlović M, Kanlić V (2022) Introdukovane vrste riba u slivovima rijeka Vrbas, Una i Sana (Republika Srpska, Bosna i Hercegovina). Treći kongres biologa Srbije, Zlatibor, 21–25. 9. 2022. Knjiga sažetaka, str. 181
Durmić-Pašić A (2008) Filogenetske i interpopulacijske relacije osobenosti mitohondrijalne DNK bh. salmonida. Prirodno-matematički fakultet Univerziteta u Sarajevu
European Union (2000) Directive 2000/60/EC of the European Parliament and of the council establishing a framework for the community action in the field of water policy. European Commission, O J Eur Commu L327(2000):1
FAO (2015) Analysis of the Fishery and Aquaculture Sector in Bosnia and Herzegovina. EU funded project "Preparation of IPARD Forest and Fisheries Sector Reviews in Bosnia and Herzegovina" - GCP/BIH/010/EC. Retrieved from: https://www.fao.org/3/au016e/au016e.pdf
FAO (2019) The State of the World's Aquatic Genetic Resources for Food and Agriculture http://www.fao.org/3/ca5256en/CA5256EN.pdf
FAO (2020a). The State of Mediterranean and Black Sea Fisheries 2020. In The State of Mediterranean and Black Sea fisheries 2020. General Fisheries Commission for the Mediterranean. https://doi.org/10.4060/cb2429en
FAO (2020b). The state of world fisheries and aquaculture 2020. Sustainability in action. Rome. https://doi.org/10.4060/ca9229en
Ficke AA, Myrick CA, Hansen LJ (2005) Potential impacts of global climate change on freshwater fisheries. WWF- World Wide Fund for Nature, Colorado. 91pp
Ficke AD, Myrick CA, Hansen LJ (2007) Potential impacts of global climate change on freshwater fisheries. Rev Fish Biol Fisheries 17:581–613
Francuski L, Ludoški J, Lukač M et al (2019) Integrative study of population structure of *Telestes dabar*, the strictly endemic cyprinid species from the Dinaric karst on the Balkan Peninsula. Eur J Wildl Res 65(66). https://doi.org/10.1007/s10344-019-1302-6
Freyhof J, Weiss S, Adrović A et al (2015) The Huchen *Hucho hucho* in the Balkan region: distribution and future impacts by hydropower development. RiverWatch & EuroNatur. 30 pp

Glamuzina B, Tutman P, Pavličević J et al. (2010) Bioraznolikost riba Hercegovine. Herceg, N. (eds) Međunarodni kolokvij "2010.- godina bioraznolikosti"; 2011, Sarajevo, Federalno ministarstvo okoliša i turizma, str. 119–135

Glamuzina B, Tutman P, Nikolić V et al (2017) Comparison of taxon-specific and taxon-generic risk screening tools to identify potentially invasive non-native fishes in the River Neretva catchment (Bosnia and Herzegovina and Croatia). River Res Appl 33(5):670–679. https://doi.org/10.1002/rra.3124

Golub D, Marić Ž, Šukalo G et al (2016) Morfološke karakteristike *Cobitis elongatoides* i *Sabanejewia balcanica* (Cobitidae) iz rijeke Suturlije, III Simpozijum biologa i ekologa Republike Srpske (SBERS 2015). Zbornik radova, SKUP 7(2):139–148

Halilović A, Adrović A (2015) Morfometrijske i merističke razlike između mužjaka i ženki *Ameiurus melas* iz akumulacionog jezara Modrac. Educa, časopis za obrazovanje, nauku i kulturu, godina VIII 8:21–40

Heckel JJ (1843) Ichthyologie [von Syrien]. In: J. von Russegger. Reisen in Europa, Asien und Africa, mit besonderer Rücksicht auf die naturwissenschaftlichen Verhältnisse der betreffenden Länder unternommen in den Jahren 1835 bis 1841, etc. Stuttgart 1(pt 2):990–1099

Heckel J, Kner R (1858) Die Süsswasserfische der Österreichischen Monarchie mit Rücksicht auf die angränzenden Länder – Monografien Vertebrata Pisces – 0001. Engelmann, Leipzig, pp 1–388

Kalamujić B (2013) Genetički diverzitet recentnih populacija *Salmo trutta* L. rijeke Neretve u odnosu na rekonstruisani simulacioni model početnog stanja. Doktorski rad. Prirodno-matematički fakultet Univerziteta u Sarajevu. COBISS.BH-ID: 20817670

Kalamujić Stroil B, Stanić M, Kalajdžić A, Pojskić N (2020) Genetic characteristics of brook trout *Salvelinus fontinalis* (MITCHILL, 1848) broodstocks from hatcheries in Bosnia and Herzegovina. Croat J Fish 78:183–194. https://doi.org/10.2478/cjf-2020-0018

Karaman S (1923) O porijeklu slatkovodnih riba našega krša. (Über die Herkunt der Süßwasserfischeunseres Karstes). Glasnik Zemaljskog muzeja u Bosni i Hercegovini XXXV. Zemaljska štamparija, Sarajevo, pp 163–176

Karaman S (1928) Beitrage zur ichthiologi von Yugoslavien. Glasnik skopskog naučnog društva 6(2):147–176. (in Serbian with German summary)

Kosorić Đ (1969) Neki podaci o rasprostranjenju i migracijama neretvanske glavatice (*Salmo marmoratus* Cuv.) u slivu Neretve. Glasnik Zemaljskog muzeja, Sarajevo 8:79–82

Kosorić Đ (1989) Endemske vrste riba i njihova zaštita u slivu Neretve. Zbornik radova sa savjetovanja o ribarstvu na hidroakumulacijama, Mostar, pp 129–132

Kosorić Đ, Vuković T, Kapetanović N et al (1983) Sastav naselja ribe rijeke Neretve u SR Bosni i Hercegovini. Godišnjak Bioloskog instituta Univerziteta u Sarajevu 36:117–128

Kottelat M, Freyhof J (2007) Handbook of European freshwater fishes. Kottelat, Cornol, Switzerland and Freyhof, Berlin

Lelo S, Hamzić A (2020) PASTRMSKI GRGEČ, *Micropterus salmoides* (La Cepède, 1802) (Chordata: Vertebrata: Actinopterygii), U BOSNI I HERCEGOVINI. Prilozi fauni Bosne i Hercegovine, UZIZAŽ 16:39–43

Mikavica D, Kosorić Đ, Savić N, Vuković D (2006) Rezultati ihtiološkog istraživanja rijeke Save u Bosni i Hercegovini od ušća Une do ušća Vrbasa. Gospodarenje ribljim resursima u ribolovnim područjima Drava-Dunav I Sava - IV Nacionalno znanstveno-stručno savjetovanje sa međunarodnim sudjelovanjem. Zbornik priopćenja. str. 53–54. Osijek. Hrvatska

Moyle PB, Cech JJ (2004) Fishes: an introduction to ichthyology, 5th edn. Prentice Hall, Emglewood Cliffs, NJ

Mrakovčić M, Brigić A, Buj I et al (2006) Crvena knjiga slatkovodnih riba Hrvatske. Državni zavod za zaštitu prirode, Republika Hrvatska, Ministarstvo kulture

Müller M, Pander J, Geist J (2018) Comprehensive analysis of >30 years of data on stream fish population trends and conservation status in Bavaria, Germany. Biol Conserv 226:311–320. https://doi.org/10.1016/j.biocon.2018.08.006

Niyoyitungiye L, Giri A, Mishra B (2019) Assessment of Physico-chemical characteristics of water at selected stations of Lake Tanganyika, Africa with special emphasis on Pisciculture purposes. Int J Basic Appl Biol 6(3):211–217

Nature Park Una –Protection Study (2019) [Park prirode Une. (n.d.) Studija zaštite – stručna osnova (2019) Republički Zavod za zaštitu kulturno- istorijskog i prirodnog nasljeđa Republike Srpske]

Official Gazette of the Federation of Bosnia and Herzegovina 44/08 [Službene novine Federacije BiH 44 (2008) Zakon o Nacionalnom parku Una]

Official Gazette of the Federation of Bosnia and Herzegovina 31/09 [Službene novine Federacije BiH 31 (2009) Pravilnik o izmjenama i dopunama Pravilnika o načinu, alatima i sredstvima kojima se obavlja ribolov]

Official Gazette of the Federation of Bosnia and Herzegovina 63/05 [Službene novine Federacije BiH 63 (2005) Pravilnik o načinu, alatima i sredstvima kojima se obavlja ribolov]

Official Gazette of the Federation of Bosnia and Herzegovina 64/04 and 20/10 [Službene novine Federacije BiH 64 (2004) i 27 (2010) Zakon o slatkovodnom ribarstvu]

Official Gazette of the Federation of Bosnia and Herzegovina 66/13 [Službene novine Federacije BiH 66 (2013) Zakon o zaštiti prirode]

Official Gazette of the Federation of Bosnia and Herzegovina 7/14 [Službene novine Federacije BiH 7 (2014) Crvena lista faune FBiH]

Official Gazette of the Brčko District of Bosnia and Herzegovina 40/07 and 19/07 [Službeni glasnik Brčko distrikta BiH 35 (2005) i 19 (2007) Zakon o slatkovodnom ribarstvu Brčko distrikta BiH]

Official Gazette of the Brčko District of Bosnia and Herzegovina 40/07 [Službeni glasnik Brčko distrikta BiH 40 (2007) Pravilnik o zaštiti riba]

Official Gazette of the Republic of Srpska 124/12 [Službeni glasnik Republike Srpske 124 (2012) Uredba o Crvenoj listi zaštićenih vrsta flore i faune Republike Srpske]

Official Gazette of the Republic of Srpska 20/14 [Službeni glasnik Republike Srpske 20 (2014) Zakon o zaštiti prirode]

Official Gazette of the Republic of Srpska 65/20 [Službeni glasnik Republike Srpske 62 (2020) Pravilnik o načinu obavljanja ribolova, lovostaju, alatima, opremi i sredstvima za ribolov]

Official Gazette of the Republic of Srpska 65/20 [Službeni glasnik Republike Srpske 65 (2020) Uredba o strogo zaštićenim i zaštićenim divljim vrstama]

Official Gazette of the Republic of Srpska 72/12 [Službeni glasnik Republike Srpske 72 (2012) Zakon o ribarstvu]

Official Gazette of the Republic of Srpska 79/19 [Službeni glasnik Republike Srpske 79 (2019) Odluka o proglašenju parka prirode Una]

Piria M, Simonović P, Kalogianni E et al (2018) Alien freshwater fish species in the Balkans—vectors and pathways of introduction. Fish Fish 19(1):138–169. https://doi.org/10.1111/faf.12242

Project Reports (2004) Management of Freshwater Fisheries on bordering rivers - a pilot study with a holistic regional approach. Aquaplan Niva, Norway

Radević M (2000) Ekološki i cenotički odnosi faune riba u srednjem i donjem toku Vrbasa i ribnjaku Bardači (monografija). PMF, Banjaluka

Radulović S, Teodorović I (2011) Ekologija i monitoring kopnenih voda. Metodološki priručnik. Prirodno-matematički fakultet Univerziteta u Novom Sadu, Departman za biologiju i ekologiju, Novi Sad

Savić N, Drinić M, Važić B, Rogić B (2013) Influence of different meal sizes on growth characteristics of young rainbow trout (*Oncorhynchus mykiss* Wal). J Agric Sci 58(3):185–193

Simić V, Pavlović S, Milošević S, Tošić J, Kljunić L (1999) Centar za očuvanje biodiverziteta akvatičnih ekosistema Srbije u ex-situ uslovima: "Akvarijum Kragujevac". Zaštita prirode 51(2):171–181

Simonović P, Povž M, Piria M et al (2014) Ichthyofauna of the River Sava System. In: Milačič M, Ščančar J, Paunović M (eds) The Sava River. The handbook of environmental chemistry, vol 31. Springer, p 490. ISBN 978-3-662-44033-9

Simonović P, Tošić A, Škraba Jurlina D et al (2017) Diversity of brown trout *Salmo cf. trutta* in the River Danube basin of Western Balkans as assessed from the structure of their mitochondrial control region haplotypes. J Ichthyol 57(4):603–616. https://doi.org/10.1134/S0032945217040154
Škraba D, Bećiraj A, Šarić I et al (2017) Haplotype diversity of brown trout (*Salmo trutta* L.) populations from Una River drainage area in Bosnia and Herzegovina: implications for conservation and fishery management. Acta Zoologica Bulgarica 69(1):25–30
Škrijelj R (2002) Populacija riba neretvanskih jezera: ihtiološka monografija. Prirodno-matematički fakultet Sarajevo, Sarajevo
Sofradžija A (2009) Slatkovodne ribe Bosne i Hercegovine. Vijeće Kongresa bošnjačkih intelektualaca Sarajevo
Steindachner F (1895) Beiträge zur Kenntniss der Süsswasserfische der Balkan-Halbinsel. k.k. Staatsdruckerei, Wien
Taler Z (1951a) Zaštita naših rijetkih i neobičnih riba. Slatkovodno ribarstvo Jugoslavije 10), god. VI:224–227
Taler Z (1951b) Podzemne ribe u našem kršu. Slatkovodno ribarstvo Jugoslavije 4), god. VI:107–109
Taler Z (1954) Rasprostranjenje i popis slatkovodnih riba Jugoslavije. Glas Prirodnog muzeja Srpske zemlje, Serija B, knjiga 5–6, Beograd. pp. 425–455
Tutman P, Glamuzina B, Dulčić J et al (2009) Endemic fish fauna of the Hutovo Blato wetland (Neretva river basin, Bosnia and Herzegovina) and their conservation status. In: Kontautas A (ed) 13th European congress of ichthyology. Abstract book. Klaipedos universitetas, Klaipeda, pp 69–69
Tutman P, Buj I, Ćaleta M et al (2017) Status and distribution of spined loaches (Cobitidae) and stone loaches (Nemacheilidae) in Bosnia and Herzegovina. Folia Zool 66(4):211–226
Tutman P, Buj I, Ćaleta M et al (2020a) Review of the lampreys (Petromyzontidae) in Bosnia and Herzegovina: a current status and geographic distribution. J Vertebr Biol 69(1):1. https://doi.org/10.25225/jvb.19046
Tutman P, Zanella D, Horvatić S et al (2020b) Freshwater gobies (Gobiidae) of Bosnia and Herzegovina: a review of the current status and distribution. J Vertebr Biol 69:4. https://doi.org/10.25225/jvb.20046
Umerfaruq HM, Solanki HA (2015) Physico-chemical parameters of water in Bibi Lake, Ahmedabad, Gujarat, India. J Pollut Effects Cont 3(2):1–5
Verma P, Chandawat D, Gupta U, Solanki HA (2012) Water quality analysis of an organically polluted lake by investigating different physical and chemical parameters. Int J Res Chem Environ 2:105–112
Vuković T (1968) Ribe Bosne i Hercegovine. Zavod za izdavanje udžbenika Sarajevo
Vuković T, Ivanović B (1971) Slatkovodne ribe Jugoslavije. B&H: Zemaljski muzej BiH, Sarajevo
Vuković D, Tursi A, Carlucci R, Dekić R (2008) Ichthyofauna of the wetland ecosystem in the Bardača area (Bosnia and Herzegovina). Ribarstvo 66(3):89–103
Law on Freshwater Fisheries of the Federation of Bosnia and Herzegovina (2021) [Zakon o slatkovodnom ribarstvu Federacije Bosne i Hercegovine. (n.d.) Retrieved from: https://fmpvs.gov.ba/wp-content/uploads/2021/11/03-PREDNACRT-Zakon-slatkovodno-ribarstvo-.pdf]
http://e-prirodafbih.ba/en/protectedspecies/
https://cin.ba/baza-zasticenih-podrucja/
https://nasljedje.org/propisi-2/
http://www.mvteo.gov.ba/Content/Read/vodni-resursi-linkovi
https://voda.ekoakcija.org/bs/brojMHEpoEntitetu
https://www.voda.ba/laboratorija-za-vode-avp-sava-podsliv-bosne-ima-najlosiji-kvalitet-voda
http://www.voders.org/upravljanje-vodama/projekti/monitoring-povr%C5%A1inskih-voda/
http://bih-chm-cbd.ba/?page_id=21

Inland Fisheries in Serbia: Historical Aspect, Fish Resources, Management, and Conservation

Vladica Simić, Miljanović Branko, Ana Petrović, Milena Radenković, Milica Stojković Piperac, Tijana Veličković, Marija Jakovljević, and Snežana Simić

Abstract The chapter presents the history of fisheries in Serbia from 1860 until today. The chapter discusses the development of society (political, social, economic) and its relationship to fisheries and fishery resources. Additionally, an analysis of the development of fisheries legislation and its practical application in commercial and recreational fisheries was carried out. The main impacts on fish stocks in the past and present are analyzed, such as the fragmentation of large rivers (Danube, Drina) by hydroelectric dams, the creation of reservoirs for the water supply on the smaller rivers, the construction of small hydropower plants on mountain rivers, land reclamation and agricultural development, deforestation, the impact of allochthonous species, pollution, overfishing, and illegal fishing. Official statistical data on the fish catches by commercial fishermen over the last 70 years in the fisheries waters of Serbia are analyzed and compared. Estimates and difficulties in estimating fishing pressure by recreational anglers are presented, as well as suggestions for overcoming these problems. The current state of fish stocks can be characterized as the following: at the limit of sustainability for parts of the Danube, Sava, and waters in some protected areas, moderately sustainable for most fishing waters, and very sustainable for some reservoirs in the Great Morava and Drina basins, a small number of salmonid waters, and protected pond ecosystems.

Supplementary Information The online version contains supplementary material available at https://doi.org/10.1007/978-3-031-36926-1_4.

V. Simić (✉) · A. Petrović · M. Radenković · T. Veličković · M. Jakovljević · S. Simić
Department of Biology and Ecology, Faculty of Science, University of Kragujevac, Kragujevac, Serbia
e-mail: vladica.simic@pmf.kg.ac.rs

M. Branko
Department of Biology and Ecology, Faculty of Sciences, University of Novi Sad, Novi Sad, Serbia

M. S. Piperac
Department of Biology and Ecology, Faculty of Sciences and Mathematics, University of Niš, Niš, Serbia

V. Simić et al. (eds.), *Ecological Sustainability of Fish Resources of Inland Waters of the Western Balkans*, Fish & Fisheries Series 43,
https://doi.org/10.1007/978-3-031-36926-1_4

Keywords Serbia · History of fisheries · Fish stocks · Fisheries legislation · Commercial fisheries · Recreational fisheries

1 Introduction

There used to be much more fish in the waters of Serbia. This thesis can often be heard in conversations with locals and fishermen and refers to the state of fishing waters in this area in the last 50–70 years. Is it the actual situation, and what are the reasons for the decrease in the ecological sustainability of the fish stock in Serbia? The answer to these questions was not easy for several reasons. For the territory of Serbia, with its difficult and often tragic history, there are very few written and, above all, precise scientific data on the state of fish and fish resources. However, it is important to point out that in the past, the fish were caught mainly with methods and tools that are prohibited by law today, including spears, dividing rivers (funnel-shaped barriers), or complete drying of river sections. In addition, fish were caught mainly in the quantity that covered the family's food needs and was more common in the diet of people who lived and traded on large rivers such as the Danube, Sava, Tisza, and Great Morava (Pančić 1860; Petrović 1998a, b, c, d[1]).

However, it turned out that for the territory of Serbia, with its difficult and often tragic history, there are very few written and, above all, precise scientific data on the state of fish and fish resources. Although they are valuable, the few published data from 160 years ago (Pančić 1860) are insufficient to scientifically assess the state of the fish stock and the causes of changes. Data from 70 years ago, when more deliberate research on fish and fish stocks in inland waters of Serbia began, are much more precise. However, there is a lack of their systematization, so their use for precise scientific analysis is very limited. Therefore, data from 30 years ago, which coincide with the past and current Fisheries Laws (Anonymous 1994, 2009, 2014), have the most scientific value for fisheries. Therefore, most of the discussion and conclusions about fish, fish stocks, and fisheries in the inland waters of Serbia are based on the data analysis from this period.

The chapter discusses the development of society (political, social, economic) and its relationship to fisheries and fishery resources. Additionally, an analysis of the development of fisheries legislation and its practical application in commercial and recreational fisheries was carried out.

[1]The original references of Petrović (1998a, b, c, d) are each from the following years: 1900, 1933, 1940, 1941.

1.1 Inland Waters of Serbia, Fishing Waters, Fishing Areas, and Current Legislation

The inland waters in Serbia belong to the basins of the Black, Adriatic, and Aegean Seas. The basin of the Black Sea is the largest and covers around 92% of the territory. All major Serbian rivers (Sava, Drina, Great Morava, and Tisza) flow into the Black Sea through the Danube. The Adriatic Basin accounts for 5% of the territory, with the most important river being the Beli Drim. The Aegean basin is the smallest, with 3%, and the most important rivers are Lepenac, Pčinja, and Dragovištica (Fig. 1).

Today, inland fishery waters in Serbia consist of 66,000 km of flowing waters (streams, rivers), floodplains, backwaters, 50 lakes, and 150 reservoirs (Marković et al. 2009). The great heterogeneity of fishing waters is due to natural and anthropogenic factors. Traditionally, the most important fishing waters include the major rivers (Danube, Sava, Drina, Great Morava, and Tisza), large marshes and floodplains in Vojvodina, and streams and rivers in the hilly and mountainous areas of central Serbia, which are parts of the sub-basins of Drina, Great Morava, and Timok rivers. Due to human activities and various needs (water supply, power generation, flood control, irrigation), artificial water bodies such as small and large reservoirs and canals have been created in Serbia at the expense of natural ecosystems (especially the large canal system in Vojvodina, Danube-Tisa-Danube (DTD), of the total length of 929 km). In addition to their main function, these artificial water bodies also represent important fishing waters. Most large reservoirs, which are important for fishing, are located in Central Serbia in the Great Morava and Drina River basins (Figs. 2, 3, 4, 5, 6, and 7).

Currently, all waters of Serbia, including fishing waters, are owned by the state (except for some smaller privately owned lakes). Serbia is a continental country, so fishing is represented only in inland waters as commercial, recreational, and sport fishing. Commercial fishing is represented only on the major rivers, the Danube, Sava, and Tisza, while recreational and sport fishing is allowed in all waters. The Ministry of Environmental Protection (https://www.ekologija.gov.rs/) manages commercial and recreational fishing on wild waters, the Ministry of Sports manages sport fishing (https://www.mos.gov.rs/), while the Ministry of Agriculture, Forestry and Water Management (http://www.minpolj.gov.rs/) regulates the cultivation of fish in fish ponds.

On the territory of Serbia, there are 17 established fishing areas. They are located in unprotected and protected areas with significant fishing waters defined within their boundaries (Fig. 8). The fishing areas are awarded to users (managers) for management for ten years through a competition announced by the Ministry.

The managers of fishing areas are legal entities and can be public state-owned enterprises or private companies that meet the conditions for managing the fishery areas according to the tender issued by the Ministry. The main conditions include that the company has experience in fisheries, employs the required number of fishery rangers with a state license, and has at least one expert with a bachelor's or master's

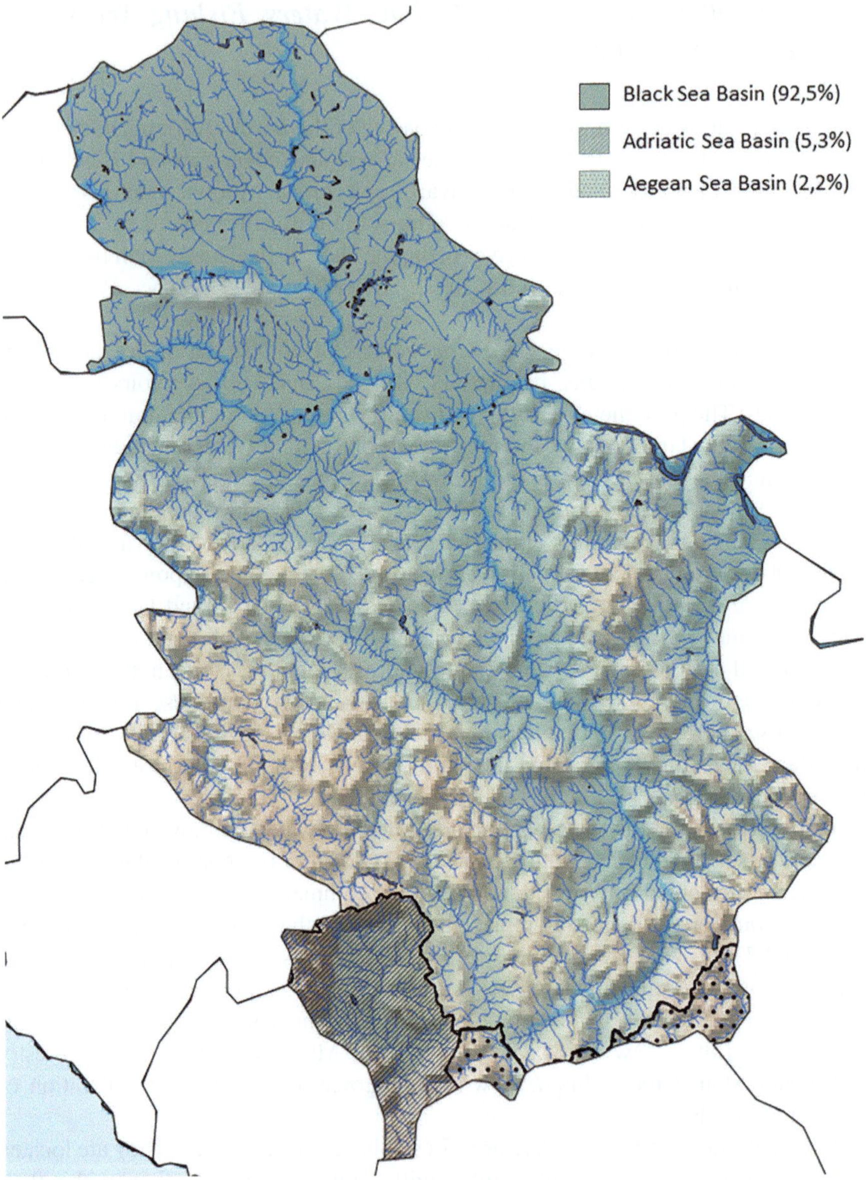

Fig. 1 Map of rivers and main drainage basins in Serbia

degree (biologist, ecologist, agricultural engineer, or veterinarian). In addition, the company must also meet the requirements for adequate equipment (vessels, all-terrain vehicles, and other equipment for fishery rangers). Fishing areas in protected natural areas are managed by a company that manages the protected natural assets (national parks, nature parks, reserves). A company managing a

Fig. 2 The natural course of the Danube River below the dam of the HPP "Đerdap 2." The picture was recorded in 2020 in Negotin, Serbia (photo by V. Simić)

Fig. 3 Drina River near Bajina Bašta, Serbia (photo by V. Simić)

Fig. 4 Great Morava at low water level during October 2017 (photo by V. Simić)

Fig. 5 Salmonid river. The Dojkinačka River in the Stara planina Nature Park in 2020 (photo by V. Simić)

Fig. 6 Gruža Reservoir during the summer of 2012 (photo by V. Simić)

Fig. 7 Zasavica River in September 2015 (photo by V. Simić)

Fig. 8 Fishing areas in Serbia: 1—FA "South Morava 1"; 2—FA "South Morava 2"; 3—FA "Rasina"; 4—FA "Ibar"; 5—FA "West Morava "; 6—FA "Kolubara"; 7—FA "Sava"; 8—FA "Srem"; 9—FA "Beograd"; 10—FA "Great Morava 2"; 11—FA "Great Morava 1"; 12—FA "Mlava"; 13—FA "Danube"; 14—FA "Timok"; 15—FA "Bačka"; 16—FA "Banat"; 17—FA "Kosovo"

fishing area in an unprotected or protected area carries out its activities following the Law on the Protection and Sustainable Use of Fish Stocks (Anonymous 2014) and regulations, of which particularly important is Regulation on Measures for the Conservation and Protection of Fish Stocks (Anonymous 2018). Fishing associations and clubs only conduct sport fishing competitions in the country and abroad, and most of them are members of the Federation of Sport Anglers of Serbia (https://

www.ribolovci.org.rs/index.php/savez-sportskih-ribolovaca-srbije) or on the territory of Vojvodina, the Federation of Sport Fishermen of Vojvodina (https://www.ssrvojvodina.org.rs/).

Managers of fishing areas sell commercial fishing licenses. A recreational fishing license is unique in Serbia. This means that a fisherman with a license obtained at his place of residence, which is within the boundaries of a specific fishing area, may fish in all other fishing areas except for fishing areas located in nature reserves. Fishing in protected areas requires the purchase of a special license that is valid only for the fishing waters of that protected area. Recreational fishing licenses are sold as daily, multi-day (valid for a maximum of seven days), or annual licenses. The Ministry sets the total price of recreational fishing licenses each year, with licenses for people with disabilities and pensioners being significantly cheaper. The managers of fishing areas set the value of the commercial fishing license for each year with the approval of the Ministry.

The manager of the fishing area is obliged by law to carry out its work based on the Fishing Area Management Program. In a public tender, the manager selects an authorized professional institution (mainly faculties and institutes dealing with fisheries biology, ichthyology, or inland water ecology) to prepare the Management Program for a period of 10 years. The law further requires that monitoring of the fish stock must be conducted every three years and that the Fishing Area Management Program may be amended and supplemented based on the monitoring. The program review is carried out by the Ministry of Environmental Protection, and the control of implementation and execution of the measures from the program is controlled by the fishery inspectors of the Republic. The manager of fishing area income is obtained by the number of sold fishing licenses, which is used for its activities (salaries of fishery rangers, experts, administration, and purchase and maintenance of equipment) as well as the improvement of the fish stock (stocking, maintenance of fish waters, training). The manager of the fishing area pays 20% of the value added tax (VAT) and 10% of the fee for using the fishing area to the state budget. In addition to the long-term program, at the end of each year, the manager of the fishing area submits a report on the work and the results obtained to the Ministry.

2 Historical Overview: Fish and Fisheries in Serbia

2.1 Fishes of Serbia

The first scientifically based work describing fish from Serbia and the adjacent countries of Bosnia, Montenegro, and Dalmatia was done by Hechel and Rudolf (1858) in the period when the Austro-Hungarian Empire held power over part of the Balkan Peninsula. Based on the results of this work, Pančić (1860) published the first study on fishes in the Serbian language under the title "Fishes in Serbia." In this work, 93 fish species are described, some of which originate from Bosnia and Dalmatia waters. This work is important because, deciphering the taxonomic synonymy, almost all fish species described at that time are still present in the waters of Serbia. However, a clear difference is observed in the spatial distribution of certain species, while a significant reduction in the range and distribution is observed for a considerable number of species. This is primarily true for all migratory anadromous species of the family Acipenseridae, as well as for species such as tench (*Tinca tinca*), crucian carp (*Carassius carassius*), true loach (*Misgurnus fossilis*), European eel (*Anguilla anguilla*), European mud-minnow (*Umbra krameri*), and huchen (*Hucho hucho*). Qualitatively, the total number of species has increased due to the invasion of allochthonous fish species (Lenhardt et al. 2011; Zorić et al. 2014) (see Chap. 13, Radenković et al. 2023).

In a relatively recent period, the fishes of Serbia were described in the works of Taler (1953), Vuković and Ivanović (1971), and somewhat later by Ristić (1977). These publications refer to the fishes of Serbian inland waters and other republics (now states) of the former Socialist Federal Republic of Yugoslavia (SFRY).

Finally, the last comprehensive publication on the fishes of Serbia appeared 21 years ago (Simonović 2001). Based on new research results, this publication needs to be revised with respect to the species occurring in the waters of Serbia. However, for the purpose of this chapter, the list of fish species according to this publication is presented (Table 1).

A more detailed consideration of the taxonomy of the ichthyofauna of Serbia is beyond this chapter's scope and will, therefore, not be discussed further. Instead, the focus is on the fish species that are important for fisheries and for which there are generally no significant changes in taxonomic status.

2.2 Fish Species Important for the Fishery in the Waters of Serbia

The fishery and economic importance of fish species have changed over time. In the work of Pančić (1860), the fish species were divided according to their taste into the most palatable (sterlet sturgeon, huchen, brown trout, pikeperch, pike, burbot, and young wels catfish), less appreciated but quite pleasant (wels catfish, European eel, lampreys, true loach, ide, starry sturgeon, barbel, and Danube barbel), the ones with a lesser appeal (Allis shad, crucian carp, carp, tench, and rudd), and so-called "white fish" (which met contains many small bones). In a later study by Hristić and Bunjevac (1991), fish divided into first class according to their economic importance for fisheries were pikeperch, Volga pikeperch, wels catfish, pike, European eel, huchen, carp, tench, sterlet sturgeon, grayling, brown trout, and Acipenseridae: beluga sturgeon, starry sturgeon, and fringerbarbel sturgeon. In addition to the mentioned autochthonous species, allochthonous species such as grass carp, silver carp, bighead carp, largemouth bass, rainbow trout, and brook trout were also considered first-class fish. The second class included asp, burbot, white bream, chub, bream, ide, nase, Prussian carp, crucian carp, and barbel. Third-class fish

Table 1 List of fish species in Serbia according to Simonović (2001)

Order number	Family/fish species valid name *according to FishBase*	English name	Literature[a]
	Petromyzontidae		
1.	*Eudontomyzon danfordi* (Regan, 1911)	Carpathian lamprey	Karaman (1974); Šorić (1992, 1998); Janković and Krpo Ćetkovic (1995)
2.	*Eudontomyzon stankokaramani* (Karaman, 1974)	Drin brook lamprey	**Karaman** (1974); **Holčik and Šorić** (2004)
3.	*Eudontomyzon mariae* (Berg, 1931)	Ukrainian brook lamprey	Janković and Krpo Ćetkovic (1995)
4.	*Eudontomyzon vladykovi* (Oliva & Zenandrea 1959)	Danubian brook lamprey	
5	*Lampetra fluviatilis* (Form. *Petromyzon fluviatilis*) (Linnaeus, 1758)	River lamprey	Pančić (1860); Janković and Krpo Ćetkovic (1995); Holčik and Šorić (2004);
6.	*Lampetra planeri* (Bloch, 1784)	European brook lamprey	Pančić (1860); Janković and Krpo Ćetkovic (1995)
7.	*Petromyzon marinus* (Linnaeus, 1758)	Sea lamprey	Pančić (1860); Janković and Krpo Ćetkovic (1995)
	Acipenseridae		
8.	*Acipenser gueldenstaedtii* (Brandt and Ratzeburg, 1833)	Danube sturgeon	Pančić (1860); Stamenković (1991); Janković and Krpo Ćetkovic (1995); Lenhardt et al. (2006, 2008, 2011); Jarić et al. (2009a, 2010)
9.	*Acipenser nudiventris* (Lovetsky, 1828)	Fringebarbel sturgeon	Pančić (1860); Ristić (1963); Janković and Krpo Ćetkovic (1995); Jarić et al. (2009a, b, 2010); Simonović et al. (2005); Lenhardt et al. (2006, 2008, 2011)
10.	*Acipenser ruthenus* (Linnaeus, 1758)	Sterlet sturgeon	Pančić (1860); Janković (1958); Ristić (1969, 1971a); Janković and Krpo Ćetkovic (1995); Kolarević (2004); Lenhardt et al. (2004, 2006, 2008, 2011); Stanić et al. (2006); Jarić et al. (2009a, 2010, 2011); Poleksić et al. (2010); Simić et al. (2014a, b); Rašković et al. (2015); Štrbac et al. (2015); **Cvijanović et al.** (2015,2017); Đikanović et al. (2015)
11.	*Acipenser stellatus* (Pallas, 1771)	Starry sturgeon	Pančić (1860); Stamenković (1991); Janković and Krpo Ćetkovic (1995); Lenhardt et al. (2006, 2011); Jarić et al. (2009a, 2010)
12.	*Acipenser sturio* (Linnaeus, 1758)	Sturgeon	Pančić (1860); Janković and Krpo Ćetkovic (1995); Jarić et al. (2009a, b, 2010); Lenhardt et al. (2011)

(continued)

Table 1 (continued)

Order number	Family/fish species valid name *according to FishBase*	English name	Literature[a]
13.	*Huso huso* (Linnaeus, 1758)	Beluga sturgeon	Pančić (1860); Stamenković (1991); Lenhardt et al. (2006, 2008, 2011); Jarić et al. (2009a, 2010)
	Salmonidae		
14.	*Oncorhynchus mykiss* (Walbaum, 1792)	Rainbow trout	Nikolić et al. (2006, 2010); Vranić and Đinović-Stojanović (2011); Vranković et al. (2021); Spasić et al. (2011); Nikčević et al. (2016)
15.	*Coregonus peled* (Gmelin, 1789)	Peled	Lenhardt et al. (2010)
16.	*Hucho hucho* (Linnaeus, 1758)	Huchen	Pančić (1860); Janković and Krpo Ćetkovic (1995); Simonović et al. (2000, 2007, 2011); **Marić et al.** (2012a, 2014); Simić et al. (2014a, b); Freyhof et al. (2015); Nikolić et al. (2016)
17.	*Salmo marmoratus* (Cuvier, 1829)	Marble trout	Pančić (1860); Janković and Krpo Ćetkovic (1995); Simonović et al. (2007)
18.	*Salmo trutta* (Linnaeus, 1758)	Trout	Pančić (1860); Janković and Krpo Ćetkovic (1995); **Marić et al.** (2004a, 2006a, 2012b, c, 2022a); **Simonović et al.** (2007); Simonović and Nikolić (2007); **Tošić et al.** (2014); Simić et al. (2014a, b); Nikolić et al. (2016); Kanjuh et al. (2021)
19.	*Salmo farioides* (Karaman, 1938)	/	**Marić et al.** (2006a, 2017)
20.	*Salmo macedonicus* (Karaman, 1924)	Macedonian trout	**Marić et al.** (2004b)
21.	*Salvelinus alpinus* (Linnaeus, 1758)	Arctic char	Janković and Krpo Ćetkovic (1995); Marić et al. (2022a)
22.	*Salvelinus fontinalis* (Mitchill, 1814)	Brook trout	Janković and Krpo Ćetkovic (1995); Marić et al. (2022a)
	Thymalidae		
23.	*Thymallus thymallus* (Linnaeus, 1758)	Grayling	Pančić (1860); **Janković** (1960, 1964, 1983a, 2010); Janković and Krpo Ćetkovic (1995); Apostolski (1978); Filipović and Janković (1978); **Marić et al.** (2012b,c); Simić et al. (2014a, b); Nikolić et al. (2016)
	Anguillidae		
24.	*Anguilla anguilla* (Linnaeus, 1758)	European eel	Pančić (1860); Janković and Krpo Ćetkovic (1995)
	Clupeidae		

(continued)

Table 1 (continued)

Order number	Family/fish species valid name *according to FishBase*	English name	Literature[a]
25.	*Alosa caspia* (Eichwald, 1838)	Caspian shad	Janković and Krpo Ćetkovic (1995)
26.	*Alosa immaculata* (Bennet, 1835)	Pontic shad	Janković and Krpo Ćetkovic (1995); Višnjić-Jeftić et al. (2009, 2010, 2013); Lenhardt et al. (2011, 2016); Smederevac Lalić et al. (2018); Đikanović et al. (2018); Tošić and Taflan (2019)
	Esocidae		
27.	*Esox lucius* (Linnaeus, 1758)	Northern pike	Pančić (1860); Budakov and Maletin (1982, 1984); Maletin and Budakov (1983); Budakov (1993a); Janković and Krpo Ćetkovic (1995); Štrbac et al. (2015); Mérő (2014); Pavlovic et al. (2015); Milošković and Simić (2015); Simić et al. (2022a)
	Umbridae		
28.	*Umbra krameri* (Walbaum, 1792)	European mud-minnow	Pančić (1860); Janković and Krpo Ćetkovic (1995); **Sekulić** (2013) **;** **Sekulić et al**. (1998a, 2013, 2015); Simić et al. (2007); **Marić et al.** (2015,2016,2019)
	Cyprinidae		
29.	*Ballerus ballerus* (Form. *Abramis ballerus*) (Linnaeus, 1758)	Zope	Pančić (1860); Grginčević (1977); Maletin and Budakov (1983); Janković and Krpo Ćetković (1995)
30.	*Abramis brama* (Linnaeus, 1758)	Freshwater bream	Pančić (1860); Janković and Krpo Ćetković (1995); Lenhardt et al. (2012); Simić et al. (2014a, b); Milošković and Simić (2015); Milošković et al. (2016); Kostić et al. (2016); Kostić-Vuković et al. (2021)
31.	*Ballerus sapa* (Form. *Abramis sapa*) (Pallas, 1814)	White-eye bream	Pančić (1860); Janković and Krpo Ćetković (1995)
32.	*Barbus barbus* (Linnaeus, 1758)	Barbel	Pančić (1860); Janković and Krpo Ćetković (1995); Sunjog et al. (2012); Simić et al. (2014a, b); Rašković et al. (2015); Milošković et al. (2016)
33.	*Barbus cyclolepis* (Heckel, 1837)	Round-scaled barbel	Pančić (1860); **Marić et al.** (2004b)
34.	*Barbus balcanicus* (Kotlík, Tsigenopoulos, Ráb, and Berrebi, 2002)	Danube barbel	Pančić (1860); Janković and Krpo Ćetković (1995); **Simonović et al.** (2018); Radojković et al. (2019); Kojadinović et al. (2020)
35.	*Carassius auratus* (Form. *Carassius auratus auratus*) (Linnaeus, 1758)	Goldfish	Pančić (1860); Budakov (1980); Žikić et al. (2000)

(continued)

Table 1 (continued)

Order number	Family/fish species valid name *according to FishBase*	English name	Literature[a]
36.	*Carassius gibelio* (Form. *Carassius auratus gibelio*) (Bloch, 1782)	Prussian carp	Budakov et al. (1979); Budakov (1980); Maletin et al. (1980); Budakov and Maletin (1984); Simić et al. (2014a, b); Milošković and Simić (2015); Milošković et al. (2022)
37.	*Carassius carassius* (Linnaeus, 1758)	Crucian carp	Pančić (1860); Janković and Krpo Ćetkovic (1995); Sekulić et al. (1998b)
38.	*Ctenopharyngodon idella* (Valenciennes, 1844)	Gras carp	Janković (1992); Janković and Krpo Ćetkovic (1995); Marković et al. (2012)
39.	*Cyprinus carpio* (Linnaeus, 1758)	Carp	Pančić (1860); Ristić (1971b); Janković (1983b); Budakov et al. (1994); Poleksić and Mitrović-Tutundžić (1994); Janković and Krpo Ćetkovic (1995); Žikić et al. (1997); Sekulić et al. (1998b); **Faster and Cakić** (1998); Ćirković et al. (2000, 2015); Marković (2010); Spasić et al. (2010); Stanković et al. (2011); Lenhardt et al. (2012); Čanak et al. (2013); Subotić et al. (2013); Simić et al. (2014a, b); Novakov et al. (2015); Štrbac et al. (2015); Rajić et al. (2016); Marković et al. (2016); Rokvić et al. (2020); Božić et al. (2021)
	Xenocyprididae		
40.	*Hypophthalmichthys molitrix* (Valenciennes, 1844)	Silver carp	Janković and Krpo Ćetkovic (1995); Lenhardt et al. (2012)
41.	*Hypophthalmichthys nobilis* (Richardson, 1844)	Bighead carp	Sekulić et al. (1998b); Marković et al. (2012)
	Gobionidae		
42.	*Romanogobio albipinnatus* (Form. *Gobio albipinnatus)* (Lukasch, 1933)	White-finned gudgeon	Šorić and Ilić (1987, 1991); Janković and Krpo Ćetkovic (1995)
43.	*Gobio obtusirostris* (Form. *Gobio gobio* (Valenciennes, 1842)	/	Pančić (1860); **Šorić and Ilić** (1987); Janković and Krpo Ćetkovic (1995)
44.	*Romanogobio kesslerii* (Dybowski, 1862)	Kessler's gudgeon	Šorić and Ilić (1991); Janković and Krpo Ćetkovic (1995)
45.	*Pseudorasbora parva* (Temminck and Schlegel, 1842)	Topmouth gudgeon	Janković and Krpo Ćetkovic (1995); Cakić et al. (2004)
46.	*Romanogobio uranoscopus* (Agassiz, 1828)	Danube longbarbel gudgeon	

(continued)

Table 1 (continued)

Order number	Family/fish species valid name *according to FishBase*	English name	Literature[a]
	Leuciscidae		
47.	*Alburnoides bipunctatus* (Bloch, 1782)	Spirlin, schneider	Pančić (1860); **Šorić and Ilić** (1985); Janković and Krpo Ćetkovic (1995); Živković and Jovanović (2011); Simić et al. (2022a)
48.	*Alburnus albidus* (Costa, 1838)	Italian bleak	Janković and Krpo Ćetkovic (1995); **Šorić** (2006); Simić et al. (2012)
49.	*Alburnus alburnus* (Linnaeus, 1758)	Bleak	Pančić (1860); Budakov and Lečić (1991); Janković and Krpo Ćetkovic (1995); Simić et al. (2016); Jovanović et al. (2018); Nikolić et al. (2022)
50.	*Alburnus scoranza* (Bonaparte, 1845)	/	**Simić et al.** (2012)
51.	*Alburnus chalcoides* (Güldenstadt, 1772)	Danube bleak	Janković and Krpo Ćetkovic (1995)
52.	*Blicca bjoerkna* (Linnaeus, 1758)	White bream	Pančić (1860); Budakov and Maletin (1984); Janković and Krpo Ćetkovic (1995); Lenhardt et al. (2012)
53.	*Chondrostoma nasus* (Linnaeus, 1758)	Nase	Pančić (1860); Janković (1965); Janković and Krpo Ćetkovic (1995); Simić et al. (2014a, b); Đikanović et al. (2016); Milošković et al. (2016)
54.	*Leuciscus aspius* (Form. *Aspius aspius*) (Linnaeus, 1758)	Aral asp	Pančić (1860); Janković and Krpo Ćetkovic (1995); Krpo Ćetković et al. (2010); Simić et al. (2022a)
55.	*Leucaspius delineatus* (Heckel, 1843)	Belica	Pančić (1860); Janković and Krpo Ćetkovic (1995)
56.	*Squalius cephalus* (Form. *Leuciscus cephalus*) (Linnaeus, 1758)	Chub	Pančić (1860); Janković and Krpo Ćetkovic (1995); Simić et al. (2014a, b); Milošković et al. (2016); **Kolarević et al.** (2018); Sunjog et al. (2019); Simić et al. (2022a)
57.	*Leuciscus idus* (Linnaeus, 1758)	Ide, orfe	Pančić (1860); Janković and Krpo Ćetkovic (1995)
58.	*Leuciscus leuciscus* (Linnaeus, 1758)	Common dace	Pančić (1860); Janković and Krpo Ćetkovic (1995)
59.	*Telestes souffia* (Risso, 1826)	Vairone	Pančić (1860); Janković and Krpo Ćetkovic (1995); **Veličkovič et al.** (2020)
60.	*Pachychilon macedonicus* (Steindachner, 1892)	/	Simić et al. (2012)
61.	*Pachychilon pictum* (Heckel & Kner, 1858)	/	**Šorić** (1979,1993); Janković and Krpo Ćetkovic (1995)
62.	*Pelecus cultratus* (Linnaeus, 1758)	Sichel	Pančić (1860); Janković and Krpo Ćetkovic (1995), Subotić et al. (2015)

(continued)

Table 1 (continued)

Order number	Family/fish species valid name *according to FishBase*	English name	Literature[a]
63.	*Phoxinus phoxinus* (Linnaeus, 1758)	Common minnow	Pančić (1860); Janković and Krpo Ćetkovic (1995); Simić et al. (2022a)
64.	*Leucos basak* (Form. *Rutilus basak)* (Heckel, 1843)	/	Janković and Krpo Ćetkovic (1995); Simić et al. (2012)
65.	*Rutilus pigus* (Lacepede, 1804)	Pigo	Janković and Krpo Ćetkovic (1995)
66.	*Rutilus virgo* (Heckel, 1852)	/	Nikolić et al. (2019)
67.	*Rutilus rutilus* (Linnaeus, 1758)	Roach	Pančić (1860); **Grginčević and Pujin** (1980); Budakov (1989); Janković and Krpo Ćetkovic (1995); Marković et al. (1996a); Sekulić et al. (1998b); Radenković et al. (2022)
68.	*Scardinius erythrophthalmus* (Linnaeus, 1758)	Rudd	Pančić (1860); Janković and Krpo Ćetkovic (1995)
69.	*Scardinius graecus* (Stephanidis, 1937)	/	Simić et al. (2012)
70.	*Scardinius knezevici* (Bianco and Kottelat, 2005)	/	**Simić et al.** (2012)
71.	*Vimba vimba* (Linnaeus, 1758)	Vimba	Janković and Krpo Ćetkovic (1995); Subotić et al. (2021)
72.	*Vimba melanops* (Heckel, 1837)	Macedonian vimba	**Simić and Šorić** (2006)
	Acheilognathidae		
73.	*Rhodeus sericeus* (Pallas, 1776)	Amur bitterling	Pančić (1860); Janković and Krpo Ćetkovic (1995)
	Tincidae		
74.	*Tinca tinca* (Linnaeus, 1758)	Tench	Pančić (1860); Janković and Krpo Ćetkovic (1995); Simić et al. (2009); Ćirković et al. (2009); Marković et al. (2010); **Lujić et al.** (2017a,b)
	Cobitidae		
75.	*Cobitis elongata* (Heckel & Kner, 1858)	Balkan loach	**Šorić** (1985); Janković and Krpo Ćetkovic (1995)
76.	*Cobitis taenia* (Linnaeus, 1758)	Spined loach	Pančić (1860); Janković and Krpo Ćetkovic (1995)
77.	*Misgurnus fossilis* (Linnaeus, 1758)	True loach	Pančić (1860); Budakov (1993b); Janković and Krpo Ćetkovic (1995)
78.	*Sabanejewia balcanica* (Filippi, 1865)	Balkan spined loach	Janković and Krpo Ćetkovic (1995); **Marić et al.** (2022b); **Marešova et al.** (2011)
79.	*Sabanejewia bulgarica*	/	Miljanović et al. (2016)
80.	*Sabanejewia romanica* (Băcescu, 1943)	/	**Marić et al.** (2022b)
	Nemacheilidae (Form. **Balitoridae)**		

(continued)

Table 1 (continued)

Order number	Family/fish species valid name *according to FishBase*	English name	Literature[a]
81.	*Barbatula barbatula* (Linnaeus, 1758)	Stone loach	Pančić (1860); Janković and Krpo Ćetkovic (1995)
82.	*Oxynoemacheilus bureschi* (Form. *Barbatula bureshi*) (Drensky, 1928)	Struma stone loach	Sipos et al. (2007)
	Siluridae		
83.	*Silurus glanis* (Linnaeus, 1758)	Wels catfish	Pančić (1860); Janković and Krpo Ćetkovic (1995); Lenhardt et al. (2012, 2021); Subotić et al. (2013); Simić et al. (2014a, b); Pavlovic et al. (2015); Milošković and Simić (2015); Milošković et al. (2016)
	Ictaluridae		
84.	*Ameiurus nebulosus* (Le Sueur, 1819)	Brown bullhead	Janković and Krpo Ćetkovic (1995)
85.	*Ameiurus melas* (Rafinesque, 1820)	Black bullhead	**Cvijanović et al.** (2005); Jaćimović et al. (2019)
	Lotidae (formerly **Gadidae**)		
86.	*Lota lota* (Linnaeus, 1758)	Burbot	Pančić (1860); Janković and Krpo Ćetkovic (1995); Smederevac Lalić et al. (2015)
	Syngnathidae		
87.	*Syngnathus abaster* (Risso, 1826)	Black-striped pipefish	Sekulić et al. (1999); Cakić et al. (2005)
	Gasterosteidae		
88.	*Gasterosteus aculeatus* (Linnaeus, 1758)	Three-spined stickleback	Pančić (1860); Janković and Krpo Ćetkovic (1995); **Cakić et al.** (2000)
89.	*Pungitius platygaster* (Kessler, 1859)	Southern ninespine stickleback	Pančić (1860); Janković and Krpo Ćetkovic (1995)
	Percidae		
90.	*Gymnocephalus baloni* (Holčik & Hensel, 1974)	Balon's ruffe	**Hegediš et al.** (1993); Janković and Krpo Ćetkovic (1995)
91.	*Gymnocephalus cernua* (Form. *Gymnocephalus cernuus*) (Linnaeus, 1758)	Ruffe	Pančić (1860); Janković and Krpo Ćetkovic (1995)
92.	*Gymnocephalus schraetser* (Linnaeus, 1758)	Schraetzer	Pančić (1860); Janković and Krpo Ćetkovic (1995)
93.	*Perca fluviatilis* (Linnaeus, 1758)	European perch	Pančić (1860); Simonović and Jovanović (1993); Simonović and Simović (1997); Pavlović et al. (2013, 2015); Subotić et al. (2015); Radenković et al. (2022)

(continued)

Table 1 (continued)

Order number	Family/fish species valid name *according to FishBase*	English name	Literature[a]
94.	*Sander lucioperca* (Linnaeus, 1758)	Pikeperch	Pančić (1860); Janković and Krpo Ćetkovic 1995); Subotić et al. (2013); Simić et al. (2014a, b); Lujić et al. (2015); Pavlovic et al. (2015); Milošković and Simić (2015); Milošković et al. (2016); Nikolić et al. (2021)
95.	*Sander volgensis* (Gmelin, 1788)	Volga pikeperch	Janković and Krpo Ćetkovic (1995)
96.	*Zingel streber* (Siebold, 1863)	Danube streber	Pančić (1860); **Šorić and Naumovski** (1991); Janković and Krpo Ćetkovic (1995)
97.	*Zingel zingel* (Linnaeus, 1766)	Zingel	**Šorić and Naumovski** (1991); Marković and Simonović (2010)
	Centrarchidae		
98.	*Lepomis gibbosus* (Linnaeus, 1758)	Pumpkinseed	Marković and Simović (1997)
99.	*Micropterus salmoides* (Lacepede, 1802)	Largemouth black bass	Maletin (1988); Marković et al. (1996b)
	Gobiidae		
100.	*Neogobius fluviatilis* (Pallas, 1811)	Monkey goby	Janković and Krpo Ćetkovic (1995); **Simonović** (1996a, 1999); **Simonović and Nikolić** (1996)
101.	*Babka gymnotrachelus* (Kessler, 1857)	Racer goby	Hegediš et al. (1991); Janković and Krpo Ćetkovic (1995); **Simonović** (1996a, 1999); Subotić et al. (2013)
102.	*Ponticola kessleri* (Form. *Neogobius kessleri*) (Gunther, 1861)	Bighead goby	Šorić and Ilić (1991); Janković and Krpo Ćetkovic (1995); **Simonović** **(1996a,b,1999)**
103.	*Neogobius melanostomus* (Pallas, 1811)	Round goby	**Nikolić and Skora** (1996); **Simonović** (1996a,b,1999); **Valković and Paunović** (1998); Simonović et al. (2001); Subotić et al. (2013)
104.	*Proterorhinus marmoratus* (Pallas, 1811)	Tubenose goby	Janković and Krpo Ćetkovic (1995); **Simonović** (1996a, 1999); Simonović and Mesaros (1998)
	Blennidae		
105.	*Salaria fluviatilis* (Asso, 1801)	Freshwater blenny	Janković and Krpo Ćetkovic (1995)
	Cottidae		
106.	*Cottus gobio* (Linnaeus, 1758)	European bullhead	Pančić (1860); Janković and Krpo Ćetkovic (1995); Bravničar et al. (2015); Simić et al. (2022a)
	Odontobutidae		
107.	*Perccottus glenii* (Dybowski, 1877)	Chinese sleeper	**Simonović et al.** (2006)

[a]Literature review of available data on fish species in Serbia (Central Balkans). Data in bold refer to taxonomic classification of species. In the rest of the text, species are referred to by their English names

(white fish) further included roach, zope, white-eye bream, rudd, vimba, and sichel. Finally, fish of the fourth class were ones considered only as food for predatory fish (bleak, gudgeon, European bitterling, and ruffe) and the so-called category of "harmful fish" (perch, black bullhead, pumpkinseed, and topmouth gudgeon) (Hristić and Bunjevac 1991).

An overview of fish species according to their economic importance today and in the past can be found in Table 2.

Compared to the period at the end of the nineteenth century and the first half of the twentieth century, the current categorization of fish according to their fishery and economic importance for recreational and commercial fishing in Serbia is qualitatively inferior. In general, all species of the family Acipenseridae have been lost for commercial fisheries today, including sterlet sturgeon as a stationary species (Danube population), followed by European eel, tench, crucian carp, true loach, Danube lamprey, herring, and sichel. Presently, as incomparably inferior substitutes for these species, allochthonous Asian species such as grass carp, silver carp, bighead carp, and Prussian carp have some commercial importance. Furthermore, the species not available for recreational fishing are marble trout, tench, crucian carp, sichel, and zingel, as well as other species of small body size used as live bait in fisheries (all species of the family Cobitidae, species of the genus *Romanogobio*). These fish species are now on the list of strictly protected species for the territory of Serbia (see Chap. 17, Galambos and Sekulić 2023).

2.3 Fisheries

Findings of fish remains, objects made of fish bones (sewing needles), and fish-shaped sculptures at the sites of Padine, Lepenski vir, and Vlasac (on the right bank of the Danube) from the period before 9500–5500 BC (Mesolithic and Early/Middle Neolithic) are evidence that people in these areas were engaged in fishing and lived from it much earlier than the historical dating of the Serbians and the state of Serbia (fourth to seventh century AD) (Živaljević 2017). Unfortunately, from the period of Roman rule, there is no written information about fishing in the area where today's Serbia is located. However, the remains of fish (beluga sturgeon and Russian sturgeon) from the site of Viminacium (Vuković 2015) indicate that fish was caught in the Danube at that time and that it was highly valued at imperial feasts in the Roman Danubian provinces (Krišpatić 1893).

There is little written information about fisheries in Serbia during the Middle Ages. It is known that the Đerdap area on the Danube was one of the best in the region and that it was under the administration of Prince Lazar (Lazar Hrebeljanović 1329–1389) from 1379 (Živaljević 2017). According to Pančić (1860), fishing in medieval Serbia (from the eleventh to the middle of the fifteenth century) was not under any legal regulations. The fish was often caught without any restrictions and by means that are illegal today (spears, various types of fish traps, and usage of quicklime and toxic plant metabolites (especially *Verbascum* spp.)). Fish was caught

Table 2 Comparative overview of fisheries and the economic importance of fish species in Serbia during the past and the present

Fish species	Fisheries and economic importance			
	Commercial fishing		Recreational fishing	
Category I				
	Past	Present	Past	Present
Pikeperch	+	+	+	+
Volga pikeperch	+	+	+	+
Wels catfish	-*+	+	+	+
Northern pike	+	+	+	+
Carp	-*+	+	+	+
Tench	-*+	+*	+	+*
Sterlet sturgeon	+***	+*	–	–
European eel	+	-*	-?	–
Grayling	+?	–	+	+
Brown trout	+?	–	+	+
Huchen	+?	–	+	+
Beluga sturgeon***, Starry sturgeon, Fringerbarbel sturgeon	–++	+*	–	–
Category II				
Aral asp, Burbot, White bream, Chub, Bream, Ide, Nase, Prussian carp, Crucian carp, Barbel	Burbot	Burbot, Nase, Bream, Barbel, Prussian carp (Vojvodina especially)	+	+
Category III				
Common roach, Zope, White-eye bream, Rudd, Vimba, Sichel	–+	Bream, Vimba, Sichel*	–+	+ (Bream)
Category IV				
	Past	Present	Past	Present
Bleak, Gudgeon, Amur biterling, Eurasian ruffe	Bleak? Danube bleak (Pančić (1860)	–	Bleak	Bleak
Category IV				
	Past	Present	Past	Present
European perch, Black bullhead, Pumpkinseed, Topmouth gudgeon	–	European perch (large specimens)	European perch (large specimens)	European perch (large specimens)
Category - other allochthonous species				
	Past	Present	Past	Present
Rainbow trout, Gras carp, Silver carp Largemouth black bass	–+**	Silver carp	–	Rainbow trout, Largemouth bass

+ significant, – not significant, –* less significant in the nineteenth century, +*significant but not available, –+ variable significant over time, *** significant for obtaining caviar mainly by the local population and fishermen who were in the service of the nobility or the clergy. The caught fish was mainly used as food. High-quality fish, usually caught and brought from the Danube, was especially present on the nobility and clergy tables. The proof of this is preserved menus from that period (Petrović 1998d) and findings of fish bones during archeological excavations of some monasteries (e.g., Studenica Monastery; Živaljević et al. 2019). According to the same source, during the Ottoman rule in Serbia (from the fifteenth to the twentieth century), the Ottomans were the most skilled and experienced fishermen on the Danube, Sava, and surrounding large ponds. They caught fish with large forged hooks ("takuni"), nets made of strong rope and large eyelets ("korane"), and fish traps made of woven willow mesh called "garde" (Fig. 9). Such tools indicate that large fish were caught and that fish populations were abundant.

Apart from fishing in the Danube, Sava, and surrounding ponds from this period, there is almost no data about fishing in other waters of Serbia. Some indirect data can be found in the travelogues of foreigners in Serbia. Thus, Kanitz (1862) stated that during his journey in central Serbia from the town of Kraljevo to the monastery Studenica at all places for rest, brown trout from nearby rivers (Brvenica and other smaller rivers) were served in large and abundant portions. Additionally, grayling from the river Studenica was served in the monastery. However, the grayling population completely disappeared from the Studenica River in the 1970s (Janković 2010). The period of struggle for independence of Serbia (1804–1878) is characterized by a relatively weak organization of fishermen. The fishing grounds on the Danube were gradually bought from the Ottoman fishermen, who left this area, and fishing was performed by fishermen (Alasi) of Serbian nationality who previously worked for Ottoman fishing masters. These fishermen joined the Fishermen's Guild (founded in 1839, made official in 1847; Bojić 2013), which lasted until the beginning of the twentieth century and later became the Belgrade Fishermen's Association (Bojić 2013; Živaljević 2017). Fishing from this period, especially on the Danube, is described in the works of the mathematician Dr. Mihailo Petrović Alas, cited earlier (Petrović 1998a, b, c, d). One of the rare works containing data on fish and fishing on other rivers of Serbia from before and after World War I, entitled "Fishing on the Great Morava," was completed in 1941 but, due to wartime circumstances, was not published until 1972 (Momirović 1972). This work shows a similar organization, techniques, and tools for fishing on the Great Morava as well as on the Danube and the Sava, but in all aspects on a much smaller scale, including the amount (biomass) of fish caught.

Analyzing the data from these works, we can indirectly conclude the state of the fish stocks on the territory of Serbia over a long period of time, practically from ancient times until the first quarter of the twentieth century. During this period, fish were caught with tools that are generally not used today and were adapted for certain types of fish, as well as for their size and the amount available for fishing. Drawings of such tools can be found in Fig. 9. Of all the tools used, the most important was the usage of the so called "garde" (and similar tools called "sup" or "seća" (Fig. 9). These tools were used on the Danube, Sava, and Great Morava rivers until 1935 when they were banned by law. From the construction, method of use, and fishing

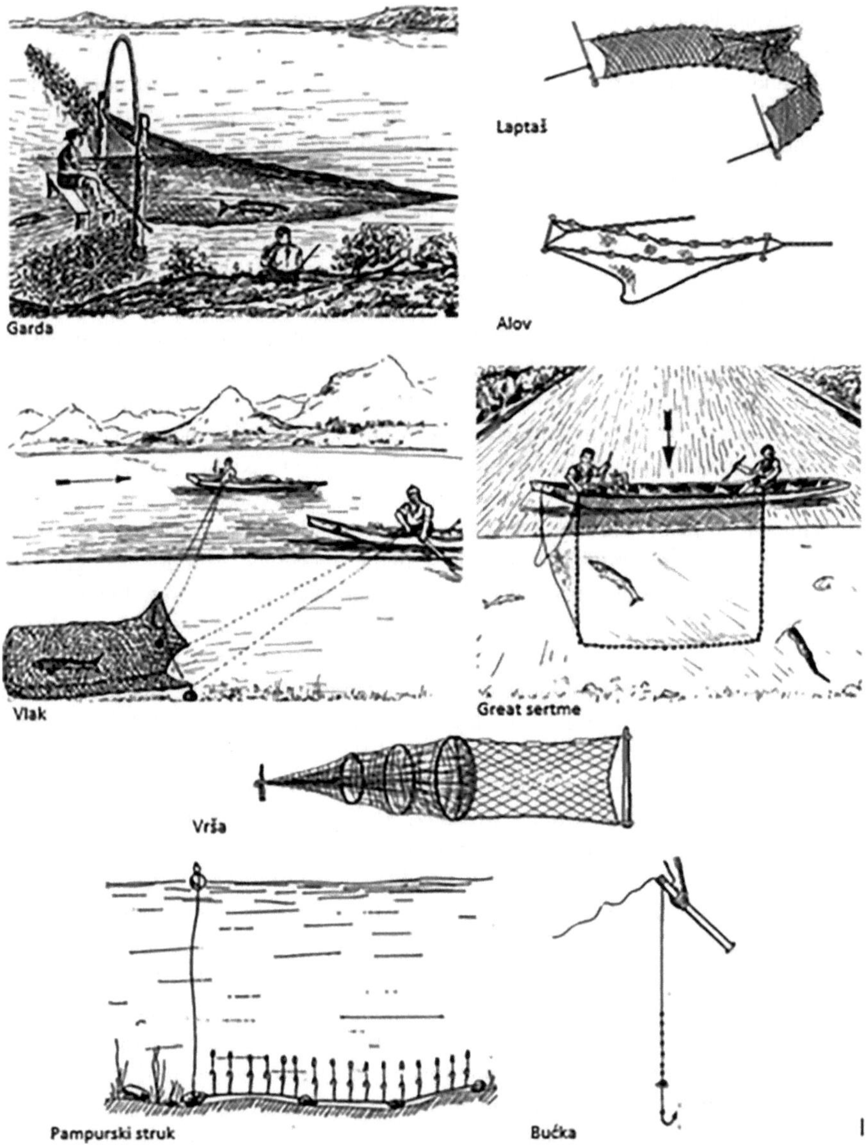

Fig. 9 Tools for commercial fishing (Source: https://www.ravnoplov.rs with the approval and courtesy of Mr. Milan Stepanović)

efficiency of these tools, it can be concluded that large, high-quality, and expensive fish were caught during this period, primarily sturgeon species (Fig. 10), followed by wels catfish (Fig. 11), carp, pikeperch, and pike.

Beluga sturgeon were hunted primarily for their caviar, which proved to be an economically very profitable food even in the post-World War II period (Fig. 12).

Fig. 10 Beluga sturgeon (weighing 135 kg) caught near Belgrade around 1910. It is being held by the head and pectoral fin by the Serbian mathematician and then a member of the Serbian Royal Academy, Mihailo Petrović Alas. (Source: https://www.ravnoplov.rs with the approval and courtesy of Mr. Milan Stepanović)

All available facts support the conclusion that in the past, for a very long period, there were significant and diverse fish stocks in the main fishing waters of Serbia, firstly in the Danube and then in the Sava, Drina, and Great Morava. However, on the other hand, during that time, there were practically no controls and restrictions on the exploitation of fish stocks.

2.4 *Legal Regulation of Fisheries in Serbia*

The organization of fisheries in Serbia can be traced through the chronological presentation of legal documents from the medieval state of Serbia to the present day.

2.4.1 The Period of the Middle Ages (1217–1459): Principality, Kingdom, Empire, Despotism

The most significant rivers, the Danube, Sava, and Drina, and their floodplains were considered general public property and a source of food for the population. There

Fig. 11 Wels catfish caught in the Danube, near Apatin, in the 1970s (Source: https://www.ravnoplov.rs/ with the approval and courtesy of Mr. Milan Stepanović)

were no restrictions and no laws regulating fishing. In the charter of Prince Lazar (1379), some places with particularly rich ichthyofauna on the Danube ("Gospođin vir" near the village of Dobra) were ceded for use by lords or monasteries (Mišić 1995a, b). It is assumed that there were abundant fish populations.

2.4.2 The Period of Complete Rule of the Ottoman Empire (1459–1804)

During this period, there was no official legal regulation and no restrictions on fishing. Moreover, the Ottomans managed all fishing places on the Danube, organized fishing, and traded fish. During this period, in the Danube, fish were also caught with "garde" (Fig. 9). A rich fish stock and a considerable amount of large fish specimens (all species of Acipenseridae, wels catfish, pikeperch, pike, carp, European eel, and bream) can be estimated from fishing gear. The fish was very cheap (Zirojević 2007).

Fig. 12 Caviar of beluga sturgeons caught in the Danube in the Đerdap gorge. This type of caviar was especially valued on the market. (Source: https://www.ravnoplov.rs with the approval and courtesy of Mr. Milan Stepanović)

2.4.3 The Period of the Struggle for Independence (1804–1878–1912): Principality of Serbia

During this time, parts of the fishing waters of the Sava and Danube rivers were leased under the so-called "Hatari." Additionally, the regulation on the payment of fishing fees of 1859/60 prescribed certain fishing conditions for the lessees. In these conditions, there are no regulations on hunting methods and restrictions on the amount of fish caught, except for the prohibition of fish poisoning. During this time, in the neighboring countries (Kingdom of Croatia, Hungary), there were already laws that contained scientifically and professionally sound measures for protecting fish stocks and sustainable fishing (Batrićević 2016).

The First Law on Fisheries in the Kingdom of Serbia (Kingdom Since 1882) (Anonymous 1898)

The first law on fishing in Serbia only applied to flowing waters. According to the law, the fishing waters of the Sava, Danube, and Drina belonged to the state, while the remaining ones to the local municipal authorities. A fishing license was introduced for each form of fishing, except when fishing with a rod with 1–2 hooks. The number of fishing tools a fisherman may use was also limited by law. Moreover, the smallest mesh diameter of fishing nets was introduced (from 10 mm for small tools used for fishing near the shore to 30 mm for large tools such as "laptaš" and "alov" used for fishing from a boat). Using tools on the whole width of the river was also forbidden. A ban on fishing during the fish spawning season was also introduced, along with a minimum size for fish below which they may not be caught. Importantly, penalties were introduced for poaching. Fishing is controlled by customs and municipal police, field guards, forest guards, and hunting guards. In addition, the law introduced the "informant," the person who reports poaching to the state authorities. These persons were entitled to financial compensation equal to 50% of the value of the fine paid by the poacher if the court proved his guilt (Anonymous 1898).

The Second Law on Fisheries of the Kingdom of Serbia (Anonymous 1911)
This law applied to flowing and standing waters, with fisheries being under the jurisdiction of the Ministry of National Economy. Provisions were introduced to encourage the construction of fishponds by private individuals, associations, and municipalities and for the leasing of certain fishing waters. Furthermore, it contained provisions for establishing fishing associations (guilds) and the Fisheries Improvement Fund. The latter, in addition to funds for fisheries improvement, provided funds for studies in the field of fisheries and fisheries-related projects (Anonymous 1911).

Measures to protect fish stocks have remained mainly the same as the previous legislation, but penalties for illegal fishing have increased.

2.4.4 The Period of Serbia under Occupation by the Austro-Hungarian Monarchy

Under the Austrian occupation of Serbia, the "Regulation on Fishing and Catching Crayfish in the Area of the Main Military Governorate in Serbia of May 2, 1917" was issued. According to this decree, fishing on the Danube, Sava, and floodplains was regulated by the military administration through military fishing stations. Fishing licenses could be acquired only to military personnel, while for civilians only with the permission of the military administration. The regulation defined fishing with a rod (hook) for the first time; i.e., fishing was defined as a sport. The Ordinance regulated the fishing time and the minimum length of the catch, which were more rigorous than in the Fishing Law of the Kingdom of Serbia (1911). The prescribed mesh sizes of nets were also increased: fishing net meshes may not be narrower than 40 mm from January 1st to July 31st and not narrower than 20 mm outside this period. The "vrška" as a fishing tool is prohibited (Fig. 9). After the end of World War I, the 1911 fishing law was reinstated in Serbia.

2.4.5 The Period after World War II

Fishing Law of the National Republic of Serbia (Anonymous 1949)
This was the first law on fishing after World War II. By this law, fish and their habitats are public property, and fisheries were under the jurisdiction of the Ministry of Agriculture. Fishing areas and enterprises were introduced, which are state economic enterprises. Additionally, the establishment of fishing associations was

especially initiated. Annual and daily fishing licenses with a hook were issued, but only to anglers that are members of fishing associations. Finally, a Fisheries Council was established to support households and associations financially (Anonymous 1949).

Fishing Law of the Socialist Republic of Serbia (Anonymous 1965)
Fishing was defined as a branch of the national economy and, in addition to fish, included the catching and breeding crayfish, mussels, and leeches.

Fishing waters were divided into closed (ponds, semi-ponds, lakes, pools, canals, and artificial lakes) and open, which includes all other waters. For the first time, the law clearly distinguished two types of fishing: commercial and sports (recreational). Commercial fishing was carried out by registered companies, fishing cooperatives whose main activity was fishing, and professional fishermen. On the other hand, sport fishing was carried out by anglers who were members of sport fishing associations belonging to the Federation of Sport Fishermen. Furthermore, fishing waters were divided into fishing areas and fishing grounds managed by local municipalities and communes. Commercial fishing was allowed only in fishing areas, while sport fishermen were also allowed to fish in fishing areas and fishing grounds. Both fishing areas and fishing grounds were used for fishing and fish farming.

Furthermore, the concept of two types of reserves was introduced. There were reserves used to exploit fish under special conditions and so-called protective reserves or natural fish spawning grounds where all forms of fishing were prohibited (mainly for a certain period). Fishing control was carried out by the fisheries inspector, whose competencies were prescribed by a particular law. Sport fishing required an annual or daily permit valid for fishing waters in the entire state (SFRY) or fishing grounds and areas in the territory of a specific municipality. The law supported and encouraged maintaining fish for stocking, which was considered a manager's duty. The law also encouraged the establishment and cooperation of fishing organizations and associations and their cooperation with scientific and research organizations. Furthermore, the law significantly regulated fines and penalties against companies that polluted fishing waters and/or otherwise destroyed fish stocks (Anonymous 1965).

Fisheries Law of the Socialist Republic of Serbia (Anonymous 1967)
It was similar to the previous one. It introduced the obligation of the organization that manages the fishing area to have a basic fisheries plan that provides all measures and actions for improving and exploiting fish stock in that area. The fisheries plan was valid for five years, and each year the manager was obliged to prepare an annual plan for the following year. According to the law, the basic fisheries plan included seven chapters. This law initiated measures for fishing certain fish species or parts of the area more than laws from previous periods (Anonymous 1967).

Law on Fisheries of the Socialist Republic of Serbia (Anonymous 1976)
It was stated that fishing waters are social property and fishing areas were managed by municipalities on whose fishing waters commercial and sport fishing could be conducted. However, this law did not apply to fishponds. The obligation of the

municipality on whose territory the fishing area was located was to introduce the estimation of maximum sustainable yield (MSY) in the fishing management plan. The law favored the production of fish for stocking and the fishing of overpopulated fish species (Anonymous 1976).

Fisheries Law of the Republic of Serbia (Anonymous 1994)
The law included fishing waters and ponds. Fisheries fell under the jurisdiction of the Ministry of Agriculture, and the Minister made all major decisions. Fishing areas could also be formed within protected areas. The manager of the area who does recreational fishing must have an expert in the field of fishing (the level of education and profession are not prescribed, but most experts come from agricultural faculties). An organization that engages in commercial fishing must have two experts, an agricultural engineering graduate, a graduate in animal husbandry, a biology graduate, or a veterinary medicine graduate. The manager limits the fishing service and must submit the so-called medium-term fishery improvement program, valid for five years. Sport fishing is based on a purchased annual license, which was valid only for the fishing area in which it was purchased. In addition to the annual license, there was the possibility of buying a daily and weekly license. An angler can practice sport fishing only if he has passed the fishing exam. The "Serbian Fishing Association" issued the fishing license after passing the exam. The manager is obliged to perform stocking of the fishing waters of the area annually. Only professional fishermen can sell their catch on the market. For the first time, the law specifies the conditions that establishments must meet for the sale of caught fish (Anonymous 1994).

Law on the Protection of Sustainable Use of Fish Stocks of Republic of the Republic of Serbia (Anonymous 2009, 2014)
According to this law, fishing is under the jurisdiction of the Ministry of Environmental Protection and is carried out in fishing waters (still and flowing waters) throughout the territory of Serbia. Fishing waters are the property of the Serbian state. The law promotes a new strategy aimed at managing fish stocks in accordance with the principle of sustainable use, which contributes to the preservation of the diversity of ichthyofauna and the ecological integrity of aquatic ecosystems. The cadaster of fishing waters is introduced and updated annually. Managers of fishing areas can be legal entities, i.e., companies or public companies. These laws abolish the possibility for fishing associations and the Association of Sport Fishermen of Serbia to be managers of fishing areas. Associations and federations can only deal with organized sport fishing competitions and fall under the jurisdiction of the Minister for Sports. Fishing areas are also formed in protected areas and are managed by the administrator of the respective protected area. Fishing areas are allocated to the managers on the basis of a public competition for ten years. After that, the manager is obliged to adopt the Fishing Area Management Program, which contains 18 chapters. The program is prepared by authorized scientific research organizations dealing with fisheries, fisheries biology, ichthyology, or freshwater ecology. The 2014 law requires that the status of the fish stock should be monitored every three years in order to correct the measures provided in the Fishing Area Management Program. A professional examination for fishkeepers will be introduced, based on

which the ministry will issue a license to the fishkeeper. The law abolished the fishing exam for recreational fishers. Commercial fishing is allowed only in the major rivers, Danube, Sava, and Tisza. Commercial fishing can be carried out by companies or entrepreneurs (professional fishermen). A professional examination for fishermen will be introduced. Recreational fishing may be conducted on the basis of an annual, daily, and multi-day license (up to seven days). With an annual license, an angler can engage in recreational fishing in all fishing areas of Serbia, except for fishing areas in nature reserves. A permit for fishing areas in protected areas can be used only for the fishing area for which it was issued. One of the important regulations that accompanied the 2009 law is the Rulebook on Categorization of Fishing Waters (Anonymous 2012). According to this rulebook, an additional license was required for fishing waters in protected areas, fishing waters inhabited by huchen, and those in which the mass percentage of brown trout and grayling exceeded 70%. The rulebook has been repealed, and the current law from 2014 provides only a single permit for all areas and specific permits for each fishing area in protected areas. The law from 2014 stipulates that the manager of the area employ at least one professional with a biology or ecology bachelor's or master's degree, but the option can also be a graduate engineer in agriculture, in animal husbandry, or a veterinarian. The 2014 law requires the managers to declare parts or all of the fishing waters in the fishing area as so-called "special fish habitats" where any form of fishing is permanently prohibited. These habitats are those where fish perform their important biological functions such as reproduction, migration, and feeding. The 2014 law provides, for the first time, that the fisheries management program must be consistent with conservation conditions prescribed by the Institute for Nature Protection. It must include a detailed work program of fisheries supervision, a program for the training of fishermen, a program for the development of fishing tourism, and an assessment of the economic viability of the fishing area. There are no provisions in the 2009/14 laws for the destruction of animals that feed on fish.

3 Fish Resources of Serbia, Current Situation and Perspective

3.1 Data Sources for Analysis

Part of the material for this section consists mainly of professional and scientific publications in the field of ichthyology and fisheries of inland waters of Serbia, available and published from 1860 to the present (see Sect. 2.1 of this Chapter). From these data, changes in the diversity of ichthyofauna and only partial quantitative changes in fish stocks can be reconstructed. The most accurate data on quantitative changes in fish stocks were obtained by analyzing technical scientific documents dealing with the assessment of fish stock status and the planning of measures for sustainable use and management, whose name has changed over time, from the “Basic Plan” (until 2009) and “Medium Term Program for Fisheries Improvement” (until 2014) to the current “Fishing Area Management Program.” These strategically important documents for the needs of managers of fisheries areas are prepared by scientific research organizations from Serbia, licensed to work in the field of fisheries, fisheries biology, ichthyology, or ecology of inland waters, and which apply scientific methods for assessing the status of fish stocks. The analysis of these scientific and research data from the management program on the state of the fish stock in the fishing areas revealed the development of important fishery parameters for the main fish species in the fishing waters of Serbia. The analysis covered the period from 2000 to 2022, which covered monitoring the evolution of the following parameters: abundance, biomass, real production, the ratio of real and potential production, the mean value of the total length of the fish stock, and the number of age classes. Based on the determined trends of the monitored parameters, the sustainability of Serbian fish stocks under the conditions of a complex of modern stressors was evaluated.

3.2 *Quantitative Data on Fish Resources in the Fishing Waters of Serbia*

Data on the amount of fish caught, both in total and by species in fishing waters, is a significant deficiency of fisheries in Serbia. This deficiency runs as a problem through all historical periods presented in the previous considerations until today (Simić et al. 2022a). In the earliest period of Serbian statehood (Serbia during the reign of the Nemanjić dynasty, later Lazarevic, Branković), there are no numerical data on the amount of fish caught by species on the territory of Serbia. During Ottoman rule, Ottoman fishermen fished mainly on the Danube, Sava, and Drina rivers and much less on the Great Morava, but there is no data on the amount of fish caught. From the period of the Principality of Serbia, there are more written indirect and direct data on fish and fish populations (Pančić 1860; Petrović 1998a, b, c, d), but again there is a lack of accurate quantitative data. In general, the following can be stated on the basis of all sources regarding fish stock and fisheries until the end of World War I.

The territory of Serbia is rich in high-quality fishery waters (clean waters, Serbia had no developed industry at that time). In the west, the Drina; in the north, the Danube, the Sava, and numerous ponds, the middle part of the Great Morava, and the Timok in the east. The abundance of high-quality water indirectly indicates the abundance of fish. Fish abundance is the only descriptive quantitative information that says something about the state of the fish stock in Serbia during this period. In contrast to fish abundance, it is assumed that it was used very little by the majority of the population on average, except in the courts of rulers and in monasteries by the clergy, where fish was used regularly. According to Pančić (1860), this is in contrast to the Orthodox faith (to which the largest percentage of the population of Serbia belongs), according to which the population is instructed to eat fish for a good part of the year—about 103 days—during fasting. At that time, Serbia had no fishponds, so only fish caught in wild waters could be used. Besides fresh fish salted fish was consumed, and much less fish in the marinade. The same source noted that there are few prohibitions in the ancient records regarding the catch of noble fish species such as brown trout, huchen, pike, and pikeperch (probably regarding *Sander lucioperca* and *Sander volgensis*), but there is no information about quantities. An indirect insight into the quantity of fish caught in the period from 1896 to 1898 is given by the statistical data of the customs offices on the Danube, through which the fish caught were exported to other countries (Petrović 1998d), referring to the part of the so-called lower Danube in the section from Kladovo to Donji Milanovac (933.8–991 rkm). According to these data, a total of 28.2 tons of large fish, or an average of 9.4 tons per year, were exported through the Donjomilanovac customs office, 1561 tons (an average of 52.04 tons) through the Tekija customs office, and 34.3 tons (an average of 11.4 tons) through the Kladovska customs office. An important detail of this data, apart from the quantity, is that it is the export of large fish. A significant part of the catch consisted of species from the Acipenseridae family, i.e., those that yield particularly valuable black caviar (called "ajvar" at that time) (Fig. 12), as well as very large specimens of beluga sturgeon (Petrović 1989d) (Fig. 10). Herring is also caught in large quantities in this part of the Danube, but the quantities are not given. Apart from the mentioned data for the lower Danube, Petrović (1989d) lists data for the Drina in 2010 and gives the most common weights of the main fish species, such as wels catfish caught up to 20 kg and huchen up to 10 kg. Additionally, pike specimens were caught up to 9 kg, carp and pikeperch up to 8 kg, and especially nase up to 3.5 kg. The list of fish from this period lacks grayling, whose population in the Drina is stable and sustainable today (Hegediš et al. 2008; Simić et al. 2014a, 2020a).

The first systematic and official data on the quantities caught by fishermen in Serbia appeared after World War II during the period of Yugoslavia. The data were first published in the Statistical Yearbook in 1951 (Smederevac Lalić et al. 2011; www.stat.gov.rs). Fishing during this period was under the supervision of the Ministry of Agriculture and referred to fish caught in wild waters and fishponds. However, the data are very sparse and refer only to commercial fisheries. Nevertheless, the data show the total number of fishermen (total and permanent), number of boats (total and motorized), and fish catch (total and only carp).

Analysis of these data is presented below.

The number of fishermen was highest in 1957, with 4468, of which 2703 were permanent. After that, the number decreased; in 1973, there were 909 total fishermen and only 461 permanent ones. From 1957 to 1973, the number of boats increased steadily, both boats and motorized ones. Fishing also increased during this period, with a total of 3585 tons of fish caught in 1957, including 1501 tons of carp, and a total of 6333 tons caught in 1973, including 3687 tons of carp.

From 1973 to 1990, the total number of fishermen stabilizes at an average of 1,100, while the total catch remains at about 6,600 tons, but the amount of carp caught decreased by an average of 900 tons. The maximum catch in this period was reached in 1974, when a total of 8,704 tons of fish were caught, of which 6,065 tons were carp.

The statistics do not cover the period from 1990 to 2006. An important reason for this is the period of disintegration of SFRY, economic, social, and political crises, and war events in this area.

In the period from 2006 to the present (Table 3), the collection and publication of statistical data on the quantities of fish caught in the fishing waters of Serbia (rivers, lakes, and canals) have improved significantly (but still insufficiently).

In contrast to the previous period, in addition to data on catches by professional fishermen, data on quantities of fish caught by sport and recreational fishermen are collected for the first time. In addition to the amount of carp, data on other fish species are also collected. Whereby the choice of other species is debatable because the data refer only to the bream as an autochthonous species but also to two allochthonous Asian species, namely the silver carp and Prussian carp. On the other hand, there is no data on the quantities of other important autochthonous fish species, such as wels catfish, pikeperch, pike, and sterlet sturgeon. The analysis of the data presented in Fig. 13 shows that the total fish catch was the highest in the period from 2009 to 2014, ranging from 3500 to 5300 tons.

An important fact that emerges from the statistical data is the greater amount of fish caught by recreational fishermen compared to the amount caught by commercial fishermen, starting in 2010, with the difference increasing steadily until today. The quantities of fish caught, including carp, bream, silver carp, and Prussian carp, are declining, according to the data after 2015, until today. The decline in carp, bream, and Prussian carp is particularly striking. Also, for these species, recreational fishermen's catches are higher than commercial fishermen's. The statistical data that have been analyzed are mostly related to the largest fishing waters, namely the Danube, Sava, and Tisza, so they do not fully reflect the state of the fish stock and fishing pressure on the whole territory of Serbia. In general and based on the comparative data shown in Fig. 14, the current quantities of fish catches in the Danube, Sava, and Tisza are several times (5–10 times) lower than in the period of the end of the nineteenth century and almost three times lower than in the second half of the twentieth century. The significant decrease is primarily due to the gradual and now complete disappearance of all sturgeon species in the fishermen's catches.

Official statistics are useful when considering the state of the fish stock globally. However, they do not reflect the closer and more realistic state of fish stock in the whole territory of Serbia outside the largest fishing waters (Danube, Sava, Tisza). Considering this fact, the data from the Program for Management of Fishing Areas of Serbia were analyzed with a special emphasis on abundance (A), biomass (B), and production (P) of the main fish species in all major fishing waters of Serbia. Together with these results, the average values of the total production of all fish species (TP), total catch (Tcatch), and average fishing pressure were calculated based on the average number of licenses sold for commercial fishing (Lic.cf) and recreational fishing (Lic. rf). The analysis covers the period from 2000 to 2022 and refers to 92% of the Serbian territory (Fig. 15a, b).

Table 3 Data on fish catch (in tons) based on the Statistical Yearbook of the Republic of Serbia (2006–2021)

	2006	2007	2008	2009	2010	2011	2012	2014	2015	2016	2017	2018	2020	2021
Total fish catch	2631	2536	3151	3845	4807	5384	4798	3591	3150	2069	2028	2083	1931	2353
Professional fishermen	1695	1447	1683	2112	2002	2260	1935	908	851	581	590	686	761	927
Sports and recreational fishermen	936	1088	1468	1732	2805	3124	2863	2683	2299	1488	1618	1397	1170	1426
Carp	204	206	289	331	494	497	469	335	312	141	148	140	169	184
Professional fishermen	132	120	200	208	198	201	182	71	58	28	23	35	52	64
Sports and recreational fishermen	72	86	89	122	296	296	286	264	254	113	125	105	117	120
Bream	359	479	431	528	399	426	505	281	247	127	109	114	154	237
Professional fishermen	259	272	215	279	163	246	226	95	72	59	40	47	61	95
Sports and recreational fishermen	101	208	216	249	238	180	279	186	176	68	69	67	93	141
Silver carp	225	121	206	244	265	233	167	105	152	159	197	222	177	207
Professional fishermen	183	104	120	223	256	218	158	94	143	119	173	169	156	184
Sports and recreational fishermen	33	16	86	21	9	15	9	11	9	40	24	53	21	24
Prussian carp	842	655	790	855	844	863	849	600	516	322	400	420	321	360
Professional fishermen	488	367	374	306	325	387	350	130	82	63	58	81	94	118
Sports and recreational fishermen	353	289	416	548	519	476	499	470	434	259	342	339	226	242

*The green color represents maximum catch. The red color represents a decrease in catches and/or minimum catches.

The analysis of the presented data shows a more or less decreasing trend of the populations of the most valuable and commercially important fish species in the main fishing waters of the Danube, Sava, and Tisza, in the following order: sterlet<carp<pike<pikeperch<wels catfish. Strong fishing pressure is also observed in these fishing waters, which are at the limit of sustainability, especially the Danube (Simić et al. 2014a).

The next feature that emerges from the analysis is the declining trend in the main fish species, such as brown trout (in all basins), huchen (Drina basin), Danube barbel (Danube, Great Morava), and chub (Great Morava basin). In the mentioned fishery waters, the populations of nase (especially in the Drina basin) and bream are ecologically sustainable (Simić et al. 2014a, 2022a). Characteristically for the Great Morava and Drina basins, populations of important fish species such as wels catfish, pikeperch, bream, and to some extent pike and carp are at a higher

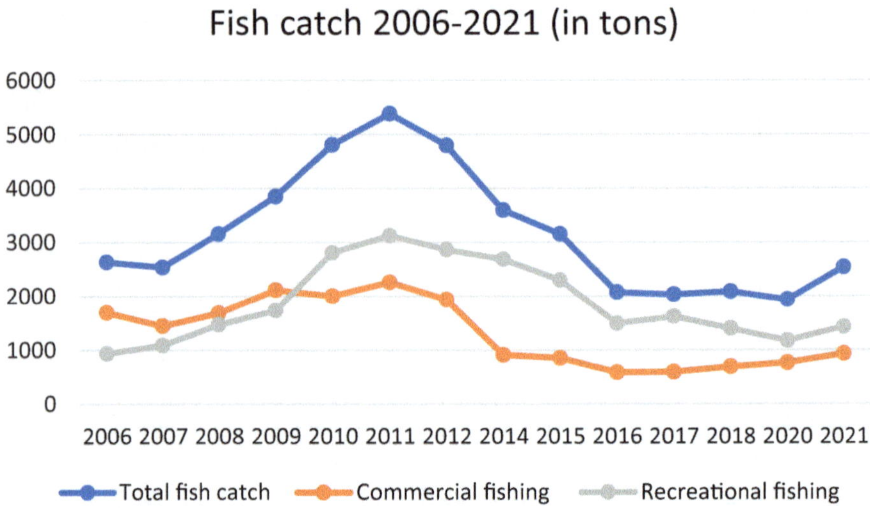

Fig. 13 Total catch, as well as catch by commercial and recreational fishermen in tons, from 2006 to 2021 based on official statistics

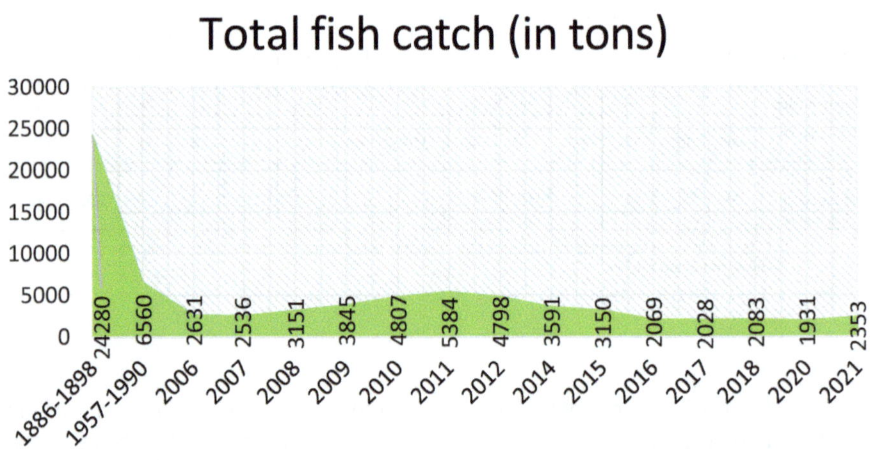

Fig. 14 Comparative overview of the total fish catches (in tons) in Serbia from 1886 to 2021

sustainability level in most reservoirs compared to the main river flows, but also to parts of the Danube and Sava rivers (Simić et al. 2022a). In most fishing waters, and especially in waters on the territory of Vojvodina, a part of the Danube in the reservoirs Đerdap I and Đerdap II (Žikić 2022), there is a trend of increasing allochthonous fish species, mainly Prussian carp, but also grass carp, bighead carp, and silver carp. These allochthonous species, especially Prussian carp, are highly valued by fishermen in Vojvodina, contrasting with their ecological status in inland

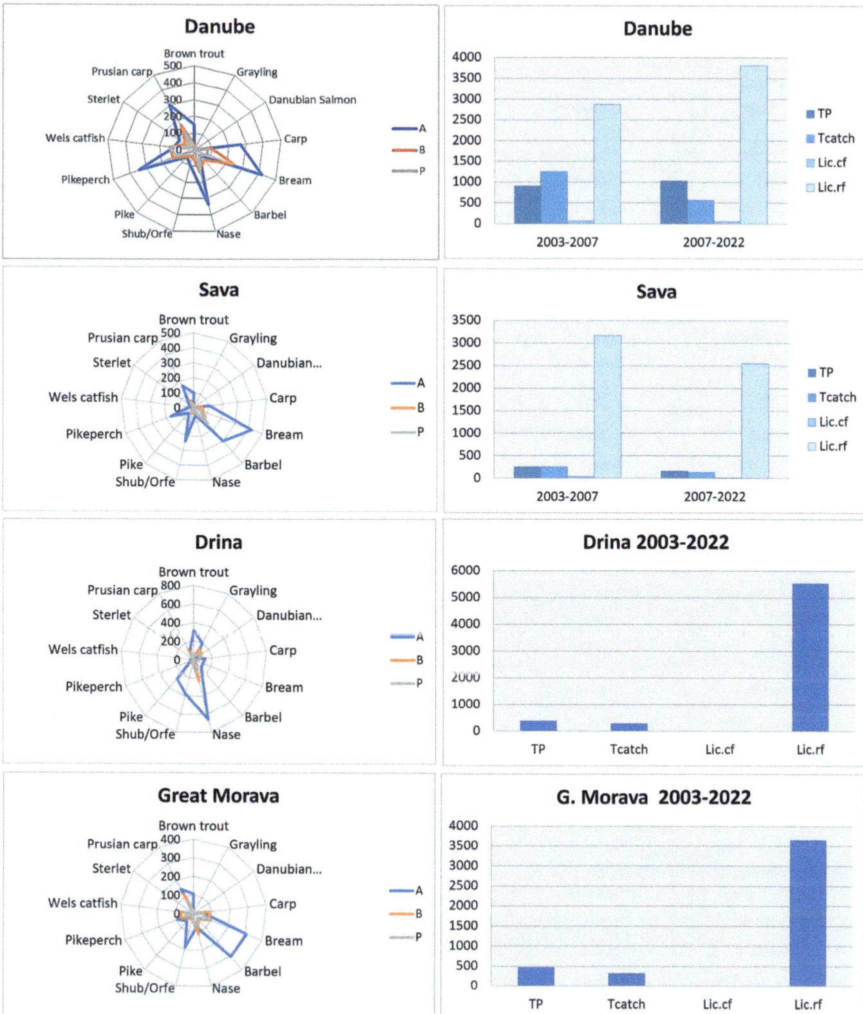

Fig. 15 (**a, b**) The state of the fish stock of the main fishing waters of Serbia is shown on the left, while the production, total catch, and fishing pressure of commercial and recreational fishermen from 2003 to 2022 are shown on the right

waters (allochthonous invasive species). Fishing pressure in the last 20 years is highest in the historically most important fishing waters such as the Upper Danube (part of the Danube from Ram, around Belgrade and upstream to the Hungarian border), the Sava, and the Tisza, where commercial fishing is present in addition to recreational fishing. In the Great Morava River basin, where there is no commercial fishing and which covers most of Serbia (about 60%), the total fishing pressure from recreational fishers is lower on average. The fish population is moderately

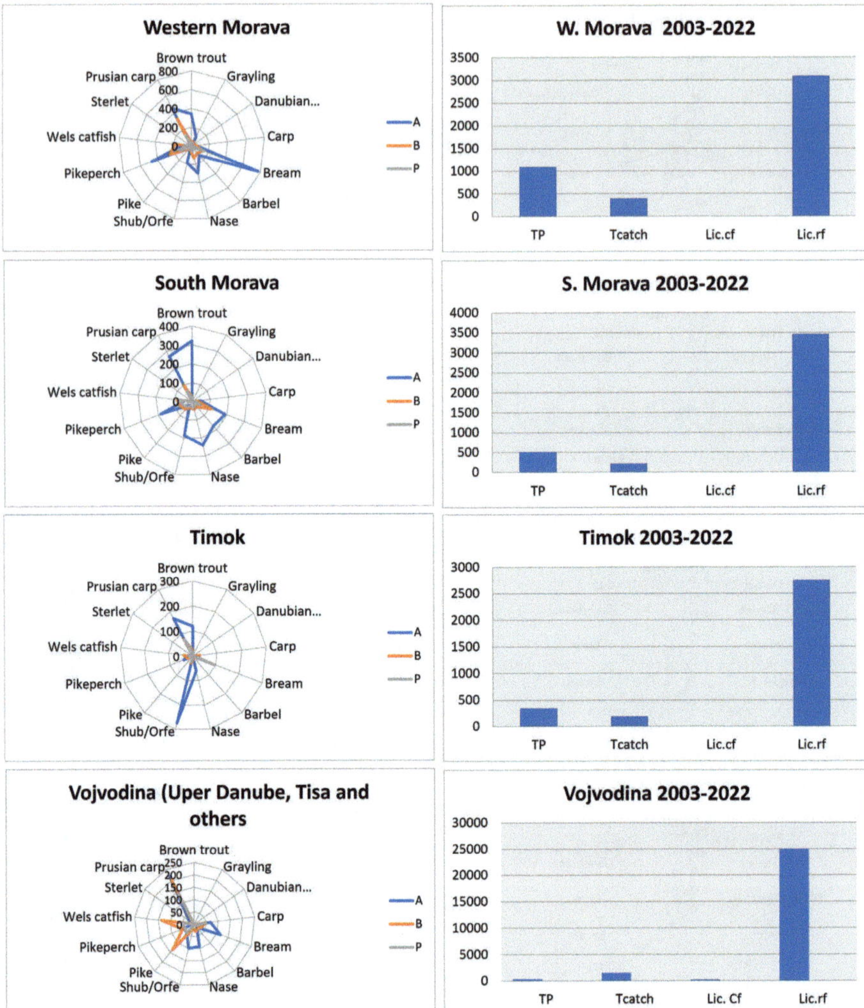

Fig. 15 (continued)

sustainable in the Great, South, and West Morava and their main tributaries, while it is highly sustainable in a large number of water assemblages in the area. The Drina has a moderate to sustainable fish population in the river itself and a very sustainable one in the reservoirs. Based on the results of the "ESE-HIPPO river basin" model (Simić et al. 2022a) in the Great Morava River basin, the small rivers of the salmonid zone have the highest sensitivity and a low level of sustainability, while the large rivers (Great, Southern, and Western Morava) have a moderate level of sustainability and a lower level of sustainability.

Analyzing the data for the average values of real production (RP), potential production (PP), and relation of the real production to potential production (% PP) of the most important fishing species, characteristic differences in these parameters can be observed in the main fishing waters of Serbia (Fig. 16a, b).

Fish species from the family Acipenseridae have undergone the greatest changes in recent times (see Chap. 10, Cvijanović et al. 2023). The damming of the Danube by the dams of the HPP "Đerdap 1" and "Đerdap 2" (Iron Gate I and II) in the area of the "Đerdap" gorge between the former SFRY and Romania in the 70 s and 80 s of the last centuries resulted in the disappearance of all migratory sturgeon species (beluga sturgeon and four other sturgeon species) in the upstream part of the Danube. In this way, Serbia lost not only the species important for commercial fishing but also the caviar industry. Caviar produced on the Kladovo fishery farm (so-called "black caviar") was particularly appreciated and protected by the law (Anonymous 1995) as food of geographical origin (Mladenović 1999). According to the records of the Kladovo fishery, the amount of sturgeon species caught (beluga sturgeon, Russian sturgeon, starry sturgeon, and sterlet sturgeon) before the construction of the power

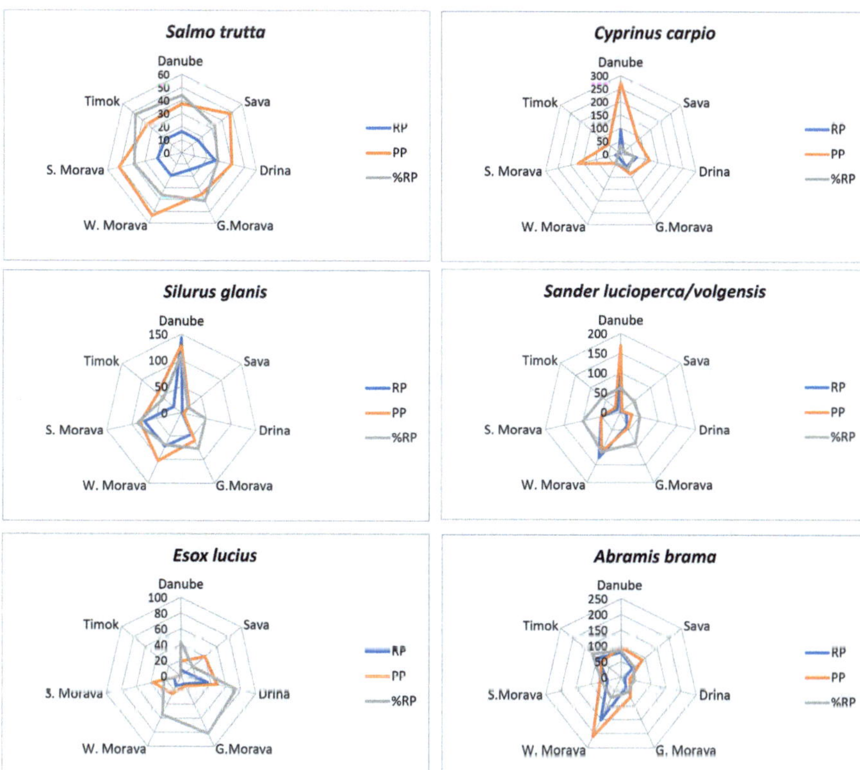

Fig. 16 (**a, b**) The most important fish species are shown in terms of real production (RP), potential production (PP), and the achieved real production in relation to the potential (%PP)

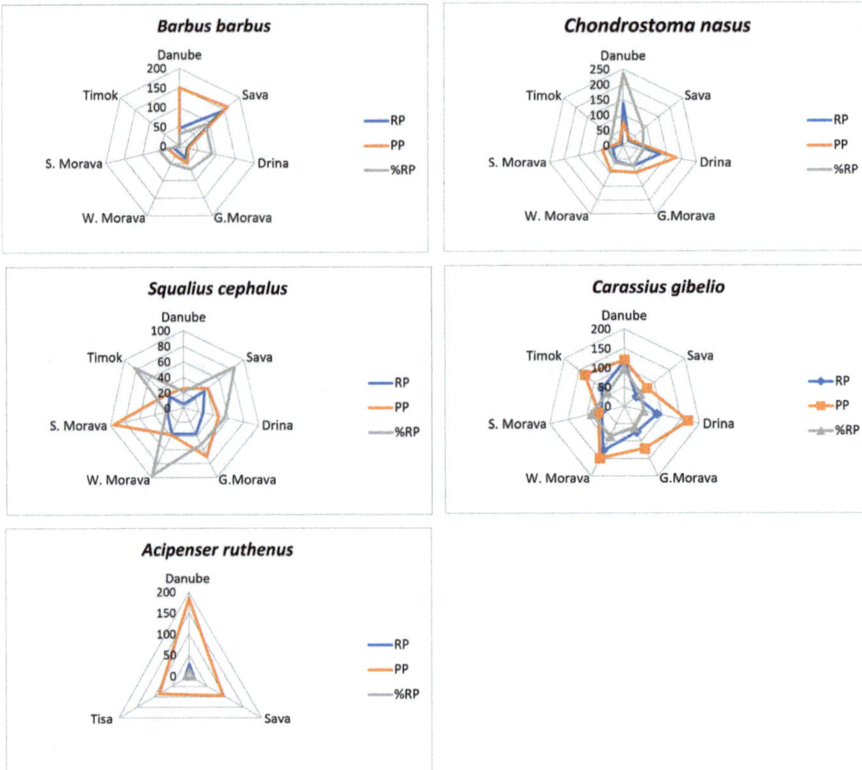

Fig. 16 (continued)

plant "Đerdap II" (before 1984) was 21 tons on average, and 15 years later it dropped to less than 3 tons per year (Janković et al. 1994; Mladenović 1999; Jarić et al. 2010). With the introduction of an intergovernmental moratorium that includes a ban on fishing migratory Acipenseridae species (beluga sturgeon and all other sturgeon species) between Romania, Bulgaria, and Serbia (2006–2012–2017), the focus of commercial fishermen in Serbia has shifted to sterlet sturgeon as the only non-migratory species (Danube population) inhabiting the Danube, Sava, Tisza, and occasionally the Great Morava. Due to overfishing, especially of mature individuals, loss of natural habitats for spawning, eutrophication, and pollution, sterlet sturgeon population in the Danube and Sava rivers has declined significantly. It is estimated that sterlet sturgeon biomass and actual production decreased by about 60% in the period 2014–2018 compared to biomass and production before this period (Janković and Krpo Ćetković 1995; Simić et al. 2020b, c, d; Žikić 2022) (Fig. 17).

Fig. 17 Juvenile sterlet sturgeon caught by fishermen in the Danube (area of the city of Belgrade) in August 2016 (photo by V. Simić)

This situation led to a complete ban on sturgeon fishing in the Danube, Sava, and Tisza rivers from 2019, based on the report of the WWF project—Life for Danube Sturgeons (Layman's Report 2020).

Brown trout is the most widespread salmonid fish in Serbia and is highly valued in recreational fisheries (Fig. 18).

However, studies clearly indicate a declining trend in populations in nearly 80% of the habitats of this fish species (Simić et al. 2014a, 2020e, f, g, h, i, j, k; 2022a, b, c, d, e, f). The most numerous populations are in the salmonid area of the Great Morava Basin and, to a lesser extent, in the Drina Basin. The rivers of the salmonid area in Serbia are less abundant in water compared to rivers in other parts of the Western Balkans (especially rivers in Bosnia and Herzegovina, Slovenia, and Montenegro) and in Western Europe. This natural characteristic of salmonid rivers is even more pronounced due to climate change. The situation is exacerbated by human activities that use the ecological services of salmonid waters for power generation, agriculture, forestry, and tourism. In general, the salmonid region in Serbia has decreased by more than 30% in the last 20 years (Simić et al. 2014a, 2022a) due to the habitat loss estimated at more than 30% (in quantitative and/or qualitative terms). Trout populations are increasingly concentrated in the upper, colder parts of rivers, i.e., the upper salmonid region. Due to spatially smaller habitats and reduced water

Fig. 18 Brown trout specimens from the Raška River (West Morava basin) caught in June 2019 (photo by V. Simić)

quantity, the percentage of trout reaching the minimum standard body length decreases SL $\geq$ 25 cm, which is allowed for fishing according to the Regulation on Measures for the Conservation and Protection of Fish Stock (Anonymous 2015). Middle and lower salmonid areas are becoming warmer on average (in 2022, summer temperatures at most sites surveyed were between 18 and 22 °C; author's measurements and observations) and more eutrophic (mass occurrence of filamentous macroalgae; author's observations). This causes the expansion of Danube barbel in salmonid regions (Fig. 19). This species is less sensitive even on the habitat fragmentation (Radojković et al. 2019).

On the other hand, the rivers (Veliki Rzav, Gradac, Đetinja, some tributaries of the Drina and Lim, rivers on the Vlasina plateau, and the right tributaries of the South Morava) where there are more numerous and fishery significant trout populations are mostly hybrid (Marić et al. 2022a). Namely, in these populations, the allochthonous Atlantic haplotype is represented to a smaller or larger percentage, originating from trout of the Atlantic and partly Adriatic genetic lineage introduced into the salmonid waters of Serbia by stocking. The introduction of non-native genes started in the middle of the last century and continued almost until the adoption of the new Law on Fish Stocks in 2014. This law explicitly prohibits the stocking of fishing waters with genetically unsuitable fish. According to the model "ES-HIPPOfish" (Simić et al. 2014a), populations of other autochthonous salmonid species, namely huchen (Fig. 20) (Marić et al. 2012a, 2014; Simonović et al. 2000, 2011) and grayling (Marić et al. 2004b, 2012b, d) in Serbian waters, also show a downward trend. Based on the same model, grayling in Serbia has medium (Drina, Lim) to low sustainable stocks (Ibar River). Additionally, grayling population in Drina is very sustainable, in Lim medium, and in Ibar has low sustainable status.

Fig. 19 Danube barbel (photo by V. Simić)

Fig. 20 Huchen (female) from the control catch during the return migration after spawning in the middle reaches of the Poblaćnica River. The breeding ground is about 40 km from the permanent habitat of the juveniles in the Lim River (photo by V. Simić)

The carp is a traditionally valued fish species in Serbia. It inhabits almost all aquatic habitats, from lowland rivers and ponds to reservoirs at altitudes of 800 to 1200 m (Vlasina Reservoir). Sustained populations exist in some protected waters of Vojvodina (Zasavica, Jegrička), in parts of the Danube and Sava, as well as canals and reservoirs in the Great Morava Basin and Vojvodina (Miljanović et al. 2018a, b, 2021a). A sharp decline in abundance, biomass, and production of the carp

Fig. 21 Carp from the Ribnica Reservoir (Drina basin) in June 2019. The population is not abundant, but mainly large specimens are caught by recreational fishermen (photo by V. Simić)

population has suffered in the course of the Great, South, and West Morava mainly due to regulation of discharge and loss of floodplains for spawning and then due to pollution and poaching (Simić et al. 2013). The highest fishing pressure on carp populations and the lowest percentage of realized production relative to potential are found for the Danube and Sava rivers (Fig. 21). In these rivers, the pressure comes from commercial and recreational fishing but also poaching. In the reservoirs in the Great Morava catchment, carp populations have been established by stocking carp fry from farms or fishponds. In Serbia, detailed genetic studies of carp populations have not been conducted, so it is not known where, to what extent, and whether native wild populations have been maintained. Moreover, carp has been listed as a vulnerable species according to the IUCN Red List (Freyhof and Kottelat 2008). Research associated with the development of the Fisheries Management Program indicates that stocking large rivers and reservoirs with carp is not producing the expected results. In general, it is difficult for carp to adapt to reservoirs with an unstable water level regime, and even when its populations adapt, they decline rapidly. Probable reasons for the rapid decline of populations include low reproductive potential of the founder population, poor conditions for reproduction, or spawning failure due to the unstable water level (Vilizzi et al. 2015).

Monitoring of the fish population of some fish waters in Serbia confirmed the absence of younger age groups even ten years after the first stocking. In addition, the low genetic diversity of the founding population and/or loss of genetic diversity may also cause the stocking failure. Poaching is also a cause of stocking failure, as increased poaching was observed after each stocking.

Besides carp, other cyprinid species that have fishing importance in the fishing waters of Serbia are bream, nase, barbel, chub, and asp. Bream is the most frequently

Fig. 22 Bream, the most common species in anglers' catches in the fishing waters of Serbia. Catch of a recreational angler in West Morava in September 2020 (photo by V. Simić)

caught species today, both by commercial and recreational anglers. Populations are still viable in the Danube, Sava, Tisza, and large reservoirs in the Great Morava River basin (Fig. 22).

Nase is often the dominant species in the Great Morava Basin, and populations with a high degree of sustainability are present in the Drina River (Hegediš 2003a; Simić et al. 2014a; Simić et al. 2020a) (Fig. 23).

There is evidence that increasing eutrophication and sea floor algal cover benefit this species by increasing the total amount of its important plant food (Fig. 24) (Gerke et al. 2021). Apart from this, dams have not yet been built on the main course of the Morava River, allowing nase to migrate freely during spawning and to spawn successfully at the mouths of larger tributaries (Simić et al. 2020d, 2022a).

The barbel (Fig. 25) is still present with relatively numerous populations in the waters of Serbia, but the ratio of real and potential production is unfavorable in the Danube, Sava, and especially Great Morava. In general, in the last 20 years, there has been a decrease in the average length of barbel caught and the average total length of the catch (Simić et al. 2013, 2014a, b, 2022a). There are many causes for this situation, but the ones that stand out include changes and destruction of habitats suitable for spawning due to exploitation of sand and gravel from rivers, changes in flow regime and water velocity, overgrowth of the bottom with algae and macro vegetation, and pollution (Mueller et al. 2018) (Fig. 26).

These factors are compounded by excessive fishing of individuals below the minimum hunting length ≥ 25 cm (SL) and poaching. The chub shows similar changes in population characteristics in Serbian waters. These changes are even more pronounced in some waters (Great Morava, South Morava, Ibar). One of the additional reasons for the decline of chub populations is probably the excessive catch of individuals below the minimum allowed length (SL ≥ 20 cm). In practice, managers of fishing areas often correct the minimum permissible length to SL ≥ 30 cm.

Fig. 23 Catch of a recreational fisherman on the Drina River in 2017. Nase (shown above) has viable populations in most of Serbia's fishing waters, while *Rutilus pigus* (shown below) is a common species in the Drina (photo by V. Simić)

Fig. 24 Traces of scraping algae from the stone during the feeding of nase in the Ibar River near Kraljevo (photo by V. Simić)

Predatory species, wels catfish, pikeperch, and pike, are among the most appreciated fish species in the waters of Serbia, because of both the quality of the meat and the price on the market. On average, the pike is less represented in the catches of professional fishermen than wels catfish and pikeperch. Pike populations are declining in large rivers, especially in the Great Morava (Fig. 27) (Simić et al. 2020d). In

Fig. 25 Barbel from the catch of recreational fishermen from West Morava in June 2018 (photo by V. Simić)

Fig. 26 The exploitation of sand and gravel from the Great Morava River (photo by V. Simić)

slow-flowing lowland rivers in Vojvodina, but also in clean mountain reservoirs such as the "Vlasina," pike reaches sustainable populations due to water quality, suitable habitat, and abundance of fish (perch, roach, Prussian carp, etc.) on which pike feed (Budakov 1993a). Overfishing of pike below the legal minimum length (SL $\geq$ 40 cm) mainly leads to declining population trends. In the Vlasina reservoir, based on monitoring in 2022, it was estimated that 35% of pike caught by

Fig. 27 Pike from Vlasina Reservoir, caught by commercial fishermen in June 2020 (photo by V. Simić)

recreational fishermen were below the minimum length and were not returned to the water (Simić et al. 2022e). Commercial fishermen on the Danube also largely do not return the specimens of pike caught in nets, and the minimum catch lengths are prescribed below.

Pikeperch and wels catfish are predatory species, mostly caught in the waters of Serbia, both because of the tasty meat without small bones and the high price on the market (especially pikeperch that costs approximately 6–8 euros per kilogram) (Fig. 28).

Wels catfish and pikeperch stocks are mostly sustainable and tend to decline gradually, especially in the Sava and Danube rivers, where they are used for commercial fishing. In the Great Morava and Drina basin reservoirs, there are sustainable and numerous wels catfish stocks. However, in some reservoirs (e.g., Vlasina, Rastovnica, and Vrutci reservoirs), wels catfish has great predatory pressure on other species, even on other predatory species (Fig. 29).

The number and biomass of species that are part of the wels catfish diet (carp, bream, etc.) decrease over time in "closed" reservoirs and generally reduce the attractiveness of fish waters.

The sustainability of pikeperch populations may be related to increasing eutrophication of water bodies and increased occurrence of prey species such as populations of resistant benthopelagic fishes like bleak *A. alburnus* and *Alburnus* spp. (Pavlovic et al. 2015), but also invasive species, mainly gobies in the Danube and Sava rivers (see Chap. 2, Piria et al. 2023). According to FishBase data, pikeperch is a highly vulnerable species to fishing. Our observations indicate that this characteristic is more pronounced in stagnant, closed waters such as reservoirs in Serbia compared to large, slow-flowing rivers. As evidence of this, we can cite the

Fig. 28 Pikeperch, a popular species on the fish market in Serbia, targeted by commercial and recreational fishermen. Pikeperch from recreational catch from Bovan Reservoir in September 2015 (photo V. Simić)

Fig. 29 Specimens of wels catfish from the Vrutci Reservoir during an experimental and selective catch in 2017 (photo by V. Simić)

example of the Bovan Reservoir on the Moravica River (South Morava river system). This reservoir was stocked with fertilized pikeperch roe (so-called artificial pikeperch nests) for several years (regularly five years), and a sustainable population was formed with a biomass of 40 kg/ha and an annual production of 33 kg/ha, which allowed an MSY of 4 tons for recreational anglers. However, by the decision of the Ministry, the accumulation was transferred to another manager of fishing area to increase economic viability. The new manager failed to organize the fishing service sufficiently efficiently, resulting in an increase in poaching with nets and hunting of bass below the minimum catch length by recreational fishers. In only two years of this uncontrolled fishing, the pikeperch stock dropped to a very low level of sustainability (biomass 5.3 kg/ha, production only 0.4 kg/ha) (Simić et al. 2020 l).

The factor of greater presence of benthopelagic fish is probably an important factor in the sustainable populations of *Leuciscus aspis* (Form. *Aspius aspius*, www.

fishbase.se) in the fish waters of Serbia. This predatory fish from the family Cyprinidae has stable populations in the main lotic fishing waters of Serbia (Danube, Sava, Great Morava) and is frequently caught by commercial and recreational fishermen. On the other hand, the asp is not considered a particularly high-quality fish and is not much appreciated by commercial fishermen. In neighboring Croatia, the asp is endangered and may not be caught (Mrakovčić et al. 2006). According to FishBase, the asp is a highly to very highly endangered species in relation to fisheries, but the consistently high proportion of catches of this species in Serbian waters by commercial and recreational fishers calls this assessment into question. Of the allochthonous species in Serbian waters (see Chap. 13, Radenković et al. 2023), the Prussian carp has the greatest fishery importance. Populations are particularly abundant and sustainable in the eutrophic waters of Vojvodina (Miljanović et al. 2018a, b, c, 2019a, b, c, d, 2021a, b, 2022; Tokodi et al. 2018), but also in certain reservoirs in the Great Morava Basin and in the Đerdap Reservoir (Simić et al. 2020d; Žikić 2022). In Vojvodina, Prussian carp is a valued fish species, and its percentage of presence and biomass reaches 70%, while in other parts of Serbia, it ranges from 30 to 50% (Miljanović et al. 2018b, Simić et al. 2022a). Among other allochthonous species, grass carp, bighead carp, and silver carp have some importance for commercial fisheries on the Danube, on the Sava, and in the Vojvodina area. Increasingly frequent catches of several age classes of bighead carp and mostly silver carp species in the Danube by commercial fishermen indicate that these Asian species are successfully spawning in the Danube and probably also in the Sava (Figs. 30 and 31) (Simić et al. 2020a, b, c; Žikić 2022). This may pose a serious threat to fish stocks in these rivers based on experience from US inland waters (Kahler et al. 2020).

Finally, largemouth bass has a limited number of habitats in Serbia (only a few reservoirs in Vojvodina and Central Serbia), but it is a highly sought-after and valued species among recreational fishermen.

4 Perspective: Management, Sustainable Use, Conservation

4.1 A Critical View of the Current Fish Stock Management Model

The good and bad sides of the current model of fish stock management on the territory of Serbia are considered on the basis of the historical changes in the model, as well as in relation to the application of the model in practice.

The current model of fish stock management in Serbia is prescribed by valid laws and regulations (see Sect. 2.3). The model has significantly changed the responsibility for fish stocks in inland waters. Until the law from 2009 (Anonymous 2009), fisheries were always under the Ministry of Agriculture and included fish stocks in wild waters and fish farming in aquacultures. With the new model (Law from 2009),

Fig. 30 Catch of bighead carp during selective fishing on Potpeć Reservoir (Drina basin) in June 2015 (photo by V. Simić)

Fig. 31 Juvenile specimens of silver carp from the Danube (photo by V. Simić)

fisheries fall under the Ministry of Environment's jurisdiction and are primarily concerned with the protection and sustainable use of fish stocks as a biological resource of inland aquatic ecosystems. This change is not only declaratory but essential, because, for the first time in the history of Serbian fisheries, protection, conservation, and sustainable use are given priority over economic exploitation, maximum production, and profit. With this change, the philosophical view of fish traders is also changing. Fish dealer managers are no longer just fish farmers and "permit sellers" but are also becoming active conservationists. In parallel with this change, the professional staff in fisheries is also gradually changing so that agricultural engineers from the field of animal husbandry are replaced by biologists and ecologists (fisheries biologists, ichthyologists, and hydroecologists). In practice, however, this conceptual shift in attitude toward fish stocks is occurring very slowly and is being embraced by managers of fishing areas, and commercial and recreational fishermen. For most managers, the most important goal is still the number of permits issued for measures to protect or improve the fish stock. Fishermen in Serbia are not less, but not much more than in other parts of the world, against measures that organize fishing in different ways (e.g., the number of fishing gears and increasing the mesh size of commercial fishing nets, total prohibition of fishing in smaller or larger areas of fishing waters, restrictions on fish stocking, fishing according to the "catch-and-release" system and the use of hooks without retrieval hooks, declaration of reserves, i.e., special habitats for fish where all forms of fishing are prohibited, reduction of catch quotas and fishing regulations, and increase in minimum catch length of fish species) (Arlinghaus et al. 2016). Commercial and recreational fishermen have the same views on the above issues only because of conflicting interests. The issue on which managers and fishermen differ is the price of fishing permits. Namely, while managers ask the state to increase the price, fishermen have serious objections to the current price and are against any further increase. We believe that the price of a fishing permit must be a secondary issue between fishermen and managers in the sustainable management of fish stocks. This issue can be resolved by directing the activities of fishermen (commercial and recreational) toward measures and actions that sustainably contribute to the conservation and enhancement of fish stocks while respecting their social and human rights, including fishing rights (Arthur et al. 2016).

On the Danube, the most important fishing water for commercial fisheries, there is intolerance between commercial fishermen and recreational fishermen, mainly in terms of high and abundant yields compared to trophies (Kolding and van Zwieten 2014), but also because of the total amount of fish available for fishing. They blame each other for the excessive fishing, especially in the parts of the Danube that pass through the big cities, especially Belgrade and Novi Sad. They do not take into account that high fishing pressure (selective and non-selective) is the main reason for low catch of commercially important and attractive species for fishing, such as pikeperch, carp, and wels catfish. The decrease in catches of high-value fish species makes commercial fishing unprofitable. It leads to a gradual decrease in the number of fishermen, and, in this respect, Serbia is no different from most EU countries (Cowx 2015). Commercial fishing is also declining due to lower demand and consumption. Almost 90% of fish demand in Serbia is covered by domestic aquaculture (mainly carp and rainbow trout) and marine fish imports. There is also a growing question about the quality of wild fish from commercial catches in Serbia, given the increasing contamination of inland waters with toxic and other pollutants (Milošković and Simić 2015; Milošković et al. 2016). This trend is consistent with Western European countries, where commercial fisheries are largely absent and recreational fishers make very limited use of the fish they catch for their diets (Cowx 2015). Compromise solutions to some contentious issues between commercial and recreational fisheries have to be made by the relevant ministry in some cases. For example, the demand of recreational fishermen to allow the "bućka" (Fig. 9), a traditional tool of commercial fishermen for catching wels catfish, to be used by recreational fishermen as well was met with very strong protests from commercial fishermen. Therefore, the responsible ministry has significantly restricted the catch of wels catfish by recreational anglers with a special regulation. The main regulation is that wels catfish may only be caught using a catch-and-release regime. Although tensions have eased, the question of effective control and compliance with this measure by recreational fishermen remains.

Training of fishermen is prescribed in the current model of fish stock management in Serbia. However, the law is vague on this issue. The manager of the fishing area is obliged to carry out the training based on the program he prepared but on a voluntary basis. The training designed in this way has no impact on the fishing population without restoring the obligation to pass the fishing exam, which was prescribed by the 1994 law and is common in many countries in the region (e.g., Croatia; see Chap. 2, Piria et al. 2023). It has been shown that education is a good form of popularization of fishing, especially among young people. If it is well organized, it may have an indirect effect on the increase in the number of anglers (Cowx 2015). In Serbia, the number of recreational anglers is gradually decreasing, especially among young people (18–35 years old), which may be a consequence of the lack of proper education. This negative trend in recreational fishing contrasts with its high social and economic value in many countries in Europe and the USA (Arlinghaus et al. 2015; Lauer and Pyron 2016).

According to the current model, commercial and recreational fishermen are required to keep records of their catches on a specially designed form and submit the annual catch report when they purchase a license for the following year. And while commercial fishermen submit more or less realistic catch records, in 90% of the cases, recreational fishermen submit completely unrealistic records that are useless for a scientific and professional assessment of fishing pressure. To improve this situation to some extent, the current law requires that the manager of the fishing area do not issue a license to the fisherman for the following year if he does not submit realistic catch records. In practice, these measures do not work because the administrator does not evaluate the reality of the data from the catch record (there is no reliable method for this) and it is not in his interest not to issue a new license to the angler, as this may discourage anglers from buying licenses. The Serbian Institute for Nature Protection's attempt to introduce an application for online catch records has not yet gained acceptance in Serbia. Overall, catch records, and thus catch statistics and realistic assessment of fishing pressure, are a weak link in Serbian fisheries and complicate the application of the ecosystem approach to fish conservation (Suuronen and Bartley 2014). However, according to this study, fisheries in the inland waters of Serbia are not too different from most other countries that have the same problem, and in general, catch statistics are an important problem in freshwater fisheries in the world (Bartley et al. 2015). One of the approaches that our team has used, in collaboration with the managers of the fishing areas, is that the recording of catches is done at least once a month during the year by an expert (biologist, ecologist) together with the fisheries service. Comparing the results obtained in this way with those obtained by analyzing the fishermen's catch records, we find that the fishing pressure of recreational fishermen is realistically lower than indicated in the data, while the fishing pressure of commercial fishermen is higher than indicated.

In the management models followed by the pre-2009 legislation, great importance is attached to the protection of aquatic habitats or parts of habitats primarily suitable for fish spawning, the so-called "natural spawning grounds." In such habitats, all types of fishing are prohibited, but usually during the fish spawning season or for a limited period of time. This option is left in the new model, but it introduced the obligation of the manager of fishing area to identify, define, declare, and permanently prohibit fishing in the fishing area in all or part of the habitat that is important in the long term for maintaining the stability and sustainability of populations of certain fish species or the ichthyocenosis. The act refers to such habitats as "special fish habitats." These habitats are similar to Freshwater Protected Areas (FPAs) in ecological and conservation terms. Their primary function is to allow fish populations or entire communities in fisheries-exploited waters to recover naturally and then, through the "spillover" effect, to increase the sustainability of fish populations and the fisheries value of habitats outside of protected areas (Hannah et al. 2019). In Serbia, special fish habitats are designated to maintain fish populations or communities with the highest ecological sustainability (maintained genetic and structural characteristics of populations, maintained diversity of communities, and maintained or intact habitat). In the future, the role of special habitats should be expanded to represent reserves for the recovery of fish populations or communities that have been affected by overfishing or some other factor that has led to a decline in the sustainability level of the fish population. Such aquatic ecological reserves, when further connected to the network, have the role of naturally maintaining fish biodiversity and population or community sustainability in a larger area (Jordan et al. 2020). The concept of aquatic ecoreserves is currently more developed for marine ecosystems and is producing positive results in conserving populations of fish species, especially those affected by overfishing (Welcomme et al. 2010). Serbia's first reserve is structurally and functionally designed according to FPAs. It was proposed for the Veliki Rzav River (West Morava Basin) to enable the recovery of the autochthonous Danubian lineage of brown trout in the middle and lower salmonid reaches of this river (Simić et al. 2021) (Fig. 32).

For the time being, managers of fishing areas are opposed to the concept of special fish habitats and strive to have as few as possible in their fishing area because they believe it will induce the reduction in the number of sold licenses. This problem is largely related to inadequate education of managers of fishing areas and fishermen. Mandatory education would allow managers and anglers to properly understand the ecological importance of special fish habitats and/or water conservation areas under the concept of FPAs, and more importantly, they will increase the sustainability of the fish stock in the long term (Hannah et al. 2019).

Apart from the income from the sale of licenses, the managers of the fishing area do not use, or use very little, other opportunities provided by the law to generate additional income from the fish stock. Fishing tourism has great potential but has

Fig. 32 Part of the middle course of the Veliki Rzav River, which is proposed for FPAs (photo by V. Simić)

very little presence in Serbia. Even if there are some forms of fishing tourism, they take place in the organization of tourism agencies, while the managers of the area have little or no benefit. The beneficiaries usually have not developed active fishing tourism programs in the fishing area they manage. A great opportunity for the development of fishing tourism exists on the Danube, Sava, and Tisza where commercial fishing still exists. Since the future of commercial fishing is questionable, fishermen could turn to traditional fishing methods and equipment, which would be part of the tourist offer with an accompanying gastronomic program (Gurung 2016). Turning fishermen to fishing tourism using traditional tools can increase their income but also their interest in conservation and richer fish stocks because the basis of fishing tourism, in addition to organization, is rich and attractive fish stocks (Bessa et al. 2017).

From the point of view of the conservation approach to fisheries, a very important innovation of the new model is the active participation of the Institute for Nature Protection of Serbia in its indirect influence on the development of the Fishing Area Management Program. This authorized state institution for nature protection prescribes the conditions that the program must meet in terms of habitat protection, fish stocks, and endangered fish species. One of the conditions is the obligation to present the comparative results of the monitoring, which verifies the state of the fish stock in the current and the previous period. In this way, incomparably greater temporal transparency on the state of the fish stock is achieved. Because the programs are available to the general public, the manager of fishing area may be subject to public criticism if monitoring indicates poor state of the fish stock or a decrease in sustainability from the previous period. The conservation approach to sustainable management of the fish stock is achieved through conditions such as the following:

- Measures to prevent small-scale fishing, selective fishing, and declines in fish community biodiversity;
- The obligation to identify and declare special fish habitats;
- Emphasize "catch-and-release" fishing regime;
- Prevent the introduction of allochthonous species;
- Explain the ecological justification for stocking (as a historically traditional and mandatory measure in earlier periods);
- Attitudes and actions toward strictly protected fish species.

The conservation approach is a contribution and strong support for the faster transition of freshwater fisheries in Serbia from convection to the ecosystem (Craig 2015).

The conservation approach is only one part and cannot be equated with the overall ecosystem approach to fisheries (Garcia et al. 2012). The ecosystem approach is largely in contrast to the conventional approach to fisheries, most of which is still supported by the current Law on the Protection and Sustainable Use of Fish Stocks in Serbia. The conventional fishery is based on the paradigm of protecting juvenile fish (on the principle that they must spawn at least once) and on strict measures on the minimum diameter of the meshes of the nets, the minimum catch length of fish (instead of the maximum length), the reduction of fishing gears, and the time limit of fishing. The principles of the ecosystem approach to fisheries are the opposite and are based on the fact that a balanced harvest (Garcia et al. 2012) provides the highest yield to the fish community with the least disturbance to the demographic structure of fish populations. In this way, the fishery focuses on the smallest but most productive components (Law et al. 2012, 2013), and larger mature fish are spared, which are the least productive but have the highest reproductive potential (Kolding et al. 2014). An ecosystem-based approach to fisheries could reconcile commercial and recreational fishermen on the Danube by implementing a balanced harvest and catch-and-release regime for large fish by recreational fishermen. From our practice and as a contribution to the support of the ecosystem approach, the high sustainability of the bream population in the "Gruža" Reservoir is interesting, despite the high and long-term fishing pressure by recreational fishers and the ubiquitous poaching. Studies of the catch structure show that the pressure is evenly distributed among all age classes during the year. In spring and early summer, recreational fishers mass hunt juvenile specimens of bream, justifying their catch on the basis that it is impossible to distinguish them from other similar species (white-eye bream and white bream) for which, unlike bream, there is no legal length limit on the catch. In early and late autumn, recreational anglers catch mainly medium-sized bream, which are also a target for poachers. In this incidental catch, large breams make up the least common catch.

The catch-and-release fishing regime is not required by law, but the Institute for Nature Protection recommends that it should be applied as widely as possible. There is a higher percentage of opposition to this fishing regulation among managers of fishing area and recreational fishermen in Serbia. This regulation is particularly opposed by anglers whose target species are from the families Cyprinidae, Percidae, and Siluridae. Fly fishermen and bait fishermen, whose target species are mainly salmonids, accept this measure much better and use it more and more. In Serbia, management programs in 95% of salmonid habitats prescribe a catch-and-release regime for brown trout and huchen. Only grayling may be fished under the "catch-and-carry" system in the Drina River in quotas that do not threaten population sustainability. Long-term implementation of the catch-and-release regime has proven to be a good measure for ecological sustainability and an attractive population of brown trout in some rivers in Serbia (Đetinja, Rasina, Gradac) (Figs. 33 and 34, Movie S1). This measure has its full significance if other measures are strictly implemented, where healthy habitat and successful natural reproduction are crucial. In addition to favorable natural conditions, successful reproduction is also highly dependent on management of fishing activities during the spawning season and the prevention of poaching (Simić et al. 2014a, 2020a, b).

A catch-and-release regime is a generally supported resource enhancement measure for recreational fisheries in fisheries biology (Cowx 2015). Potential consequences of this regulation include increased mortality due to injury, severe stress, and animal welfare rights (Cooke and Schramm 2007). Our experience working with managers of fishing areas and fishermen suggests that this measure has a more humane character when the requirement to use hooks without retrieval hooks is added. However, there are also opinions that fishing is a human companion activity that is primarily about catching fish for food and supports fishing under the "catch-and-carry" system. A potential solution must be offered by science that considers all modern factors affecting ecosystems, fish stocks, and freshwater in general (Bănăduc et al. 2022). The ecosystem approach to fisheries can enable catch-and-carry fisheries, but only if the consistent implementation of measures that result from scientific

Fig. 33 Đetinja River (West Morava basin). A catch-and-release fishing regime with a well-organized fishkeeping service enables a sustainable brown trout population

Fig. 34 Đetinja River, where the lighter part of the bottom has been cleaned and prepared for brown trout spawning

research, such as absolute compliance by fishermen with the prescribed regime and catch quotas, is designed according to the principle of balanced harvest. Training anglers can be very important in applying the ecosystem approach to fisheries. Other alternative but highly humane forms of recreational fishing, such as "fishing with an underwater camera," are currently very little represented in Serbia. Underwater cameras are used in recreational fishing only as a tool for more successful "trophy fishing," but not for "trophy fishing photography" as the only form of fishing. Underwater cameras are increasingly being used in scientific research, and the recording techniques, methods, and equipment are also applicable to recreational fishing (Struthers et al. 2015).

Developments in molecular biology, conservation genetics, etc., have provided new insights that allow a better understanding of the ecological effects of stocking on autochthonous fish populations. Stocking, as one of the most important measures for fish stock improvement and enrichment, is represented throughout the history of Serbian fisheries legislation and practice. In general, stocking is a very widely used measure to improve freshwater fisheries in the world (Cowx 1994, 1998; Daupagne et al. 2021). On the other hand, research has shown that stocking can negatively impact the genetic diversity of autochthonous populations and reduce their adaptive potential (Rhymer and Simberloff 1996; Storfer 1999). Studies on the effects of stocking some French lakes have shown that this measure does not increase populations or the ratio of large to small fish (Daupagne et al. 2021). Stocking is one of the measures most supported by managers of fishing areas and fishermen in Serbia. Fishing area managers support stocking as a measure to enrich and improve the fish stock but also as a marketing measure aimed at increasing the attractiveness of the fishing water and, thus, the number of licenses sold. On the other hand, fishermen generally demand and positively evaluate stocking because they consider it a concern of managers of fishing areas for the fish stock, but also as an acceptable compensation for the fishing license price. From a professional point of view, stocking fishing waters was a widespread (almost obligatory) measure until the 2009 law, which also allowed the stocking with allochthonous fish species. From today's scientific point of view, stocking fishing waters with allochthonous fish species is ecologically completely wrong and prohibited by the new law. In Serbia, inland waters were stocked with allochthonous species after World War II. Large newly created reservoirs were stocked with Asian planktivorous and herbivorous fish (silver carp and bighead carp, grass carp) to slow down the process of eutrophication. Today, conflicting opinions prevail in scientific circles about the importance and role of silver carp and bighead carp in controlling eutrophication and the occurrence of cyanobacterial blooms (see Chap. 11, Simić et al. 2023). According to recent studies by Wang et al. (2022), positive results are obtained in controlling cyanobacterial blooms in Lake Qiandao in China when these species are introduced and act together. On the other hand, studies from areas where these two species are not native (Germany, France) indicate that they do not affect the eutrophication process (Domaizon and Devaux 1999; Radke and Kahl 2002). However, the invasive nature of these species in waters outside their historical ranges poses a serious threat to native biodiversity and fish stocks (Kahler et al. 2020). Of the salmonid species, rainbow trout were stocked most frequently, with brown trout and brook trout stocked to a lesser extent. Also, at the beginning of the 1950s, the newly created Vlasina Reservoir was stocked with a larger amount of Ohrid trout (Janković and Raspopović 1960), which, according to research, adapted well and showed better growth on average (see Chap. 9, Schöffmann and Marić 2023). Stocking of allochthonous salmonids, mainly rainbow trout, was neither massive nor long-term, so there are no established sustainable populations in the waters of Serbia, as is the case in Slovenia (see Chap. 1, Pliberšek and Tavčar 2023). A much more serious problem in the salmonid waters of Serbia is the long-term stocking of trout fry of the Atlantic lineage. Trout of the Atlantic genetic lineage was largely brought from ponds in Bosnia in the early 1950s and placed in two hatcheries in Serbia for spawning. Stocking trout waters with trout of the Atlantic haplotype was carried out regularly, often with a large amount of juvenile fish. In the last 15 years, serious genetic studies have been conducted on wild populations of brown trout in the waters of Serbia. The results indicated significant genetic contamination of the autochthonous Danube lineage with allochthonous genes of the Atlantic lineage and the existence of hybrid populations. The Center for Fisheries and Conservation of Inland Water Biodiversity Faculty of Science Kragujevac (CERIKOB) and the Faculty of Biology in Belgrade have conducted and continue to research on the genetic diversity of brown trout (Marić et al. 2022; Marić et al. 2006a, b, 2022a; Tošić et al. 2014, 2016; Simonović et al. 2017; Škraba Jurlina et al. 2020) in the waters of Serbia. Recent studies indicate that in the Danube drainage alone, an Atlantic brown trout was heavily introduced in about 7% of the rivers (Veličković et al. 2019). In the Reprocenter "Braduljica," genotyping of the broodstock of brown trout was also carried out, and all individuals with the Atlantic haplotype were removed from the further production process. In the newly established brown trout reprocenter on the Tolišnica River, hatchery broodstock was established in respect of the autochthony of the rivers intended for stocking. Stocking is based on the principle of returning the young fish to the same habitats from which the parents originated. It is also planned to capture new females or collect sperm from males every year, cryopreserve them, and use them for artificial insemination and reproduction when needed. Constant monitoring and measurement of the genetic diversity of the broodstock should prevent its decline in captive populations.

Despite this approach based on scientific facts, stocking salmonid waters in Serbia will be the last resort in the future; that is, it will be carried out only if all other measures do not lead to natural recovery and the formation of sustainable autochthonous populations. The fact that the new approach to stocking salmonid waters does not meet with too much resistance from the managers of fishing waters along the fishermen is encouraging. Fishermen's demands for stocking cyprinid waters remain strong. In accordance with the Fishing Area Management Programs, the managers are stocking mainly reservoirs in the Great Morava River basin (excluding the Great Morava River itself) and stagnant and slow-flowing waters in the Vojvodina area. Stocking consists mainly of carp, bream, and pikeperch, significantly less of pike and other fish species. Water reservoirs with muddy bottoms are not stocked with carp, since carp feeds on the bottom and promotes eutrophication by allowing the release of nitrogen and phosphorus from the sediment into the water. The most common fish species used to stock reservoirs for water supply is pikeperch (usually in the form of fertilized roe called "pikeperch nest"). Feeding pikeperch on zooplanktivorous fish (bleak, roach, juvenile bream, and juveniles of other species) allows faster growth and development of zooplankton, which more effectively reduces the density of phytoplankton and thus indirectly slows the process of eutrophication (Sundblad et al. 2020). This model of reservoir stocking (based on the food chain) strengthens not only the ecological but also the economic services of the ecosystem. The attractiveness of reservoirs to anglers increases due to perch stocking, which increases the sale of fishing licenses and equipment, and the slowing of the eutrophication process decreases the cost of drinking water production (Bauer et al. 2018). The fish community, regardless of how much its composition has been formed to reduce eutrophication in reservoirs for water supply, cannot prevent it, especially the occurrence of cyanobacterial "blooms" if there is another factor that accelerates this process, such as a large influx of nutrients or organic material into tributaries or due to heavy erosion from the watershed after heavy rains (e.g., the Barje, Vrutci, and Ćelije reservoirs). In general, Serbia lags behind other Western Balkan countries, especially Slovenia, in terms of stocking, especially in the last two decades (see Chap. 1, Pliberšek and Tavčar 2023). Apart from the reduction in the amount of fish stocking due to the possible ecological consequences this may have on the fish population, the problem is also the lack of quantity and quality of fish for stocking. On the territory of Serbia, the production of salmonid fish (mainly brown trout, much less grayling, and huchen) for stocking, which lasted for several decades, was achieved by only one hatchery, the fish hatchery "Braduljica" (initially owned by a fishing association and now a private company). For many years in this facility,

Fig. 35 Modern equipment for artificial spawning of brown trout (reprocenter on the Tolišnica river), December 2021 (photo by V. Simić)

juvenile fish were bred from fish with a dominant Atlantic haplotype. Only with the recent opening of the stocking center for brown trout on the river "Tolišnica," owned by SE "Srbijašuma," the conditions for further stocking in accordance with science have been created (Fig. 35).

Production of other fish species, such as carp, bream, and pikeperch, was carried out in several ponds in Vojvodina, but only one pond had long-term continuity and is the only one active today. The low level of stocking leads to a revolt of groups or individual recreational fishermen who resort to independent unplanned stockings. For example, wels catfish, an apex predator, and highly resilient fish species, has been introduced into reservoirs in Serbia in an unplanned manner, which can significantly affect the structure and composition of the fish community and significantly reduce the abundance of other important fish species (e.g., Vlasina Reservoir) (Simić et al. 2022e). In recent years, there have also been cases of unplanned introduction of huchen into fishery waters south of their range (Nišava River, Mlava River) (Simić et al. 2020e, l). The effects of such unplanned introduction on the fish population of rivers where the juveniles did not survive (or survived only for a very long period of time) have not yet been studied.

Commercial fishermen are vigorously resisting legislative changes that would restrict fishing gear and the mesh size of nets and self-tailing gear. According to the law, commercial fishing in Serbia is allowed with nets (set nets, trawls) with a mesh size of at least 50 mm during summer and 40 mm during winter season. In the past, fishing with nets with a mesh diameter of 40 mm was allowed for a longer period of time. Today, fishermen mostly use nets with the minimum permitted mesh diameter because it allows them to achieve a catch that is economically viable for them. Commercial fishermen on the Danube blame the low numbers of large fish on recreational fishermen who target and selectively hunt large specimens and generally disagree with the catch-and-release regulation. This situation on the Danube is largely a consequence of the prevailing and long-term conventional approach to fisheries in Serbia. In the coming period and given the increasing climatic and other threats to fisheries and inland aquatic ecosystems as a whole, it is necessary to change the approach to fisheries from a convective one to an ecosystemic one or another approach that will be the result of new scientific research.

5 Factors Affecting the Sustainability of the Serbian Fish Stock and Perspectives

5.1 Changes and Destruction of Habitats

Habitat loss in inland waters is greater than the losses and changes that occur in other types of ecosystems (Freyhof et al. 2015; Cowx 2015; Craig 2015; Bănăduc et al. 2022). Moreover, Serbia's more significant economic development in the last 15 years has caused greater anthropogenic pressure on aquatic habitats and fish waters. During this period, the greatest changes and loss of habitat have occurred in fishery waters populated predominantly by salmonids, where more than 100 (110) small hydropower plants (SHPP) have been built in the last ten years (with more than 800 planned; Ristić et al. 2018; Weiss et al. 2018). In the planning of SHPP and in the construction itself, economic interest (electricity generation) completely dominated over ecological interest, and the fish population came last in this chain. Most SHPP were built in ecologically important habitats with autochthonous and often rare or unique haplotypes. A large number of SHPP were also built in protected areas (national parks, nature parks, reserves) (Fig. 36). They also resulted in the loss of biomass and production of brown trout populations in some rivers (Vlasina, Rupska, Jelašnička, Tripušnica) estimated at over 85% (Simić et al. 2018a, b; Simonović 2020; Simić et al. 2022a). Due to the protests of environmental activists and scientific evidence of the serious consequences of SHPP on the fish population of mountain rivers, the construction of SHPP in protected areas was prohibited by law and further construction on other rivers was suspended, but without guarantees from the state that it will not continue.

Before the recent phase of construction of SHPP, there was a phase of construction of dams for large hydroelectric power plants on the major rivers in Serbia that include the Danube (Iron Gates I and II), Drina (Bajina Bašta, Zvornik), Lim (Potpeć), Uvac (Radoinja, Zlatar, Sjenička), West Morava (Međuvršje), and Ibar (Gazivode). Due to the double damming of the Danube, Serbian fisheries have lost migratory sturgeon species and the highly valued caviar industry (see Chap. 10, Cvijanović et al. 2023). In addition to sturgeon species, eel and herring populations have also declined dramatically.

Impoundment of other major rivers in Serbia (Drina, Lim, Uvac) led to the fragmentation of populations of grayling and huchen. According to Freyhof et al. (2015), huchen habitats in the Danube basin would decrease in the coming period due to the planned damming of rivers by 1000 km. In addition to damming rivers and

Fig. 36 The small hydroelectric power plant on the Samokovska River (Kopaonik National Park). This river is the largest in the National Park and had a sustainable population of brown trout until the small hydroelectric power plant was constructed. The river's upper course has also been affected by organic pollution from the wastewater of the winter tourist complex (photo by V. Simić)

creating reservoirs to generate electricity, more than 150 dams were built on Serbia's rivers in the postwar period, and an equal number of small (100), medium (34), and large (26) artificial lakes (reservoirs) were created. The purposes of these reservoirs are most diverse, and they include water supply, flood control, irrigation, fishing, recreation, and tourism (Radojević et al. 2008). Regardless of the purpose, water recharge has negative ecological consequences for the native fish communities and loss of aquatic biodiversity due to habitat and population fragmentation, disruption of migration routes, habitat change from rheophilic to limnophilic, or complete loss

Fig. 37 Vlasina Reservoir (1200 masl). The reservoir submerged the former high-altitude valley and caused the decline of autochthonous populations of fish and other organisms. From the perspective of recreational fishing, it is significant due to the sustainable populations of pike, catfish, bream, carp, chub, and trout, which were mostly formed by planned stocking, with the exception of catfish populations that were introduced unplanned by recreational fishermen (photo by V. Simić)

of habitat. A good example of the complete loss of a specific and unique habitat is the flooding of a large raised bog ("Vlasinsko blato") on the Vlasina River and the formation of a large reservoir (for the needs of the Serbian electricity industry) (Fig. 37). Due to the flooding of the Tresava River, there was a drastic decrease in the populations of autochthonous fish species such as brown trout, Danube barbel, and common minnow (Marić et al. 2022a).

The reservoir submerged the former high-altitude valley and caused the decline of autochthonous populations of fish and other organisms. From the point of view of recreational fishing, it is significant due to the sustainable populations of pike, wels catfish, bream, carp, chub, and brown trout, which were mostly formed by planned stocking, except for wels catfish populations (introduced unplanned by recreational fishermen).

From the point of view of fishing in reservoirs (as new habitats for fish), a rich and fishery-important fish stock has formed through stocking or spontaneously, especially in the reservoirs of the Great Morava basin (Simić et al. 2022a). The productivity and biomass of fish species important for fisheries (pikeperch, bream) in some reservoirs is higher than the biomass and production in the Danube and Sava rivers. In some eutrophic reservoirs (Gruža, Ćelije, Bovan), the fish stock has the capacity for balanced commercial fishing, but it is currently prohibited by law in Serbia.

In addition to the fragmentation of dammed rivers, changes and loss of habitat for fish and fisheries also occur due to the regulation of river flows, resulting in a significant loss of floodplains. For example, the Great Morava has been shortened by 80 km due to meander correction (Gavrilović and Dukić 2014). The loss of floodplains in the lower reaches of the river resulted in the disappearance of some fish species important for fisheries, such as crucian carp and tench, as well as a significant decline in carp and bream populations (Simić et al. 2014a, 2022a). During 2021 and 2022, the West Morava and South Morava rivers' lengths were regulated over almost 90 km due to the construction of new highways. In addition to the long-term changes in river habitats, the short-term changes that occur due to sudden fluctuations in water levels at the time of fish spawning (especially phytophilous species, carp, bream, and tench) are very unfavorable for the fish population of reservoirs intended for power generation. Due to rapid water withdrawal, fertilized eggs or juveniles and often adult fish remain trapped in muddy lagoons, ponds, and bays and die before they can find their way to deeper water. This phenomenon occurs almost regularly on the Danube in the coastal sections of the large Đerdap reservoirs, but also on smaller reservoirs in the Great Morava (Vlasina, Zavoj Reservoirs) and Drina (Zlatar, Sjenička reservoirs) basins. Among the habitat changes that have a negative impact on the fish population is severe eutrophication, which causes an overgrowth of planktonic algae, the so-called "water bloom," within cyanobacteria being particularly dangerous for fish and other organisms (Đorđević et al. 2015; Simić et al. 2018a, b; also see Chap. 11, Simić et al. 2023). The massive development of macrovegetation in some reservoirs, especially in the Danube-Tisa-Danube Canal System (DTD) in the Vojvodina region, leads to the disappearance of large specimens of quality fish species (carp, bream, wels catfish) and the number of small specimens of invasive species such as Prussian carp, black and brown bullhead, and pumpkinseed (Miljanović et al. 2018a, 2018b, 2019b; Simić et al. 2020c). In this monograph, an entire chapter is devoted to invasive fish species and their impact on fish stocks in Serbia (see Chap. 13, Radenković et al. 2023).

Anglers often have conflicting interests and views regarding the scientific views of allochthonous and allochthonous invasive fish species (Lenhardt et al. 2011, 2016). Because anglers' education is ineffective in Serbia, recreational anglers have difficulty accepting common views about allochthonous fish species. For example, it is unclear to them why allochthonous fish species cannot be used in angling waters. This is especially true for grass carp and largemouth bass, as these species are attractive to recreational anglers. Recreational anglers are bothered by invasive allochthonous species of average small growth, such as brown and black bullhead, pumpkinseed, and all species of gobies, primarily because of interference with angling. In fact, due to dense populations, especially in reservoirs and large rivers, these fish often attack bait intended for other fish, which reduces the efficiency of the catch for fishermen but also the relaxing effect of fishing (Fig. 38).

Commercial fishermen do not mind silver carp and bighead carp, because these species sell well in the fish market. Prussian carp is the most accepted allochthonous invasive species in Serbia, especially in the fishing waters of Vojvodina, both among fishermen and recreational fishermen. As highly resistant and adaptable species, allochthonous invasive species benefit from warmer waters due to climate change, increased eutrophication, reduction in water quality, habitat degradation, and

Fig. 38 Fry of black bullhead in the Gruža Reservoir (photo by V. Simić)

unstable structure of the autochthonous fish community, and therefore constantly pose a serious threat to biodiversity and fish populations (Zorić et al. 2014) from unlimited fishing of allochthonous fish species and take active measures to prevent their spread. Managers of fishing areas generally take selective fishing measures. These measures are generally expensive and insufficiently efficient and produce short-term results or even have the opposite effect. Managers are more likely to accept them and implement them as a marketing measure that positively affects increased license purchases by recreational anglers, especially if the measures aim to catch black and brown bullhead and/or pumpkinseed. Experience with allochthonous species from other countries varies. However, selective commercial fishing and the use of these species for human or animal consumption, as well as processing them into sardines and fish pies, is an increasingly common method of controlling their populations. This process can also have adverse effects, especially when a profitable market is created, and the economic interest of the community requires that the species be kept permanently or even spread to other areas (Nuńez et al. 2012).

Pollution of inland waters is generally a very pronounced risk factor for fish stocks, not only at local but also at global scales (Bănăduc et al. 2022; also see Chap. 12, Milošković and Kojadinović 2023). Historically, pollution of Serbian inland waters has been positively correlated with economic development. From World War II to 1990, rapid industrialization and intensive agriculture increased pollution. The war events after 1990 and the cessation of industrial and agricultural work led to a significant recovery of aquatic ecosystems. However, with Serbia's renewed rapid economic growth in the last 15 years, the situation has deteriorated again. Serbia suffers from an unacceptably low number of treatment plants for wastewater from urban settlements and industry. All major cities in Serbia

(Belgrade, Novi Sad, Niš) do not have centralized wastewater treatment plants, so all wastewater (municipal water, industrial water) is discharged into the major rivers, Danube, Sava, and Great Morava. Moreover, the existing plants are mostly not fully functional (e.g., the one in Kragujevac). This problem is being solved only slowly, despite the sharp increase in the number of city inhabitants. Agriculture in Vojvodina and Great Morava valley is intensive and represents the largest source of phosphorus and nitrogen, which are responsible for the severe eutrophication of inland waters. Eutrophication of standing waters is a very pronounced process in Serbia and is exacerbated by climate changes that cause greater warming of water, low water levels, and/or strong torrents that wash nutrients and organic matter from the soil and introduce them into aquatic ecosystems. In addition to nutrients, substances such as heavy metals, pesticides, polychlorinated biphenyls, and microplastics are also introduced into aquatic ecosystems along with wastewater. These substances usually have a cumulative effect, and they accumulate in various tissues of fish and reducing their safe use in human and animal nutrition: plastic waste (bottles, bags, packaging) is becoming an increasingly serious problem for the inland waters of Serbia (Fig. 39).

Plastics that turn into micro- and nano-plastics through mechanical decomposition accumulate in the intestinal tracts and fish tissues (Rios et al. 2010; Dehaut et al. 2019).

Our investigations of several reservoirs (Gruža, Bovan, and Ćelije) for water supply indicate the presence of plastic in the intestinal tract of the bleak, which was used as a potential indicator (Nikolić et al. 2022). In the dry years of 2021/22, numerous accidents occurred in Serbia due to the discharge of wastewater from industry and sewage into rivers, resulting in massive fish kills. These accidents are becoming more frequent, and the low water level of rivers exacerbates their harmful effects in the summer months.

Serbia has a negative birth rate, according to the latest census of 2022. The distribution and density of the population are very uneven. In general, the total population is decreasing rapidly, especially in the southern, eastern, and northeastern parts of the country, as the number and density are increasing in the cities and especially in the capital (Belgrade), where more than one-third of the total population of Serbia now lives. The population density is not in accordance with the condition and abundance of fish stock in the fishing waters/areas of Serbia. The "Belgrade" fishing area, which includes parts of the Danube, Sava, and DTD canal systems with high population density, has a large number of recreational fishermen and commercial fishermen. Fishing pressure on the fishing waters is high, and estimates (see Discussion) indicate that the status of fish stocks is at the limit of ecological sustainability with a high risk of quickly becoming unsustainable (Simić et al. 2014a, 2020c, 2022a). In contrast, the parts of the Danube in the eastern part of Serbia (lower Danube), which are part of the "Danube" fishery area, have diverse and sustainable fish stock. At the same time, the population is incomparably smaller compared to the fishery area "Belgrade," and therefore has a significantly lower number of recreational and active professional fishermen. However, fishing pressure in the area has increased and is disproportionate to the number of local fishermen.

Fig. 39 Plastic waste on the bank of the Raška River (West Morava Basin) (photo by V. Simić)

This phenomenon is a consequence of some shortcomings of the model of individual licenses on the territory of Serbia. Namely, because of the diverse and attractive fish stock, recreational fishermen from all over Serbia come here to fish; however, they have obtained a license in their place of residence, which belongs to another fishing area. This situation leads to the fact that the fishery area "Danube" is not economically sustainable and managers cannot effectively implement measures for the protection and conservation of the fish stock, so the opportunities for poaching have increased. The fishery area "Danube" is a border fishing water with Romania, further increasing fishing pressure and the risk of overfishing.

In addition to current population densities, seasonal increases in density in fishing areas and some accompanying human activities also threaten fish populations. This influence is particularly pronounced during the warm season at attractive fishing waters in unprotected areas (Gruža Reservoir, Bovan Reservoir, Danube, Sava, Tisa), but also in protected areas (Vlasinsko Reservoir, Međuvršje Reservoir, inland waters of Kopaonik National Park and Zlatibor Nature Park) (Fig. 40).

The problem arises from the lack of regional and local development plans and the appearance of illegal constructions of a large number of weekend houses or fishing camps on the banks of reservoirs and rivers (Drina, Danube). Most of these facilities are owned by recreational fishermen who actively fish during the fishing season. These unplanned settlements and camps lack adequate sewerage and wastewater storage and are a significant source of eutrophication and water pollution, negatively affecting fish populations. Mass winter tourism in Kopaonik National Park is accompanied by growing infrastructure (roads, trails, cable cars), accommodation facilities (hotels, private houses for rent, etc.), and economic activities such as SHPP construction. All these activities put much pressure on the mountain streams within

Fig. 40 Waste left by recreational fishermen on the Bojnik Reservoir on the Pusta River (photo by V. Simić)

the park. The brown trout population in the National Park "Kopaonik" has decreased by about 60% in the last 20 years (Janković et al. 2005; Simić et al. 2020k, 2022g).

According to the data from the medium-term fisheries management programs, overfishing is constant in the part of the Sava River and the part of the Danube River called the "upper Danube" from 1075 km upstream (the major cities of Smederevo, Belgrade, and Novi Sad are located in this part). Overfishing in these parts of the Sava and Danube rivers is caused by strong pressure from commercial fishermen (Hegediš 2003b; Nikolić and Simonović 2003; Nikolić et al. 2003a, b, c, d, e). In these areas, the fish stock has not yet fully recovered. However, the number of professional fishermen has decreased by 70%, and the number of recreational fishermen has increased by less than 20% (Simić et al. 2020c). Illegal fishing (poaching) is still a significant factor in fishing pressure in Serbian fishing waters, which is difficult to control. It is estimated that poaching was highest during the period of economic crisis from 1990 to 2000. During this period, fish and fish stocks were considered an additional food source for the population. This "social" status of fishing in wild waters is still largely maintained today. It is estimated that poaching in Serbia is high (although a slight decrease has been observed in the last 2–3 years) and accounts for between 20 and 50% of the total fishing pressure. The estimation was made on the basis of the number of filed misdemeanor and criminal complaints, as well as seized nets and other fishing equipment (handmade equipment for electrofishing, self-catching nets, fish traps, etc.). The rate of poaching is high everywhere, but it is noted that it is higher in economically less developed communities, where the fish thieves are mostly people with lower economic and social status and use the caught fish as food. However, there are also fishing boats with modern equipment and much better economic status. In addition to the low ecological awareness and low morale of poachers, the weak and inadequately organized (and sometimes corrupt) fishing service of fishing area managers is the cause of widespread poaching. A major problem is the number of fish wardens, especially in large fishing areas where they are insufficient for effective guarding. Managers of fishing areas strive to employ a minimum number of fish wardens to ensure their own economic livelihood. Poached fish is sold on the "black" market, usually to restaurants or private individuals. Restaurants often state that the poached fish is legal and was purchased from commercial fishermen or registered fish farms (e.g., sterlet sturgeon, see Chap. 10, Cvijanović et al. 2023). Interestingly, some restaurants offer fish on their menus from local waters where commercial fishing is not allowed. In the restaurant on the Drina River, nase is offered, but also huchen, which is found in all waters of Serbia where it is caught and released. In addition to the lake "Gruža," the restaurant menu offers the "Gružan pikeperch," although according to the law it cannot be bought in the lake, as only recreational fishing is allowed, and recreational fishermen do not have the right to sell their catch. From Gazivode Reservoir, bleak is illegally fished with nets and thus ends up in restaurants near Ibar River, where it can be ordered as a specialty. In addition to poaching with illegal methods and tools, noncompliance with fishing regulations and quotas (fishing during the spawning season, catching fish under the permitted length, etc.) is also a major problem and an important factor that can have a negative impact on the fish population.

6 Conclusion

Fish stocks in the inland waters of Serbia have changed according to the intensity and nature of human activities. In the distant past, available data indicated diverse and sustainable fish stocks in the main fishing waters, Danube, Sava, Drina, and Great Morava. The sustainable trend lasted until the beginning of the intensive economic development of Serbia in the 1970s. From that time, a decrease in sustainability was recorded in the main fishing waters to a low level of sustainability in parts of the Danube (upper Danube), Sava, eutrophic waters of Vojvodina (Palić, Ludaš, some parts of the DTD channel), and salmonid waters in the Drina and Great Morava basins. The medium level of fish stock sustainability was recorded in the Lower Danube, Great Morava, Drina, Tisza, and their larger tributaries. The highest degree of fish stock sustainability in Serbia is in the reservoirs in the Great Morava and Drina basins. However, it should be emphasized that reservoirs are very sensitive and volatile, which may be the reason for the rapid decline in fish stock sustainability in these catchments. The reasons for declining fish stock sustainability in Serbia are not different from those in other parts of Europe and the world. Fragmentation of rivers by dams for the needs of large and small hydropower plants, regulation and melioration of river courses, eutrophication, pollution with organic and toxic substances and microplastics, and invasion of invasive fish and overfishing, together with the current climate changes, are the main factors for the decline in the ecological status of fish stocks (Stojković et al. 2013, 2014). In the last ten years, the fragmentation of salmonid rivers due to the construction of SHPP has led to the collapse of brown trout populations in numerous mountain rivers in Serbia.

Apart from the abovementioned factors, fish stock management had and continues to have a major impact on the state of the fish stock in the fishing waters of Serbia. Notwithstanding the change in strategy from a purely commercial fishery aimed at maximum production to a sustainable approach promoted by the current law from 2014, fishing and fisheries in Serbia have, in practice, essentially maintained a conventional approach. It includes numerous restrictions on the minimum length of the catch, the diameter of the meshes of nets, and the use of tools. In the main fishing waters, the Danube and the Upper Danube, the Sava, and the Tisza, in the period before the law's adoption from 2009, it was higher than production and was considered unsustainable. After 2009, the number of professional fishermen on the Danube, Sava, and Tisza decreased by more than 50% on average, while the number of recreational fishermen increased by about 20%. The current status of the fish stock in these historically important rivers differs only slightly; on the Upper Danube, it is at the limit of sustainability, and on the Lower Danube, it can be considered moderate to sustainable, although sturgeon species are no longer present. For further development of fisheries in Serbia in wild waters, a clear and realistic economic valorization (value) of fish resources by the state is necessary. Economic value is an assessment of all ecological services that fish resources can provide for the prosperity of the community. The principles of sustainable management of fish resources must be the basis for economic valorization. Economic valorization by the state should enhance the importance and prestige of fisheries and clearly define their place in society. The future of commercial fisheries in the inland waters of Serbia is questionable, especially if they continue to rely on the convection approach. Turning to the ecosystem approach, with a clear tendency to return to traditional methods and tools with small quotas that meet the needs of angling tourism, is a direction that can enable the survival and economic viability of commercial fisheries on the Danube, Sava, and Tisza rivers. Recreational fishing has a future, but it is necessary to change the legislation in many points. In this regard, better-organized training and education of the anglers are particularly important. With trained fishermen, implementing measures and strategies aimed at sustainable management of the fish stock is more successful. The problem of recording catches, especially by recreational fishermen, is one that needs to be solved more efficiently soon. In addition to fishermen, managers of fishing area administrative staff must also be trained and reorganized, especially with regard to the importance and good organization of the fishing service. Developing the strategy and practice of fisheries in Serbia should be focused on an ecosystem approach. The introduction of a balanced fishing regime is based on an accurate scientific assessment of the production and ecological potential of the aquatic ecosystem and the introduction of the optimum sustainable yield (OSY) instead of the maximum sustainable yield (MSY). The ecosystem approach is also accompanied by the need to establish a network of Freshwater Protected Areas (FPAs) or "Aquatic Protected Areas," which would represent areas without any human impact or fishing. These areas would have a strong long-term impact on the sustainability of fish stocks beyond their boundaries through the so-called "spillover effect." Special fish habitats currently required by law should be enhanced using the model of FPAs, which takes into account their spatial and temporal variability.

In addition to FPAs for the conservation of important fish habitats and fish stocks, increasing pressure on Serbian inland waters also requires the establishment of a network of NATURA and EMERALD areas for the conservation of endangered fish species. Serbia is actively involved in these processes.

Acknowledgments The investigation was supported by Grant (Agreement No. 451-03-68/2022-14/200122) funded by the Serbian Ministry of Science, Technological Development and Innovation). The authors express their gratitude to Andrija Dumanović for his help in figures design and graphic image processing and Milan Stepanović for the permission to download photos from the Ravnoplov website (https://www.ravnoplov.rs/lov-na-dunavske-kitove/).

References

Anonymous (1898) Fisheries law. "Officiale Gazette of the Kingdom of Serbia" No. 169/1898 (*In Serbian*)

Anonymous (1911) Fisheries law. "Officiale Gazette of the Kingdom of Serbia" No. 17/1911 (*In Serbian*)

Anonymous (1949) Fisheries law. "Officiale Gazette of NRS" No. 12/1949 (*In Serbian*)

Anonymous (1965) Fisheries law. "Officiale Gazette of SRS" No. 35/1965 (*In Serbian*)

Anonymous (1967) Fisheries law. "Officiale Gazette of SRS" No. 17/1967 (*In Serbian*)

Anonymous (1976) Fisheries law. "Officiale Gazette of SRS" No. 8/1976 (*In Serbian*)

Anonymous (1994) Fisheries law. "Officiale Gazette of SRJ" No. 35/1994 (*In Serbian*)

Anonymous (1995) Law on Geographical Indications. "Officiale Gazette of SRJ" No. 15/95 (*In Serbian*)

Anonymous (2009) Law on Protection and Sustainable Use of Fish Stocks. "Officiale Gazette of RS" No. 36/09 (*In Serbian*)

Anonymous (2012) Regulation on the categorization of fishing waters. "Officiale Gazette of RS" No. 10/2012 (*In Serbian*)

Anonymous (2014) Law on protection and sustainable use of fish stocks. "Officiale Gazette of RS" No. 128/2014, 95/2018 (*In Serbian*)

Anonymous (2018) Regulation on measures for the conservation and protection of fish stocks. "Officiale Gazette of RS" No. 56/15, 94/18 (*In Serbian*)

Apostolski K (1978) Prvo veštačko mrešćenje lipljena. Ribarstvo Jugoslavije, Zagreb pp. 33–39

Arlinghaus R, Cooke S, Cowx S (2015) Explaining participation rates in recreational fishing across industrialized countries. Fish Manag Ecol 22:71–77. https://doi.org/10.1111/fme.12075

Arlinghaus R, Lorenzen K, Johnson M, Cooke S, Cowx I (2016) Management of freshwater fisheries: addressing habitat, people and fishes. In: Craig J (ed) Freshwater fisheries ecology. Wiley Blackwell, pp 557–580

Arthur R, Friend R, Bene C (2016) Social benefits from inland fisheries; implications for a people-centered response to management and governance challenges. In: Craig J (ed) Freshwater fisheries ecology. Wiley Blackwell, pp 500–512

Bănăduc D, Simić V, Cianfaglione K, Barinova S, Afanasyev S, Öktener A, McCall G, Simić S, Curtean-Bănăduc A (2022) Freshwater as a sustainable resource and generator of secondary resources in the 21st century: stressors, threats, risks, management and protection strategies, and conservation approaches. Int J Environ Res Public Health 19(24):16570. https://doi.org/10.3390/ijerph192416570

Bartley M, De Graf G, Valbo-Jorgenson J, Mammula G (2015) Inland capture fisheries: status of data issues. Fish Manag Ecol 22:71–77. https://doi.org/10.1111/fme.12104

Batrićević A (2016) Zaštita životinja u srpskoj pravnoj istoriji. Institut za kriminološka i sociološka istraživanja, Beograd

Bauer B, Meier M, Casini M, Hoff A, Margoński P, Orio A, Saraiva S, Steenbeek J, Tomczak M (2018) Reducing eutrophication increases spatial extent of communities supporting commercial fisheries: a model case study. ICES J Mar Sci 75:1306–1317. https://doi.org/10.1093/icesjms/fsy003

Bessa E, Silva B, Sabino J (2017) Impact of fish tourism. In: Blumstein D et al (eds) Ecotourism's promise and peril. Springer International Publishing

Bojić D (2013) Dunavski ribari sa Dorćola i porodica Bogdanović. Istorijski muzej Srbije, Beograd

Božić G, Rašković B, Stanković M, Poleksić V, Marković Z (2021) Effects of different feeds on growth performance parameters, histology of liver, distal intestine, and erythrocytes morphology of common carp (*Cyprinus carpio* L.). Biologia 76:3769–3779. https://doi.org/10.1007/s11756-021-00882-y

Bravničar J, Palandačić A, Marić S, Šanda R, Snoj A (2015) Phylogeny of European Bullhead (*Cottus* sp.) in Western Balkans. Front. Mar. Sci. Conference Abstract: XV European congress of ichthyology, 7–11 September 2015, Porto, Portugal

Budakov Lj (1980) Variranje morfoloskih karaktera karasa - *Carassius auratus* L. u nekim lokalitetima SAP Vojvodine. Doktorska disertacija, Univerzitet u Novom Sadu

Budakov Lj (1989) Rast bodorke - *Rutilus rutilus* L. (1758), (Cyprinidae, Pisces). 20. Annual Meeting of Tisza Research Committe, Szeged

Budakov L (1993a) Characteristics of feeding and growth of pike perch and pike in the protected part of river Begej. Tiscia 27:53–56

Budakov Lj (1993b) *Misgurnus fossilis* L. (1758) on the list of Red Book of Serbia. Book of Abstracta and Programme 12, Thessaloniki

Budakov L, Lečić B (1991) Prilog ocenjivanju kvaliteta vode Palićkog jezera na osnovu tempa porasta vrste *Alburnus alburnus*. Zastita voda 91:143–147

Budakov Lj, Maletin S (1982). Tempo porasta riba kao parametar kvaliteta vode. "Zaštita '82", Ohrid

Budakov Lj, Maletin S (1984) Dynamics of longitudinal growth and body mass of *Esox lucius* L., *Blicca bjoerkna* L., and *Carassius aurata gibelio* Bloch in the Tisa. Tiscia (Szeged). XIX:173–178

Budakov Lj, Pujin V, Maletin S, Mućenski V (1979) Wachstum der Silberkarausche (*Carassius auratus gibelio* BLOCH, in der Donau und einigen Nebenflüssen in der Sozialistischen Autonomen Prowinz Wojwodina. XXI Arbeitstagung der IAD, Novi Sad, 202–08

Budakov L, Branković D, Gajin S (1994) Comparative investigation between phytoplankton and the intestinal contents of carp. Tiscia 28:47–51

Cakić P, Lenhardt M, Petrović Z (2000) The first record of *Gasterosteus aculeatus* L. 1758 (Pisces: Gasterosteidae) in the Yugoslav section of Danube. Ichtyologia 32(1):79–82

Cakić P, Lenhardt M, Kolarević J, Micković B, Hegediš A (2004) Distribution of the Asiatic cyprinid *Pseudorasbora parva* in Serbia and Montenegro. J Fish Biol 65(5):1431–1434. https://doi.org/10.1111/j.0022-1112.2004.00525.x

Cakić P, Lenhardt M, Mićković D, Sekulić N, Budakov L (2005) Biometric analysis of *Syngnathus abaster* populations. J Fish Biol 60(6):1562–1569. https://doi.org/10.1006/jfbi.2002.2011

Čanak S, Vasiljević Z, Radivojević D, Ivanović S, Marković T (2013) Influence of the carp pond model on the amount of investments for its construction. The first international symposium on agricultural engineering, Belgrade – Zemun, Serbia, 4–6 October 2013, p 2/20–2/33

Ćirković M, Jeremić S, Ćirković D (2000) Intensive farming of carpfish species. Contemporary fishery of Yugoslavia. Monograph. 4th Yugoslav symposium "fishery in Yugoslavia". Vršac, p 56

Ćirković M, Marković G, Simić V, Maletin S, Milošević N, Momirov D (2009) Reintrodukcija i repopulacija linjaka (*Tinca tinca* L.) u ribnjačke sisteme i otvorene vode. IV International conference "Fishery", Faculty of Agriculture, 27–29 May 2009, Belgrade, Zemun

Ćirković M, Ljubojević D, Novakov N, Đorđević V (2015) Gajenje i kvalitet mesa šaranskih riba. Monografija, Naučni institut za veterinarstvo, Novi Sad

Cooke S, Schramm L (2007) Catch-and-release science its application to conservation management of recreational fisheries. Fish Manag Ecol 14(2):73–79. https://doi.org/10.1111/j.1365-2400.2007.00527.x

Cowx I (1994) Stocking strategies. Fish Manag Ecol 1(1):15–30

Cowx I (1998) Stocking and introduction of fish. Fishing New Books, Blackwell Science, Oxford

Cowx I (2015) Characterisation of inland fisheries in Europe. Fish Manag Europe 22:78–87. https://doi.org/10.1111/fme.12105

Craig J (2015) Freshwater fisheries ecology. Introduction. In: Craig J (ed) Freshwater fisheries ecology. Wiley Blaskwell

Cvijanović G, Lenhardt M, Hegediš A (2005) The first record of black bullhead *Ameiurus melas* (Pisces, Ictaluridae) in Serbian waters. Arch Biol Sci 57(4):307–308. https://doi.org/10.2298/ABS0504307C

Cvijanović G, Adnađević T, Lenhardt M, Marić S (2015) New data on Sterlet (*Acipenser ruthenus* L.) genetic diversity in the middle and lower Danube sections, based on mitochondrial Dna analyses. Genetika-Belgrade 47(3):1051–1062. https://doi.org/10.2298/GENSR1503051C

Cvijanović G, Adnađević T, Jarić I, Lenhardt M, Marić S (2017) Genetic analysis of sterlet (*Acipenser ruthenus* L.) populations in the middle and lower Danube sections. North-Western. J Zool 13(1):34–43

Cvijanović G, Đikanović V, Galambos L, Pengal P, Smederevac-Lalić M (2023) Past and future of sturgeon species (Acipenseridae) in Western Balkans – case for permanent conservation or sustainable management. In: Simić V, Simic S, Pešić V (eds) Ecological sustainability of fish resources of inland waters of the Western Balkans in the Anthropocene. Springer, Cham

Daupagne L, Rolan-Meynard M, Logez M, Argiller C (2021) Effect of fish stocking and fishing pressure on fish community structures in French lakes. Fish Manag Ecol 28(4):317–327. https://doi.org/10.1111/fme.12476

Dehaut A, Hermabessiere L, Duflos G (2019) Current frontiers and recommendations for the study of microplastics. Trends Anal Chem 116:346–359. https://doi.org/10.1016/j.trac.2018.11.011

Đikanović V, Skorić S, Lenhardt M, Smederevac-Lalić M, Višnjić-Jeftić Ž, Spasić S, Micković B (2015) Review of sterlet (*Acipenser ruthenus* L. 1758) (Actinopterygii: Acipenseridae) feeding habits in the River Danube, 1694-852 river km) J Nat Hist 49(5–8):411–417. https://doi.org/10.1080/00222933.2013.877991

Đikanović V, Skorić S, Jarić I, Lenhardt M (2016) Age-specific metal and accumulation patterns in different tissues of nase (*Chondrostoma nasus*) from the Medjuvršje reservoir. Sci Total Environ 566-567:185–190. https://doi.org/10.1016/j.scitotenv.2016.05.072

Đikanović V, Tošić K, Lenhardt M (2018) A review of the diet and intestinal parasites of Pontic shad (*Alosa immaculata* Bennett, 1835) in the Danube River. 3rd International Congress on Applied Ichthyology & Aquatic Environment, 8-11 November 2018, Volos, Greece

Domaizon I, Devaux J (1999) Experimental study of the impact of silver carp on plankton Communities of eutrophic Villerest reservoir (France). Aquat Ecol 33:193–204. https://doi.org/10.1023/A:1009984713183

Đorđević N, Simić S, Ćirić A (2015) First identification of the cylindrospermopsin (CYN) - producing cyanobacterium *Cylindrospermopsis raciborskii* (Woloszyńska) Seenayya & Subba Raju in Serbia. Fresenius Environ Bull 24(11a):3736–3742

Faster S, Cakić P (1998) Karyotype characteristics of *Cyprinus carpio* L. (Pisces) individuals caught in the Danube by Zemun town [Serbia, Yugoslavia]. Acta Veterinaria (Yugoslavia)

Filipović D, Janković D (1978) Odnos sastava faune dna ishrane riba u brdskim tekućicama Istočne Srbije. Ichthyologia 10:29–40

Freyhof J, Kottelat M (2008) *Cyprinus carpio*. The IUCN red list of threatened species 2008: e. T6181A12559362

Freyhof J, Weiss S, Adrović A, Ćaleta M, Duplić A, Hrašovec B, Kalamujić B, Marčić Z, Milošević D, Mrakovčić M, Mrdak D, Piria M, Schwarz U, Simonović P, Šljuka S, Tomljanović T, Zabric D (2015) The Huchen *Hucho hucho* in the Balkan region: distribution and future impacts by hydropower development. RiverWatch & EuroNatur 30

Galambos L, Sekulić N (2023) Future perspective of sustainable development of freshwater fisheries and conservation of threatened fish species, crustaceans and molluscs. In: Simić V, Simic S, Pešić V (eds) Ecological sustainability of fish resources of inland waters of the Western Balkans in the Anthropocene. Springer, Cham

Garcia S, Kolding J, Rice J, Rochet M, Zhou S, Arimoto T, Beyer J, Borges L, Bundy A, Dunn D, Fulton E, Hall M, Heino M, Law R, Makino M, Rijnsdorp A, Simard F, Smith A (2012) Reconsidering the consequences of selective fisheries. Science 335(6072):1045–1047. https://doi.org/10.1126/science.1214594

Gavrilović Lj, Dukić D (2014) Reke Srbije.. Zavod za udžbenike Beograd

Gerke M, Hűbner D, Schneider J, Winkelann C (2021) Can top-down effect of cypriniform fish be used to mitigate eutrophication effect in medium-sized European rivers? Sci Total Environ 755: 142547. https://doi.org/10.1016/j.scitotenv.2020.142547

Grginčević M (1977) Ekologija i problem hibridizacije vrste *Abramis ballerus* L. u nekim vodotocima Vojvodine. Doktorska disertacija, Univerzitet u Novom Sadu

Grginčević M, Pujin V (1980) Prilog poznavanju taksonomskih karaktera vrste *Rutilus rutilus* L. u nekim vodotocima Vojvodine. — IV Simpozijum biosistematicara Jugoslavie, Đerdap

Gurung TB (2016) Enabling water–energy–food–nexus: a new approach for sustainable agriculture and food security in mountainous landlocked countries. J Nep Agric Res Counc 2:46–50. https://doi.org/10.3126/jnarc.v2i0.1621

Hannah L, Costello C, Elliot V, Owashi B, Nam S, Oyanedel R, Chea R, Vibol O, Phen C, McDonald G (2019) Designing freshwater protected areas (FPAs) for indiscriminate fisheries. Ecol Model 393:127–134. https://doi.org/10.1016/j.ecolmodel.2018.12.006

Hechel E, Rudolf K (1858) Die Süsswasserfische der Österreichischen Monarchie mit Rücksicht auf die angränzenden Länder. Monografien Vertebrata Pisces 0001:1–88

Hegediš A (2003a) Srednjoročni program unapređenja ribarstva na ribarskom području "Kolubara" za period 2003–2007. IMI, Beogad

Hegediš A (2003b) Srednjoročni program unapređenja ribarstva na "Savskom jezeru" za period 2003–2007. JP "Ada Ciganlija", Beograd

Hegediš A, Nikčević M, Mićković B, Janković D, Anđus R (1991) Discovery of the goby *Neogobius gymnotrachelus* in Yugoslav fresh waters. Arch Biol Sci 43:39–40

Hegediš A, Cakić P, Mićković B, Nikčević M, Anđus R (1993) *Gymnocephalus baloni* Holčik and Hensel, 1974 – a new percid in Yugoslav fresh waters. Arch Biol Sci 45:35–36

Hegediš A, Nikčević M, Mićković B (2008) Srednjoročni program za unapređenje ribarstva na ribarskom području "Srbija-Zapad" za period 2008–2012. IMI, Beograd

Holčik J, Šorić V (2004) Redescription of *Eudontomyzon stankokaramani* (Petromyzontes, Petromyzontidae) – a little known lamprey from the Drin River drainage, Adriatic Sea basin. Folia Zoologica 53(4):399–410

Hristić Đ, Bunjevac I (1991) Gajenje slatkovodnih riba. Građevinska knjiga Beograd

Jaćimović M, Lenhardt M, Krpo Ćetković J, Jarić I, Gačić Z, Hegediš A (2019) Boom-bust like dynamics of invasive black bullhead (*Ameiurus melas*) in Lake Sava (Serbia). Fish Manag Ecol 26(2):153–164. https://doi.org/10.1111/fme.12335

Janković D (1958) Ekologija dunavske kečige (*Acipenser ruthenus* L.). II izdanje, Biološki institut N.R. Srbije

Janković D (1960) Sistematika i ekologija lipljena Jugoslavije. Posebna izdanja 2, Biološki institut i Stručno udruženje za unapređenje slatkovodnog ribarstva Jugoslavije, Beograd

Janković D (1964) Synopsis of biological data on European grayling *Thymallus thymallus* L. FAO Fisheries Division, p 49

Janković D (1965) Uzroci širenja areala skobalja (*Chondrostoma nasus* L.) u reci Studenici. Archives of Biological Sciences 173–185
Janković D (1983a) Specifičnost ihtiofaune reke Vape i njenih pritoka. II Simpozijum o Fauni u Srbiji. Zbornik naučnih adova, Beograd, pp. 121–123.
Janković D (1983b) Ishrana sarana (*Cyprinus carpio* L.) u Skadarskom jezeru. Cmogorska akademija nauka i umetnosti, Naučni skupovi 9:211–229
Janković D (1992) Natural reproduction of herbivorous fish *Ctenopharyngodon idella* and *Hypophthalmichthys molitrix* in the Djerdap accumulations (Danube river, Yugoslavia). Ichthyology (Yugoslavia) 24(1):57–59
Janković D (2010) Serbia – Part of the South zone of the range of distribution of the European grayling (*Thymallus thymallu*s L.). Arch Biol Sci 62(1):115–121. https://doi.org/10.2298/ABS1001115J
Janković D, Krpo Ćetkovic J (1995) Diverzitet slatkovodnih riba (Osteichthyes) i kolousta (Cephalaspidomorpha) Jugoslavije sa pregledom vrsta od međunarodnog znacaja. In: Stevanovic V, Vasic V (eds) Biodiverzitet Jugoslavije. Bioloski fakultet i Ekolibri, Belgrade, pp 425–445
Janković D, Raspopović M (1960) Ohridska pastrmka (*Salmo letnica typicus* K.) pod promenjenim uslovima u Vlasinskom baražnom jezeru. Arch Biol Sci 12:117–122
Janković D, Krpo J, Lenhardt M, Hegediš A (1994) Ecological characteristics of the Danubian fish species. In: Janković D, Jovičić M (eds) The Danube in Yugoslavia. Institute for biological research „Siniša Stanković", Belgrade; Institute for development of water resources "Jaroslav Černi", Belgrade; Commission of the European communities Brussels, Belgium; Ministry of Science and technology of Republic of Serbia, Belgrade pp. 159–173.
Janković D, Simić V, Anđelković D (2005) Srednjoročni plan za unapređenje ribarstva Nacionalnog parka Kopaonik. JP „Nacionalni park Kopaonik "(*In Serbian*).
Jarić I, Lenhardt M, Cvijanović G, Ebenhard T (2009a) Population viability analysis and potential of its application to Danube sturgeons. Arch Biol Sci 61(1):123–128. https://doi.org/10.2298/ABS0901123J
Jarić I, Lenhardt M, Cvijanović G, Ebenhard T (2009b) *Acipenser sturio* and *Acipenser nudiventris* in the Danube – extant or extinct? J Appl Ichthyol 137–141. https://doi.org/10.1111/j.1439-0426.2009.01227.x
Jarić I, Ebenhard T, Lenhardt M (2010) Population viability analysis of the Danube sturgeon populations in a Vortex simulation model. Rev Fish Biol Fish 20:219–237. https://doi.org/10.1007/s11160-009-9151-0
Jarić I, Višnjić–Jevtić Ž, Cvijanović G, Gačić Z, Jovanović Lj, Skorić S, Lenhardt M (2011) Determination of differential heavy metal and trace element accumulation in liver, gills, intestine and muscle of sterlet *(Acipenser ruthenus*) from the Danube River in Serbia by ICP-OES. Microchem J 98:77–81 https://doi.org/10.1016/j.microc.2010.11.008
Jordan S, Chakona A, Colff D (2020) Protected areas and Endemic Freshwater Fishes of the Cape Fold Ecoregion: Missing the Boat for Fish Conservation. Front Environ Sci. https://doi.org/10.3389/fenvs.2020.502042
Jovanović J, Kolarević S, Milošković A, Radojković N, Simić V, Dojčinović B, Kračun Kolarević M, Paunović M, Kostić J, Sunjog K, Timilijić J, Đorđević J, Gačić Z, Žegura B, Vuković Gačić B (2018) Evaluation of genotoxic potential in the Velika Morava basin in vitro and in situ. Sci Total Environ 15(621):1289–1299. https://doi.org/10.1016/j.scitotenv.2017.10.099
Kahler J, Wuyu Liu R, Newcomb J, Herbst S, Gore L (2020) Public risk perceptions associated with Asian carp introduction and corresponding response action. Manag Biol Invasions 11:80–95. https://doi.org/10.3391/mbi.2020.11.1.06
Kanitz F (1862) Serbien byzantinsche Monumente. Hof—und Staatsdrucherei, Vien
Kanjuh T, Tomić S, Marić A, Škraba Jurlina D, Nikolić V, Simonović P (2021) Trout *Salmo* spp. (Salmoniformes: Salmonidae) Molecular Diversity in Streams on the Southern Slopes of the Stara Planina Mts. in Serbia. Acta Zool Bulg 73(3):425–429

Karaman MS (1974) *Eudontomyzon vladykovi stankokaramani* n.ssp. a new subspecies of lamprey from tributaries of the Ohrid – Drim – Skadar system in West Balkan peninsula. Folia Balcanica 3:1–14

Kojadinović N, Marinović Z, Veličković T, Milošković A, Jakovljević M, Horvath A, Simić V (2020) Cryopreservation of Danube barbel *Barbus balcanicus* sperm and its effects on sperm subpopulation structure. Arch Biol Sci 72(4):525–534. https://doi.org/10.2298/ABS200831046K

Kolarević J (2004) Populaciono-ekološka studija kečige (*Acipenser ruthenus* L.) u vodama Dunava na području Beograda, Magistarski rad, Biološki Fakultet, Univerzitet u Beogradu.

Kolarević S, Sunjog K, Kračun-Kolarević K, Kostić-Vuković J, Jovanović J, Simonović P, Simić V, Piria M, Gacić Z, Lenhardt M, Paunović M, Vuković-Gacić B (2018) The Genetic Variability (RAPD) and Genotoxicity In Vivo (Alkaline and Fpg-Modified Comet Assay) in Chub (*Squalius cephalus*): The Sava River Case Study. Int J Environ Res 12(5):703–712. https://doi.org/10.1007/s41742-018-0127-6

Kolding J, van Zwieten PAM (2014) Sustainable fishing of inland waters. J Limnol 73:132–148. https://doi.org/10.4081/jlimnol.2014.818

Kolding J, Medard M, Mkumbo O, Van Zwieten PAM (2014) Status, trends and management of the Lake Victoria fisheries. Inland fisheries evolution and management—case studies from four continents. FAO Fisheries and Aquaculture Technical Paper 579:675.

Kostić Vuković J, Kolarević S, Kračun Kolarević M, Višnjić Jevtić Ž, Rašković B, Poleksić V, Gačić Z, Lenhardt M, Vuković Gačić B (2021) Temporal variation of biomarkers in common bream *Abramis brama* (L., 1758) exposed to untreated municipal wastewater in the Danube River in Belgrade, Serbia. Environ Monit Assess 193(465). https://doi.org/10.1007/s10661-021-09232-6

Kostić J, Kolarević S, Kračun–Kolarević M, Aborgiba M, Gačić Z, Lenhardt M, Vuković-Gačić B (2016) Genotoxicity assessment of the Danube River using tissues of freshwater bream (*Abramis brama*). Environ Sci Pollut Res 23(20):20783–20795 https://doi.org/10.1007/s11356-016-7213-0

Krišpatić M (1893) Ribe. Prirodoslovne i kulturne crtice. Tisak Karla Albrechta, Zagreb

Krpo Ćetković J, Hegediš A, Lenhardt M (2010) Diet and growth of asp, *Aspius aspius* (Linnaeus, 1758), in the Danube River near the confluence with the Sava River (Serbia). J Appl Ichthyol 513–521. https://doi.org/10.1111/j.1439-0426.2010.01456.x

Lauer T, Pyron M (2016) Freshwater fisheries of the United States. In: Craig FJ (ed) Freshwater Fisheries Ecology. Wiley, pp 166–180

Law R, Plank J, Kolding J (2012) On balanced exploitation of marine ecosystems: result from dynamic size spectra. J Mar Sci 69(4):602–614. https://doi.org/10.1093/icesjms/fss031

Law R, Kolding J, Plank J (2013) Squaring the circle: reconciling fishing and conservation of aquatic ecosystems. Fish Fish:160–174. https://doi.org/10.1111/faf.12056

Layman's Report (2020) Sustainable protection of Lower Danube Sturgeons by preventing and counteracting poaching an illegal wildlife trade, WWF.

Lenhardt M, Kolarević J, Jarić I, Cvijanović G, Poleksić V, Micković B, Gačić Z, Cakić P, Nikčević M (2004) Assessment concepts for river ecosystems characterization based on sterlet (*Acipenser ruthenus* L.) population research, Proceedings, Fifth International Symposium on Ecohydraulics "Aquatic habitats: Analysis & Restoration", 12–17 September 2004, Madrid, Spain, pp. 153 156.

Lenhardt M, Jarić I, Kalauzi A, Cvijanović G (2006) Assessment of extinction risk and reasons for decline in sturgeon. Biodivers Conserv 15:1967–1976. https://doi.org/10.1007/s10531-005-4317-0

Lenhardt M, Jarić I, Cvijanović G, Smederevac Lalić M (2008) The key threats to sturgeons and measures for their protection in the lower Danube Region. In: Lagutov V (ed) Rescue of sturgeon species in the Ural River basin. NATO Science for Peace Security Series C: Environ mental Security, Springer, Dordrecht

Lenhardt M, Marković G, Hegediš A, Maletin S, Ćirković M, Marković Z (2010) Non-native and translocated fish species in Serbia and their impact on the native ichthyofauna. Rev Fish Biol Fish 21:407–421. https://doi.org/10.1007/s11160-010-9180-8

Lenhardt M, Višnjić Jeftić Ž, Navodaru I, Jarić I, Vassilev M, Gačić Z, Nikčević M (2011) Fish stock management cooperation in the lower Danube region: a case study of sturgeons and Pontic shad. In: Lagutov V (ed) Environmental security in watersheds: The Sea of Azov. NATO Science for Peace Security Series C: Environmental Security. Springer, Dordrecht

Lenhardt M, Jarić I, Višnjić Jeftić Ž, Skorić S, Gačić Z, Pucar M, Hegediš A (2012) Concentrations of 17 elements in muscle, gills, liver and gonads of five economically important fish species from the Danube River. Knowl Manag Aquat Ecosyst 407:02. https://doi.org/10.1051/kmae/2012028

Lenhardt M, Navodaru I, Vassilev M, Kalauzi A, Regner S, Višnjić Jeftić Ž, Tošić K, Smederevac Lalić M (2016) Model of the Pontic shad *Alosa immaculata* (Bennett, 1835) and anchovy *Engraulis encrasicolus* (Linnaeus, 1758) catch in the Danube River and Black Sea for the period 1920–2008. Acta Zool Bulg 68(4):557–561

Lenhardt M, Smederevac Lalić M, Spasić S, Hont S, Paraschiv M, Iani M, Nikčević M, Klimley P (2021) Seasonal changes in depth position and temperature of European catfish (*Silurus glanis*) tracked by acoustic telemetry in the Danube River. Int Rev Hydrobiol 106(3–4):191–201. https://doi.org/10.1002/iroh.202002049

Lujić J, Matavulj M, Poleksić V, Rašković B, Marinović Z, Kostić D, Miljanović B (2015) Gill reaction to pollutants from the Tamiš River in three freshwater fish species, *Esox lucius* L. 1758, *Sander lucioperca* (L. 1758) and *Silurus glanis* L. 1758: a comparative study. Anatomia Histologia Embryologia 44:128–137. https://doi.org/10.1111/ahe.12119

Lujić J, Kohlmann K, Kersten P, Marinović Z, Ćirković M, Simić V (2017a) Phylogeographic Identification of Tench *Tinca tinca* (L.,1758) (Actinopterygii: Cyprinidae) from the Northern Balkans and Adjacent Regions and its Implications for Conservation. Zool Stud 56(3). https://doi.org/10.6620/ZS.2017.56-03

Lujić J, Bernath G, Marinović Z, Radojković N, Simić V, Ćirković M, Urbanyi B, Horvath A (2017b) Fertilizing capacity and motility of tench *Tinca tinca* (L., 1758) sperm following cryopreservation. Aquac Res 48(1):102–110. https://doi.org/10.1111/are.12865

Maletin S (1988) The largemouth black bass (*Micropterus salmoides* Lacepede, 1802 Pisces, Centrarchidae) presence in the open waters of Vojvodina. The Proceeding of scientific meeting "Minerali, stijene, izumrli i živi svijet" Sarajevo, pp. 537–539.

Maletin S, Budakov Lj (1983) Growth of some species in Tisa river. Tiscia (Szeged). XVIII:113–117.

Maletin S, Pujin V, Budakov Lj (1980) Variranje morfoloških karaktera *Carassius auratus gibelio* Bloch (1783) (Cyprinidae) u nekim vodama Vojvodine. IV simpozijum biosistematicara Jugoslavie, Đerdap.

Marešova E, Delić A, Kostov V, Marić S, Mendel J, Šanda R (2011) Genetic diversity of *Sabanejewia balcanica* (Actinopterygii: Cobitidae) in the western Balkans and comparison with other regions. Folia Zool 60(4):335–342. https://doi.org/10.25225/fozo.v60.i4.a11.2011

Marić S, Nikolić V, Simonović P (2004a) Pilot study on the morphological identity of wild brown trout (*Salmo trutta*) stocks in the streams of the Danube river basin. Folia Zool 53(4):411–416

Marić S, Hegediš A, Nikolić V, Simonović P (2004b) Conservation status of two Eastern Balkan endemic fish species in Serbia and a proposal for their protection. Acta Zool Bulg 56(2): 213–222

Marić S, Sušnik S, Simonovic P, Snoj A (2006a) Phylogeographic study of brown trout from Serbia, based on mitochondrial DNA control region analysis. Genet Sel Evol 38:411–430. https://doi.org/10.1186/1297-9686-38-4-411

Marić S, Snoj A, Nikolić V, Simonović P (2006b) Genetic differentiation of trout (*Salmo* spp.) populations in Serbia ascertained using RFLP technique on PCR amplified control region of mitochondrial DNA. Acta Vet Beograd 56(5–6):423–430. https://doi.org/10.2298/AVB0606423M

Marić S, Razpet A, Nikolić V, Snoj A, Simonović P (2012a) Genetic diversity of the huchen (*Hucho hucho*) in Serbia. II International Hucho Symposium: Species of the Genus Hucho Gunther, 1866: Population Status, Conservation, Biology, Ecology, Genetics and Culture. 9–12 September 2012, Lopuszna, Poland.

Marić S, Kalamujić B, Snoj A, Razpet A, Lukić Bilela L, Pojskic N, Susnik–Bajec S (2012b) Genetic variation of European grayling (*Thymallus thymallus*) populations in the Western Balkans. Hydrobiologia 691(1):225–237 https://doi.org/10.1007/s10750-012-1076-2

Marić S, Nikolić V, Tošić A, Simonović P (2012c) Record of the brown trout *Salmo trutta* L., 1758 in the main riverbed of the Serbian part of the Danube River. J Appl Ichthyol 28(1):135–137. https://doi.org/10.1111/j.1439-0426.2011.01881.x

Marić S, Razpet A, Nikolić V, Simonović P (2012d) Genetic differentiation of European grayling (*Thymallus thymallus*) populations in Serbia, based on mitochondrial and nuclear DNA analyses. Genet Sel Evol 43(2). https://doi.org/10.1186/1297-9686-43-2

Marić S, Razpet A, Nikolić V, Snoj A, Simonović P (2014) Analysis of Genetic Structure of Huchen (*Hucho hucho*) in Serbia Inferred from Mitochondrial and Nuclear DNA. Acta Vet Belgrade 64(2):236–244. https://doi.org/10.2478/acve-2014-0022

Marić S, Snoj A, Sekulić N, Krpo Ćetković J, Šanda R, Jojić V (2015) Genetic and morphological variability of the European mudminnow *Umbra krameri* (Teleostei, Umbridae) in Serbia and Bosnia and Herzegovina, a basis for future conservation activities. J Fish Biol 86:1534–1548. https://doi.org/10.1111/jfb.12657

Marić S, Stanković D, Wanzenböck J, Šanda R, Erős T, Takács P, Specziár A, Sekulić N, Bănăduc D, Ćaleta M, Trombitsky I, Galambos L, Sipos S, Snoj A (2016) Phylogeography and population genetics of the European mudminnow (*Umbra krameri*) with a time-calibrated phylogeny for the family Umbridae. Hydrobiologia 792:151–168. https://doi.org/10.1007/s10750 016 3051 9

Marić S, Sušnik Bajec S, Schöffmann J, Kostov V, Snoj A (2017) Phylogeography of stream dwelling trout in the Republic of Macedonia and a molecular genetic basis for revision of the taxonomy proposed by S. Karaman. Hydrobiologia 785(1):249–260. https://doi.org/10.1007/s10750-016-2930-4

Marić S, Stanković D, Šanda R, Ćaleta M, Čolić S, Šukalo G, Snoj A (2019) Genetic characterisation of European mudminnow (*Umbra krameri*) populations from the Sava River system. Knowl Manag Aquat Ecosyst 420:46. https://doi.org/10.1051/kmae/2019035

Marić A, Špelić I, Radočaj T, Vidović Z, Kanjuh T, Vilizzi L, Piria M, Nikolić V, Škraba Jurlina D, Mrdak D, Simonović P (2022) Changing climate may mitigate the invasiveness risk of non-native salmonids in the Danube and Adriatic basins of the Balkan Peninsula (south-eastern Europe). In: Giannetto D, Piria M, Tarkan AS, Zięba G (eds) Recent advancements in the risk screening of freshwater and terrestrial non-native species, NeoBiota, vol 76. Pensoft, Sofia, pp 135–161. https://doi.org/10.3897/neobiota.76.82964

Marić S, Stanković D, Susnik Bajec S, Vukić J, Šanda R, Stefanov T, Nikolić D, Snoj A (2022a) Perils of brown trout (*Salmo* spp.) mitigation-driven translocations: a case study from the Vlasina Plateau, Southeast Serbia. Biol Invasions 24:999–1016. https://doi.org/10.1007/s10530-021-02688-0

Marić S, Bănăduc D, Gajić Đ, Šanda R, Veličković T (2022b) *Sabanejewia romanica* (Băcescu, 1943) (Actinopterygii: Cobitidae), a new species for the ichthyofauna of Serbia. Acta Zool Bulg 74(3):369–377

Marković Z (2010) Šaran – gajenje u ribnjacima i kaveznim sistemima. Izdanje prof. dr Zoran Marković, Beograd

Marković G, Simonović P (2010) Endangerment and conservation status of zingel (*Zingel zingel* L. 1766, Percidae, PISCES). Acta Agric Serb XV(29):103–105

Marković G, Simović S (1997) Pumpkinseed (*Lepomis gibbosus* L. 1758) – unwelcome member to the ichthyofauna of the Yugoslav reservoirs. The proceeding of symposium "Ribarstvo Jugoslavije" 123–128 pp. (*In Serbian with English summary*).

Marković G, Simonović P, Simović S (1996a) Linear growth and some ecological characteristics of roach (*Rutilus rutilus* L. 1758) in Zapadna Morava River (Serbia, Yugoslavia). Hydrobiology-Sofia, p 40.

Marković G, Simović S, Vasiljević V (1996b) Possible outcome of largemouth bass (*Micropterus salmoides* Lacepede 1802) propagation in water courses and ponds of Yugoslavia. Poljoprivreda i šumarstvo, Podgorica 42:101–107. (*In Serbian with English summary*)

Marković Z, Poleksić V, Živić I, Stanković M (2009) State of the art of fishery in Serbia. Conference proceedings. IV international Conference "Fishery" Faculty of Agriculture. Belgrade, Serbia, pp. 30–38.

Marković G, Ćirković M, Maletin S, Milošević N (2010) A contribution to the data on tench (*Tinca tinca* L., Cyprinidae, Pisces) distribution in Serbia. Zbornik Matice Srpske za prirodne nauke, Novi Sad 118:127–142

Marković G, Ćirković M, Maletin S (2012) The role of allochthonous (non-native) fish species in Serbian aquaculture. Cent Eur Agric 13(3)

Marković Z, Stanković M, Rašković B, Dulić Z, Živić I, Poleksić V (2016) Comparative analysis of using cereal grains and compound feed in semi-intensive common carp pond production. Aquac Int 24:1699–1723. https://doi.org/10.1007/s10499-016-0076-z

Mérő TO (2014) Diet in pike (*Esox lucius*) in northwestern Vojvodina (Serbia). Nat Croat 23(1): 27–34

Miljanović B, Jurca T, Vukov D, Simeunović J, Mijić I, Pogrmić S, Živković M, Novković M, Bajić A, Šipoš Š, Pankov N (2016) Program upravljanja ribarskim područjem "Karaš-Nera" za period 2016–2025. PMF Novi Sad (*In Serbian*).

Miljanović B, Simeunović J, Jurca T, Pogrmić S, Mijić Oljačić I, Bajić A, Živković M, Šipoš Š, Pankov N (2018a) Program upravljanja ribarskim područjem PP „Palić "za period 2018–2027. PMF Novi Sad (*In Serbian*).

Miljanović B, Simeunović J, Jurca T, Pogrmić S, Mijić Oljačić I, Bajić A, Živković M, Šipoš Š, Pankov N (2018b) Program upravljanja ribarskim područjem u okviru SRP "Ludaško jezero" za period 2018–2027. PMF Novi Sad (*In Serbian*).

Miljanović B, Simeunović J, Jurca T, Pogrmić S, Mijić Oljačić I, Bajić A, Živković M, Šipoš Š, Pankov N (2018c) Program upravljanja ribarskim područjem u zaštićenom području park prirode "Ponjavica" za period 2018–2027.PMF Novi Sad (*In Serbian*).

Miljanović B, Simeunović J, Jurca T, Pogrmić S, Mijić Oljačić I, Bajić A, Živković M, Šipoš Š, Pankov N (2019a) Program upravljanja ribolovnom vodom "Bačkotoplosko doline" za period 2018–2027. PMF Novi Sad (*In Serbian*).

Miljanović B, Simeunović J, Jurca T, Pogrmić S, Mijić Oljačić I, Bajić A, Živković M, Šipoš Š, Pankov N (2019b) Program upravljanja ribarskim područjem "Stara Tisa" – kod Bisernog ostrva za period 2020–2029. PMF Novi Sad (*In Serbian*).

Miljanović B, Simeunović J, Jurca T, Pogrmić S, Mijić Oljačić I, Bajić A, Živković M, Šipoš Š, Pankov N (2019c) Program upravljanja ribarskim područjem u okviru SRP "Kraljevac" za period 2020–2029. PMF Novi Sad (*In Serbian*).

Miljanović B, Simeunović J, Jurca T, Pogrmić S, Mijić Oljačić I, Bajić A, Živković M, Šipoš Š, Pankov N (2019d) Program upravljanja ribarskim područjem parka prirode „Jegrička "za period 2020–2029. PMF Novi Sad (*In Serbian*).

Miljanović B, Simeunović J, Jurca T, Pogrmić S, Mijić Oljačić I, Bajić A, Živković M, Pankov N (2021a) Program upravljanja ribarskim područjem u okviru granica SRP "Gornje Podunavlje" za period 2022–2031. PMF Novi Sad (*In Serbian*).

Miljanović B, Jurca T, Pogrmić S, Mijić Oljačić I, Bajić A, Pankov N (2021b) Program upravljanja ribarskim područjem "Mali Begej" za period 2021–2030. PMF Novi Sad (*In Serbian*).

Miljanović B, Simeunović J, Jurca T, Pogrmić S, Mijić Oljačić I, Bajić A, Živković M, Pankov N (2022) Program upravljanja ribarskim područjem u SRP "Karađorđevo" za period 2023–2032. PMF Novi Sad (*In Serbian*).

Milošković A, Kojadinović N (2023) Pollution of fishing waters of the Western Balkan – potentially toxic elements and their impact on the ecological sustainability of fish resources. In:

Simić V, Simic S, Pešić V (eds) Ecological sustainability of fish resources of inland waters of the Western Balkans in the Anthropocene. Springer, Cham

Milošković A, Simić V (2015) Arsenic and Other Trace Elements in Five Edible Fish Species in Relation to Fish Size and Weight and Potential Health Risks for Human Consumption. Pol J Environ Stud 24(1):199–206. https://doi.org/10.15244/pjoes/24929

Milošković A, Dojčinović B, Kovačević S, Radojković N, Radenković M, Milošević Đ, Simić V (2016) Spatial monitoring of heavy metals in the inland waters of Serbia: a multispecies approach based on commercial fish. Environ Sci Pollut Res 23(10):9918–9933. https://doi.org/10.1007/s11356-016-6207-2

Milošković A, Stojković Piperac M, Kojadinović N, Radenković M, Đuretanović S, Čerba D, Milošević Đ, Simić V (2022) Potentially toxic elements in invasive fish species Prussian carp (*Carassius gibelio*) from different freshwater ecosystems and human exposure assessment. Environ Sci Pollut Res 29(19):29152–29164. https://doi.org/10.1007/s11356-021-17865-w

Mišić S (1995a) Lov u srednjovekovnoj Srbiji, Istorijski glasnik, br. 1–2/1995, 51–56 (*In Serbian*).

Mišić S (1995b) Lov u srednjovekovnoj Srbiji, Istorijski glasnik, br. 1–2/1995, 65 (*In Serbian*).

Mladenović S (1999) Kladovski kavijar. Ribarsko gazdinstvo "Đerdap" Kladovo, Srbija.

Momirović P. (1972). Ribolov na Velikoj Moravi. Novi Sad.

Mrakovčić M, Brigić A, Buj I, Ćaleta M, Mustafić P, Zanella D (2006) Red Book of Freshwater Fish of Croatia. Ministry of Culture, State of Institute for Nature Protection, Zagreb, Croatia.

Mueller M, Pander J, Geist J (2018) Comprehensive analysis of <30 years of data on streams fish population trend and conservation status in Bavaria, Germany. Biol Conserv 226:311–320. https://doi.org/10.1016/j.biocon.2018.08.006

Nikčević M, Skorić S, Cvijanović G, Mićković B, Hegediš A (2016) First record of smoltified rainbow trout *Oncorhynchus mykiss* (Walbaum, 1792) in the main riverbed of the Serbian part of the Danube River. J Appl Ichthyol 32(6). https://doi.org/10.1111/jai.13230

Nikolić V, Simonović P (2003) Program ribarstva na ribarskom području "Nacionalnog Parka Đerdap" od 1040 km do 942 km za period 2006–2010. Biološki Fakultet Beograd (In Serbian).

Nikolić V, Skora K (1996) Vertebral number in Ponto-Caspian gobies: phylogenetic relevance. J Fish Biol 49:1027–1029

Nikolić V, Marić S, Hegediš A, Simonović P (2003a) Program unapređenja ribarstva na delu ribarskog područja „Sava 1 "od 95 km do 121 km za period 2004–2007. Biološki Fakultet Beograd *(In Serbian)*.

Nikolić V, Marić S, Hegediš A, Simonović P (2003b) Program unapređenja ribarstva na delu ribarskog područja „Sava 2 "od 62 km do 0 km za period 2004–2007. Biološki Fakultet Beograd (In Serbian).

Nikolić V, Marić S, Simonović P (2003c) Program unapređenja ribarstva na delu ribarskog područja "Dunav 2" od 1297 km do 1233 km za period 2004–2007. Biološki Fakultet Beograd (In Serbian).

Nikolić V, Marić S, Hegediš A, Simonović P (2003d) Program unapređenja ribarstva na delu ribarskog područja "Dunav 3" od 1187 km do 1112 km za period 2004–2007. Biološki Fakultet Beograd (In Serbian).

Nikolić V, Marić S, Hegediš A, Simonović P (2003e) Program unapređenja ribarstva na delu ribarskog područja "Dunav 4" od 942 km do 845,5 km za period 2004–2007. Biološki Fakultet Beograd (In Serbian).

Nikolić V, Simonović P, Marić S (2006) Occurrence of *Chilodonella hexasticha* (Ciliophora, Protista) on farmed rainbow trout (*Oncorhynchus mykiss*) throughoutteh season. Acta Vet Beograd 56(1):55–61

Nikolić V, Đikanović V, Paunović M, Marić S, Simonović P (2010) Parasitic infections in pond-reared rainbow trout Oncorhynchus mykiss in eastern Serbia. In: The 1st International Symposium of Fisheries and Fishing Tourism, Center for Fisheries "Neretva". Boracko Lake, Bosnia and Herzegovina, Konjic

Nikolić V, Marić S, Škraba D, Tošić A, Mrdak D, Simonović P (2016) First Record of Ectobiont Community on Wild Salmonids in Serbia. Int J Innov Stud Aquat Biol Fish 2(1):25–28. https://doi.org/10.20431/2455-7670.0201002

Nikolić D, Cvijanović G, Smederevac Lalić M, Skorić S, Hegediš A, Jovičić K, Krpo Ćetković J (2019) Length–weight relationship and condition factor of the Cactus roach (*Rutilus virgo*) in the Perućac reservoir (Serbia). International Conference Adriatic Biodiversity Protection – AdriBioPro2019, 7–10 April 2019, Kotor, Montenegro.

Nikolić D, Skorić S, Poleksić V, Rašković B (2021) Sex-specific elemental accumulation and histopathology of pikeperch (*Sander lucioperca*) from Garaši reservoir (Serbia) with human health risk assessment. Environ Sci Pollut Res 28(53700–53):711. https://doi.org/10.1007/s11356-021-14526-w

Nikolić M, Milošković A, Jakovljević M, Radenković M, Veličković T, Đuretanović S, Kojadinović N, Nikolić M, Simić V (2022) The first observation of the presence of microplastics in wild common bleak (*Alburnus alburnus* L.) and standardization of extraction protocols. Kragujevac J Sci 44:267–282. https://doi.org/10.5937/KgJSci2244267N

Novakov N, Ćirković M, Kartalović B, Pelić M, Matić D, Ljubojević D, Božić B (2015) Infestation of *Larnaea cyprinacea* (Crustacea: Cepopoda) in different categories of common carp (*Cyprinus carpio*) reared in Serbia. Arc Vet Med 8(2):8–11. https://doi.org/10.46784/e-avm.v8i2.110

Nuñez M, Kuebbing S, Dimarco R, Simberloff D (2012) Invasive Species: to eat or not to eat, that is the question. Conserv Lett:334–341. https://doi.org/10.1111/j.1755-263X.2012.00250.x

Pančić J (1860) Ribe u Srbiji. Glasnik društva srbske slovesnosti XII, Beograd.

Pavlović M, Paunović M, Simić V (2013) Feeding of Eurasian perch (*Perca fluviatilis* L.) in three reservoirs in Serbia. Water Res Manag 3(4):41–46

Pavlovic M, Simonovic P, Stojkovic M, Simic V (2015) Analysis of diet of piscivorous fishes in Bovan, Gruza and Sumarice Reservoir, Serbia. Iran J Fish Sci 14(4):908–923

Petrović M (1998a) [ориг. 1900]. "О нашим риболовима на Доњем Дунаву", у Д. Трифуновић (ур.) Рибарство. Сабрана дела Михаила Петровића. Београд: Завод за уџбенике и наставна средства: 273–305 (In Serbian).

Petrović M (1998b) [ориг. 1933]. "Риболови у Тимочкој Крајини", у Д. Трифуновић (ур.) Рибарство. Сабрана дела Михаила Петровића. Београд. Завод за уџбенике и наставна средства: 312–324 (*In Serbian*).

Petrović M (1998c) [ориг. 1940]. "Београд негдашњи центар великог рибарства", у Д. Трифуновић (ур.) Рибарство. Сабрана дела Михаила Петровића. Београд: Завод за уџбенике и наставна средства: 10–174 *(In Serbian)*.

Petrović M (1998d) [ориг. 1941]. "Ђердапски риболови у прошлости и садашњости", у Д. Трифуновић (ур.) Рибарство. Сабрана дела Михаила Петровића. Београд: Завод за уџбенике и наставна средства: 175–270. 345 p (In Serbian).

Piria M, Suić J, Špelić I (2023) Inland fisheries in Croatia: historical aspects, fish resources, management and conservation. In: Simić V, Simic S, Pešić V (eds) Ecological sustainability of fish resources of inland waters of the Western Balkans in the Anthropocene. Springer, Cham

Pliberšek K, Tavčar T (2023) Fish resources of inland waters and fisheries in Slovenia, management, sustainability and conservation. In: Simić V, Simic S, Pešić V (eds) Ecological sustainability of fish resources of inland waters of the Western Balkans in the Anthropocene. Springer, Cham

Poleksić V, Mitrović-Tutundžić V (1994) Fish gills as a monitor of sublethal and chronic effects of pollution. Water Res 23(3):341–344

Poleksić V, Lenhardt M, Jarić I, Djordjević D, Gačić Z, Cvijanović G, Rašković B (2010) Liver, gills and skin histopathology and heavy metal content of the Danube sterlet (*Acipenser ruthenus* Linnaeus, 1758). Environ Toxicol Chem 29(3):515–521. https://doi.org/10.1002/etc.82

Radenković M, Stojković Piperac M, Milošković A, Kojadinović N, Đuretanović S, Veličković T, Jakovljević M, Nikolić M, Simić V (2022) Diet seasonality and food overlap of *Perca fluviatilis* (Actinopterygii: Perciformes: Percidae) and *Rutilus rutilus* (Actinopterygii: Cypriniformes:

Cyprinidae) juveniles: A case study on Bovan Reservoir, Serbia. Acta Ichthyol Piscat 52(1): 77–90. https://doi.org/10.3897/aiep.52.78215
Radenković M, Zorić K, Petrović A (2023) Invasive macroinvertebrate and fish species and their impact on fish resources: a case study on fishing waters of Serbia. In: Simić V, Simic S, Pešić V (eds) Ecological sustainability of fish resources of inland waters of the Western Balkans in the Anthropocene. Springer, Cham
Radke R, Kahl U (2002) Effects of a filter–feeding fish [silver carp, Hypophthalmichthys molitrix (Val.)] on phyto- and zooplankton in a mesotrophic reservoir: results from an enclosure experiment. Freshw Biol:2337–2344. https://doi.org/10.1046/j.1365-2427.2002.00993.x
Radojević I, Stefanović D, Čomić L, Ostojić A (2008) Information system of lakes and reservoirs of the Serbia – SeLaR. Kragujevac J Sci 30:81–98
Radojković N, Marinović Z, Milošković A, Radenković M, Đuretanović S, Lujić J, Simić V (2019) Effects of Stream Damming on Morphological Variability of Fish: Case Study on Large Spot Barbell *Barbus balcanicus*. Turk J Fish Aquat Sci 19(3):231–239. https://doi.org/10.4194/1303-2712-v19_3_06
Rajić Z, Vignjević Đorđević N, Čanak S (2016) Production and economic results of intensive carp (*Cyprinus carpio*) farming in Serbia. Econ Agric 4:1445. https://doi.org/10.5937/ekoPolj1604445R
Rašković B, Poleksić V, Višnjić–Jeftić Ž, Skorić S, Gačić Z, Đikanović V, Jarić I, Lenhardt M (2015) Use of histopathology and elemental accumulation in different organs of two benthophagous fish species as indicators of river pollution. Environ Toxicol 30(10): 1153–1161 https://doi.org/10.1002/tox.21988
Rhymer J, Simberloff D (1996) Extinction by hybridization and introgression. Annu Rev Ecol Evol Syst 27:83–109
Rios L, Jones P, Moore C, Narayan U (2010) Quantitation of persistent organic pollutants adsorbed on plastic debris from the Northern Pacific Gyre's "eastern garbage patch". J Environ Manage 12:2226–2236. https://doi.org/10.1039/c0em00239a
Ristić M (1963) Ribarsko biološka studija Acipenseridae-a jugoslovenskog dela donjeg Dunava. Ribarstvo Jugoslavije 2:48–58
Ristić M (1969) The specifity of sterlet (*Acipenser ruthenus* L.) growth in the Danube, Sava and Tisa river. Ribarstvo Jugoslavije 2:21–31
Ristić M (1971a) Prilog objašnjenju postojanja kratkorilih i dugorilih kečiga (*Acipenser ruthenus* L. morpha kamensis et typica Lovetzky 1834, Berg L.S.1932–1934) kao posebnih formi u Dunavu. Savi i Tisi. Ribarstvo Jugoslavije 26:25–28
Ristić M (1971b) Biologija, morfoloske karakteristike, rastenje, plodnost i migracija dunavskog šarana. Ribarstvo Jugoslavije 5:93–107
Ristić M (1977) Ribe i ribolov u slatkim vodama. Nolit, Beograd
Ristić R, Malušević I, Polovina S, Milčanović V, Radić B (2018) Male hidroelektrane derivacionog tipa: beznačajna energetska korist i nemerljiva ekološka šteta. Vodoprivreda 50(294–6):311–317. (*In Serbian with English summary*)
Rokvić N, Aksentijević K, Kureljušić J, Vasiljević M, Todorović N, Zdravković N, Stojanac N (2020) Occurrence and transfer of mycotoxins from ingredients to fish feed and fish meat of common carp (*Cyprinus carpio*) in Serbia. World Mycotoxin J 13(4):545–552. https://doi.org/10.3920/WMJ2020.2580
Schöffmann J, Marić M (2023) Salmonid fish species – opportunities for sustainable use under multiple pressures and current climatic change. In: Simić V, Simic S, Pešić V (eds) Ecological sustainability of fish resources of inland waters of the Western Balkans in the Anthropocene. Springer, Cham
Sekulić N (2013) Ekološke karakteristike i morfološko-genetička diferencijacija populacija crnke (*Umbra krameri* Walbaum, 1792) sa područja Bačke. Doktorska disertacija, Univerzitet u Beogradu, Biološki fakultet, Mačve i Semberije

Sekulić N, Budakov L, Branković D (1998a) Distribution of the European mudminnow *Umbra krameri* (Umbridae) in Serbia. Ital J Zool 65:381–382. https://doi.org/10.1080/11250009809386851

Sekulić N, Budakov L, Branković D (1998b) Comparison of feeding spectra of some exotic and native fishes in Vojvodina (Serbia). Ital J Zool 65:377–380. https://doi.org/10.1080/11250009809386850

Sekulić N, Cakić P, Lenhardt M, Vučić D, Budakov L (1999) Short-snouted pipefish *Syngnathus abaster* (Acanthopterygii: Syngnathidae) in Yugoslav section of the Danube. Ichthyologia 3: 79–82

Sekulić N, Marić S, Galambos L, Radošević D, Krpo Ćetković J (2013) New distribution data and population structure of the European mudminnow *Umbra krameri* in Serbia and Bosnia and Herzegovina. J Fish Biol 3(3):659–666. https://doi.org/10.1111/jfb.12183

Sekulić N, Krpo Ćetković J, Jojić V, Galambos L, Radošević D, Snoj A, Šanda R, Marić S (2015) Biological and ecological characteristics of the European mudminnow *Umbra krameri* as a basis for in-situ and ex-situ conservation. VII International conference" Water & Fish", Faculty of Agriculture, 10–12 June 2015, Belgrade Zemun, Serbia.

Simić V, Šorić V (2006) New data of ichthyofauna of Serbia. Arch Biol Sci 58(3):29P–30P

Simić V, Simić S, Paunović M, Petrović A, Stanković M (2007) Neke ugrožene vrste u specijalnom rezervatu prirode „Zasavica" (*Umbra krameri*, Pisces i *Batrachospermum gelatinosum*, Rhodophyta). Zbornik. Naučno-stručni skup Zasavica 2007 sa međunarodnim učešćem, Pokret gorana Sremska Mitrovica pp. 99–107.

Simić V, Simić S, Ćirković M, Pantović N (2009) Preliminary results of the research of the population of tench (*Tinca tinca*) in the water ecosystems of Serbia. IV International conference" Fishery", Faculty of Agriculture, 27–29 May 2009, Belgrade, Zemun, Serbia.

Simić V, Simić S, Paunović M, Simonović P, Radojković N, Petrović A (2012) *Scardinius knezevici* Bianco & Kottelat, 2005 and *Alburnus scoranza* Bonaparte, 1845: New species of ichthyofauna of Serbia and the Danube basin. Arch Biol Sci. https://doi.org/10.2298/ABS1203981S

Simić V, Lujić J, Kostić D, Ćirković M, Bjelić-Čabrilo O, Simić S, Marković G (2013) Diversity characteristics of the fish species important for fishery in the waters of Serbia. Bulg J Agric Sci 19(S1):77–87

Simić V, Simić S, Stojković-Piperac M, Petrović A, Milošević Đ (2014a) Commercial fish species of inland waters: a model for sustainability assessment and management. Sci Total Environ 497–498C:642–650. https://doi.org/10.1016/j.scitotenv.2014.07.092

Simić V, Petrović A, Boris E, Dimović D, Makovinska J, Karadžić B, Paunović M (2014b) Indicative Status Assessment, Biodiversity Conservation, and Protected Areas within the Sava River Basin. In: Milačić R, Ščančar J, Paunović M (eds) The Sava River. Springer, pp 453–500

Simić V, Simić S, Paunović M, Radojković N, Petrović A, Talevski T, Milošević Đ (2016) The Alburnus benthopelagic fish species of the Western Balkan Peninsula: An assessment of their sustainable use. Sci Total Environ 540:410–417. https://doi.org/10.1016/j.scitotenv.2015.07.030

Simić S, Đorđević N, Mitrović A (2018a) Preliminary assessment of the degree of vulnerability and health risk in some fishing waters based on cyanobacteria in 2017. VIII International conference "Water & Fish" Faculty of Agriculture, Belgrade-Zemun, Serbia p 394.

Simić V, Petrović A, Radenković M (2018b) Izveštaj o proceni uticaja MHE na biodiverzitet makrobeskičmenjaka i riba reka Vlasine, Jošanice, Prištavice i Gračanice. Univerzitet u Kragujevcu, Prirodno–matematički fakultet (In Serbian).

Simić V, Simić S, Petrović A (2020a) Izmene i dopune programa upravljanja ribarskim područjem "Kolubara" za period 2017–2026. "Cassini Fiching" d.o.o. Arilje (In Serbian).

Simić V, Simić S, Petrović A, Đoković M (2020b) Izmene i dopune Programa upravljanja ribarskim područjem „Dunav "(2017–2026). JP "Srbijašume" Beograd (In Serbian).

Simić V, Simić S, Petrović A, Đoković M (2020c) Izmene i dopune Programa upravljanja ribarskim područjem "Beograd" (2017–2026). JP "Srbijašume" Beograd (In Serbian).

Simić V, Simić S, Petrović A (2020d) Izmene i dopune programa upravljanja ribarskim područjem "Velika Morava" za period 2017–2026. Balkan eko tim d.o.o. Prijepolje (In Serbian).
Simić V, Simić S, Petrović A, Đoković M (2020e) Izmene i dopune Programa upravljanja ribarskim područjem "Mlava" (2017–2026). JP "Srbijašume" Beograd (In Serbian).
Simić V, Simić S, Petrović A (2020f) Izmene i dopune programa upravljanja ribarskim područjem "Južna Morava 2" za period 2017–2026. Asocijacija "Veternica-Vlasina" d.o.o. Leskovac (In Serbian).
Simić V, Simić S, Petrović A (2020g) Izmene i dopune programa upravljanja ribarskim područjem "Timok" za period 2017–2026. Zajednica ribolovaca "Timočka krajina" d.o.o. Zaječar (In Serbian).
Simić V, Simić S, Petrović A (2020h) Izmene i dopune programa upravljanja ribarskim područjem „Ibar "za period 2017–2026. „Plus sport "d.o.o. Adrani, Kraljevo (In Serbian).
Simić V, Simić S, Petrović A (2020i) Izmene i dopune programa upravljanja ribarskim područjem „Radan "za period 2017–2026. JP „Stbijašume "Beograd (In Serbian).
Simić V, Simić S, Petrović A (2020j) Izmene i dopune programa upravljanja ribarskim područjem „Zlatibor "za period 2017–2026. JP „Srbijašume "Beograd (In Serbian).
Simić V, Simić S, Petrović A (2020k) Izmene i dopune programa upravljanja ribarskim područjem „Kopaonik "za period 2017–2026. JP „Nacionalni park Kopaonik "Kopaonik (In Serbian).
Simić V, Simić S, Petrović A (2020l) Izmene i dopune programa upravljanja ribarskim područjem „Južna Morava 1 "za period 2017–2026. Južna Morava 2. d.o.o. Niš (In Serbian).
Simić V, Simić S, Petrović A, Mitrović A, Veličković T, Simović P (2021) Ekološke vrednosti reke Veliki Rzav značajne za forniranje rečnog ekološkog rezervata (RER-VR). Water Logist. d.o.o. and PMF Kragujevac (In Serbian).
Simić V, Bănăduc D, Curtean Bănăduc A, Petrović A, Veličković T, Stojković Piperac M, Simić S (2022a) Assessment of the ecological sustainability of river basins based on the modified the ESHIPPOfish model on the example of the Velika Morava basin (Serbia, Central Balkans). Front Environ Sci 10:952692. https://doi.org/10.3389/fenvs.2022.952692
Simić V, Simić S, Petrović A (2022b) Izmene i dopune programa upravljanja ribarskim područjem „Zapadna Morava "za period 2017–2026. Balkan eko tim d.o.o. Prijepolje (In Serbian).
Simić V, Simić S, Petrović A (2022c) Izmene i dopune programa upravljanja ribarskim područjem „Stara planina "za period 2017–2026. JP „Srbijašume "Beograd (In Serbian).
Simić V, Simić S, Petrović A (2022d) Izmene i dopune programa upravljanja ribarskim područjem „Golija "za period 2017–2026. JP „Srbijašume "Beograd (In Serbian).
Simić V, Simić S, Petrović A (2022e) Izmene i dopune programa upravljanja ribarskim područjem „PIO Vlasina "za period 2017–2026. TOOS Surdulica (In Serbian).
Simić V, Simić S, Petrović A (2022f) Izmene i dopune programa upravljanja ribarskim područjem „Dolina Pčinje "za period 2017–2026. Pravoslavna Eparhija Vranjska, Vranje (In Serbian).
Simić V, Simić S, Petrović A (2022g) Izmene i dopune programa upravljanja ribarskim područjem „Kopaonik "za period 2017–2026. JP „Nacionalni Park Kopaonik "(*In Serbian*).
Simić S, Đorđević N, Tokodi N, Drobac Backović D, Marinović Z (2023) Eutrophication of fishing waters and the influence of cyanobacterial occurrence and blooming on fish resources – case studies in Serbia. In: Simić V, Simic S, Pešić V (eds) Ecological sustainability of fish resources of inland waters of the Western Balkans in the Anthropocene. Springer, Cham
Simonović P (1996a) Karyological relationships between gobies (Gobiidae: Perciformes) of Euro-Mediterranean and Ponto-Caspian districts. Ichthyologia 28:25–31
Simonović P (1996b) Cranial osteology of the bighead goby *Neogobius kessleri* from the rivers Danube and Sava (Serbia, Yugoslavia). Ital J Zool 63:65–72
Simonović P (1999) Phylogenetic relationships of Ponto Caspian gobies and their relationship to the Atlantic Mediterranean Gobiinae (Gobiidae, Perciformes). J Fish Biol 54(3):533–555. https://doi.org/10.1111/j.1095-8649.1999.tb00634.x
Simonović P (2001) Ribe Srbije [Fish of Serbia]. NNK International, Belgrade. 247 pp
Simonović P (2020) The impact of small derivate hydropower plants on fish species and other aquatic organisms in the mountain river ecosystems of the Republic of Serbia. In: Anđelković M

(ed) Environmental impact of small hydropower plants. Serbian Academy of Sciences and Arts, p 37

Simonović P, Jovanović V (1993) Morphology of the Eurasian perch (*Perca fluviatilis*) from two different localities of the Danube basin in Serbia. 1: Similarities and differences of juveniles. Folia Zool 423:257–268

Simonović P, Mesaroš G (1998) Phenetic relationship between gobies (Gobiidae) based on traits of external morphology. Acta Biologica Yugoslavica–Ichthyologia Belgrade 30:43–50

Simonović P, Nikolić V (1996) Cranial osteology of the sand goby *Neogobius fluviatilis* (Pallas, 1881) from the River Sava (Serbia, Yugoslavia). Acta Biol Szeged 41:45–55.

Simonović P, Nikolić V (2007) Density-dependence of growth characteristics and maturation in stream-dwelling resident brown trout, *Salmo trutta*, in Serbia. Fish Manag Ecol 14:1–6. https://doi.org/10.1111/j.1365-2400.2006.00517.x

Simonović P, Simović S (1997) Growth-in-length of Eurasian perch (*Perca fluviatilis* L.) in Gruzansko [water] reservoir [Serbia, Yugoslavia] before and after its formation. Ichthyologia (Yugoslavia).

Simonović P, Marić S, Nikolić V (2000) Growth characteristics of huchen *Hucho hucho* (L.) from Rivers Drina, Una, and Sana. Acta Biologica Yugoslavica – Ekologija Belgrade 35:123–126

Simonović P, Paunović M, Popović S (2001) Morphology, Feeding, and Reproduction of the Round Goby, *Neogobius melanostomus* (Pallas), in the Danube River Basin, Yugoslavia. J Gt Lakes Res 27(3):281–289. https://doi.org/10.1016/S0380-1330(01)70643-0

Simonović P, Budakov L, Nikolić V, Marić S (2005) Recent record of the ship sturgeon *Acipenser nudiventris* in the middle Danube (Serbia). Biologia 60(2):1–2

Simonović P, Marić S, Nikolić V (2006) Records of Amur sleeper *Perccottus glenii* (Odontobutidae) in Serbia and its recent status. Arch Biol Sci 58(1):7P–8P. https://doi.org/10.2298/ABS0601001S

Simonović P, Marić S, Nikolić V (2007) Trout *Salmo* spp. complex in Serbia and adjacent regions of the western Balkans: reconstruction of evolutionary history from external morphology. J Fish Biol 70:359–380. https://doi.org/10.1111/j.1095-8649.2007.01516.x

Simonović P, Nikolić V, Tošić A, Marić S (2011) Length-weight relationship in adult huchen *Hucho hucho* (L., 1758) from Drina River, Serbia. Biologia 66(1):156–159. https://doi.org/10.2478/s11756-010-0135-2

Simonović P, Tošić A, Škraba-Jurlina D, Nikolić V, Piria M, Tomljanović T, Šprem N, Mrdak D, Milošević D, Bećiraj A, Dekić R, Povž M (2017) Diversity of brown trout *Salmo* cf. *trutta* in the River Danube basin of western Balkans as assessed from the structure of their mitochondrial control region haplotypes. J Appl Ichthyol 57:603–616. https://doi.org/10.1134/S0032945217040154

Simonović P, Marić S, Tošić A, Škraba Jurlina D, Nikolić V (2018) Morphological and Molecular Differentiation between Rheophilic Barbels *Barbus* spp. (Cyprinidae) from Headwaters at the Divide between Drainage Areas of the Danube, Vardar and Struma Rivers, Central Balkans. Acta Zool Bulg 70(2):153–164

Sipos S, Kostov V, Miljanović B (2007) The first record of *Barbatula bureschi* Drensky, 1928, (fam. Balitoridae) in Serbia. Pisces Hungarici 2.

Škraba Jurlina D, Marić A, Mrdak D, Kanjuh T, Špelić I, Nikolić V, Piria M, Simonović P (2020) Alternative life-history in native trout (*Salmo* spp.) suppresses the invasive effect of alien trout strains introduced into streams in the western part of the Balkans. Front Ecol Evol 8:188. https://doi.org/10.3389/fevo.2020.00188

Smederevac Lalić M, Višnjić Jeftić Ž, Pucar M, Mićković B, Skorić S, Nikčević M, Hegediš A (2011) Fishing circumstances on the Danube in Serbia. Water Res Manag 1(4):45–49

Smederevac Lalić M, Skorić S, Višnjić Jeftić Ž, Đikanović V, Micković B (2015) Growth and Weight-Length Relationship of Burbot *Lota lota* (L.) (Lotidae) in the Danube River at Backa Palanka (Serbia). Acta Zool Bulg 67(1):97–103

Smederevac Lalić M, Kalauzi A, Regner S, Navodaru I, Višnjić Jeftić Ž, Gačić Z, Lenhardt M (2018) Analysis and forecast of Pontic shad (*Alosa immaculata*) catch in the Danube River. Iran J Fish Sci 17(3):443–457. https://doi.org/10.22092/IJFS.2018.116611

Šorić V (1979) Morfološko-ekološka studija populacija balkanskog endemita *Pachychilon pictum* (Heckel i Kner, 1858). Doktorska disertacija, Univerzitet u Kragujevcu.

Šorić V (1985) *Cobitis elongata*, Heckel & Kner 1858 u ihtiofauni SR Srbije. Ichthyologia 17(1): 29–39

Šorić V (1992) The second contribution to the knowledge of the Metohija potamologic system. Ichthyologia 24:33–42

Šorić V (1993) Osteology and taxonomic status of the genus *Pachychilon* Steind, 1882 (Pisces, Cyprinidae). Bios (Macedonia, Greece) 1(2):49–67

Šorić V (1998) New data on the taxonomy and distribution of lampreys (Petromyzontidae) in southern Europe. Ichthyologia 30:71–82

Šorić V (2006) Redescription and some ecological characteristics of *Alburnus arborella* (Bonapartae, 1841). Kragujevac J Sci 28:83–92

Šorić V, Ilić K (1985) Systematical and ecological characteristics of *Alburnoides bipunctatus* (Bloch) in some waters of Yugoslavia. Ichthyologia 17:47–58

Šorić V, Ilić K (1987) The species of the genus Gobio I. *Gobio gobio* in the Velika Morava system. Ichthyologia 19:53–68

Šorić V, Ilić K (1991) The species of the genus Gobio III. *Gobio albipinnatus vladykovi* and *Gobio kessleri banaticus*. Ichthyologia 23:79–90

Šorić V, Naumovski M (1991) The species of the genus *Zingel* in the waters of Yugoslavia. Ichthyologia 23:73–78

Spasić M, Poleksić V, Stanković M, Dulić Z, Rašković B, Živić I, Ćirić M, Relić R, Vukojević D, Bošković D, Marković Z (2010) Selective breeding programme of common carp (*Cyprinus carpio* L.) in Serbia: Preliminary results. Journal of. Agric Sci 55(3):243–251

Spasić M, Poleksić V, Stanković M, Rašković B et al. (2011) Establishing selective breeding program for rainbow trout (*Oncorhynchus mykiss* Walbaum) in Serbia. Conference Proceedings. 5th International Conference Aquaculture and Fishery, Belgrade (Serbia).

Stamenković S (1991) Evolution des captures de trois especes d'esturgeons *Huso huso*, *Acipenser gueldenstaedtii*, et *Acipenser stellatus* dans la partie Yougoslave du Danube. In: Williot, P. (ed.): Acipenser. Cemagret Publ. (In French).

Stanić S, Andrić N, Zorić S, Grubor–Lajsić G, Kovačević R (2006) Assessing pollution in the Danube River near Novi Sad (Serbia) using several biomarkers in sterlet (*Acipenser ruthenus* L.). Ecotoxicol Environ Saf 65: 395–402 https://doi.org/10.1016/j.ecoenv.2005.08.005

Stanković M, Dulić Z, Marković Z (2011) Protein sources and their significance in carp (*Cyprinus carpio* L.) nutrition. Journal of. Agric Sci 56(1):75–86. https://doi.org/10.2298/JAS1101075S

Stojković M, Simić V, Milošević Đ, Mančev D, Penczak T (2013) Visualization of fish community distribution patterns using the self-organizing map: A case study of the Great Morava River system (Serbia). Ecol Modell 248:20–29. https://doi.org/10.1016/j.ecolmodel.2012.09.014

Stojković M, Milošević Đ, Simić S, Simić V (2014) Using a Fish-Based Model to Assess the Ecological Status of Lotic Systems in Serbia. Water Resour Manag 28(13):4615–4629. https://doi.org/10.1007/s11269-014-0762-4

Storfer A (1999) Gene flow and endangered species translocations: a topic revisited. Center for Ecology, Evolution and Behavior and T.H. Morgan School of Biological Sciences, University of Kentucky, Lexington, KY 40506–0225, USA.

Štrbac S, Kašanin Grubin M, Jovančićević B, Simonović P (2015) Bioaccumulation of Heavy Metals and Microelements in Silver Bream (*Brama brama* L.), Northern Pike (*Esox lucius* L.), Sterlet (*Acipenser ruthenus* L.), and Common Carp (*Cyprinus carpio* L.) From Tisza River, Serbia. J Toxicol Environ Health - A 78:663–665. https://doi.org/10.1080/15287394.2015.1023406

Struthers D, Danylshuk A, Wilson A, Cooke S (2015) Action Cameras; Bringing Aquatic and Fisheries Research into View. Fisheries 40(10):503–511. https://doi.org/10.1080/03632415.2015.1082472

Subotić S, Jeftić V, Spasić S, Hegediš A, Krpo–Ćetković J, Lenhardt M (2013) Distribution and accumulation of elements (As, Cu, Fe, Hg, Mn, and Zn) in tissues of fish species from different trophic levels in the Danube River at the confluence with the Sava River (Serbia). Environ Sci Pollut Res 20(8):5309–5317 https://doi.org/10.1007/s11356-013-1522-3

Subotić S, Višnjić–Jeftić Ž, Spasić S, Hegediš A, Krpo-Ćetković J, Lenhardt M (2015) Concentrations of 18 elements in muscle, liver, gills, and gonads of sichel (*Pelecus cultratus*), ruffe (*Gymnocephalus cernua*), and European perch (*Perca fluviatilis*) in the Danube River near Belgrade (Serbia). Water Air Soil Pollut 226:287https://doi.org/10.1007/s11270-015-2544-x

Subotić S, Višnjić Jeftić Ž, Bojović S, Đikanović V, Krpo Ćetković J, Lenhardt M (2021) Seasonal variations of macro-, micro-, and toxic elements in tissues of vimba bream (*Vimba vimba*) from the Danube River near Belgrade, Serbia. Environ Sci Pollut Res 28:63087–63101. https://doi.org/10.1007/s11356-021-15073-0

Sundblad G, Bergström L, Söderqvist T, Bergström U (2020) Predicting the effects of eutrophication mitigation on predatory fish biomass and the value of recreational fisheries. Ambio 49: 1090–1099. https://doi.org/10.1007/s13280-019-01263-1

Sunjog K, Gačić Z, Kolarević S, Višnjić Jeftić Z, Jarić I, Knežević Vukčević J, Vuković Gačić B, Lenhardt M (2012) Heavy metal accumulation and the genotoxicity in barbel (*Barbus barbus*) as indicators of the Danube River pollution. Sci World J 2012:351074. https://doi.org/10.1100/2012/351074

Sunjog K, Kolarević S, Kračun-Kolarević M, Višnjić-Jeftić Ž, Gačić Z, Lenhardt M, Vuković-Gačić B (2019) Seasonal variation in metal concentration in various tissues of the European chub (*Squalius cephalus* L.). Environ Sci Pollut Res 26(9):9232–9243. https://doi.org/10.1007/s11356-019-04274-3

Suuronen P, Bartley M (2014) Challenges in managing inland fisheries – using the ecosystem approach. Boreal Environ Res 19:245–255

Taler Z (1953) Rasprostranjenje i popis slatkovodnih riba Jugoslavije. Glasnik Prir. muz. Srpske zemlje. Ser B 5–6:425–455

Tokodi N, Drobac D, Jussi M, Marinović Z, Vazić T, Simeunović J, Lazić G, Petrović T, Vuković-Gačić B, Sunjog K, Kolarević S, Kračun-Kolarević M, Subakov-Simić G, Miljanović B, Svirčev Z (2018) Cyanobacterial effects in lake Ludoš, Serbia - Is preservation of a degraded aquatic ecosystem justified? Sci Total Environ 635:1047–1062. https://doi.org/10.1016/j.scitotenv.2018.04.177

Tošić K, Taflan E (2019) Observation on morphological color changes in Pontic shad (*Alosa immaculata* Bennett 1835) during spawning migration in the Danube, vol 24. Scientific Annales of the Danube Delta Institute, Tulcea

Tošić A, Škraba D, Nikolić V, Mrdak D, Simonović P (2014) New Mitochondrial DNA Haplotype of Brown Trout *Salmo trutta* L. from Crni Timok Drainage Area in Serbia. Turk J Fish Aquat Sci 14:37–42. https://doi.org/10.4194/1303-2712-v14_1_05

Tošić A, Škraba D, Nikolić V, Atlagić JC, Mrdak D, Simonović P (2016) Haplotype diversity of brown trout *Salmo trutta* (L.) in the broader Iron Gate area. Turk J Zool 40:655–662

Valković B, Paunović M (1998) Round goby *Neogobius melanostomus*, a new Ponto-Caspian element for Yugoslavia. Folia Zool Brno 47:305–312

Veličković T, Stanković D, Snoj A, Simić V, Vukić J, Šanda R, Barcytė D, Marić S (2019) Genetic structure of brown trout (*Salmo trutta* L.) in the Danube drainage in Serbia, inferred from microsatellite DNA. International symposium "Advances in the population ecology of stream salmonids - V". Granada, Spain.

Veličkovič T, Simić V, Šanda R, Radenković M, Milošković A, Radojković N, Marić S (2020) New record of a population of *Telestes souffia* (Risso, 1827) (Actinopterygii: Cyprinidae) in Serbia. Acta Zool Bulg 72(1):13–20

Vilizzi L, Ekmekçi G, Tarkan S, Jackson J (2015) Growth of common carp *Cyprinus carpio* in Anatolia (Turkey), with a comparison to native and invasive areas worldwide. Ecol Freshw Fish 24:166–180. https://doi.org/10.1111/eff.12141

Višnjić Jeftić Ž, Lenhardt M, Navodaru I, Hegediš A, Gačić Z, Nikčević M (2009) Reproducibility of age determination by scale and vertebra in Pontic shad (*Alosa pontica* Eichwald, 1838), from the Danube. Arch Biol Sci 61(2):337–342. https://doi.org/10.2298/ABS0902337V

Višnjić Jeftić Ž, Jarić I, Lj J, Skorić S, Smederevac Lalić M, Nikčević M, Lenhardt M (2010) Heavy metal and trace element accumulation in muscle, liver and gills of the Pontic shad (*Alosa immaculata* Bennett 1835) from the Danube River (Serbia). Microchem J 95(2):341–344. https://doi.org/10.1016/j.microc.2010.02.004

Višnjić-Jeftić Ž, Lenhardt M, Vukov T, Gačić Z, Skorić S, Smederevac–Lalić M, Nikčević M (2013) The geometric morphometrics and condition of Pontic shad, *Alosa immaculata* (Pisces: Clupeidae) migrants to the Danube River. J Nat Hist 47:1121–1128 https://doi.org/10.1080/00222933.2012.752048

Vranić D, Đinović-Stojanović SA (2011) Rainbow trout (*Oncorhynchus mykiss*) from aquaculture – meat quality and importance in the diet. Meat Technol 52(1):122–133

Vranković J, Stanković M, Marković Z (2021) Levels of antioxidant enzyme activities in cultured rainbow trout (*Oncorhynchus mykiss*) fed with different diet compositions. Bull Eur Assoc Fish Pathol 4(4):135–145

Vuković S (2015) Životinje u rimskin amfiteatrima. Studija slučaja amfiteatra u Viminacijumu. Doktorska disertacija, Filozovski fakultet, Univerziteta u Beogradu.

Vuković T, Ivanović B (1971) Slatkovodne ribe Jugoslavije. Zemaljski muzej BiH, Sarajevo

Wang Z, Wang Q, Wang J, Wei H, Qian J, Zhang Y, Feng K, Chen Q, Yuan J, Liu J, Li D (2022) Evaluation of the control effect of bighead carp an silver carp on cyanobacterial blooms based on the analysis of differences an algal digestion process. J Clean Prod 375:134106

Weiss S, Apostolou A, Đug S, Marčić Z, Mušović M, Oikonomou A, Shumka S, Škrijelj R, Simonović P, Vesnić A (2018) Endangered fish species in Balkan rivers: Their distributions and threats from hydropower development. Riverwatch Euro Natur 162

Welcomme R, Cowx I, Coates D, Bene C, Funge-Smith S, Hals A, Lorenzen K (2010) Inland capture fisheries. Philos Trans R Soc B Biol Sci 1554:2881–2896

Žikić G (2022) Godišnji program upravljanja ribarskim područjem „Đerdap "za 2022. godinu. JP „Nacionalni park Đerdap "Donji Milanovac (In Serbian).

Žikić V, Štajn A, Ognjanović B, Pavlović S, Saičić Z (1997) Activities of Superoxide Dismutase and Catalase in Erythrocytes and Transaminases in the Plasma of Carps (*Cyprinus carpio* L.) Exposed to Cadmium. Physiol Res 46:391–396

Žikić R, Štajn A, Pavlović S, Ognjanović B, Saićić Z (2000) Activities of superoxide dismutase and Catalase in erythrocytes and plasma transaminases of goldfish (*Carassius auratus gibelio* Bloch.) exposed to cadmium. Physiol Res 50:105–111

Zirojević O (2007) Srbija pod turskom vlašću 1459–1804. Srpski genealoški centar, Beograd

Živaljević R (2017) Ribolov na Đerdapu u ranom holocenu (10.-6. milienijum pre.n.e). Univerzitet u Beogradu, Filozovski fakultet. Doktorska disertacija (In Serbian).

Živaljević I, Marković N, Maksimović M (2019) Food worthy of kings and saints: Fish consumption in the medieval monastery Studenica (Serbia). Anthropozoologica 54(16):179–201. https://doi.org/10.5252/anthropozoologica2019v54a16

Živković D, Jovanović B (2011) Spatial morphometric plasticity of spirlin *Alburnoides bipunctulus* (Bloch, 1782) phenotype from the Nišava River, Serbia, Danube basin. Biologica Nyssana 2(1): 67–77

Zorić K, Simonović P, Đikanović V, Marković V, Nikolić V, Simić V, Paunović M (2014) Checklist of non-indigenous fish species of the river Danube. Arch Biol Sci 66:629 639. https://doi.org/10.2298/ABS1402629Z

Links

https://www.ekologija.gov.rs
https://www.fishbase.se
https://www.minpolj.gov.rs
https://www.mos.gov.rs
https://www.ribolovci.org.rs/index.php/savez-sportskih-ribolovaca-srbije
https://www.ssrvojvodina.org.rs/
https://www.stat.gov.rs

The Freshwater Fish Resources and Freshwater Fisheries of Montenegro, Management, Sustainability, and Conservation: With a Special Perspective on Lake Skadar

Danilo Mrdak and Dragana Milošević

Abstract The wealth of fish and fishery potential represent one of the most important biological and economical resources in Montenegro. Lake Skadar is the largest Balkan lake, a region widely recognized as a European biodiversity hotspot, one of the key biodiversity areas in the Mediterranean, and the most important fisheries region of Montenegro. Even though fishing in the continental part of Montenegro is still relatively undeveloped, this activity has great importance for the local population of Lake Skadar in terms of tradition, culture, and economic development. This chapter provides an overview of the fish resources and continental fisheries in Montenegro both throughout history and in recent times, highlighting the main limitations and threats, as well as the potential for its further development and sustainable use in the future.

Keywords Tradition · Fishing gear · Commercial fishery · Aquaculture · Lake Skadar

1 Introductions

The area of the continental part of Montenegro is 13.812 km^2 (Radulović et al. 2015). There are two hydrographic basins in the territory of Montenegro: the Black Sea basin and the Adriatic Sea basin. Of the total area of Montenegro, the Adriatic basin comprises 47.8%, and the Black Sea basin 52.2% (Fig. 1). The watershed between the two basins passes through the mountains of Golija, Vojnik, Lola, Sinjajevina, and Žijovo. The larger rivers of the Adriatic basin are the Morača (97 km) with its tributaries the Zeta (60 km) and the Cijeva (33 km in Montenegro)

D. Mrdak · D. Milošević (✉)
Faculty of Science and Mathematics, Department of Biology, University of Montenegro, Podgorica, Montenegro
e-mail: dragana.mi@ucg.ac.me

V. Simić et al. (eds.), *Ecological Sustainability of Fish Resources of Inland Waters of the Western Balkans*, Fish & Fisheries Series 43,
https://doi.org/10.1007/978-3-031-36926-1_5

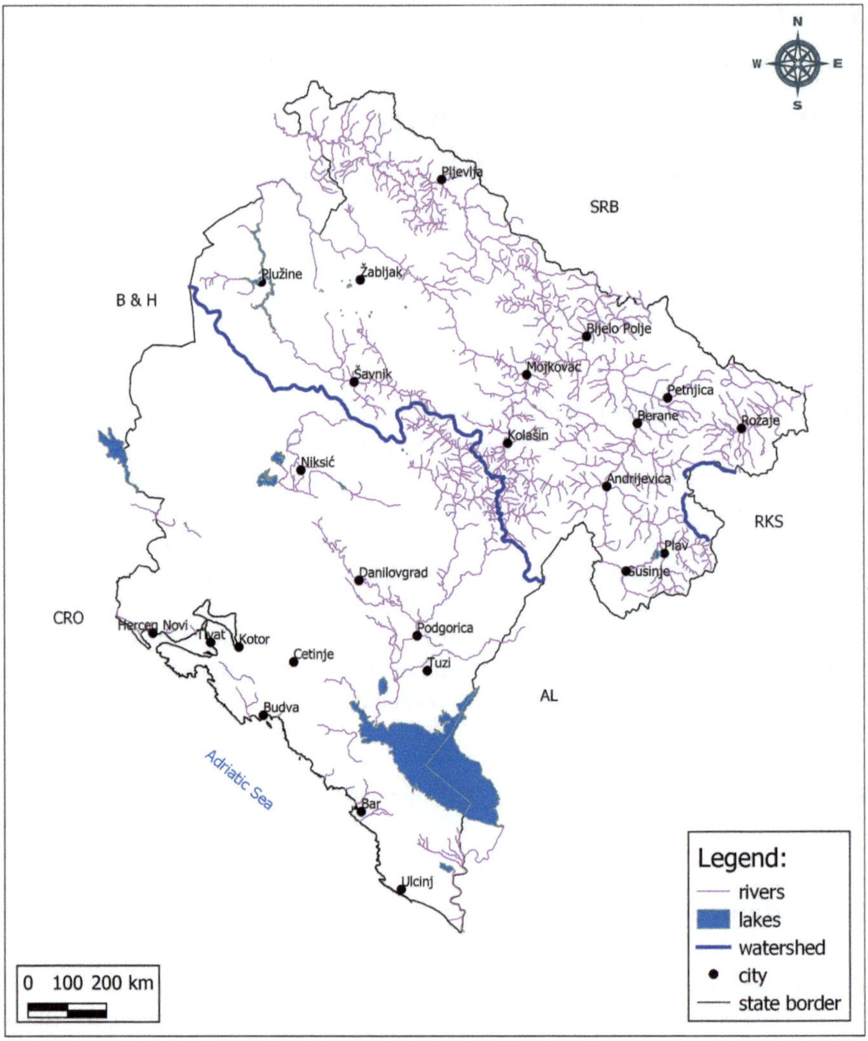

Fig. 1 Inland waters of Montenegro (personal map)

as well as the Bojana (44 km), which represents the connection with the Adriatic Sea (Radulović et al. 2015). The rivers of the Black Sea basin are more numerous and richer in water. The river network consists of the Ibar (25 km in Montenegro), the Lim (86 km in Montenegro), the Piva (82 km) which is mostly submerged, the Tara (141 km), and the Ćehotina (100 km in Montenegro) (Radulović et al. 2015). In the southern part of Montenegro, there are two larger lakes: Lake Skadar and Lake Šasko, while smaller lakes, which are more numerous and glacial in origin, are located in the northern, mountainous, part of Montenegro. The most important fishing area within the territory of Montenegro is Lake Skadar. The catchment area of the lake covers a total of 5631 km^2 with about 81% falling within the territory of Montenegro, and the remainder in the territory of Albania (Radulović et al. 2015). It is a transborder lake, with a surface area fluctuating from 350 to 500 km^2, depending on water levels. The exact origin of the lake remains unknown, though it is likely that it is a result of tectonic processes during the Pleistocene (Stanković 1926). It is a shallow lake with a mean depth of around 5 m, while the maximum depth (without the sublacustrine springs) varies seasonally between 8.3 and 10.4 m (Lasca et al. 1981).

Ichthyofauna represent one of four indicators that determine the ecological status of lakes according to the EU Water Framework Directive. The Directive requires that good ecological status, i.e. the status of the fish fauna (and other biota) under nearly undisturbed conditions, has to be maintained or restored, and that specific measures are to be taken to fulfil this requirement. Freshwater fishing is an economic branch that features valuable but underutilized resources, especially if we take into account the undoubted advantages of fishing in relation to other spheres of agricultural, as well as the significant economic potential that it offers. Although fisheries represent a small segment of Montenegro's economy, the existence of rivers and lakes characterized by relatively preserved ecosystems, with extraordinary fishery (commercial, sport, and recreational fishing) and aquaculture potential, are real prerequisites for improving the fishing sector in Montenegro. Due to the significant number of natural resources available in Montenegro, all types of freshwater fishing are possible: commercial fishing, sport fishing, "fly fishing", and aquaculture. According to the Law on Freshwater Fisheries and Aquaculture (Official Gazette of the Republic of Montenegro 17/18), fishing waters, fishing areas, and fishing grounds where sport, recreational, and commercial fishing can be carried out are defined. There are indications and concerns that the fish stock is being overexploited and that fish populations are under various pressures due to illegal fishing, the introduction of invasive species, and environmental pollution; however, there is no systematic research or constant monitoring of fish populations. The only recent project that was implemented in Montenegro, and whose focus was the state of the fish stocks and fisheries of Lake Skadar, was the project "Conservation and Sustainable Use of Biodiversity on Lake Prespa, Ohrid and Skadar Lake (CSBL)" which was implemented by GIZ (Deutsche Gesellschaft für Internationale Zusammenarbeit). During the three annual fishing seasons during the implementation of the project (2013–2016), fish sampling was performed based on European standard EN 14757. During the 3 years of the project, a total of 18 species were found. The report on the project states that the multi-mesh gill net fishing method works reasonably well to obtain an overview about species composition, abundance, biomass, and size structure in the Montenegrin context.

The data presented in this chapter provide the main information and identifies the knowledge gaps about the continental fish resources in Montenegro, both throughout history and in recent times, as well as offering insights into the current conditions of the main commercial fish species, which is an important identifying factor in planning the next steps towards sustainable fishing in the country.

2 Fish Resources and Fisheries

Within the territory of Montenegro, 89 species of freshwater fish have been registered so far across both basins (i.e. the Adriatic Sea basin and the Black Sea basin) (Marić 2019). This number also includes 14 species mentioned in the literature, although some of them were not registered in the latest species lists (Marić 2019; Marić and Milošević 2011) which means that the ichthyofauna of Montenegro currently numbers 75 species. Of this number, 62 species are autochthonous and 14 allochthonous. The autochthonous ichthyofauna of the Adriatic basin consists of 43 species, while 19 of them are found in the Black Sea basin. There are 14 non-native species in the Adriatic basin, and 7 in the Black Sea basin (Marić 2019). Seven species are endemic: *Salmo obtusirostris* Hadžišće, 1962—the soft muzzled trout, *Gobio skadarensis* Karaman, S., 1936, *Barbatula zetensis* (Šorić 2000), *Knipowitschia montenegrina* (Kovačić and Šanda 2007), *Pomatoschistus montenegrensis* (Miller and Šanda 2008), and *Rutilus albus* (Marić 2010). In the wider area of the Adriatic basin or in the southeastern Adriatic ecoregion, there are 18 endemic species that have been registered in Montenegro. According to the classification of the International Union for Conservation of Nature, IUCN, in 2020, the Skadar nase (*Chondrostoma scodrense*) was classified as extinct—EX. Two sturgeon species, *Acipenser naccarii* Bonaparte, 1836, and *Acipenser sturio* Linnaeus, 1758, are classified as critically endangered—CR (www.iucn.org). However, these two species of sturgeon in the territory of Montenegro have not been registered for more than 30 years and in terms of the waters of Montenegro they are considered extinct in nature—EW (Marić 2019). The *Hucho hucho* is endangered throughout its range—EN (Kottelat and Freyhof 2007; www.iucn.org). The *Salmo obtusirostris*, *Salmo marmoratus*, and *Anguilla anguilla* are critically endangered—CR www.iucn.org (Marić 2019). Three species are protected by national legislation: *Acipenser naccarii* Bonaparte, 1836, *Acipenser sturio* Linnaeus, 1758, and *Salmo zetensis* Hadžišće, 1962. All of the above gives an indication of the importance of Montenegro for the ichthyofauna in both drainage basins. Being the southernmost and, due to high mountains, one of the sourcing regions of the Black Sea drainage basin, it is kind of refuge area for the one of the flagship species and, at the same time, the most endangered European fish species—the river huchen (*Hucho hucho*). Together with other Balkan rivers in the Adriatic and Ionian drainage basin, it is a place of unique evolution which has driven a large number of endemic fish species.

The ancient state of Zeta (later Montenegro) was situated in the southeastern part of the former Yugoslavia, close to the Albanian border. Over time, it lost its coastal regions and the greater part of Lake Skadar to Venetian conquests in the fifteenth century and Turkish invasions in the sixteenth century. However, the Montenegrin people, and its military, never ceased to fight in order to reclaim the lost areas of its national territory. Finally, at the end of the Russian-Turkish War of 1876–78, in which Montenegro also took part, the old longing of the Montenegrin people was realized. By the determination of Article 28 of the Treaty of Berlin in 1878, Montenegro regained the territory of its coastal region, extending from the mouth of the small River Željeznica (near Bar) up to the mouth of River Bojana, including the southern part of the western shore of the same river and approximately the western shore of Lake Skadar (Franetović 1960).

In terms of both biodiversity and fishery, Lake Skadar is the richest and most important area in Montenegro. The largest number of published research papers from the area of Montenegro is precisely related to the area of Lake Skadar and its biodiversity. It is the largest Balkan lake and a region widely recognized both as a European biodiversity hotspot (Gaston and David 1994) and one of the key biodiversity areas in the Mediterranean (Darwall et al. 2014). The catchment area of Lake Skadar is located between 18°41′ and 19°47′ East and 42°58′ and 40°10’ North, in a karst area in the outer part of the southeastern Dinaric Alps. Lake Skadar's main lake inlet is the River Morača (with average discharge of 202 m^3/s), while the River Bojana is the main Lake outlet (with average discharge of 310 m^3/s). The rest (110 m^3/s) comes by precipitation, smaller tributaries, and underwater karst springs called "vrulja" placed around the Lake. Lake Skadar is relatively shallow, with a mean depth of 5.01 m, but there are locations where the Lake depth reaches 60 m in sublacustric springs called "oka" (Raduško oko). The Lake shoreline (including islands) is 207 km long. The River Bojana (the main outlet) is a wide lowland river, 40 km in length, forming a huge estuary delta on the Adriatic coast. The Lake is situated in the Mediterranean climate zone with an average number of 130 "summer days" (days with a temperature equal to or higher than 25 °C). The mean value for precipitation is 2250 mm and there are no "winter days" (days with a temperature equal to or below 0 °C). The southern and south-western shores (as well as narrow north-eastern part towards Rijeka Crnojevića) feature steep mountains and hills slopes, while the northern and north-eastern shore is a wide lowland area (with swamps, flooded meadows, lowland forests, and so on). Due to this climate, as well as its hydrology, geomorphology, and dramatic geological history, this Lake is characterized by significant biodiversity, including numerous endemic and relict species of flora and fauna. A specific characteristic of Lake Skadar is its many sublacustric springs that are deeper than 30 m, called "oka" by the locals. The existence of sublacustric springs in the Lake is of vital importance for fish, especially for the bleak which is the dominant fish species whose seasonal intensive fishing in these springs has been considered rational fishing management in the biological and economic terms (Drecun and Ristić 1964). According to its nutrient content, Lake Skadar is classified as oligotrophic to mesotrophic (Rakočević and Rakaj 2015). Lake Skadar is a unique ecosystem, as it provides excellent conditions for its diverse ichthyofauna, since it provides habitats for both cold water (e.g. salmonids) and warm water species (e.g. cyprinids). The ichthyofauna of the lake is composed of a relatively high number of species of freshwater and marine origin, some of which, however, only periodically inhabit the lake. The freshwater ichthyofauna of Lake Skadar consists of 59 species; 44 of them are native to the watershed and 15 have been introduced. Seven species are endemic (Marić 2019). The Lake is also a spawning ground and habitat for larval and juvenile fish, respectively, for several catadromous and anadromous (e.g. *Alosa fallax*, *Anguilla anguilla*, *Acipenser* spp.)

as well as certain estuarine fish species (*Mugil cephalus*, *Dicentrarchus labrax*, and *Platichthys flesus*).

Such significant fish species diversity is a consequence of a diverse combination of ecological factors and the presence of various biotopes. Thanks to its shallowness, almost the whole water masses are in a photic zone and the lake waters are rich in oxygen. Although the lake temperature is very variable (8–24 °C) there are locations where the temperature of the deeper part is pretty much constant (in the Lake "oka") and provides an environment with relatively constant thermic conditions in different seasons of the year, and this is used by different fish species (e.g. lake brown trout and marble trout during the summer or bleak during winter months).

The lake offers diverse habitats ranging from extensive fields composed of submersed and/or floating plants (such as *Potamogeton* sp., yellow water lilies *Nuphar lutea*, or water chestnut *Trapa natans*) to reed (*Phragmites* sp.) belts of various sizes and also areas of sparse aquatic shoreline vegetation (Ristić and Vizi 1981; Bajković et al. 2017). Aquatic vegetation plays an important role in the life cycle of many fish (it provides food for young fish and shelter from predators and physical disturbances). In addition, the lake forms several habitats that are important for fish or their developing stages (eggs or larvae). The northern part of the lake, which is widely covered by macrophytes, are important spawning and nursery grounds, e.g., for carp (*C. carpio*) and other cyprinids (*Carassius* sp., *Scardinius* sp.) (Bajković et al. 2017).

The human population of the relatively wide shore zone of Lake Skadar has been engaged in fishing since ancient times. By constantly using the lake as a rich source of food, fishing has become an integral part of the traditional life of the people of this region and the production of fishing tools, their use, and the method of catching and preparing fish became important parts of its culture. Fishing methods were developed in parallel with the improvement in fishing gears. The use of fishing tools made of wood dates back to the beginning of the nineteenth century, after which the use of more efficient net-type fishing tools became more and more intensive. In the earliest times, the fishing tools on Lake Skadar have been made from materials offered by the natural environment. Shoreline vegetation provided the basis for making wooden tools. For example, a tool made of willow wood called "vrša" (a type of fyke net trap) is considered the oldest tool for fishing bleak on the lake. Among the tools made of metal, hooks and "ošća" (a type of spear with several metal cogs with the joint base on the top) were used. Their production and application were constantly improved, in contrast to the "vrša", which remained unchanged. In the first half of the nineteenth century, beach seines—"veliki grib"—and other fishing nets began to be used, and because of their effectiveness they gradually suppressed the use of the "vrša". According to Lopičić (1931)), these large nets (beach seines—"grib") can simultaneously catch up to 50 tonnes small fish, alongside eel, carp, nase, and several big trout they called "mladica". In December 1891, 30 tonnes of fish were caught at once and the fishermen were taking them out from water for three days (Rovinski 1897). The large-scale catch of fish encouraged domestic crafts to produce nets from silk thread. The silkworm was common in the villages Bobija, Drušići, Dodoši, Meterizi, Rvašima, Dujevi, and Prevlaci, among others (Radusinović 1983).

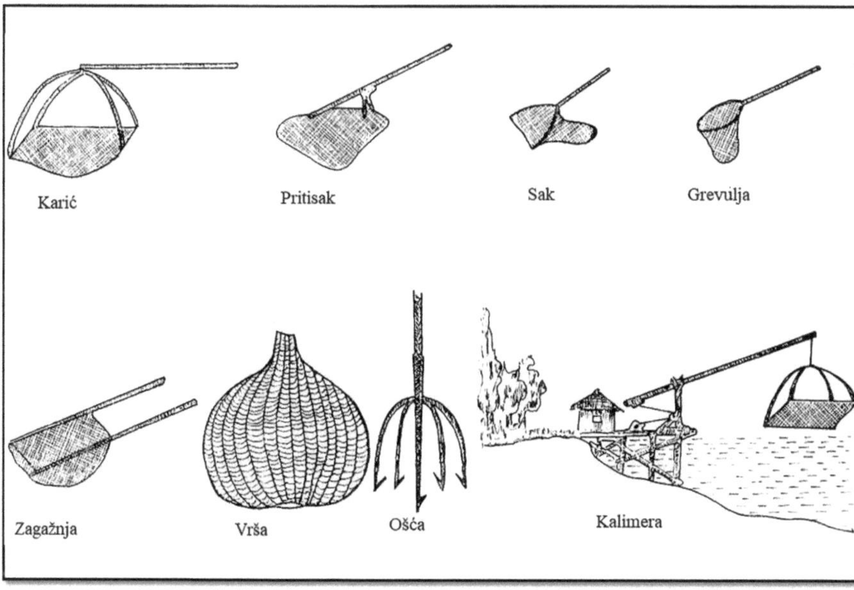

Fig. 2 Traditional fishing tools used in the past on Lake Skadar. According to Radusinović 1983

In the earliest time of fishing development on Lake Skadar, the following fishing tools were used: the "vrša", the "vršica", the "koš", the "košić", the "koca", the "daljan", the "kalimera", the "pari", the "ošća", the "karić", the "pritisak", the "sak", the "grevulja", and the "zagažnja" (Fig. 2). In the twentieth century, other fishing gear were in use: the "sačmarica" or "samolovka", the "veliki grib"—bleak beach seine, the "mali grib"—nase beach seine, and different bottom or pelagic gill nets. Depending on the target species, the dimensions of the gill net and mesh size vary. They are named according to the target species. For example, the "ukljevna mreža" is a short net 100 m in length and 3.5 m in height, with a mesh size of 15–15.5 mm (knot to knot). It is used for bleak fishing and could be benthic or pelagic, depending on the fishing location (for lake "oka", a pelagic bleak net is used, while for the shorelines away from the lake "oka" fishermen use benthic bleak nets); the "krapnjača" is a net with a total length of 50–70 m, which is 3.5–4 m high, with a mesh size of 50–55 mm (knot to knot) used for carp, trout, and other larger species; the "skobaljna mreža"—50 m long, 2 m in height, with a mesh size of 38–40 mm (knot to knot) is used for nase and chub fishing during spring and autumn; and the "pastrvna mreža"—80 m long and 3–3.5 m in height with a mesh size of 60–65 mm (knot to knot) is used for trout fishing during spring and autumn.

The following are descriptions of some of the traditional fishing tools used in the past. Some of them are still used today (Fig. 2):

"Vrša"—a fyke net. This pear-shaped net is woven from thin willow twigs. It consists of two openings: the lower one for the fish to enter and the upper one for taking it out. The total length is 1–1.20 m, and it has a net opening of 0.50–0.60 m.

"Koš"—This tool is woven from wicker. It is mostly used for catching chub, as well as smaller carp and European eel. The total length is 1 m and the width is 80 cm.

"Ošća"—Technically this is a spear made of a long wood stick with the metal made into five sharp metal teeth which start from one base. The middle tooth, called "kalauz", is the longest. This tool was mainly used for carp fishing at night during the spawning period. In recent times, its use has been prohibited.

"Pari"—they are made of rope 150 to 200 m long with hooks attached to it with a lateral line (longline fishing gear type). Smaller fish such as roach are placed on the hooks. It is used to catch carp and eel.

"Karić"—A grid squared shape, with dimensions of 3x3 m woven of cotton thread. It is attached at the ends with two cross-bent and connected sticks.

"Pritisak"—Composed of a smaller net called "kuljar" tied to a frame 1.5–2 m in diameter. A handle is attached to the frame both directly (the end of handle at one end of the frame) and via a V-shaped connector (the connector is attached approximately midway along the handle and the other end of the frame).

"Sak"—The frame is in the shape of a triangle with two curved sides (similar to a Reuleaux triangle). A pleated net is tied to the frame to form a "kuljar" about 1 m long. A handle is attached to the frame at the point where the two curved sides meet. This type of net is designed to enclose a funnel of a dammed stream and catch fish carried by the current. This net is mainly used to catch common nase—*Chondrostoma nasus*.

"Grevulja"—An elliptical shaped frame with a handle. A dense net, forming a "kuljar", is tied to the frame.

"Zagažnja"—A small net, 1.5 by 1.5 m in size. The ends of the net are folded and tied to two parallel handles in such a way that the net forms a half cylindrical shape. The net derives its name from the term "zagaziti", meaning to step into (as stepping into water is required to use this net).

"Veliki grib"—bleak beach seine—This is the largest net which is mainly used for bleak fishing. The total length is 200–250 m and 16 m in height in the middle of the net, decreasing gradually towards the end of the wings to 2–3 m. This net consists of long wings, with a mesh size of up to 26 mm, with 2 thick sticks at their end to which two ropes are attached and of the cod-end—sak, mesh size 18–22 mm. About 200 fishermen handle the net at the same time (Radusinović 1983) (Fig. 3).

"Kalimera"—a big net (30 x 30 m) posed horizontally in water and which harvest bleak during extraction in a vertical column above it. They are still used nowadays, particularly in Raduš Bay, during the winter period (December, January, and February) when industrial fishing of bleak is performed by the concession holders (Fig. 4).

Fig. 3 "Veliki grib"—bleak beach seine, Raduš 1974 (Photo Ondrej Vizi)

Fig. 4 A kalimera net at Raduš Bay during winter bleak fishing

Many technological advances in fishing gear have been incorporated by the fishermen on Lake Skadar since 1947. Nylon thread has been shown to be the best for eel fishing. The catch rate is much higher, whereas the loss of hooks is decreased

Fig. 5 Fyke net for eel sampling on Lake Skadar (Photo: Jelena Brnović)

to a minimum. Economic analysis demonstrates the trawl lines made of nylon thread are more profitable (Drecun 1956). By 1958, nylon gill nets had almost completely replaced cotton nets and with these changes the efficiency of capture by gill nets increased (Hewson 1951; Berts 1961; Högman 1973). A technique of fishing by trawl in the open part of the lake was described by Drecun (1955)). During the 22 efforts, nine economically important fish species were registered. Of particular note were the catch of the sturgeon (31 kg) and a catch of the bleak during August (10 kg). To catch the European eel in Lake Skadar the fishermen use the longlines and/or fyke nets (Fig. 5). The handmade fyke (Hegediš et al. 2005) was used for sampling glass eel from the River Bojana. The nets consist of 1-mm holes, three chambers, and a total length of 3 m, with a height of 0.4 m and wings of 2 metres at a 90-degree angle. The mouth of the trap encompasses an area of 1.2 m^2 (Fig. 6).

Recently, through the project "Conservation and sustainable use of Biodiversity of Lake Prespa, Ohrid and Shodra/Skadar—CSBL" which was implemented during 2013–2018 for the first time on Lake Skadar, the EU standard researching nets (multi-mesh size gill nets—MMG) EN 14757 were applied. The standard benthic MMG nets are 30 m long, 1.5 m high, and made of 12 different panels with different mesh size diameter ("knot-knot" measures). The mesh size ranged between 5 and 55 mm in a geometric row. Nowadays, all types of fishing gears that are allowed to be used in inland fisheries in Montenegro, as well as their characteristics, are detailed in the relevant by-law (Anonymous 2018).

In the past when travelling around Lake Skadar fishermen traditionally used "čun"—small crafts hollowed from one piece of oak trunk (Fig. 7). Nowadays, fishermen use similar boat with the same name "čun" but built either of beech planks

Fig. 6 Fyke net for sampling glass eels from the small streams of the River Bojana

or made of metal plates. The length of a "čun" is 7 m, its width is 1 m, and it has a height of 0.5 m. One of several types of fishing boats used at Lake Skadar was a "lađa"—a boat of varying lengths (10–12 m, 16–17 m, or up to 24 m) and varying heights above the water level (0.7–1 m). The boat can repel waves owing to the upper half of the bow being raised above the water level. The boat itself is made of chestnut and mulberry wood (chestnut and mulberry planks and mulberry ribs). Small outboard motors came into prominent use around 1962. With outboard motors came increased mobility, and consequently decreased physical effort, which raised the pressure on the fish populations with more fishermen fishing a wider area.

The first data about fish from Montenegro appeared in the second half of the nineteenth century. The authors of those papers were mostly foreign researchers and the papers were most often in the form of short reports providing information documenting the presence of a taxon and list of the species (Heckel and Kner 1858; Steindachner 1883). At the beginning of the twentieth century, there are the first professional and scientific works based on more detailed ichthyological research (Karaman 1923; Karaman 1933; Ivanović 1962; Ivanović et al. 1983; Janković 1971; Janković 1975; Vuković 1961; Vuković 1965; Šorić 1980; Šorić 1982; Marić 1995). The first information about the fishery in Montenegro can be found

Fig. 7 "Čun" on the Rijeka Crnojevića (Photo: Mijat Božović)

in the paper by Andrija Jovićević from 1909 in his publications Lake Skadar and Fishing on it (Jovićević 1909). In this publication, the author gives the list of 17 fish species presented in the Lake as well as data on the organization of fishing and the gears in use. The first fishery related research started with Đorđije Drecun, who, in addition to examining fishing practices, was also involved in ichthyology research. In the period from 1951 to 1980, he published over 40 scientific papers that are still cited and represent valuable historical and scientific data for all future generations. Some aspects of the organization, development, and limitations of fishing in Montenegro were the subject of papers (Kapac 1949a; Kapac 1949; Lopičić 1953; Stanković 1929; Stanković 1939). Comprehensive data about the ichthyofauna of freshwater fishes can be found in publications (Ivanović 1973; Marić and Milošević 2011; Marić 2019). Up to 2010, 300 published ichthyologic papers were listed in the Bibliography of Fisheries and Ichthyological Research in the Freshwaters of Montenegro (Marić 2010). Neither comprehensive monitoring of fish stocks nor the estimation of maximum sustainable yield on Lake Skadar has ever been performed in the past. Only recently, through the CSBL projects in period 2013–2018, has the monitoring of fish populations been established but, unfortunately, that process did not continue beyond the end of this project. Most scientific papers on Lake Skadar fish focused primarily on fish taxonomy and the ecology of individual species (Bianco and Kottelat 2005; Kovačić and Šanda 2007; Krivokapić 1992; Krivokapić 1994; Krivokapić 2005; Marić 1995, 1998, 2002; Miller and Šanda 2008; Milošević et al. 2011, 2012; Milošević and Marić 2012; Milošević and Mrdak 2016; Mrdak et al. 2017; Marić et al. 2022). According to Anonymous (2018) fishing on the Lake can be commercial, sport-recreational, scientific research, rehabilitation, and selective.

3 Commercial Fishing

Lakes Skadar and Šasko, as well as the River Bojana, represent the most important and only areas where commercial freshwater fishing occurs in Montenegro. According to Marić (2018), economically important species in the area of Lake Skadar include bleak, carp, chub, flounder, bream, sable, eel, charadan, yellow bream, white bream, twaite shad, sea bass, and perch. In the same source, it was stated that in the past the annual catch on Lake Skadar amounted to 1500–2000 tonnes, while today it amounts to less than 1/3 of that catch.

The Fishery Cooperative of Rijeka Crnojevića started bleak fishing on 22 November 1948 (Kapac 1949a). According to the results of fisheries for period 1947–1960, the dominant fish were of the family Cyprinidae, which formed 87% of the total catch. The most important fish species was bleak (54.1%), carp (19.9%), nase (8.8%), and common rudd 1.0% of the total catch. The same results show the productivity of Lake Skadar, which at 50 kg/ha is close to some eutrophic areas of the Baltic Sea (Drecun and Miranović 1962). For the period 1947–1960, contribution of the most important fish species (Alburnus, Cyprinus, Chondrostoma, and Alosa) fell below 50%. Simultaneously, the catch of scarp fish has increased with less desirable fish (Rutilus and Pachychilon) dominating in the catch (Stein et al. 1981). The population of Chondrostoma species has disappeared from the lake (Drecun 1985). In 1951, the total catch amount for Lake Skadar was 1,026,033 kg of fish, of which 500 t was bleak, 220 t was carp, then nase, mugil, skadar shad, eel, trout, and so on (Drecun 1952a). One of the largest registered catches was on 10 March 1949, at the fishing area called Ploča. 150,000 kg of fish were caught in one effort, of which over 80% were bleak (Drecun and Miranović 1962). According to Drecun (1961), the data show a bleak catch in a quantity of 700–800 t per year (Fig. 8).

According to the 1947–1977 statistics, the annual catch which averaged 500 tonnes decreased to only 100 tonnes in later years. Such a sudden drop resulted from certain ecological changes caused by excessive fishing and the pollution of the lake with waste material. One author suggested a fishing ban of bleak for a three-year period (Drecun 1985).

In the period 1950–1954, 6000–28,000 kg of twaite shad (*Alosa fallax*) were caught from Lake Skadar (Drecun 1955). All of the above is shown in Fig. 9, and it is easily noticeable that each time the total catch reached or exceeded the sum of 1000 tonnes we were faced with a decrease in the total catch over the subsequent few years. The most dramatic decrease happened in period of 1961–1968 due to the overfishing of bleak and carp, which are two most commercially important species in

Fig. 8 The bleak catch at Bazagur 1977 (Photo Ondrej Vizi)

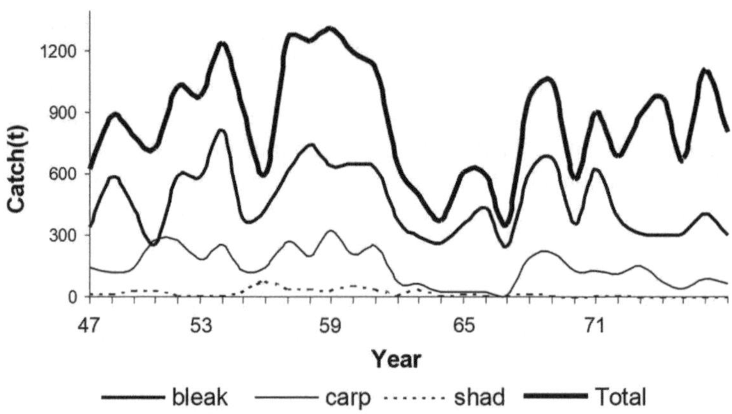

Fig. 9 Catch statistic on Lake Skadar for the period 1947–1976 (Stein et al. 1981)

the Lake. The River Bojana represent a significant source of high-quality glass eel, which can produce between 320 and 427 kg annually (Hegediš et al. 2005). It has been estimated that, until the recent introduction of a closure to the eel fishing season, no less than 50 tonnes of eel were caught (legally and illegally) from Lake Skadar every year, including a third of the total yellow eel catch (Simon et al. 2020).

Regarding fishing pressure, in 2018 74 vessels used for commercial fishing operated on Lake Skadar, while in 2019 there were 97 commercial fishing vessels (Official report of "Lake Skadar" National Park). The number of commercial fishing

Table 1 Number and type of fishing licences issued for the 2019–2022 period

		2019	2020	2021	2022[a]	Average 2019–2021
Skadar Lake recreational fishing licence	Daily licence	329	435	191	32[a]	318
	Yearly licence	68	350	145		188
	TOTAL	397	785	336	32[a]	506
Other inland water body recreational fishing licence	Daily licence	1135	663	851	137[a]	883
	Yearly licence	3069	2648	2468	1996[a]	2728
	TOTAL	4204	3311	3319	2133[a]	3611
TOTAL Recreational fishing licences		4601	4096	3655	2165[a]	4117
Commercial fishing licences (Skadar Lake)		312	367	274	306[a]	318

[a]Not a final figure since 2022 is not finished

licences has varied over the last few years. For instance in the 2021–2022 fishing season, 306 licences were issued, in the 2020–2021 season—274, in the 2019–2020 season—367, and in 2018–2019, 312 licences were issued (Table 1). The licences last for one year and they are not related to a calendar year but rather cover one year starting from the date of issue; e.g. if the fisherman receives his licence on 15th March 2022, it will be valid until 15th March 2023. There are six types of licences: (a) a licence for so-called "carp" nets for up to 10 nets, (b) licences for so-called "bleak" nets, (c) licences for so-called "nase" nets, (d) licences for so-called "mullet" nets, (e) licences for fyke nets and longlines for eel fishing, and (f) licences for longlines for carp fishing with up to 500 hooks. One fisherman can have more than one commercial fishing licence, so the number of sold licences does not match with the number of fishermen. Depending on year, Lake Skadar Lake has had no fewer than 160 and not more than 240 commercial fishermen.

One additional problem on Lake Skadar is poaching or IUU fishing (illegal, unreported, or unregulated fishing). Although reporting each catch is mandatory for all fishermen in Montenegro, neither the commercial nor the recreational fishermen fulfil this obligation at all times. Therefore, almost all the commercial fishing on Lake Skadar is unreported even though all the commercial fishing is legal and respects the regulation in terms of the fishing ban season and regulations related to fishing gears. However, since that there is strong and constant marked demand for Lake fish and that fishing is a kind of tradition in the local communities around the Lake, there are lot of poaching and poachers on Lake Skadar. The Lake itself is wide with many islands, with bays, and with dense coastal forests and shrubbery coverage making poaching and hiding relatively easy. We estimate that at least 35% of the catch on Lake Skadar comes from IUU fishing activities (fishing during the fishing ban season, fishing with illegal gears, and fishing without any kind of licence).

Fig. 10 A zig-zag fish trap on the River Bojana (Buna)

As we said, log books are defined in the Law on Freshwater Fishing and Aquaculture, for both commercial and recreational fishing, but many fishermen are not consistent in their record keeping. Therefore, the precise catch statistics are still unknown, both for commercial and for recreational fishery. The data we cited above are the result of the well-organized and controlled fishing processes on Lake Skadar which occurred during the Communist period (1945–1989). Some species that in the past had significant commercial potential, such as grey mullet, Lake Skadar rudd, twite shad, nase, and chub, no longer even play a similar role in contemporary fishing. The reasons for such a situation are most probably numerous, but all of them have one common thing. All are caused by human activity. During the 1990s, Montenegro was in a huge economic crisis, especially in the first half of that decade, due to UN sanctions. The whole economy collapsed and the population survive mainly on natural resources and on the so-called "black" and "grey" economy, based on the smuggling of goods from other countries. In that period, the fish of Lake Skadar was the best source of fresh, healthy, and what was most important, almost free protein. In that period, there was almost no legal fishing and the Lake Skadar fish populations were under enormous and uncontrolled fishing pressure. Moreover, Lake Skadar was the main smuggling route for activities to and from neighbouring Albania since this is a transborder water body. In that period, pollution was also permanent due to the intensive traffic through the Lake and the fact that people mainly smuggle oil and gas with a lot of leaking of these products into Lake waters. In addition to that, communal and industrial wastewaters entered the Lake Skadar drainage area without any processing and ended up the Lake. An additional problem which is present even today is the existence of traps in the form of zig-zag partitions in the corners on which nets are mounted for catching fish that migrate to or from Lake Skadar. These barriers are located on the River Bojana, downstream from the town of Skadar (in Albania), and are placed along the entire width of the river course (Fig. 10).

In this way, the upstream and downstream migration of anadromous and catadromous fish species, which occurs during higher water levels in spring and late autumn are prevented, and the majority of migratory fish finish as part of the catch. Only during extremely high water levels, when the water level of the River Bojana is above these fishing tools, can some of the fish migrate into or out of the Lake, but even during this events migration is hindered. A special problem is the fact that it is not known how this fishing toll is operating. What is obvious is that either the fish species that would be expected to migrate from the sea to the lake are completely absent or their number has been drastically reduced. Therefore, it is necessary to form a joint commission (between Albania and Montenegro) that would prescribe the way of operation and control of this zig-zag fishing gear, thus solving the problem of ensuring the passage of the River Bojana for upstream migration. Without regulation of this operation and without a controlling mechanism, the legal protection of eels and the fishing ban season on this fish species in Montenegro during its migration into the sea is worthless. Since Lake Skadar is one of the most important and largest habitats for eels in the whole eastern Mediterranean, the basic implementation of the so-called Eel Directive (EC No 1100/2007) and allowing the migration of eels to and from the sea is urgently required.

4 Recreational Fishing

As Montenegro is a mountainous country, it is also the most southern watershed among the Black Sea and the Adriatic drainage areas. For both watersheds, Montenegro is a sourcing region and consequently the majority of inland waterbodies, rivers, and lakes are cold water in terms of their ecology (except the furthest downstream part of the River Morača and the whole watercourse of the River Bojana). So, it is not surprising that all the rivers and lakes are inhabited by salmonid fish fauna and cyprinids common to middle river parts (nase, barbel minnow, and chub). In the Black Sea drainage area, river huchen (*Hucho hucho*), brown trout (*Salmo trutta*), and grayling (*Thymallus thymallus*) are dominant, while in the Adriatic Sea drainage there are soft muzzle trout (*Salmo obtusirostris*), marble trout (*Salmo marmoratus*), and Adriatic brown trout (*Salmo farioides*). Recreational fishing in Montenegro has a long tradition it is not a kind of "noble" sport activity, but is rather spread among all social categories no matter the income and education level. The total length of large rivers and their tributaries in Montenegro is about 1700 km or about 2100 ha of water surface (Burić 2010). Their special value is their good water quality, so they also represent a rich potential resource for both recreational fishing and aquaculture. The surface of natural mountain lakes inhabited by fish covers a total of 547 ha (Lake Plav—250 ha, the Black Lake—63 ha, and Lake Biograd—43 ha). The surface of the main accumulation in Montenegro is about

3090 ha: Lake Pivsko 1250 ha, Lake Slano 900 ha, Lake Krupac 570 ha, Lake Otilovići 270 ha, and Lake Liverovići 100 ha.

Recreational fishing occurs on all inland waterbodies within Montenegro and fishermen are mainly oriented towards salmonid fish species. Only lately have some recreational fishermen oriented towards cyprinid species such are carp, nase, chub, rudd, and barble. Recreational fishing of cyprinid species mainly occurs in Lake Skadar, as well as in the lower parts of the Rivers Morača and Lim, the River Bojana, Lake Šasko, and several accumulations such as Slano, Krupac, Otilovići, Liverovići, and Borovičko. Fishermen are organized into Sport Fisherman Clubs and the Clubs mainly represent recreational fishermen and are organized on a municipality basis—one Club for one municipality even though there are some municipalities with more than one Club (e.g. the Podgorica municipality).

Sport Fisherman Clubs are the concessioners of the fish populations within the exact municipality administrative borders, and they are in charge of protecting the fish population, organization and controlling fishing activities, fighting against poachers, the legal processing of poachers, issuing licences, the organization of cups and competition, and the enforcement of fishing bans and restrictions.

According to data held by the Ministry of Agriculture, Forestry and Water Management (of the Government of Montenegro) from the Register of Issued Fishing Licenses in 2019, there were 4601 licences issued for recreational fishing of which 397 licences were issued for Lake Skadar, while 4204 licences were issued for other waterbodies in Montenegro (Table 1). In the following years, we had a slight decrease in the number of issued licences, but the average number for 2019–2021 period was about 4117 licences (Table 1).

One recreational fisherman can have one yearly licence issued by their local Club (of which this fisherman has membership) and many other daily permissions for fishing either on Lake Skadar or on waterbodies under the management of other Clubs. Therefore, the number of issued licences does not match directly with the number of active recreational fishermen. The number of yearly licences issued for waterbodies other than Lake Skadar is the number which indicates the number of recreational fishermen (Table 1). In addition to that, if we consider the number of yearly licenses issued on Lake Skadar and the fact that there are recreational fishermen who are not member of any Club, so they fish only on the basis of daily permits, we can conclude that the number of recreational fishermen in Montenegro is unlikely to be below 3500.

The sector of recreational fishery, the organization defined by the latest law (Anonymous 2018), showed that it is not sustainable, due to several factors of which we highlight the following: a) the complete lack of financial sustainability due to the low number of Club members; b) the conflict of interest due to the obligation to organize fishing activity, protect the fish population, and engage in law enforcement as well as the processing of law offender towards legal institutions; c) we do not have any clear idea of the total yearly catch of recreational fisherman; and d) the overlap between sport and fishery jurisdiction. Regarding everything previously stressed, a new Law on Freshwater Fishing and Aquaculture is in the final drafting phase at the moment.

5 Aquaculture, Stocking, and Fish Processing

Aquaculture aims to achieve intensive production and high yields of fish on smaller water surfaces, using appropriate technical and technological solutions. The practice of fish farming in the territory of Montenegro originated in 1951. The first hatchery facility with a capacity of about 250 thousand eggs was founded near the Vidrovan, in Nikšić (Drecun 1951). The first fish stocking within the territory of Montenegro was done in 1900. The Durmitor mountain lakes were stocked with fish from the River Bukovica near Šavnik (Drecun 1952b). The first stocking of rainbow trout was made in 1951 in the upper stream of the River Zeta (Drecun 1952c). The first stock artificial lakes (Lakes Krupac and Liverovići) from the central part of Montenegro, supplying them with rainbow trout and brook trout, was done in 1957, while Lake Zabojsko was stocked with trout fingerling in 1958 (Drecun 1960; Drecun 1958–1959). The first and the most important trout fish farm in Montenegro is the "Morača" fish farm near Podgorica, and its characteristics, capacity, and facilities have been described in detail (Drecun 1961).

Montenegro has great natural potential for a range of aquaculture: the farming of trout species in basins (both stocking fish and fish for market) in the continental part of the country; the farming of carp species in basins—in the sub-Mediterranean and Mediterranean areas; and cage systems type fish farms—in reservoirs or in brackish waters for species that tolerate conditions of variable water salinity. In order to improve and develop the profitability of freshwater aquaculture, it is necessary to take advantage of the favourable environmental conditions and improve the work of the existing ones, as well as to establish new fish farms, taking care to preserve the existing quality of the water. In addition, in order to increase production, it is necessary to modernize the existing farms, as well as to strengthen their competitiveness and efficiency, while respecting the highest standards in environmental protection and human health. Nowadays, freshwater aquaculture is based on the production of cold water fish species, with rainbow trout (*Oncorhynchus mykiss*) being by far the most widely produced. According to the official data of MONSTAT, in 2022 there were 19 fish farms with a total area of 2.1 ha as well as 5 cage systems. In this sector in 2022, 100 workers were employed (FAO 2022). Fish farms which farm trout species are separate in three categories: large farms are those with an area covering more than 3000 m^2, medium from 1000 to 3000 m^2, and small fish farms up to 1000 m^2. In Montenegro there are two farms that are categorized as large, and 5 belonging to the category of medium farms, while the remaining 18 are in the category of small fish farms (Marković and Pešić 2021). The quantity of farmed trout in fish farms for the period 2014–2021 is presented in Fig. 11. Only one carp farm has ever been established in Montenegro, in 1974. It produced 100 tonnes of farmed carp per year, but unfortunately, 12 years later in 1986, it was closed due to low market demand. Nowadays, carp farming in Montenegro does not exist (Marković and Pešić 2021).

There are several reservoirs in Montenegro: "Krupac", "Slano", "Liverovići", "Grahovo", "Piva", and "Otilovići". These reservoirs belong to the type of salmonid

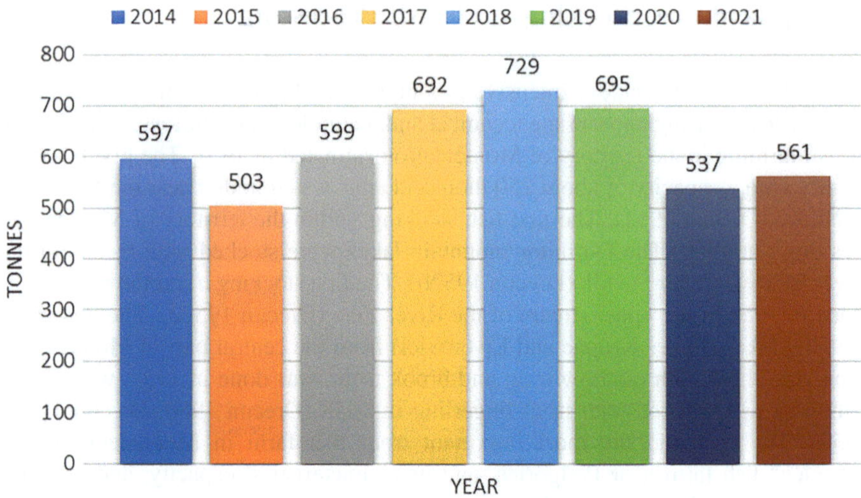

Fig. 11 Quantity of farmed trout in trout farms in Montenegro, 2014–2021 (MONSTAT)

waters, but they do not provide favourable conditions for the natural spawning of trout. After construction, they were stocked with trout species, so they are also used for sport fishing. However, the greatest potential of these reservoirs is in the development of aquaculture, which, unfortunately, has not yet been fully exploited.

According to the Law on Freshwater Fishing (Anonymous 2018) and Aquaculture, the rivers, natural lakes, and reservoirs in Montenegro can be stocked only with the autochthonous fish species and lineages. The "Breznica" fish farm, operated by the "Lipljen" Sport Fishing Club from Pljevlja, is the only farm which produces brown trout of the right phylogenetic lineage for stocking the waterbodies in the Black Sea drainage area of Montenegro. Technically, it is hatchery with an annual production capacity of up to 300,000 fingerlings and with small basins for the so-called parental fish from the wild.

Historically, along with the improvement in both tools and methods of fishing, the catch of fish has been constantly increasing. Over time, the need to preserve fish as a food source arose. The other reason for fish processing is that some of the fish species are present in the catch in huge amounts only during certain restricted and relatively short periods (e.g. bleak). Considering that fact, the methods of preserving bleak (salting and drying) are very old and practised over centuries up to the present day. The methods of preserving bleak first included its salting and drying in simple houses that were located near the fishing areas. There were over 150 such simple houses in just three fishing areas; for example, at Ploča there were 59, at Bazagur there were more than 40, and near Prevlaka there were 53 (Rovinski 1897). One of the most primitive ways of preserving bleak was cold-drying with smoke, which was then sold under the name "saraga". Cleaned and salted fish were placed on willow branches and dried over the smoke of the fire. In addition to that, part of the fish was also dried in the sun. According to historical data, in 1302 bleak was sent to Apulia and in 1312 it was sold dry in Drač (Durres; Albania) (Franetović 1960). In addition, records show that bleak from Lake Skadar was exported and used in the Studenica Monastery (in Serbia) during the holidays (Jiriček 1923). Apart from this method of fish processing and its permanent use in the area of Lake Skadar, bleak was also used for other purposes. For example, artificial pearls were made from the scales of this fish species. Bartolomeo Biasoletto first gave information about the use of bleak scales for the making of artificial pearls in 1841 (Franetović 1960). The first attempts to obtain silver from the scales of bleak at Rijeka Crnojevića were made in 1908.

The canning factory "Ribarstvo" for the processing of fish was established in 1927 at Rijeka Crnojevića. The annual capacity of this factory was planned for the processing of 950 tonnes of fresh fish, of which 700 tonnes was intended for processing into canned fish, 120 tonnes as cold smoked fish, and 130 tonnes for the processing of fish flour. In 1958, 329 tonnes of canned fish and 226 tonnes of other processed products were produced. Carp, bleak, eel, and twaite shad were mainly used for canning. The factory also had a plant for the production of fish flour and oil as well as fish scales. The discarded fish was mainly used as raw material for fish flour, while bleak was used for scale production. Fish flour was widely used in the nutrition of domestic animals, and fish oil as an industrial material (Drecun 1962). In 1928, 20,000 kg of fish scales were produced in the factory (Jovićević 1928). From this data, we can easily calculate the amount of bleak harvested in 1928. Knowing that bleak scales hardly represent 3% of bleak total mass we can calculate that in 1928 it was harvested at least 666 tonnes. Later, the production in the factory was extended to the processing of sea fish, sardines, and mackerel. Up to 360 people were employed in production. At the time when the factory was privatized in 1995, 176 people were employed and they produced 20,000 cans of sardines per day, while with the new equipment, about 60 workers were able to produce around 40,000 cans per one day. The factory ran into financial problems and is currently not operational. According to the data provided by the Administration for Food Safety, Veterinary and Phytosanitary Affairs, two processing companies operate in Montenegro: "Zeta Fish" and "MM Ribarstvo". "MM Ribarstvo" is located in Cetinje and is registered as a limited liability company. The company is registered for processing and preserving fish, crustaceans, and molluscs with a capacity of up to 50,000 cans per year. The plant has recently started its operations and currently produces only one product—cans of hot smoked carp, but the diversification of their production is planned, to include the processing and preservation of other species of fish from Lake Skadar. "MM Ribarstvo" is a family business, where several members of the household work, with the occasional hiring of workers for fish cleaning and dressing. All the raw materials are supplied from Lake Skadar, and imports are not planned because the owners want to keep the traditional and typical product of Montenegrin origin. The company "Zeta" in Podgorica is engaged in smoking and drying fish. The company processes freshwater fish from the domestic market and sea fish from imports. These include carp, Californian trout, bleak, tuna, mackerel, and salmon. The daily production level is 300–500 kg of fish. 8 workers are permanently employed by the company. The products are obtained by a clean ecological process, without using preservatives. The company also has a HACCP (Hazard Analysis Critical Control Point) certificate (Brdar 2015).

5.1 Management

The governance of the fishery sector is under the jurisdiction of the Ministry of Agriculture, Forestry and Water Management, and specifically the Directorate for Fisheries. The Ministry and the Directorate for Fisheries are in charge of the drafting and implementation of fisheries policy as well as the drafting and proposing of regulations and the preparation of the annual budget. The Directorate for Fisheries consists of three administrative units: the Department for the Management of Resources and the Fishing Fleet, the Department for Monitoring, and the Department for Structural Measures, State Support, EU funds and Fisheries and Aquaculture Market Management. Within the Department of Monitoring is the Office for the Information Systems in Fisheries, which provides technical assistance through the establishment and maintaining the necessary databases. Two key tasks of this Office include the maintenance of the Montenegrin Register of Fishing Vessels and the maintenance of the Fishery Information System.

It has been estimated that, until the recent introduction of the closure to the eel fishing season, no less than 50 tonnes of eel were caught (legally and illegally) on Lake Skadar every year, including a third of the yellow eel catch (Simon et al. 2020). The basic objective of the fisheries statistics is to collect data on catches of marine and freshwater fish. In addition to this, MONSTAT has the task to collect data on the number of staff employed, numbers of fishing vessels, types and numbers of fishing equipment, the production and catch of fish, the production of spawn and fish fry, and feed consumption at farming sites. The data on freshwater and marine fisheries are collected from the direct reports of fishermen, while from 2017, the data collection method has been replaced and nowadays is based on an administrative source, in that the data are obtained from the Ministry of Agriculture and Water Management. The information on marine and freshwater fisheries collected through reporting method is taken from the accounting records of the companies engaged in fishing as a main or secondary activity. So far, the method of assessment has been applied due to the absence of an alternative data source. Pursuant to the Law on Marine Fisheries and Mariculture (Anonymous 2009, 2011, 2015), the Ministry of Agriculture, Forestry, and Water Management collects data on the catch and landing of fish. Such data collected by the Ministry of Agriculture, Forestry and Water Management are stored in electronic catch logbooks that are directly filled by fishermen.

Environmental protection is, according to law which governs this area, under the jurisdiction of the Environmental Protection Agency of Montenegro, which is responsible for administrative and professional tasks in this area.

CETI (the Centre for Ecotoxicological Research of Montenegro) was founded by the Decision of the Government of Montenegro on 20th December 1996. It started its operations in 1998, when it got all the necessary certificates and authorizations. In 2012, on the basis of the Law on Business Organizations of the Government of Montenegro, the Public Institution "Centre for Ecotoxicological Researches of Montenegro" was transformed into a limited liability company owned by the Government of Montenegro. CETI's main activities are technical examinations and analyses, which include, among others, ecotoxicological research on the quality of all segments of the environment; analysis of the safety of food and health safety of general use items, dosimetry, quality assurance, and quality control; toxicological analyses of biological material; and the production of toxicology studies, analyses, and programmes for the needs of the public authorities, scientific and research institutions, companies, and other entities.

The Diagnostic Veterinary Laboratory, a public institution, was established by the Decision of the Government of the Republic of Montenegro on 7th October 2004 and officially started work on 1st March 2005. It is the only veterinary laboratory in Montenegro. The laboratory carries out specialist diagnostic and research veterinary activities in Montenegro for the purpose of the protection and improvement of animal health, the protection of animals against infectious diseases, the detection and diagnosis of animal diseases, pathomorphological tests, bacteriological and mycological checks on materials, microbiological checks on food of animal origin, parasitological checks (among others, on animals and fish), serological testing, molecular methods, virology analyses (infectious pancreatic necrosis in trout, among other viral diseases), and various other testing and research activities.

The University of Montenegro and its Institute for Marine Biology as well as the Department of Biology at the Faculty of Science and Mathematics are the main scientific advisory institution for fisheries in Montenegro. The Institute for Marine Biology is oriented towards marine fisheries, while the Department of Biology at the Faculty of Science and Mathematics takes an advisory role in relation to freshwater fisheries. The advisory role is based on their scientific research and experience. The Ministry of Agriculture, Forestry and Water Management officially asks for the opinion in relation to any regulation changes (by-laws) or requests from fishermen and/or their organization, while representatives of the faculty are members of all the working group that are related to freshwater fisheries and aquaculture. The Faculty of Science and Mathematics is responsible for collecting biological data related to freshwater fish.

A number of laws, by-laws, and regulations directly or indirectly regulate the use and conservation of freshwater fish resources in Montenegro. The main law is as follows:

- The Law on Freshwater Fisheries and Aquaculture (Anonymous 2018); The Law on Freshwater Fisheries governs the use, protection, conservation, and catch of fish and other living aquatic organisms in inland waters, the requirements for aquaculture, and other issues of relevance for freshwater fisheries.

Other legislation governing areas related to fisheries include:

- The Law on Food Safety (Anonymous 2015).

This law determines the requirements for food and feed safety (including fish and other aquatic organisms) and regulates the marketing, labelling, presentation and promotion of food and feed (including fish and other aquatic organisms), materials, as well as packaging and items that come into contact with food or feed, food additives, fast frozen food, genetically modified food and feed, water needs, traceability, the recall of unsafe food and feed (including fish and other aquatic organisms), the authorization and licensing of facilities, and so on.

- The Law on Animal Welfare Protection (Anonymous 2008).

This law deals with the issues of the protection of farmed fish and other aquatic organisms during the period of production, the construction and maintenance of farms and farming sites, nutrition, watering, the movement of animals, procedures in cases of disease or injury, protection during transportation, and conditions for keeping animal.

- The Law on Waters (2018). This law regulates the legal status and manner of the integrated management of water, waterways and coastal land and water facilities, the conditions and manner of conducting water-related activities, and other issues of importance for waters and water estate management.
- The Veterinary Law (Anonymous 2016). This law governs animal diseases; the prevention of occurrence, control, and eradication of infectious animal diseases; preventive veterinary measures; the mandatory scope of animal health protection; animal welfare protection; the financing of animal health protection; the organization of legal persons for carrying out veterinary activities; and ensuring the exercise of activities of public interest and other matters of relevance for the carrying out of veterinary activities.
- The Law on Alien and Invasive Alien Species of Plants, Animals, and Fungi (Anonymous 2019. This law governs the manner of the prevention of the introduction and spread of alien and invasive alien species of plants (including aquatic plants), animals (including fish and other aquatic animals), and fungi, to mitigate and minimize the harmful effects on biodiversity, ecosystem services, and/or human health.
- The Law on the Environment (Anonymous 2016). This law governs the principles of environmental protection and sustainable development, instruments, and environmental protection measures and other issues of relevance to the environment.

5.2 *By-laws*

The following by-laws and regulations have been adopted:

- The Rulebook on the Form, Content, and Manner of issuing Commercial Fishing Permits ("Official Gazette of Montenegro", No. 8/19)

- The Rulebook on the Form, Content, and Manner of issuing permits for Sports and Recreational Fishing and the Quantity of Fish that may be Caught ("Official Gazette of Montenegro", No. 11/19)
- The Rulebook on the Content of the Annual Plan for the Protection, Preservation, and Development of Fish and other Aquatic Organisms ("Official Gazette of Montenegro", No. 8/19)
- The Rulebook on the Content of the Fishery Master Plan ("Official Gazette of Montenegro", No. 1/19)
- The Rulebook on the List of Alien and Locally Absent Species of Fish and other Aquatic Organisms in Aquaculture ("Official Gazette of Montenegro", No. 61/18)
- The Rulebook on the Form of the Permit and the Content of the Aquaculture Permit Register ("Official Gazette of Montenegro", No. 61/18)
- The Rulebook on the Values of Compensation for the Damage Caused to Fish and other Aquatic Organisms ("Official Gazette of Montenegro", No. 76/18)
- The Rulebook on the Designation of Fishing Grounds ("Official Gazette of Montenegro", No. 78/18, 36/20)
- The Rulebook on the Content of the Protocol on Controls carried out, Uniform Appearance, Forms, Content, and Manner of Issuing Official Identification Documents for Fishing Grounds Rangers, and the Manner of Keeping Records on Official Identification Documents of Fishing Grounds Rangers ("Official Gazette of Montenegro", No. 85/18)
- The Rulebook on the Type of Fishing Equipment and Gear, the Technical Characteristics, the Manner of Marking and the Quantity of Fishing Equipment and Gear used in Commercial Fishing, and the Number of Permits that may be Issued for a Specific Fishing Area ("Official Gazette of Montenegro", No. 1/19, 37/20)
- The Rulebook on the Type, Technical Characteristics, and Quantity of Fishing Equipment and Gear that may be used in Sports and Recreational Fishing ("Official Gazette of Montenegro", No. 1/19)
- The Order on Closed Fishing Areas, Closed Seasons, and the Minimum Sizes of Fish and other Aquatic Organisms below which Catch is prohibited ("Official Gazette of Montenegro", No. 37/20, 9/21)
- The Decision on the Amount of the Fee for Issuing Permits for Commercial Fishing and Aquaculture ("Official Gazette of Montenegro", No. 70/19)
- The Decision on the Designation of the Fishing Area in the Fishery Waters of the Lake Skadar National Park ("Official Gazette of Montenegro", No. 10/19)
- The Decision on the Designation of Protected Fishing Grounds ("Official Gazette of Montenegro", No. 12/19)

5.3 The Management of Lake Skadar and Fishery Legislation in Montenegro

In 1983, the Montenegrin part of Lake Skadar and the surrounding areas were declared a National Park and, since then, it has been managed by the Public Enterprise for National Parks (PENP) of Montenegro. The mandate, structure, and procedures of the National Parks management body are laid out in the Law on National Parks (Official Gazette 56/09), which is a *Lex specialis*.

- The Order on Fishing Prohibitions, Restrictions and Measures for the Protection of Fish Stocks (Official Gazette 21/12) sets restrictions in terms of closing periods (fishing bans), fishing gear and methods, minimum fish dimensions, and catch limits for different fish species in the lake and its associated waters. It also addresses measures for the control of fish diseases.
- The Rules of Conduct within the Biogradska Gora, Lovćen, Durmitor, and Lake Skadar National Parks (Official Gazette 32/05) endorsed by the Management Board of the PENP identify the mode of use of National Park assets, which also concerns the taking of fish.
- The Decision on the Modes and Conditions for Commercial Fishing on Lake Skadar issued by the Management Board of the PENP defines the conditions for fishing (the purchase of licenses, the allowed fishing gear, and the permitted fishing areas) in the lake.
- The Decision on the Fees and Modes of Payment for Using the Goods of the National Parks, Carrying out Activities, and Offering Services (Official Gazette No. 31/02) includes information about the fees related to fishing in National Parks. The Decision was prepared by the Management Board of the PENP.
- The Decree to Prohibit the Use of Vessels with an Engine Power more than 4.5 HP by Physical Persons on Lake Skadar/Shkodra (Official Gazette of the RM, No. 9/86).

6 Sustainability and Conservation

The freshwater ichthyofauna of Montenegro represents an important resource both in terms of biodiversity, considering the uniqueness of the fish fauna with numerous freshwater endemic species, and in terms of fisheries. The existing resources are renewable, but they are highly degraded, so they cannot be used currently to their full capacity. Therefore, it is necessary to undertake systematic activities towards the recovery of freshwater fishing resources in order to enable their sustainable use, and thus their preservation. Currently, the most significant problem in this area is large-scale poaching by all possible illegal means (electro aggregates for catching fish, underwater harpoons, or illegal nets), as well as fishing and poaching during the prescribed fishing bans, which makes it impossible for fish to spawn and, therefore,

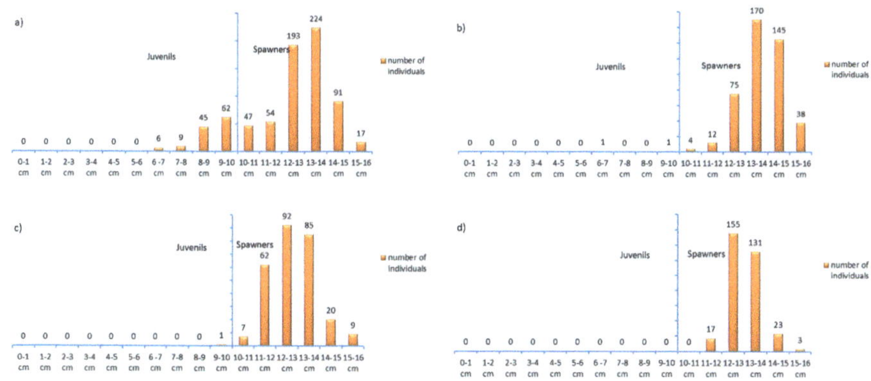

Fig. 12 Length frequency of bleak from the sample: (**a**) nets from the 0–3 m depth stratum; (**b**) nets from the 3–6 m depth stratum; (**c**) kalimera nets from the Raduš concessionary area, November 2013; (**d**) kalimera net from the Raduš concessionary area, January 2014 (Mrdak et al. 2017a)

for any kind of recovery of these resources. On most watercourses, we are very close to a situation where the only possible and effective measure will be the imposition of a total ban for two or three years in order for the fish populations to recover and for the establishment of sustainable management of this resource to be considered. As Lake Skadar is the only freshwater body where commercial fishing takes place, in order to ensure the sustainability of fishing on this Lake it is necessary to determine the amount of fish in it, the annual yield by species, and the maximum sustainable catch and, in accordance with this, to precisely determine the maximum number of fishermen who can fish commercially on this Lake. All of this has not been done so far. One additional problem is the quantity of allochthons fish species that were introduced in the last three decades of the twentieth century and whose population number has risen, so that they have become competitors for food and space, further threatening the autochthonous species.

As we have mentioned previously, the only regular and standardized monitoring of fish fauna related to fisheries was done within the "Conservation and Sustainable Use of Biodiversity on Lake Prespa, Lake Ohrid and Lake Skadar (CSBL)"—project which was implemented by the GIZ (Deutsche Gesellschaft für Internationale Zusammenarbeit). For the purpose of that research, standardized MMG nets were in use and research covered all three depth zones of Lake Skadar; 0–3 m, 3–6 m, and more than 6 m (in 2014 and 2014) as well as the Raduš Bay concessionary area. In 2013, the data showed a decrease or absence of the juveniles of bleak in both strata and the Raduš Bay concessionary area during the winter shoaling period (Mrdak et al. 2017a) (Fig. 12).

This suggests that in previous years overfishing events took place and that bleak fishing will be faced with a decrease in upcoming years (Mrdak et al. 2017a). In the subsequent year, the situation with bleak changed dramatically (Fig. 13).

In 2014, in both strata, juveniles were almost completely absent, while in 2015, juveniles finally appear in the samples but the abundance of adults obviously

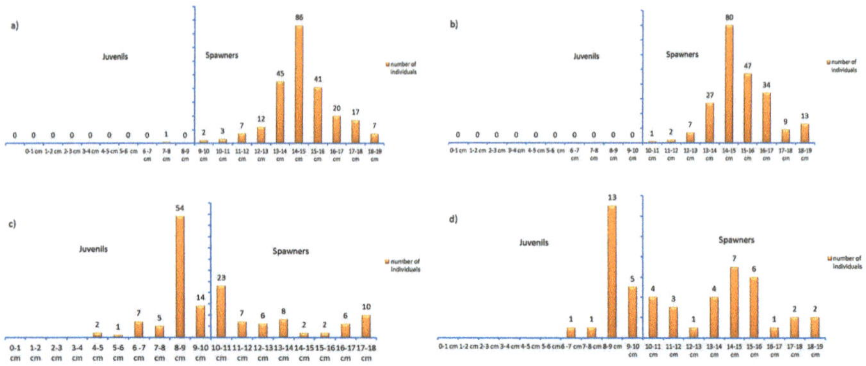

Fig. 13 Length frequency of bleak from the sample: (**a**) nets from the 0–3 m depth stratum for 2014; (**b**) nets from the 3–6 m depth stratum for 2014; (**c**) nets from the 0–3 m depth stratum for 2015; (**d**) nets from the 3–6 m depth stratum for 2014 (Mrdak et al. 2017a)

dropped (Fig. 13). The decrease in relative biomass (WPUE, weight per unit of effort) and relative abundance (NPUE, number of individuals per unit of effort) was more than obvious in the subsequent years, 2014 and 2015, as is shown in Fig. 14.

The most dramatic reduction of both relative biomass and relative abundance occurred in the 0–3 m stratum where an approximately tenfold decrease in 2015 was detected, comparing to the reference year of 2013 (Fig. 14). For the 3–6 m stratum, the decrease was also significant, being almost sevenfold for NPUE and almost ninefold for the WPUE, comparing 2015 to the reference year of 2013 (Fig. 14). All of this was also noticed by the professional fishermen of Lake Skadar and the absence of bleak from fish markets lasted until the early winter of 2017, while the first significant catch of bleak occurred in the late autumn of 2018 and again in the early winter of 2019 (personal observation).

Bleak from Lake Skadar is short-living species which becomes fertile by the end of the first year of their life cycle (Ivanović 1964; Ružić 2018). It lives up to 5 years and feeds on zooplankton, mainly on Cladocera and Copepods but also on Chironomid larva and pupa, as well as on other insect larva and adults that drop or land on the water surface (Ivanović 1965). It is a highly mobile species which, according to Ivanović (1964), undertakes several migrations in Lake Skadar during one year cycle: a) a winter migration which bleak undertakes with the first cold and windy days of late autumn or early winter and shoal in wintering areas of lake bays with sublacustric springs or in river inlets; b) a spring migration which bleak undertakes in late March and early April, so that the individuals spread all over the lake open water and start feeding intensively; c) a spawning migration towards the spawning ground near the lake shores, where spawning occurs on coastal rocks, gravel, pebble, and sand; d) a summer migration that occurs after the spawn, where the bleak migrates towards the deep central areas of the lake being scattered and intensive feeding; and e) an autumn migration that occurs after the summer hot period, when the bleak migrates from the deep central lake areas towards coastal area and its wintering habitats.

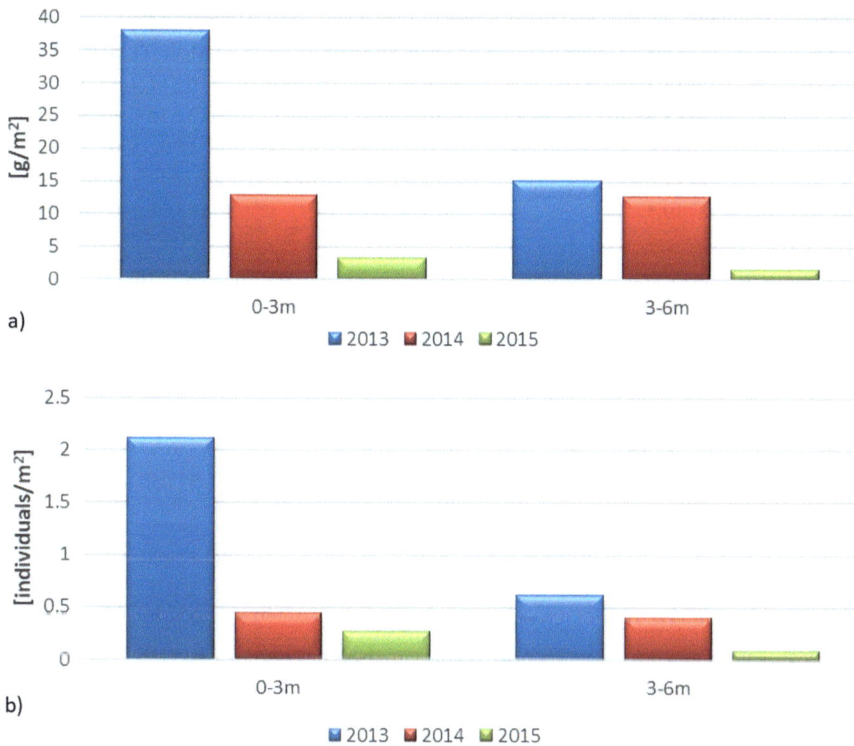

Fig. 14 Drop in bleak relative biomass (**a**) and relative abundance (**b**) per strata in three consecutive years (2013, 2014, and 2015) (Ivanović 1965)

Regarding the fact that bleak has portion spawning and that for it to spawn it lays its eggs in shallow coastal zone up to 30 cm depth, but most commonly in the contact zone between water and land (Ivanović 1964), it is clear that fertilized eggs are at risk of emerging from the water due to water level decreasing. One of the adaptations to this risk is the bleak portion spawn which means that one female bleak releases eggs several times during one spawning season, while the second adaptation is short incubation time which lasts only 5 days (Ivanović 1964). Karaman (1924) stated that bleak in Lake Skadar spawns in June, while Nedeljković (1959) wrote that bleak normally spawns in the first half of spring, up until May, while Ivanović (1964) concluded that the time interval of bleak spawn in Lake Skadar is prolonged and starts from the second half of April and ends by the second half of July. The temperature of the water is the main trigger for spawning, and the optimal temperature for spawning is 20–23 °C, while sudden cooling stops the bleak spawn (Ivanović 1964). An additional threat to fertilized eggs are strong winds which form waves on the lake shore, making eggs periodically emerge so that they decay. All this implies that the management of bleak population in Lake Skadar should be done with spatial care. There is a need for the monitoring not only of the bleak population but also bleak spawn.

Table 2 Maximum, minimal, and mean values of total length (TL) (Ivanović 1965)

Cohort	Max TL [cm]	Min TL [cm]	Mean TL and SD [cm]
0+	11.1	6.5	9.514 ± 1.0024
1+	15.2	10.6	13.536 ± 0.8399
2+	16.4	13.9	15.02 ± 0.5137
3+	17.5	16.0	16.644 ± 0.5434
4+	19.5	18.5	19.067 ± 0.5131

Table 3 Maximum, minimal, and mean values of total weight (TW) (Ivanović 1965)

Cohort	Max TW [g]	Min TW [g]	Mean TW and SD [g]
0+	9.77	2.41	6.079 ± 1.9070
1+	31.62	7.69	20.374 ± 4.2901
2+	37.99	20.40	28.208 ± 3.7986
3+	53.35	30.74	39.97 ± 6.6284
4+	56.94	49.93	53.650 ± 3.5247

Table 4 Growth of bleak from Lake Skadar in percentages of TL (total length) and TW (total weight)

	1+	2+	3+	4+
Yearly growth in TL (%)	42.27	10.96	10.81	14.56
Yearly growth in TW (%)	235.2	38.45	41.70	34.23

Ružić (2018) gives several population traits of Lake Skadar bleak. They pointed out that in winter 2017/2018 individuals from the 1+ cohort were dominant, making up 59.06%, suggesting population recovery and a rising bleak abundance in the lake. They found out that 0+ cohort represented only 16.13% in the total sample, explaining that with the low efficiency of spawn in 2017 as well as with the different habitat presential of juveniles comparing with the adults. Of total sample (Ružić 2018) reports 50.4% were females and 43.4% were males, while only 6.2% of all individuals were immature.

In terms of first maturity, it is experimentally demonstrated that it happens in the first year of the lifespan of the bleak, and the differentiated ovary was noticed for individuals with a minimal TL of 8.9 cm and a TW of 4.37 g. Differenced testicles were noticed for individuals with a minimal TL of 8.4 cm and a TW of 4,21 g (Ružić 2018).

In the following tables (Tables 2 and 3), data about the maximum, minimal, and mean values of total length (TL) and total weight (TW) for all cohorts are given.

Ružić (2018) showed that bleak from Lake Skadar, which was expected, has the fastest growth in the first year, while they also showed that bleak slows down in the second year and has a similar growth of TL in the second and the third year of their lifespan, while in the fourth year it is slightly larger (Table 3). As for growth in terms of TW, we have a similar pattern only with higher values in the first year (Table 4).

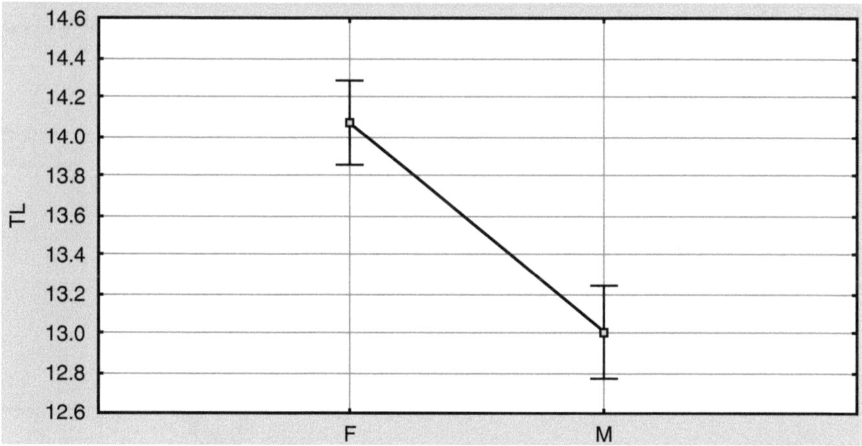

Fig. 15 ANOVA analysis showing the significant differences in TL (total length) among bleak females (F) and males (M) from Lake Skadar, F (1374) = 42.710, $p = 0.0000$; 95% confidence intervals

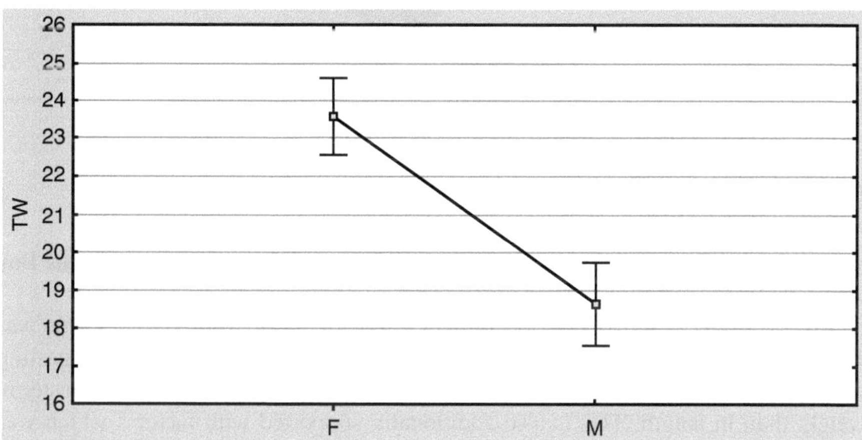

Fig. 16 ANOVA analysis showing the significant differences in TW (total weight) among bleak females (F) and males (M) from Lake Skadar, F (1374) = 42,374, $p = 0.0000$; 95% confidence intervals

Ružić (2018) also showed that there is also a significant difference in terms of TL and TW of females and males, suggesting a larger size of females within the bleak population from Lake Skadar (Figs. 15 and 16).

As is true of all fish species with short lifespans, Lake Skadar bleak is characterized by high mortality among all cohorts, with the highest value at the end of the second year (89.77%) and at the end of the fourth year (Table 5) (Ružić 2018). Ivanović (1968) also found similar values, while Coble and Knežević (1981)

Table 5 Mortality (q_x) and survival (p_x) rate of bleak from Lake Skadar [88] * high mortality value is due to low number of 0+ individuals in sample

Cohort (x)	Mortally rate (q_x)	Survival rate (p_x)
0+	2.552*	–*
1+	0.630	0.369
2+	0.897	0.102
3+	0.666	0.333
4+	1	0

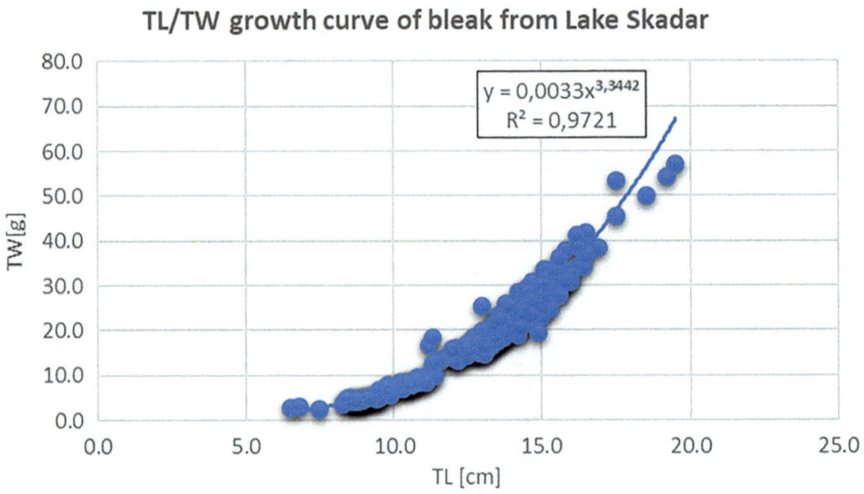

Fig. 17 TL/TW relation and growth curve for bleak from Lake Skadar. Ružić (2018)

reported mortality within the range of 83%–94%, with the exception of Raduš Bay where they reported overall mortality of 68%.

Ružić (2018) reported that the growth curve of bleak from Lake Skadar was TW = 0.0033 TL = 3.344 (Fig. 17). The allometry factor b was 3.344 suggesting positive allometry growth or, in other words, bleak from Lake Skadar Lake faster in weight than in length. This is also additionally supported with factor a which was low (0.0033) but again indicating that bleak from Lake Skadar grows slightly faster in weight than in length.

Tsagarakis et al. (2015) in their comparison of the TW/TL relationship and growth curve of three species from the genus Alburnus (*A. chalcoides*, *A. mossulensis*, and *A. tarichi*) from Turkey give the following values: *A. chalcoides* a = 0.0079 and b = 3.051; *A. mossulensis* a = 0.0061 and b = 3.060; and *A. tarichi* a = 0.0220 and b = 2.653. For the species *A. chalcoides* and *A. mossulensis*, the values are similar to those of the bleak from Lake Skadar, while for *A. tarichi* there was negative allometric growth. It is also noticeable that the value of allometry factor b is higher for bleak from Lake Skadar than in those species that were investigated by Tsagarakis et al. (2015). It seems that this could be a consequence of the excellent environmental and/or trophic conditions for bleak in Lake Skadar, which means that this species invests less energy in feeding and other activities in the Lake. Another explanation could be that this is a kind of biological characteristic of the bleak population of Lake Skadar.

By total catch, the second most important species which inhabits Lake Skadar is the common carp, *Cyprinus carpio*. The total annual catch of common carp on Lake Skadar reaches up to 300 tonnes (Stein et al. 1981). Since the market price of the common carp is more than three times higher than bleak, this species is the most important for the fisherman, since it brings much more income. Its fishing season does not depend on the wintering shoaling period (as is the case for bleak), and except during the fishing ban period, the season is year-round. Bearing this in mind, it is not surprising that the common carp is the number one target species for all professional and recreational fishermen from Lake Skadar.

Although it is the most important fish species, in terms of fishing, the common carp from Lake Skadar has in fact not been widely investigated over recent years. Having said that, data about population composition (age and TL), growth and mortality (Drecun and Ristić 1972; Janković 1971; Coble and Knežević 1981), feeding preferences (Janković 1983; Stein et al. 1975), and carp reproduction (Janković 1971) have emerged. In their paper, Drecun and Ristić (1972) gave data about the sample composition, while Coble and Knežević (1981) estimated the total annual carp mortality from data published by them at 44%, with the instantaneous mortality rate for the 4+ to 10+ age group being 58.2% (Coble and Knežević 1981). They also used data published by Janković (1975) and calculated the annual mortality rate for carp from Lake Skadar at 41%, while based on the same data, the instantaneous mortality rate for the 4+ to 12+ age cohorts was calculated at 52.6%. Based on data taken from their own sample, Coble and Knežević (1981) reported the total annual mortality of carp from Lake Skadar at 52%, while based on the same sample set, the instantaneous mortality rate for the 2+ to 9+ age cohorts was 72.7%. Based on this data, it seems that half of the adult carp from Lake Skadar died every year. This data corresponds with some similar research done by Kolaneci et al. (2010a), b) since they reported 71% annual mortality for the common carp calculated on the basis of a sample from the Albanian part of Lake Skadar. Karataş et al. (2007) reported total annual mortality of 64% for the common carp population from Lake Almus Dam (Tokat—Turkey). Based on data published by Coble and Knežević (1981), we calculated and construct the Von Bertalanffy growth curve (Fig. 18).

The calculated $\boldsymbol{K}$ value was 0.2489, while growth performance index was $\boldsymbol{\Phi}$" = 3.10 indicating excellent growth performance, since the higher growth performance index of common carp is mostly related to some eutrophic lakes from temperate region (Cochrane 1985; Sharma 1987; Li et al. 1990; Bajer et al. 2009; Oyugi et al. 2011; Vilizzi et al. 2014). Kolaneci et al. (2010) reported a significantly higher growth performance index for the common carp from Lake Skadar and then we calculated based on the dataset from Coble and Knežević (1981). Regarding food items of common carp from Lake Skadar, Janković (1983) reported differences in food preferences between juveniles and adults. The common carp fry mostly feeds on the zooplankton Rotatoria, Cladocera, and Copepoda during the spring and summer (Janković 1983). Juveniles of more than 6 months, in addition to zooplankton, feed on Chyronomidae, Ephemeroptera, and Trichoptera maggots with a

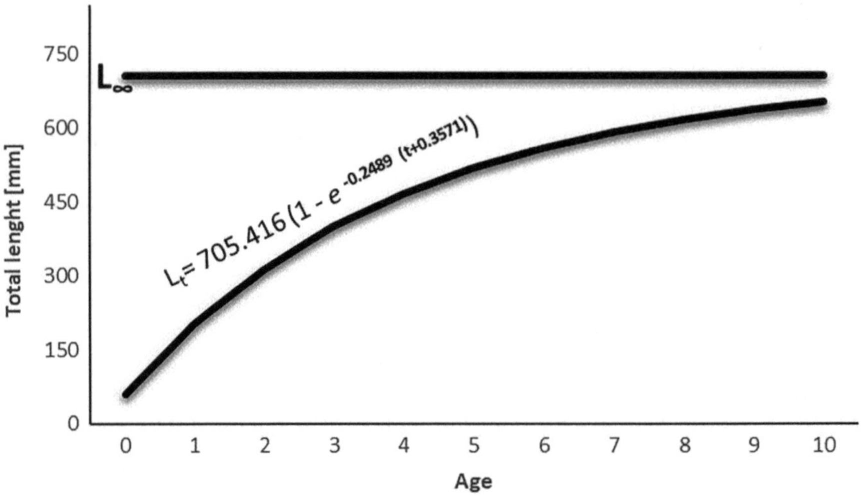

Fig. 18 Relation between age (years) and body length (cm) for common carp (*Cyprinus carpio*) from Lake Skadar, calculated using the Von Bertalanffy growth model (L∞—maximal expected length—length infinity, Lt—expected total length) based on data from Coble and Knežević (1981)

predominance of Chyronomidea. Similar food items were found in juveniles of a year old, while the older adult fish (two years and older) feed on various items, mostly on plants, mussel, and snails, but also on isopoda crustaceans as well as on insect maggots (Janković 1983). Stein et al. (1975) reported that more than 30% of all carp from Lake Skadar in their stomach contained benthic molluscs and within this group Valvata snail was the preferred prey with an electivity index five times that of any other species of mollusc. They suggested that the carp in Lake Skadar choose Valvata for following four reasons: (I) Valvata has a thin shell, (II) it has a turbinate physical shape, (III) a large proportion of the shell is occupied, and (IV) it has a high level of high nutrition value (Stein et al. 1975). The data of the length–weight relationship (LWR) and the condition factor of the carp in Lake Skadar during the spawning period were given by Milošević and Marić (2012). The value of exponent "b" of the LWR in this study was generally equal to or less than 3 (2.72 to 3.05). The LWR in five groups (juveniles, mature males, mature females, males, and females after spawning) of carp showed negative allometric and isometric growth patterns (juveniles = 3.02, mature females = 3.05). Estimates of the average condition factor ranged from 1.12, as shown by the male after spawning, to 1.35 as shown by the mature female. Significant differences were found between juveniles and males and females after spawning and between males and females after spawning. There was no difference in the condition factor between mature males and females. The mature females gonadosomatic index (GSI) ranged from 8.1 to 24.5. The coefficient of correlation between K and GSI was not significant in mature females. No significant differences were found in the ratio between sexes from the hypothetical distribution of 1:1 ($p > 0.05$) (Milošević and Marić 2012).

In the last five decades, the number of allochthonous species in ecosystems worldwide has increased exponentially (Hulme 2009; Vilà et al. 2010). The invasion of alien species is nowadays considered to be one of the main threats to biodiversity and the second leading cause of animal extinctions (MEA 2005; Clavero and García-Berthou 2005). The introduction of freshwater fish has increased substantially during recent decades, mainly due to the negative socio-economic impact of globalization on native freshwater biodiversity (Cucherousset and Olden 2011; Gozlan et al. 2010). The negative impact of alien species on native species includes predation, trophic competition, behavioural interference, hybridization, the spread of novel parasites and diseases, alterations to food webs, and the modification of biochemical cycles (Cucherousset and Olden 2011; Arthington and Lloyd 1989; Mills et al. 2004; Keller and Brown 2008). Some recent surveys from Balkan countries have revealed that 15%–23% of their fish fauna is alien (Barbieri et al. 2015; Piria et al. 2016a; b; Simon et al. 2013). Invasive species have a negative impact on fishing conditions, effecting the autochthonous fish fauna. Their invasion could cause a decrease in the abundance of domestic species and consequently a decrease in their annual catch. Among all allochthons species which are detected within the freshwater ecosystems in Montenegro, two of them have reached high levels of abundance and could have or already have had a negative impact on fishery: perch (*Perca fluviatilis*) and gibel carp (*Carassius gibelio*).

Gibel carp was detected in Montenegro for the first time in 1975 in Lake Skadar, while in 1979 this species was detected in Lake Šasko, which is indirectly connected to Lake Skadar through the River Bojana (Vuković et al. 1975, Knežević 1979]. After that some of the first research on gibel carp biology was done by Marić (1998) in his PhD thesis, where he elaborated on the adaptation of gibel carp from Lake Skadar. Marić (1998) also gave the morphological and some of ideo-ecological characteristics such as their nutrition, growth, reproduction, fertility, maturing, and so on. He concluded that gibel carp has been adapted well in Lake Skadar and that it represents one of most abundant fish species in this Lake.

Marić (1998) showed that research on the food items of the introduced gibel carp in Lake Skadar has revealed its intensive and versatile nutrition during the entire year. The Lake Skadar gibel carp population belongs to the euryphagous or omnivorous group with a significant participation of detritus (30–90%) in its diet. There were some records of certain selectivity, while differences as regards their age have not been found (Marić 2000). The diet of Skadar's population differs significantly from the diet of populations from continental waterbodies and that there is a significant competition for food with Lake Skadar autochthonous cyprinids, especially with the common carp (Marić 2000).

The gibel carp from Lake Skadar becomes mature at the age of one year (to 25% of specimens) at a size exceeding 100 mm of TL, while at the age of two years, all the specimens were mature (Marić 2002). This is in line with the invasive species strategy, to accomplish acclimatization and to establish its populations in a new environment so they "invest" energy in early reproduction rather than in fast growing. Marić (2002) also stated that with the specimens which undergo sexual maturation in the first year, the ovogenesis terminates later (May–June) than with

Table 6 Mean (±SD) relative abundance (NPUE, individuals m − 2) and mean (±SD) biomass (WPUE, g m − 2) of perch and native dominant species (October–November of 2013, 2014, and 2015, 0–3 m and 3 m + depth strata) (Mrdak et al. 2017)

Depth	Unit	*Perca fluviatilis*	*Alburnus scoranza*	*Rutilus prespensis*	Other 14 species pooled
0–3 m	NPUE	0.712 ± 0.548	2.117 ± 1.236	1.172 ± 0.470	0.169 ± 0.098
	WPUE	32.02 ± 19.30	18.12 ± 17.90	21.68 ± 8.99	4.62 ± 1.76
3 m+	NPUE	0.454 ± 0.405	0.621 ± 0.762	0.576 ± 0.428	0.046 ± 0.052
	WPUE	18.27 ± 10.92	7.39 ± 6.13	17.99 ± 8.43	2.18 ± 1.92

older ones (March–April). The spawning of gibel carp in Lake Skadar starts in March, and is at its most intensive in the April–May period which overlaps with the spawn of other cyprinid species from the lake (Marić 2003). The population of this gibel carp from Lake Skadar is unisexual, composed only of females, and it has been experimentally proved that the following species participate in the initiation of egg cell division: *Cyprinus carpio*, *Leuciscus cephalus*, *Scardinius knezevici*, *Rutilus prespensis*, and *Pachychilon pictum*, while the embryonic development of fertile egg lasts around 100 degrees C/days (Marić 2002). This fact also showed the invasion strategy of this species and it explains establishment of the high abundance of gibel carp only several years after the first finding. Marić (2004) stated that gibel carp reached its maximum abundance in Lake Skadar some 10 years after its introduction. Nowadays, the abundance of gibel carp is at a much lower level than it was during the 1990s, and it seems that some of the intrapopulation and/or interpopulation abundance regulation mechanisms have regulated the numbers of the present population (Mrdak et al. 2017b).

The perch was first noticed in Lake Skadar in 1979 (Knežević and Marić 1979). Until recently, there had been no investigation on its dispersal in the lake, its growth, its stock development, or the effects on local autochthonous fish (Mrdak et al. 2017). Data based on three consecutive years of monitoring of Lake Skadar fish fauna (with 17 fish species from Lake Skadar within the sample) revealed that perch had the third-highest relative abundance (NPUE) values, and with regard to its relative biomass (WPUE), perch was by far the most dominant species, followed by *Alburnus scoranza* and *Rutilus prespensis* (see Table 6) (Mrdak et al. 2017).

Mrdak et al. (2017) found a lack of significant variation in WPUE, in contrast to NPUE and body size TL, suggesting that the perch population of Lake Skadar is relatively stable in terms of biomass (Table 7). A negative correlation was also revealed between the NPUE and TL values in the relatively constant biomass over the three years of the research (Mrdak et al. 2017). It was suggested that it could be that the perch in Lake Skadar has already reached its biomass maximum for the existing ecological conditions (e.g. the available food and space resources) given the lack of abundant and effective predators or any serious fishing pressure (Mrdak et al. 2017). They suggested that in such conditions, the dominant regulatory mechanism in the perch population of Lake Skadar could be the intraspecies competition for limited lake resources (Mrdak et al. 2017).

Table 7 Results of Fisher LSD test for WPUE (g/m^2), NPUE (individuals/m^2), and TL (cm) of perch (DF-90, *-significance level $p < 0.05$) (Mrdak et al. 2017)

WPUE					
	Infra-littoral 2013	Littoral 2013	Infra-littoral 2014	Littoral 2014	Infra-littoral 2015
Littoral 2013	0.009*				
Infra-littoral 2014	0.901	0.101			
Littoral 2014	0.001*	0.995	0.013*		
Infra-littoral 2015	0.554	0.434	0.984	0.129	
Littoral 2015	0.238	0.506	0.849	0.116	0.999
NPUE					
Littoral 2013	0.000*				
Infra-littoral 2014	0.000*	0.999			
Littoral 2014	0.000*	0.998	0.977		
Infra-littoral 2015	0.346	0.013*	0.058	0.001*	
Littoral 2015	0.344	0.000*	0.007*	0.000*	0.999
TL					
Littoral 2013	0.449				
Infra-littoral 2014	0.092	0.901			
Littoral 2014	0.000*	0.000*	0.004*		
Infra-littoral 2015	0.039*	0.000*	0.000*	0.000*	
Littoral 2015	0.004*	0.000*	0.000*	0.000*	0.99

The perch population of Lake Skadar mainly consists of younger fish, and according to Claessen et al. (2000), we could mark this population as "stunted", but Mrdak et al. (2017) stated the fact that perch from Lake Skadar showed high growth rates (Fig. 19) drawing us away from such a conclusion.

Mrdak et al. (2017) report on the high mortality occurring in the third year of life, a relatively small number of age classes, and no individuals older than 5+ in perch from Lake Skadar (Table 8). Heibo et al. (2005) reported on a situation where there is a lack of older perch individuals in southern populations, a reduction in juvenile and adult mortality rates at higher latitudes, and the increase of the perch lifespan with latitude. Heibo et al. (2005) also reported about a tendency towards an increase in growth performance Φ' and in growth parameter k values from north to south, which is in line with the high k and Φ' (2500) values Mrdak et al. (2017) found in the perch population in Lake Skadar.

In Mrdak et al. (2017), it was concluded that perch seems to have successfully colonized Lake Skadar and that their population traits (notably their abundance, biomass, and growth performance) indicate that this species is now naturalized and

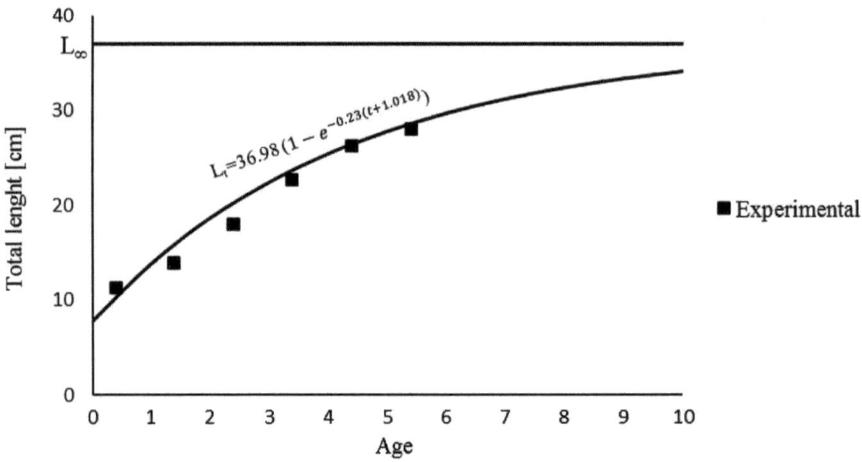

Fig. 19 Relation between age (years) and body length (cm) for the perch (*Perca fluviatilis*) of Lake Skadar calculated by the Von Bertalanffy growth model (Experimental—data obtained from samples, mean values of length of each cohort, Lt—expected total length) (Mrdak et al. 2017)

Table 8 Mortality (qx) and survival (px) rates of perch (*P. fluviatilis*) population in Lake Skadar (Mrdak et al. 2017)

Age	q_x	p_x
0	0.031	0.962
1	0.191	0.809
2	0.894	0.106
3	0.615	0.385
4	0.6	0.4
5	1	0

doing well there. The environmental conditions in Lake Skadar seem to favour a life history strategy of short lifespan, fast growth, and early maturity as the most beneficial adaptation to this Mediterranean environment (Mrdak et al. 2017).

The European eel is the third most important fish species in terms of fisheries in Lake Skadar. It is the most valuable catch on Lake Skadar, due to the high market price of this species, which varies from 12 to 18 EUR/kg, depending on the time of year and the size of the fish. It inhabits all the watercourses of the Adriatic drainage basin in Montenegro, but main fishing area for this species is Lake Skadar. This fish species is also part of the recreational fisherman catch in almost all the rivers that flow into Lake Skadar, but the main catch, more than 90%, occurs in the lake itself. It is poorly researched and only recently have some investigation been carried out on adults through the Conservation and Sustainable Use of Biodiversity on Lake Prespa, Lake Ohrid, and Lake Skadar (CSBL III project funded by Deutsche Gesellschaft für Internationale Zusammenarbeit (GIZ) GmbH).

Since the European eel has a well-known and complicated life history and because of the fact that its population in freshwater bodies naturally depends on the number of individuals (glass eels) which enter from the sea into the freshwater systems, we can state that the situation with the transfer of glass eels and the accessibility of adult habitats in Montenegro is still good. The River Bojana, which connects most of the Adriatic drainage waters in Montenegro, still has unobstructed estuary into the Adriatic Sea, and it is still freely accessible to young European eels to pass through. Even the fishing traps situated on the River Bojana downstream from city of Sakar do not represent a problem to young eels, since the mesh size allows them to migrate towards Lake Skadar via other Montenegrin rivers. Regarding the glass eel migration, there are data from the past, and in 1998, it was estimated that the glass eel migration season started in November, most intense in the period of February–April, and an end coming in April–May (Hegediš 2007). The migrating glass eels had a mean total body length of 57 mm and a mean body weight if 0.171 g (Hegediš 2007). The mean biometric index of these glass eels was 0.7051, while the predominant pigment stages of the migrating glass eels were VIA1, VIA2, and VIA3. They estimated that between 320 kg and 427 kg of glass eel migrated annually and entered the River Bojana from the Adriatic Sea (Hegediš 2007).

The body length of the European eels examined from Lake Skadar ranged from 35.0 cm to 47.5 cm for males and from 38 cm to 75.0 cm for females, while the percentage of silver eels in the sample from Lake Skadar was 95%, while 5% of the fish were yellow eels (Simon et al. 2020). The same authors report that the mean body length of the silver eels from this lake was 40.4 cm for males and 57.0 cm for females, which corresponds to a mean body weight of 107 g for males and 356 g for females. The age of the eels from Lake Skadar ranged from 5 to 11 years old for males and 5 to 15 years old for females, while the overall mean annual length increment during the years of life in Lake Skadar was higher in females (6.3 cm per year) as compared to males (4.9 cm per year) (Simon et al. 2020). The current growth of European eels in the eel management unit of the Drin river system is reflected by the von Bertalanffy growth formula with $L_t = 111.3(1 - e\text{-}^{0.989(t + 0.073)})$ for females and with $Lt = 60.3(1 - e^{-0.903(t + 0.107)})$ for males (Fig. 20) (Simon et al. 2020).

Simon et al. (2020) give the length–weight relationship calculated on the basis of the 172 eels of the EMU (Eel Management Unit) Drin River (Lakes Ohrid and Skadar) system that were sampled and investigated in the autumn and winter of 2019. The current length–weight relationship of eels in the eel management unit Drin River system is reflected by the formula W = 0.0017 L2.9946 for males and W = 0.0007 L3.2384 for females (Fig. 21) (Simon et al. 2020). Research on the relationship between the Fulton condition factor and the total length indicates that European eels increased in length, proving that Lake Skadar, the River Bojana, and the River Sutorina represent good habitats for eels (Piria et al. 2016a; b). This statement was confirmed by the isometric growth (b = 2.956) of the same specimens obtained in the analysis of their length—weight relationship (Milošević and Mrdak 2016).

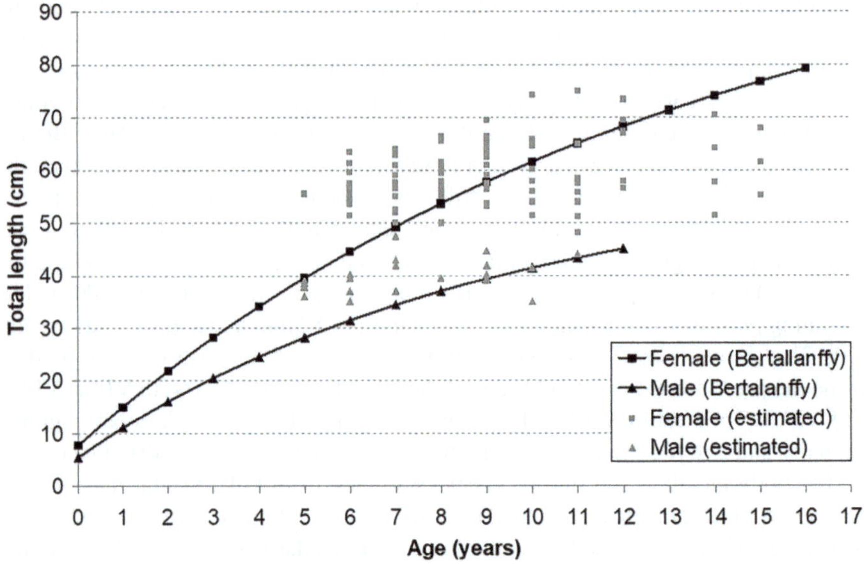

Fig. 20 Estimated age and von Bertalanffy growth curves of (■) female (n = 89) and (▲) male (n = 21) silver European eels from the River Drin catchment area (Simon et al. 2020)

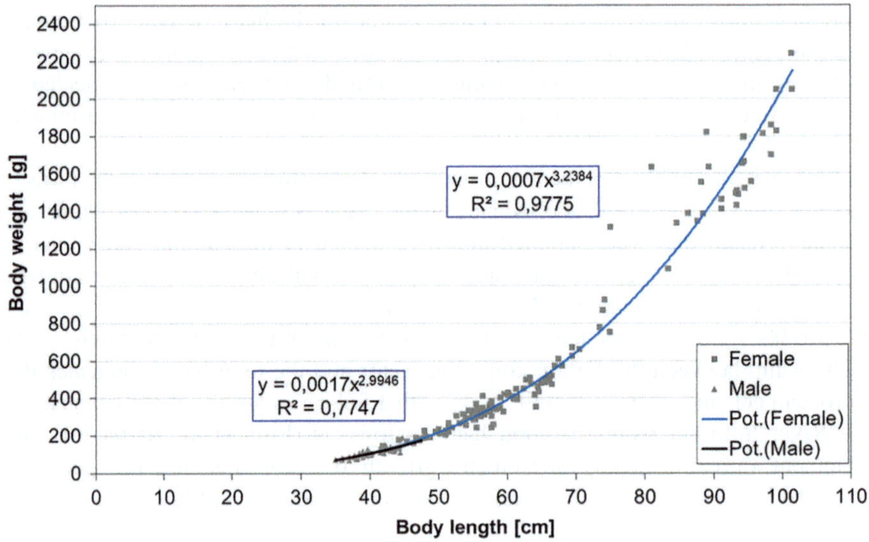

Fig. 21 Correlation between the length and weight of (■) female (n = 143) and (▲) male (n = 29) eels from the EMU Drin river system (Simon et al. 2020)

7 Towards Sustainable Fishing

The establishment of a sustainable freshwater fishery concept in Montenegro requires the improvement of the fishery sector and a number of specific activities:

- The establishment of the regular annual monitoring of fish fauna in Lake Skadar, based on previous experience and focused on economically important fish species.
- Surveys of the fish population parameters within the freshwater waterbodies and the improvement of decision-making process which should be based on biological data.
- The improvement or drafting of a new Law on Freshwater Fisheries and Aquaculture towards sustainability, taking into consideration all the characteristics and problems of the Montenegro freshwater fishery sector. The improvement of the existing law or the drafting of a new one has to be in harmony with the real problems we have faced in Montenegro over the last three decades.
- The improvement mechanisms for guarding and protecting waters and the implementation of zero tolerance towards poaching.
- The establishment of faster communication channels between professional fishermen, scientists, and fishery administration, in order to have the faster sharing of information and a faster decision-making process.
- The harmonization of fishery regulations on both sides of Lake Skadar (i.e. in both Montenegro and Albania) as well as the establishment of a bilateral scientific/administrative body for the regulation/management of fishing and fish stocks in Lake Skadar.
- Continuing with the diversification of catch in freshwater fishing on Lake Skadar and with the reduction of fishing pressure on the most important species.
- The improvement of law enforcement, especially those which tackle the monitoring and control of the fish market and preventing the trade or restaurant selling of endangered and rare fish species, as well as the common fish species during the fishing ban period (e.g. the common carp).
- Prohibit gravel and sand extraction from freshwater waterbodies.
- Provide for the free flowing of rivers in all places where it is possible and remove all obstacles on the river courses that were created by an action of an individual or a company (during the exploitation of gravel or other actions) or that are the result of industrial water catchment that is no longer used.
- When stocking, it is necessary to take into account the origin and quality of the fry. This means obtaining the consent from the competent scientific and research institutions on the genetic matching of the fry with the indigenous population of the area.
- The creation of a better "ecosystem" for the fish processing sector, which will lead to economic improvement and make fishing more sustainable, since it will generate a larger income on the basis of the same or an even smaller amount of basic resources, while allowing for the employment of more people.

- Bilaterally regulate the operation of fishing traps downstream of Lake Skadar on the River Bojana, which currently prevents both the upstream and downstream migration of fish.
- Improve the work of the angler and fisherman associations in Montenegro, as the only legal entities related to fishing recognized by the current law.
- Clarify the distinction between types of fishing and sport activities in the fishing sector, especially in relation to recreational fishery.
- Until the drafting and adoption of the new law, ensure the strict enforcement of the current Law on Freshwater Fishing (Official Gazette of Montenegro, No. 17/2018 of 20 March 2018) as well as making the most urgent amendments and drafting all the necessary by-laws.
- The production and enforcement of management plans related to allochthon fish species, in order to reduce their negative impact on autochthonous fish fauna.

Acknowledgments We dedicate this chapter to the late Dr. Đorđije Drecun who established modern fishery science in Montenegro. As young researchers, we had the opportunity to meet him personally and, alongside the valuable bibliography and many framing objects which Dr. Drecun left behind him, we will always remember his positive attitude and great love of, and dedication to, fish and fisheries.

References

Arthington AH, Lloyd LN (1989) Introduced poeciliids in Australia and New Zealand. In: Meffe GK, Snelson FF (eds) Ecology and evolution of livebearing fishes (Poeciliidae). Prentice Hall, Englewood Cliffs, NJ, pp 333–348

Bajer PG, Sullivan G, Sorensen PW (2009) Effects of a rapidly increasing population of common carp on vegetative cover and waterfowl in a recently restored Midwestern shallow lake. Hydrobiologia 632:235–245

Bajković I, Hadžiablahović S, Cakovic D, Medenica B, Zaimi K, Elbasani O, Peci D, Simixhiu V, Zennaro B (2017) Shorezone functionality Skadar/Shkodra Lake – implementing the EU water framework directive in South-Eastern Europe. Technical report. Deutsche Gesellschaft für Internationale Zusammenarbeit (GIZ), Bonn, Eschborn. Pegi Sh. P.K. Book Publishers, Tirana, p 83

Barbieri R, Zogaris S, Kalogianni E, Stoumboudi M, Chatzinikolaou Y, Giakoumi S, Economou A N (2015) Freshwater fishes and lampreys of Greece: an annotated checklist. Monographs on marine sciences no. 8. Athens, Greece: Hellenic Center for Marine Research. ISBN: 978-960-9798-06-8, p 96

Berts AJ (1961) Selectivity and efficiency of experimental gill nets in South Bay and Georgian Bay of Lake Huron. Trans Am Fish Soc 90:413–418

Bianco PG, Kottelat M (2005) *Scardinius knezevici*, a new species of Rudd from Lake Skadar, Montenegro Teleostei: Cyprinidae. Ichthyol Explor Freshw 16(3):231–238

Brdar R (2015) Prerada ribe kroz proces dimljenja, procjena ukupne količine nastalog otpada, nutritivna vrijednost otpada i mogućnosti njegove upotrebe radi smanjenja opterećenja životne sredine. Magistarski rad. Prirodno-matematički fakultet. Univerzitet Crne Gore. 63 pp.

Burić M (2010) Atlas of waters of Montenegro, Izdanja Leksikografskog centra, 1, Crnogorska akademija nauka i umjetnosti, Podgorica

Claessen D, de Roos AM, Persson L (2000) Dwarfs and giants: cannibalism and competition in size-structured populations. Am Nat 155:219–237

Clavero M, García-Berthou E (2005) Invasive species are a leading cause of anima extinctions. Trends Ecol Evol 20:110

Coble DW, Knežević B (1981) Growth and mortality of bleak (*Alburnus alburnus alborella*) and carp (*Cyprinus carpio*) in Lake Skadar. In: Karaman G, Beeton A (eds) The biota and limnology of Lake Skadar. Univerzitet 'Veljko Vlahović', Institut za biološka i medicinska istraživanja u SRCG, Biološki zavod Titograd, Yugoslavia; Smithsonian Institution. Center for Great Lakes Studies University of Wisconsin-Milwaukee, U.S.A, Washington, DC, pp 311–316

Cochrane KL (1985). The population dynamics and sustainable yield of the major fish species in Hartbeespoort Dam. PhD Thesis, University of the Witwatersrand, South Africa. 514 pp.

Cucherousset J, Olden JD (2011) Ecological impacts of non-native freshwater fishes. Fisheries 36: 215–230

Darwall W, Carrizo S, Numa C, Barrios V, Freyhof J, Smith K (2014) Freshwater key biodiversity areas in the Mediterranean basin hotspot: informing species conservation and development planning in freshwater ecosystems. IUCN, Cambridge/Malaga. x + 86 p

Drecun Đ (1951) About a significant anniversary in the fishery in the republic of Montenegro. Ribarstvo Jugoslavije 6(11):249–250

Drecun Đ (1952a) Achievements of freshwater fisheries in republic of Montenegro in 1951. Ribarstvo Jugoslavije 7(1–2):4–6

Drecun Đ (1952b) Durmitor's lakes. Ribarstvo Jugoslavije 7(7):128

Drecun Đ (1952c) One year results of stocked rainbow trout stock in the upper stream of the Zeta River. Ribarstvo Jugoslavije 7(6):105–106

Drecun D (1955) Experimental fishing by trawl nets at Skadar Lake. Naša poljoprivreda 1(4):63–70

Drecun Đ (1956) The results of experiment with trawl lines made from nylon thread for eel fishing at Skadar Lake. Naša poljoprivreda 2(2):81–86

Drecun Đ (1958–1959) Stocking of Zabojsko Lake. Lov i ribolov, 3: 17

Drecun Đ (1960) Results of stocking of artificial lakes in Nikšić. Ribarstvo Jugoslavije 15(4):80–82

Drecun Đ (1961) Experimental trout pond "Morača" at Titograd. Lov I ribolov 2–3:16–19

Drecun Đ (1962) The fishing on Skadar Lake from 1947–1960. Hydrobiologia Montenegrina, Tom I, No 10

Drecun Đ (1985) The decrease of the population of skadar bleak (*Alburnus albidus alborella*) in the Lake Skadar. Agrosaznanje 2:15–28

Drecun Đ, Miranović M (1962) The fishing on Lake Skadar from 1947 to 1960. Hydrobiologia Montenegrina 1(10):1–19

Drecun Đ, Ristić M (1964) Sublacustric spring "oka" and their character in connection with the fishing in Skadar Lake. Hydrobiologia Montenegrina 2(5):1–21

Drecun Đ, Ristić DM (1972) Biology, morphological characteristics and growth of Skadar Lake Carp. Ribarstvo Jugoslavije 27(2):21–42

FAO (2022) Montenegro. Text by Krasic, M., Mandic, M.. Fisheries and Aquaculture Division [online]. Rome. https://www.fao.org/fishery/en/countrysector/me/en?lang=en

Franetović DB (1960) Historija pomorstva i ribarstva Crne Gore do 1918. Istorijski Institut Republike Crne Gore, Titograd, Godine

Gaston KJ, David R (1994) Hotspots across Europe. Biodivers Lett 2(4):108–116

Gozlan RE, Britton JR, Cowx I, Copp GH (2010) Current knowledge on non native freshwater fish introductions. J Fish Biol 76:751–786

Heckel J, Kner R (1858) Die Süsswasserfische der Östereichischen Monarchie mit Rücksichtauf der angrenzenden Länder. Engelmann, Leipzig, 388 p

Hegediš A (2007) Migration and characteristics of glass eel (*Anguilla anguilla*) as limitation factors for fish farming. PhD dissertation. University in Novi Sad (Serbia), Faculty of Agriculture

Hegediš A, Kalauzi A, Mićković B, Nikčević M, Anđus KR (2005) Modeling of the migration of the European glass eel. Ann N Y Acad Sci 1048:85–91. https://doi.org/10.1196/annals.1342.008

Heibo E, Magnhagen C, Vøllestad A (2005) Latitudinal variation in life-history traits in Eurasian perch. Ecology 86(12):3377–3386. https://doi.org/10.1890/04-1620
Hewson LC (1951) A comparison of nylon and cotton gill nets in the Lake Winnipeg winter fishery. Can Fish Cult 11:7–9
Högman WJ (1973) The relative efficiency of nylon gill nets after the transition from cotton nets in a multispecies fishery. Trans Am Fish Soc 102:778–785
Hulme PE (2009) Trade, transport and trouble: managing invasive species pathways in an era of globalization. J Appl Ecol 46:10–18
Ivanović B (1962) The longitudinal succession of fishing villages in the lower part of the Morača river in the summer period. Naša poljoprivreda i šumarstvo 8(4):23–35
Ivanović B (1964) The ecology of Skadar Lake bleak *Alburnus albidus alborella* (Filipi). PhD thesis. Biological Institute in Sarajevo, Sarajevo
Ivanović B (1965) Feeding and food items of Skadar Lake bleak. Agric For Podgorica 11(3–4): 49–61
Ivanović B (1968) Ecology of *Alburnus albidus alborella* (Filippi). Godišnjak Biološkog instituta Univerziteta u Sarajevu 21:5–70
Ivanović B (1973) Ichthyofauna of Skadar Lake. Institut za biološka i medicinska istraživanja. 146 pp. Titograd
Ivanović B, Knežević B, Vuković T (1983) The population of fishes in Lake Skadar and their protection. Zbornik radova CANU, Radovi sa simpozijuma, Titograd 30. i 31.10. i 1.11., 9: 123–128
Janković D (1971) Reproduction of carp (*Cyprinus carpio*, L) from Lake Skadar. Arhiv bioloških nauka 23(1–2):71–90
Janković D (1975) The effect of fisheries exploitation on the age composition of carps in Lake Skadar. Verh Internat Verein Limnol 19(4):2589–2592
Janković D (1983) Nutrition of carp (*Cyprinus carpio* L.) on Skadar Lake. CANU – Zbornik radova sa simpozijuma Skadarsko jezero, Titograd 30. i 31.10. i 1.11., 9: 211–229
Jiriček K (1923) Istorija Srba II. Beograd. C 36:59
Jovićević A (1909) Skadar lake and its fishery. Srpski etnografski zbornik 13:157–259
Jovićević A (1928) Ribolov na Skadarskom jezeru. Zapisi III, Pp, p 38
Kapac E (1949) Two days with fisherman on Lake Skadar. Ribarstvo Jugoslavije 4(1–2):12–14
Kapac E (1949a) The freshwater fishery of Montenegro. Ribarstvo Jugoslavije 4(1–2):5–8
Karaman S (1923) Uber die Herkunft der Susswasserfische unseres Karstes. Glasnik zemaljskog muzeja za Bosnu i Hercegovin 35:163–176
Karaman S (1924). Pisces Macedoniae. Split
Karaman S (1933) Prilozi ihtiologiji Crne Gore. Ribarski List, God. VIII. Br, 9 i 10
Karataş M, Çiçek E, Başusta A, Başusta N (2007) Age, growth and mortality of common carp (*Cyprinus carpio* Linneaus, 1758) population in Almus Dam Lake (Tokat-Turkey). J Appl Biol Sci 1(3):81–85
Keller K, Brown C (2008) Behavioural interactions between the introduced plague minnow, *Gambusia holbrooki* and the vulnerable native Australian rainbow fish, *Rhadinocentrus ornatus*, under experimental conditions. J Fish Biol 73:1714–1724
Knežević B (1979) Pojava *Carassius auratus gibelio* (Bloch, 1783) i *Perca fluviatilis* (Linnaeus, 1758) u Šaskom jezeru kod Ulcinja. Poljoprivreda i šumarstvo 25(2):101–107
Knežević B, Marić D (1979) *Perca fluviatilis* Linnaeus, 1758. (Percidae, Pisces). New species for the Yugoslav part of Lake Skadar. Glasnik Republičkog zavoda za zaštitu prirode-Prirodnjačkog muzeja 12:177–180
Kolaneci V, Kuka B, Boçari A, Sadikaj R (2010a) Population parameters of carp (*C. carpio*) in Shkodra Lake. Proceedings, international scientific conference "Skadar Lake – current status and the future" 19–21 June 2010. Podgorica-Shkodër, Montenegro-Albania 2:67–72
Kolaneci V, Shegani Ç, Nikolla M, Kuka B (2010b) Some biological properties of two cyprinid fishes living in Shkodra Lake. Balwois 25:1–16

Kovačić M, Šanda R (2007) A new species of Knipowitschia (Perciformes: Gobiidae) from southern Montenegro. J Nati Mus (Prague), Nat Hist Ser 179(5):81–89
Kottelat M, Freyhof J (2007) Handbook of European Freshwater Fishes. Kottelat, Carnol, Switzeraland and Freyhof. Berlin, Germany. 646 pp
Krivokapić M (1992) Population structure of *Leuciscus souffia montenegrinus* (Vuković, 1963) endemic subspecies from the Morača River. Glasnik republičkog zavoda za zaštitu prirode – Prirodnjačkog muzeja 25:83–91
Krivokapić M (1994) Distribution and morphometric characteristics of *Leuciscus souffia montenegrinus* Vuković, 1963 (Osteichthyies, Cyprinidae). Glasnik odeljenja prirodnih nauka CANU 11:143–166
Krivokapić M (2005) Population structure of white chub (*Leuciscus cephalus albus)* (Pisces, Cyprinidae) from Skadar Lake, by weight group. Natura Montenegrina 4:55–60
Lasca N, Radulović V, Ristić R, Cherkauer D (1981) Geology, hydrology, climate and bathymetry of Lake Skadar. In: Karaman G (ed), The biota and limnology of Lake Skadar. Univerzitet Veljko Vlahović, Titograd; Smithsonian Institution, Washington; Center for Great Lakes Studies, University of Wisconsin, Milwaukee, pp. 17–38
Law on alien and invasive alien species of plants, animals, and fungi (2019) "Official Gazette of Montenegro", No. 18/2019
Law on Environment (2016) "Official Gazette of Montenegro", No. 52/16
Law on Food Safety (2015) "Official Gazette of Montenegro", No. 57/2015
Law on Marine Fisheries and Mariculture (2009) "Official Gazette of Montenegro", No. 56/09
Law on Marine Fisheries and Mariculture (2011) "Official Gazette of Montenegro", 40/11
Law on Marine Fisheries and Mariculture (2015) "Official Gazette of Montenegro", other regulation, 47/15
Li S, Zhou B, Lin Q (1990) The yield and growth of major fish species in a large Chinese reservoir. Asian Fish Sci 3(2):185–196
Lopičić S (1931) The fishery in Ceklin, in pictures. Šematizam Zetske banovine, 84 pp
Lopičić P (1953) The fishery in Ceklin. Ribarstvo Jugoslavije 8(5):97–101
Marić D (1995) Endemic fish species of Montenegro. Biol Conserv 72:187–194
Marić D (1998) Adaptation of introduced German carp (*Carassius auratus gibelio* Bolch, 1783) in Skadar Lake. PhD thesis, pp. 268. Kragujevac
Marić D (2000) Feeding of *Carassius auratus gibelio* (Bloch) in Skadar Lake (Montenegro) and Competitive relations with autochthonous cyprinid species. CANU, Glasnik odjeljenja prirodnih Nauka 13:237–258
Marić D (2002) Sexual maturation and sexual cycle of gonads in introduced Gold fish (*Carassius auratus gibelio*, Bloch) in Skadar Lake. Natura Montenegrina 1:109–124
Marić D (2003) Spawning and embryonic development of introduced Goldfish (*Carassius auratus gibelio*, Bloch) in Skadar Lake. Natura Montenegrina 2:45–55
Marić D (2004) Distribution and abundance of introduced German Carp *Carassius auratus gibelio* (Bloch) in the Lake Skadar in period 1972–1992. Glasnik Republičkog zavoda za zaštitu prirode 27–28:113–126
Marić D (2010) Bibliography on fishery-Ichthyological investigation in freshwaters of Montenegro (150 years of investigation). Natural History Museum of Montenegro. 140 pp. ISBN 978-86-908729-3-0
Marić D (2018) The Ichthyofauna of Lake Skadar/Shkodra: diversity, economic significance, condition, and conservation status. In: The handbook of environmental chemistry. Springer, Berlin, Heidelberg
Marić D (2019) Fauna of freshwater fish (Osteichthyes) of Montenegro. Montenegrin academy of science and art. Special editions (Monographies and Studies), vol. 149, 419 p. ISBN: 978-86-7215-453-5
Marić D, Milošević D (2011) Katalog slatkovodnih riba (Osteichthyes) Crne Gore (ISBN 978-86-7215-270-8). Crnogorska akademija nauka i umjetnosti. Katalozi 5, Knjiga 4. Podgorica. pp. 114

Marić D, Burzanović K, Marić S (2022) Variability of length-weight relationship and condition factor of the European eel (*Anguilla anguilla* L.) – case study from the Lake Skadar (Montenegro). Agric For 68(2):175–191. https://doi.org/10.17707/AgricultForest.68.2.13

Marković Z, Pešić A (2021) Fishery sector study for the Ipard programme. Prepared by Seedev. 117 pp.

MEA (Millennium Ecosystem Assessment) (2005) Ecosystems and human wellbeing: biodiversity synthesis. World Resources Institute, Washington, DC

Miller PJ, Šanda R (2008) A new west Balkanian send-goby (Teleostei: Gobiidae). J Fish Biol 72: 259–270

Mills MD, Rader RB, Belk MC (2004) Complex interactions between native and invasive fish: the simultaneous effects of multiple negative interactions. Oecologia 141:713–721

Milošević D, Marić D (2012) Length-weight relationship and condition factor of *Cyprinus carpio* from Skadar Lake (Montenegro) during spawning period. Agric For 52(1–4):53–60

Milošević D, Mrdak D (2016) Length-weight relationship of nine fish species from Skadar Lake (Adriatic catchment area of Montenegro). J Appl Ichthyol 32:1331–1333

Milošević D, Winkler AK, Marić D, Weiss S (2011) Genotipic and phenotypic evaluation of Rutilus sp. From Skadar, Ohrid and Prespa lakes support revision of endemic as well as taxonomic status of several taxa. J Fish Biol, 79(5):1094–1110

Milošević D, Pešić V, Petrović D, Pavićević A, Marić D (2012) Length-weight relationship and condition factor of two sympatric Rutilus (Rafinesque, 1820) species from Lake Skadar (Montenegro). Arch Biol Sci 64(3):991–994

Mrdak D, Pietrock M, Brämick U, Simonović P, Milošević D (2017a) Population traits and colonization success of non-native Eurasian perch (*Perca fluviatilis*) 35 years after its first appearance in the Mediterranean Lake Skadar. Environ Biol Fish 101:417

Mrdak D, Milošević D, Despotović V (2017b) Pietrock, M., Ritterbusch, D., Brämick, U. (eds) Fish and Fisheries Skadar/Shkodra Lake. Deutsche Gesellschaft für Internationale Zusammenarbeit (GIZ) GmbH

Nedeljković R (1959) The Skadar Lake: study of organic production in karst lake. Special issue of Biological Institute, Belgrade

Oyugi DO, Cucherousset J, Ntiba MJ, Kisia SM, Harper DM, Britton JR (2011) Life history traits of an equatorial common carp Cyprinus carpio population in relation to thermal influences on invasive populations. Fish Res 110(1):92–97

Piria M, Milošević D, Šprem N, Mrdak D, Tomljanović T, Matulić D, Treer T (2016a) Condition of European eel from the Adriatic basin of Croatia and Montenegro. 51st Croatian and 11th International Symposium on Agriculture, February 15–18, 2016 Opatija, Croatia, Proceedings, 270–273

Piria M, Povž M, Vilizzi L, Zanella D, Simonović P, Copp GH (2016b) Risk screening of non-native freshwater fishes in Croatia and Slovenia using FISK (fish invasiveness screening kit). Fish Manag Ecol 23:21–31

Radulović MM, Radulović M, Stevanović Z, Sekulić G, Radulović V, Burić M, Novaković D, Vako E, Blagojević M, Dević N, Radojević D (2015) Hydrogeology of the Skadar Lake basin (Southeast Dinarides) with an assessment of considerable subterranean inflow. Environ Earth Sci 74:71–82

Radusinović SP (1983) Basic traits of traditional fishing on Lake Skadar. In proceedings Lake Skutari. Montenegrin Acad Sci Arts 9:646–665

Rakočević J, Rakaj M (2015) Phytoplankton of Lake Skadar, final report. In: Peveling R, Brämick U, Densky H, Parr B, Pietrock M (eds) Initial Characterization of Lakes Prespa, Ohrid and Shkodra/Skadar. Deutsche Gesellschaft für Internationale Zusammenarbeit (GIZ). GmbH, Eschborn

Ristić J, Vizi O (1981) Synoptic survey of the dominant macrophytes in Lake Skadar. In: Karaman GS, Beeton AM (eds) Smithsonian Institution, Center for Great Lakes Studies, University of Wisconsin The biota and limnology of Lake Skadar, Titograd, Washington D.C.,

MilwaukeeUniverzitet "Veljko Vlahović, Institut za biološka i medicinska istraživanja u SRCG, Biološki zavod
Rovinski P (1897) Cernogorija v jejo proslom I nastajašćem. II, 1, Sankpetersburg. p: 718–736 Rutilus sp. From Skadar, Ohrid and Prespa lakes support revision of endemic as well as taxonomic status of several taxa. J Fish Biol 79(5):1094–1110
Ružić N (2018) Age population structure, yearly growth, mortality and fecundity of bleak (*Alburnus scoranza* Bonaparte, 1845) from Skadar Lake. Master thesis. University of Montenegro, Faculty of Science and Mathematics, Podgorica
Sharma VK (1987) The biology and fishery of *Cyprinus carpio* Linn. From the Gobindsagar reservoir, Himachal Pradesh, India. Res Bull Panjab Univ Sci 38:171–173. Thesis Abstract (PhD)
Simon J, Hala E, Fladung E, Mrdak D, Milošević-Malidžan D, Ilik-Boeva D, Pietrock M, Simonović P, Tošić A, Vassilev M, Apostolou A, Mrdak D, Ristovska M, Copp GH (2013) Risk identification of non-native freshwater fishes in four countries of the Balkans Region using FISK. Mediterr Mar Sci 14:369–376
Simon J, Hala E, Fladung E, Mrdak D, Milošević-Malidžan D, Ilik-Boeva D, Pietrock M, Kapedani R, Brämick U, Spirkovski Z, Goertler A, Cani E (2020) Eel management plan for the Drin/Drim River Basin (draft). Deutsche Gesellschaf fuer Internationale Zusammenarbeit (GIZ) Through the CSBL III Project (Conservation and Sustainable Use of Biodiversity of Big Balkan Lakes)
Šorić V (1980) Comparative analysis of the morphological characters of *Alburnus alburnus alborella* (De Filippi, 1844) from the White Drim and Lake Skutari. Zvornik radova 1:171–180
Šorić V (1982) Individual growth in *Pachychilon pictum*, West-Balcanic endemic fish species in the Lake Skadar and Ohrid. Ichthyologia 14(1):87–93
Stanković, S. (1926). O tercijernim reliktima u fauni Triklada Skadarskog jezera. Glas geografskog društva, 12, Beograd. (in Serbian)
Stanković S (1929) Fish productivity in South-Balkan lakes. Glasnik geografskog društva, 5 pp 22, Belgrade
Stanković S (1939) On our lake fishing. Ribarstvo Jugoslavije 2:3–12
Stein RA, Kitcheel JF, Knežević B (1975) Selective predation by carp (*Cyprinus carpio* L.) on benthic molluscs in Skadar Lake, Yugoslavia. J Fish Biol 7:391–399
Stein RA, Mecom JO, Ivanović B (1981) Commercial exploitation of fish stocks in Lake Skadar/Shkodra, Yugoslavia, 1947–1973. Biol Conserv 8:1–18
Steindachner F (1883) Ichthyologische Birtage XII. Sitz-Ber Math Naturwiss Cl Acad Wiss Wien, Abt 1(86):61–82
The Law on Animal Welfare Protection (2008) "Official Gazette of Montenegro", No. 14/2008
The Law on Freshwater Fisheries and Aquaculture (2018) "Official Gazette of Montenegro", No. 17/18.23
The Law on Waters (2018) "Official Gazette of Montenegro", No. 084/18
Tsagarakis K, Başusta N, Başusta A, Biandolino F, Bostanci D, Buz K, Djodjo Z, Dulčić J, Gokoglu M, Gücü A. C, Machias A, Maravelias C. D, Özvarol Y, Polat N, Prato E, Vasilakopoulos P, Yedier S (2015) New fisheries-related data from the Mediterranean Sea. Medit Mar Sci, 16: 3. pp. 703–713
Veterinary Law (2016) "Official Gazette of Montenegro", 52/16
Vila M, Basnou C, Pyšek P, Josefsson M, Genovesi P, Gollasch S, Partners D (2010) How well do we understand the impacts of alien species on ecosystem services? A pan-European, cross-taxa assessment. Front Ecol Environ 8:135–144
Vilizzi L, Güler Ekmekçi F, Tarkan AS, Jackson ZJ (2014) Growth of common carp *Cyprinus carpio* in Anatolia (Turkey), with a comparison to native and invasive areas worldwide. Ecol Freshw Fish:1–16
Vuković T (1961) Populations and spawning of *Alosa fallax nilotica* (Geoffroy) in the waters of the Neretva and Lake Skadar. Godišnjak Biološkog instituta Univerziteta u Sarajevu, 15

Vuković T (1965) *Leuciscus souffia montenegrinus* n. spp. (Cyprinidae, Pisces) eine neue Unterart aus dem Flusses Morača. Bulliten Scientifique, Conseil de l'' Academie de la RSF de Yugoslavie, Section A, Sciences Naturalles Tehniques et Medical 10:217–220
Vuković T, Kažić D, Knežević B (1975) *Carassius auratus* (Linnaeus, 1758) (Pisces Cyprinidae) new species for the Yugoslav part of Lake Skadar. Bull Sci Yugosl 20:1293

Fish and Fisheries of the Republic of North Macedonia, Current Situation, and its Perspective

Trajče Talevski, Blagoja Trajčeski, Marina Talevska, and Orhideja Tasevska

Abstract Freshwaters in the Republic of North Macedonia belong to two ecoregions: ecoregion 420 (Southeast Adriatic basin, Prespa, and Ohrid lakes) and ecoregion 422 (Vardar basin with Dojran lake).

Despite being landlocked, the Republic of North Macedonia has three large natural lakes (Ohrid, Prespa, and Dojran), more artificial lakes (man-made lakes), and river basins.

Eighty species represent the fish fauna in the Republic of North Macedonia. Sixty-one fish species which belong to 15 families have been described in large natural lakes (Prespa, Ohrid, and Dojran lakes).

Fish populations in the Republic of North Macedonia can be divided into native, non-native, and invasive species. For example, in Lake Ohrid, 22 natives and seven introduced fish species were described; in Lake Prespa, 11 native and 12 introduced fish species were described; and in Lake Dojran, 15 native and 7 introduced fish species were described.

The IUCN Red List is widely considered the most objective and authoritative system for classifying the species in terms of the risk of extinction. According to the IUCN Red List, fish (total 80) from the Republic of North Macedonia belong to the following categories: LC—35 fish species, CR—3 species, EN—2 species, VU—9 species, NT—1, species, NE—1 species, DD—10 species, IN—18 species, and 1 species which is not categorized by the IUCN Red List.

T. Talevski (✉) · B. Trajčeski
Department of Cyprinid Fish, PSI Hydrobiological Institute, Ohrid, Republic of North Macedonia
e-mail: trajce.talevski@hio.edu.mk

M. Talevska
Department of Hydrobotany, PSI Hydrobiological Institute, Ohrid, Republic of North Macedonia

O. Tasevska
Department of Zooplankton, PSI Hydrobiological Institute, Ohrid, Republic of North Macedonia

V. Simić et al. (eds.), *Ecological Sustainability of Fish Resources of Inland Waters of the Western Balkans*, Fish & Fisheries Series 43,
https://doi.org/10.1007/978-3-031-36926-1_6

From the statistical data on fishing in the Republic of North Macedonia, it can be concluded that the catch has been declining in recent decades, with a significant change in the qualitative and quantitative composition of fish.

This condition is the result of changes in water quality, macrophyte vegetation, planktonic communities, and other biological components.

Keywords Freshwater · Ecoregions · Lakes · Endemism · Fish species · Fishing

1 Introduction

Freshwater fish spend part or all their lives in freshwater, such as rivers and lakes, with less than 1.05% salinity. They comprise 50% of all fish species described worldwide and represent an essential part of global biodiversity (Loh et al., 2005). Fish are part of food chain dynamics, nutrient cycling, and ecosystem resilience.

Fish play an essential role in nutrient cycles because they store many ecosystem nutrients in their tissues, transport nutrients further than other aquatic animals, and excrete nutrients in a dissolved form readily available to primary producers in nutrient recycling. The ichthyofauna of the freshwaters in the western and central parts of the Balkan Peninsula undoubtedly arose in this form and composition after the last Pleistocene glaciation (Stankovic, 1960).

The ichthyofauna of the freshwaters in the western and central parts of the Balkan Peninsula undoubtedly originated in this form and composition after the last Pleistocene glaciation (Stankovic, 1960).

Freshwater fish and aquatic ecosystems have many more unknown parts, and developing a fish resource utilization strategy is necessary.

Freshwater fish and fisheries (including other freshwater organisms (Lynch et al., 2017) and the aquatic ecosystems they rely on (Lynch et al., 2017; Tickner et al., 2020) are prominent examples of overlooked natural resources that make significant contributions to the Sustainable Development Goals, SDGs). Their fundamental characteristics (e.g., high biodiversity, diverse fisheries, and definitions of sustainability) pose challenges in grasping their value and adequately representing them in policy decisions. However, given the prevalence of freshwater fisheries and fisheries and their importance in many regions of the world (Dickens et al., 2020), it is worthwhile to seek ways to address these challenges and identify opportunities for increasing recognition of their values (Lynch et al., 2017).

The autochthonous freshwater ichthyofauna in the western and central parts of the Balkan Peninsula belongs to the Central European and Southern European fauna. Within the southern European fauna, representatives of the western Balkans (including the Adriatic and Ionian sea basins) are the richest part with endemic species (Stankovic, 1960).

Fish and fishery have always been essential in Lake Ohrid, Prespa, and Lake Dojran (Republic of North Macedonia). Nowadays, fishing is still of high importance to fishers and their families living in neighboring cities and villages around lakes in the Republic of Albania and the Republic of North Macedonia. However,

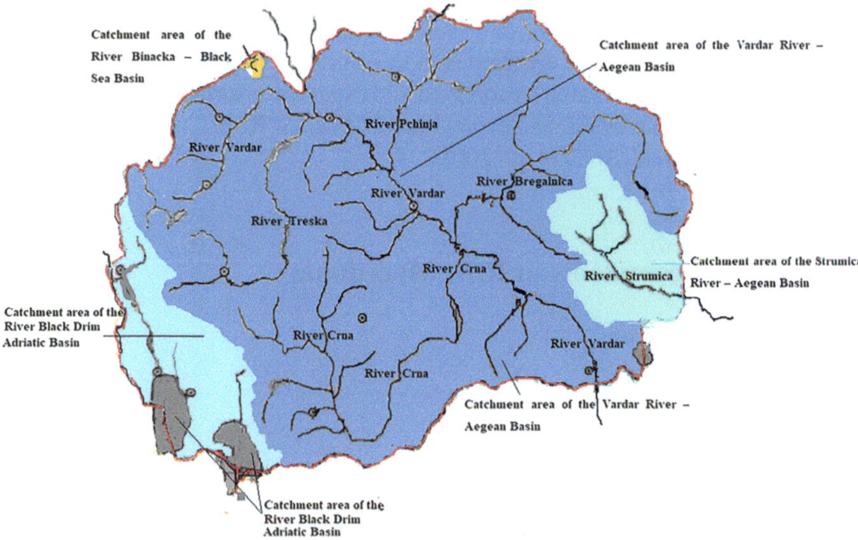

Fig. 1 Drainage basins in the Republic of North Macedonia: Aegean, Adriatic, and Black Sea basins (original, Talevski T. 2023)

the current state of the lake's fish stock is very different from the past because fish are exposed to many stressors such as pollution and habitat destruction, unsustainable fishing, invasive species and others (Spirkovski et al., 2001; Watzin et al., 2002; Talevski, 1996), hunting of endemic species during spawning (Talevska, 2002; Talevski & Talevska, 2008; Talevska & Talevski, 2009; Talevski et al., 2009a; Talevski et al., 2009b; Talevski et al., 2010a; Talevska and Talevski, 2018a).

There are 1100 significant water sources in the Republic of North Macedonia. The rivers flow into three different basins: the Aegean, the Adriatic, and the Black Sea basin (Talevski & Talevska, 2010) (Fig. 1). The most significant river basin is the Aegean Basin, while the largest river is the Vardar River.

Black Drim River flows out of Lake Ohrid near Struga and forms the Adriatic Basin. Despite being landlocked, North Macedonia has three major lakes (Ohrid, Prespa, and Dojran), about thirty artificial lakes, and about 50 lakes.

Since 1906, the hypothesis that the waters of Lake Prespa flow into Lake Ohrid through the karst massif of Galicica has been known (Cvijić, 1906). The plausibility of this hypothesis was proven by isotope-based tests in 1992 (Anovski et al., 1992) and 2001 (Anovski et al., 2001).

Most underground aquifers have been discovered in the Galicica limestones, and the water flows into Lake Ohrid through numerous springs. It is estimated that about 45% of the water flowing out of Prespa Lake reaches Ohrid through several underwater springs. In comparison, the remaining 55% comes from surface springs, the most famous of which are the springs at Sveti Naum in the Republic of North Macedonia and Tushemisht in the Republic of Albania. Lake Prespa and Lake Ohrid represent a coherent complex water system with hydrogeographical and ecological characteristics.

There are 426 freshwater ecoregions in the world, of which Macedonian freshwater belongs to two ecoregions: ecoregion 420 (Southeast Adriatic), which includes Lake Prespa and Lake Ohrid, and ecoregion 422 (Vardar River), which includes the Dojran Lake supplied (Abell et al., 2008).

2 Aquatic Ecosystems in R.N. Macedonia

2.1 Natural Lakes

2.1.1 Lake Ohrid

Lake Ohrid (Fig. 2) stretches along the mountainous border between the southwestern Republic of North Macedonia and the eastern Republic of Albania. Lake Ohrid is the deepest lake in the Balkans, with a maximum depth of 288 meters and an average depth of 155 meters. It covers an area of 358 km^2, and is 30.4 km long and 14.8 km

Fig. 2 Map of lakes Ohrid, Prespa, and Dojran, rivers, and artificial lakes (original, Talevski T. 2023)

wide. It has a coastline of 87.53 km, of which 56.02 km belongs to the Republic of North Macedonia and 31.51 km to the Republic of Albania. The lake is fed primarily by underground springs on the east coast but also by rivers and rainfall. Over 20% of the lake's water comes from the nearby Prespa Lake, which is 150 meters higher than Lake Ohrid (Matzinger et al., 2007; Matzinger et al., 2006; Albrecht & Wilke, 2008).

Similar to Lake Baikal and Lake Tanganyika, Lake Ohrid hosts endemic species covering the entire food chain, from phytoplankton and sessile algae (20 species; e.g., *Cyclotella fottii*) to plant species (2 species; e.g., *Chara ohridana*), zooplankton (5 species; e.g., *Cyclops ochridanus*), carp fish (8 species; e.g., *Pachychilon pictus*) to predatory fish (2 species of trout; *Ohrid trout*, *Salmo letnica* and "Belvica," *Acantholingua ohridana*), and finally its diverse endemic bottom fauna (176 species, e.g., *Ochridagammarus solidus*), with considerable endemism in crustaceans, mollusks, sponges, and planarians (Stankovic, 1960; Riis & Hawes, 2002).

Three cities lie on the shores of the lake: Ohrid and Struga in the Republic of North Macedonia and Pogradec in Albania. There are also several fishing villages on either side.

2.1.2 Lake Prespa

Lake Prespa (Fig. 2) has a tectonic origin in the Pliocene and fills the deepest part of the valley of the same name. It lies at the crossroads of three large geological masses: a granitic massif to the east, a karst massif from Galicica to the west, and Suva Gora to the south (Matzinger et al., 2007). The tectonic valley lies between the Baba mountains (2601 m) to the east, the Galicica (2288 m) and Petrino mountains to the west, and the Suva Gora mountains (1857 m) to the south (Matzinger et al., 2007).

It consists of Great Prespa Lake and Small Prespa Lake. Almost 65% of the Greater Prespa Lake is in Macedonia, 17% of its southern part is in Greece, and its southeastern part (18%) is in the Republic of Albania.

Lake Prespa is a transboundary lake shared by three countries: the Republic of Albania, Greece, and North Macedonia. It is located in the southwestern part of the Republic of North Macedonia with an area of 274 km^2, a maximum length of about 28 km, a maximum width of about 17 km, and a depth of 54 meters. The lake draws its water from underground springs and rivers: Istocka, Golema, Pretorska, and the Brajcinska river in the Macedonian part of the watershed and the Stara river in the Greek territory.

Both lakes (Great Prespa Lake and Small Prespa lake) are important in terms of biodiversity.

The total area of the basin is 1200 km^2. Prespa Lake has no surface spills. The underground stream Zavir is the main outlet of the lake (Fig. 3), which supplies Lake Ohrid through the springs in Sveti Naum and Tushemisht (Chavkalovski, 1997; Chavkalovski, 2000).

During the 1970s, water was diverted from the Devol River into the lake for later irrigation. Since the 1990s, the water level has declined due to the multiple uses for

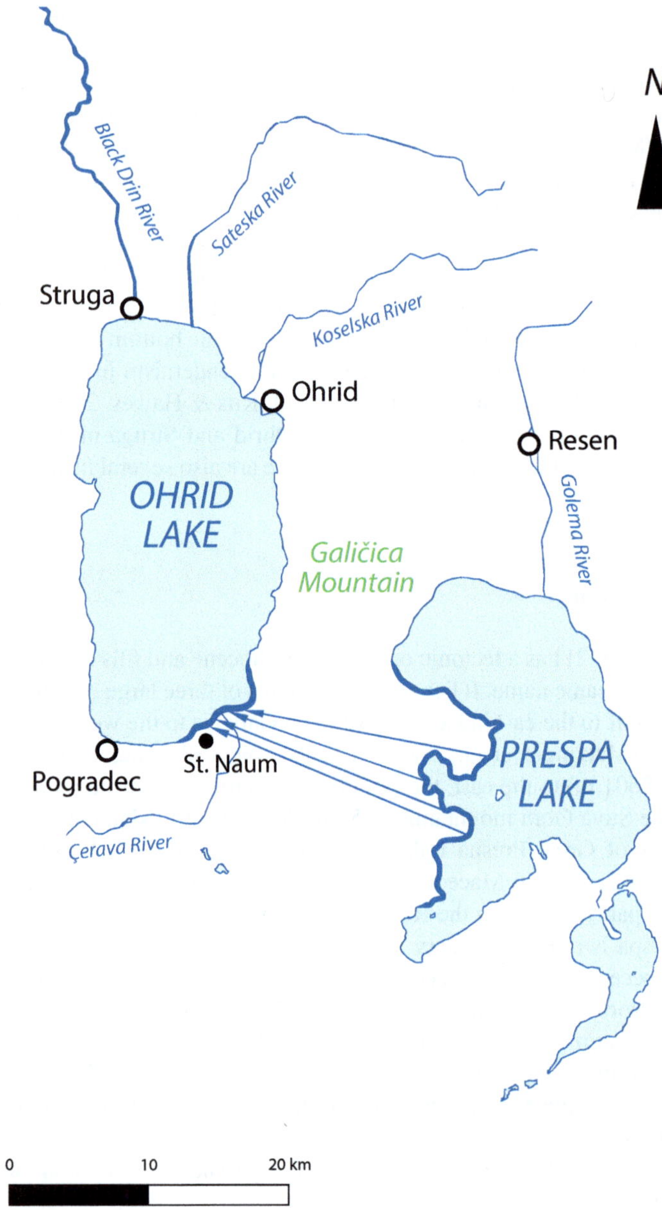

Fig. 3 Communication between Lake Prespa and Lake Ohrid under Galicica Mountain (original, Talevski T. 2023)

irrigation and changes in the local climate (Chavkalovski, 1997; Chavkalovski, 2000).

In aquatic ecosystems, water level fluctuations cause disturbances in the ecosystem (Watzin et al., 2002; Avital & Sarig, 1990; Wantzen et al., 2008; Strayer & Findlay, 2010a).

Seasonal and interannual water level fluctuations affect the survival of flora and fauna in the littoral zone (Avital & Sarig, 1990; Wantzen et al., 2008; Strayer & Findlay, 2010a), as well as fish spawning grounds (Riis & Hawes, 2002). Sediment drainage and subsequent drying stimulate macrophyte species' proliferation and improve nutrient cycling (Baldwin & Mitchell, 2000). Through the interaction of direct (e.g., physiological stress) and indirect (e.g., habitat alteration) mechanisms, water level fluctuations create temporal and spatial heterogeneity that structures littoral zone communities.

Natural water level fluctuations are critical to ecosystem structure and function (Winfield, 2004; Peters & Lodge, 2009; Strayer & Findlay, 2010b).

2.1.3 Lake Dojran

Dojran Lake (Fig. 2) is the third largest tectonic lake, a relict of the Mediterranean Sea, the Natural Monument, the Balkan Green Belt, CORINE, and Ramsar site. It is a transboundary lake shared by the Republic of North Macedonia and Greece.

Lake Dojran is located in the southeastern part of the Republic of North Macedonia and covers an area of 43.1 km^2, with a maximum depth of 10 m and a width of 7.1 km (Bonacci & Popovska, 2006; Bonacci et al., 2014; Löffler, 2004).

The lake is home to several endemic species and subspecies of fish, invertebrates, and many protected bird species. The lake and surrounding marshes are essential habitats for migratory birds. It has rich biodiversity and significant habitat value; serious water quality problems are associated with falling water levels and eutrophication.

3 Rivers in the Republic of North Macedonia

In the Republic of North Macedonia, there are also essential river basins for fisheries: the Black Drim river basin, the Treska river basin, the Strumica river basin, the Pinja river basin, the Vardar river basin (upper stream, middle stream, and lower stream), Bregalnica river basin, and Crna river basin (Petruseski & Markoski, 2016). With a length of 388 km, the Vardar River is the largest river flowing through the Republic of North Macedonia and one of the largest in the Balkans. Its catchment area covers an area of around 20,500 km. The main tributaries of the Vardar are the Bregalnica, Treska, Crna, and Pinja rivers (Fig. 2). The Bregalnica river is the second largest in the Republic of North Macedonia with a length of 211.5 km. The Treska River is a tributary of the Vardar River and is 132 km long. Finally, the Crna River is the most prominent right tributary of the Vardar River, with a length of 207 km.

4 Artificial Lakes in the Republic of North Macedonia

In the Republic of North Macedonia, there are many artificial lakes: artificial lake Berovsko (Ratevo), artificial lake Vodocha, artificial lake Glaznja, artificial lake Globochica, artificial lake Gratche, artificial lake Debar (Shpilje), artificial lake Kalimanci, artificial lake Kozjak, artificial lake Lipkovo, artificial lake Mavrovsko, artificial lake Mantovo, artificial lake Matka, artificial lake Mladost, artificial lake Prilepsko, artificial lake Slatino, artificial lake Strezevo, artificial lake Tikvesh, artificial lake Turija, and others (Table 1).

Building of artificial lakes is often a resource to improve fish production. Still, they can converse themselves on freshwater systems by altering river flow, changing fish habitats, and blocking fish migration routes.

With the construction of several high artificial lakes on the Drim River: artificial lakes Globochica and Spilje, which are about 100 meters high, the communication with the Adriatic Sea, and with it the contact of the eel, was permanently broken with the sea. Nevertheless, the excellent communication between the Adriatic Sea and Lake Ohrid is evidenced by the fact that *Alossa falax nilotica* was caught in Lake Ohrid (Tocko, 1959).

5 Fish Species in the Republic of North Macedonia

Freshwater ecosystems in the Republic of North Macedonia contain about 80 fish species from 17 families. The family Petromyzonidae consists of 2 species, Acipenseridae—1 species, Anguillidae—1 species, Clupeidae—1 species, Cyprinidae—46 species, Cobitidae—5 species, Nemacheilidae—3 species, Ictaluridae—1 species, Siluridae—1 species, Esocidae—1 species, Salmonidae—11 species, Poecilidae—1 species, Cottidae—1 species, Centrarchidae—1 species, Percidae—2 species, Blenniidae—1 species, and Gobiidae—1 species (Talevski & Talevska, 2010; Talevski et al., 2010b; Milošević & Talevski, 2015; Talevski & Milosevic, 2016; Talevski & Talevska, 2018; Talevski et al., 2017; Spirkovski et al., 2007; Talevski et al., 2010c; Kottelat & Freyhof, 2007; Stefanoic, 1948).

The fish populations in the Republic of North Macedonia are divided into native, non-native, and invasive species.

Native species are species that originated and evolved in their surrounding habitat and have adapted to life in that particular environment.

Non-native species originated somewhere other than their current location and have been introduced to the area where it now lives (also called exotic species).

Table 1 Artificial lakes (dams) in the Republic of North Macedonia

Artificial lake	Location	Dimensions	Depth	Used for
Artificial lake Berovsko	Eastern Macedonia	~ 1 km^2	–	Recreation fishing
Artificial lake Vodocha	Southeastern Macedonia	1.94 km^2	42 meters	Recreation fishing, rainbow trout aquaculture
Artificial lake Glaznja	North Macedonia	96.5 ha	–	Recreation fishing
Artificial lake Globochica	Western Macedonia	2.69 km^2	–	Recreation fishing, carp aquaculture
Artificial lake Gratche	Eastern Macedonia	0.19 km^2	29 meters	Recreation fishing, rainbow trout aquaculture
Artificial lake Debar (Shpilje)	Western Macedonia	13.2 km^2	92 meters	Recreation fishing, rainbow trout aquaculture
Artificial lake Kalimanci	Eastern Macedonia	4.23 km^2	80 meters	Recreation fishing, rainbow trout aquaculture
Artificial lake Kozjak	Western Macedonia	32 km length and 400 meters wide	130 meters	Recreation fishing, carp aquaculture
Artificial lake Lipkovsko	North Macedonia	1480 meters length and 120 meters wide	–	Recreation fishing
Artificial lake Mavrovsko	Western Macedonia	10 km length and 5 km wide	50 meters	Recreation fishing, trout aquaculture
Artificial lake Mantovo	Eastern Macedonia	5.5 km length and 0.8 km wide	40 meters	Recreation fishing, rainbow trout aquaculture
Artificial lake Matka	North Macedonia	~2.5 km^2	–	Recreation fishing
Artificial lake Mladost	Central Macedonia	~ 1 km^2	–	Recreation fishing
Artificial lake Prilepsko	South Macedonia	0.54 km^2	28 meters	Recreation fishing
Artificial lake Slatino	Southwestern Macedonia	420 ha	–	Recreation fishing
Artificial lake Strezevo	South Macedonia	7.5 km^2	–	Recreation fishing
Artificial lake Tikvesh	Central Macedonia	14 km^2	105 meters	Recreation fishing, carp aquaculture
Artificial lake Turija	Southeastern Macedonia	0.16 km^2	–	Recreation fishing

Invasive species is a species of plant or animal that outcompetes other species, causing damage to an ecosystem.

22 native and 7 introduced fish species have been described in Lake Ohrid; 11 native and 12 introduced fish species are described in Lake Prespa; 15 native and 7 introduced fish species are described in Lake Dojran (Table 2).

Table 2 Conservation status of native fish species in the Republic of North Macedonia according to Kottelat and Freyhof (2007)

FAMILY	GENUS	SPECIES	FAMILY/SPECIES	Adriatic		Aegen		IUCN
				Native	Introduced	Native	Introduced	
I			PETROMYZONIDAE					
	1.	1.	*Eudontomyzon stankokaramani* Karaman, 1974	n				LC
		2.	*Eudontomyzon mariae* (Berg, 1931)			n		LC
II			ACIPENSERIDAE					
	1.	3.	*Acipenser gueldenstaedtii* (Brandt & Ratzeburg, 1833)					CR
III			ANGUILLIDAE					
	1.	4.	*Anguilla anguilla* (Linnaeus, 1758)	n		n		CR
IV			CLUPEIDAE					
	1.	5.	*Alosa falax* (La Capède, 1803)		i			
V			CYPRINIDAE					
	1.	6.	*Alburnoides bipunctatus* (Bloch, 1782)			n		LC
		7.	*Alburnoides ohridanus* (Karaman, 1928)	n				VU
		8.	*Alburnoides prespensis* (Karaman, 1924)	n				VU
	2.	9.	*Alburnus belvica* (Karaman, 1924)	n				VU
		10.	*Alburnus macedonicus* (Karaman, 1928)			n		CR
		11.	*Alburnus scoranza* (Bonaparte, 1845)	n				LC
		12.	*Alburnus thessalicus* (Stephanidis, 1950)			n		LC
	3.	13.	*Barbus balcanicus* (Kotlik, Tsigenopoulos, Rab, & Berrebi, 2002)			n		LC
		14.	*Barbus macedonicus* (Karaman, 1928)			n		DD
		15.	*Barbus prespensis* (Karaman, 1924)	n				LC
		16.	*Barbus rebeli* (Koller, 1925)	n				LC
		17.	*Barbus strumicae* (Karaman, 1955)			n		LC
	4.	18.	*Carassius carassius* (Linnaeus, 1758)				i	
		19.	*Carassius gibelio* (Bloch, 1782)		i		i	

	5.	**20.**	***Chondrostoma ohridanus* (Karaman, 1924)**	**n**				**NE**
		21.	*Chondrostoma prespense* (Karaman, 1924)	n				VU
		22.	*Chondrostoma vardarense* (Karaman, 1928)			n		NT
	6.	23.	*Ctenopharyngodon idella* (Valenciennes 1844)		i			
	7.	24.	*Cyprinus carpio* (Linnaeus, 1758)	n		n		LC
	8.	25.	*Gobio bulgaricus* (Drensky, 1926)			n		LC
		26.	*Gobio gobio* (Linnaeus, 1758)	n				?
		27.	*Gobio ohridanus* (Karaman, 1924)	n				VU
	9.	28.	*Gymnocephalus cernua* (Linnaeus, 1758)				i	
	10.	29.	*Hypophthalmichthys molitrix* (Valenciennes, 1844)	i			i	
	11.	30.	*Pachychilon macedonicum* (Steindachner, 1892)			n		DD
		31.	*Pachychilon pictum* (Heckel et Kner, 1858)	n				LC
	12.	32.	*Parabramis pekinensis* (Basilewsky, 1855)		i			
	13.	33.	*Pelasgus minutus* (Karaman, 1924)	n				DD
		34.	*Pelasgus prespensis* (Karaman, 1924)	n				EN
	14.	35.	*Phoxinus limaireul* (Schinz, 1840)	n				LC
		36.	*Phoxinus phoxinus* (Linnaeus, 1758)	n				LC
	15.	37.	*Pseudorasbora parva* (Temminck & Schlegel, 1846)		i		i	
	16.	38.	*Rhodeus amarus* (Bloch, 1782)		i			
		39.	*Rhodeus meridionalis* (Karaman, 1924)				i	
	17.	40.	*Romanogobio elimeius* (Kattoulas, Stephanidis, & Economidis, 1973)			n		LC
	18.	41.	*Rutilus ohridanus* (Karaman, 1924)	n				LC
		42.	*Rutilus prespensis* (Karaman, 1924)	n				VU
		43.	*Rutilus rutilus* (Linnaeus, 1758)			n		LC
		44.	*Rutilus karamani* (Fowler, 1977)	n				DD
	19.	45.	*Scardinius erythrophthalmus* (Linnaeus, 1758)			n		LC
		46.	*Scardinius knezevici* (Bianco & Kottelat, 2005)	n				LC

(continued)

Table 2 (continued)

FAMILY	GENUS	SPECIES	FAMILY/SPECIES	Adriatic		Aegen		IUCN
				Native	Introduced	Native	Introduced	
	20.	47.	*Squalius prespensis* (Fowler, 1977)	n				LC
		48.	*Squalius squalus* (Bonaparte, 1837)	n				LC
		49.	*Squalius vardarensis* (Karaman, 1928)			n		LC
	21.	50.	*Tinca tinca* (Linnaeus, 1758)				i	
	22.	51.	*Vimba melanops* (Heckel, 1837)			i		
VI			COBITIDAE					
	1.	52.	*Cobitis meridionalis* (Karaman, 1924)	n				VU
		53.	*Cobitis ohridana* (Karaman, 1928)	n				LC
		54.	*Cobitis vardarensis* (Karaman, 1928)			n		LC
		55.	*Cobitis strumicae* (Karaman, 1955)			n		LC
	2.	56.	*Sabanejewia balcanica* (Karaman, 1922)			n		LC
VII			NEMACHEILIDAE					
	1.	57.	*Barbatula barbatula* (Linnaeus, 1758)			n		LC
		58.	*Barbatula sturanyi* (Steindachner, 1892)	n				LC
	2.	59.	*Oxynoemacheilus bureschi* (Drensky, 1928)			n		LC
VIII			ICTALURIDAE					
	1.	60.	*Ameiurus nebulosus* (Lesuer, 1819)				i	
IX			SILURIDAE					
	1.	61.	*Silurus glanis* (Linnaeus, 1758)		i	n		LC
X			ESOCIDAE					
	1.	62.	*Esox lucius* (Linnaeus, 1758)				i	
XI			SALMONIDAE					
	1.	63.	*Oncorhynchus mykiss* (Walbaum, 1792)		i		i	
	2.	64.	*Salmo aphelios* (Kottelat, 1997)	n				DD

		65.	*Salmo balcanicus* (Karaman, 1927)	n				DD
		66.	*Salmo farioides* (Karaman, 1938)	n				DD
		67.	*Salmo letnica* (Karaman, 1924)	n				DD
		68.	*Salmo lumi* (Poljakov, Filipi, & Basho 1958)	n				DD
		69.	*Salmo macedonicus* (Karaman, 1924)			n		DD
		70.	*Salmo marmoratus* (Cuvier, 1829)	n				LC
		71.	*Salmo ohridanus* (Steindachner, 1892)	n				VU
		72.	*Salmo pelagonicus* (Karaman, 1938)			n		VU
		73.	*Salmo pestericus* (Karaman, 1938)			n		EN
XII			POECILIDAE					
	1.	74.	*Gambusia holbrooki* (Girard, 1859)		i		i	
XIII			COTTIDAE					
	1.	75.	*Cottus gobio* (Linnaeus, 1758)				i	
XIV			CENTRARHIDAE					
	1.	76.	*Lepomis gibbosus* (Linnaeus, 1758)		i		i	
XV			PERCIDAE					
	1.	77.	*Perca fluviatilis* (Linnaeus, 1758)			n		LC
	2.	78.	*Zingel balcanicus* (Karaman, 1936)			n		DD
XVI			BLENNIDAE					
	1.	79.	*Salaria fluviatilis* (Asso, 1801)			n	/	LC
XVII			GOBIIDAE					
	1.	80.	*Economidichthys pygmaeus* (Holly, 1929)	n				LC

The species *Chondrostoma ohridanus* Karaman, 1924, occurring in Lake Ohrid is not categorized (Kottelat & Freyhof, 2007; The IUCN Red List Categories and Criteria, n.d.).

The presence of endemic fish characterizes the fish fauna in the Republic of North Macedonia. In fact, 7 fish species (33.3%) are endemic to Lake Ohrid: *Salmo ohridanus* (Steindachner, 1892); *Salmo aphelios* (Kottelat, 1997); *Salmo balcanicus* (Karaman, 1928); *Salmo letnica* Karaman, 1924; *Salmo lumi* (Poljakov, Filip, & Basho, 1958); *Barbatula sturanyi* (Steindachner, 1892); and *Gobio ohridanus* (Karaman, 1924).

Salmo balcanicus is one of four species of the Ohrid trout endemic to Lake Ohrid, along with *Salmo letnica*, *Salmo aphelios*, and *Salmo lumi*. The different forms are thought to be different species distinguished by their breeding season and habitat, making them reproductively isolated. *Salmo balcanicus* spawns specifically from October to January at the lake's outlet, the Black Drin river, at the northwestern end of the lake near the Macedonian-Albanian border area.

9 species of fish (82%) are endemic to Lake Prespa: Prespa trout (*Salmo peristericus*), Prespa spirlin (*Alburnoides prespensis*), Prespa bleak (*Alburnus belvica*), Prespa barbel (*Barbus prespensis*), Prespa nase (*Chonrostma prespense*), Prespa spined loach (*Cobitis meridionalis*), Prespa minnow (*Pelasgus prespensis*), Prespa roach (*Rutilus prespensis*), and Prespa Chub (*Squalius prespensis*).

7 species of fish are endemic to Lake Dojran (Table 2): Macedonian bleak (*Alburnus macedonicus*), Danube barbel (*Barbus balcanicus*), Macedonian roach (*Pachychilon macedonicum*), Vardar Chub (*Squalius vardarensis*), Vardar spined loach (*Cobitis vardarensis*), Golden spined loach (*Sabanejewia balcanica*), and Vardar streber (*Zingel balcanicus*) (Talevski & Milosevic, 2016; Talevski & Talevska, 2018).

6 The Conservation Status of the Fish in the Republic of North Macedonia

The IUCN Red List (The IUCN Red List Categories and Criteria, n.d.) is widely recognized as the most objective and authoritative system for grading species according to their risk of extinction. It is based on precise criteria relevant to all fish species and all regions of the world. In addition, the IUCN Red List aims to convey to the public the urgency of species conservation and help the international community to reduce species extinction. In the IUCN Red List, fish species are classified into nine groups determined by criteria such as rate of decline, population size, area of geographic distribution, and degree of population and distribution fragmentation.

Categories included are EX—Extinct, EW—Extinct in the Wild, CR—Critically Endangered, EN—Endangered, VU—Vulnerable, LR/cd—Lower Risk/conservation dependent, NT—Near Threatened (includes LR/nt—Lower Risk/near threatened), LC—Least Concern (includes LR/ LC—Lower Risk, most minor concern), DD—Data Deficient, has not yet been evaluated against the criteria, and NE—Not Evaluated—has not yet been assessed against the requirements.

According to the IUCN Red List, fish species from the Republic of North Macedonia (80 in total) belong to the following categories: Least Concern (LC)—35 fish species, Critically Endangered (CR)—3 species, Endangered (EN)—2 species, Vulnerable (VU)—9 species, Near Threatened (NT)—1 species, Not Evaluated against the criteria (NE)—1 species, and Data Deficient (DD)—10 species. In addition, 18 species of fish are introduced into the aquatic ecosystems of the Republic of North Macedonia and excluded from the list of endangered species (Table 2) and evaluated.

The conservation status of the fish from the Republic of North Macedonia is now being researched at the tremendous natural lakes: Lake Ohrid (Talevski et al., 2010b; Talevski & Milosevic, 2016; Stefanoic, 1948), Lake Prespa (Milošević & Talevski, 2015; Talevski & Milosevic, 2016; Stefanoic, 1948), and Lake Dojran (Talevski & Milosevic, 2016; Talevski & Talevska, 2018) (Table 2). Also, threats and dangers of commercial fishing in the great natural lakes are studied for the native species (Talevski et al., 2010b; Milošević & Talevski, 2015).

According to Kotellat and Freyhof (2007), Lake Ohrid listed on the IUCN Red List, *Anguilla anguilla* (Linnaeus, 1758) is a Critically Endangered (CR) species, while *Salmo aphelios* (Kottelat, 1997) is potential Endangered species (EN). *Alburnoides ohridanus* (Karaman, 1928), *Cyprinus carpio* (Linnaeus, 1758), *Gobio ohridanus* (Karaman, 1924), *Salmo ohridanus* (Steindachner, 1892), and *Scardinius knezevici* (Bianco & Kottelat, 2005) are Vulnerable (VU) species. *Eudontomyzon stankokaramani, Alburnus scoranza* (Heckel et Kner, 1858), *Barbatula sturanyi* (Steindachner, 1892), *Barbus rebeli* (Köller, 1926), *Chondrostoma ohridanus* (Karaman, 1924), *Cobitis ohridana* (Karaman, 1928), *Pachychilon pictum* (Heckel et Kner, 1858), *Phoxinus limaireul* (Schinz, 1840), *Rutilus karamani* (Karaman, 1924), *Rutilus ohridanus* (Karaman, 1924), and *Squalius squalus* (Bonaparte, 1837) are Least Concerns (LC). *Pelasgus minutus* (Karaman, 1924), *Salmo balcanicus* (Karaman, 1928), *Salmo letnica* (Karaman, 1924), and *Salmo lumi* (Poljakov, Filip, & Basho, 1958) are Data Deficient (DD) species (Table 2).

According to the IUCN Red List of Ichthyofauna in Lake Prespa (Table 2), 1 species (*Anguilla anguilla* (Linnaeus, 1758) is Critically Endangered (CR), 2 species are EN (Endangered species), and 8 species are Vulnerable (VU), while in Lake Dojran, 2 species: *Anguilla anguilla* (Linnaeus, 1758) and *Alburnus macedonicus* (Karaman, 1928) are Critically Endangered (CR), 1 species is Vulnerable (VU), 10 species are Least Concern (LC), and 1 species is Data Deficient (DD).

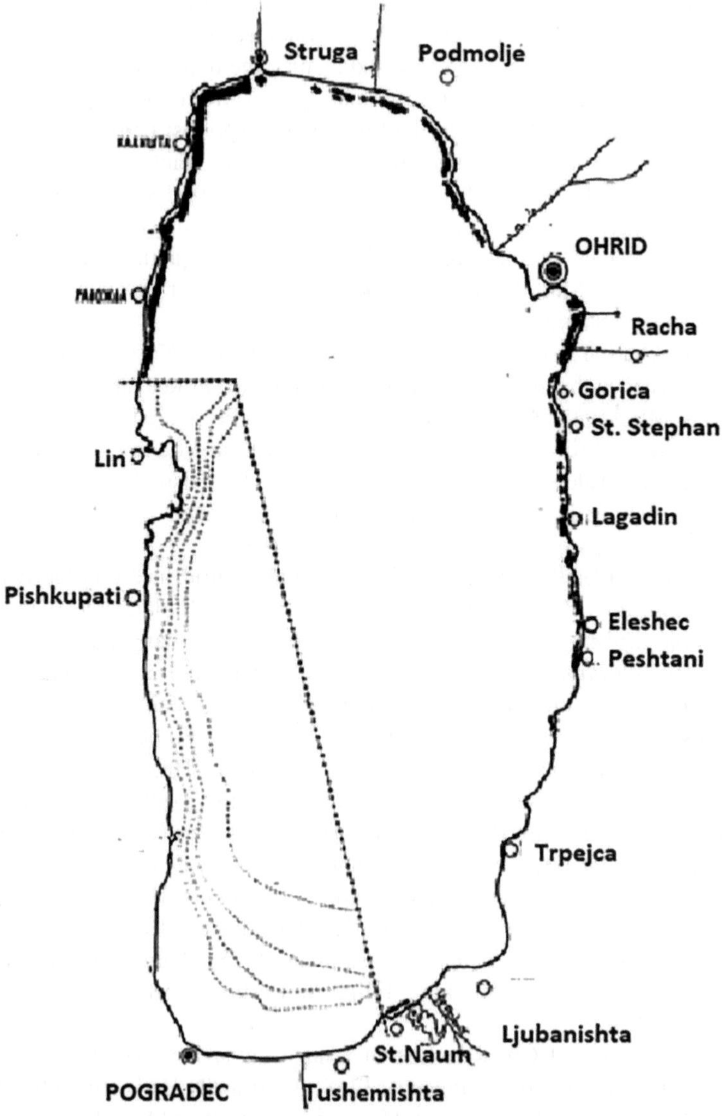

Fig. 4 Belt of Reed, *Phragmites australis* from Lake Ohrid (Talevska, 1996)

7 Valuable Fish Habitats

7.1 Macrophyte Vegetation as Habitat and Spawning Ground of Fish

In Lakes Ohrid, Prespa, and Dojran, three types of macrophyte vegetation are usually discernible from the shore to deeper water: emerged, floating, and submerged. The dominant plant species in lakes is the reed *Phragmites australis* (Cav.) Trin. Ex Steud. which form the natural discontinuous belt (Figs. 4 and 6). The other emergent plants (*Typha latifolia* L., *Typha angustifolia* L., and *Schoenoplectus lacustris* (L.) Palla) are rarely present. Floating plants are rare, and submerged macrophytes in lakes are *Potamogeton perfoliatus* L., *Stuckenia pectinata* Gmel., *Potamogeton lucens* L., *Potamogeton pusillus* L., *Zannichellia palustris* L., *Myriophyllum spicatum* L., *Myriophyllum verticillatum* L., *Ceratophyllum demersum* L., *Ceratophyllum submersum* L., *Vallisneria spiralis* L., *Najas marina* L., *Najas minor* All., *Chara globularis* Thuiller 1799, *Chara vulgaris* L., *Nitella opaca* (Bruz.) Ag., and others (Talevski & Talevska, 2018; Talevska & Talevski, 2019a; Talevska & Talevski, 2019b; Talevska, 2003; Talevska, 2009; Talevska & Talevski, 2015; Spirkovski et al., 2002). In lakes, the green alga *Cladophora* sp. adhere to sand, gravel, rocks, or plant organs. Macrophytes provide food, habitat, shelter, and spawning grounds for many carp fish and play an important role in protection from their predators (Spirkovski et al., 2001; Talevska, 2002; Talevski & Talevska, 2008; Talevska & Talevski, 2009; Talevski et al., 2009a; Talevski et al., 2009b; Talevska & Talevski, 2010), (Talevska & Talevski, 2018a; Talevska & Talevski, 2018b; Talevska, 1996).

The fish from Macedonian aquatic ecosystems can be divided into four groups depending on the spawning substrate (Tocko, 1987):

(a) psamophyl—on fine sand substrate
(b) Phytophil—on macrophyte vegetation
(c) Litophil—on gravel and rock substrate
(d) Fluviatile—spawning in the tributaries

Therefore, according to (Tocko, 1987), the following spawning grounds can be identified:

1. Shallow sandy bottom with underwater vegetation. This species is the most common, except on the rocky parts of the lake shore and bottom. In the past, before the lakes were mined, it was mainly grayling and chub that came to the beaches during the spawning season (in very shallow water and even on dry land). However, both species have migrated to the deeper parts of the lakes (quite far from shore) for spawning where submerged vegetation occurs due to lowering of the water level (Tocko, 1987).
2. Shallow sand bottom with emerged vegetation (reed belts). This type—the reed belts—concerns mainly the common carp (*Cyprinus carpio*) and part of the native

Table 3 Spawning periods, substrate, and regions of Ohrid trout forms

Fish species	Spawning period	Spawning substrate and region
Salmo balcanicus	January–March	Between stones, gravel, and sand/littoral and sublittoral (east and west coast)
Salmo letnica	December–January	Fine sand/littoral and south coast and formerly in River Crni Drim
Salmo lumi	November–January	Sheltered part of the tributaries
Salmo aphelios	July–August	Rocky bottom/sublittoral around amount sublacustrine springs with a temperature of 10.5 °C (Gradishte–Veljapesh). East coast

roach population. With the changes in the lake's water level, parts of these reed belts have gone ashore and other parts have extended further into the lakes. It is worth mentioning that the carp spawning can also take place outside the reed belts and in deeper waters (Tocko, 1987).

3. Rocky parts of the bottom, like spawning ground for fish, are the smallest part of the lake bottom. With the drop in water levels, the presence of this type of substrate in the lakes changes significantly. This mainly affects the endemic fish species, especially the barbel population, since their eggs, in contrast to the other carp fish, do not stick to the substrate (Tocko, 1987).

   There are two important spawning seasons for fish in Lake Ohrid. On the one hand, the winter period, in which mainly the Ohrid trout (Table 3) and Belvica spawn, and on the other hand, the spring-summer period, in which the carp did the same thing.

Cyprinid fish species from lakes in the Republic of North Macedonia are mostly phytophilic species in relation to the substrate on which they lay their eggs, since they lay their eggs on plants (Spirkovski et al., 2001; Talevska, 2002; Talevski & Talevska, 2008). The various carp fish spawn in a certain month in the spring-summer period in other regions of Lake Ohrid and on different substrates. Any change in the qualitative and quantitative composition of macrophytic vegetation inevitably leads to changes in spawning grounds. Spawning in the same place under new conditions inevitably leads to less efficient natural spawning: less developed juvenile fish and therefore fewer adult fish (Talevska & Talevski, 2019a; Talevska, 1996). In Lake Ohrid, carp spawn in the reed belts of *Phragmites australis* (Talevska & Talevski, 2019b) (Fig. 4) and in the belt of Perfoliate pondweed, *Potamogeton perfoliatus* (Grillas, 2011) (Fig. 5).

Previous research has shown that fish (including endemic and commercial species) spawn in many (if not all) areas along coasts (Talevska, 2002; Talevski & Talevska, 2008; Talevska & Talevski, 2009; Talevski et al., 2009a; Talevski et al., 2009b; Talevska & Talevski, 2010), (Talevska & Talevski, 2018a; Talevska & Talevski, 2018b; Talevska, 1996). Nevertheless, certain zones in the lake are of particular importance for the local fish fauna, as certain sites appear to be excellent spawning sites or provide shelter from predators due to the composition of the

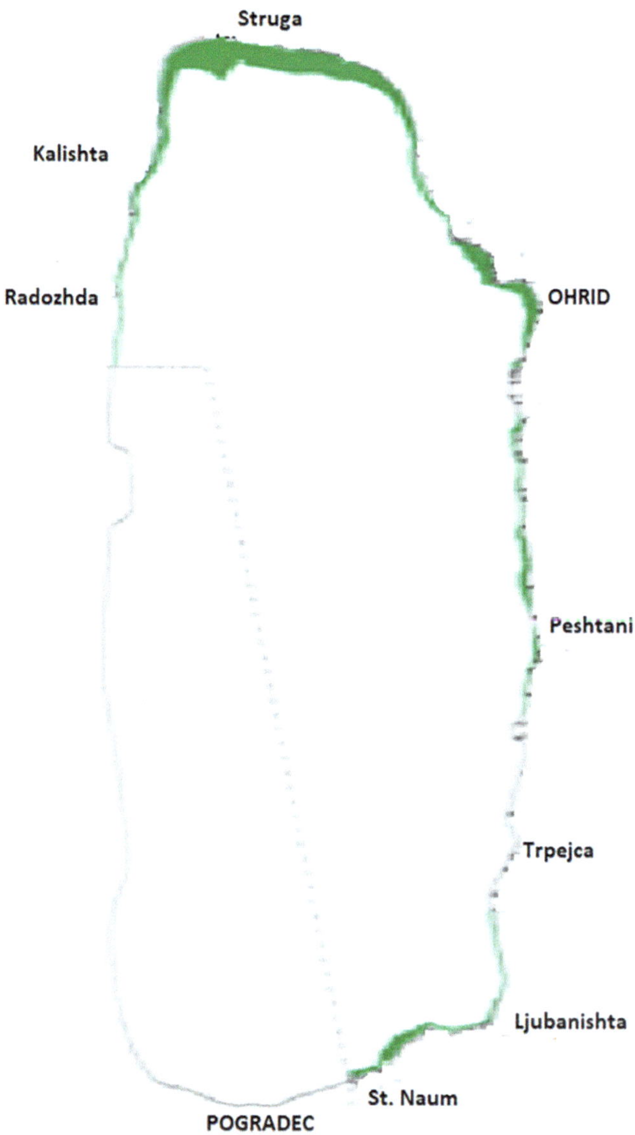

Fig. 5 Belt of Perfoliate pondweed, *Potamogeton perfoliatus* from Lake Ohrid (Talevska, 2005)

macrophyte vegetation. The type of substrate (soil) and the composition of macrophyte vegetation in Lake Ohrid (Figs. 4 and 5) provide favorable spawning sites for *Alburnoides ohridanus*, *Alburnus scoranza, Barbus rebeli, Chondrostoma ohridanus, Cyprinus carpio, Gobio ohridanus, Pachychilon pictum, Pelasgus minutus, Phoxinus limaireul, Rutilus ohridanus, Scardinius knezevici*, and *Squalius squalus* (Talevska & Talevski, 2010; Talevska & Talevski, 2018a).

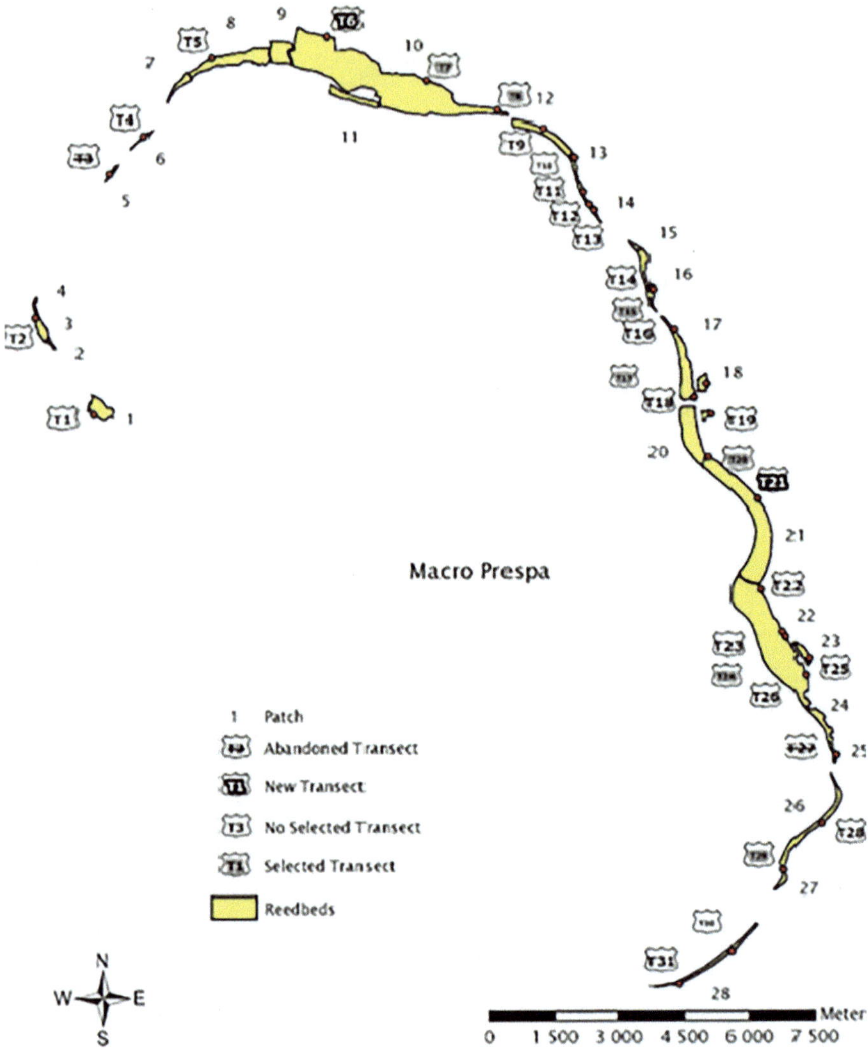

Fig. 6 Reed belt from Lake Prespa, Macedonian coastline (Grillas, 2011)

The spawning season for most fish species in Lake Prespa is from April to June, with the exception of salmonids (trout) which spawn in the connected rivers from November to March. Another exception is the lien pumpkin seed (*Lepomis gibbosus*), which spawns twice a year—in spring and autumn. For reproduction, the fish prefer different habitats within the lake. For example, lithophile species (like spirlin) lay their eggs on gravel and rocky substrates, while phytophiles (like carp) require submerged vegetation for spawning (Talevski et al., 2009a).

There are 11 native fish species in Lake Prespa, of which *Phragmites australis* (Talevski et al., 2009a) (Fig. 6) live and spawn in the reed belt.

The following fish species are found in Lake Dojran: *Rutilus rutilus*, *Cyprinus carpio*, *Leucaspius delineatus*, *Scardinius erythrophthalmus*, *Tinca tinca*, *Alburnus alburnus*, *Rhodeus amarus*, *Pachychilon macedonicum*, *Gobio gobio*, *Barbus macedonicus*, *Carassius gibelio*, *Cobitis vardarensis*, *Silurus glanis*, *Anguilla anguilla*, *Perca fluviatilis*, *Salaria fluviatilis*, and *Sabanejewia doiranica (Talevski & Milosevic,* 2016*; Talevski & Talevska,* 2018*)* (Table 3).

8 Fishing Tools in the Republic of North Macedonia

Fishing tools used on lakes in the Republic of North Macedonia have certain adaptations to the fish species and the characteristics of the lakes.

In the past, the fishing tools used on Lake Ohrid were made from natural materials. It differed according to its dimensions, the places where it was used, what species of fish was caught, and others.

9 Fishing Tools in Lake Ohrid

According to Stankovic, 1960 (Stankovic, 1960), the fishing techniques on Lake Ohrid and the fishing tool used since ancient times are not only adapted to the ecological characteristics of the target fish species but also to the characteristics of the lake such as the size and depth of the lake and the underwater relief of the lake. Fishing on Lake Ohrid is a pro using extensive fishing tools that experienced fishermen use. It is clearly shore fishing in the littoral and sublittoral area of the lake, no deeper than 80 meters. Fishing tools used in Lake Ohrid are grouped into the following categories: seine nets (vlak), standing gill nets, angling with fish hooks, and fishing installations called "daljani."

Fishing with nets is the most significant and complex. This mainly applies to fishing with vlak, the seine nets, the most effective type of nets used for mass fishing on the lake. Regardless of the differences in their construction (length and height of the net and mesh size), all Vlak types are made of cotton thread in an effectively rectangular shape with narrowed edges and a cone bag in the middle. The top and bottom edges of the net are reinforced with ropes that are placed along the entire length of the net. On the bottom rope, weights are placed at different distances (e.g., rocks), which helps for adequate placement of the rope on the lake floor, and floats are attached to the upper rope. On both ends of the net long ropes are attached, with the help of which it is pulled toward the shore. The net is set up with the use of a boat in the form of a semicircle, after which it is dragged and pulled out either on the shore or another boat.

There are more seines used in Lake Ohrid and each one is used for fishing specific species of fish (Megdanski Vlak, Vlachka, Grunchar, Vlache, Peshtanski Vlak, Sachma (Fig. 10), and others) (Stankovic, 1960). Megdanski vlak (Stankovic,

Fig. 7 Peshtanski vlak (Photo German Nikola)

1960) is the largest seine net used on Lake Ohrid. Its wings are longer than 150 m and its height reaches from 7 to 11 m. This type of seine net's maximum height is at the cone bag's level and amounts to 11 m. The wings have mesh sizes with larger dimensions, while the cone bag is made of a denser net with a mesh size of 25 mm and a length of 16 m. This type of seine is mainly used for carp fishing and partly for fishing other cyprinid fish species. It is mainly used during the winter season of the year and spring, especially in May, when the lake's macrophytes are not yet overgrown making it easier to pull the net on the bottom of the lake. In summer, this net is only used in a few places on the lake shores, where macrophytes are less common.

Vlachka (Stankovic, 1960) is a seine used mainly for fishing Belvica (*Salmo ohridanus*). It differs from Megdanski Vlak in that it has a shorter length and height and a smaller mesh size.

Grunchar (Stankovic, 1960) is a seine similar in length to the vlachka but slightly larger. This seine is used to catch cyprinids, mainly the Ohrid roach (*Rutilus ohridanus*). The name Grunchar was derived from the genus name of the Ohrid roach—grunec (Stankovic, 1960).

The smallest of all the seines used in Lake Ohrid is Vlache, a seine with a mesh size of no more than 10–12 mm, except for the ends of the wings where the mesh size

Fig. 8 Complete fishing crew for Ohrid chun fishing (Foto German Nikola)

is slightly larger. The Vlache is about 70 m long and 9 m high. It is used for fishing for bleak and other smaller species of fish.

The most complicated method of construction and use of the autochthonous seine used in Lake Ohrid is the so-called Peshtanski Vlak (Figs. 7 and 9). Named after the village of Peshtani on the east shore of the lake. Nevertheless, it is used in various fishing grounds on the other side of the lake. This type differs from all other seine nets in that it is placed at greater depths, up to 80 m in the subtidal area. The Seine net is not pulled onto the bank, although fishermen pull the net from the bank. The entire maneuver of setting and lifting the net is carried out by boats (Fig. 8), always on the border of the littoral zone (18–20 m depth). This seine is mainly used during the summer season and would be totally useless in the littoral area overgrown with macrophytes. The build of this seine net is very complicated. The basic shape of the net is in the form of an ellipse. The wing tips have a height of 16.2 m, while in the middle, where the cone bag is located, the height is about 39.6 m. The total length of the net is about 100.8 m. In the structure of the wings and the cone bag, there are different parts with different mesh sizes. At the beginning of the wings there is a part of the net called Vezgja with a height of 16.2 m and a mesh size of 2 cm. This part protects the nets from tearing during the dragging, and the outer part carries a rope (Staliga) on which the main dragging rope is attached. After this part, toward the middle of the net, there are initially two fields, one with a mesh size of 30 mm, a height of approx. 21.6 m, and a width of 32 meshes, and the other with a mesh size of 45 mm, the height of about 23.4 m, and the width of 22 meshes. This part is followed by 10 pieces with larger mesh sizes of 65 mm, with a width of 32 meshes and a cascade height of 23.4–36 m, with the last 4 pieces having a maximum height of

36 m. After these parts comes the center of the net, which is the most complicated of all the other parts mentioned before. From each side of the middle and toward the bottom and top of the net are the following parts: 1. small tail, 10.8 m high, 22 mesh wide, 45 mm in size; 2. large tail with a height of 12.6 m, a width of 22 meshes with a size of 45 mm; third in the center with a height of 18 m, a width of 22 meshes with a size of 45 mm (this part of the mesh is called a Stoche).

Between these parts, toward the top and bottom of the net and partially entering the cone bag, are the following parts: 1. "eleven" with a height of 19.8 m, a width of 22 meshes with a size of 45 mm, toward placed the "little tail"; 2. "thirteen" with a height of 23.4 m, a width of 32 meshes with a size of 35 mm, placed toward the "big tail"; 3. two so-called "bindermini," with a height of 12.6 m, a width of 32 meshes, and a size of 35 mm, placed in the "middle." The cone bag is made up of the following parts: 1. five so-called "kaculi," each with a width of 32 meshes with a size of 20 mm, the first four with a height of 5.4 m, and the fifth with a height of 18 m; 2. five so-called "tepini" with a height of 3.6 m, a width of 42 meshes, and a size of 15 mm; 3. the so-called "dipche," which is the bottom of the cone bag, with a height of 1.8 m, built into the "tepini" with a mesh size of 15 mm. The top and bottom edges of the net are secured with ropes made up of different parts ("Isina," "Big Ise," "Small Ise"). The top and bottom edges of the net are secured with ropes made up of different parts (Isina, Big Ise, Small Ise). The top rope carries 25 floats along its length, of which the middle one is the largest and is called Mana. The lower rope has 2 kg rocks attached, called Shenci, and the middle one is called Mantash. The rocks attached to the Staliga ropes are called Petli.

Peshtanski vlak (Fig. 8) is unique with its intricate construction, and this form of netting is rarely seen in the seine used in other lakes in Europe and on the seas. This seine is a unique type used in Lake Ohrid, specialized for use in the deeper parts of the lake where vegetation is absent and where the summer form of the Ohrid trout lives. The enormous height of the net makes it difficult for the fish to escape. The maneuver with this seine net is complicated, and the dragging is exhausting because the net is dragged on a steep bottom; because of its height, it has to fight water resistance. The dragging is performed from the shore, and the ropes go over two boats to reach it. The boats, during the dragging, are gradually close to each other. This maneuver is used to close the net, after which the net is lifted on the boats. Because of the structure and the method of using, this type of seine can be used only on areas of the lake with an adequate bottom (Stankovic, 1960).

Another unique type of seine used in Lake Ohrid is the so-called "kluchot." This seine net with a characteristic structure is only used on Lake Ohrid. This net gives the impression of a single-winged seine net without the cone bag. The net is 32.4 m long and only 3.6 m high at the leading edge. The back edge is 12.6 m in height. Two metal circles are attached to this edge, with a rope running past them. Attached to the end of the rope is a rock called "Gla." Pulling the rope closes the top and bottom edges of the net, causing the net to fold itself and form a pouch. This type of seine is used for fishing adult trout at their spawning grounds. The maneuver to use this net begins with placing the front end of the net in the water and turning the boat in a circle around the spawning ground until the back end of the net reaches the front end.

Fig. 9 Cleaning and repairing fishing nets (Foto German Nikola)

After the net has been set, the ropes are pulled in. This seine is particularly effective for catching trout during the spawning season (Stankovic, 1960).

Standing gill nets are also very effective fishing tools. In contrast to the seine, these nets are placed vertically at a certain depth, where they rest in the form of a long net wall, blocking the way for the fish. The fish will get caught in the meshes when trying to pass through them. Standing gillnets are in the form of massive elongated rectangles. Rope weights are attached to the lower rope and a multitude of small floats are attached to the upper rope. Gill nets are made from cotton, silk, and, more recently, nylon threads. The nets are put into the water at dusk and lifted at dawn.

There are different types of standing nets. "Mreza" is a standing net with a mesh size of up to 45 mm, 2 m high, and 40 m long. It is mainly used for fishing trout in winter season. "Mreica" is a standing gill-net with a mesh size of up to 30 mm. It is

Fig. 10 Fishing by Sachma (Foto Radovan Dimitrievski)

used for fishing Belvica and lies at a depth of 50 m. “Plashicarnik” is a standing net up to 2 m with a mesh size up to 15 mm and is used for fishing for bleak. Finally, “Gruneshnik” is a vertical net used for fishing roaches, and “Kraparnik” is used for fishing carp. “Kraparnik” has the enormous mesh sizes of all standing nets used in Lake Ohrid.

Triple standing nets are rarely used in Lake Ohrid, although they are very effective. This standing net is a triple wall net where the two outer parts have enormous mesh sizes and the middle one has a mesh size of 30–35 mm.

After each fishing all fishing tools are cleaned, dried, repaired, and prepared for the next fishing (Figs. 9 and 10).

As a professional fishing type, angling is uncommon on Lake Ohrid. However, angling is used for fishing eel, chub, and other fish species.

An ancient type of fishing used on Lake Ohrid is a permanent installation called “daljani.” Such fishing installations are found throughout the Balkan lakes and are based on the labyrinth principle with fishing chambers known as “kotec.” The chambers are of rods and canes, shaped like a heart, with a narrow opening. When the fish enters the narrow opening, it is trapped and cannot escape. All parts of the installations, which can be very complicated, are designed to guide the fish toward the fishing chamber. In other words, these installations are automatic fishing tools placed in fish movement and migration areas.

Fig. 11 Model of fishing installation—“daljan” (Foto Radovan Dimitrievski)

The most famous installations on Lake Ohrid were located on river Drim near Struga (Fig. 11). They were built with vertically and densely placed rods. These installations were used for fishing eels. The eel from the lake toward the river Drin reaches the installation’s wall. Looking for a passage, it enters the channels that have closed ends. From there, the eels are forced to join the fishing chamber, where they remain trapped.

9.1 *Fishing Tools in Lake Prespa*

Similar fishing tools are used on Lake Prespa with specific adjustments to the fish and the characteristics of the lake (Vlak, Grunchar, Vlache, and others). Also, in this lake, seine nets are dragged by fishermen from lake depths to the shore with ropes.

9.2 *Fishing Tools in Lake Dojran*

9.2.1 Fishing in Mandra

Dojran Lake is known for the traditional way of fishing with the help of birds (cormorants) in reed fences called mandras. This is one of the few places in the world where fish are caught using birds (the wings of which have previously been cut off, the so-called argat). The residents of Dojran are particularly proud of their fishing tradition.

Mandra is a body of water near the coast with a fisherman's house surrounded by reeds on three sides. Preparations for fishing in this old traditional way start in October and last until March. In October, both sides of the fisherman's house are fenced with reeds, and the side facing Lake Dojran remains open.

In November, when the cormorants come to the lake in search of food, they approach the fence and chase the fish in that direction. When the fishermen find that there is enough fish, they build the third part of the dairy. Over time, the enclosure narrows and gets closer to shore, creating a small area called the manger from which the fish are drawn (Apostolski, 1968).

10 Commercial and Recreational Fisheries

Commercial and recreational fisheries are essential sources of protein and contribute significant economic benefits to local and national economies (Arlinghaus et al., 2019; Cowx, 1994, 2002; Cowx & Portocarrero, 2011; Cowx et al., 2004, 2010). Commercial fishing was done explicitly to catch fish products for sale (Smith et al., 2011). In recent years, commercial fishing has repeatedly been identified as the main reason for the global decline in fish stocks (Talevska & Talevski, 2018b; Smith et al., 2015; Hilborn, 1976; Hilborn & Stokes, 2010; Hilborn et al., 2003; Hilborn, 2010). Global catch trends in freshwater fisheries are generally on the rise.

The main commercial fisheries are subject to intensive exploitation and can deplete fish stocks. Consequently, new slow-growing fish species that previously had no commercial importance or those at a lower trophic level are being sought (Hilborn et al., 2012; Hilborn et al., 2014; Pauly, 1988; Pauly et al., 2000).

Although there are assumptions, limitations, and cautions about this data, the message is clear that commercial fishing can deplete fishery resources. While opportunities to target some stocks remain, commercial catches continue to decline despite increased efforts (Pauly et al., 2000). Aquaculture proliferation is also a worrying trend and potentially unsustainable. Aquaculture relies heavily on feeds formulated from fish protein sources, putting further pressure on dwindling wild populations (Reynolds et al., 2001).

Recreational fisheries are typically those where fishing is undertaken by individuals for sport and recreation, with a possible secondary objective of catching fish for personal consumption (Naylor et al., 2000). Sometimes this definition is extended to include the sale of excess catches to offset costs (Cowx, 1994). Despite the perception that recreational fishing is benign, global participation is huge. Worldwide, recreational fishing is now highly developed and practiced by many people, mainly for pleasure but also for income generation and to supplement food supplies. The basic principles of recreational fishing are high effort and low effort (Pitcher & Hollingworth, 2002), while in commercial fishing they are high effort and low effort. This is relevant to understanding the characteristics of the fisheries and potential management and conservation strategies.

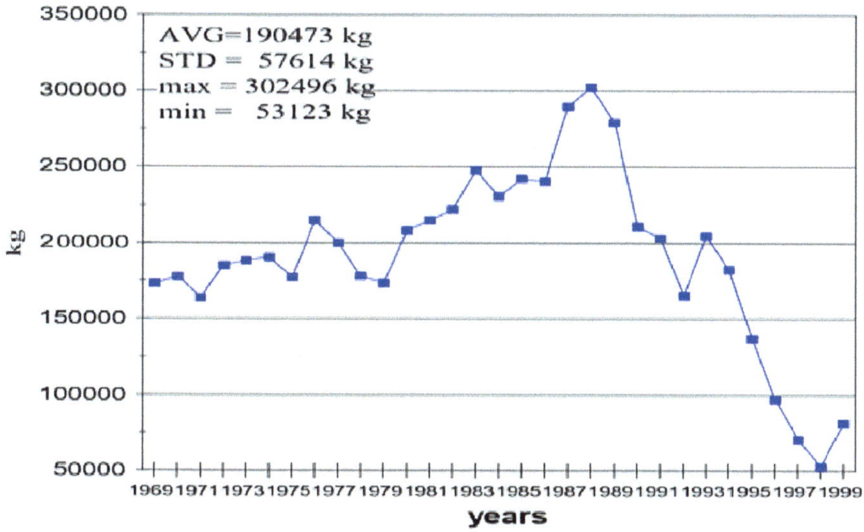

Fig. 12 Changes in the total annual catch from commercial fishing in the Macedonian part of Lake Ohrid for the period 1969–1999

10.1 Lake Ohrid

From 1945 to 1994, commercial and recreational fishing in Lake Ohrid was organized and operated by two state-owned companies, each responsible for a part of the lake, as the Macedonian lake district was divided into the Ohrid and Struga fishing sub-basins. These companies employed professional fishermen who were licensed by the state after passing fishing exams at the Ohrid Hydrobiological Institute, the authorized fisheries board. The main target of recreational fishing was the Ohrid trout and fishing was done from boats (4–5 m length) using nylon ropes with spinner hooks. At times, more than 2000 licenses were issued daily by the Recreational Fisheries Association. The catch per person (or permit) was limited to four specimens of Ohrid trout per day, but this limit was often far exceeded. In some years, the catches of recreational fisheries accounted for more than 50% of the annual income of commercial fishermen. However, these figures have never been included in fisheries statistics. In 1994, both state-owned fisheries were privatized. The total annual catches of the commercial fisheries are shown in Fig. 12. The composition of fish catches differed between the first half and the end of the twentieth century (Figs. 13 and 14), indicating a shift from salmonid to cyprinid species.

In 2002 and 2003, however, professional fishermen again mainly fished for trout.

In April 2004, the state government withdrew the concession for the fish stocks of Lake Ohrid because of unsustainable fishing. As a result, a total fishing moratorium was imposed until September 2012. Only fishing for scientific reasons (collection of Lake Ohrid trout spawn for artificial breeding) was carried out during this period.

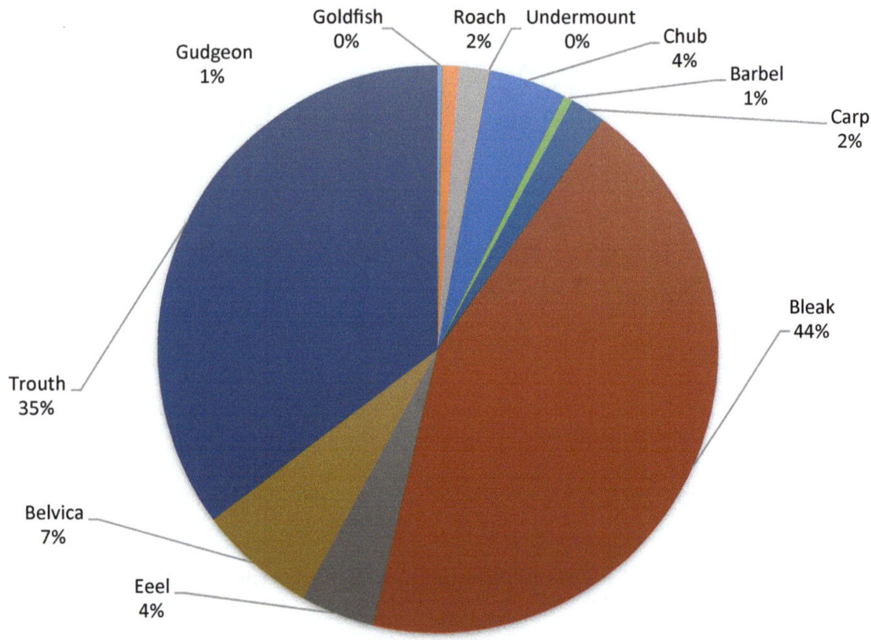

Fig. 13 Percentage participation of economically significant fish species in the total annual catch in Lake Ohrid in the period 1969–1999

After 8.5 years of total fishing moratorium on Lake Ohrid, a new concession was tendered and signed in September 2012. The current concessionaire currently employs 47 fishermen. Fig. 13 reflects the species composition of the total catch of the new concession period. It is a period of adjustment of the company to the special conditions in the lake, training of the fishermen, getting to know the habitats of the fish, and a relatively short fishing period.

As in the Republic of Albania, stock restocking measures with fish of different development stages were carried out in the Republic of North Macedonia. Nowadays, restocking of the Macedonian part of Lake Ohrid is usually done with juvenile fish that are nine months old. The number of hired individuals was determined year by year. According to the published data on fisheries statistics for the period 1969/2001 (Talevska & Talevski, 2018b), the fishing catch for the Macedonian part of Lake Ohrid amounts to an average of 185 tons with a maximum of 302 tons in 1988 and a minimum of 53 tons in 1998. The main stock consisted of Ohrid trout and Belvica (both endemic species) and bleak. Trout and Belvica contributed 42% of the total catch for that period (trout 35.6%, Belvica 6.8%)—again giving the lake the attribute of trout lake—while bleak was represented at 43.7%. Unfortunately, from 1999 to 2001, the two salmonids were present in the catch together with a reduced proportion of 19% for trout and 13.5% for belvica, while the proportion of bleak (Cyprinidae) on the other hand was increased to 55% (Fig. 16). Data from 2003 show enormous changes in trout population composition with respect to disturbance age and sex classes, as well as changes in spawning ecology (Talevska & Talevski,

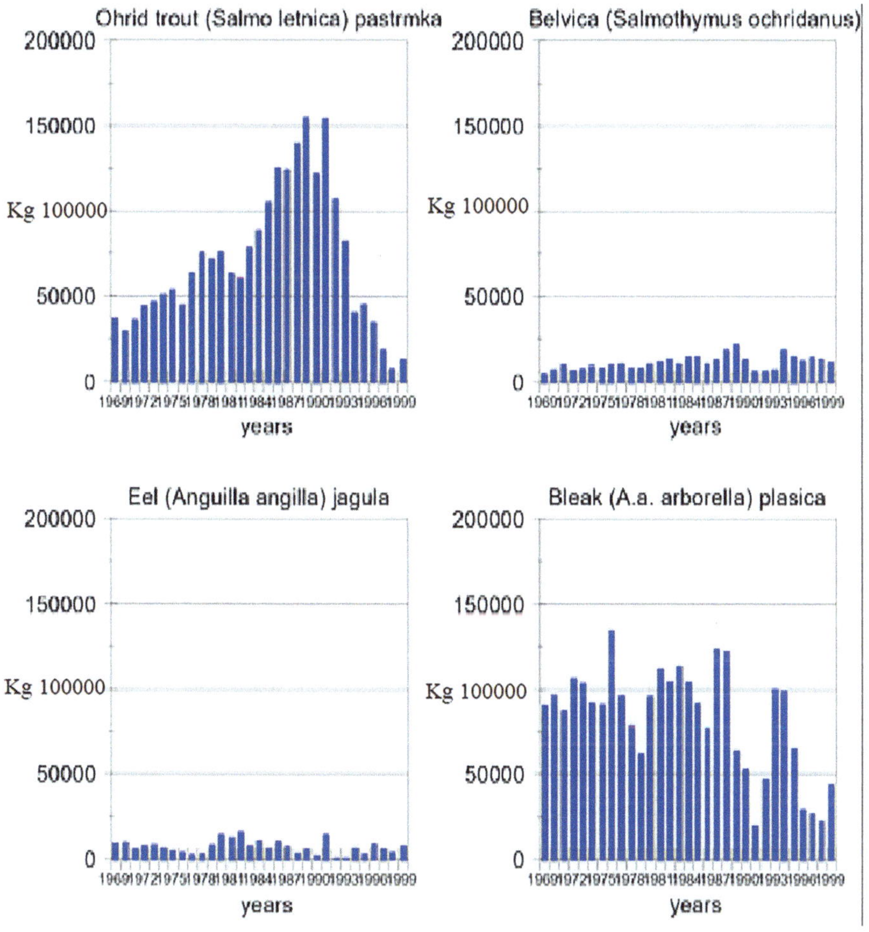

Fig. 14 Total annual catch by species over a more extended period in Lake Ohrid (Republic of North Macedonia part)

2018b). During the spawning season of trout in winter around a decade ago, there were designated zones in the upper littoral at water depths of 0.5–3 m, where trout mated and spawned. In these areas, the trout nest density was 5–8 per 10 meters. Today it is almost impossible to find nests in the upper littoral; Trout spawning takes place at greater depths. Overfishing appears to be the main cause of the decline in the trout population. However, the adverse impact of effluents leading to the destruction of natural fish breeding grounds and reducing the abundance of organisms on which autochthonous fish species feed plays a significant role (Pitcher & Hollingworth, 2002). Controls on the number and size of fish must be carried out and coordinated on both sides of the lake (Republic of North Macedonia and Republic of Albania). Because the fish in the lake are a single, connected population, they must be managed collectively, with similar requirements in neighboring countries, the Republic of North Macedonia and the Republic of Albania.

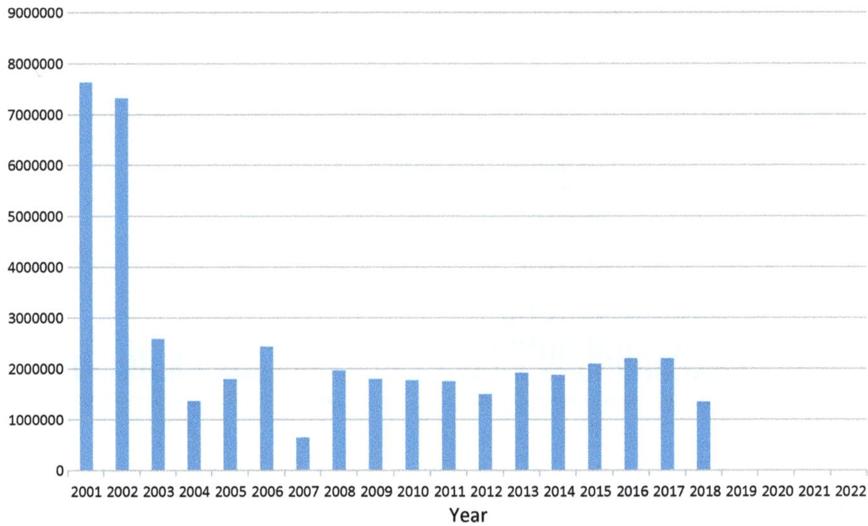

Fig. 15 Stocking of Lake Ohrid with juvenile Ohrid trout in the period 2001–2022

Table 4 Minimum legal size of catch for some commercial species at Lake Ohrid

Common name	Scientific name	Republic of North Macedonia	Republic Albania
Ohrid trout	*Salmo letnica*	35 cm	32 cm
Ohrid belvica	*Salmo ohridanus*	22 cm	30 cm
Carp	*Cyprinus carpio*	40 cm	30 cm
Bleak	*Alburnus scoranza*	12 cm	10 cm
Roach roach	*Rutilus ohridanus*	–	12 cm
Nase	*Chondrostoma ohridanus*	25 cm	–
Chub	*Squalius squalius*	30 cm	15 cm
Rudd	*Scardinius knezevici*	20 cm	12 cm
Moranec	*Pachychilon pictum*	10 cm	–
Eel	*Anguilla anguilla*	60 cm	25 cm

The socioeconomic pressures that have led to overfishing have hit trout harder than other fish stocks due to the greater demand and higher economic value of this fish. Another unfavorable moment is the presence of several introduced allochthonous fish species, such as gambusia, goldfish, rainbow trout, sunfish, and bitterlings, which through competitive relationships have an adverse effect on the lake's endemic fish fauna.

Stocking of Lake Ohrid with juvenile Ohrid trout in the period 2001–2022 is presented in Fig. 15.

The minimum legal catches for some commercial species at Lake Ohrid in the Republic of North Macedonia part and Albanian parts of Lake Ohrid are given in Table 4.

Table 5 Fishing ban due to spawning of fish species according to Republic of North Macedonian legislation

Common name	Scientific name R.N. Macedonia	
Ohrid trout	*Salmo letnica*	1st December–20th March
Ohrid belvica	*Salmo ohridanus*	1st November–31st January
Carp	*Cyprinus carpio*	20th May–19th June
Chub	*Squalius squalus*	1st May–31st May
Nase	*Chondrostoma ohridanus*	15th April–15th May
Rudd	*Scardinius knezevici*	1st May–31st May
Barbel	*Barbus rebeli*	20th May–19th June
Bleak	*Alburnus scoranza*	20th April–15th May

The catch of eels in the waters of Lake Ohrid has dropped drastically from ten tonnes to less than one ton in recent times.

10.2 Stocking for Fisheries

Fish stocking supplies wild stocks with fish from hatcheries in order to establish new fisheries, strengthen endangered or overfished native populations, or support recreational fishing. This global management practice has existed for over a century (Naylor et al., 2000; Pitcher & Hollingworth, 2002). Most stocking is done with native species.

In the past (from 1935 to the present), Lake Ohrid was stocked with young trout of different ages and in different quantities. The seasonal fishery for salmonid and cyprinid species in the Republic of North Macedonia part of Lake Ohrid includes 30 days of spawning, which can vary from year to year. However, it must be within the specified time frame as shown in Table 5.

Hydrological changes and the associated loss of habitats threaten the functioning of ecosystems and biodiversity in lakes (Pereira & Hansen, 2003). Altered water level regimes involve frequent, extreme fluctuations, and water level stabilization, creating novel environments around.

10.2.1 Lake Prespa

Based on various data, eight species of fish are endemic to the Prespa Lake basin and one to the Balkans. Seven are classified as vulnerable or endangered (threatened or critically endangered).

Figure 16 shows the total participation of individual fish species in the total annual catch in period 1946–2007 (Fig. 17).

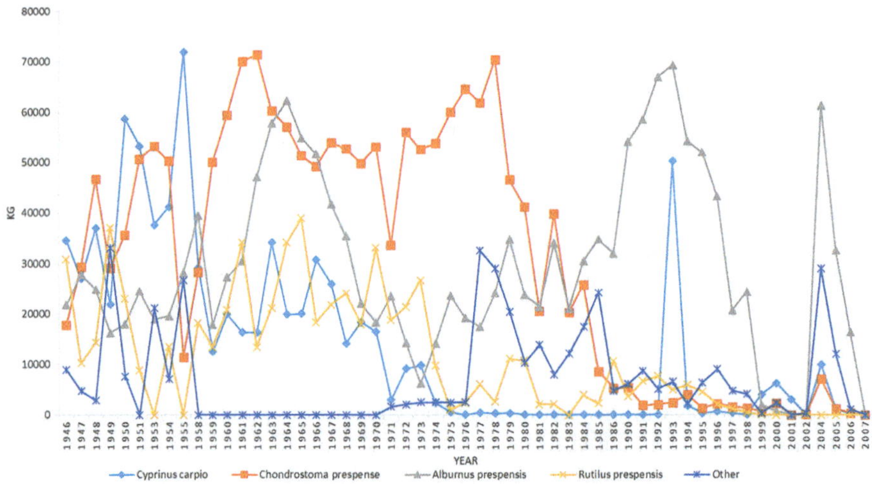

Fig. 16 Participation of individual fish species in the total annual catch in the period 1946–2007

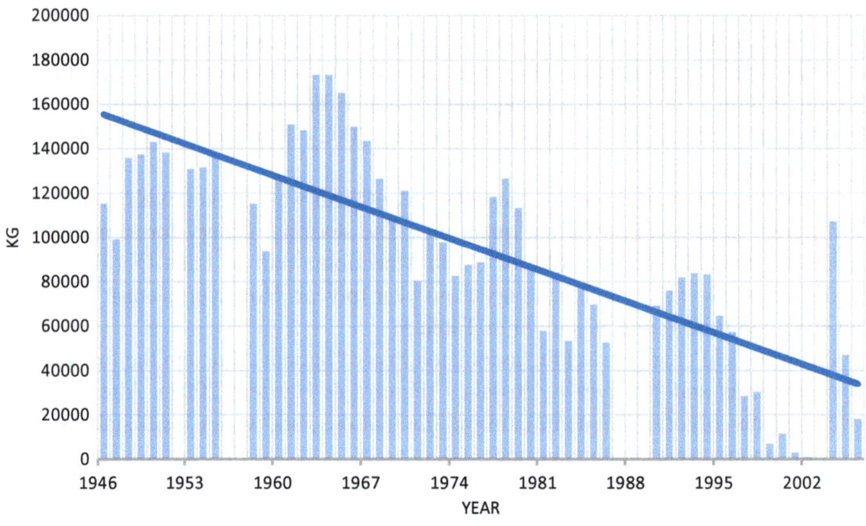

Fig. 17 Total annual catch over a longer period (1946–2002) and catch regression curve

From figures 16 and 17, it can be concluded that the catch of certain types of fish has been constantly decreasing in recent years as a result of the increased negative conditions for the fish populations that are present in the lake.

The first specimens of *Lepomis gibbosus* were caught in the mid-1990s, which have had a significant impact on the fish stocks of Lake Prespa in the years that followed and up to the present day.

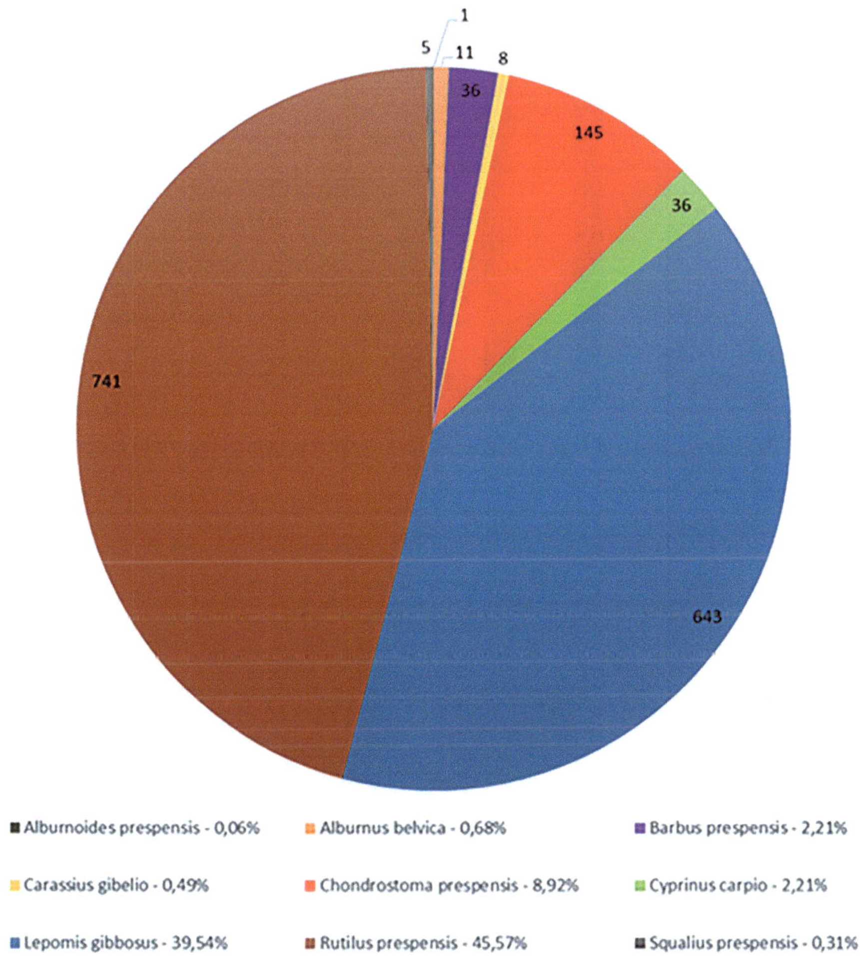

Fig. 18 Percentage participation of fish species in experimental catches from Lake Prespa

In Fig. 18, the percentage participation of individual fish species in the experimental catch from Prespa Lake is shown. A survey conducted in the fall of 2021 shows that the Lepomis gibbosus population accounts for a large percentage of the total trial catch (Trajchevski & Talevski, 2022).

The analyses of the length classes of Lepomis gibbosus obtained from the experimental catches from Prespa Lake (Fig. 19) indicate that individuals with a length of 10 and 11 cm dominate.

The development of the length-to-weight ratio of individuals of *Lepomis gibbosus* shows their excellent condition (Fig. 20).

The mesotrophic to eutrophic character of the Prespa Lakes allows for relatively high fish production. As a result, the annual catch has varied from 173 tons annually

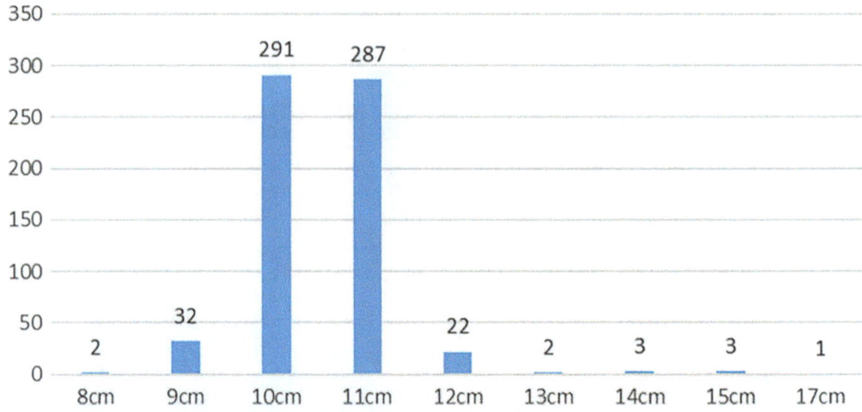

Fig. 19 The length classes of *Lepomis gibbosus* in the experimental catches from Prespa Lake

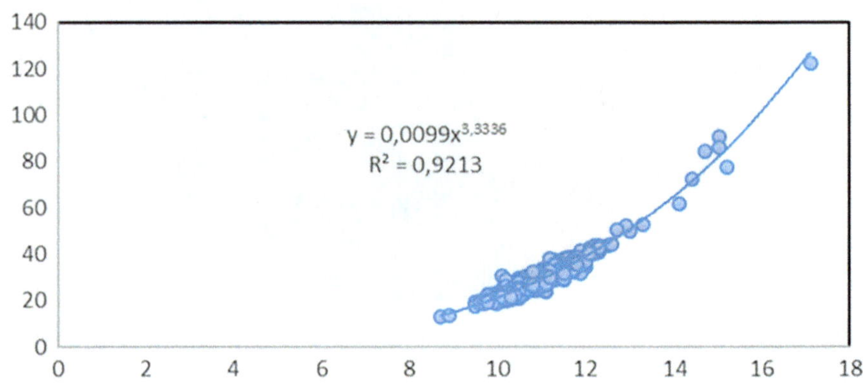

Fig. 20 The length–weight ratio of the *Lepomis gibbosus* population

in the 1960s to only about 20 tons toward the end of the last decade. Fish prices have risen steadily in the past. Since 2013 there have been concessionaires managing the fish stocks of Lake Prespa. Around 45 commercial fishermen work for the concessionaire on the Republic of North Macedonia part of Lake Prespa.

The fishing statistics for Lake Prespa go back to 1946. Depending on market demand and the fishing gear used, different species predominate in the annual catches. In the aftermath of World War II, fishermen used trawls and seine. In the early 1960s, purse-seine nets were introduced as a new fishing gear, which led to a higher proportion of bleaks in the annual catch at 55 tons per year.

The fishing sector represents a significant part of local income generation. About 40 licensed fishermen earn their income from fish from the lake. Another 50–60 people without a license catch fish from the lake to eat or sell it.

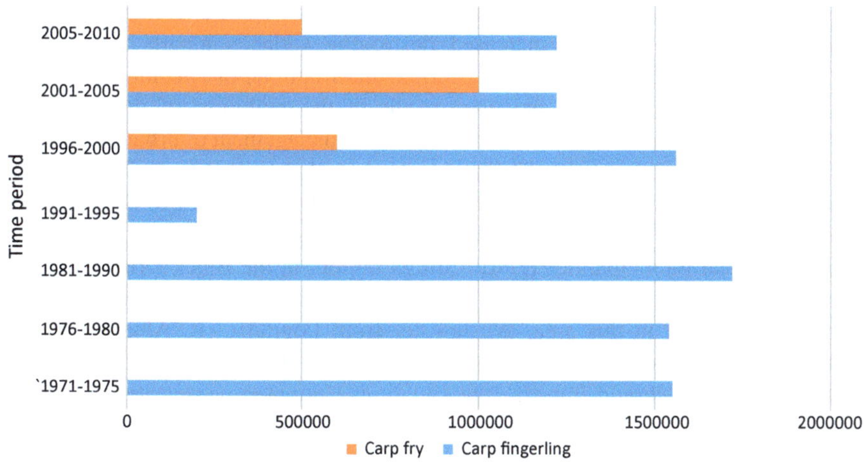

Fig. 21 Stocking of Lake Prespa with juvenile carp in the period 1971–2010

From 1971, Lake Prespa was stocked only with autochthonous juvenile carp from the state hatchery in the village of Zvezda (Republic of Albania). From 1990, only Makro Prespa was stocked with juvenile carp fish and carp flies (Fig. 21).

11 Pressures, Transboundary Impacts, and Responses

Lake Ohrid is a unique body of water that is a biodiversity hotspot of international importance and considered a natural heritage. Consequently, the entire lake (and adjacent areas) is worthy of protection from anthropogenic disturbances such as environmental degradation and pollution. It is one of the deepest and oldest lakes in Europe with a unique aquatic ecosystem of global importance with many endemic species.

On the Macedonian side, Lake Ohrid was declared a World Heritage Site by UNESCO in 1979, and in 1980, the site was expanded to include the cultural-historical area of Ohrid. In 2014, the cross-border Ohrid-Prespa Reserve between the Republic of Albania and the Republic of North Macedonia was added to the UNESCO World Network of Biosphere Reserves. The Albanian side of Lake Ohrid was also declared a UNESCO World Heritage Site in 2019.

11.1 The Main Threats to the Lake Ohrid

Human interventions have changed the hydrological regime of Lake Ohrid. Diverting the Sateska River into the lake increased its catchment area, agricultural runoff, and sediment input. Due to unsustainable forest management and subsequent erosion, sediment loads have also increased, destroying wetlands in parts of the lake in both countries (Republic of North Macedonia and Republic of Albania).

Unsustainable agricultural practices result in the pollution of the lake with pesticides and nutrients. The lack of or insufficient management of urban wastewater and sewage runoff plays an equally important role in the nutrient load of the lake and exerts a slight pressure on the karstic origin of Lakes Prespa and Lake Ohrid. On the Macedonian side of the lake, where a sewage system was built collecting sewage from the coastal municipalities (Ohrid-Struga region, in the Black Drin basin), the municipalities were reduced from municipal sewage. The construction of further plants was planned in this area. Wastewater from Pogradec (R. Albania) mainly contributes to the phosphorus and organic pollution of the lake. New collection and treatment facilities should improve the situation. Untreated wastewater discharges from industrial activities in Pogradec (food processing, textile, metal and wood processing, and other light industries) are also considered a significant source of pollution.

11.2 Threats to Fish Stocks from Lake Ohrid

Due to inadequate protection and unsustainable fishing practices, the lake's two endemic freshwater fish species (Ohrid trout and whiting) have been overfished in recent years and their stocks are at imminent risk of extinction. The uncontrolled destruction (degradation) of the reed belt in the two neighboring countries leads to the deterioration of the fish habitats and thus to the disturbance of the natural state of the fish populations, which threatens the spawning and wintering grounds of some fish species. In the destroyed reed beds, Ohrid carp and other carp-like fish left their natural habitats and natural spawning grounds. All of this leads to a reduction in the overall biodiversity in Lake Ohrid. Also, the presence of introduced (non-native) fish species negatively affects the native fish populations in the lake.

For these reasons, fisheries master plans have been drawn up in both countries (Republic of North Macedonia and Republic of Albania). Specific measures have been proposed to harmonize fisheries legislation, as well as measures such as the modernization of stock hatcheries and the strengthening of fisheries control bodies, which will implement specific measures to combat illegal fishing.

11.3 Lake Prespa

Because of its wealth of rare fauna and flora, Lake Prespa was declared a Ramsar site in 1999. In addition, the Prespa area was declared a transnational park in 2000. In 2014, the Ohrid-Prespa Transboundary Reserve between Albania and North Macedonia was included in the UNESCO World Network of Biosphere Reserves.

11.4 The Main Threats to the Lake Prespa

The main threats to the ecosystem of Lake Prespa and in particular to the biodiversity of the region include:

- Nutrient (nitrogen and phosphorus) stress leads to eutrophication and low levels of dissolved oxygen. The main sources of pollution are inadequate treatment of sewage from human settlements and improper use of fertilizers on agricultural land.
- Loss of water level in Makro Prespa leading to changes in coastal habitats.
- Reducing the use of lake water by agriculture. Sediment transport results from improper land management (agriculture and forestry), periodic flooding, and changes in river regimes.
- Pollution with hazardous substances (accumulations in water and sediments) and improper use of agrochemicals.

11.5 Threats to Fish Stocks from Lake Prespa

Significant factors in the reduction of native fish populations are the reduction in the water level of the lake (Fig. 22), pollution of the lake water with excess organic nutrients, causing a decrease in the level of dissolved oxygen, the toxic effects of the use of agrochemicals, insecticides, and pesticides, fisheries with poor regulation and

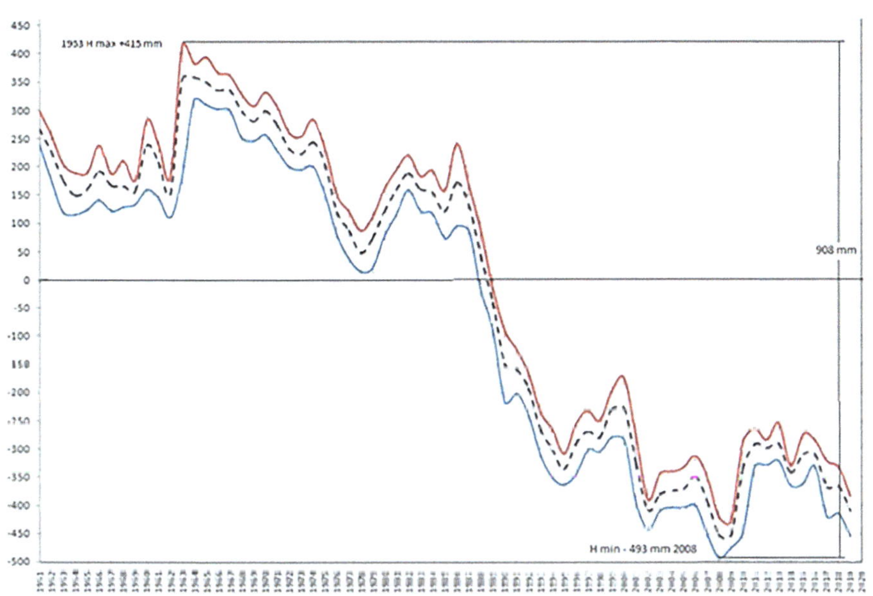

Fig. 22 The level of Lake Prespa in the period 1951–2020

enforcement, use of illegal nets, loss of habitat and spawning grounds (with destruction of the reed belt), and competition from introduced fish species.

The population of the introduced non-native fish species *Lepomis gibbosus* (about 15 years ago) competes with the native species for food and habitat, and their population is increasing. Therefore, it can disturb the biodiversity of the lake. The decline of native fish populations in Lake Prespa leads to a potential loss of income for fishermen and a loss of biodiversity.

11.6 Lake Dojran

Lake Dojran has been declared a Natural Monument and Ramsar Site as it is rich in endemic species and an essential habitat for migratory birds. Lake Dojran is rich in fish biomass and species. The annual catch was about 500,000 kg, i.e., 150 kg/ha (production of 178 kg/ha in Lake Dojran due to abundance of food, the highest freshwater production in Europe).

The eutrophic nature of the lake, declining levels, and changes in water quality have significant impacts on species of living organisms (including fish) and their habitats. Figures 23 and 24 show the total catch of fish, the reduction of which is directly related to the reduction of the water level in the Dojran Lake.

The southeastern part of the Republic of North Macedonia was strongly influenced by climate changes in the last quarter of the twentieth century. Frequent, long-lasting droughts (with high temperatures in summer) and excessive consumption of water for irrigation of agricultural lands in the two neighboring countries led to a drastic drop in the water level in the lake (Fig. 25).

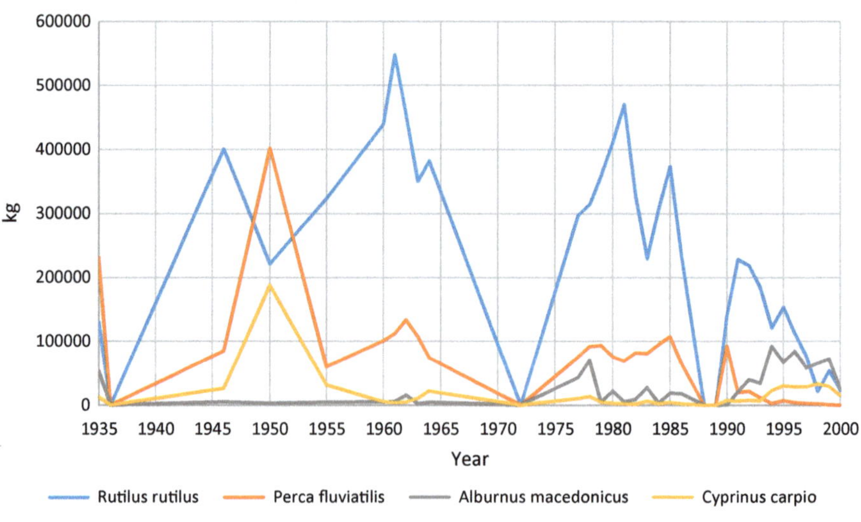

Fig. 23 Participation of individual economically important fish species in the annual catch from Lake Dojran

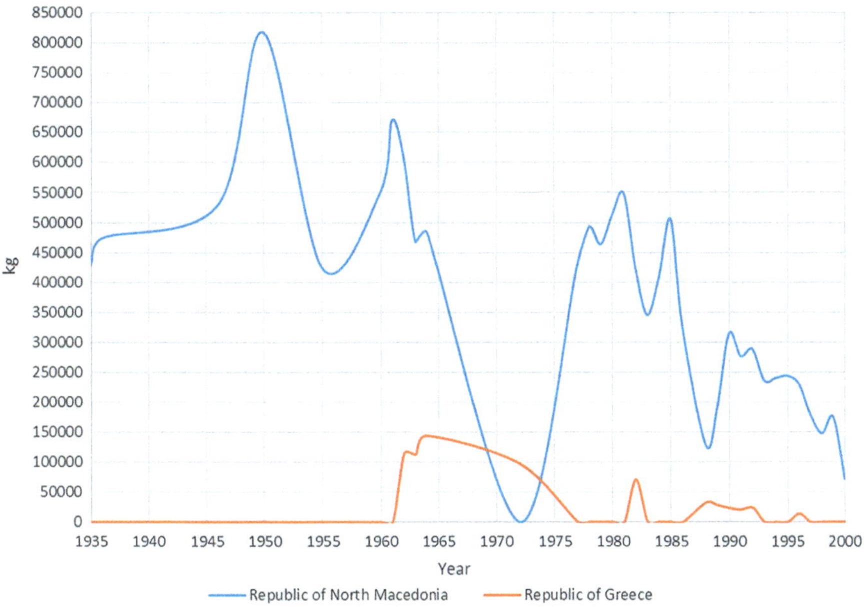

Fig. 24 Total annual catch from Lake Dojran in period 1935–2000 (Republic of North Macedonia and Republic of Greece)

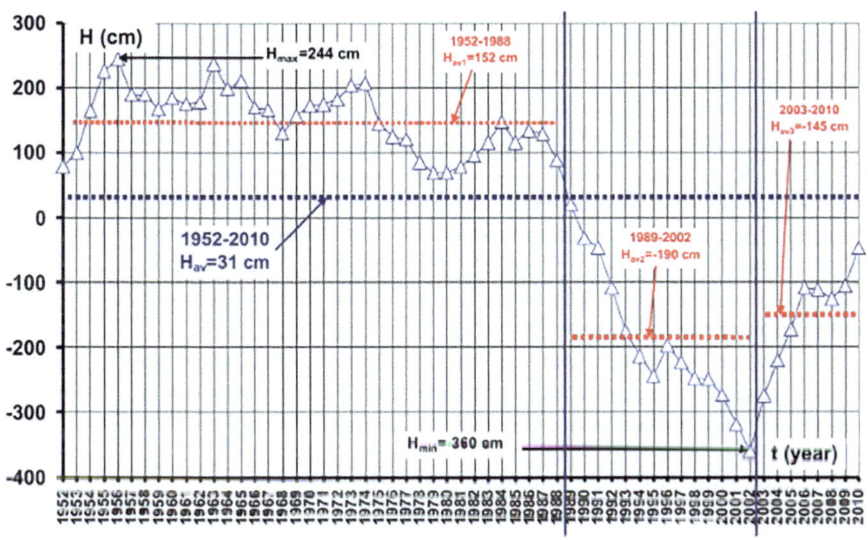

Fig. 25 The level of Lake Dojran in the period 1952–2010

11.7 The Main Threats to the Lake Dojran

Lake Dojran is (since the early 1990s) under the influence of water quantity reduction and water quality reduction due to activities in the two countries, the Republic of North Macedonia and the Republic of Greece, such as water abstraction and removal of municipal sewage.

The situation was worsened by low rainfall in 1989–1993 and high evaporation rates in the lake basin. In the last 20 years, the level of the lake has continuously decreased also due to the increasing Greek abstraction, mainly for irrigation. The most extreme drop in water level and volume occurred from 1988 to 2000.

The main threats to Lake Dojran include a decline in the water level in Lake Dojran, point sources of pollution (domestic and industrial effluents), diffuse sources (agriculture, solid effluents, untreated domestic effluents, etc.) and agrochemicals, and cleaning sprayers and chemical waste containers. The other threats are climate change, land-use change (intensive agriculture), sediment erosion, urbanization and the lack of integrated land-use planning, tourism—rising numbers with insufficient infrastructure, and others. Its effects can be felt in both countries.

11.8 Threats to Fish Stocks from Lake Dojran

However, declining water levels (below the allowable hydrobiological minimum) and deteriorating water quality have drastically affected fish stocks and some spawning grounds. Eutrophication is a phenomenon that has serious consequences for fish. This is because algal blooms, which often follow eutrophication, block light from reaching deeper water layers, preventing aquatic plants from photosynthesizing and providing oxygen. Intense algal blooms cause the death of various aquatic organisms, including fish, which sink to the bottom and begin to decompose. Their microbial decomposition consumes oxygen, leading to further oxygen deficit that can lead to a hypoxic environment that limits the survival of sensitive species, especially fish. The decrease in water level and water quality leads to changes in the composition of the fish population.

Other factors affecting declines in native fish populations include the toxic effects of agrochemical, insecticide, and pesticide use, fishing with poor regulation and enforcement, use of wild nets, habitat and spawning loss (with the destruction of the reed belt), and competition with introduced fish species.

The population of the introduced non-native fish species *Carassius gibelio* competes with the native species for food and habitat and may disrupt the lake's biodiversity (Talevski & Milosevic, 2016; Talevski & Talevska, 2018).

At the same time, the occurrence of autochthonous (native) economically important species in the catch has been significantly reduced, and invasive species such as the Prussian carp are more numerous. Lake Dojran used to be one of the most productive lakes in Europe and fishing still plays a vital role in the local economy.

50% of the total annual fish catch in Republic of North Macedonia used to come from Dojran.

12 Fish Stocks in Artificial Lakes

In the period after 2005 until a few years ago, in the fisheries management plan for the artificial lakes Tikvesh and Shpilje, it was planned to give these two artificial lakes a concession for commercial fishing. However, this attempt turned out to be relatively unsuccessful, since the artificial lakes Shpilje and Tikvesh were simultaneously used for aquaculture and the production of carp in fish farms, and the artificial lake Shpilje was used for aquaculture for the production of trout and carp in fish farms. Therefore, commercial fishing in these artificial lakes has been abandoned. In addition, recreational fishermen and landowners receiving aquaculture concessions have had many negative reactions.

13 Recreational Fishing

Recreational fishing is practiced on purely natural and artificial lakes and on almost all major rivers and their catchment areas, where different species of fish are caught depending on the season and the place where recreational fishing is practiced.

The number of recreational fishermen is about 3200 organized in about 40 fishing associations. But, unfortunately, there are no accurate statistics on the number of species and other characteristics of the fish specimens caught.

13.1 Natural Lakes

The most important species of fish for recreational fishing on Lake Ohrid are *Salmo letnica, Cyprinus carpio*, *Rutilus ohridanus*, *Squalius squalus*, *Barbus rebeli*, etc. In earlier times, the Ohrid whitefish and the Ohrid trout were catheterized with a fishing tool, a blinker, pulled from a boat.

The main fish species for recreational fishing at Lake Prespa are *Cyprinus carpio*, *Alburnus belvica, Rutilus prespensis*, *Squalius prespensis*, *Barbus prespensis,* etc.

The main fish species for recreational fishing at Lake Dojran are *Alburnus macedonicus*, *Barbus balcanicus*, *Carassius gibelio*, *Cyprinus carpio, Pachychilon macedonicum*, *Rutilus rutilus*, *Scardinius erythrophthalmus*, *Squalius vardarensis*, *Tinca tinca*, *Anguilla anguilla*, *Silurus glanis*, and *Perca fluviatilis*.

13.2 Artificial Lakes

Recreational fishermen in the artificial lake Strezevo catch the following species of fish: *Salmo pelagonicus*, *Salmo letnica*, *Oncorhynchus mykiss*, *Alburnus alburnus*, *Carassius carassius*, *Carassius gibelio*, *Chondrostoma vardarense*, *Hypophthamichthys molitrix*, *Hypophthamichthys nobilis*, *Rutilus rutilus*, *Squalius vardarensis*, *Tinca tinca*, *Anguilla anguilla*, *Silurus glanis*, *Esox lucius*, and *Perca fluviatilis*.

Recreational fishermen in the artificial lake Mladost catch the following species of fish: *Alburnus alburnus*, *Carassius carassius*, *Carassius gibelio*, *Ctenopharyngodon idella*, *Cyprinus carpio*, *Hypophthalmichthys molitrix*, *Hypophthalmichthys nobilis*, *Rutilus rutilus*, *Squalius vardarensis*, *Tinca tinca*, *Silurus glanis*, *Esox lucius*, *Perca fluviatilis*, and *Stizostedion lucioperca*.

Recreational fishermen in the artificial lake Mavrovica catch the following species of fish: *Alburnus alburnus*, *Barbus balcanicus*, *Carassius carassius*, *Carassius gibelio*, *Ctenopharyngodon idella*, *Cyprinus carpio*, *Hypophthalmichthys molitrix*, *Hypophthalmichthys nobilis*, *Rutilus rutilus*, *Scardinius erythrophthalmus*, *Squalius vardarensis*, *Tinca tinca*, *Silurus glanis*, and *Perca fluviatilis*.

Recreational fishermen in artificial lake Lipkovo catch the following fish: *Salmo macedonicus*, *Barbus balcanicus*, *Carassius gibelio*, *Chondrostoma vardarense*, *Cyprinus carpio*, *Rutilus rutilus*, *Scardinius erythrophthalmus*, *Squalius vardarensis*, *Tinca tinca*, *Vimba melanops*, and *Anguilla anguilla*.

Recreational fishermen in the artificial lake Glaznja catch the following fish: *Salmo macedonicus*, *Alburnus alburnus*, *Barbus balcanicus*, *Carassius gibelio*, *Cyprinus carpio*, *Squalius vardarensis*, and *Anguilla anguilla*.

Recreational fishermen in artificial lake Ratevo catch the following species of fish: *Salmo macedonicus*, *Alburnus alburnus*, *Barbus balcanicus*, *Carassius carassius*, *Chondrostoma vardarense*, *Cyprinus carpio*, *Squalius vardarensis*, *Tinca tinca*, and *Anguilla anguilla*.

Recreational fishermen in the artificial lake Paljurci catch the following fish: *Alburnus alburnus*, *Carassius gibelio*, *Cyprinus carpio*, *Rutilus rutilus*, *Scardinius erythrophthalmus*, *Squalius vardarensis*, *Tinca tinca*, *Silurus glanis*, and *Perca fluviatilis*.

Recreational fishermen in the artificial lake Kozjak catch the following fish: *Salmo macedonicus*, *Oncorhynchus mykiss*, *Alburnus alburnus*, *Barbus balcanicus*, *Carassius gibelio*, *Chondrostoma vardarense*, *Cyprinus carpio*, *Rutilus rutilus*, *Squalius vardarensis*, *Vimba melanops*, *Anguilla anguilla*, *Silurus glanis*, and *Esox lucius*.

Recreational fishermen in the artificial lake Mantovo catch the following fish: *Alburnus alburnus*, *Barbus balcanicus*, *Carassius gibelio*, *Chondrostoma vardarense*, *Ctenopharyngodon idella*, *Cyprinus carpio*, *Rutilus rutilus*, *Scardinius erythrophthalmus*, *Squalius vardarense*, *Tinca tinca*, *Silurus glanis*, and *Perca fluviatilis*.

Recreational fishermen in the artificial lake Turija catch the following species of fish: *Barbus macedonicus, Carassius gibelio, Cyprinus carpio, Rutilus rutilus, Squalius vardarense*, *Silurus glanis,* and *Perca fluviatilis.*

Recreational fishermen in the artificial lake Vodocha catch the following species of fish: *Barbus macedonicus, Carassius gibelio, Cyprinus carpio, Rutilus rutilus, Squalius vardarense, Silurus glanis*, and *Perca fluviatilis.*

Recreational fishermen in the artificial Kalimanci lake catch the following species of fish: *Salmo macedonicus, Alburnus alburnus, Carassius gibelio, Chondrostoma vardarense, Cyprinus carpio, Rutilus rutilus, Squalius vardarensis, Tinca tinca, Anguilla anguilla*, and *Silurus glanis.*

13.3 Rivers

Recreational fishermen in the Treska catch the following fish: *Salmo macedonicus, Oncorhynchus mykiss, Alburnus alburnus, Barbus balcanicus, Carassius gibelio, Chondrostoma vardarense, Cyprinus carpio, Squalius vardarensis, Vimba melanops,* and *Anguilla anguilla.*

Recreational fishermen in the Strumica River catch the following species of fish: *Alburnus alburnus, Barbus strumicae, Carassius gibelio, Chondrostoma nasus, Cyprinus carpio, Gobio bulgaricus, Squalius vardarensis,* and *Esox lucius.*

Recreational fishermen in the Pchinja River catch the following species of fish: *Salmo macedonicus, Alburnus alburnus, Barbus balcanicus, Carassius gibelio, Chondrostoma vardarense, Cyprinus carpio, Gobio bulgaricus, Pachychilon macedonicum, Squalius vardarense, Vimba melanops, Silurus glanis, Esox lucius,* and *Perca fluviatilis.*

Recreational fishermen in the upper part of the Vardar River catch the following species of fish: *Salmo farioides, Salmo macedonicus, Salvelinus fontinalis, Oncorhynchus mykiss, Alburnus alburnus, Barbus balcanicus, Carassius gibelio, Chondrostoma vardarense, Pachychilon macedonicum, Squalius vardarensis, Vimba melanops*, and *Anguilla anguilla.*

Recreational fishermen in the middle part of the Vardar River catch the following species of fish: *Salmo macedonicus, Oncorhynchus mykiss, Alburnus alburnus, Barbus balcanicus, Barbus macedonicus, Carassius carassius, Carassius gibelio, Chondrostoma vardarense, Cyprinus carpio, Pachychilon macedonicum, Rutilus rutilus, Scardinius erythrophthalmus, Squalius vardarense, Tinca tinca, Vimba melanops, Anguilla anguilla, Silurus glanis,* and *Perca fluviatilis.*

Recreational fishermen in the lower part of the Vardar River catch the following species of fish: *Salmo macedonicus, Oncorhynchus mykiss, Alburnus alburnus, Barbus balcanicus, Carassius carassius, Carassius gibelio, Chondrostoma vardarense, Cyprinus carpio, Pachychilon macedonicum, Rutilus rutilus, Scardinius erythrophthalmus, Squalius vardarensis, Tinca tinca, Vimba melanops, Anguilla anguilla, Silurus glanis*, and *Perca fluviatilis.*

Recreational fishermen in the Bregalnica river catch the following fish: *Salmo macedonicus, Oncorhynchus mykiss, Alburnus alburnus, Barbus balcanicus, Carassius gibelio, Chondrostoma vardarense, Cyprinus carpio, Pachychilon macedonicum, Squalius vardarensis, Vimba melanops, Anguilla anguilla*, and *Silurus glanis.*

Recreational fishermen in the Crna Reka River catch the following species of fish: *Salmo pelagonicus, Oncorhynchus mykiss, Alburnus alburnus, Barbus balcanicus, Barbus macedonicus, Carassius carassius, Carassius gibelio, Chondrostoma vardarensis, Cyprinus carpio, Pachychilon macedonicum, Rutilus rutilus, Scardinius erythrophthalmus, Squalius vardarensis, Tinca tinca, Vimba melanops, Silurus glanis, Esox lucius*, and *Perca fluviatilis.*

14 Aquaculture

According to the method of fish farming, ponds, canals, and cages are primarily used as facilities. In the past, many warm-water ponds have been decommissioned or production reduced.

Due to this situation in aquaculture, the fish production in the Republic of North Macedonia is currently not enough for the required amount of fish for the Macedonian market, so the country has been forced to import fish in bulk from abroad to meet the missing amount of fish in the market; according to data of the State Statistics Service, the total catch in 2021 was 2,171,537 kg, which is 4.1% higher compared to 2020. 65.6% of the total catch in 2021 are cold-water fish; the share of warm-water fish is 34.4%. Compared to 2020, cold-water fish catches increased by 1.8% and warm-water fish catches increased by 7.9%. Fish production and fishing is based on aquaculture production (92.9%), commercial fishing (5.8%), and recreational fishing, which participates with only 1.3%.

In 2021 compared to 2020, an increase in fishing was observed for all. In 2021, registered reproduction centers produced 1815 million roe and young with a total weight of 132,403 kg. However, the production of juvenile roe in a controlled environment in 2021 decreased by 14.2%, while juvenile weight increased by 21.2%. The rainbow trout is mainly grown as a juvenile of the fish species.

15 Legislative (Laws)

In 2007, the existing Republic of North Macedonian Fisheries Law (1993) was replaced by the Law on Fisheries and Aquaculture (LFA, Official Gazette 7/2008 of 15.01.2008). This law has several amendments (Official Gazette of Republic of North Macedonia 67/2010, 47/2011, and 53/2011; Official Gazette of Republic of North Macedonia 95/2012; Official Gazette of Republic of North Macedonia

164/2013; Official Gazette 116/2014; Official Gazette 154/2015 and 193/2015; and Official Gazette 39/2016).

The following documents are complementary to the Law on Fishery and Aquaculture:

- Law on the protection of Ohrid, Prespa, and Dojran Lakes (Official Gazette 45/1977 of 09.09.1977). This law has four amendments (Official Gazette 08/1980, 51/1988, 10/1990, and 62/1993).
- Law on nature protection (Official Gazette 67/2004 of 04.10.2004). This law has five amendments (Official Gazette 14/2006, 84/2007, 35/2010, 47/2011, and 148/2011).
- Law for the environment (Official Gazette 53/2005 of 05.07.2005). This law has seven amendments (Official Gazette 81/2005, 24/2007, 159/2008, 83/2009, 48/2010, 124/2010, and 51/2011).

The first Lake Ohrid Fisheries Master Plan for the period 2011–2016 was published in the Official Gazette 145/2011, while the current plan (for the period 2017–2022) was made publicly available in the Official Gazette 55/2017,

The following regulations are of importance to Macedonian fishery and have to be considered too:

– "Regulation on the Form, Content and the Way of Performing Evidence of Fish Production as for the amount of the Sold Fish per Species" (2008)
– "Regulation for Performing the Fish Guarding Service, the Form and the Content of the Fish Guardian Legitimation, as the Way of its Issuing and Withdrawing" (2008)
– "Regulation of the Content of the Program for Examining, the Form and Content of the Certificate, as the cost for Issuing Certificate for Commercial Fishery" (2008)
– "Regulation on the Form and the Content of the Evidence Formulary in the Fishing Regions" (2008)
– "Regulation of the Content of the Fishery Master Plan" (2008)
– "Regulation of the Content of the Annual Plan for Protection and Exploitation of the Fish and the Content of the Annual Report of Realization of the Plan" (2008)
– "Regulation on the Technical Requirements for the Landing Sites" (2008)
– "Regulation on the Quality, Size and Weight, as also the Way of Declaring the Fish for Traffic Market"(2008)
– "Regulation on the Way of Marking of the Boats and Tagging and Evidencing of the Fishing Gear" (2008)
– "Regulation on the Form and the Content of the Document for the Origin of the Fish and the Way of its Issuing and Fulfilling" (2010)
– "Regulation on the Way of Issuing Licenses for Recreational Fishing, the Required Documentation for Issuing, the Form and Content of the Evidence Formulary, the Way of Evidencing and Delivering the Data" (2010)
– "Regulation on the Form and the Content of the Legitimation for Recreational Fishing and the Way of its Issuing" (2010)

- “Regulation on the Allowed Fishing Gears and Equipment and their Use for Commercial and recreational Fishing” (2011)
- “Regulation on the Length of the Fish below which it Cannot be Taken by Commercial and recreational fishing” (2011)
- “Regulation on the Quality, Size and Weight, as also the Way of Declaring the Fish for Traffic Market”(2013)
- “Regulation for Amendments of Regulation on the Allowed Fishing Gears and Equipment and their Use for Commercial and Recreational Fishing” (2013)
- “Regulation for Changes of the Length of the Fish below which it Cannot be Taken by Commercial and Recreational Fishing” (2013)

16 Conclusions

1. Freshwater in the Republic of North Macedonia is one of two ecoregions (out of 426 ecoregions worldwide) classified as ecoregion 420 (the catchment area of the southeastern Adriatic basin, which includes Lake Prespa and Lake Ohrid) and ecoregion 422 (the Vardar basin, which includes Lake Dojran covered).
2. The fish populations in the Republic of North Macedonia can be divided into native, non-native, and invasive species.
3. On the territory of the Republic of North Macedonia, 80 species of fish belong to the families: Petromyzonidae, Acipenseridae, Anguillidae, Clupeidae, Cyprinidae, Cobitidae, Nemacheilidae, Ictaluridae, Siluridae, Esocidae, Poecilidae, Salmonidae, Gobiidae, etc.
4. The fish fauna in the Republic of North Macedonia is characterized by endemism (a total of 23 endemic fish species).
5. According to the IUCN Red List, fish species (80 in total) from the Republic of North Macedonia belong to the following categories: Least Concern (LC)—35 fish species, Critically Endangered (CR)—3 species, Endangered (EN)—2 species, Vulnerable (VU)—9 species, Near Threatened (NT)—1 species, Not assessed according to criteria (NE)—1 species, Data Deficient (DD)—10 species. Therefore, eight fish species have been introduced into the aquatic ecosystems of the Republic of North Macedonia and have been omitted and assessed from the endangered species list.
6. Return of the Satesca river to the old trough.
7. Reducing the use of groundwater and rivers in the Prespa watershed for irrigation.
8. Continuous operation of the Djawato hydro system to supply clean water to Lake Dojran to maintain the water level in the lake.
9. Completion of the ring construction of the collector system for Ohrid and Struga and construction of sewage treatment plants in rural settlements, the water from which flows directly into Lake Ohrid.
10. Reconstruction of the Prespa Lake Wastewater Treatment Plant.
11. Modernization and reconstruction of the purification system of Lake Dojran.

12. Reducing the use of artificial fertilizers, pesticides, and herbicides in the watersheds of Lake Ohrid, Prespa, and Dojran.
13. Continuous fishing with fish regeneration from trout and carp in natural and artificial lakes and rivers in the Republic of Macedonia through a program prepared by an appropriate scientific institution.
14. Fishing with older rejuvenation.
15. Protection and restoration of fish spawning grounds in aquatic ecosystems.
16. Respect for professional and recreational fishermen in accordance with the legal norms provided for by law (spawning season, established catch quotas, fishing).
17. Strict compliance with quotas for endemic species in aquatic ecosystems.
18. Maintaining optimal water levels in concessionaire pools and lakes for electric irrigation at a level that does not affect the ecology of flora and fauna.
19. Ban on the construction of mini-hydroelectric power plants on rivers that are habitats of endemic and endangered species that prevent fish migration.
20. Implementation of continuous monitoring of natural lakes and other important aquatic ecosystems in the Republic of North Macedonia.

Acknowledgments We are grateful to the Historical Archives of the Republic of North Macedonia, Department in Ohrid, for the donated photos from the German Nikola Fund on fishing tools and their use on Lake Ohrid in the 1930s.

We also express our gratitude to Radovan Dimitrievski, professional photographer, videographer, and director of the show Fisherman's Truth, for providing photos from his archives.

References

Abell R, Thieme ML et al. (2008) Freshwater ecoregions of the world: A new map of biogeographic units for freshwater biodiversity conservation. BioScience 58(5), 403–414 https://www.researchgate.net/publication/232677514_Freshwater_Ecoregions_of_the_World_A_New_Map_of_Biogeographic_Units_for_Freshwater_Biodiversity_Conservation. Accessed Sep 19 2022

Albrecht C, Wilke T (2008) Ancient Lake Ohrid: biodiversity and evolution. Hydrobiologia 615: 103–140

Anovski B, Andonovski B, Minceva B (1992) Study the hydrological relationship between Lake Ohrid and Prespa. Proceedings of Symposium on Isotope Techniques in Water Resources Development, IAEA, Vienna, Austria, March 1991

Anovski TM, Milevski KJ, Ristevski P, Stamos A (2001) Hydrological aspects and water balance of Prespa Lakes. In Anovski, T. (ed.), Progress in the study of Prespa Lake using nuclear and related techniques. Project report, IAEA regional project RER/8/008, ISBN 9989-650-21-7, Skopje, Macedonia, 53–66

Apostolski KM (1968) Contribution to the fish populations study in Lake Dojran. Izdanija Institut de Pisciculture de la RS Macedoine 4:1–30. (in Macedonian with English abstract)

Arlinghaus R, Abbott JK, Fenichel EP, Carpenter SR, Hunt LM, Alós J et al (2019) Governing the recreational dimension of global fisheries. Proc Natl Acad Sci U S A 116:5209–5213. https://doi.org/10.1073/pnas.1902796116

Avital G, Sarig G (1990, January) Effects of water level fluctuation on the structure and function of the Littoral zone. Large Lakes. https://doi.org/10.1007/978-3-642-84077-7_8

Baldwin DS, Mitchell AM (2000) The effects of drying and re-flooding on the sediment and soil nutrient dynamics of lowland river-floodplain systems: a synthesis. Regul Rivers Res Manag 16: 457–467

Bonacci O, Popovska C (2006) Dojran Lake. Hrvatska Vodoprivreda 158:14–21

Bonacci O, Popovska C, Geshovska V (2014) Analysis of transboundary Dojran Lake mean annual water level changes. In: NATO science for peace and security series-C: environmental security, vol 73. Springer, Cham, p 3177

Chavkalovski I (1997) Hydrology of Prespa Lake. Proceedings of the international symposium "toward integrated conservation and sustainable development of transboundary macro and micro Prespa lakes", Korcha, Albania, pp 9–14

Chavkalovski I (2000) Anthropogenic influence on the denivelation of Macro and Micro Prespa Lake. Proceedings of the International Symposium "Sustainable Development of Prespa Region", Oteshevo, pp 250–257

Cowx IG (1994) Stocking strategies. Fish Manage Ecol 1:15–30

Cowx IG (2002) Recreational fishing. In: Handbook of fish biology and fisheries, vol 2. Blackwell Science, Oxford, pp 367–390

Cowx IG, Portocarrero AM (2011) Paradigm shifts in fish conservation: moving to the ecosystem services concept. J Fish Biol 79:1663–1680

Cowx IG, Almeida O, Bene C et al (2004) Value of river fisheries. In: Sustaining livelihoods and biodiversity in the new millennium. Proceedings of the 2nd large Rivers symposium, vol 1. Mekong River Commission, Phnom Penh, Cambodia, pp 1–20

Cowx IG, Arlinghaus R, Cooke SJ (2010) Harmonizing recreational fisheries and conservation objectives for aquatic biodiversity in inland waters. J Fish Biol 76:2194–2215

Cvijić J (1906) Osnove za geografiju i geologiju Makedonije i stare Srbije. Srpska kraljevska akademija, Beograd 190:1272

Dickens C, McCartney M, Tickner D, Harrison IJ, Pacheco P, Ndhlovu B (2020) Evaluating the global state of ecosystems and natural resources: within and beyond the SDGs. Sustainability 12:7381. https://doi.org/10.3390/SU12187381

Grillas P (2011) Pilot application of the transboundary monitoring system for the Prespa Park: aquatic vegetation (Reedbed) monitoring, final report, Society for the Protection of Prespa - tour du Valat, Agios Germanos"

Hilborn R (1976) Optimal exploitation of multiple stocks by a common fishery: a new methodology. J Fish Res Board Can 33(1):1–5

Hilborn R (2010) Pretty Good Yield and exploited fishes. Mar Pol 34:193–196

Hilborn R, Stokes K (2010) Defining overfished stocks: have we lost the plot? Fisheries 35:113–120

Hilborn R, Quinn TP, Schindler DE, Rogers DE (2003) Biocomplexity and fisheries sustainability. Proc Natl Acad Sci U S A 100:6564–6568

Hilborn R, Stewart IJ, Branch TA, Jensen OP (2012) Defining trade-offs among conservation of species diversity abundances, profitability, and food security in the California current bottom-trawl fishery. Conserv Biol 26:257–266

Hilborn R, Hively DJ, Jensen OP, Branch TA (2014) The dynamics of fish populations at low abundance and prospects for rebuilding and recovery. ICES J Mar Sci 71(8):2141–2151

Kottelat M, Freyhof J (2007) Handbook of European freshwater fishes. Ichthyol Res 55:99. https://doi.org/10.1007/s10228-007-0012-3

Löffler H (2004) The origin of lake basins. In: O'Sullivan PE, Reynolds CS (eds) The lakes handbook, vol 1. Blackwell, Malden, pp 8–60

Loh J, Green RE, Ricketts T, Lamoreux J, Jenkins M, Kapos V et al (2005) The living planet index: using species population time series to track trends in biodiversity. Phil Trans R Soc B 360:289–295. https://doi.org/10.1098/RSTB.2004.1584

Lynch AJ, Cowx IG, Fluet-Chouinard E, Glaser SM, Phang SC, Beard TD et al (2017) Inland fisheries - invisible but integral to the UN sustainable development agenda for ending poverty by 2030. Glob Environ Change 47:167–173

Matzinger A, Jordanoski M, Veljanoska-Sarafiloska E, Sturm M, Muller B, Wuest A (2006) Is Lake Prespa jeopardizing the ecosystem of ancient Lake Ohrid? Hydrobiologia 553:89–109

Matzinger A, Schmid M, Veljanoska-Sarafiloska E, Patceva S, Guseska D, Wagner B, Muller B, Sturm M, Wuest A (2007) Eutrophication of ancient Lake Ohrid: global warming amplifies detrimental effects of increased nutrient inputs. Limnol Oceanogr 52:338–353

Milošević D, Talevski T (2015) Conservation status of native species in Natural Lakes of Drim system (Prespa, Ohrid and Skadar Lake) and dangers of commercial fishing Bulg. J Agric Sci Supplement 1, 21:61–67

Naylor R, Goldburg R, Primavera J et al (2000) Effect of aquaculture on world fish supplies. Nature 405:1017–1024. https://doi.org/10.1038/35016500

Pauly D (1988) Fisheries research and the demersal fisheries of Southeast Asia. In: Gulland JA (ed) Fish population dynamics. John Wiley and Sons, New York, pp 329–348

Pauly D, Christensen V, Walters CJ (2000) Ecopath, Ecosim, and Ecospace as tools for evaluating ecosystem impact of fisheries. ICES J Mar Sci 57:697–706

Pereira DL, Hansen MJ (2003) A perspective on challenges to recreational fisheries management: summary of the symposium on active management of recreational fisheries. N Am J Fish Manag 23:1276–1282

Peters J, Lodge DM (2009) Invasive species policy at the regional level: a multiple weak links problem. Fisheries 34:373–380

Petruseski I, Markoski B (2016) Rekite vo Republika Makedonija

Pitcher TJ, Hollingworth CE (2002) Fishing for fun: where's the catch? In: Pitcher TJ, Hollingworth Cs (eds) Recreational fisheries: ecological, economic, and social evaluation. Blackwell Science, Oxford, pp 1–16

Reynolds JD, Jennings S, Dulvy NK (2001) Life histories of fishes and population responses to exploitation. In: Reynolds JD, Mace GM, Redford KH, Robinson JG (eds) Conservation of exploited species. Cambridge University Press, Cambridge (United Kingdom), pp 147–168

Riis T, Hawes I (2002) Relationships between water level fluctuations and vegetation diversity in shallow water of New Zealand lakes. Aquatic Botany 74:133–148. https://doi.org/10.1016/S0304-3770(02)00074-8. Corpus ID: 83562068

Smith AD, Brown CJ, Bulman CM, Fulton E, Johnson P, Kaplan IC, Lozano-Montes H, Mackinson S, Marzloff M, Shannon LJ (2011) Impacts of fishing low-trophic level species on marine ecosystems. Science 333:1147–1150

Smith MD, Fulton EA, Day RW (2015) An investigation into fisheries interaction effects using Atlantis. ICES J Mar Sci 72(1):275–283

Spirkovski Z, Talevski T, Ilic D (2001) Lake Ohrid fish - spawning grounds. In: Lake Ohrid conservation project, Lake Ohrid monitoring program. 2nd midterm report. Hydrobiological Institute, Ohrid, pp 129–149

Spirkovski Z, Talevski T, Stojanovski S, Jordanova V (2002) Lake Ohrid fish in Lymnological Investigations of Lake Ohrid, Vol. 1 & 2, – Special editions: PSI Hydrobiological Institute, Ohrid. 129–214

Spirkovski Z, Talevski T, Ilik-Boeva D, Kostoski G, Sand-lund OT (2007) A summary of the environmental and socio-economic characteristics of the Crna Reka (Crna River) watershed, Macedonia. - NINA Report 293. 37 pp + 12 Annexes

Stankovic S (1960) The Balkan Lake Ohrid and its living world. Monographiae Biologiae Vol. IX. Uitgeverij Dr. W. Junk. Den Haag, 357 p

Stefanoic D (1948) Rasna i ekoloska ispitivanjana Ohridskim salmonidama slika

Strayer D, Findlay S (2010a) Ecology of freshwater shore zones environmental science. Aquat Sci 72:127. https://doi.org/10.1007/s00027-010-0128-9. Corpus ID: 16690527

Strayer DL, Findlay SEG (2010b) Ecology of freshwater shore zones. Aquat Sci 72(2):127–163. https://doi.org/10.1007/s00027-010-0128-9. Journal of Great Lakes Research

Talevska M (1996) Biomass, production and main nutrients in the reed, *Phragmites communis* TRIN. From Lake Ohrid. Institute of Biology, Faculty of Natural Sciences and Mathematics, Skopje, University "St. Kyril and Methodius"

Talevska M (2002) Investigation of cyprinid spawning grounds: condition for sustainable fish management of transboundary Lake Ohrid. Second International Conference "Sustainable management of Transboundary Waters in Europe". Miedzyzdroje, Poland, pp 389–396
Talevska M (2003) Biodiversity of macrophytes from Lake Ohrid. The 32nd Annual Conference of Yugoslav Water Pollution Control Society "Water 2003". Zlatibor, pp 127–132
Talevska M (2005) Biomass, production and main nutrients in the red head grass, Potamogeton perfoliatus L. from Lake Ohrid. - Ph. thesis in Biology, Institute of Biology, Faculty of Natural Sciences and Mathematics, Skopje, University "St. Kyril and Methodius"
Talevska M (2009) Review of vascular macrophytes in Lake Ohrid, Lake Prespa and Lake Dojran. 5th Balkan botanical congress, Belgrade. Serbia. Book of abstracts: 99
Talevska M, Talevski T (2009) Macrophyte vegetation, slope of littoral and tip of soil in particular regions of Lake Ohrid (important factors for fish spawning grounds). COMBAFF- I st conference on conservation and Management of Balkan Freshwater Fishes. 20–24 May, Ohrid. R. Macedonia. Abstract book p. 25. ISNN 1409-9373.Ohrid. R. Macedonia
Talevska M, Talevski T (2010) Macrophyte vegetation, Typ of soil and slope of Littoral of north-eastern Littoral of Lake Ohrid - important factors for cyprinid fish spawning grounds, I Symposium of Ecologists of the Republic Serbian. Banja Luka. The book of abstracts and programme
Talevska M, Talevski T (2015) Qualitative composition of macrophyte vegetation and cyprinid fauna from Lake Ohrid. 6th International Conference "Aquaculture and Water Ecosystems". Plovdiv. Bulgarian J Agric Sci 21(Supplement 1):68–75
Talevska M, Talevski T (2018a) Macrophyte vegetation - habitat and spawning ground of some cyprinid fish from Lake Ohrid, 8th international conference "Water &Fish" - Conference Proceedings 453–458
Talevska M, Talevski T (2018b) Macrophyte vegetation - habitat and spawning ground of some cyprinid fish from Lake Ohrid". VIIIth International Conference Water & Fish. Belgrade. Republic of Serbia. Conference proceedings: 453–458
Talevska M, Talevski T (2019a) Researches on reed belt from Lake Prespa as habitat and spawning ground for cyprinid fishes. Third International Conference on Zoology, Zoonoses and Epidemiology. Hisar, Bulgaria. Book of abstracts: 125
Talevska M, Talevski T (2019b) Researches on belts of reed and pondweed from Lake Ohrid (Republic of North Macedonia). International conference "Lakes & Reservoirs: hot spot and topics in limnology". Mikorzyn, Poland. Book of abstracts: 110
Talevski T (1996) Rhodeus sericeus amarus (Bloch, 1782) нов вид во алохтоната ихтиофауна од Охридското Езеро I Конгрес на биолози на Македонија (со меѓународно учество), 146, Охрид
Talevski T, Milosevic D (2016) Biodiversity and conservation status of the fish in the great natural lakes (Dojran Lake, Prespa Lake and Ohrid Lake) of the Macedonia impact factor IF-0.532. Acta zoologica bulgarica
Talevski T, Talevska M (2008) The researches of cyprinid spawning grounds in north and north-western part of Lake Ohrid. III symposium of ecologists of the Republic of Montenegro. The book of abstracts and programme pp 53, 8–12. October. Bijela, Herceg Novi. ISBN 978-86-908743-2-3; COBISS.CG-ID 12971024
Talevski T, Talevska A (2010) Comparative analysis of fish biodiversity of the Adriatic and the Aegean basin in the Republic of Macedonia, IV International Symposium of Ecologists of Montenegro, 06–10.10.2010 Budva, Monte Negro The book of Abstracts and programme pp 78–79
Talevski T, Talevska M (2018) Biodiversity and conservation status of the fish and qualitative composition of the macrophytic vegetation in the Lake Dojran, Republic of Macedonia, 8th International Conference "Water &Fish" - Conference Proceedings 448–452
Talevski T, Talevska M, Milosevic D, Talevska A (2009a) Cyprinids spawning grounds and macrophyte vegetation in Lake Prespa, GEC16 abstract book: pp 62–64

Talevski T, Talevska M, Milosevic D, Talevska A (2009b) Cyprinids spawning grounds and macrophyte vegetation in Lake Ohrid, GEC16 abstract book: pp 67–69

Talevski T, Milosevic D, Talevska A (2010a) Anthropogenic influence and conservation status of autochthonous fish fauna from Lake Ohrid, conference of water observation and information system for decision support, Balwois 2010 http://balwois.com/balwois/administration/full_paper/ffp-587.pdf 43

Talevski T, Milosevic D, Talevska A (2010b) Anthropogenic influence and conservation status of autochthonous fish fauna from Lake Prespa, conference of water observation and information system for decision support, Balwois 2010, http://balwois.com/balwois/administration/full_paper/ffp-1588.pdf

Talevski T, Talevska M, Milosevic D, Talevska Lesoska A (2010c) Anthropogenic Influence on Ichthyofauna and Macrophyte Diversity in the Black Drim Ecosystem May 2010 Conference: BALWOIS 2010

Talevski T, Trajceski B, Talevska M, Leshoski J, Milosevic D (2017) Qualitative composition of fish population of the Crna River and reservoirs Streževo and Tikves" (135). 7th International Conference of Ecosystems (ICE 2017).Tirana. R.Albania. Abstract book (Essays on Ecosystem and Environmental Research), pp78–79. isbn: 978-9928-4248-8-4

The IUCN Red List Categories and Criteria. https://www.iucnredlist.org/

Tickner D, Opperman JJ, Abell R, Acreman M, Arthington AH, Bunn SE et al (2020) Bending the curve of global freshwater biodiversity loss: an emergency recovery plan. Bioscience 70:330–342. https://doi.org/10.1093/biosci/biaa002

Tocko MN (1959) Alosa fallax nilotica (Geoffrey) vo Ohridskoto Ezero. Zbornik na rabotite, Hidrobioloski zavod, Ohrid, God. VII. br.9: 1-3

Tocko MN (1987) The development and nutrition of the young cyprinids of Lake Ohrid, part I special edition Hydrobiological Institute Ohrid 1-165 Alosa fallax nilotica (Geoffrey) vo Ohridskoto Ezero. Zbornik na rabotite,Hidrobioloski zavod, Ohrid, God. VII. br.9: 1-3

Trajchevski B, Talevski T, (2022) Length-weight relationships of five native endemic and three alien introduced fish species from Lake Prespa, Alblakes 4 - International Conference on "Building Resilience of water ecosystems through scientific knowledge"

Wantzen KM, Rothaupt K-O, Mortl M, Cantonati M, Toth L, Fischer P (2008) Eco-logical effects of water-level fluctuations in lakes: an urgent issue. Hydrobiologia 613:151

Watzin M C, Puka V, Naumoski T B, eds. (2002) Lake Ohrid and its watershed, State of the Environment Report. Lake Ohrid Conservation Project. Tirana, R.Albania and Ohrid, Macedonia

Winfield J (2004, May) Fish in the littoral zone: ecology, threats and management Limnologica 34, 1–2, 124–131 https://doi.org/10.1016/S0075-9511(04)80031-8. Get rights and content 457–467

Fish Resources of Inland Waters and Fisheries in Albania, Management, Sustainability, and Conservation

Spase Shumka, Radek Šanda, Dhimitër Dhora, and Linda Grapci-Kotori

Abstract In the current contexts of development and state of the Anthropocene, ecological sustainability of freshwater ecosystems cannot be dissociated from socio-economic challenges and therefore entails policy framework, planning, and management. The general purpose of this chapter is to underline the fish resources of inland waters and fisheries in Albania, the key issues related to fishery, and associated problems within management, sustainability, and conservation. The total fish caught covering all fishing categories for the year 2020 was assessed at value of 16,892 tonnes, while the fisheries of inland waters at 2844 tonnes. In all planning and strategic documents, the inland fisheries are recognized as important to food security, livelihoods, and human well-being, but there is a poor integration in local planning and development practices. An ecosystem-based approach may help that inland fisheries be better evaluated and recognized in broader natural and water resource planning, and decision-making frameworks, enhancing their value and sustainability for the future. This chapter and current results presented here show that inland waters resources are subject to numerous anthropogenic interventions.

Keywords Freshwater ecosystems · Management · Fisheries · Conservation · Vulnerable stocks · Overuse

This designation is without prejudice to positions on status and it is in line with UNSCR 1244 and the ICJ Opinion on the Kosovo declaration of independence.

S. Shumka (✉)
Department of Biotechnology and Food, Faculty of Biotechnology and Food, Agricultural University of Tirana, Tirana, Albania
e-mail: s.shumka@ubt.edu.al

R. Šanda
Department of Zoology, National Museum, Prague, Czech Republic

D. Dhora
University of Shkodra "Luigj Gurakuqi", Shkodë, Albania

L. Grapci-Kotori
Linda Grapci-Kotori University of Prishtina "H. Prishtina", Prishtina, Kosovo

V. Simić et al. (eds.), *Ecological Sustainability of Fish Resources of Inland Waters of the Western Balkans*, Fish & Fisheries Series 43,
https://doi.org/10.1007/978-3-031-36926-1_7

1 Introduction

Albania is located in the western part of the Balkan Peninsula and covers landscapes from high mountains to the sea coast, and is rich in water resources including lakes, rivers, springs, and lagoons, with a high quantity of available water. The aquatic ecosystems are shelter for unique flora and fauna making many zones a biodiversity hot spot in the Mediterranean region (Freyhof and Brooks 2011; Freyhof 2012; Weiss et al. 2018; Shumka et al. 2020a, b). However, at present, many habitats are seriously endangered. The transformation of the Albanian economy to an open market situation during the past decade caused a significant damage to the natural resources of the country, mainly to natural waters, most of them exposed and unprotected (Cullaja et al. 2005; Shumka et al. 2020a, b). Besides development, urban wastewater and solids are directly affecting the rivers and lake ecosystems.

Albanian territory covers about 65% of a total watershed area of 43,905 km^2 of Drini River with an average height of the hydrographic territory of about 700 m above the sea level (Lazaj and Xhelilaj 2017). There are more than 152 rivers and streams in Albania, of which the largest areas are Buna (41 km), Drini (285 km), Mati (115 km), Ishmi (74 km), Erzeni (109 km), Shkumbini (181 km), Semani (281 km), and Vjosa (272 km), which run southeast to northwest toward the Adriatic coast (Kabo 1991). Albanian rivers are characterized by a high flow rate: the total annual mean flow is 1308 m^3s^{-1}, which corresponds to an annual water volume of $42.25 \times 10^9 m^3$ of which 30% belongs to subterranean waters. This accounts for more than 13,000 m^3 per capita per year (Cullaja et al. 2005). The rivers, fed mainly by precipitation (69%), show a typical Mediterranean regime, with seasonal variation in the flow rate, and high flow from October to May (Cullaja et al. 2005; Shumka et al. 2018). There are several main contemporary independent river and lake systems in Albania (Fig. 1). Listed from north to south they are as follows: Ohrid-Drin-Skadar system (including River Buna) (A), rivers Mat (B), Ishëm (C), Erzen (D), Shkumbin (E), Seman (composed of two big inflows—Devoll and Osum) (F), and Vjosë (Aoos in Greece) (G), several short rivers flowing from the Cikes mountains to the southernmost part of the Adriatic Sea and the northernmost part of the Ionian Sea (H), area around Butrint Lagoon (rivers Bistrica and Pavllo) (I), and Lake Prespa (K). Most of the above-mentioned lakes and rivers (A–G) belong to the Adriatic catchment and the southernmost area (I) to the Ionian Sea slope. Area H lies on the border of the Adriatic and Ionian Seas. Whereas there is no surface outflow of water from Prespa Lakes, there are underground connections with Lake Ohrid (Amataj et al. 2007). Only a very small area in the northernmost part of Albania in the Albanian Alps belongs to the Danube basin (J) (Dill 1993).

Discharge of all Albanian rivers is seasonally highly variable, being sometimes more than ten times less in summer than in winter. The beds of the main rivers are usually very wide, as a great amount of gravel and pebbles is deposited around the flow itself (Dill 1993).

The basic biological objective of fisheries management in Albania is to control the amount of fish that can be caught (Weiss et al. 2018; Shumka et al. 2020a, b),

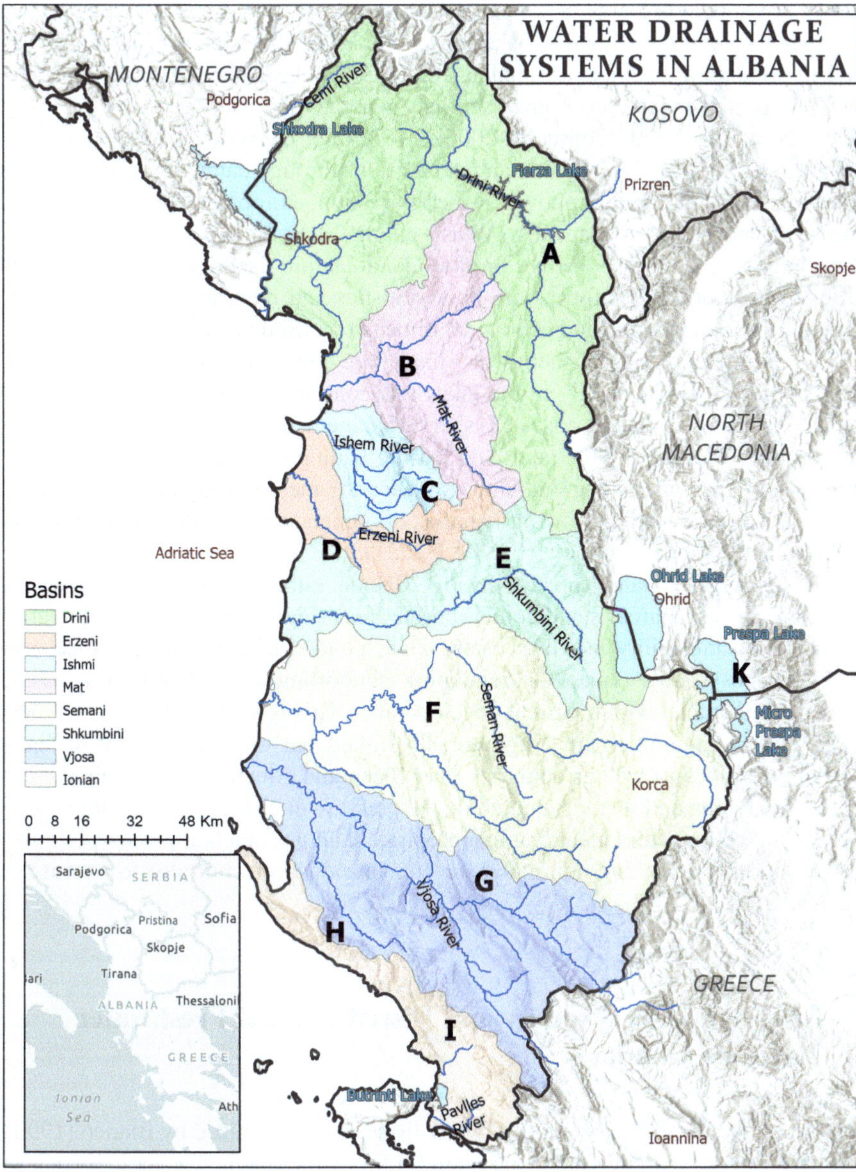

Fig. 1 Water drainage systems in Albania

following principles of sustainability from a renewable resource (Kolding and Zwieten 2014), while in this context, sustainable means that a fishery can be carried out indefinitely without hampering the resources' reproductive capacity. The trends of commercial inland capture fisheries, particularly in lakes and rivers, are in continuous decline (Weiss et al. 2018), similar to what is happening across almost all of Europe (Cowx 2015). So, the experiences and case studies have revealed that similarly to commercial marine fisheries, exploitation of freshwater fish stocks can lead to population declines (Arlinghaus et al. 2016) and alter structural and functional properties and the ecosystem services generated by fishes (Allan et al. 2005; Lewin et al. 2006; Dudgeon et al. 2006). Freshwater ecosystems of Albania have also been strongly affected by human interventions through damming, water habitats diversions, agriculture, hydropower, water extraction, eutrophication, pollution, and release of non-native organisms (Weiss et al. 2018; Shumka et al. 2020a, b; Catsadorakis et al. 1996). Further on the inland capture fisheries around the world suffer from poor catch information, poor statistics and reporting, and a poor integration of socioeconomic and ecological dimensions (Shumka et al. 2020a, b).

The commercial inland fisheries in Albania are considered of significant importance and are focused mainly in the lakes and coastal lagoons of the Adriatic and Ionian Sea basins. At the local scale Fishery Management Organizations (FMOs) are responsible for managing the lakes and leasing fishing rights to fishermen, particularly to commercial fishing individuals, who are members of the organization. Following official statistics inland fisheries do generate a considerable income for the local and national economy (INSTAT 2021).

All significant inland ecosystems of Albania either are crucial part of the protected areas system or stand at their buffer areas (Fig. 2). This needs to integrate ecological sustainability into a protected areas context of national ecological network and ecosystem services (i.e., regulating, supporting, and cultural services) and is even more crucial within current developments. This is also due to the catchment continuum; communities are longitudinally linked but often owned by different legislative entities as well as countries (Der Sleen and Albert 2021). While there is a lack of planning and sectors integration, the pressure on fisheries sustainability is in essence linked to factors including inappropriate land use practices, water pollution, habitat modifications, and non-native species introductions, but also to overuse of fish resources.

2 Brief Notes on Diversity and Distribution of Freshwater Fishes in Albania

The distinctiveness of Albanian freshwater fishes was recognized by Bianco (1990), who defined specific ichthyological region for this area. The freshwater fish fauna of Albania represents a specific value of national heritage, particularly due to its diversity and high degree of endemicity. Primarily, this is due to its complex geological and climatic features and position within the Adriatic and Ionian Sea on the west and Balkan continental areas on the other direction that has allowed different colonization from outside the area and long periods of speciation.

The knowledge about the diversity and distribution of freshwater fishes in the Mediterranean has recently increased considerably. As of July 2022, the freshwater

Fig. 2 (**a**) Lakes Ohrid and Prespa, a view from Galichica Mountain; (**b**) Reservoir of Koman; (**c**) Lake Shkodra; (**d**) River Black Drini; and (**e**) River Vjosa (Author of photos: S. Shumka)

fish fauna of Albania is known to consist of 101 species belonging to 25 families. The actual number of inland water fish species is not easily determined, as various marine species are often encountered in the lower and middle sections of water courses, especially in transitional waters, while new alien species are becoming established in the natural ecosystems. Further on, taxonomic changes are taking place at a rapid pace, with new species being described, numerous of former synonyms being reinstated, and former "subspecies" validated to species rank. Within these circumstances and dynamics there are permanent changes in country checklist of freshwater fish species. This also applies to the knowledge of Albanian freshwater fishes. The primary past sources of information were the books of Poljakov et al. (1958), which included 36 freshwater species (Rakaj 1995) and where 77 species were mentioned. The difference between the coverage in these two publications is probably due in part to inclusion of newly introduced species (Dhora 2015), but more so by changes in the taxonomic status of many species. Moreover, the data about the freshwater fishes of Albania increased considerably in the twenty-first century. The number of new species was described from this area (Bianco and Kottelat 2005; Economidis 2005; Kovačić and Šanda 2007; Miller and Šanda 2008; Zupančič et al. 2010; Bogutskaya et al. 2010), though most of the species were described fromtransboundary water bodies. The updated checklists of Dhora (2015) included altogether 120 species, thought almost 20 of them were considered doubtful. Detailed field-based studies of specific drainages provided additional valuable information on the freshwater ichthyofauna of Albania (Shumka et al. 2020a, b; Shumka and Apostolou 2018; Grapci-Kotori et al. 2020). Furthermore, numerous studies focusing on specific groups (Böhlen et al. 2006; Mendel et al. 2008; Šanda et al. 2008; Buj et al. 2010; Marková et al. 2010; Kalous et al. 2013; Palandačić et al. 2015; Stierandová et al. 2016; Tutman et al. 2017; Vukić et al. 2019; Schönhuth et al. 2018; Bartáková et al. 2019; Viñuela Rodríguez et al. 2020; Barbieri et al. 2000; Böhlen et al. 2003; Marić and Radujković 2009), focused on the European (Kottelat and Freyhof 2007) or peri-Mediterranean region (Böhlen et al. 2003; Marić and Radujković 2009), helped to determine the systematic position of a number of populations of Albanian freshwater fishes. Based on a revision of the available resources, we counted 101 freshwater fish species belonging to 25 families (Table 1). Out of them 76 species are considered native freshwater fishes, while 16 are amphidromous species regularly entering freshwaters, and 25 are introduced species (Table 1).

Regarding the species richness, the highest number of species inhabits Drini basin (including the lakes Ohrid and Skadar) (Table 1), followed by Vjosa basin, while the least number of species is found in the Danubian basin, presented only very marginally in Albanian, moreover in high altitude.

From bio-geographic point of view, most of the Albanian region formed a specific district (delimited by the Drini basin on the north and Vjosa in the south), called Albanian, south Adriatic, or southeastern Adriatic by different authors (Bianco 1990; Snoj et al. 2009; Perea et al. 2010; Geiger et al. 2014; Zogaris et al. 2009; Oiknomou et al. 2014; Zogaris and Economou 2017; Economidis and Daoulas 2003; Albrecht and Wilke 2008). From this area, Prespa Lake is sometimes delimited as a specific region (Perea et al. 2010) (Fig. 3).

3 Significant Fish Resources Water Bodies in Albania

3.1 Lake Ohrid

The oldest lake in Europe is Lake Ohrid, which has tectonic origins and is thought to be two to three million years old (Matzinger et al. 2006). The lake is located between Albania and North Macedonia. Its catchment area (combined with Lake Prespa) comprises about 2600 km^2(Spirkovski 2004), and the lake's surface is about 358 km^2. Lake Ohrid is fed primarily by spring water from the two main surface springs at the southern shores: St. Naum's in North Macedonia and Tushemisht in Albania. There are also a lot of sub-lacustrine (underwater) springs along the eastern

Table 1 Fish species presence in different water basins of Albania

No	Species	English name	Categories	Drini Basin	Vermoshi	Prespa	Mati	Ishmi	Erzeni	Shkumbini	Semani	Vjosa	Ionian Streams	Pavllo
1	*Acipenser naccarii* Bonaparte, 1836	Adriatic sturgeon	N	+								+		
2	*Acipenser sturio* Linnaeus, 1758	Atlantic sturgeon	N	+								+		
3	*Alburnoides devolli* Bogutskaya et al. 2010	Devolli spirlin	N								+			
4	*Alburnoides fangfangae* Bogutskaya et al. 2010	Osumi spirlin	N								+			
5	*Alburnoides ohridanus* (Karaman, 1928)	Ohrid spirlin	N	+			+	+	+					
6	*Alburnoides prespensis* (Karaman, 1928)	Prespa spirlin	N			+				+	+	+	+	
7	*Alburnus belvica* Karaman, 1924	Prespa bleak	N			+								
8	*Alburnus scoranza* Bonaparte, 1845	Skadar bleak	N	+			+					+		

(continued)

Table 1 (continued)

No	Species	English name	Categories	Drini Basin	Vermoshi	Prespa	Mati	Ishmi	Erzeni	Shkumbini	Semani	Vjosa	Ionian Streams	Pavllo
9	*Alosa fallax* (La Cepède, 1803)	Twait shad	N	+								+		
10	*Alosa* sp. Skadar	Skadar shad	N	+										
11	*Ameiurus nebulosus* (Lesueur, 1819)	Brown bullhead	A	+										
12	*Anguilla anguilla* (Linnaeus, 1758)	European eel	N	+		+	+	+	+	+	+	+	+	+
13	*Atherina boyeri* Risso, 1810	Sand smelt	N	+			+				+	+	+	+
14	*Abramis brama* (Linnaeus, 1758)	Freshwater bream	A							+	+			
15	*Aphanius fasciatus* (Valenciennes, 1821)	Mediterranean toothcarp	N	+			+				+	+		+
16	*Barbatula sturanyi* (Steindachner, 1892)	Brook loach	N	+										
17	*Barbatula zetensis* (Soric, 2001)	Zeta stone loach	N	+										

18	*Barbus balcanicus* (Kotlík, Tsigenopoulos, Ráb, & Berrebi, 2002)		N	+	+									
19	*Barbus peloponnesius* (Valenciennes, 1842)	Peloponnese barbel	N										+	
20	*Barbus prespensis* (Karaman, 1924)	Prespa barbel	N			+				+	+	+	+	
21	*Barbus rebeli* (Köller, 1926)	Balkan barbel	N				+	+	+					
22	*Barbus* sp. Drini		N	+										
23	*Carassius auratus* (Linnaeus, 1758)	Goldfish	A	+						+	+			
24	*Carassius carassius* (Linnaeus, 1758)	Crucian carp	A	+						+	+			
25	*Carassius gibelio* (Bloch, 1782)	Prussian carp	A	+		+		+	+	+	+	+		
26	*Carassius langsdorfii* (Temminck & Schlegel, 1846)	Ginbuna carp	A	+							+			

(continued)

Table 1 (continued)

No	Species	English name	Categories	Drini Basin	Vermoshi	Prespa	Mati	Ishmi	Erzeni	Shkumbini	Semani	Vjosa	Ionian Streams	Pavllo
27	*Caspiomyzon graecus* (Renaud & Economidis, 2010)	Epirus brook lamprey	N											+
28	*Chelon aurata* (Risso, 1810)	Golden grey mullet	N	+								+		
29	*Chelon ramada* (Risso, 1827)	Thinlip grey mullet	N	+								+		+
30	*Chelon saliens* (Risso, 1810)	Sharpnose grey mullet	N	+								+		
31	*Oedalechilus labeo* (Cuvier, 1829)	Boxlip grey mullet	N	+								+		
32	*Chondrostoma scodrense* Elvira, 1987	Skadar nase	N	+										
33	*Chondrostoma ohridanum* Karaman, 1924	Ohrid nase	N	+								+		
34	*Chondrostoma prespense* Karaman, 1924	Prespa nase	N			+								
35	*Cobitis meridionalis* Karaman, 1924	Prespa spined loach	N			+								
36	*Cobitis ohridana* Karaman, 1928	Ohrid spined loach	N	+			+	+	+	+	+	+		

37	*Cottus gobio* Linnaeus, 1758	European bullhead	N		+									
38	*Citharus linguatula* (Linnaeus, 1758)	Atlantic spotted flounder	N	+										
39	*Ctenopharyngodon idella* (Valenciennes 1844)	Grass carp	A	+		+				+	+			
40	*Cyprinus carpio* Linnaeus, 1758	Carp	N	+		+	+	+	+	+	+	+		+
41	*Dicentrarchus labrax* (Linnaeus, 1758)	Sea bass	N	+								+		
42	*Economidichthys pygmaeus* (Holly, 1929)	Western Greece goby	N	+										
43	*Eudontomyzon stankokaramani* Karaman, 1974	Drini brook lamprey	N	+										
44	*Gambusia holbrooki* Girard, 1859	Mosquitofish	A	+		+	+	+	+	+	+	+		
45	*Gasterosteus aculeatus* Linnaeus, 1758	Western three-spined stickleback	N	+									+	
46	*Gobio ohridanus* Karaman 1924	Ohrid gudgeon	N	+										

(continued)

Table 1 (continued)

No	Species	English name	Categories	Drini Basin	Vermoshi	Prespa	Mati	Ishmi	Erzeni	Shkumbini	Semani	Vjosa	Ionian Streams	Pavllo
47	*Gobio skadarensis* Karaman, 1936	Skadar gudgeon	N	+			+	+	+	+	+	+		
48	*Hypophthalmichthys molitrix* (Valenciennes, 1844)	Silver carp	A	+		+					+			
49	*Hypophthalmichthys nobilis* (Richardson, 1845)	Bighead carp	A	+							+			
50	*Huso huso* (Linnaeus, 1758)	Beluga	A	+										
51	*Knipowitschia goerneri* Ahnelt, 1991		N											+
52	*Knipowitschia montenegrina* Kovačić and Šanda 2007		N	+										
53	*Lampetra soljani* Tutman et al. 2017		N											
54	*Lepomis gibbosus* (Linnaeus, 1758)	Pumpkinseed	A	+		+	+	+	+	+	+	+		

55	*Luciobarbus albanicus* (Steindachner, 1870)	Strossidi	N									+		
56	*Megalobrama terminalis* (Richardson, 1846)	Black Amur bream	A	+							+			
57	*Mugil cephalus* Linnaeus, 1758	Striped mullet	N	+								+	+	+
58	*Mylopharyngodon piceus* (Richardson, 1846)	Black carp	A	+										
59	*Oreochromis niloticus* (Linnaeus, 1758)	Nile tilapia	A					+						
60	*Oncorhynchus mykiss* (Walbaum, 1792)	Rainbow trout	A	+			+		+	+		+		
61	*Oxynoemacheilus pindus* (Economidis, 2005)	Pindus stone loach	N						+	+	+	+		
62	*Pachychilon pictum* (Heckel et Kner, 1858)	Ohrid roach	N	+			+	+	+	+	+	+		+
63	*Parabramis pekinensis* (Basilewsky, 1855)	White amur bream	A			+					+			

(continued)

Table 1 (continued)

No	Species	English name	Categories	Drini Basin	Vermoshi	Prespa	Mati	Ishmi	Erzeni	Shkumbini	Semani	Vjosa	Ionian Streams	Pavllo
64	*Pelasgus minutus* (Karaman, 1924)	Ohrid minnow	N	+										
65	*Pelasgus prespensis* (Karaman, 1924)	Prespa minnow	N			+					+			
66	*Pelasgus thesproticus* (Stephanidis, 1939)	Epiros minnow	N									+		+
67	*Perca fluviatilis* Linnaeus, 1758	European perch	A	+										
68	*Petromyzon marinus* Linnaeus, 1758	Atlantic sea lamprey	N	+								+		
69	*Phoxinus karsticus* Bianco & De Bonis 2015		N	+										
70	*Phoxinus* sp. Drini		N	+										
71	*Platichthys flesus* (Linnaeus, 1758)	European flounder	N	+								+		

72	*Pomatoschistus montenegrensis* Miller & Sanda, 2008	Skadar goby	N	+										
73	*Pseudorasbora parva* (Temminck & Schlegel, 1845)	Topmouth gudgeon	A	+		+	+	+	+	+	+	+		
74	*Rhodeus amarus* (Bloch, 1782)	Bitterling	A	+		+					+			
75	*Rutilus albus* Marić, 2010	White Skadar roach	N	+										
76	*Rutilus ohridanus* (Karaman, 1924)	Ohrid roach	N	+										
77	*Rutilus prespensis* (Karaman, 1924)	Prespa roach	N	+		+								
78	*Sabanejewia balcanica* (Karaman, 1922)	Balkan golden loach	N	+	+									
79	*Ichthyocoris (Salaria) fluviatilis* (Asso, 1801)	Freshwater blenny	N	+			+					+	+	+

(continued)

Table 1 (continued)

No	Species	English name	Categories	Drini Basin	Vermoshi	Prespa	Mati	Ishmi	Erzeni	Shkumbini	Semani	Vjosa	Ionian Streams	Pavllo
80	*Salmo aphelios* Kottelat, 1997		N	+										
81	*Salmo balcanicus* (Karaman, 1927)		N	+										
82	*Salmo farioides* Karaman, 1938	Adriatic brown trout	N	+			+			+		+		
83	*Salmo labrax* Pallas, 1814	Danubian brown trout	N		+									
84	*Salmo letnica* (Karaman, 1924)	Ohrid trout	N	+		+								
85	*Salmo lumi* Poljakov, Filipi & Basho, 1958		N	+										
86	*Salmo marmoratus* Cuvier, 1829	Marble trout	N	+										
87	*Salmo obtusirostris* (Heckel, 1852)	Softmouth trout	N	+										
88	*Salmo ohridanus* Steindachner, 1892	Belvica	N	+										

89	*Salmo peristericus* Karaman, 1938	Prespa trout	N			+								
90	*Sander lucioperca* (Linnaeus, 1758)	Pike-perch	A	+										
91	*Scardinius knezevici* Bianco & Kottelat, 2005	Skadar rudd	N	+										
92	*Silurus glanis* (Linnaeus, 1758)	Wels catfish	A			+								
93	*Squalius pamvoticus* (Stephanidis, 1939)	Pamvotis chub	N											+
94	*Squalius platyceps* (Zupančič, Marić, Naseka, & Bogutskaya, 2010)	Skadar chub	N	+			+	+	+	+	+	+		
95	*Squalius prespensis* (Fowler, 1977)	Prespa chub	N			+								
96	*Syngnathus abaster* Risso, 1827	Black-striped pipefish	N	+									+	+

(continued)

Table 1 (continued)

No	Species	English name	Categories	Drini Basin	Vermoshi	Prespa	Mati	Ishmi	Erzeni	Shkumbini	Semani	Vjosa	Ionian Streams	Pavllo
97	*Telestes pleurobipunctatus* (Stephanidis, 1939)	Greek riffle dace	N										+	+
98	*Telestes montenigrinus* (Vuković, 1963)	Montenegro riffle dace	N	+										
99	*Thymallus thymallus* (Linnaeus, 1758)	European grayling	A	+										
100	*Tinca tinca* (Linnaeus, 1758)	Tench	A	+		+					+			
101	*Valencia letourneuxi* (Sauvage, 1880)	Corfu toothcarp	N											+
				73	4	22	17	13	14	20	29	33	10	15

Fig. 3 Fish species of conservation interest. (**a**) *Cobitis meridionalis*; (**b**) *Barbus perspensis*; (**c**) *Oxynoemacheilus pindus*; (**d**). *Salmo farioides*; (**e**) *Eudontomyzon stankokaramani*; (**f**) *Ichthyocoris (Salaria) fluviatilis*; and (**g**) *Valencia letourneuxi* (Author of photos: S. Shumka)

and, to some extent, western shores. The surface and underwater springs in the south (and east) draw their water from the neighboring Lake Prespa.

Fish and fishery have always played an important role in the Lake Ohrid region. Nowadays, fishing is still of high importance to fishers and their families living in neighboring cities and villages in Albania and North Macedonia. The fish stocks in the lake are currently in a significantly different state than they were in the past because they are exposed to a variety of stresses, including invasive species, unsustainable fishing practices, environmental contamination, and habitat deterioration (Talevski et al. 2009).

The fish fauna of Lake Ohrid is represented by 20 autochthonous species from five families: Salmonidae (5), Cyprinidae (2), Leuciscidae (11), Cobitidae (1), Nemacheilidae (1), and Anguillidae (1) and 6 allochthonous species (belonging to four families). With the exception of eel, none of them is a migratory species. The high level of endemism present in Lake Ohrid, which also applies to fish species, is a distinctive feature (Albrecht and Wilke 2008). According to Talevski et al. (2009) there are seven endemic fish species in the lake, namely *Salmo ohridanus*; *S. aphelios*; *S. balcanicus*; *S. letnica*; *S. lumi*; *Barbatula sturanyi* and *Gobio ohridanus*. Furthermore, occasional occurrence of *Alosa fallax* has been recorded by Tocko (GIZ 2017) and the finding of a single Drin brook lamprey (*Eudontomyzon stankokaramani*) has been reported by (GIZ 2017a).

In Lake Ohrid, fishing has been performed by local people from both riparian countries for many years. At present, eleven fish species are commercially used, of which Ohrid trout (*Salmo letnica*), belvica (*Salmo ohridanus*), eel (*Anguilla anguilla*), carp (*Cyprinus carpio*), and bleak (*Alburnus scoranza*) are of special interest. Of these named species, the bleak is of lowest commercial value.

During the communist regime (before 1990), the Albanian fishery at Lake Ohrid was managed by a state enterprise which employed about 35 fishers. After the collapse of the political system, there were an estimated 800 legal and illegal fishers on the lake (Levkov et al. 2007; Löffler et al. 1998). During those times, the legal fishers were organized into different groups or cooperatives. Every licensed fisher has been a member of a Fisheries Management Organization (FMO) since 2002. This organization is charged with managing a landing site and taking part in the co-management of fisheries resources. The FMO can only issue 110 licenses total.

3.2 Lake Prespa

The Prespa Lakes Basin is a high-altitude system (850–2600 m) with a catchment area of over 2500 km^2. It includes Lakes Macro and Micro Prespa and covers parts of the territories of Albania, North Macedonia, and Greece. Lake Macro Prespa is a subtropical dimictic lake. According to GIZ (2017a) and Crivelli et al. (1997) it has a surface area of currently about 254 km^2 with a maximum water depth of 48 m (mean 14 m). Micro Prespa is of much smaller size (surface area 47 km^2) and has a maximum water depth of 9 m. Prespa Lake has four tributaries: Agios Germanos (Greek part), Brajcinska, Kranska, and Golema Reka (North Macedonian part). At the Albanian territory, there are no perennial streams feeding the lake.

Since the late twentieth century, the Prespa Lakes environment has been susceptible to excessive nutrient inputs as a result of intensive agricultural activities, which has led to dramatic overabundant plant growth (eutrophication) (Spirkovski 2004; GIZ 2017a). The trophic state of the lake developed from oligotrophic in the mid-1970s to mesotrophic in the following two decades and to eutrophic status (Crivelli et al. 1997) from the turn of the century. Changes in trophic conditions are demonstrated by physicochemical water parameters and the development of phytoplankton communities (Šanda et al. 2008). Reduced water levels due to excessive lake water use for agriculture at the same time as an increase in fertilizer loading have also significantly altered the lake's environmental characteristics. For instance, water transparency has significantly decreased since a few decades ago (Löffler et al. 1998; Crivelli et al. 1997).

Twenty-five fish taxa have been identified in the Prespa Lakes. With the exception of catadromous European eel, *Anguilla anguilla*, none of them is a migratory species. The lake has a high percentage of endemic species, albeit it has recently been discovered that several of the supposedly isolated species also live in other drainages. Thus, the strict endemics are Prespa bleak (*Alburnus belvica*), Prespa nase (*Chondrostoma prespensis*), Prespa trout (*Salmo peristericus*), Prespa chub

(*Squalius prespensis*), and Prespa loach (*Cobitis meridionalis*), which were yet not reported outside the lake.

Out of determined 13 native fishes, 8 species are endemic: Prespa spirlin (*Alburnoides prespensis*), Prespa bleak (*Alburnus belvica*), Prespa barbell (*Barbus prespensis*), Prespa nase (*Chondrostoma prespensis*), Prespa minnow (*Pelasgus prespensis*), Prespa roach (*Rutilus prespensis*), Prespa trout (*Salmo peristericus*), and Prespa chub (*Squalius prespensis*). At first sight, the proportion of endemism in the fish populations of the Prespa Lakes seems remarkable. It should be mentioned however, that, according to Catsadorakis et al. (2018) and Economidis (1992), the taxonomic position of a number of taxa occurring in the Prespa Lakes remains doubtful. At present, only the barbel would appear to be undoubtedly endemic to (Micro and Macro) Prespa Lakes (Economidis 1992). Prespa barbel also presents species with some economic importance in the Prespa watershed.

According to Crivelli et al. (1997), two endemic species (Prespa barbel and Prespa trout) are classified as "endangered," which is an important criterion used for the identification of priority species for conservation of animals in the Prespa region. Both species are listed as "vulnerable" species on the IUCN Red List of Threatened Animals (Globally threatened species and Regional-European threatened species) (Freyhof and Brooks 2011; Freyhof 2012). Further, Prespa barbel is present in all three countries (Macedonia, Albania, Greece) sharing Prespa Lake, which is important for transboundary collaboration in terms of species conservation. Pelister stream trout (*Salmo peristericus*), on the contrary, can be only found in Macedonian and Greek rivers (River Brajcinska and its tributaries, Agios Germanos stream, and others) of the Prespa basin.

3.3 Lake Skadar/Shkodra

On the Balkan Peninsula, Lake Skadar/Shkodra is a sizable fluvial lake that is shared by Albania and Montenegro (GIZ 2017b). Its drainage basin is situated along the Montenegrin/Albanian border and has a surface area of roughly 5500 km^2 (Stierandová et al. 2016).. The lake itself is located in the southern-eastern Dinaric Alps' outer karstic region. Lake Skadar/Shkodra is the largest of the Balkan lakes, with a seasonally variable surface size of between 370 and 600 km^2. From immemorial time, the lake has been of great importance to the people in the region, providing food and income and serving as transportation route (Hassert 1892). The lake is home to numerous animal and plant species, many of which are rare or threatened. Moreover, many taxa of the occurring flora and fauna are endemic to this area (Hassert 1892; Pešić and Glöer 2013; Shumka 2019). Due to its high importance for species conservation, the lake was included in the Ramsar list as a wetland of international importance in 1995 (Montenegro) and 2005 (Albania).

The ichthyofauna of the lake is composed of a relatively high number of species of freshwater and marine origin, some of which, however, only periodically inhabit the lake (Marić and Milošević 2011). The number of fish species in the lake system

(Lake Skadar/Shkodra and adjoining rivers) increased over the last 50 years because of uncontrolled stocking of allochthonous fish species (*Hypophthalmichthys molitrix, Thymallus thymallus*, *Carassius gibelio*, *Perca fluviatilis*) which mainly originate from the Black Sea drainage area. Besides those introductions, however, the lake system is inhabited by several sub-endemic fish species such as *Salmo obtusirostris* and endemic including *Barbatula zetensis*, *Chondrostoma skadarensis*, *Gobio skadarensis*, *Knipowitschia montenegrina*, *Pomatoschistus montenegrensis*, and *Rutilus albus* of the families Salmonidae, Cyprinidae, Gobiidae, and Cobitidae (Marić 1995). Similarly, some further fishes (*Acipenser naccarii*, *Acipenser sturio*, *Alosa fallax*, *Rhodeus amarus*, *Telestes montenigrinus*, and *Salmo marmoratus*) deserve special protection for conservation purposes and have been specified in the Natura 2000 annexes (Shumka 2019). Finally, the lake is also spawning place and habitat of larval and juvenile fish, respectively, of several catadromous and anadromous (e.g., *Alosa fallax*, *Anguilla anguilla*, *Acipenser* spp.) as well as estuarine fish species (*Mugil cephalus*, *Dicentrarchus labrax*, and *Platichthys flesus*).

According to Marić and Milošević (2011), there are roughly 50 species registered in Lake Skadar/Shkodra, of which about one-fifth are introduced.

The data of total lakes commercial catches in Albania for the period 1947–2020 (in tonnes) show a significant decrease in the period after the year 1990 (Fig. 4).

3.4 Coastal Lagoons

The Albanian costal lagoons represent the most sensitive and important environment of the Albanian coast. These lagoons are areas of multiple ecological and economic values and uses, as they provide fish and wildlife habitats, support complex food web, absorb water to reduce flooding and damage from storms, provide erosion control, improve the quality of water, and in particularly provide open space and aesthetic value. It is estimated that coastal lagoons, as specific ecosystems between the land and the sea, are the most productive water ecosystems. The ecological and economic importance of transitional water biological reaches has been well understood (Veiga et al. 2010; Russi et al. 2012; Bon et al. 2018), while there is little knowledge about the temporal and spatial patterns of fish populations, and their importance as nursery grounds for species of both commercial and conservationist interest (Russi et al. 2012; Elliott et al. 2007). To that fact, this study aims to assess the value of the different lagoon habitat types by determining spatial patterns of habitat use by fish assemblages.

The main eight coastal lagoons are situated on the Albanian coastline with a length of about 430 km that contains two geographic entities: the Adriatic and the Ionian Sea coastal areas. Seven of them (from north to south; Viluni, Kune, Vain, Patok, Karavasta, Narta, and Orikumi) are situated in the Adriatic Sea coastline and only one (Butrint Lagoon) is situated in Ionian Sea coastline. The traditional management approach in the past mostly based on the consumptive use of wild renewable resources (to produce fish and to create opportunities for seasonal hunting

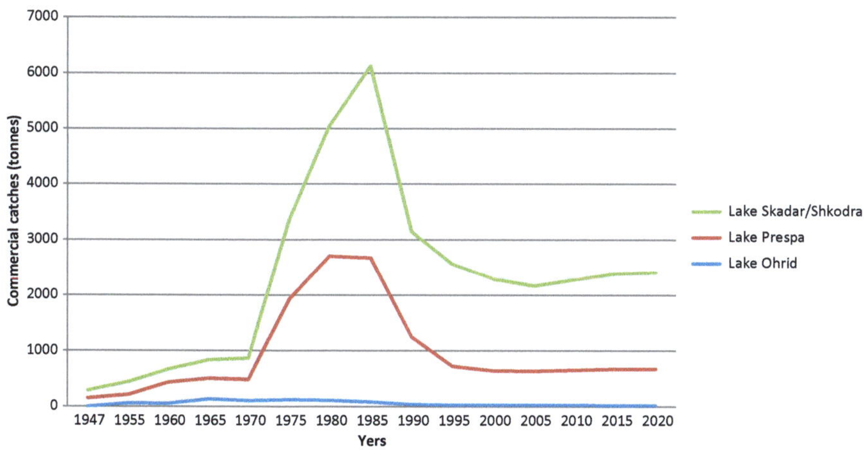

Fig. 4 Total lakes commercial catches in Albania for the period 1947–2020 (in tones)

of aquatic birds) has evolved today toward models integrating the concept of sustainability, i.e., management models in which biodiversity conservation and economic activities must be correctly included.

There is a lack of data regarding the present knowledge on the water quality and the water volume-reaching lagoon from its watershed and fewer data regarding water quality of the lagoons (as described above for each lagoon). However, in general, terrestrial inputs and inputs of water originating from the watersheds of lagoon systems threaten today the integrity of these systems by accelerating and degrading their biological production. In such a case of the Albanian lagoons, particularly for Karavasta and Narta lagoon there is no doubt that freshwater input is needed in order to avoid strong salinization (saltiness), and to favor fish recruitment, but this water has to be clean and unpolluted with nutrients in order to be beneficial for the system.

The fish community of Albanian coastal lagoons is typical of Mediterranean lagoons with two groups of species: the migratory fish and the sedentary fish. *Sparidae, Mugilidae, Moronidae, Soleidae, Anguillidae,* and *Belonidae* are the main groups of migratory species, and *Gobiidae, Cyprinodontidae, Atherinidae,* and *Syngnthidae* are the main groups with sedentary species, although some species of these groups are also migratory (Shumka 2021, 2022; FAO 2012). As described for each lagoon, there are a few differences on the fish species, and main fish species are: *Sparus auratus, Mugil* spp. *(Mugil cephalus, M. ramada, M. saliens, Chelon labrosus), Anguilla anguilla, Dicentrarchus labrax, Solea vulgaris, Atherina boyeri,* and *Syngnathus* spp.

In Table 2, the Karavasta lagoon shows an annual fish production of about 50 tonnes, Butrinti about 58 t/year, Narta lagoon about 18 t/year; Vain about 54 t/year, Kune 20 t/year; Orikumi lagoon 14 t/year, Viluni 16 t/year, and Patok lagoon 23 t/year (FAO 2012).

This decline first of all is caused by the eutrophication process (i.e., Karavasta lagoon) and illegal fishing (especially with dynamite and pesticides) within lagoon

Table 2 Data on fish production of the main Albanian coastal Lagoons (year 2018)

Lagoon	Licensed fishermen	Target Species	Production (tonnes/ year)	Productivity (kg/ha year)	Type of fishing	CPUE (Catch per unite effort)
Karavasta	35	Seabass, bream mullet, eel	50	12.8	Fish wair, nets fyke nets	40–60
Butrint	20	Seabass, bream mullet, eel	58	36.2	Fish wair, nets fyke nets, hooks	23–40
Narta	25	Seabass, bream mullet, eel	18	6.7	Fish wair, nets fyke nets, hooks	54
Orikum	12	Seabass, bream mullet, eel	14	87	Fish wair, selective fyke nets	4
Kune	13	Seabass, bream mullet, eel	20	80	Nets trawls fyke nets	13
Vain	20	Seabass, bream mullet, eel	54	63.5	Fish wair, nets fyke nets, trawls	27
Viluni	13	Seabass, bream mullet, eel	16	57	Nets fyke nets	27.3
Patok	12	Seabass, bream mullet, eel	23	88.4	Nets fyke nets	12.8

and in the coastal area. More threatened lagoons are the large lagoons, such as Karavasta, Butrinti, Narta, and Vain.

Fishing practices in lagoon are practiced within the lagoon and in the channel connecting lagoon with the sea. Different kinds of fishing gears are used within lagoon like fyke nets for catching all fish species, especially for eels; gill and trammel nets for mullet; sea bass and gilthead bream and hooks for eels; and traps for mullet. In the channels, fishing is practiced in fish barrier located in the canal connecting the lagoon to the sea. At the moment only in Karavasta, Narta, Butrinti, Vain, and Orikum lagoon, there are fish traps.

The main problem that the lagoon fishermen face is the decrease in catches. Shumka (2021) elaborated the data collected from a study dedicated to Lagoon of Karavasta which represents the major lagoon where fisheries' exploitation occurs and revealed interesting facts concerning their dynamics, structure, functioning, and importance. The data show a clear trend of steadily decreasing landings in lagoon, albeit with different patterns, with drastic changes occurring during short time periods and increased inter-annual fluctuations (Fig. 5). Among the other a significant decrease has been recorded for the eel catch (Figs. 6 and 7).

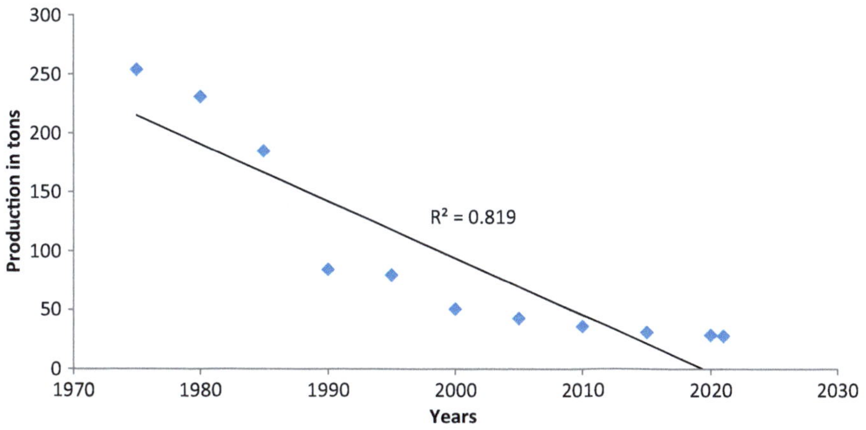

Fig. 5 Long-term trend of the annual yield of Karavasta lagoons in the Adriatic coast (Municipality of Divjaka)

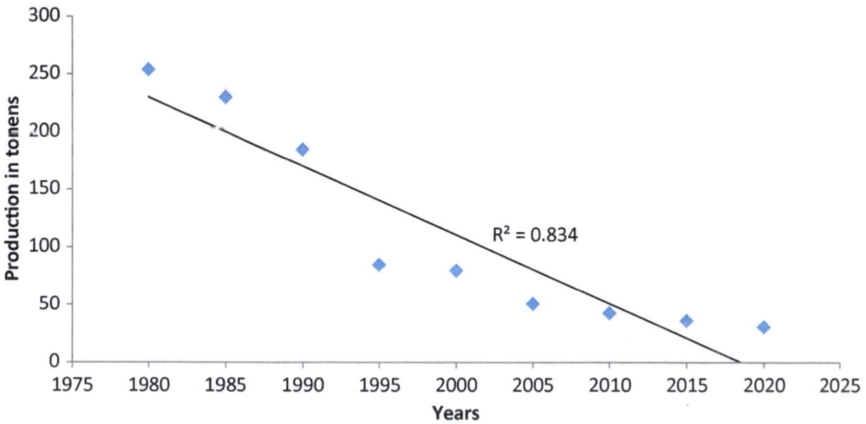

Fig. 6 The trend line of fish catches reduction. Karavasta Lagoon presenting period of years 1980–2021 in the Adriatic coast (Municipality of Divjaka)

Other significant problems faced by the Albanian lagoons are pollution generated by agricultural activities, residential sewage, and the shortage of freshwater input into the lagoon, the draining, and the poor infrastructure. Erosion is also a potential problem in several coastal areas, since the strips of sand that separate many coastal lagoons are diminishing.

Fig. 7 Commercially important species caught in Karavasta lagoon. (**a**) Eel (*Anguilla anguilla*); (**b**) gilthead seabream (*Sparus aurata*); (**c**) sand smelt (*Atherina boyeri*); (**d**) thin lip grey mullet (*Liza ramada*); (**e**) gudgen (*Gobius* sp.); (**f**) common sole (*Solea vulgaris*); (**g**) flathead grey mullet (*Mugil cephalus*); (**h**) European sea bass (*Dicentrarchus labrax*); (**i**) karamote prawn (*Melicertus kerathurus*); (**j**) garfish (*Belone belone*) (Author of photos: S. Shumka)

4 Commercial Fisheries Syntheses

As explained above, there are eight main lagoons with a total area of 10,000 hectares. In 2020, the catches in inland water amounted to 2844 tonnes (Table 3), while that of lagoons at 79 tonnes reflecting a significant reduction compared to 2017 when the same amounted to 599 tonnes (FAO 2012). In terms of inland water, fishing takes place mostly in Shkodra, Ohrid, Great Prespa, and Small Prespa. Further on, there are five large artificial lakes, Fierza, Koman, Vau i Dejes, and Ulza and many other man-made water bodies.

The local experience in aquaculture, using mainly semi-intensive practices, began in the early 1960s. Until the 1990s, it involved the production of carp family species, reaching a total of 25 hatcheries which covered a surface of 800 ha. Given the country's abundant water resources, climate, biological potential, and sociopolitical factors, the prospects for developing this important sector for the Albanian economy were high (Shumka 2022; FAO 2012). The economic directives of the earlier political system had encouraged polyculture of carps family species in all Albanian inland waters as a cheap and easy way of achieving food security.

Table 3 Data on fish catch categories in Albania for the period of 2016–2020 (the % calculation is based on the data of the last year, i.e., 2020)

	Description	2016	2017	2018	2019	2020	%
	Fishing categories						
I	Total fishing (1 + 2 + 3 + 4)	7884	8289	8629	8707	7808	
1	Marine	4646	4609	5537	5499	4521	26.76
2	Coastal line	952	1074	315	342	365	2.16
3	Coastal lagoons	598	599	350	94	79	0.47
4	Inland waters	1688	2007	2427	2772	2844	16.83
II	Aquaculture	3200	4000	5138	5229	8799	52.09
III	Mytilus galloprovincialis	1450	430	1108	1075	285	1.69
	Total fish caught (I + II + III)	**12,534**	**12,719**	**14,875**	**15,011**	**16,892**	**100**

Trout farming is concentrated in the southeast of Albania, where about 60 companies have their facilities. Cultivation of the endemic Ohrid trout (Salmo letnica) is restricted to the area near lake Ohrid. A few hatcheries are involved in the cultivation of common carp (Cyprinus carpio) and Chinese carps such as grass carp (Ctenopharyngodon idellus), bighead carp (Hypophthalmichthys nobilis), and silver carp (Hypophthalmichthys molitrix).

5 National Fisheries Management Policy and Obstacles

Fisheries management is the process by which trustworthy information is used to accomplish management goals and operational objectives specified for fisheries resources (Welcomme 2001), while achievement of management goals (Anonymous 2007) is linked with three major pillars, namely sustainable use of freshwater fish stocks, conservation of freshwater biodiversity, and equitable sharing of benefits among stakeholders with a focus on local communities.

The Fisheries Strategy of Albania highlights a series of priorities for every key aspect of the sector. Implementation has been slow mainly due to limited budgets and capacity to adopt the most appropriate management practices. The Law on Fisheries establishes institutional arrangements for sector governance. The law has provisions for a Central Consultative Commission for Fisheries and Aquaculture (CCCFA) at national level and for Fishery Management Organization (FMO) at local level. The Fishery Inspectorate conducts inspections to ensure that the Law on Fisheries and pertinent regulatory provisions is being followed. Fishing in protected regions, in particular estuary areas, and natural or man-made lagoons requires taking precautions. 1% of the territorial waters are covered by this management choice.

Numerous statutes and by-laws make up the regulatory framework for the fisheries and aquaculture industries. Albania is in the process of applying to be an EU candidate country, and as a result, it is also preparing to align its laws with the *acquis communautaire* of the EU. Several by-laws that incorporate some of the tenets of the Common Fishery Policy into Albanian law have been enacted. The legislation also includes the core values of the FAO's Code of Conduct for Responsible Fisheries, and work has already started to establish the Fishery Management Organization for inland and marine waterways. The main law regulating this sector is Law no 64/12 of 2012 "On Fishery." By-laws in perfect conformity with EU Regulations and GFCM Recommendations have advanced the fisheries legislation. The 2016 Albanian fisheries plan places a strong emphasis on excellent governance, sustainability, a competitive fleet with resource capacity in balance, and compliance with pertinent EU regulations.

Decision No. 701, dated 12.10.2016 "On the approval of the National Fisheries Strategy" 2016–2021," Sect. 1.6 of the Strategy refers to issues of coastal Lagoons, where the references on average CPUE reduction from 5.6 mt (*metric ton-a unit of weight equal to 1000 kilograms*) in 2002 to 1.7 mt in 2013 are noted, while the average production mention is 36 kg/ha.

In Sect. 3.4 (Fishery in coastal lagoons and freshwaters), referring to Strategy objectives, it is underlined that fishery has to be based on knowledge CPUE (*assuming that has to be determined in advance*). The document in the same section notes that "Amount of fish catch is proportional to maximum sustainable production and availability of fish stocks."

Law No. 80/2017: "For Some Additions and Amendments to Law No. 64/2012, "On Fishing," Amended, reflects the EU documents listed in this survey as follows: (i) Council Regulation (EC) no. 1224/2009, dated November 20, 2009, "On the establishment of a community control system for ensuring compliance with the rules of the common fisheries policy, amending previous Regulations (EC); (ii) Regulation (EU) no. 1380/2013 of the European Parliament and of the Council, dated December 11, 2013, "On a common policy of fisheries, amending Council Regulations (EC); (iii) Regulation (EU) 2015/2102 of the European Parliament and of the Council, dated October 28, 2015, amending Regulation (EU) no. 1343/2011 on certain provisions on fisheries in the field of the GFCM (General Fisheries Commission for the Mediterranean); and (iv) Commission Implementing Decision (EU) 2016/1251, dated July 12, 2016, "On the approval of a multi-annual program of Union for the collection, management and use of data in the fisheries and aquaculture sector for the period 2017–2019;

Order Nr. 536, dated 27.9.2019 [52] "On the approval of the European eel Management Plan (*Anguilla anguilla*) in Albania for the period 2019-2024" issued by MARD (Ministry of Agriculture and Rural development). This extremely important regulation since European eel comprises the main fished species within Karavasta lagoon. This management plan is in approximation and implementation of the Council Regulation of September 18, 2007, no. 1100/2007, laying down measures for the reconstruction of the European eel stock. This order in its introductory section underlines the fact that the European eel has been classified since 2009 as a "Critically Endangered Species—CR" based on the opinion of the International Maritime Exploration Council (ICES), which in the 2000s considered that the biological safety limits for this stock had been exceeded. The European Union, with the adoption of EC Regulation 1100/2007, has obliged the member states that use the eel stock to reduce the fishing effort by 25% and to draw up a management plan for the eel. Furthermore, since 2009 the European eel is also included in Annex II of the International Trade Convention on Endangered Species of Wild Fauna and Flora (CITES), which deals with species that are not necessarily endangered, but for the trading of which a check is planned, in order to ensure that they are not subject to exploitation incompatible with their survival. It is worth to mention that, besides that this is a management plan, there are no scaled-up specific measures that will ensure the improvement of eel status, but mostly counting the already existing data and regulations (e.g., statistics, seasons of opening and closing of channels, etc.)

6 Threats

Similarly to the rest of the region and world where there are hardly any large rivers systems left in which fish do not face a multitude of threats (Freyhof 2012; Weiss et al. 2018; Lazaj and Xhelilaj 2017), severe and persistent pollution continues to be a major problem in many rivers and includes pesticides and fertilizers running off from agricultural land, mercury from gold mining in headwaters, microplastics, and pollutants that affect the sex ratio in fish populations (Freyhof 2012). Loss, or severe reduction, of river connectivity is another major threat.

There are many anthropogenic threats to populations of freshwater fishes in freshwater basins. The primary threat is (i) the construction of hydropower plants that are leading to direct effects on fish life patterns (Weiss et al. 2018; Lazaj and Xhelilaj 2017). In Albania, river and stream barriers range from large hydroelectric dams (started to be installed mostly in 70′) and low-head run-of-the-river dams to small pond dams, alteration, and in-penstock diversions, etc. Their impact on fish populations depends on size, fish passage facilities, fish species, and operation practices that in every case are of high concern. After the legislative measures and the amendments of the existing ones, regarding the concessions, from 2007 (Anonymous 2007), private–public partnership was considered as a great opportunity for the sector of hydropower. Therefore, the interest for the hydropower sector grew rapidly, by providing enormous benefits for the private sector and at the same time meeting people's needs for electricity. Back then, the environmental and social impacts of hydropower plants (HPPs) into the ecosystems and biodiversity were unknown.

Recent studies reveal important data on the environmental and social negative impacts of HPPs on the environment's quality (Freyhof 2012; Weiss et al. 2018; Lazaj and Xhelilaj 2017). These reports confirm a critical situation related to rare and endangered plants and animals (fish, amphibians and reptiles, birds, and mammals), as well as rare and endangered habitats. Most of the threatened reported species belong to the Albanian Red List (2013), and/or IUCN endangered species. Dams inevitably alter thousands of square miles of waterways, affecting fish and wildlife, forests, farms, the water quality, and, of course, human population.

Table 4 The number of HPPs for each of the basins

Basins	Existing	In Construction	Planned	In total
Drini	16	3	140	159
Ishem/Erzen	1	0	10	11
Mat	2	0	57	59
Seman	6	5	68	79
Shkumbin	7	7	65	79
Vjosa	12	12	21	45
TOTAL	44	27	361	432

Currently, the surveys identify and visualize 432 HPPs in Albania, 361 of which are planned to be developed, 27 are in the constructing process, and 44 currently exist (Table 4) (Lazaj and Xhelilaj 2017).

Other threats include (i) improper water management (water quantity and quality). One of the biggest challenges to all ecosystems is water pollution. Industrial, urban, and agricultural waste are the sources of pollution, while one of the most detrimental effects of water contamination is eutrophication. It is brought on by the introduction of non-native species and the large-scale discharge of nutrient-rich substances into water systems, such as nitrogen and phosphorus. (ii) One of the biggest risks to all ecosystems and the nearby fish they support is the presence of non-native "invaders." Through restocking, they are unintentionally or purposely introduced into the aquatic environments. In total, there are 27 non-native species identified, although only nine of them have established populations. (iii) The failure to protect crucial fish spawning grounds. It is clear that when spawning areas in natural ecosystems become overrun with obstructions, fish cannot reproduce there. (iv) Pressure from unsustainable fishing methods. The use of "non-friendly fishing methods and tools," fishing during spawning periods, etc., are examples of inappropriate fishing techniques. There is a good chance that fish resources are being overused because there are no statistics on fisheries or information on fishermen (both legal and illegal). High fishing pressure which is frequently made worse by insufficient regulation enforcement (such as the inability to stop poaching). (v) Monitoring and research shortages could all be contributing factors. Lack of data on fish and fishing effort as well as knowledge gaps has not been documented. Besides the requirements of the WFD for considering fish as a key component in water monitoring, this is almost not covered within national monitoring programs mostly due to lack of resources.

7 Concluding Remarks

A particular feature of inland fishery in Albania is that it is characterized by a multi-use level at the major freshwater ecosystems (including all large lakes and coastal lagoons) where fish resources' exploitation is overlapped with ecological conservation approaches (protected areas, Ramsar site, Biosphere Reserve, etc.) and sectors such as agriculture and tourism development.

Due to this interaction among economic and ecological systems, management and sustainable utilization of inland fishery resources is an issue of sectoral integration, planning, and cooperation among different authorities and end users. This also includes water management, flood control, hydropower, and navigation. Further on, a serious obstacle lies in unclear competences of the local authorities and legal and institutional overlaps. In the process of design and management planning, there is an urgent need for particular fisheries management interventions such as harvest regulations or the type and amount of stocking and the fishery manager must ensure to be well represented in all decisions.

Based on the identified threats, it seems that inland fishery remains marginalized due to countries giving priority to marine fishery and aquaculture, and it suffers political and administrative support. Therefore, an integrated planning and management approach that integrates current inland fisheries within the broader scope of aquatic ecosystem management is often needed for sustainable inland fisheries. Further studies may enable definitions of clear elements of an integrative approach to inland fisheries planning and management.

References

Albrecht C, Wilke T (2008) Ancient Lake Ohrid: biodiversity and evolution. Hydrobiologia 615: 103–140. https://doi.org/10.1007/978-4020-9582-59

Allan JD, Abell R, Hogan Z, Revenga C, Taylor BW, Welcomme RL, Winemiller K (2005) Overfishing of inland waters. BioScience 55:1041–1051. https://doi.org/10.1641/0006-3568

Amataj S, Anovski T, Benischke R, Eftimi R, Gourcy LL, Kola L, Leontiadis I, Micevski E, Stamos A, Zoto J (2007) Tracer methods used to verify the hypothesis of Cvijic about the underground connection between Prespa and Ohrid Lake. Environ Geol 51:749–753. https://doi.org/10.1007/s00254-006-0388-9

Anonymous (2007) Official Gazette, 2007. 1 Law No. 9663 dated on 18/12/2006 "On concessions"; Council of Ministers Decision No. 27, dated on 19/1/2007 "On approval of rules for evaluation and concession procedures", changed with MCD No. 87 dated on 16.01.2008; · Council of Ministers Decision No.150, dated on 22/03/2007 "For the organization and function of Concession's Treated Agency" (ATRAKO); Minister's decision No.536, dated on 27/07/2007 "On regulations approval for the administration of the documents and requests for concessionary agreements and "Bonus evaluation criteria"

Arlinghaus R, Lorenzen K, Johnson BM, Cooke SJ, Cowx IG (2016) Management of freshwater fisheries: addressing habitat, people and fishes. In: Craig JF (ed) Freshwater fisheries ecology, 1st edn. John Wiley & Sons, Ltd.. Published 2016 by John Wiley & Sons, Ltd, pp 557–579

Barbieri R, Daoulas C, Psarras T, Stoumboudi MT, Economou A (2000) The biology and ecology of *Valencia letourneuxi* Sauvage 1880 (Valenciidae) - Prospects for conservation. Mediterranean Marine Science 1(2):75–90

Bartáková V, Bryja J, Šanda R, Bektas Y, Stefanov T, Choleva L, Smitha C, Reichard M (2019) High cryptic diversity of bitterling fish in the southern West Palearctic. Mol Phylogenet Evol 133:1–11

Bianco PG (1990) Potential role of the paleohistory of the Mediterranean and Paratethys basins on the early dispersal of Euro-Mediterranean freshwater fishes. Ichthyol Explor Freshw 1:167–184

Bianco PG, Kottelat M (2005) *Scardinius knezevici*, a new species of rudd from Lake Skadar, Montenegro (Teleostei: Cyprinidae). Ichthyol Explor Freshw 16(3):231–238

Bogutskaya NG, Zupančič P, Naseka AM (2010) Two new species of freshwater fishes of the genus *Alburnoides*, *A. fangfangae* and *A. devolli* (Actinopterygii: Cyprinidae), from the Adriatic Sea basin in Albania. Proc Zool Institute RAS 314(4):448–468

Böhlen J, Šlechtová V, Šanda R, Kalous L, Freyhof J, Vukić J, Mrdak D (2003) *Cobitis ohridanus* and *Barbatula zetensis* in the River Morača basin, Montenegro: distribution, habitat, population structure and conservation needs. Folia biologica (Krakow) 51(Suppl):147–153

Böhlen J, Šlechtová V, Bogutskaya N, Freyhof J (2006) Across Siberia and over Europe: Phylogenetic relationships of the freshwater fish genus *Rhodeus* in Europe and the phylogenetic position of *R. sericeus* from the River Amur. Molecular Phylogenetics and Evolution 40:856–865. https://doi.org/10.1016/j.ympev.2006.04.020

Bon N, Jjosipovič D, Kanjir U (2018) Linking geomorphological and demographic movements: The case of Southern Albania. ApplGeogr 100:55–67. https://doi.org/10.1016/j.apgeog.2018.09.002

Buj I, Vukić J, Šanda R, Perea S, Ćaleta M, Marčić Z, Bogut I, Povž M, Mrakovčić M (2010) Morphological comparison of bleaks (*Alburnus*, Cyprinidae) from the Adriatic Basin with the description of a new species. Folia Zoologica 59(2):129–141

Catsadorakis G, Malakou M, Crivelli AJ (1996) The Prespa barbel Barbus prespensis, Karaman 1924 in the Prespa lakes basin, north-western Greece. Tour du Valat, Arles, 79 p

Catsadorakis G, Papadopoulou E, Petrakos M, Koutseri I (2018) Status of fisheries at Megali Prespa Lake and Mikri Prespa Lake, Greece, Based on a census of fishermen's opinions. Environ Ecol Res 6(6):583–592. https://doi.org/10.13189/eer.2018.060608

Cowx IG (2015) Characterisation of inland fisheries in Europe. Fish Manage Ecol 22(1):78–87. https://doi.org/10.1111/fme.12105

Crivelli AJ, Catsadorakis G, Malakou M, Rosecchi E (1997) Fish and fisheries of the Prespa lakes. Hydrobiologia 351:107–125. https://doi.org/10.1023/A:1003064509018

Cullaja A, Hasko A, Miho A, Schanzd F, Brandle H, Bachofenf R (2005) The quality of Albanian natural waters and the human impact. Environ Int 31:131–146. https://doi.org/10.1016/j.envint.2004.06.008

der Sleen V, Albert A (2021) Patterns in freshwater fish diversity. Reference Module in Earth Systems and Environmental Sciences. https://doi.org/10.1016/B9778-0-12-819168-8.00056-6

Dhora D (2015) Te dhena me te plota per peshqit alien te Shqiperise. Bul. Shk., Seria Shk. Nat. 65: 94–102

Dill WA (1993) Inland fisheries of Europe. EIFAC Technical Paper. No. 52 Suppl. Rome, FAO, p 281

Dudgeon D, Arthington AH, Gessner MO, Kawabata ZI, Knowler DJ, Leveque C, Naiman RJ, Prieur-Richard AH, Soto D, Stiassny ML, Sullivan CA (2006) Freshwater biodiversity: importance, threats, status and conservation challenges. Biol Rev 81:163–182. https://doi.org/10.1017/S1464793105006950

Economidis PS (1992) Fish. In: The red data book of threatened vertebrates of Greece (M. Karandinos, E. Legakis, eds.), H.Z.S. & H.O.S., Athens, 43–81

Economidis PS (2005) *Barbatula pindus*, a new species of stone loach from Greece (Teleostei: Balitoridae). Ichthyol Explor Freshw 16(1):67–74

Economidis PS, Daoulas C (2003) *Barbus prespensis* Karaman, 1924. In: Banarescu PM, Bogutskaya NG (eds) The Freshwater Fishes of Europe, *Cyprinidae* 5/II, Part II, *Barbus*. Aula–Verlag, Wiebelsheim, Germany, pp 365–376

Elliott M, Whitfield AK, Potter IC, Blaber SJM, Cyrus DP, Nordlie FG, Harrison TD (2007) The guild approach to categorize estuarine fish assemblages: a global review. Fish Fisheries 8:241–268. https://doi.org/10.1111/j.1467-2679.00253.x

FAO (2012) National statistics division. http://faostat3.fao.org/home/E

Freyhof J (2012) Threatened freshwater fishes and mollusks of the Balkan, Potential impacts of hydropower projects. Report ECA Watch Austria and EuroNatur, 81 p

Freyhof J, Brooks E (2011) European red list of freshwater fishes. Publications Office of the European Union, Luxembourg

Geiger MF, Herder F, Monaghan MT, Almada V, Barbieri R, Bariche M, Berrebi P, Bohlen J, Casal-Lopez M, Delmastro GB, GPJ D, Dettai A, Doadrio I, Kalogianni E, Kärst H, Kottelat M, Kovačić M, Laporte M, Lorenzoni M, Marčić Z, Özulug M, Perdices A, Perea S, Persat H, Porcelotti S, Puzzi C, Robalo J, Šanda R, Schneider M, Šlechtová V, Stumboudi M, Walter S, Freyhof J (2014) Spatial heterogeneity in the Mediterranean Biodiversity Hotspot affects barcoding accuracy of its freshwater fishes. Mol Ecol Resour 14(6):1210–1221. https://doi.org/10.1111/1755-0998.12257

GIZ (2017) Fish and Fisheries Ohrid Lake. Implementing the EU Water Framework Directive in South-Eastern Europe. Tirana, p 83

GIZ (2017a) Fish and Fisheries Prespa Lake. Implementing the EU Water Framework Directive in South-Eastern Europe, Tirana, p 87

GIZ (2017b) Fish and Fisheries Skadar/Shkodra Lake. Implementing the EU Water Framework Directive in South-Eastern Europe, Tirana, p. 88

Grapci-Kotori L, Vavalidis T, Zogaris D, Šanda R, Vukić J, Gedi D, Ibrahimi H, Bilalli A, Zogaris S (2020) Fish distribution patterns in the White Drin (Drini i Bardhë) river, Kosovo. Knowl Manage Aquat Ecosyst 421(29)

Hassert K (1892) Der Scutarisee. IV. Globus 62:87–91

INSTAT (2021) Albania in figures - Agriculture and Fishery (http://instat.gov.al/)

Kabo M (1991) Physical Geography of Albania, vols. I and II. Academy of Sciences, Geographic Center; Tirana (in Albanian)

Kalous L, Rylková K, Bohlen J, Šanda R, Petrtyl M (2013) New mtDNA data reveal a wide distribution of the Japanese ginbuna (Carassius langsdorfii; Cyprinidae) in Europe. J Fish Biol 82:703–770

Kolding J, Zwieten PA (2014) Sustainable fishing of inland waters. J Limnol 73:128–144. https://doi.org/10.4081/jlimnol.2014.818

Kottelat M, Freyhof J (2007) Handbook of European freshwater fishes. Kottelat, Cornol, and Freyhof, Berlin, p 646. https://doi.org/10.1643/OT-08-098a.1

Kovačić M, Šanda R (2007) A new species of *Knipowitschia* (Perciformes: Gobiidae) from southern Montenegro. J Nat Museum (Prague), Nat Hist Series 176(5):81–89

Lazaj L, Xhelilaj R (2017) Mapping of hydropower plant in Albania, using Geographic Information System. Tirana, Report 39 p

Levkov Z, Blanko S, Krstic S, Nakov T, Ector L (2007) Ecology of benthic diatoms from Lake Macro Prespa (Macedonia). Algol Stud 124:71–83. https://doi.org/10.1127/1864-1318/2007/0124-0071

Lewin WC, Arlinghaus R, Mehner T (2006) Documented and potential biological impacts of recreational fishing: insights for management and conservation. Rev Fish Sci 14:305–367. https://doi.org/10.1080/10641260600886455

Löffler H, Schiller E, Kusel E, Kraill H (1998) Prespa Lake, a European natural monument, endangered by irrigation and eutrophication? Hydrobiologia 384:69–74. https://doi.org/10.1023/A:1003336027258

Marić D, Milošević D (2011) Catalog of freshwater fishes (Osteichthyes) of Montenegro. Montenegrin Academy of Sciences and Arts 5 (4), Podgorica, 114

Marić D, Radujković B (2009) *Rutilus ohridanus* (Karaman, 1924) (Teleostei, Cyprinidae) from Ohrid, Prespa and Skadar lakes. Natura Montenegrina 8(3):137–150

Marková S, Šanda R, Crivelli A, Shumka S, Wilson IF, Vukić J, Berrebi P, Kotlík P (2010) Nuclear and mitochondrial DNA sequence data reveal the evolutionary history of *Barbus* (Cyprinidae) in the ancient lake systems of the Balkans. Mol Phylogenet Evol 55:488–500. https://doi.org/10.1016/j.ympev.2010.01.030

Matzinger A, Jordanoski M, Veljanoska-Sarafiloska E, Sturm M, Müller B, Wüest A (2006) Is Prespa Lake jeopardizing the ecosystem of ancient Lake Ohrid? Hydrobiologia 553:89–109. https://doi.org/10.1007/s10750-005-6427-9

Mendel J, Lusk S, Vasil'eva ED, Vasil'ev VP, Lusková V, Erk'akan F, Ruchin A, Koščo J, Vetešník L, Halačka K, Šanda R, Pashkov AN, Reshetnikov SI (2008) Molecular phylogeny of the genus Gobio Cuvier, 1816 (Teleostei: Cyprinidae) and its contribution to taxonomy. Mol Phylogenet Evol 47:1061–1075. https://doi.org/10.1016/j.ympev.2008.03.005

Miller PJ, Šanda R (2008) A new west Balkanian sand-goby (Teleostei: Gobiidae). J Fish Biol 72(1):259–270

Oiknomou A, Lperieur F, Leonardos ID (2014) Biogeography of freshwater fishes of the Balkan Peninsula, Hydrobiologia, 738(1). https://doi.org/10.10507/s10750-014-1930-5

Palandačić A, Bravničar J, Zupančič P, Šanda R, Snoj A (2015) Molecular data suggest a multispecies complex of Phoxinus (Cyprinidae) in the Western Balkan Peninsula. Mol Phylogenet Evol 92:118–123

Perea S, Bohme M, Zupančič P, Freyhof J, Šanda R, Ozulug M, Abdoli A, Doadrio I (2010) Phylogenetic relationships and biogeographical patterns in Circum-Mediterranean Subfamily Leuciscinae (Teleostei, Cyprinidae) inferred from both mitochondrial and nuclear data. BMC Evol Biol 10(265):1–27. https://doi.org/10.1186/1471-2148-10-265

Pešić V, Glöer P (2013) A new freshwater snail genus (Hydrobiidae, Gastropoda) from Montenegro, with a discussion on gastropod diversity and endemism in Skadar Lake. ZooKeys 281:69–90. https://doi.org/10.3897/zookeys.281.4409

Poljakov GD, Filipi N, Basho K, Hysenaj A (1958) Pesquit e Shqiperise (Fishes of Albania). Mihal Duri, Tirana

Rakaj N (1995) Iktiofauna e Shqiperise (Ichthyofauna of Albania). Shtëpia Botuese "Libri Universitar", Tirana

Russi D, Ten Brink P, Farmer A, Badura T, Coates D, Forster J, Kumar R, Davidson N (2012) The economics of ecosystems and biodiversity for water and wetlands (Final Consultation Draft), a contribution to CBD COP2011. Ramsar

Šanda R, Vukić J, Choleva L, Křížek J, Šedivá A, Shumka S, Wilson IF (2008) Distribution of loach fishes (Cobitidae, Nemacheilidae) in Albania, with genetic analysis of populations of *Cobitis ohridana*. Folia Zoologica 57(1–2):42–50

Schönhuth S, Vukić J, Šanda R, Yang L, Mayden RL (2018) Phylogenetic relationships and classification of the Holarctic family Leuciscidae (Cypriniformes: Cyprinoidei). Mol Phylogenet Evol 127:781–799. https://doi.org/10.1016/j.ympev.2018.06.026

Shumka S (2019) Natura 2000 potential sites in Albania-a contribution of fish species of community interest. Proceedings Book of the 3rd International Symposium for Environmental Science and Engineering Research, 25–27 May 2019, Konya, Turkey

Shumka S (2021) Survey of fishes and fishery of the Lagoon of Karavasta. JICA, Albania, p 66

Shumka S (2022) Mark-Recapture mullets survey in Lagoon of Karavasta. JICA, Albania, p 31

Shumka S, Apostolou A (2018) Current Knowledge on the Status of the Most Common Non-indigenous Fish Species in the Transboundary Greater Prespa Lake (Albanian Side). Acta Zoologica Bulgarica 70(2):203–209

Shumka S, Meulenbroek P, Schiemer F, Šanda R (2018) Fishes of the River Vjosa – an annotated Checklist. Acta ZooBot Austria 155:163–176

Shumka S, Kalogianni E, Šanda R, Vukić J, Shumka L, Zimmerman B (2020a) Ecological particularities of the critically endangered killifish Valencia letourneuxi and its spring-fed habitats: a long-lost endemic species of south Albania. Knowl Manage Aquat Ecosyst 421(45):1–10. https://doi.org/10.1051/kmae/2020036

Shumka S, Shumka L, Trajce K, Ceci S (2020b) First record of the Western Greece goby - *Economidichthys pygmaeus* (Holly, 1929), in Greater Prespa Lake (Albania). Ecologica Montenegrina 35:78–81. https://doi.org/10.37828/em.2020.35.6

Snoj A, Maric, S, Berrebi P, Crivelli AJ, Shumka S, Susnik S (2009) Genetic architecture of trout from Albania as revealed by mtDNA control region variation. Genetics Selection Evolution, 41: 22. ISSN (Print Edition): 0999-193X ISSN (Electronic Edition): 1297–9686

Spirkovski Z (2004) Changes in the spawning ecology of the Lake Ohrid trout, *Salmo letnica* (Karaman). Proceedings of the 2nd Congress of Ecologists of Macedonia, 205–209

Stierandová S, Vukić J, Vasil'eva ED, Zogaris S, Shumka S, Halačka K, Vetesník L, Švátora M, Nowak M, Stefanov T, Koščo J (2016) A multilocus assessment of nuclear and mitochondrial sequence data elucidates phylogenetic relationships among European spirlins (Alburnoides, Cyprinidae). Mol Phylogenet Evol 94:479–491

Talevski T, Milosevic D, Maric D, Petrovic D, Talevska M, Talevska A (2009) Biodiversity of ichthyofauna from Lake Prespa, Lake Ohrid and Lake Skadar. Biotechnol Biotechnol Equip 23: 400–404. https://doi.org/10.1080/13102818.2009.10818449

Tutman P, Buj I, Ćaleta M, Hamzić A, Korjenić E, Adrović A, Glamuzina B (2017) Status and distribution of spined loaches (Cobitidae) and stone loaches (Nemacheilidae) in Bosnia and Herzegovina. Folia Zoologica 66:211–226. https://doi.org/10.25225/fozo.v66.i4.a2.2017

Veiga P, Vieira L, Bexiga C, Erzini K (2010) Structure and temporal variations of fish assemblages of the Castro Marim salt marsh, southern Portugal. Estuar Coast Shelf Sci 70:27–38. https://doi.org/10.1016/j.ecss.2016.07.009

Viñuela Rodríguez N, Šanda R, Zogaris S, Vukić J (2020) Distribution and genetic diversity of two species of *Pelasgus minnows* (Leuciscidae) in southern Greece. Knowl Manage Aquat Ecosyst 421:27. https://doi.org/10.1051/kmae/2020019

Vukić J, Eliášová K, Marić D, Šanda R (2019) Occurrence of alien spirlin (Alburnoides sp.) in the Neretva river basin. Knowl Manage Aquat Ecosyst 420:15. https://doi.org/10.1051/kmae/2019007

Weiss S, Apostolou A, Đug S, Marčić Z, Mušović M, Oikonomou A, Shumka S, Škrijelj R, Simonović P, Vesnić A, Zabric D (2018) Endangered Fish Species in Balkan Rivers: their distributions and threats from hydropower development. Riverwatch and EuroNatur, 162 p

Welcomme RL (2001) Inland fisheries. Ecology and management. Fishing News Books, Blackwell Science, Oxford

Zogaris S, Economou AN (2017) The biogeographic characteristics of the river basins of Greece. The handbook of environmental chemistry. https://doi.org/10.10007/698.2017.475

Zogaris S, Economou AN, Dimopoulos P (2009) Ecoregions in the southern Balkans: should they be revised? Environ Manag 43:682–697. https://doi.org/10.1007/s00267-008-9243-y

Zupančič P, Marić D, Naseka AM, Bogutskaya NG (2010) *Squalius platyceps*, a new species of fish (Actinopterygii: Cyprinidae) from the Skadar Lake basin. Zoosystematica Rossica 19(1):154–167

Part II
Some Fish Species of the Western Balkans Important for Fisheries and Conservation

Summary

This part contains chapters presenting the results of the status of populations of fish species that are most exploited in the past and today, whether they are attractive for fishing or economically important. In addition to fish, a chapter is devoted to decapod crustaceans. In the Western Balkans, crayfish are exploited by the local population to a very limited extent. However, they are hunted and exported to other European countries, especially Scandinavian countries. A special chapter is devoted to migratory sturgeon species, the most threatened fish species in the Western Balkans today, with a detailed discussion on the possibilities of conservation and restoration of populations, including sterlet sturgeon as the only stationary species. Trout, both endemic and non-endemic, are particularly threatened by the increasingly intensive economic expansion of the Western Balkan countries and the use of the ecological services of inland waters for other needs, such as the production of electricity and healthy drinking water.

Freshwater Crayfish of the Western Balkans: Is it Possible to Use them Sustainably or Do they Need Prompt Conservation Actions?

Simona Đuretanović, Milka Rajković, and Ivana Maguire

Abstract The extinction of species and biodiversity decline are the most severe global consequences of environmental threats in the Anthropocene era. Freshwater ecosystems are the most endangered and invertebrates residing in them are the most vulnerable, particularly freshwater crayfish. The freshwater crayfish are significant components of biodiversity, are indicators of good water quality, are also known as keystone species in food webs, and have an important ecological role in the proper functioning of a freshwater ecosystem. The Western Balkans freshwater ecosystems are home to four indigenous species, *Austropotamobius torrentium*, *A. pallipes*, *Astacus astacus*, and *Pontastacus leptodactylus*. Three of them are listed in the IUCN Red List of Threatened Species as Endangered (*A. pallipes*), Vulnerable (*A. astacus*), or Data Deficient (*A. torrentium*), while *P. leptodactylus* is categorized as "Least Concern," and it is, along with *A. astacus*, a species grown in astaciculture. This chapter aims to present an overview of threats and problems that crayfish of the Western Balkans face, with special reference to the sustainability of already threatened populations.

Keywords Aquaculture · Astaciculture · Crayfish · Conservation

Supplementary Information The online version contains supplementary material available at https://doi.org/10.1007/978-3-031-36926-1_8.

S. Đuretanović (✉)
Department of Biology and Ecology, Faculty of Science, University of Kragujevac, Kragujevac, Serbia
e-mail: simona.djuretanovic@pmf.kg.ac.rs

M. Rajković
Environment Protection Agency of Montenegro, Podgorica, Montenegro

I. Maguire
Department of Biology, Faculty of Science, University of Zagreb, Zagreb, Croatia

V. Simić et al. (eds.), *Ecological Sustainability of Fish Resources of Inland Waters of the Western Balkans*, Fish & Fisheries Series 43,
https://doi.org/10.1007/978-3-031-36926-1_8

1 Introduction

Human activities have profoundly transformed the face of the Earth, and the processes of urbanization and industrialization along with demographic pressures are considered critical to environmental issues such as biodiversity loss, climate change, and the ecological health of the natural environments (Perdikaris et al. 2012; Lewis and Maslin 2015). The negative effects are projected to increase exponentially in the coming decades and centuries (Hofman and Rick 2018), with biotechnology having immense potential to cause additional environmental changes (Goudie and Viles 2013). The main concerns are pollution, habitat degradation through physical alteration and fragmentation, and the extinction of wild species; however, some of the greatest threats come from biological invasions (Hofman and Rick 2018). Invasive species are threatening biodiversity worldwide, particularly in aquatic ecosystems, and are among the most relevant causes of extinction (Souty-Grosset et al. 2016).

Freshwater ecosystems are among the most diverse ecosystems, highly complex and productive (Balian et al. 2008). At the same time, those ecosystems are heavily altered, degraded, polluted, and overexploited in terms of water abstraction, but also in terms of wild and commercial species overexploitation, with strong negative impacts of non-native invasive species (Geist 2011). Jussila et al. (2021) emphasize that freshwaters exhibit a high degree of endemism and extinction rates due to their insular nature and frequently small size. Further, they observed that the limited dispersal ability of many freshwater species commonly results in their adaptation to narrow habitat conditions, and if these are altered, extinction is a frequent consequence. Therefore, residents of freshwater ecosystems are under greater threat of range reduction and extinction compared to terrestrial taxa. Native freshwater invertebrates are particularly imperiled, especially freshwater mussels and crayfish. Concurrently, among the most successful freshwater ecosystem invaders are non-native crayfish (Hudina et al. 2016; Žganec et al. 2018), which can displace native crayfish through direct competition and as vectors of diseases such as the crayfish plague (Roessink et al. 2022).

2 The Importance of Crayfish

Freshwater crayfish are a highly diverse and very important group of aquatic invertebrates, and among the most important aquatic taxa in terms of ecosystem integrity retention (Alvanou et al. 2022). Their value to humans is unquestionable in many aspects, from culture to economy. They are physiologically, ecologically, and behaviorally remarkable animals residing in a wide range of habitats (Kubec et al. 2019).

Indigenous freshwater crayfish are known as keystone species (Nyström et al. 1999; Dorn and Wojdak 2004; Kouba et al. 2014), as well as ecosystem engineers, considering their prominent impact on the physical structure of their environment (Füreder and Reynolds 2003; Pârvulescu et al. 2011; Kouba et al. 2016). These largest mobile freshwater invertebrates are also indicators of good water quality (Reynolds et al. 2013; Lette 2020). Moreover, they are recognized as umbrella species for the conservation of freshwater communities (Kuklina et al. 2013). Crayfish can directly affect ecosystem processes, species abundance, and diversity (Pyšek and Richardson 2010). Their removal/disappearance from or their addition to a community may lead to cascading effects on the food webs as well as ecosystem services, which could finally lead to negative consequences that affect humankind (Gherardi 2011).

2.1 Distribution of Crayfish and the Main Threats

Freshwater crayfish have a global distribution, with the North American cambarids forming the most diverse group in the world with 428 described species, while European freshwaters are inhabited by only one family (Astacidae) that includes five indigenous species (ICS) (Kawai and Crandall 2016). Apart from ICS, currently in Europe, there are fifteen non-indigenous crayfish species (NICS), meaning that ICS are outnumbered by NICS (Kouba et al. 2014; Kotovska et al. 2016).

The Western Balkans freshwater ecosystems are inhabited by four ICS: the stone crayfish *Austropotamobius torrentium* (Schrank, 1803), the white-clawed crayfish *A. pallipes* (Lereboullet, 1858), the noble crayfish *Astacus astacus* (Linnaeus, 1758), and the narrow-clawed crayfish *Pontastacus leptodactylus* (Eschscholtz, 1823). Also, even five NICS originating from North America and Australia occur in the area (Maguire and Klobučar 2003; Pavlović et al. 2006; Maguire et al. 2008, 2011a, b, 2018; Simić et al. 2008; Hudina et al. 2009, 2011a, 2013, 2017; Jaklič and Vrezec 2011; Lipták et al. 2013; Samardžić et al. 2014; Jussila et al. 2015; Rebrina et al. 2015; Govedič 2017; Trožić-Borovac et al. 2019; Dragičević et al. 2020, 2021a; Zorić et al. 2020a, b; Žganec et al. 2020; Dobrović et al. 2021; Panjković et al. 2021): the spiny-cheek crayfish *Faxonius limosus* (Rafinesque, 1817), the red swamp crayfish *Procambarus clarkii* (Girard, 1852), the marbled *P. virginalis* (Lyko, 2017), the signal crayfish *Pacifastacus leniusculus* (Dana, 1852), and the Australian red claw crayfish *Cherax quadricarinatus* (Von Martens, 1868).

Freshwater crayfish have been recognized as taxa in need of conservation attention worldwide, with 32% of species listed as globally threatened by the IUCN (Richman et al. 2015; Bland 2017). According to IUCN Red List for European ICS, declines between 50% and 80% have been observed for the white-clawed crayfish (Füreder et al. 2010a), and between 50% and 70% for the noble crayfish (Edsman et al. 2010). Therefore, they are listed as endangered and vulnerable, respectively. On the other hand, the narrow-clawed crayfish has been assessed as Least Concern, owing to the majority of its population appearing to be relatively stable even though it has suffered significant declines both in the past and currently (Gherardi and

Souty-Grosset 2017). The stone crayfish has been assessed as Data Deficient (Füreder et al. 2010b), even though the application of IUCN criteria in some countries revealed its status as a vulnerable or endangered species (Simić et al. 2008; Gottstein et al. 2011; Đuretanović et al. 2017; Anonymous 2002, 2004, 2010a, b). Consequently, ICS are protected by laws in many European countries, and according to the official data, local authorities report them as highly endangered or threatened at local levels. Also, they are protected by international legislation (Appendix III of the Bern Convention, Annex II and V of the EU Habitats Directive (92/43/EEC)).

The introduction of NICS to Europe was initiated by the need to replace ICS, which were decimated by the crayfish plague, in fishery and aquaculture (Jussila et al. 2021). Initially a "good idea," it soon turned into a disaster because NICS soon turned out to be invasive species that outcompete ICS (Jussila et al. 2021). Nowadays, ICS are seriously devastated, among others due to lethal disease (i.e., crayfish plague) introduced by asymptomatic NICS that are relatively resistant to it, initially imported from North America during the late nineteenth century (Perdikaris et al. 2012; Jussila et al. 2021). The causative agent of the crayfish plague, pathogenic oomycete *Aphanomyces astaci* Schikora, 1906, was introduced into European freshwaters on the unknown American NICS and triggered a grave epidemic that decimates ICS populations since 1860. *Aphanomyces astaci* is listed among the "100 of the world's worst invasive alien species" (Holdich et al. 2009; Lowe et al. 2000), indicating that crayfish plague is one of the most devastating wildlife diseases known. The infection first appeared in Northern Italy and quickly spread over Continental Europe, then to Finland, and Sweden, and in the 1970s and 1980s, it was detected in the Iberian Peninsula, Norway, Ireland, and the UK (Jernelöv 2017; Jussila et al. 2021). Crayfish plague did not omit the area of the Western Balkans (Vrezec et al. 2013; Jussila et al. 2015, 2017; Maguire et al. 2016; Mrugała et al. 2017; Panteleit et al. 2018; Pavić et al. 2020, 2021; Dragičević et al. 2021b; Dobrović et al. 2022), but there is a lack of information for Albania, the Republic of North Macedonia, and Montenegro.

Apart from NICS and their pathogens, European ICS populations are also greatly negatively affected by habitat loss and degradation, overharvesting, translocation, and mixing of the populations. The best example of this is the state of the noble crayfish populations (Schrimpf et al. 2014; Gross et al. 2021). Overall, human activities, in the form of translocations and re/introductions of both ICS and NICS, have led to the degradation of this species' natural genetic structure and their mixing and loss across Europe (Souty-Grosset and Reynolds 2009; Schrimpf et al. 2011, 2014; Gross et al. 2013, 2017, 2021; Makkonen et al. 2015).

Overexploitation or overharvesting has not been implicated as a single cause of the decline of ICS; however, these factors have been listed as contributing factors in several cases (Taylor 2001). Overexploitation or overharvesting is not pronounced in the Western Balkans because people in the area rarely or not at all have a tradition of consuming crayfish in their diet. Also, established astaciculture does not exist.

The latest research by Hossain et al. (2018) predicted that 87% of freshwater crayfish species are highly sensitive to climate change, primarily due to habitat specialization, 35% have low adaptive capacity, and 57% are highly exposed. When it comes to the Western Balkans, substantial reductions of suitable habitats for the stone and the noble crayfish are predicted under different climate change scenarios (Lovrenčić et al. 2022a, b). Predictions are that 80% of the currently suitable habitat for the stone crayfish will be lost, and only 13% of the current potential distribution will remain stable for the noble crayfish.

2.1.1 Literature Review

In order to have insights into the publication on freshwater crayfish in the area of the Western Balkans, we searched Google scholar and the Scopus database. The analysis included relevant publications, not restricted to any journals, but restricted to the period from 2000 to 2022. We searched literature by crossing the following keywords: Crayfish AND Slovenia, Crayfish AND Croatia, Crayfish AND Bosnia and Herzegovina, Crayfish AND Serbia, Crayfish AND Montenegro, Crayfish AND Albania, and Crayfish AND Republic of North Macedonia.

Croatia leads in the number of publications (84), compared to Serbia (24), Slovenia (19), Bosnia and Herzegovina (10), Montenegro (7), Albania (4), and the Republic of North Macedonia (3) (Table 3 in the appendix).

The authors published the largest number of papers in the field of invasive species research (30 publications), distribution (21 publications), genetics (18 publications), and morphometry (14 publications), conservation (12 publications), crayfish plague and biology (11 publications each), physiology (9 publications), ectosymbiosis (6 publications), diversity (5 publications), and ecotoxicology (4 publications).

It is worth mentioning that during the literature review no paper was found regarding astaciculture.

2.2 *Crayfish through Human History*

The first reference to the crayfish in scientific literature is in Aristotle's *Historia Animalium* (Greek: Τῶν περὶ τὰ ζῷα ἱστοριῶν) (Chidester 1912). Since that time, they have been the subject of zoological research and have often been used as model organisms. One of the pages in *Theatrum Sanitatis*[1] features a miniature showing one of the first crayfish parties (Fig. 1) ever recorded in human history (Gherardi 2011). Freshwater crayfish have left a cultural mark almost everywhere. They appear in dictums, fables, and stories, names of villages and rivers, and mythology, and are

[1] ***Tuqwīm as-Siḥḥa*** (تقويم الصحة Maintenance of Health) is originally an eleventh century Arab medical treatise by Arabian doctor Ibn Butlan of Baghdad (i.e., Ububchasym de Baldach). In the West, the work is known by the Latinized name ***Theatrum Sanitatis***. The Latin manuscript is based on the medical insights of the Arabian doctor and adorned with 208 high-quality, artistically appealing illustrations.

Fig. 1 A miniature from medical treatise *Theatrum Sanitatis* represents a crayfish party (adopt from Gherardi 2011)

part of folklore and superstitions, especially among the Scandinavian peoples (Swahn 2004).

Consumption is the most obvious direct use of crayfish, and they have been known to mankind as a source of animal protein since prehistoric times (Alvanou et al. 2022). According to Patoka et al. (2016), the earliest archaeological evidence of crayfish used as a food source was reported in Central Europe (Bohemia and Moravia region), dated to the Neolithic, 4774–4500 BCE. The species was identified as *Astacus astacus* (Patoka et al. 2014), i.e., the noble crayfish, the most prized ICS in Europe (Đuretanović 2022). Several crayfish species have gained regional, national, and even international acclaim (Meyer-Arendt and Henderson 2020). Freshwater crayfish have been known and recognized in local communities on the European continent as a delicious treat but also as an item of cultural heritage (Perdikaris et al. 2012; Kouba 2011) (Rotten Carrot Production (2022). They have had significant value in both fishermen's folklore and as a part of traditional cuisine/gastronomy (Füreder 2015), and are commonly exploited as luxury food that commands a high price in Western markets (Holdich 1993).

Their culinary progress began in the monasteries during fasting periods, and from there, the eating of crayfish spread to wider circles (Swahn 2004). From the plates of royals and aristocrats, in the seventeenth century, freshwater crayfish arrived on the plates of common people who began to consume and appreciate them as a delicacy (Gherardi 2011; Patoka et al. 2014).

The tradition of crayfish parties "kräftskiva" in Fennoscandia started in the eighteenth century as an upper-class custom but has spread more widely, and even today, these parties are popular and especially dear to the Scandinavians, where they gather and celebrate good summer harvests (Svanberg and Locker 2020).

Folks describe the crayfish as a mini lobster since their deliciously succulent meat tastes similar. Crayfish meat has high nutritional parameters due to favorable fatty acid and amino acid profiles, as well as balanced mineral content, and it is a better source of Ca, K, Mg, Na, P, and Cu than meat from livestock animals, so it can be used as an alternative in the human diet (Śmietana et al. 2020). Crayfish muscles are characterized by high levels of proteins (reaching 13.24% of the dry matter) and low levels of lipids (less than 0.8% of the dry matter) (Gherardi 2011).

According to Swahn (2004), Scandinavians consider the noble crayfish aromatically far superior to its European relatives, so in time, crayfishing became an important export business. That business has become so big that it has led to the overexploitation of wild populations, even their extermination.

3 The Astaciculture: Freshwater Crayfish Farming

Aquaculture has been practiced for thousands of years. It is the fastest-growing food production sector (Brummett and Beveridge 2015) that plays an important role in addressing nutrition and food security, especially for the poor in developing countries (Rath 2018). Also, aquaculture represents a powerful tool for meeting the demands of both human consumption and conservation (Kouba 2011). Global production of aquatic animals was estimated at 178 million tonnes in 2020, a slight decrease from the record of 179 million tonnes in 2018 (FAO 2022).

Among the various aquatic organisms that are cultivated or farmed worldwide, the majority belongs to fish, followed by bivalves, while crustaceans including decapods fall in the third place (Alvanou et al. 2022). Crayfish farming is considered to be relatively lucrative, given the high market price and the demand both locally and for export (Wang et al. 2014). Commercial crayfish aquaculture is typically based on an extensive or semi-extensive production approach. The North American red swamp crayfish is one of the world's most widely introduced freshwater species for aquaculture purposes, especially due to its high economic importance (Loureiro et al. 2015). It is cultured in the USA and China and forms the bulk of farm-raised and wild-captured crayfish globally. In Europe, this species arrived via Spain, where it was legally introduced for aquaculture purposes in 1973 (Souty-Grosset et al. 2016). Moreover, another North American species (*P. leniusculus*) and two European species (*A. astacus* and *P. leptodactylus*) constitute the main cultured species from the family Astacidae. They are grown in small facilities throughout Europe and parts of Asia, while four native species of Oceania (Parastacidae) are cultured in their native ranges and were introduced for aquaculture in several locations around the globe (McClain 2020).

Europeans always prefer and value more ICS, such as the noble crayfish and narrow-clawed crayfish, than NICS, such as red swamp crayfish (Holdich et al. 2006). This is the reason why Europeans pay more to have the noble crayfish on their plate. D'Agaro et al. (1999) indicated that Italians pay even double for the noble crayfish than for other indigenous or non-indigenous species.

Until the 90s of the last century, the farming of freshwater crayfish was mainly carried out in the southern states of the USA, in Australia, and in Europe, while at least equivalent amounts have been harvested from the wild in North America, China, Australia, Kenya, Turkey, and Europe (Holdich 1993). During the last few decades, international trade in crayfish has greatly increased, largely as a result of rising farm production in PR China (Wickins and Lee 2008; Wang et al. 2014). There has been a significant increase in global crayfish production from just over 720,000 tonnes in 2015, to more than 1.7 million tonnes in 2018 (Table 1). The observed rise is primarily a consequence of red swamp crayfish farming in China (Cai et al. 2020).

According to data from the Food and Agriculture Organization of the United Nations (FAO 2020), PR China predominates (95.74%) in the global freshwater aquaculture production, and became a major exporter of crayfish aquaculture products in the world with 1,638,662 tonnes per year (Table 1). Chinese and American crayfish production is based mainly on the red swamp crayfish.

Genus *Cherax* contains several species that have been introduced around the world for aquaculture purposes, while they are intensively cultured within Australia (Taylor 2001).

The group "freshwater crustaceans," including freshwater crayfish, was the 13th largest group in the world aquaculture in 2018, with 1,711,635 tonnes of crayfish produced in aquaculture (FAO 2020). However, in terms of farmgate value, crayfish came in the fourth place accounting for 5.49% of the world aquaculture farmgate value in 2018, following salmons, trouts, and smelts (8.65%), marine shrimps and prawns (14.6%), and carps, barbels, and other cyprinids (23.67%) (Fig. 2).

The official international statistics (i.e., FAO) underestimates crayfish production (Ackefors 2000), since significant quantities of crayfish are obtained from commercial harvesting from wild populations and recreational catches (Holdich 1993; Policar and Kozák 2005).

Data collected from different sources provide different quantities of harvested crayfish from the wild and from astaciculture (Tables 1 and 2). The discrepancy is possible as a consequence of the time when data were collected (Policar and Kozák (2015) versus FAO (2020)).

Table 1 Global crayfish aquaculture production in 2018

Crayfish farming country, 2018	Crayfish aquaculture production, 2018 (tonnes)	Share of world crayfish aquaculture (%)	Species item (ranked by production tonnage)
1. China	1,638,662	95.74	Red swamp crayfish (*Procambarus clarkii*)
2. USA	72,682	4.25	Red swamp crayfish (*Procambarus clarkii*)
3. Australia	166	0.01	1. Smooth marron crayfish (*Cherax cainii*; 66 tones, 40%); 2. Yabby crayfish (*Cherax destructor*; 51 tones, 31%); 3. Red claw crayfish (*Cherax quadricarinatus*; 49 tones, 29%)
4. Malaysia	83	<0.01	Red claw crayfish (*Cherax quadricarinatus*)
5. Iran (Islamic Republic of)	25	<0.01	Narrow-clawed crayfish (*Pontastacus leptodactylus*)
6. Singapore	6	<0.01	Yabby crayfish (*Cherax destructor*)
7. Bulgaria	5	<0.01	Narrow-clawed crayfish (*Pontastacus leptodactylus*)
8. South Africa	4	<0.01	Smooth marron crayfish (*Cherax cainii*)
9. Sweden	1	<0.01	Euro-American crayfishes nei[a] (Astacidae, Cambaridae)
10. Estonia	0.58	<0.01	Noble crayfish (*Astacus astacus*)
11. Barbados	0.50	<0.01	Red claw crayfish (*Cherax quadricarinatus*)
12. Poland	0.34	<0.01	Euro-American crayfishes nei[a] (Astacidae, Cambaridae)
World	1,711,635	100.00	

Data source: FAO Fishery and Aquaculture Statistics. Global aquaculture production 1950–2018 (FishstatJ). July 2022. Available at www.fao.org/fishery/statistics/software/fishstatj/en
[a]Not elsewhere included

3.1 *Crayfish Farming in Europe in Comparison to the Western Balkans*

Commercial harvesting of crayfish became a big industry in the nineteenth century since the demands for crayfish increased (Policar and Kozák 2005). Originally

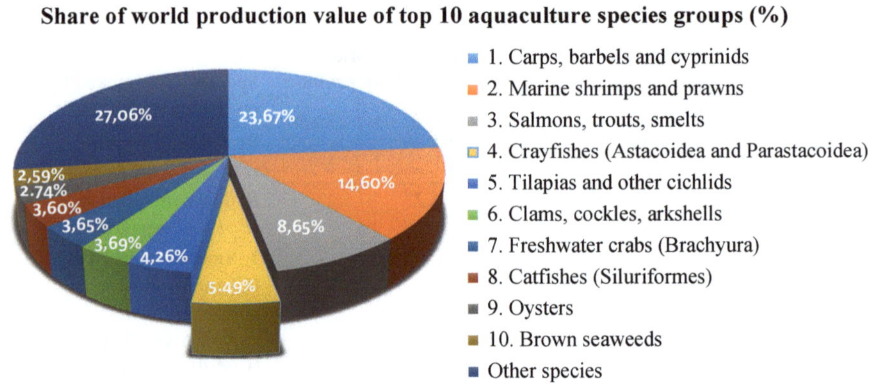

Fig. 2 Share of the top 10 aquaculture species groups by production value (Data source: https://www.fao.org/3/ca9383en/CA9383EN.pdf)

Europeans cultivated ICS, but after the wild populations of ICS were devastated by the crayfish plague, they began to introduce NICS as a substitution for devastated ICS populations and started to grow them in astaciculture.

Europe was self-sufficient in crayfish production before the onset of the crayfish plague that started spreading in the mid-1800s. The original, pre-crayfish plague, harvest was based mainly on the noble crayfish and the white-clawed crayfish, and countries such as Finland (Westman 1991) and Sweden (Gydemo 1989) were main exporters.

Nowadays, crayfish production in Europe is mainly based on the harvesting of the wild population (96%) (D'Agaro et al. 1999), while a small percentage is obtained from commercial aquaculture (astaciculture) and recreational capture for domestic consumption (Table 2). The harvesting of the wild noble crayfish is carried out mainly in the western, central, and northern Europe. Sweden and Finland are highly important producers of crayfish and harvest the noble crayfish and signal crayfish mainly through commercial catching/trapping (Policar and Kozák 2015) (Table 2).

In 1988, Finland was the largest producer of the noble crayfish, providing over 50% of the total European catch, i.e., 3.5 million crayfish or about 140 tonnes (Reynolds and Souty-Grosset 2011).

Turkey was a major exporter of the narrow-clawed crayfish to Europe until the outbreak of the crayfish plague in 1984 (Harlıoğlu 2008). After 1986 crayfish production declined dramatically (from 5000 tonnes to 200 tonnes) in most Turkish lakes; however, there has been an increase in the wild catch in recent years (Mazlum 2007).

Out of 12 countries worldwide where crayfish are bred in astaciculture, in Europe they are farmed in Bulgaria, Sweden, Estonia, and Poland (FAO 2020). Once grown in 13 European countries (Westman et al. 1990), nowadays the noble crayfish is farmed only in Estonia (0.58 tone) according to FAO data (2020). Bulgarians are breeding the narrow-clawed crayfish (5 tones), while Swedes and Poles are breeding

Table 2 Marketable crayfish species in Europe; amounts obtained from wild capture and astaciculture

Country	Marketable species	Wild capture (tones/annually)	Astaciculture (tones/annually)
Spain	*Procambarus clarkii*	1500	
	Pacifastacus leniusculus		0.25
Sweden	*Astacus astacus*	50–200	
	Pacifastacus leniusculus	265	
Finland	*Astacus astacus*	65–80	
	Pacifastacus leniusculus	200–260	
Norway	*Astacus astacus*	12–15	
Ukraine	*Astacus astacus*	2–11	
	Pontastacus leptodactylus	Unknown amount	
Bulgaria	*Pontastacus leptodactylus*	20	
Estonia	*Astacus astacus*	1–3	10–15
Great Britain	*Pontastacus leptodactylus*	Unknown amount	
	Pacifastacus leniusculus	8	
France	*Astacus astacus*		Unknown amount
	Pontastacus leptodactylus		
	Pacifastacus leniusculus		
Italy	*Astacus astacus*		30–40
	Pontastacus leptodactylus		
	Procambarus clarkii		
	Cherax destructor		
Germany	*Astacus astacus*	10	
Lithuania	*Astacus astacus*	15–20	
Montenegro	*Astacus astacus*	5	
Greece	*Astacus astacus*	5	
EU part of Russia	*Pontastacus leptodactylus*	Unknown amount	
Southern Portugal	*Procambarus clarkii*	Unknown amount	

Data source: Policar and Kozák (2015)

European and American crayfish (Astacidae, Cambaridae) (1 tone and 0.34 tonne, respectively) (FAO 2020).

There are no official data on astaciculture in the Western Balkans. The only mentions found in the literature most likely concern wild harvesting. Holdich (1993) mentioned Yugoslavia, along with Spain, Britain, and Greece, as the main European suppliers of crayfish to France in 1991. Figures on the crayfish trade in Italy in 1995 reported 959 tonnes of crayfish imported from Macedonia, Greece, New Guinea, and Australia (D'Agaro et al. 1999). The noble crayfish exploitation was also recorded in the Zeta River system (Montenegro) by companies from Sweden, so Rajković (2004) and Rajković et al. (2006) warned of the pressure on the populations in the country, and expressed the need for adequate protection. In spite of their recommendation, during the following years, Policar and Kozák (2015) noted that Montenegro produces up to five tonnes of marketable noble crayfish through wild populations harvesting (Table 2). Even today, in exclusive restaurants in Montenegro, it is possible to find noble crayfish on the menu that come from wild populations. For this reason, work is underway to amend the Law on Protection, which will define measures to protect indigenous crayfish species and completely ban their harvesting (personal communication M. Rajković).

As mentioned above, due to a lack of crayfish consuming habits and undeveloped astaciculture, neither the noble crayfish nor narrow-clawed crayfish has been of commercial interest in the Western Balkan. Sporadic studies were conducted on the possibilities of freshwater crayfish breeding (Klobučar and Maguire 2011), and there were also attempts at crayfish breeding in astaciculture (personal communication M. Ćirković), but they did not materialize.

Through personal communication with local people, they frequently mentioned that earlier (40–50 years ago), when they were children, they caught crayfish in the rivers and ate them. Additionally, the fishermen commonly use crayfish as a bait for fishing different fish species. However, completely unaware of the consequences of translocation, they transfer crayfish (mostly the noble crayfish) from one ecosystem to another.

Even though crayfish are a valuable source of food with high nutritional value, and could be a precious source for growing human population, one should also consider their role in the freshwater ecosystems, their evolutionary history, and diversity. The Balkan Peninsula is a hotspot of ICS diversity on European level, still guarding their unique and original pattern. Knowing the level of threats that ICS on the Balkans, and globally, face today, harvesting wild populations or growing them in astaciculture for consumption should be extremely carefully inspected/ considered. In order to conserve their diversity, astaciculture should be focused/ dedicated on *ex situ* conservation projects. Such projects, if carefully planned and executed, could help (along with habitat protection/restorations) preserving unique ICS diversity and enable them to survive for next generations.

4 Conclusion

Interest in freshwater crayfish as culinary items within the past century has led to the research and development of culture methods that have inspired aquaculture production and commerce (McClain 2020). The commercial culture of crayfish is carried out in a wide range of countries across the continents with the aim of providing stable and high-quality marketable crayfish production for consumption. However, this industry has never been developed in the area of the Western Balkans—what is on one hand a drawback, but on the other an advantage, because it was frequently shown that astaciculture of NICS caused a lot of havoc across Europe (Jussila et al. 2021). Therefore, the introduction of non-indigenous species for aquaculture purposes should be highly scrutinized and discouraged in most cases (McClain 2020).

On the other hand, if astaciculture is focused on ICS through carefully planned conservation projects, it can significantly support the conservation, protection, and future spreading of native crayfish species in open waters (Kozák et al. 2011).

Acknowledgments The investigation was supported by a grant (Agreement No. 451-03-68/2022-14/200122) funded by the Serbian Ministry of Education, Science and Technological Development.

Appendix

Table 3 List of published papers in the Western Balkans countries

ALBANIA
1. Đuretanović S, Jaklič M, Milošković A, Radojković N, Radenković M, Simić V, Maguire I (2017) Morphometric variations among *Astacus astacus* populations from different regions of the Balkan Peninsula. Zoomorphology 136: 19–27. doi: 10.1007/s00435-016-0331-x 2. Gross R, Lovrenčić L, Jelić M, Grandjean F, Đuretanović S, Simić V, Burimski O, Bonassin L, Groza MI, Maguire I (2021) Genetic diversity and structure of the noble crayfish populations in the Balkan Peninsula revealed by mitochondrial and microsatellite DNA markers. PeerJ, 9, p e11838 3. Mrugała A, Šanda R, Shumka S, Vukić J (2017) Filling the blank spot: first report on the freshwater crayfish distribution in Albania. Knowledge & Management of Aquatic Ecosystems, (418), p 34 4. Subchev MA (2011) First record of Branchiobdella Odier, 1823 (Annelida: Clitellata) in Albania and an overview of the geographic distribution of Branchiobdella hexodonta Gruber, 1882 in Europe. Acta Zool Bulg, 63(1), pp 109–112
BOSNIA AND HERZEGOVINA
1. Afanasyev S, Lietytska O, Golub O, Iarochevitch O, Mudra K, Marushevska O (2021) New Findings of White Clawed Crayfish, *Austropotamobius pallipes* (Decapoda, Astacidae), and Peculiarities of its Spatial Distribution in Neretvica (Bosnia and Herzegovina). Zoodiversity, 55(4). 2. Mrugała A, Šanda R, Petrusek A, Marić D, Vukić J (2017) Recent acute crayfish mortality reveals *Aphanomyces astaci* presence in Bosnia and Herzegovina. Journal of invertebrate pathology, 150, pp 73–75 3. Roljić R, Nikolić V, Savić N (2020a) Morphological variability and sexual dimorphism of stone crayfish *Austropotamobius torrentium* from the Marjanovića Stream (Bosnia and Herzegovina). Kragujevac Journal of Science, (42), pp 143–150 4. Roljić R, Nikolić V, Savić N (2020b) Morphological variability and sexual dimorphism of noble crayfish *Astacus astacus* from the Balkana Lake. Bulletin of Natural Sciences Research, 10(2), pp 7–11 5. Roljić R, Nikolić V, Savić N (2020c) Investigations of variability of morphometric characteristics in stone crayfish in order to preserve autochthonous genome. Quality of Life: A Multi-Disciplinary Journal of Food Science, Environmental Science & Public Health, 11 6. Roljić R, Savić N, Nikolić V (2022d) First Insight into Distribution and Ecology of *Austropotamobius torrentium* (Schrank, 1803) (Decapoda: Astacidae) in the North-Western Bosnia and Herzegovina. Acta Zoologica Bulgarica 74(2): 265–271 7. Trožić-Borovac S (2011) Freshwater crayfish in Bosnia and Herzegovina: the first report on their distribution. Knowledge and Management of Aquatic Ecosystems, (401), p 26 8. Trožić-Borovac S (2012a) Distribucija bjelonogog raka *Austropotamobius pallipes* (Lereboullet, 1858) species complex (Astacoidea Latreille, 1802; Astacidae Latreille, 1802) u Bosni i Hercegovine. In "Struktura i dinamika ekosistema Dinarida-stanje, mogućnosti i perspektive. Posebno izdanje CXLIX. Akademija nauka Bosne i Hercegovine. Sarajevo, 23, pp 153–166 9. Trožić-Borovac S, Macanović A, Škrijelj R (2012b) The morfometrics characterics and condition index of *Austropotamobius pallipes* in the Neretva river basin. Radovi Šumarskog fakulteta Univerziteta u Sarajevu, 42(2), pp 13–30 10. Trožić-Borovac S, Škrijelj R, Vesnić A, Đug S, Mušović A, Šljuka S, Borovac B, Gajević M (2019) Negative effects of introducing allochthonous species *Pacifastacus leniusculus* (Dana, 1852) into aquatic ecosystems of Bosnia and Herzegovina. Paper presented in the 3rd Symposium of Freshwater Biology, eds M. Ivkovic, I. Stankovic, R. Matonickin Kepcija, and R. Gracan (Zagreb: Croatian Association of Freshwater Ecologists), pp 33–33

(continued)

Table 3 (continued)

CROATIA
1. Bačnik K, Kutnjak D, Černi S, Bielen A, Hudina S (2021) Virome analysis of signal crayfish (*Pacifastacus leniusculus*) along its invasion range reveals diverse and divergent RNA viruses. Viruses, 13(11), p 2259
2. Bekavac A, Beck A, Dragičević P, Dragun Z, Maguire I, Ivanković D, Fiket Ž, Gračan R, Hudina S (2021) Disturbance in invasion? Idiopathic necrotizing hepatopancreatitis in the signal crayfish *Pacifastacus leniusculus* (Dana, 1852) in Croatia. Journal of Fish diseases 45, 261–276
3. Bielen A, Vladušić T, Kuharić N, Hudina S, Šver L, Likić S (2014) First evidence of the presence of Multixenobiotic Resistance Mechanism activity in freshwater invasive species, signal crayfish *Pacifastacus leniusculus* (Dana, 1852) Periodicum biologorum 116 (3), 267–273
4. Boštjančić Lj L, Bonassin L, Anušić L, Lovrenčić L, Besendorfer V, Maguire I, Grandjean F, Austin C M, Greve C, Hamadou A B, Mlinarec J (2021) The *Pontastacus leptodactylus* (Astacidae) Repeatome Provides Insight Into Genome Evolution and Reveals Remarkable Diversity of Satellite DNA. Frontiers in Genetics 11: 1820 doi: 10.3389/fgene.2020.611745
5. Dakić L, Maguire I (2016) Year cycle and morphometrical characteristics of *Austropotamobius torrentium* from two karstic rivers in Croatia. Natura Croatica: periodicum Musei historiae naturalis Croatici 25, 1: 87–100
6. Danilović M, Maguire I, Füreder L (2022) Overlooked keystone species in conservation plans of fluvial ecosystems in Southeast Europe: a review of native freshwater crayfish species. Knowl. Manag. Aquat. Ecosyst. 423: 21, https://doi.org/10.1051/kmae/2022016
7. Dobrović A, Maguire I, Boban M, Grbin D, Hudina S (2021) Reproduction dynamics of the marbled crayfish *Procambarus virginalis* Lyko, 2017 from an anthropogenic lake in continental Croatia. Aquatic Invasions 16, 482–498, doi.org/10.3391/ai.2021.16.3.06
8. Dobrović A, Geček S, Klanjšček T, Haberle I, Dragičević P, Pavić D, Petelinec A, Boštjančić LL, Bonassin L, Theissinger K, Hudina S (2022) Recurring infection by crayfish plague pathogen only marginally affects survival and growth of marbled crayfish. NeoBiota 77, 155–177
9. Dragičević P, Faller M, Kutleša P, Hudina S (2020) Update on the signal crayfish, *Pacifastacus leniusculus* (Dana, 1852) range expansion in Croatia: a 10-year report. BioInvasions Record, 9(4)
10. Dragičević P, Bielen A, Petrić I, Vuk M, Žučko J, Hudina S (2021a) Microbiome of the successful freshwater invader, the signal crayfish, and its changes along the invasion range. Microbiology spectrum, 9(2), pp e00389–21
11. Dragičević P, Grbin D, Maguire I, Blažević SA, Abramović L, Tarandek A, Hudina S (2021b) Immune response in crayfish is species-specific and exhibits changes along invasion range of a successful invader. MDPI Biology, 10(11), p 1102
12. Dražina T, Korša A, Špoljar M, Maguire I, Klobučar GIV (2018) Epifauna of native and alien freshwater crayfish species (Crustacea: Decapoda): a host-specific community? Freshwater Science, 37(3): 593–604. doi: 10.1086/698764
13. Faller M, Maguire I, Klobučar G (2006) Annual activity of the noble crayfish (*Astacus astacus*) in the Orljava River (Croatia). Bulletin Français de la Pêche et de la Pisciculture 383: 23–39
14. Gottstein S, Kerovec M, Maguire I, Bukvić I (1999) Ecological notes on *Austropotamobius pallipes italicus* (Faxon, 1914) (Decapoda, Astacidae) in the karstic spring of Neretva. Proceedings of the 12th Symposium of the International Association of Astacology: 620–628
15. Gottstein S, Žganec K, Maguire I, Kerovec M, Jalžić B (2007) Viši rakovi slatkih i bočatih voda porječja rijeke Krke. Rijeka Krka i Nacionalni park "Krka" – Prirodna i kulturna baština, zaštita i održivi razvitak, Marguš D (ured). Javna ustanova Nacionalni park "Krka", Šibenik, 421–430 (In Croatian)
16. Gottstein S, Hudina S, Lucić A, Maguire I, Ternjej I, Žganec K (2011) Crveni popis rakova (Crustacea) slatkih i bočatih voda Hrvatske. Zagreb: Državni zavod za zaštitu prirode. (Red list of freshwater and brackish Crustacea in Croatia; in Croatian)
17. Gross R, Lovrenčić L, Jelić M, Grandjean F, Đuretanović S, Simić V, Burlinski O,

(continued)

Table 3 (continued)

Bonassin L, Groza MI, Maguire I (2021) Genetic diversity and structure of the noble crayfish populations in the Balkan Peninsula revealed by mitochondrial and microsatellite DNA markers. PeerJ 9:e11838, doi.org/10.7717/peerj.11838
18. Hudina S, Maguire I, Galić N, Đurić P (2004) Dinamika populacije riječnoga raka u Nacionalnome parku "Paklenica". Paklenički zbornik vol. 2, Lukač G. (ured). Javna ustanova Nacionalni park "Paklenica", Starigrad-Paklenica, 89–94 (In Croatian)
19. Hudina S, Maguire I, Klobučar GIV (2008) Spatial dynamics of the noble crayfish (*Astacus astacus* L.) in the Paklenica National Park. Knowledge and Management of Aquatic Ecosystems 388: 01p12, doi: 10.1051/kmae:2008001
20. Hudina S, Faller M, Lucić A, Klobučar G, Maguire I (2009) Distribution and dispersal of two invasive crayfish species in the Drava River basin, Croatia. Knowledge and Management of Aquatic Ecosystems 394–395: 09p11, doi: 10.1051/kmae/2009023
21. Hudina S, Lucić A, Žganec K, Janković S (2011a) Characteristics and movement patterns of a recently established invasive *Pacifastacus leniusculus* population in the river Mura, Croatia. Knowledge and management of aquatic ecosystems, (403), p 07
22. Hudina S, Janković S, Lucić A, Žganec K (2011b) The status of *Astacus astacus* in the northernmost part of Croatia (Međimurje County) in the face of invasion by *Pacifastacus leniusculus* (Crustacea: Astacidae) Lauterbornia 72, 31–44
23. Hudina S, Hock K, Žganec K, Lucić A (2012) Changes in population characteristics and structure of the signal crayfish at the edge of its invasive range in a European river. Annales de Limnologie – International Journal of Limnology 48 (1), 3–11
24. Hudina S, Žganec K, Lucić A, Trgovčić K, Maguire I (2013) Recent invasion of the karstic river systems in Croatia through illegal introductions of the signal crayfish. Freshwater Crayfish 19(1): 21–27
25. Hudina S, Hock K, Radović A, Klobučar G, Petković J, Jelić M, Maguire I (2016) Species-specific differences in dynamics of agonistic interactions may contribute to the competitive advantage of the invasive signal crayfish (*Pacifastacus leniusculus*) over the native narrow-clawed crayfish (*Astacus leptodactylus*). Marine & Freshwater Behaviour & Physiology 49 (3): 147–157
26. Hudina S, Kutleša P, Trgovčić K, Duplić A (2017) Dynamics of range expansion of the signal crayfish (*Pacifastacus leniusculus*) in a recently invaded region in Croatia. Aquatic Invasions, 12(1)
27. Jelić M, Klobučar GIV, Grandjean F, Puillandre N, Franjević D, Futo M, Amouret J, Maguire I (2016) Insights into the molecular phylogeny and historical biogeography of the white-clawed crayfish (Decapoda, Astacidae). Molecular Phylogenetics and Evolution 103: 26–40
28. Jussila J, Maguire I, Kokko H, Tiitinen V, Makkonen J (2020) Narrow-clawed crayfish in Finland: *Aphanomyces astaci* resistance and genetic relationship to other selected European and Asian populations. Knowl. Manag. Aquat. Ecosyst., 421, 30 DOI 10.1051/kmae/2020022
29. Klobučar GIV, Maguire I, Gottstein S, Gelder S (2006) Occurrence of Branchiobdellida (Annelida: Clitellata) on the freshwater crayfish in Croatia. Annales de Limnologie – International Journal of Limnology 42, 4: 251–260
30. Klobučar GIV, Malev O, Šrut M, Štambuk A, Lorenzon S, Cvetkovi, Ferrero EA, Maguire I (2012) Genotoxicity monitoring of freshwater environments using caged crayfish (*Astacus leptodactylus*). Chemosphere 87: 62–67
31. Klobučar G, Podnar M, Jelić M, Franjević D, Faller M, Štambuk A, Gottstein S, Simić V, Maguire I (2013) Role of the Dinaric Karst (western Balkans) in shaping the phylogeographic structure of the threatened crayfish *Austropotamobius torrentium*. Freshwater Biology 58: 1089–1105
32. Lovrenčić L, Maguire I (2021) Gap analysis revealed a moderate efficiency of protected areas for the conservation of the endangered noble crayfish in Croatia. Natura Croatica 30(2): 529–536. doi: 10.20302/NC.2021.30

(continued)

Table 3 (continued)

33. Lovrenčić L, Temunović M, Maguire I (2020a) How well does NATURA 2000 protect threatened stone crayfish in Croatia?. Natura Croatica 29(2): 241–253. https://doi.org/10.20302/NC.2020.29.30

34. Lovrenčić L, Bonassin L, Boštjančić Lj L, Podnar M, Jelić M, Klobučar G, Jaklič M, Stamenković V, Hinić J, Maguire I (2020b) New insights into the genetic diversity of the stone crayfish: taxonomic and conservation implications. BMC Evolutionary Biology 20, 146

35. Lovrenčić L, Pavić V, Majnarić S, Abramović L, Jelić M, Maguire I (2020c) Morphological diversity of the stone crayfish – traditional and geometric morphometric approach. Knowl. Manag. Aquat. Ecosyst. 421: 1. https://doi.org/10.1051/kmae/2019042

36. Lovrenčić L, Temunović M, Bonassin L, Grandjean F, Austin CM, Maguire I (2022a) Climate change threatens unique genetic diversity within the Balkan biodiversity hotspot – The case of the endangered stone crayfish. Global Ecology and Conservation 39, e02301. https://doi.org/10.1016/j.gecco.2022.e02301

37. Lovrenčić L, Temunović M, Gross R, Grgurev M, Maguire I (2022b) Integrating population genetics and species distribution modelling to guide conservation of the noble crayfish, *Astacus astacus*, in Croatia. Scientific Reports 12, 2040. doi.org/10.1038/s41598-022-06027-8

38. Lovrenčić L, Ferrón HG, Grbin D, Maguire I (2022c) Insight into the noble crayfish morphological diversity: a geometric morphometric approach. Knowl. Manag. Aquat. Ecosyst. 423, 9. https://doi.org/10.1051/kmae/2022006

39. Lucić A, Erben R, Maguire I (2004) Occurrence of the pathogen *Psorospermium haeckeli* in Astacid populations in Croatia. Bulletin Français de la Pêche et de la Pisciculture 372–373: 375–383

40. Lucić A, Hudina S, Faller M, Cerjanec D (2012) A comparative study of the physiological condition of native and invasive crayfish in Croatian rivers. Biologia, 67(1), pp 172–179

41. Lucić A, Paunović M, Tomović J, Kovačević S, Zorić K, Simić V, Atanacković A, Marković V, Kračun-Kolarević M, Hudina S, Lajtner J (2015) Aquatic macroinvertebrates of the Sava River. In: Milačič R, Ščančar J, Paunović M (eds) The Sava River, Springer, Berlin, pp 335–359

42. Maguire I. (2009) Die Flussekrebse Osteuropas. U: Flusskrebse Biologie–Ökologie–Gefährdung, Füreder L (ured). Folio-Verlag Veröffentlichungen des Naturmuseums Südtirol, 6, Innsbruck. 92–97

43. Maguire I (2010) Slatkovodni rakovi – Priručnik za inventarizaciju i praćenje stanja. Državni zavod za zaštitu prirode, Zagreb. 4 (In Croatian)

44. Maguire I (2013) Nacionalni programi za praćenje stanja očuvanosti vrsta u Hrvatskoj – Plemeniti ili riječni rak (*Astacus astacus*). Državni zavod za zaštitu prirode. 28 p (In Croatian)

45. Maguire I (2014a) Nacionalni programi za praćenje stanja očuvanosti vrsta u Hrvatskoj – Potočni rak ili rak kamenjar (*Austropotamobius torrentium* (Schrank, 1803)). Državni zavod za zaštitu prirode, 30 p (In Croatian)

46. Maguire I (2014b) Nacionalni programi za praćenje stanja očuvanosti vrsta u Hrvatskoj – Bjelonogi ili primorski rak (*Austropotamobius pallipes* (Lereboullet, 1858)). Državni zavod za zaštitu prirode, 28 p (In Croatian)

47. Maguire I, Klobučar GIV (2003) Appearance of *Orconectes limosus* in Croatia. Crayfish news 25, 3: 7

48. Maguire I, Gottstein-Matočec S (2004) The distribution pattern of freshwater crayfish in Croatia. Crustaceana 77, 1: 25–47

49. Maguire I, Jelić M (2010) Distribucija rakova iz porodice Astacidae na području Banovine. U: Zrinska gora regionalni park prirode, Bučar M (ured). Grafički zavod Hrvatske, Zagreb. 212–216

50. Maguire I, Dakić L (2011a) Comparative analyses of *Astacus leptodactylus* morphological characteristics from Croatia and Armenia. Biologia, Section zoology 66, 3. 491–498. doi: 10.2478/s11756-011-0045-y

51. Maguire I, Klobučar G (2011b) Size structure, maturity size, growth and condition index of

(continued)

Table 3 (continued)

stone crayfish (*Austropotamobius torrentium*) in North-West Croatia. Knowledge and Management of Aquatic Ecosystems 401. doi: 10.1051/kmae/2011026
52. Maguire I, Erben R, Klobučar GIV, Lajtner J (2002) Year cycle of *Austropotamobius torrentium* (Schrank). Bulletin Français de la Pêche et de la Pisciculture 367: 943–957
53. Maguire I, Klobučar GIV, Gottstein S, Erben R (2003a) Distribution of *Austropotamobius pallipes* (Lereboullet) in Croatia and notes on its morphology. Bulletin Français de la Pêche et de la Pisciculture 370–371: 57–71
54. Maguire I, Klobučar G, Lucić A, Štefok B (2003b) Nalaz vrste *Astacus leptodactylus* na području Parka prirode Lonjsko polje. Bilten Parka prirode "Lonjsko polje" 5 (1–2): 10–13 (In Croatian)
55. Maguire I, Hudina S, Erben R (2004a) Estimation of noble crayfish (*Astacus astacus*) population size in the Velika Paklenica Stream (Croatia). Bulletin Français de la Pêche et de la Pisciculture 372-373: 353–366
56. Maguire I, Hudina S, Klobučar G, Štambuk A. (2004b) Godišnji ciklus riječnog raka (*Astacus astacus* L.) u potoku Velika paklenica. Paklenički zbornik vol. 2, Lukač G. (ured). Javna ustanova Nacionalni park "Paklenica", Starigrad-Paklenica, 81–88 (In Croatian)
57. Maguire I, Klobučar G, Erbene R. (2005) The relationship between female size and egg size in freshwater crayfish *Austropotamobius torrentium*. Bulletin Français de la Pêche et de la Pisciculture 376-377: 777–785
58. Maguire I, Hudina S, Faller M, Lucić A, Erben R (2006a) Morphometric characteristics of noble crayfish (*Astacus astacus*) in Croatia. Proceedings of the 15th Symposium of the International Association of Astacology: 324–331
59. Maguire I, Klobučar G, Faller M, Machino Y, Kučinić M, Žužul M (2006b) Updates on the distribution of the white-clawed and the narrow-clawed crayfish in Croatia. Crayfish news 28, 1: 4–5
60. Maguire I, Gottstein Matočec S, Klobučar G, Erben R, Kerovec M (2007) Rasprostanjenost slatkovodnih rakova reda Decapoda na području Istre, Kvarnera i Gorskog kotara – prošlost i sadašnjost. Prirodoslovna istraživanja riječkog područja II, Arko–Pijevac M, Surina B (ured). Prirodoslovni muzej Rijeka, Rijeka, 191–198 (In Croatian)
61. Maguire I, Klobučar G, Marčić Z, Zanella D (2008) The first record of *Pacifastacus leniusculus* in Croatia. Crayfish news 30, 4: 4
62. Maguire I, Lucić A, Schletterer M, Sint D, Erben R, Füreder L (2010a) Fecundity of indigenous stone crayfish populations: Implications for crayfish conservation in Austria and Croatia. Freshwater crayfish 17: 103–107
63. Maguire I, Cerjanec D, Lucić A (2010b) Life history of *Astacus leptodactylus* in two Croatian rivers. Freshwater crayfish 17: 255–260
64. Maguire I, Klobučar G, Posavec Z, Hudina S (2011a) Update on the *Pacifastacus leniusculus* spread in Croatia. Crayfish news 33: 5–5
65. Maguire I, Jelić M, Klobučar G (2011b) Update on the distribution of freshwater crayfish in Croatia. Knowledge and Management of Aquatic Ecosystems 401. doi: 10.1051/kmae/2011026
66. Maguire I, Špoljarić I, Klobučar G (2013a) The indigenous crayfish of Plitvice Lakes National Park, Croatia. Freshwater Crayfish 19, 1: 91–96
67. Maguire I, Špelić I, Jelić M, Klobučar G (2013b) Is it Possible to Detect Narrow-clawed and Noble Crayfish Probable Hybrids Using Multivariate Discriminant Analysis of Morphometric Data? Freshwater crayfish 19, 2: 219–227
68. Maguire I, Podnar M, Jelić M, Štambuk A, Schrimpf A, Schulz H, Klobučar G (2014) Two distinct evolutionary lineages of the *Astacus leptodactylus* species–complex (Decapoda : Astacidae) inferred by phylogenetic analyses. Invertebrate Systematics 28 (2): 117–123
69. Maguire I, Jelić M, Klobučar G, Delpy M, Delaunay C, Grandjean F (2016) Prevalence of the pathogen *Aphanomyces astaci* in freshwater crayfish populations in Croatia. Diseases of aquatic organisms 118: 45–53

(continued)

Table 3 (continued)

70. Maguire I, Marn N, Klobučar G (2017) Morphological evidence for hidden diversity in the threatened stone crayfish *Austropotamobius torrentium* (Schrank, 1803) (Decapoda: Astacoidea: Astacidae) in Croatia. Journal of Crustacean Biology 37(1): 7–15. doi:10.1093/jcbiol/ruw012

71. Maguire I, Klobučar G, Žganec K, Jelić M, Lucić A, Hudina S (2018) Recent changes in distribution pattern of freshwater crayfish in Croatia – threats and perspectives. Knowledge and Management of Aquatic Ecosystems, 419: 2. doi: 10.1051/kmae/2017053

72. Malev O, Šrut M, Lorenzon S, Maguire I, Štambuk A, Ferrero EA, Klobučar GIV (2010) Genotoxic, physiological and immunological effects caused by temperature increase, hypoxia or food deprivation in freshwater crayfish *Astacus leptodactylus*. Comparative Biochemistry and Physiology – Part C: Toxicology & Pharmacology 152: 433–443

73. Mijošek T, Jelić M, Mijošek V, Maguire I (2017) Molecular and morphometric characterisation of the invasive signal crayfish populations in Croatia. Limnologica 63: 107–118. doi:10.1016/j.limno.2017.02.003

74. Mlinarec J, Mužić M, Pavlica M, Šrut M, Klobučar G, Maguire I (2011) Comparative karyotype investigations in the European crayfish *Astacus astacus* and *A. leptodactylus*. Crustaceana 84 (12–13): 1497–1510

75. Mlinarec J, Porupski I, Maguire I, Klobučar G (2016) Comparative karyotype investigations in the white-clawed crayfish (*Austropotamobius pallipes* species complex) and stone crayfish (*A. torrentium*) (Decapoda, Astacidae). Journal of crustacean biology 36 (1): 87–93

76. Njegovan V, Orešković M, Hudina S, Maguire I (2017) Crayfish fauna of the Maksimir lakes in the urban area of Zagreb. Freshwater crayfish 23(1): 1–11 https://doi.org/10.5869/fc.2017.v23-1.XXX

77. Pavić D, Čanković M, Petrić IS, Makkonen J, Hudina S, Maguire I, Vladušić T, Šver L, Hrašćan R, Orlić K, Dragičević P, Bielen A (2020) Non-destructive method for detecting *Aphanomyces astaci*, the causative agent of crayfish plague, on the individual level. Journal of Invertebrate Pathology 169: 107274. https://doi.org/10.1016/j.jip.2019.107274

78. Pavić D, Bielen A, Hudina S, Špoljarić I, Grandjean F, Jussila J, Maguire I (2021) Distribution of *Aphanomyces astaci*, the causative agent of crayfish plague, in the Plitvice Lakes national park, Croatia. Bionivasion records 10 (3), 654–668. doi.org/10.3391/bir.2021.10.3.16

79. Rebrina F, Skejo J, Lucić A, Hudina S (2015) Trait variability of the signal crayfish (*Pacifastacus leniusculus*) in a recently invaded region reflects potential benefits and trade-offs during dispersal. Aquatic Invasions 10 (1)

80. Samardžić M, Lucić A, Maguire I, Hudina S (2014) The First Record of the Marbled Crayfish (*Procambarus fallax* (Hagen, 1870) f. virginalis) in Croatia. Crayfish News 36, 4: 4–4

81. Schrimpf A, Theissinger K, Dahlem J, Maguire I, Pârvulescu L, Schulz H, Schulz R (2014) Phylogeography of noble crayfish (Astacus astacus) reveals multiple refugia. Freshwater biology 59: 761–776

82. Šarić I, Klobučar G, Podnar M, Štambuk A, Maguire I (2018) Molecular phylogeny of branchiobdellidans (Annelida: Clitellata) and their host–epibiont association with *Austropotamobius* freshwater crayfish. Invertebrate Systematics, 32(1): 55–68

83. Topić Popović N, Sauerborn Klobučar R, Maguire I, Strunjak-Perović I, Kazazić S, Barišić J, Jadan M, Klobučar GIV, Čož–Rakovac R (2014) High-throughput discrimination of bacteria isolated from *Astacus astacus* and *A. leptodactylus*. Knowledge and Management of Aquatic Ecosystems 413: 04. doi: 10.1051/kmae/2014005

84. Žganec K, Lajtner J, Ćuk R, Crnčan P, Pušić I, Atanacković A, Kralj T, Valić D, Jelić M, Maguire I (2020) Alien macroinvertebrates in Croatian freshwaters. Aquatic Invasions 15, 4, 593–615

MONTENEGRO

1. Marić D, Rajković M (2004) Competitors and predators the cause of crayfish (*Astacus astacus* L.) numerosity decrease in the waters of Nikšić region. Natura Montenegrina, 3, pp 101–111

2. Marić D, Stanić M (2021) Bioaccumulation of 18 trace metals in muscle and exoskeleton in the

(continued)

Table 3 (continued)

noble crayfish (*Astacus astacus* L.) in the River Zeta (Montenegro). Poljoprivreda i Sumarstvo, 67(3), pp 31–42 3. Petrovic A, Rajkovic M, Simic S, Maguire I, Simic V (2013) Importance of genetic characteristics in the conservation and management of crayfish in Serbia and Montenegro. Bulgarian Journal of Agricultural Science, 19(5), pp 1093–1104 4. Rajković M, Simić V, Petrović A (2006) Length-weight gain of European crayfish *Astacus astacus* (L.) in the area of the upper course of the Zeta River, Montenegro. Archives of Biological Sciences, 58(4), pp 233–238 5. Rajkovic M, Petrovic A, Maguire I, Simic V, Simic S, Paunovic M (2012) Discovery of a new population of the species complex of the white-clawed crayfish, *Austropotamobius palipes/italicus* (Decapoda, Astacidae) in Montenegro, range extension, endangerment, and conservation. Crustaceana, 85(3), p 333 6. Simić V, Paunović M, Petrović A, Rajković M (2008) Crayfish of Serbia and Montenegro – the population status and the level of endangerment. Crustaceana, 81(10), pp 1153–1176 7. Simić V, Maguire I, Rajković M, Petrović A (2015) Conservation strategy for the endangered crayfish species of the family Astacidae: the ESHIPPO crayfish model. Hydrobiologia, 760(1), pp 1–13
REPUBLIC OF NORTH MACEDONIA
1. Albrecht C, Wilke T (2008) Ancient Lake Ohrid: biodiversity and evolution. Hydrobiologia, 615(1), pp 103–140 2. Lovrenčić L, Bonassin L, Boštjančić LL, Podnar M, Jelić M, Klobučar G, Jaklič M, Slavevska-Stamenković V, Hinić J, Maguire I (2020b) New insights into the genetic diversity of the stone crayfish: taxonomic and conservation implications. BMC evolutionary biology, 20(1), pp 1–20 3. Panjković B, Rat M, Mihajlović S, Galambos L, Kiš A, Puzović S, Nadaždin B, Šeat J, Vukajlović F, Tot I, Đapić M (2021) Invasive Alien Species in the Balkan Peninsula. Invasive Alien Species: Observations and Issues from Around the World, 3, pp 42–87
SERBIA
1. Borković SS, Pavlović SZ, Kovačević TB, Štajn AŠ, Petrović VM, Saičić ZS (2008) Antioxidant defence enzyme activities in hepatopancreas, gills and muscle of Spiny cheek crayfish (*Orconectes limosus*) from the River Danube. Comparative Biochemistry and Physiology Part C: Toxicology & Pharmacology, 147(1), pp 122–128 2. Đuretanović S, Jaklič M, Milošković A, Radojković N, Radenković M, Simić V, Maguire I (2017) Morphometric variations among *Astacus astacus* populations from different regions of the Balkan Peninsula. Zoomorphology, 136(1), pp 19–27 3. Đuretanović SR, Veličković TZ, Milošković AM, Radenković MD, Nikolić MD, Maguire I, Simić VM (2021) Preliminary results regarding phylogeny of the noble crayfish (Decapoda, Astacidae, *Astacus astacus*) in Serbia. Paper presented at the 1st International Conference on Chemo and BioInformatics, Kragujevac, Serbia, pp 222–225 4. Gross R, Lovrenčić L, Jelić M, Grandjean F, Đuretanović S, Simić V, Burimski O, Bonassin L, Groza MI, Maguire I (2021) Genetic diversity and structure of the noble crayfish populations in the Balkan Peninsula revealed by mitochondrial and microsatellite DNA markers. PeerJ, 9, p e11838 5. Kovačević TB, Borković SS, Pavlović SZ, Radojičić R, Saičić ZS (2006) The concentrations of antioxidant compounds in the hepatopancreas, the gills and muscle of some freshwater crayfish species. Acta Biologica Hungarica, 57(4), pp 449–458 6. Kovačević TB, Borković SS, Pavlović SZ, Despotović SG, Saičić ZS (2008) Glutathione as a suitable biomarker in hepatopancreas, gills and muscle of three freshwater crayfish species. Archives of Biological Sciences, 60(1), pp 59–66 7. Kovacevic S, Radojkovic N, Miloskovic A, Radenkovic M, Simic S, Cirkovic M, Simic V (2015) The possibility of conservation and sustainable use of noble crayfish *Astacus astacus* in

(continued)

Table 3 (continued)

Serbia. Paper presented at 7th International conference “Water & Fish”, Belgrade, Serbia; pp 328–332
8. Lipták B, Vitázková B, Stloukal E (2013) First record of the spinycheek crayfish (*Orconectes limosus*) in the Serbo-Romanian Tamiš River. Freshwater Crayfish, 19(2), pp 229–232
9. Lucić A, Paunović M, Tomović J, Kovačević S, Zorić K, Simić V, Atanacković A, Marković V, Kračun-Kolarević M, Hudina S, Lajtner J (2015) Aquatic macroinvertebrates of the Sava River. In: Milačič R, Ščančar J, Paunović M (eds) The Sava River, Springer, Berlin, pp 335–359
10. Marković V, Zorić K, Ilić M, Marinković N, Đuretanović S, Miloŝković A, Radojković N (2017) A contribution to the knowledge on the distribution of native crayfish *Austropotamobius torrentium* (Schrank, 1803) in Serbia. Paper presented at the XXV International Conference on Ecological Truth" Eco-Ist'17" (pp 96–99), Bor, University of Belgrade, Technical Faculty in Bor
11. Marković V, Zorić K, Ilić M, Đuretanović S, Smederevac–Lalić M (2022) Stone crayfish *Austropotamobius torrentium* (Schrank, 1803) in south-eastern Serbia: Children knowledge survey. Paper presented at the 29th International Conference Ecological Truth and Environmental Research – EcoTER'22, Sokobanja, Serbia, pp 290–294
12. Milošević SM, Živić NV, Milosavljević MV, Jakšić TR (2016) The Activity of Antioxidant Defence Enzymes (Glutathione Peroxidase, Glutathione Reductase and Phase II Biotransformation Enzyme Glutathione-S-Transferase) in Some Tissues of Stone Crayfish (Austropotamobius torrentium Shrank) from Krajkovacka River. Water Research and Management, 6, (3), 39–42
13. Panteleit J, Keller NS, Diéguez-Uribeondo J, Makkonen J, Martin-Torrijos L, Patrulea V, Pîrvu M, Preda C, Schrimpf A, Pârvulescu L (2018) Hidden sites in the distribution of the crayfish plague pathogen *Aphanomyces astaci* in Eastern Europe. Relicts of genetic groups from older outbreaks?. Journal of Invertebrate Pathology, 157, pp 117–124
14. Pavlović S, Milošević S, Borković S, Simić V, Paunović M, Žikić R, Saičić Z (2006) A report of *Orconectes* (*faxonius*) *limosus* (Rafinesque, 1817) [Crustacea: Decapoda: Astacidea: Cambaridae: *Orconectes*: subgenus *Faxonius*] in the Serbian part of the river Danube. Biotechnology & Biotechnological Equipment, 20(1), pp 53–56
15. Petrović A, Rajković M, Simić S, Maguire I, Simić V (2013) Importance of genetic characteristics in the conservation and management of crayfish in Serbia and Montenegro. Bulg. J. Agric. Sci., 19, 1093–1104
16. Simić V, Simić S, Petrović A, Paunović M, Tomović J (2000) The Strategy and The Results of The Ex Situ Conservation of The Hydro-Bionates in The Aquarium of The Institute For Biology and Ecology, Kragujevac-Serbia. Paper presented at the BALWOIS 2010, Ohrid, Republic of Macedonia, 25–29 May 2010
17. Simić V, Paunović M, Petrović A, Rajković M (2008) Crayfish of Serbia and Montenegro – the population status and the level of endangerment. Crustaceana, 81(10), pp 1153–1176
18. Simić V, Maguire I, Rajković M, Petrović A (2015) Conservation strategy for the endangered crayfish species of the family Astacidae: the ESHIPPO crayfish model. Hydrobiologia, 760(1), pp 1–13
19. Zorić K, Ilić M, Marinković N, Đuretanović S, Miloŝković A, Radojković N, Marković V, (2018) New records of native crayfish *Austropotamobius torrentium* (Schrank, 1803) in Serbia. Paper presented at the 8th International Conference, Faculty of Agriculture, University of Belgrade, Serbia, 13-15 June 2018, pp 418–422
20. Zorić K, Atanacković A, Ilić M, Csányi B, Paunović M (2020a) The spiny-cheek crayfish *Faxonius limosus* (Rafinesque, 1817) (Decapoda: Cambaridae) invades new areas in Serbian inland waters. Acta Zoologica Bulgarica, 72(4), pp 623–627
21. Zorić K, Atanacković A, Tomović J, Vasiljević B, Tubić B, Paunović M (2020b) Diversity of Alien Macroinvertebrate Species in Serbian Waters. Water, 12(12), p 3521
22. Živić NV, Atanacković A, Milošević S, Milosavljević M (2014) The Distribution of Astacidae (Decapoda) Fauna in Kosovo and Metohija, Serbia. Water Research and Management.
23. Živić I, Stojanović K, Marković Z (2022a) Springs and headwater streams in Serbia: the

(continued)

Table 3 (continued)

hidden diversity and ecology of aquatic invertebrates. In: Pešić V, Milošević D, Miliša M (eds) Small Water Bodies of the Western Balkans. Springer, Cham, pp 189–210 24. Živić I, Božanić M, Todorović N, Stojanović K (2022b) Branchiobdella Odier, 1823 (Annelida: Branchiobdellea), a New Genus for the Fauna of Serbia. Acta Zoologica Bulgarica, 74(2), pp 325–327
SLOVENIA
1. Bedjanič M (2004) Novi podatki o razširjenosti raka navadnega koščaka *Austropotamobius torrentium* (Schrank, 1803) v Sloveniji (Crustacea: Decapoda). Natura Sloveniae, 6(1), pp 25–33 (In Slovenian) 2. Danilović M, Maguire I, Füreder L (2022) Overlooked keystone species in conservation plans of fluvial ecosystems in Southeast Europe: a review of native freshwater crayfish species. Knowledge & Management of Aquatic Ecosystems, (423), p 21 3. Đuretanović S, Jaklič M, Milošković A, Radojković N, Radenković M, Simić V, Maguire I (2017) Morphometric variations among *Astacus astacus* populations from different regions of the Balkan Peninsula. Zoomorphology 136: 19–27. doi: 10.1007/s00435-016-0331-x 4. Gelder SR, Delmastro GB, Ferraguti M (1994) A report on branchiobdellidans (Annelida: Clitellata) and a taxonomic key to the species in northern Italy, including the first record of Cambarincola mesochoreus on the introduced American red swamp crayfish. 61: 2, 179–183. doi: 10.1080/11250009409355880 5. Govedič M (2017) First record of the spiny-cheek crayfish (*Orconectes limosus*) in Slovenia – 300 km upstream from its known distribution in the Drava River. Knowledge & Management of Aquatic Ecosystems, (418), p 7 6. Gross R, Lovrenčić L, Jelić M, Grandjean F, Đuretanović S, Simić V, Burimski O, Bonassin L, Groza MI, Maguire I (2021) Genetic diversity and structure of the noble crayfish populations in the Balkan Peninsula revealed by mitochondrial and microsatellite DNA markers. PeerJ, 9, p e11838 7. Lucić A, Paunović M, Tomović J, Kovačević S, Zorić K, Simić V, Atanacković A, Marković V, Kračun–Kolarević M, Hudina S, Lajtner J (2015) Aquatic macroinvertebrates of the Sava River. In: Milačič R, Ščančar J, Paunović M (eds) The Sava River, Springer, Berlin, pp 335–359 8. Machino Y (1997) Crayfish of the upper Soca and upper Sava rivers, Slovenia. Bulletin Français de la Pêche et de la Pisciculture, (347), pp 721–729 9. Maguire I, Jelić M, Klobučar G, Delpy M, Delaunay C, Grandjean F (2016) Prevalence of the pathogen *Aphanomyces astaci* in freshwater crayfish populations in Croatia. Diseases of Aquatic Organisms, 118(1), pp 45–53 10. Jaklič M, Vrezec A (2011) The first tropical alien crayfish species in European waters: the redclaw *Cherax quadricarinatus* (Von Martens, 1868) (Decapoda, Parastacidae). Crustaceana, 84(5–6), pp 651–665 11. Jaklič M, Simčič T, Vrezec A (2015) Comparison of thermal niche between native and non-native crayfish species. Journal of Experimental Biology, 213, pp 881–893 12. Jussila J, Vrezec A, Makkonen J, Kortet R, Kokko H (2015) Invasive crayfish and their invasive diseases in Europe with the focus on the virulence evolution of the crayfish plague. Biological invasions in changing ecosystems: vectors, ecological impacts, management and predictions. De Gruyter Open, Berlin, pp 183–211 13. Jussila J, Vrezec A, Jaklič T, Kukkonen H, Makkonen J, Kokko H (2017) *Aphanomyces astaci* isolate from latently infected stone crayfish (*Austropotamobius torrentium*) population is virulent. Journal of invertebrate pathology, 149, pp 15–20 14. Simčič T, Pajk F, Vrezec A, Brancelj A (2012a) Size scaling of whole-body metabolic activity in the noble crayfish (*Astacus astacus*) estimated from measurements on a single leg. Freshwater Biology, 57(1), pp 39–48 15. Simčič T, Pajk F, Brancelj A, Vrezec A (2012b) Preliminary multispecies test of a model for non-lethal estimation of metabolic activity in freshwater crayfish. Acta Biol. Slov, 55, pp 15–27

(continued)

Table 3 (continued)

16. Simčič T, Pajk F, Jaklič M, Brancelj A, Vrezec A (2014) The thermal tolerance of crayfish could be estimated from respiratory electron transport system activity. Journal of thermal biology, 41, pp 21–30
17. Smital T, Ahel M (2015) Ecotoxicological characterization of the Sava River: biomarker responses and biological assays. The Sava River, pp 177–200
18. Trontelj P, Machino Y, Sket B (2005) Phylogenetic and phylogeographic relationships in the crayfish genus *Austropotamobius* inferred from mitochondrial COI gene sequences. MPE, 34, 212–226
19. Vrezec A, Ocepek M, Jenčič V (2013) *Aphanomyces astaci* in wild crayfish populations in Slovenia: first report of persistent infection in a stone crayfish *Austropotamobius torrentium* population. Diseases of aquatic organisms, 103(2), pp 157–169

References

Ackefors HE (2000) Freshwater crayfish farming technology in the 1990s: a European and global perspective. Fish Fish 1(4):337–359

Afanasyev S, Lietytska O, Golub O, Iarochevitch O, Mudra K, Marushevska O (2021) New findings of white clawed crayfish, *Austropotamobius pallipes* (Decapoda, Astacidae), and peculiarities of its spatial distribution in Neretvica (Bosnia and Herzegovina). Zoodiversity 55(4):285

Albrecht C, Wilke T (2008) Ancient Lake Ohrid: biodiversity and evolution. Hydrobiologia 615(1): 103–140

Alvanou MV, Papadopoulos DK, Lattos A, Georgoulis I, Feidantsis K, Apostolidis AP, Michaelidis B, Giantsis IA (2022) Biology, distribution, conservation status and stocking perspective of freshwater crayfish in Greece: an updated review, vol 53. Aquaculture Research, p 5115

Anonymous (2002) Pravilnik o uvrstitvi ogroženih rastlinskih in živalskih vrst v rdeč i seznam. Uradni list Republike Slovenije, 82/2002: 8893–8975, 42/2010:5990–5993 (in Slovenian)

Anonymous (2004) Uredba o zavarovanih prostoživeč ih živalskih vrstah v Sloveniji. Uradni list Republike Slovenije, 46/2004: 5963–6017 in 109/2004, 84/2005, 115/2007, 96/2008, 36/2009, 102/2011 (in Slovenian)

Anonymous (2010a) Правилник о проглашењу и заштити строго заштићених и заштићених дивљих врста биљака, животиња и гљива: 5/2010–46, 47/2011–134, 32/2016–59, 98/2016–97. Службени гласник RS 5/10 (in Serbian)

Anonymous (2010b) Regulation on proclamation and protection of strictly protected and protected plant, animal and fungi species. Appendix 2. Protected species. Official Gazette RS 5/10 (in Serbian)

Bačnik K, Kutnjak D, Černi S, Bielen A, Hudina S (2021) Virome analysis of signal crayfish (Pacifastacus leniusculus) along its invasion range reveals diverse and divergent RNA viruses. Viruses 13(11):2259

Balian EV, Lévêque C, Segers H, Martens K (2008) The freshwater animal diversity assessment: an overview of the results. Hydrobiologia 595:627–637

Bedjanič M (2004) Novi podatki o razširjenosti raka navadnega koščaka Austropotamobius torrentium (Schrank, 1803) v Sloveniji (Crustacea: Decapoda). Natura Sloveniae 6(1):25–33. (In Slovenian)

Bekavac A, Beck A, Dragičević P, Dragun Z, Maguire I, Ivanković D, Fiket Ž, Gračan R, Hudina S (2021) Disturbance in invasion? Idiopathic necrotizing hepatopancreatitis in the signal crayfish *Pacifastacus leniusculus* (Dana, 1852) in Croatia. J Fish Dis 45:261–276

Bielen A, Vladušić T, Kuharić N, Hudina S, Šver L, Likić S (2014) First evidence of the presence of multixenobiotic resistance mechanism activity in freshwater invasive species, signal crayfish *Pacifastacus leniusculus* (Dana, 1852). Period Biol 116(3):267–273

Bland LM (2017) Global correlates of extinction risk in freshwater crayfish. Anim Conserv 20(6): 532–542

Borković SS, Pavlović SZ, Kovačević TB, Štajn AŠ, Petrović VM, Saičić ZS (2008) Antioxidant defence enzyme activities in hepatopancreas, gills and muscle of spiny cheek crayfish (*Orconectes limosus*) from the river Danube. Comparative Biochemistry and Physiology Part C: Toxicology & Pharmacology 147(1):122–128

Boštjančić Lj L, Bonassin L, Anušić L, Lovrenčić L, Besendorfer V, Maguire I, Grandjean F, Austin CM, Greve C, Hamadou AB, Mlinarec J (2021, 1820) The *Pontastacus leptodactylus* (Astacidae) Repeatome provides insight into genome evolution and reveals remarkable diversity of satellite DNA. Front Genet 11. https://doi.org/10.3389/fgene.2020.611745

Brummett RE, Beveridge MCM (2015) Aquaculture. In: Craig JF (ed) Freshwater fisheries ecology. Wiley Blackwell, Oxford, pp 580–593

Cai J, Galli G, Zhou X (2020) Top 10 species groups in global aquaculture 2018. FAO WAPI fact sheets. Rome, FAO Fisheries and Aquaculture Department, 4 p. (also available at http://www.fao.org/3/ca9383en/ca9383en.pdf)

Chidester FE (1912) The biology of the crayfish. Am Nat 46(545):279–293

D'Agaro E, De Luise G, Lanari D (1999) The current status of crayfish farming in Italy. Freshwater Crayfish 12:506–517

Dakić L, Maguire I (2016) Year cycle and morphometrical characteristics of *Austropotamobius torrentium* from two karstic rivers in Croatia. Natura Croatica: periodicum Musei historiae naturalis Croatici 25(1):87–100

Danilović M, Maguire I, Füreder L (2022) Overlooked keystone species in conservation plans of fluvial ecosystems in Southeast Europe: a review of native freshwater crayfish species. Knowl Manag Aquat Ecosyst 423:21. https://doi.org/10.1051/kmae/2022016

Dobrović A, Maguire I, Boban M, Grbin D, Hudina S (2021) Reproduction dynamics of the marbled crayfish *Procambarus virginalis* Lyko, 2017 from an anthropogenic lake in continental Croatia. Aquat Invasions 16:482–498. https://doi.org/10.3391/ai.2021.16.3.06

Dobrović A, Geček S, Klanjšček T, Haberle I, Dragičević P, Pavić D, Petelinec A, Boštjančić LL, Bonassin L, Theissinger K, Hudina S (2022) Recurring infection by crayfish plague pathogen only marginally affects survival and growth of marbled crayfish. NeoBiota 77:155–177

Dorn NJ, Wojdak JM (2004) The role of omnivorous crayfish in littoral communities. Oecologia 140(1):150–159

Dragičević P, Faller M, Kutleša P, Hudina S (2020) Update on the signal crayfish, *Pacifastacus leniusculus* (Dana, 1852) range expansion in Croatia: a 10-year report. BioInvasions Rec 9(4): 793

Dragičević P, Bielen A, Petrić I, Vuk M, Žučko J, Hudina S (2021a) Microbiome of the successful freshwater invader, the signal crayfish, and its changes along the invasion range. Microbiol Spectr 9(2):e00389–e00321

Dragičević P, Grbin D, Maguire I, Blažević SA, Abramović L, Tarandek A, Hudina S (2021b) Immune response in crayfish is species-specific and exhibits changes along invasion range of a successful invader. Biology 10(11):1102

Dražina T, Korša A, Špoljar M, Maguire I, Klobučar GIV (2018) Epifauna of native and alien freshwater crayfish species (Crustacea: Decapoda): a host-specific community? Freshw Sci 37(3):593–604. https://doi.org/10.1086/698764

Đuretanović S (2022) The noble crayfish – a short video. In: YouTube. https://www.youtube.com/watch?v=ZQLGa9_qWI4

Đuretanović S, Jaklič M, Milošković A, Radojković N, Radenković M, Simić V, Maguire I (2017) Morphometric variations among *Astacus astacus* populations from different regions of the Balkan Peninsula. Zoomorphology 136(1):19–27

Đuretanović SR, Veličković TZ, Milošković AM, Radenković MD, Nikolić MD, Maguire I, Simić VM (2021) Preliminary results regarding phylogeny of the noble crayfish (Decapoda, Astacidae, *Astacus astacus*) in Serbia. Paper presented at the 1st International Conference on Chemo and BioInformatics, Kragujevac, Serbia, pp 222–225

Edsman L, Füreder L, Gherardi F, Souty-Grosset C (2010) *Astacus astacus*. The IUCN Red List Of Threatened Species 2010:e.T2191A9338388. https://doi.org/10.2305/IUCN.UK.2010-3.RLTS.T2191A9338388.en. Accessed on 19 September 2022

Faller M, Maguire I, Klobučar G (2006) Annual activity of the noble crayfish (*Astacus astacus*) in the Orljava River (Croatia). Bulletin Français de la Pêche et de la Pisciculture 383:23–39

FAO (2020) Fishery and aquaculture statistics. Global aquaculture production 1950–2018 (FishStatJ). In: FAO fisheries and aquaculture department [online]. Rome, Downloaded September 2022. www.fao.org/fishery/statistics/software/fishstatj/en

FAO (2022) The state of world fisheries and aquaculture 2022. Towards blue transformation. FAO, Rome. https://doi.org/10.4060/cc0461en

Füreder L (2015) Crayfish in Europe: biogeography, ecology and conservation. In: Freshwater crayfish. CRC Press, pp 604–637

Füreder L, Reynolds JD (2003) Is *Austropotamobius pallipes* a good bioindicator? Bulletin Français de la Pêche et de la Pisciculture 370–371:157–163

Füreder L, Gherardi F, Holdich D, Reynolds J, Sibley P, Souty-Grosset C (2010a) *Austropotamobius pallipes*. The IUCN Red List Of Threatened Species 2010: e.T2430A9438817. https://doi.org/10.2305/IUCN.UK.2010-3.RLTS.T2430A9438817.en. Accessed on 19 September 2022

Füreder L, Gherardi F, Souty-Grosset C (2010b) *Austropotamobius torrentium* (errata version published in 2017). The IUCN Red List of Threatened Species 2010: e.T2431A121724677. Accessed on 23 September 2022

Geist J (2011) Integrative freshwater ecology and biodiversity conservation. Ecol Indic 11(6): 1507–1516

Gelder SR, Delmastro GB, Ferraguti M (1994) A report on branchiobdellidans (Annelida: Clitellata) and a taxonomic key to the species in northern Italy, including the first record of *Cambarincola mesochoreus* on the introduced American red swamp crayfish. Bollettino di Zoologia 61(2):179–183. https://doi.org/10.1080/11250009409355880

Gherardi F (2011) Towards a sustainable human use of freshwater crayfish (Crustacea, Decapoda, Astacidea). Knowl Manag Aquat Ecosyst 401:02

Gherardi F, Souty-Grosset C (2017) *Pontastacus leptodactylus* (amended version of 2016 assessment). The IUCN Red List of Threatened Species 2017: e.T153745A120103207. https://doi.org/10.2305/IUCN.UK.20173.RLTS.T153745A120103207.en. Accessed on 23 September 2022

Gottstein S, Kerovec M, Maguire I, Bukvić I (1999) Ecological notes on *Austropotamobius pallipes italicus* (Faxon, 1914) (Decapoda, Astacidae) in the karstic spring of Neretva. Proceedings of the 12th symposium of the International Association of Astacology, pp 620–628

Gottstein S, Žganec K, Maguire I, Kerovec M, Jalžić B (2007) Viši rakovi slatkih i bočatih voda porječja rijeke Krke. Rijeka Krka i Nacionalni park "Krka" – Prirodna i kulturna baština, zaštita i održivi razvitak, Marguš D (ured). Javna ustanova Nacionalni park "Krka", Šibenik, pp 421–430

Gottstein S, Hudina S, Lucić A, Maguire I, Ternjej I, Žganec K (2011) Crveni popis rakova (Crustacea) slatkih i bočatih voda Hrvatske. Zagreb: Državni zavod za zaštitu prirode. (Red list of freshwater and brackish Crustacea in Croatia; in Croatian)

Goudie AS, Viles HA (2013) The earth transformed: an introduction to human impacts on the environment. John Wiley & Sons, Hoboken

Govedič M (2017) First record of the spiny-cheek crayfish (*Orconectes limosus*) in Slovenia–300 km upstream from its known distribution in the Drava River. Knowl Manage Aquat Ecosyst 418:7

Gross R, Palm S, Koiv K, Prestegaard T, Jussila J, Paaver T, Geist J, Kokko H, Karjalainen A, Edsman L (2013) Microsatellite markers reveal clear geographic structuring among threatened noble crayfish (*Astacus astacus*) populations in northern and Central Europe. Conserv Genet 14(4):809–821

Gross R, Kõiv K, Pukk L, Kaldre K (2017) Development and characterization of novel tetranucleotide microsatellite markers in the noble crayfish (*Astacus astacus*) suitable for highly multiplexing and for detecting hybrids between the noble crayfish and narrow-clawed crayfish (*A. leptodactylus*). Aquaculture 472:50–56

Gross R, Lovrenčić L, Jelić M, Grandjean F, Đuretanović S, Simić V, Burimski O, Bonassin L, Groza MI, Maguire I (2021) Genetic diversity and structure of the noble crayfish populations in the Balkan Peninsula revealed by mitochondrial and microsatellite DNA markers. PeerJ 9: e11838

Gydemo R (1989) Studies on reproduction and growth in the noble crayfish. Doctoral dissertation, Stockholm University, Sweden

Harlioğlu MM (2008) The harvest of the freshwater crayfish *Astacus leptodactylus* Eschscholtz in Turkey: harvest history, impact of crayfish plague, and present distribution of harvested populations. Aquac Int 16(4):351–360

Hofman CA, Rick TC (2018) Ancient biological invasions and Island ecosystems: tracking translocations of wild plants and animals. J Archaeol Res 26(1):65–115

Holdich DM (1993) A review of astaciculture: freshwater crayfish farming. Aquat Living Resour 6(4):307–317

Holdich D, Haffner P, Noël P (2006) Species files. In: Souty-Grosset C, Holdich D, Noel PY, Reynolds JD, Haffner P (eds) Atlas of crayfish in Europe, Muséum national d'Histoire naturelle, vol 64. Paris Patrimoines Naturels, Paris, pp 50–129

Holdich DM, Reynolds JD, Souty-Grosset C, Sibley PJ (2009) A review of the ever increasing threat to European crayfish from non-indigenous crayfish species. Knowl Manag Aquat Ecosyst 394–395:11

Hossain MA, Lahoz-Monfort JJ, Burgman MA, Böhm M, Kujala H, Bland LM (2018) Assessing the vulnerability of freshwater crayfish to climate change. Divers Distrib 24(12):1830–1843

Hudina S, Maguire I, Galić N, Đurić P (2004) Dinamika populacije riječnoga raka u Nacionalnome parku "Paklenica". Paklenički zbornik vol. 2, Lukač G. (ured). Javna ustanova Nacionalni park "Paklenica", Starigrad-Paklenica, 89–94

Hudina S, Maguire I, Klobučar GIV (2008) Spatial dynamics of the noble crayfish (*Astacus astacus* L.) in the Paklenica National Park. Knowl Manag Aquat Ecosyst 388:01p12. https://doi.org/10.1051/kmae:2008001

Hudina S, Faller M, Lucić A, Klobučar G, Maguire I (2009) Distribution and dispersal of two invasive crayfish species in the Drava River basin, Croatia. Knowl Manag Aquat Ecosyst 394–395:09p11. https://doi.org/10.1051/kmae/2009023

Hudina S, Lucić A, Žganec K, Janković S (2011a) Characteristics and movement patterns of a recently established invasive *Pacifastacus leniusculus* population in the river Mura, Croatia. Knowl Manag Aquat Ecosyst 403:07

Hudina S, Janković S, Lucić A, Žganec K (2011b) The status of *Astacus astacus* in the northernmost part of Croatia (Međimurje County) in the face of invasion by *Pacifastacus leniusculus* (Crustacea: Astacidae). Lauterbornia 72:31–44

Hudina S, Hock K, Žganec K, Lucić A (2012) Changes in population characteristics and structure of the signal crayfish at the edge of its invasive range in a European river. Ann Limnol Int J Limnol 48(1):3–11

Hudina S, Žganec K, Lucić A, Trgovčić K, Maguire I (2013) Recent invasion of the karstic river systems in Croatia through illegal introductions of the signal crayfish. Freshwater Crayfish 19(1):21–27

Hudina S, Hock K, Radović A, Klobučar G, Petković J, Jelić M, Maguire I (2016) Species-specific differences in dynamics of agonistic interactions may contribute to the competitive advantage of the invasive signal crayfish (*Pacifastacus leniusculus*) over the native narrow-clawed crayfish (*Astacus leptodactylus*). Mar Freshw Behav Physiol 49(3):147–157

Hudina S, Kutleša P, Trgovčić K, Duplić A (2017) Dynamics of range expansion of the signal crayfish (*Pacifastacus leniusculus*) in a recently invaded region in Croatia. Aquat Invasions 12(1):67

Jaklič M, Vrezec A (2011) The first tropical alien crayfish species in European waters: the redclaw *Cherax quadricarinatus* (Von Martens, 1868) (Decapoda, Parastacidae). Crustaceana 84(5–6): 651–665

Jaklič M, Simčič T, Vrezec A (2015) Comparison of thermal niche between native and non-native crayfish species. J Exp Biol 213:881–893

Jelić M, Klobučar GIV, Grandjean F, Puillandre N, Franjević D, Futo M, Amouret J, Maguire I (2016) Insights into the molecular phylogeny and historical biogeography of the white-clawed crayfish (Decapoda, Astacidae). Mol Phylogenet Evol 103:26–40

Jernelöv A (2017) Crayfish plague in Europe. In: The long-term fate of invasive species. Springer, Cham. https://doi.org/10.1007/978-3-319-55396-2_7

Jussila J, Vrezec A, Makkonen J, Kortet R, Kokko H (2015) Invasive crayfish and their invasive diseases in Europe with the focus on the virulence evolution of the crayfish plague. Biological invasions in changing ecosystems: vectors, ecological impacts, management and predictions. De Gruyter Open, Berlin, pp 183–211

Jussila J, Vrezec A, Jaklič T, Kukkonen H, Makkonen J, Kokko H (2017) *Aphanomyces astaci* isolate from latently infected stone crayfish (*Austropotamobius torrentium*) population is virulent. J Invertebr Pathol 149:15–20

Jussila J, Maguire I, Kokko H, Tiitinen V, Makkonen J (2020) Narrow-clawed crayfish in Finland: *Aphanomyces astaci* resistance and genetic relationship to other selected European and Asian populations. Knowl Manag Aquat Ecosyst 421:30

Jussila J, Edsman L, Maguire I, Diéguez-Uribeondo J, Theissinger K (2021) Money kills native ecosystems: European crayfish as an example. Front Ecol Evol 9:476

Kawai T, Crandall KA (2016) Global diversity and conservation of freshwater crayfish (Crustacea: Decapoda: Astacoidea). In: A global overview of the conservation of freshwater decapod crustaceans. Springer, Cham, pp 65–114

Klobučar G, Maguire I (2011) Studija izvodivosti uzgajališta slatkovodnih rakova vrste *Astacus astacus* i *Austropotamobius pallipes*. In Croatian

Klobučar GIV, Maguire I, Gottstein S, Gelder S (2006) Occurrence of Branchiobdellida (Annelida: Clitellata) on the freshwater crayfish in Croatia. Annales de Limnologie – international. J Limnol 42, 4:251–260

Klobučar GIV, Malev O, Šrut M, Štambuk A, Lorenzon S, Cvetkovi Z, Ferrero EA, Maguire I (2012) Genotoxicity monitoring of freshwater environments using caged crayfish (*Astacus leptodactylus*). Chemosphere 87:62–67

Klobučar G, Podnar M, Jelić M, Franjević D, Faller M, Štambuk A, Gottstein S, Simić V, Maguire I (2013) Role of the Dinaric karst (western Balkans) in shaping the phylogeographic structure of the threatened crayfish *Austropotamobius torrentium*. Freshw Biol 58:1089–1105

Kotovska G, Khrystenko D, Patoka J, Kouba A (2016) East European crayfish stocks at risk: arrival of non-indigenous crayfish species. Knowl Manage Aquat Ecosyst 417:37

Kouba A (2011) Intensification of juvenile crayfish culture. Doctoral dissertation, Faculty of Fisheries and Protection of Waters, University of South Bohemia České Budějovice

Kouba A, Petrusek A, Kozák P (2014) Continental-wide distribution of crayfish species in Europe: update and maps. Knowl Manag Aquat Ecosyst 413:05

Kouba A, Tíkal J, Císař P, Veselý L, Fořt M, Příborský J, Patoka J, Buřič M (2016) The significance of droughts for hyporheic dwellers: evidence from freshwater crayfish. Sci Rep 6(1):1–7

Kovačević TB, Borković SS, Pavlović SZ, Radojičić R, Saičić ZS (2006) The concentrations of antioxidant compounds in the hepatopancreas, the gills and muscle of some freshwater crayfish species. Acta Biol Hung 57(4):449–458

Kovačević TB, Borković SS, Pavlović SZ, Despotović SG, Saičić ZS (2008) Glutathione as a suitable biomarker in hepatopancreas, gills and muscle of three freshwater crayfish species. Arch Biol Sci 60(1):59–66

Kovacevic S, Radojkovic N, Miloskovic A, Radenkovic M, Simic S, Cirkovic M, Simic V (2015) The possibility of conservation and sustainable use of noble crayfish *Astacus astacus* in Serbia. Paper presented at 7th international conference "Water & Fish", Belgrade, Serbia, pp 328–332

Kozák P, Füreder L, Kouba A, Reynolds J, Souty-Grosset C (2011) Current conservation strategies for European crayfish. Knowl Manag Aquat Ecosyst, (401), p 01

Kubec J, Kouba A, Buřič M (2019) Communication, behaviour, and decision making in crayfish: a review. Zool Anz 278:28–37

Kuklina I, Kouba A, Kozák P (2013) Real-time monitoring of water quality using fish and crayfish as bioindicators: a review. Environ Monit Assess 185(6):5043–5053

Lette ED (2020) Modernising conservation through technology: a metabolomic investigation of a critically endangered freshwater crayfish. Doctoral dissertation, Edith Cowan University

Lewis SL, Maslin MA (2015) Defining the Anthropocene. Nature 519(7542):171–180

Lipták B, Vitázková B, Stloukal E (2013) First record of the spinycheek crayfish (*Orconectes limosus*) in the Serbo-Romanian Tamiš River. Freshwater Crayfish 19(2):229–232

Loureiro TG, Anastácio PMSG, Araujo PB, Souty-Grosset C, Almerão MP (2015) Red swamp crayfish: biology, ecology and invasion – an overview. Nauplius 23:1–19

Lovrenčić L, Maguire I (2021) Gap analysis revealed a moderate efficiency of protected areas for the conservation of the endangered noble crayfish in Croatia. Natura Croatica 30(2):529–536. https://doi.org/10.20302/NC.2021.30

Lovrenčić L, Temunović M, Maguire I (2020a) How well does NATURA 2000 protect threatened stone crayfish in Croatia? Natura Croatica 29(2):241–253. https://doi.org/10.20302/NC.2020.29.30

Lovrenčić L, Bonassin L, Boštjančić Lj L, Podnar M, Jelić M, Klobučar G, Jaklič M, Stamenković V, Hinić J, Maguire I (2020b) New insights into the genetic diversity of the stone crayfish: taxonomic and conservation implications. BMC Evol Biol 20:146

Lovrenčić L, Pavić V, Majnarić S, Abramović L, Jelić M, Maguire I (2020c) Morphological diversity of the stone crayfish – traditional and geometric morphometric approach. Knowl Manag Aquat Ecosyst 421:1. https://doi.org/10.1051/kmae/2019042

Lovrenčić L, Temunović M, Bonassin L, Grandjean F, Austin CM, Maguire I (2022a) Climate change threatens unique genetic diversity within the Balkan biodiversity hotspot – the case of the endangered stone crayfish. Glob Ecol Conserv 39:e02301

Lovrenčić L, Temunović M, Gross R, Grgurev M, Maguire I (2022b) Integrating population genetics and species distribution modelling to guide conservation of the noble crayfish, *Astacus astacus*, in Croatia. Sci Rep 12(1):1–17

Lovrenčić L, Ferrón HG, Grbin D, Maguire I (2022c) Insight into the noble crayfish morphological diversity: a geometric morphometric approach. Knowl Manag Aquat Ecosyst 423:9. https://doi.org/10.1051/kmae/2022006

Lowe S, Browne M, Boudjelas S, De Poorter M (2000) 100 of the World's worst invasive alien species: a selection from the global invasive species database. (The invasive species specialist group (ISSG) a specialist group of the species survival commission (SSC) of the world conservation union (IUCN), 2000)

Lucić A, Erben R, Maguire I (2004) Occurrence of the pathogen *Psorospermium haeckeli* in Astacid populations in Croatia. Bulletin Français de la Pêche et de la Pisciculture 372–373: 375–383

Lucić A, Hudina S, Faller M, Cerjanec D (2012) A comparative study of the physiological condition of native and invasive crayfish in Croatian rivers. Biologia 67(1):172–179

Lucić A, Paunović M, Tomović J, Kovačević S, Zorić K, Simić V, Atanacković A, Marković V, Kračun-Kolarević M, Hudina S, Lajtner J (2015) Aquatic macroinvertebrates of the Sava River. In: Milačič R, Ščančar J, Paunović M (eds) The Sava River. Springer, Berlin, pp 335–359

Machino Y (1997) Crayfish of the upper Soca and upper Sava rivers, Slovenia. Bulletin Français de la Pêche et de la Pisciculture 347:721–729
Maguire I (2009) Die Flussekrebse Osteuropas. U: Flusskrebse Biologie–Ökologie–Gefährdung, Füreder L (ured). Folio-Verlag Veröffentlichungen des Naturmuseums Südtirol, 6, Innsbruck. 92–97
Maguire I (2010) Slatkovodni rakovi – Priručnik za inventarizaciju i praćenje stanja. Državni zavod za zaštitu prirode, Zagreb. 40 (In Croatian)
Maguire I (2013) Nacionalni programi za praćenje stanja očuvanosti vrsta u Hrvatskoj – Plemeniti ili riječni rak (*Astacus astacus*). Državni zavod za zaštitu prirode. 28 p (In Croatian)
Maguire I (2014a) Nacionalni programi za praćenje stanja očuvanosti vrsta u Hrvatskoj – Potočni rak ili rak kamenjar (*Austropotamobius torrentium* (Schrank, 1803)). Državni zavod za zaštitu prirode. 30 p (In Croatian)
Maguire I (2014b) Nacionalni programi za praćenje stanja očuvanosti vrsta u Hrvatskoj – Bjelonogi ili primorski rak (*Austropotamobius pallipes* (Lereboullet, 1858)). Državni zavod za zaštitu prirode. 28 p (In Croatian)
Maguire I, Dakić L (2011) Comparative analyses *of Astacus leptodactylus* morphological characteristics from Croatia and Armenia. Biologia, Section Zool 66(3):491–498. https://doi.org/10.2478/s11756-011-0045-y
Maguire I, Gottstein-Matočec S (2004) The distribution pattern of freshwater crayfish in Croatia. Crustaceana 77(1):25–47
Maguire I, Jelić M (2010) Distribucija rakova iz porodice Astacidae na području Banovine. U: Zrinska gora regionalni park prirode, Bučar M (ured). Grafički zavod Hrvatske, Zagreb. 212–216 (In Croatian)
Maguire I, Klobučar GIV (2003) Appearance of *Orconectes limosus* in Croatia. Crayfish news 25(3):7
Maguire I, Klobučar G (2011) Size structure, maturity size, growth and condition index of stone crayfish (*Austropotamobius torrentium*) in North-West Croatia. Knowl Manag Aquat Ecosyst 401. doi: https://doi.org/10.1051/kmae/2011026
Maguire I, Erben R, Klobučar GIV, Lajtner J (2002) Year cycle of *Austropotamobius torrentium* (Schrank). Bulletin Français de la Pêche et de la Pisciculture 367:943–957
Maguire I, Klobučar GIV, Gottstein S, Erben R (2003a) Distribution of *Austropotamobius pallipes* (Lereboullet) in Croatia and notes on its morphology. Bulletin Français de la Pêche et de la Pisciculture 370–371:57–71
Maguire I, Klobučar G, Lucić A, Štefok B (2003b) Nalaz vrste *Astacus leptodactylus* na području Parka prirode Lonjsko polje. Bilten Parka prirode "Lonjsko polje" 5(1–2):10–13
Maguire I, Hudina S, Erben R (2004a) Estimation of noble crayfish (*Astacus astacus*) population size in the Velika Paklenica stream (Croatia). Bulletin Français de la Pêche et de la Pisciculture 372–373:353–366
Maguire I, Hudina S, Klobučar G, Štambuk A. (2004b) Godišnji ciklus riječnog raka (*Astacus astacus* L.) u potoku Velika paklenica. Paklenički zbornik vol. 2, Lukač G. (ured). Javna ustanova Nacionalni park "Paklenica", Starigrad-Paklenica, 81–88 (In Croatia)
Maguire I, Klobučar G, Erbene R (2005) The relationship between female size and egg size in freshwater crayfish *Austropotamobius torrentium.* Bulletin Français de la Pêche et de la Pisciculture 376 377:777 785
Maguire I, Hudina S, Faller M, Lucić A, Erben R (2006a) Morphometric characteristics of noble crayfish (*Astacus astacus*) in Croatia. Proceedings of the 15th symposium of the International Association of Astacology, pp 324–331
Maguire I, Klobučar G, Faller M, Machino Y, Kučinić M, Žužul M (2006b) Updates on the distribution of the white-clawed and the narrow-clawed crayfish in Croatia. Crayfish news 28(1):4–5
Maguire I, Gottstein-Matočec S, Klobučar G, Erben R, Kerovec M (2007) Rasprostanjenost slatkovodnih rakova reda Decapoda na području Istre, Kvarnera i Gorskog kotara – prošlost i

sadašnjost. Prirodoslovna istraživanja riječkog područja II, Arko–Pijevac M, Surina B (ured). Prirodoslovni muzej Rijeka, Rijeka, pp 191–198

Maguire I, Klobučar G, Marčić Z, Zanella D (2008) The first record of *Pacifastacus leniusculus* in Croatia. Crayfish News 30(4):4

Maguire I, Lucić A, Schletterer M, Sint D, Erben R, Füreder L (2010a) Fecundity of indigenous stone crayfish populations: implications for crayfish conservation in Austria and Croatia. Freshwater crayfish 17:103–107

Maguire I, Cerjanec D, Lucić A (2010b) Life history of *Astacus leptodactylus* in two Croatian rivers. Freshwater Crayfish 17:255–260

Maguire I, Klobučar G, Posavec Z, Hudina S (2011a) Update on the *Pacifastacus leniusculus* spread in Croatia. Crayfish News 33:5–5

Maguire I, Jelić M, Klobučar G (2011b) Update on the distribution of freshwater crayfish in Croatia. Knowl Manag Aquat Ecosyst 401. https://doi.org/10.1051/kmae/2011026

Maguire I, Špoljarić I, Klobučar G (2013a) The indigenous crayfish of Plitvice Lakes National Park, Croatia. Freshwater Crayfish 19(1):91–96

Maguire I, Špelić I, Jelić M, Klobučar G (2013b) Is it possible to detect narrow-clawed and Noble crayfish probable hybrids using multivariate discriminant analysis of morphometric data? Freshwater Crayfish 19(2):219–227

Maguire I, Podnar M, Jelić M, Štambuk A, Schrimpf A, Schulz H, Klobučar G (2014) Two distinct evolutionary lineages of the *Astacus leptodactylus* species–complex (Decapoda: Astacidae) inferred by phylogenetic analyses. Invertebr Syst 28(2):117–123

Maguire I, Jelić M, Klobučar G, Delpy M, Delaunay C, Grandjean F (2016) Prevalence of the pathogen *Aphanomyces astaci* in freshwater crayfish populations in Croatia. Dis Aquat Org 118(1):45–53

Maguire I, Marn N, Klobučar G (2017) Morphological evidence for hidden diversity in the threatened stone crayfish *Austropotamobius torrentium* (Schrank, 1803) (Decapoda: Astacoidea: Astacidae) in Croatia. J Crustac Biol 37(1):7–15. https://doi.org/10.1093/jcbiol/ruw012

Maguire I, Klobučar G, Žganec K, Jelić M, Lucić A, Hudina S (2018) Recent changes in distribution pattern of freshwater crayfish in Croatia – threats and perspectives. Knowl Manag Aquat Ecosyst 419:2. https://doi.org/10.1051/kmae/2017053

Makkonen J, Kokko H, Jussila J (2015) Mitochondrial cytochrome oxidase I gene analysis indicates a restricted genetic background in Finnish noble crayfish (*Astacus astacus*) stocks. Knowl Manag Aquat Ecosyst 416:21

Malev O, Šrut M, Lorenzon S, Maguire I, Štambuk A, Ferrero EA, Klobučar GIV (2010) Genotoxic, physiological and immunological effects caused by temperature increase, hypoxia or food deprivation in freshwater crayfish *Astacus leptodactylus*. Comp Biochem Physiol Part C: Toxicol Pharmacol 152:433–443

Marić D, Rajković M (2004) Competitors and predators the cause of crayfish (*Astacus astacus* L.) numerosity decrease in the waters of Nikšić region. Natura Montenegrina 3:101–111

Marić D, Stanić M (2021) Bioaccumulation of 18 trace metals in muscle and exoskeleton in the noble crayfish (*Astacus astacus* L.) in the river zeta (Montenegro). Poljoprivreda i Sumarstvo 67(3):31–42

Marković V, Zorić K, Ilić M, Marinković N, Đuretanović S, Milošković A, Radojković N (2017) A contribution to the knowledge on the distribution of native crayfish *Austropotamobius torrentium* (Schrank, 1803) in Serbia. Paper presented at the XXV international conference on ecological truth Eco-Ist'17″ (pp. 96–99), Bor, University of Belgrade, Technical Faculty in Bor

Marković V, Zorić K, Ilić M, Đuretanović S, Smederevac-Lalić M (2022) Stone crayfish *Austropotamobius torrentium* (Schrank, 1803) in south-eastern Serbia: Children knowledge survey. Paper presented at the 29th International Conference Ecological Truth and Environmental Research – EcoTER'22, Sokobanja, Serbia, pp 290–294

Mazlum Y (2007) Stocking density affects the growth, survival, and cheliped injuries of third instars of narrow-clawed crayfish, *Astacus leptodactylus* Eschscholtz, 1823 juveniles. Crustaceana 80(7):803–815

McClain WR (2020) Crayfish aquaculture. Fish Aquac 9:259

Meyer-Arendt KJ, Henderson ML (2020) Crawfish and crawfish festivals in the Pacific northwest. Yearbook Assoc Pac Coast Geog 82(82):38–67

Mijošek T, Jelić M, Mijošek V, Maguire I (2017) Molecular and morphometric characterisation of the invasive signal crayfish populations in Croatia. Limnologica 63:107–118. https://doi.org/10.1016/j.limno.2017.02.003

Milošević SM, Živić NV, Milosavljević MV, Jakšić TR (2016) The activity of antioxidant Defence enzymes (glutathione peroxidase, glutathione reductase and phase II biotransformation enzyme glutathione-S-transferase) in some tissues of stone crayfish (*Austropotamobius torrentium* shrank) from Krajkovacka River. Water Res Manag 6(3):39–42

Mlinarec J, Mužić M, Pavlica M, Šrut M, Klobučar G, Maguire I (2011) Comparative karyotype investigations in the European crayfish *Astacus astacus* and *A. leptodactylus*. Crustaceana 84(12–13):1497–1510

Mlinarec J, Porupski I, Maguire I, Klobučar G (2016) Comparative karyotype investigations in the white-clawed crayfish (*Austropotamobius pallipes* species complex) and stone crayfish (*A. torrentium*) (Decapoda, Astacidae). J Crustac Biol 36(1):87–93

Mrugała A, Šanda R, Petrusek A, Marić D, Vukić J (2017) Recent acute crayfish mortality reveals *Aphanomyces astaci* presence in Bosnia and Herzegovina. J Invertebr Pathol 150:73–75

Njegovan V, Orešković M, Hudina S, Maguire I (2017) Crayfish fauna of the Maksimir lakes in the urban area of Zagreb. Freshwater Crayfish 23(1):1–11. https://doi.org/10.5869/fc.2017.v23-1.XXX

Nyström P, Brönmark C, Granéli W (1999) Influence of an exotic and a native crayfish species on a littoral benthic community. Oikos 85:545–553

Panjković B, Rat M, Mihajlović S, Galambos L, Kiš A, Puzović S, Nadaždin B, Šeat J, Vukajlović F, Tot I, Đapić M (2021) Invasive alien species in the Balkan Peninsula. Invasive Alien Species: Observations and Issues from Around the World 3:42–87

Panteleit J, Keller NS, Diéguez-Uribeondo J, Makkonen J, Martín-Torrijos L, Patrulea V, Pîrvu M, Preda C, Schrimpf A, Pârvulescu L (2018) Hidden sites in the distribution of the crayfish plague pathogen *Aphanomyces astaci* in Eastern Europe: relicts of genetic groups from older outbreaks? J Invertebr Pathol 157:117–124

Pârvulescu L, Pacioglu O, Hamchevici C (2011) The assessment of the habitat and water quality requirements of the stone crayfish (*Austropotamobius torrentium*) and noble crayfish (*Astacus astacus*) species in the rivers from the Anina Mountains (SW Romania). Knowl Manag Aquat Ecosyst 401:03

Patoka J, Fišáková MN, Kalous L, Škrdla P, Kuča M (2014) Earliest evidence for human consumption of crayfish. Crustaceana 87(13):1578–1585

Patoka J, Kocánová B, Kalous L (2016) Crayfish in Czech cultural space: the longest documented relationship between humans and crayfish in Europe. Knowl Manag Aquat Ecosyst 417:5

Pavić D, Čanković M, Petrić I, Makkonen J, Hudina S, Maguire I, Vladušić T, Šver L, Hrašćan R, Orlić K, Dragičević P (2020) Non-destructive method for detecting *Aphanomyces astaci*, the causative agent of crayfish plague, on the individual level. J Invertebr Pathol 169:107274

Pavić D, Bielen A, Hudina S, Špoljarić I, Grandjean F, Jussila J, Maguire I (2021) Distribution of *Aphanomyces astaci* Schikora, 1906, the causative agent of crayfish plague, in the Plitvice Lakes National Park, Croatia. BioInvasions Rec 10:654–668

Pavlović S, Milošević S, Borković S, Simić V, Paunović M, Žikić R, Saičić Z (2006) A report of *Orconectes (faxonius) limosus* (Rafinesque, 1817) [Crustacea: Decapoda: Astacidea: Cambaridae: Orconectes: subgenus Faxonius] in the Serbian part of the river Danube. Biotechnol Biotechnological Equipment 20(1):53–56

Perdikaris C, Kozák P, Kouba A, Konstantinidis E, Paschos I (2012) Socio-economic drivers and non-indigenous freshwater crayfish species in Europe. Knowl Manag Aquat Ecosyst 404:01

Petrovic A, Rajkovic M, Simic S, Maguire I, Simic V (2013) Importance of genetic characteristics in the conservation and management of crayfish in Serbia and Montenegro. Bulgarian J Agr Sci 19(5):1093–1104

Policar T, Kozák P (2005) Comparison of trap and baited stick catch efficiency for noble crayfish (*Astacus astacus* L.) in the course of the growing season. Bulletin Français de la Pêche et de la Pisciculture 376–377:675–686

Policar T, Kozák P (2015) Production and culture of crayfish. Crayfish biology and culture. Vodňany (Czech Republic): University of South Bohemia in České Budějovice, Faculty of Fisheries and Protection of waters, pp 293–363

Pyšek P, Richardson DM (2010) Invasive species, environmental change and management, and health. Annu Rev Environ Resour 35:25–55

Rajković M (2004) Optimalni ekološki uslovi za razvoj riječnog raka (*Astacus astacus* L.) u vodenim ekosistemima na području gornjeg toka rijeke Zete. Specijalistički rad Univerzitet u Kragujevcu (In Serbian)

Rajković M, Simić V, Petrović A (2006) Length-weight gain of European crayfish *Astacus astacus* (L.) in the area of the upper course of the Zeta River, Montenegro. Arch Biol Sci 58(4):233–238

Rajkovic M, Petrovic A, Maguire I, Simic V, Simic S, Paunovic M (2012) Discovery of a new population of the species complex of the white-clawed crayfish, *Austropotamobius pallipes/italicus* (Decapoda, Astacidae) in Montenegro, range extension, endangerment, and conservation. Crustaceana 85(3):333

Rath RK (2018) Freshwater aquaculture, 3rd Revised and Enlarged edition. New Delhi, Scientific Publishers

Rebrina F, Skejo J, Lucić A, Hudina S (2015) Trait variability of the signal crayfish (*Pacifastacus leniusculus*) in a recently invaded region reflects potential benefits and trade-offs during dispersal. Aquat Invasions 10(1):41

Reynolds J, Souty-Grosset C (2011) Management of freshwater biodiversity: crayfish as bioindicators. Cambridge University Press, Cambridge

Reynolds J, Souty-Grosset C, Richardson A (2013) Ecological roles of crayfish in freshwater and terrestrial habitats. Freshwater Crayfish 19(2):197–218

Richman NI, Böhm M, Adams SB, Alvarez F, Bergey EA, Bunn JJ, Burnham Q, Cordeiro J, Coughran J, Crandall KA, Dawkins KL (2015) Multiple drivers of decline in the global status of freshwater crayfish (Decapoda: Astacidea). Philos Trans R Soc B: Biol Sci 370(1662):20140060

Roessink I, Van der Zon KA, De Reus SR, Peeters ET (2022) Native European crayfish *Astacus astacus* competitive in staged confrontation with the invasive crayfish *Faxonius limosus* and *Procambarus acutus*. PLoS One 17(1):e0263133

Roljić R, Nikolić V, Savić N (2020a) Morphological variability and sexual dimorphism of stone crayfish *Austropotamobius torrentium* from the Marjanovića stream (Bosnia and Herzegovina). Kragujevac J Sci 42:143–150

Roljić R, Nikolić V, Savić N (2020b) Morphological variability and sexual dimorphism of noble crayfish *Astacus astacus* from the Balkana Lake. Bull Natural Sci Res 10(2):7–11

Roljić R, Nikolić V, Savić N (2020c) Investigations of variability of morphometric characteristics in stone crayfish in order to preserve autochthonous genome. Quality of Life: A Multi-Disciplinary Journal of Food Science, Environmental Science & Public Health, 11

Roljić R, Savić N, Nikolić V (2022) First Insight into Distribution and Ecology of *Austropotamobius torrentium* (Schrank, 1803) (Decapoda: Astacidae) in the North-Western Bosnia and Herzegovina. Acta Zoologica Bulgarica 74(2):265–271

Rotten Carrot Production (2022) The crayfish tale – a short & entertaining educational film. In: YouTube. https://www.youtube.com/watch?v=oZLIaCJOwPo

Samardžić M, Lucić A, Maguire I, Hudina S (2014) The First Record of the Marbled Crayfish (*Procambarus fallax* (Hagen, 1870) *f. virginalis*) in Croatia. Crayfish News 36, 4:4–4

Šarić I, Klobučar G, Podnar M, Štambuk A, Maguire I (2018) Molecular phylogeny of branchiobdellidans (Annelida: Clitellata) and their host-epibiont association with *Austropotamobius* freshwater crayfish. Invertebr Syst 32(1):55–68

Schrimpf A, Schulz HK, Theissinger K, Pârvulescu L, Schulz R (2011) The first large-scale genetic analysis of the vulnerable noble crayfish *Astacus astacus* reveals low haplotype diversity in central European populations. Knowl Manag Aquat Ecosyst 401:35

Schrimpf A, Theissinger K, Dahlem J, Maguire I, Parvulescu L, Schulz HK, Schulz R (2014) Phylogeography of noble crayfish (*Astacus astacus*) reveals multiple refugia. Freshw Biol 59(4):761–776

Simčič T, Pajk F, Vrezec A, Brancelj A (2012a) Size scaling of whole-body metabolic activity in the noble crayfish (*Astacus astacus*) estimated from measurements on a single leg. Freshw Biol 57(1):39–48

Simčič T, Pajk F, Brancelj A, Vrezec A (2012b) Preliminary multispecies test of a model for non-lethal estimation of metabolic activity in freshwater crayfish. Acta Biol Slov 55:15–27

Simčič T, Pajk F, Jaklič M, Brancelj A, Vrezec A (2014) The thermal tolerance of crayfish could be estimated from respiratory electron transport system activity. J Therm Biol 41:21–30

Simić V, Simić S, Petrović A, Paunović M, Tomović J (2000) The strategy and the results of the ex situ conservation of the hydro-bionates in the aquarium of the institute for biology and ecology, Kragujevac – Serbia. Paper presented at the BALWOIS 2010, Ohrid, Republic of Macedonia, 25–29 May 2010

Simić V, Paunović M, Petrović A, Rajković M (2008) Crayfish of Serbia and Montenegro – the population status and the level of endangerment. Crustaceana 81(10):1153–1176

Simić V, Maguire I, Rajković M, Petrović A (2015) Conservation strategy for the endangered crayfish species of the family Astacidae: the ESHIPPO crayfish model. Hydrobiologia 760(1): 1–13

Śmietana N, Panicz R, Sobczak M, Śmietana P, Nędzarek A (2020) Spiny-Cheek Crayfish, *Faxonius limosus* (Rafinesque, 1817), as an alternative food source. Animals 11(1):59

Smital T, Ahel M (2015) Ecotoxicological characterization of the Sava River: biomarker responses and biological assays. In: Milačič R, Ščančar J, Paunović M (eds) The Sava River. Springer, Berlin, pp 177–200

Souty-Grosset C, Reynolds JD (2009) Current ideas on methodological approaches in European crayfish conservation and restocking procedures. Knowl Manag Aquat Ecosyst 394–395:01

Souty-Grosset C, Anastacio PM, Aquiloni L, Banha F, Choquer J, Chucholl C, Tricarico E (2016) The red swamp crayfish *Procambarus clarkii* in Europe: impacts on aquatic ecosystems and human well-being. Limnologica 58:78–93

Subchev MA (2011) First record of Branchiobdella Odier, 1823 (Annelida: Clitellata) in Albania and an overview of the geographic distribution of *Branchiobdella hexodonta* Gruber, 1882 in Europe. Acta Zool Bulg 63(1):109–112

Svanberg I, Locker A (2020) Ethnoichthyology of freshwater fish in Europe: a review of vanishing traditional fisheries and their cultural significance in changing landscapes from the later medieval period with a focus on northern Europe. J Ethnobiol Ethnomed 16:68

Swahn JÖ (2004) The cultural history of crayfish. Bulletin Français de la Pęche et de la Pisciculture 372–73:243–251

Taylor CA (2001) Taxonomy and conservation of native crayfish stocks. Biology of freshwater crayfish

Topić Popović N, Sauerborn Klobučar R, Maguire I, Strunjak-Perović I, Kazazić S, Barišić J, Jadan M, Klobučar GIV, Čož-Rakovac R (2014) High-throughput discrimination of bacteria isolated from *Astacus astacus* and *A. leptodactylus*. Knowl Manag Aquat Ecosyst 413:04. https://doi.org/10.1051/kmae/2014005

Trontelj P, Machino Y, Sket B (2005) Phylogenetic and phylogeographic relationships in the crayfish genus *Austropotamobius* inferred from mitochondrial COI gene sequences. MPE 34: 212–226

Trožić-Borovac S (2011) Freshwater crayfish in Bosnia and Herzegovina: the first report on their distribution. Knowl Manag Aquat Ecosyst 401:26

Trožić-Borovac S (2012) Distribucija bjelonogog raka *Austropotamobius pallipes* (Lereboullet, 1858) species complex (Astacoidea Latreille, 1802; Astacidae Latreille, 1802) u Bosni i

Hercegovine. In "Struktura i dinamika ekosistema Dinarida-stanje, mogućnosti i perspektive. Posebno izdanje CXLIX. Akademija nauka Bosne i Hercegovine Sarajevo, 23, 153–166

Trožić-Borovac S, Macanović A, Škrijelj R (2012) The morfometrics characterics and condition index of *Austropotamobius pallipes* in the Neretva river basin. Radovi Šumarskog fakulteta Univerziteta u Sarajevu 42(2):13–30

Trožić-Borovac SR, Škrijelj A, Vesnić S, Đug A, Mušović S, Šljuka B, Borovac M, Gajević (2019) Negative effects of introducing allochthonous species *Pacifastacus leniusculus* (Dana, 1852) into aquatic ecosystems of Bosnia and Herzegovina. In: Ivkovic M, Stankovic I, Matonickin-Kepcija R, Gracan R (eds) Book of abstracts – 3rd symposium of freshwater biology (Zagreb: Croatian Association of Freshwater Ecologists), pp. 33–33

Vrezec A, Ocepek M, Jenčič V (2013) *Aphanomyces astaci* in wild crayfish populations in Slovenia: first report of persistent infection in a stone crayfish *Austropotamobius torrentium* population. Dis Aquat Org 103(2):157–169

Wang Q, Cheng L, Liu J, Li Z, Xie S, De Silva SS (2014) Freshwater aquaculture in Pr China ends and prospects. Rev Aquac 7(4):283–302. https://doi.org/10.1111/raq.12086

Westman K (1991) The crayfish fishery in Finland – its past, present and future. Finn Fish Res 12: 187–216

Westman K, Pursiainen M, Westman P (1990) Status of crayfish stock, fisheries, diseases and culture in Europe. Report of the FA0 European Inland Fisheries Advisory Commission (EIFAC) Working Party of Crayfish. Helsinki, 206 p

Wickins JF, Lee DOC (2008) Crustacean farming: ranching and culture. John Wiley & Sons, Hoboken

Žganec K, Ćuk R, Tomović J, Lajtner J, Gottstein S, Kovačević S, Hudina S, Lucić A, Mirt M, Simić V, Simčič T (2018) The longitudinal pattern of crustacean (Peracarida, malacostraca) assemblages in a large south European river: bank reinforcement structures as stepping stones of invasion. In: Annales de Limnologie – international journal of limnology, vol 54. EDP Sciences, p 15

Žganec K, Lajtner J, Ćuk R, Crnčan P, Pušić I, Atanacković A, Kralj T, Valić D, Jelić M, Maguire I (2020) Alien macroinvertebrates in Croatian freshwaters. Aquat Invasions 15(4):593–615

Živić NV, Atanacković A, Milošević S, Milosavljević M (2014) The distribution of Astacidae (Decapoda) Fauna in Kosovo and Metohija, Serbia. Water Research and Management

Živić I, Stojanović K, Marković Z (2022a) Springs and headwater streams in Serbia: the hidden diversity and ecology of aquatic invertebrates. In: Pešić V, Milošević D, Miliša M (eds) Small water bodies of the Western Balkans. Springer, Cham, pp 189–210

Živić I, Božanić M, Todorović N, Stojanović K (2022b) Branchiobdella Odier, 1823 (Annelida: Branchiobdellea), a new genus for the Fauna of Serbia. Acta Zoologica Bulgarica 74(2): 325–327

Zorić K, Ilić M, Marinković N, Đuretanović S, Milošković A, Radojković N, Marković V, (2018) New records of native crayfish *Austropotamobius torrentium* (Schrank, 1803) in Serbia. Paper presented at the 8th international conference, Faculty of Agriculture, University of Belgrade, Serbia, 13–15 June 2018, pp 418–422

Zorić K, Atanacković A, Ilić M, Csányi B, Paunović M (2020a) The spiny-cheek crayfish *Faxonius limosus* (Rafinesque, 1817) (Decapoda: Cambaridae) invades new areas in Serbian inland waters. Acta Zoologica Bulgarica 72(4):623–627

Zorić K, Atanacković A, Tomović J, Vasiljević B, Tubić B, Paunović M (2020b) Diversity of alien macroinvertebrate species in Serbian waters. Water 12(12):3521

Salmonid Fish Species: Opportunities for Sustainable Use under Multiple Pressures and Current Climatic Change

Johannes Schöffmann and Saša Marić

Abstract The chapter covers the current status of salmonid fishes in the Western Balkans in the Adriatic, Aegean, and Danube drainage basins on the territories of Slovenia, Croatia, Bosnia and Herzegovina, Montenegro, Serbia, and North Macedonia. Salmonids (family Salmonidae) are freshwater and anadromous fish native to the northern hemisphere. The family is composed of three subfamilies: Salmoninae, Thymallinae, and Coregoninae. Only species of the families Salmoninae and Thymallinae occur naturally in the inland waters of the Western Balkans; these species do not normally migrate to the sea in this area. Of the Salmoninae, in addition to the Danube salmon (*Hucho hucho*), a large number of trout species (*Salmo* spp.) can be found here, which is hard to find anywhere else in their distribution range. Two species of the Thymallinae are represented, the European grayling (*Thymallus thymallus*), which has its southernmost distribution limit in this region, and the Adriatic grayling (*Thymallus aeliani*) from the northern Adriatic basin. In addition, exotic salmonid species—e.g. rainbow trout (*Oncorhynchus mykiss*), brook char (*Salvelinus fontinalis*), and lake char (*Salvelinus umbla*)—have been introduced and naturalized in a number of water bodies.

In the inland waters of the Western Balkans, salmonids play an important economic and social role in both commercial and recreational fisheries. Unfortunately, this has long since led to the fact that many of the stocks have been severely depleted, in some cases almost destroyed, by the pressure of fishing. To counteract this, stocking with hatchery fish is common in many places. However, stocking with non-native brown trout has often resulted in genetic introgression into the autochthonous populations of *Salmo* trouts and loss of their genetic integrity. Other threats to the survival of salmonid populations include water pollution from agriculture and sewage, as well as dam construction, which has particularly increased recently. The

J. Schöffmann
St. Veit/Glan, Austria

S. Marić (✉)
Institute of Zoology, Faculty of Biology, University of Belgrade, Belgrade, Serbia
e-mail: sasa@bio.bg.ac.rs

V. Simić et al. (eds.), *Ecological Sustainability of Fish Resources of Inland Waters of the Western Balkans*, Fish & Fisheries Series 43,
https://doi.org/10.1007/978-3-031-36926-1_9

increased demand for drinking water as a result of the growing tourism industry poses an additional threat to aquatic ecosystems and could be exacerbated by climate change.

Keywords Salmonids · Western Balkans · Current state · Threats · Conservation

1 Commercial and Social Importance of Salmonid Species

Brown trout (*Salmo* spp.) are important for sport fishing throughout the Balkans (Simonović et al. 2007; Snoj et al. 2009; Marić et al. 2022), as are softmouth trout and marble trout where they are still plentiful. Some river sections of the Neretva River in Bosnia-Herzegovina still have fairly good softmouth trout stocks, at least so far (Glamuzina et al. 2018). Based on the relevant literature (e.g. Povž et al. 1996; Turin et al. 2006) and the authors' own observations, it can be assumed that the largest stock of genetically pure marble trout is found in the Soča River drainage in Slovenia. In the rivers of the Danube basin, grayling and huchen, which are particularly appreciated by anglers, also play an important role (Marić et al. 2009, 2011, 2012; Veličković et al. 2023) (Fig. 1). The non-native rainbow trout has been established in several waters and now represents a large part of the catch in some places. This species is also the main source for edible fish production in aquaculture, while brown trout—predominantly of the Atlantic and Danubian strains—are mainly farmed for stocking purposes (Marić et al. 2010). Another alien salmonid sometimes kept in aquaculture is the brook char. Where these fish are kept in floating cages, such as in hydropower reservoirs along the Neretva or in Lake Skadar, they sporadically find their way into open water (Schöffmann et al. 2019). In general, however, the brook char can only be found in a few rivers and lakes of the Western Balkans and is therefore of little importance for fishing. The lake char, which originates from Central Europe, has become naturalized in several high mountain lakes, most of which were previously fishless. They, too, are only of local importance for anglers.

Native trout, caught by legal anglers or illegal fishermen, who often use casting nets or other prohibited fishing gear, are sometimes sold to local restaurants in some areas, e.g. Bosnia-Herzegovina, Montenegro, Albania (Figs. 2 and 3). In Bosnia-Herzegovina and Serbia, in rare cases, grayling or huchen are even offered in restaurants. A significant commercial fishery for salmonids is actually only practiced in Lake Ohrid in North Macedonia and Albania.

2 Salmonid Species Native to the Western Balkans

Most of the salmonid taxa found in the inland waters of the Western Balkan belong to the genus *Salmo*. In the entire natural range of this genus, which stretches across Europe, Northwest Africa, Asia Minor to Central Asia, there is no other region with

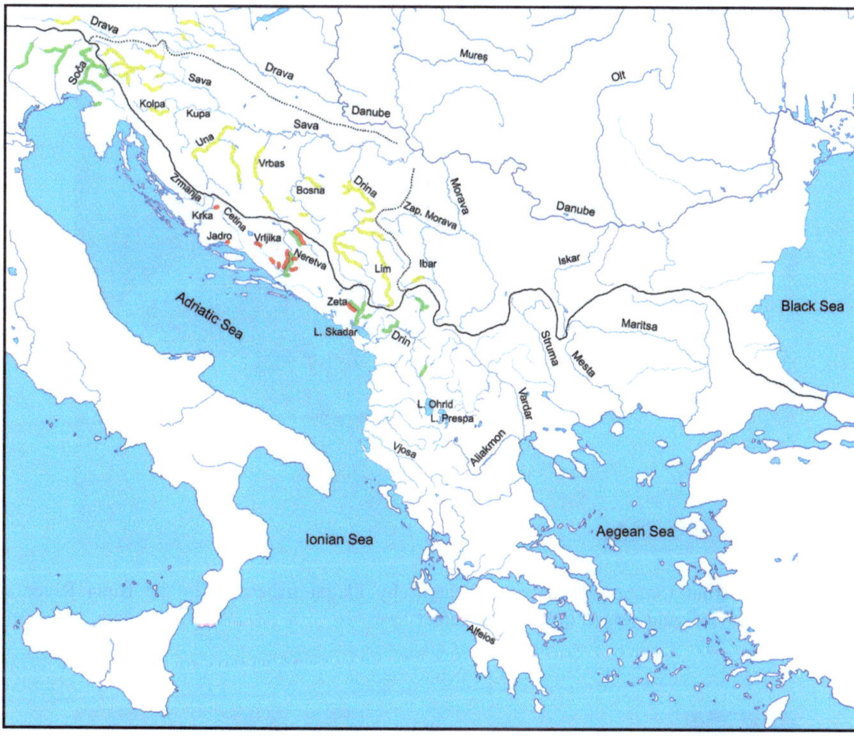

Fig. 1 Hydrographic map of the Balkan Peninsula—distribution of four of the most endangered salmonid species. Green: *Salmo marmoratus*; red: *Salmo obtusirostris*; yellow: *Hucho hucho* and (largely) *Thymallus thymallus*. Black line: drainage divide between Mediterranean and Black Sea basins. Dotted line: distribution boundary between *Salmo taleri* (Sava drainage) and *Salmo labrax* (rest of the Danube basin)

such a large diversity of trout forms and species. This remarkable diversity is reflected in the large number of species and subspecies that have been described there since the middle of the nineteenth century.

2.1 *Brown Trout* (Salmo *spp.)*

Various, often controversial, taxa have been described for the fish commonly referred to as brown trout. The largest number of endemic trout species are known from the catchment area of the Adriatic Sea. The most important of these, as far as considered as valid, are *Salmo farioides* (Karaman, 1938) and *S. montenigrinus* (Karaman, 1933) for river and stream-dwelling brown trout; *S. aphelios* (Kottelat, 1997); *S. balcanicus* (Karaman, 1927); *S. letnica* (Karaman, 1924); *S. lumi* (Poljakov et al. 1958) from Lake Ohrid and *S. peristericus* (Karaman, 1938) from Lake Prespa and its tributaries. Two questionable species from the Adriatic basin

Fig. 2 Native trout (*Salmo obtusirostris*) caught by illegal fishermen in the Buna River in Bosnia-Herzegovina, and served in local restaurant

Fig. 3 Native trout (*Salmo farioides*) caught by illegal fishermen in the upper part of Valbona River in Albania, and served in local restaurant

Fig. 4 Brown trout (*Salmo farioides*), ca. 32 cm TL; Shala River, Drin drainage, Albania. Photo by Johannes Schöffmann

should also be mentioned: *S. zrmanjaensis* (Karaman, 1938) from the Zrmanja River, Croatia, and *S. visovacensis* (Taler, 1950) from Lake Visovac of the lower Krka drainage, Croatia.

From the other large Mediterranean catchment area, that of the Aegean, the following species are known: *S. macedonicus* (Karaman, 1924) and *S. pelagonicus* (Karaman, 1938), both from the Vardar drainage in North Macedonia and Greece.

Two species have been described from the Danube basin: *S. labrax* (Pallas, 1814), Black Sea basin of Europe, except Sava River drainage; *S. taleri* (Karaman, 1933), upper Zeta River in Montenegro and the Sava River drainage.

All native brown trout from the Mediterranean catchment areas can be assigned to three phylogenetic lineages, Adriatic (AD), Mediterranean (ME), and *marmoratus* (MA), while those from the Danube basin belong to the Danubian (DA) lineage (Bernatchez et al. 1992; Bernatchez 2001). The southernmost occurrence of native brown trout on the Balkan Peninsula is found in the Alfeios River drainage on the Peloponnese, whereby Greece does not belong to the areas to be treated here.

Some details on the main brown trout species of the Western Balkans:

Salmo farioides (Karaman, 1938) (Fig. 4). The distribution area of this stream-dwelling brown trout extends over the Mediterranean (Adriatic Sea, Ionian Sea) of the Balkan Peninsula from the Zrmanja River, Croatia, in the north to the Peloponnese, Greece, in the south. Karaman (1938) described this trout, listing a series of Adriatic tributaries as the type locality: the Krka, Neretva, and from the Drim drainage the Radika in North Macedonia and the Bistrica in Serbia, in addition to the small tributaries of Lake Ohrid and two tributaries of Lake Skadar (Cijevna/Cem and Rijeka Crnojevića). Later, other authors extended the distribution to include southwestern Greece (Delling 2003; Kottelat and Freyhof 2007). As shown by mtDNA analysis, a core range of *S. farioides* (Neretva and Drim-Skadar river systems) is strongly associated with the distribution of a distinct haplogroup designated as the Balkan cluster (Snoj et al. 2009). Karaman (1938) described the Zrmanja trout as a new subspecies, *S. farioides zrmanjaensis* (Fig. 5), which was later often considered a separate species, and Taler (1950) described the trout from Lake Visovac as *S. visovacensis*. According to the recent nDNA analysis, both taxa

Fig. 5 Brown trout (*Salmo farioides zrmanjaensis*), ca. 35 cm TL; Zrmanja River, Croatia. Photo by Johannes Schöffmann

Fig. 6 Brown trout (*Salmo macedonicus*), ca. 24 cm TL; Belica River, upper Vardar drainage, North Macedonia. Photo by Johannes Schöffmann

represent monophyletic subdivisions standing outside the so-called Adriatic brown trout clade or Balkan cluster (Pustovrh et al. 2014).

Salmo montenigrinus (Karaman, 1933). Karaman first described this species from the Morača (tributary of Lake Skadar) in Montenegro and suspected a close relationship to softmouth trout due to the high body, blunt snout, and short maxilla. Based on morphological differences, in 1957 Karaman classified the brown trout from the lower Radika as *S. montenigrinus* and from the upper Radika as *S. farioides*. According to later authors, *S. montenigrinus* occurs also in the Neretva drainage, although there it may actually be a hybrid form of brown trout and softmouth trout (Delling 2003; Razpet et al. 2007).

Salmo macedonicus (Karaman, 1924) (Fig. 6). This trout was described from the upper Vardar drainage in North Macedonia and Serbia and its occurrence was also suggested in the Struma and Mesta catchments of Serbia and Bulgaria. Based on genetic data, the trout from the Došnica, a tributary of the lower reaches of the Vardar in North Macedonia, can also be assigned to this species (Marić et al. 2017).

Salmo pelagonicus (Karaman, 1938; Fig. 7) was described from three streams emptying into the Crna Reka, a right tributary to the lower reaches of the Vardar in North Macedonia. Later investigations also identified this species in three tributaries

Fig. 7 Brown trout (*Salmo pelagonicus*), ca. 28 cm TL; Bela Reka, tributary of Crna Reka, lower Vardar drainage, North Macedonia. Photo by Johannes Schöffmann

of the Sakoulevas River (Eleška Reka, in Macedonian), which flows from Greece into the Crna Reka (Apostolidis et al. 2008; Schöffmann 2016). Some authors also suspect the species in the upper Aliakmon drainage in northern Greece (Kottelat and Freyhof 2007).

Salmo peristericus (Karaman, 1938). The Prespa trout was originally described as a new subspecies, *S. macedonicus peristericus*. The description refers to four specimens from Great Prespa Lake and one specimen from the Brajčinska River, which flows from the Baba Mountains east of the lake. Today, the Prespa trout can only be found sporadically in the Great Prespa Lake, but occurs, albeit in small populations, in two other tributaries in North Macedonia (Leva Reka, Kranska Reka) and in a stream in Greece (Agios Germanos). The latter originally flowed into the Little Prespa Lake and the trout probably only colonized the stream after an artificial connection to the Great Prespa Lake was established (Berrebi et al. 2013).

Salmo taleri (Karaman, 1932) was described from the upper reaches of the Zeta River in Montenegro. The river belongs to the Adriatic catchment. The upper and the lower sections of the river are connected by subterranean channels. Based on morphological features, Karaman concluded that the upper Zeta brown trout belongs to the Danubian strain and extended the distribution of *S. taleri* to the entire Sava drainage (Fig. 8). Whether the trout arrived in the upper Zeta from undocumented translocation or ancient connections with the Drina River system in the Danube basin has not been clarified. No comprehensive genetic data are available for *S. taleri* (Schöffmann et al. 2019). However, the reconstruction of the evolutionary history from external morphology suggests a clear distinction between *S. taleri* from the western Danube drainage and brown trout of the southern Danube drainage, the Velika Morava (Great Morava) River system (Simonović et al. 2007). There are also a few lacustrian brown trout populations in the range of *S. taleri* (Fig. 9), such as in Lake Bohinj (Bohinjsko jezero) and Lake Bled (Blejsko jezero) in Slovenia and in Lake Plav (Plavsko jezero) in Montenegro. In Lake Bled, specimens of lake trout with a weight of about 10 kg and a length of up to one metre are regularly caught by fishermen. The Plavsko lake trout, known by the name "blatnjača" (mud fish), shows

Fig. 8 Brown trout (*Salmo taleri*), ca. 34 cm TL; Mala Pišnica, upper Sava drainage, Slovenia. Photo by Johannes Schöffmann

Fig. 9 Lacustrian Brown trout (*Salmo taleri*), ca. 61 cm TL; Jesenice, Sava Dolinka system, Slovenia. Photo by Tine Pokorn

a considerable increase in length and weight, especially during the first five years of life, and can reach a weight of up to 12 kg. Due to its geographic and ecological isolation, this lake trout has developed particular morphological characteristics that distinguish it from other brown trout populations in the region (Janković 1963).

Salmo labrax (Pallas, 1814) was first described with the type locality of Sevastopol and the river Biyuk-ozen on the Crimean Peninsula. Later many authors used this name, sometimes as subspecies from *S. trutta*, for resident and anadromous brown trout from the entire Danube basin (Fig. 10). According to mtDNA haplotypes, all trout populations native to the southeast European Black Sea basin are thought to cluster into the Danubian phylogenetic lineage. However, some haplotypes from the lower Danube drainage and southern Black Sea basin (northern Turkey) belong to a specific haplogroup (DA-BS) that differs substantially from the prevailing subclade of the Danubian lineage (Da-ES), consisting of haplotypes found in most of the Danube drainages and in the Caspian and Aral Sea basins (Kohout et al. 2013; Veličković et al. 2023). In addition, increased sampling, particularly in Serbia, led to the identification of a new haplogroup (DA-INT) occupying an intermediate position with respect to DA-ES and

Fig. 10 Brown trout (*Salmo labrax*), ca. 25 cm TL; Lukovska River, Southern Morava drainage, Serbia. Photo by Saša Marić

Fig. 11 Marble trout (*Salmo marmoratus*), Soča River, Slovenia. Underwater photo by Benedikt Schöffmann

DA-BS haplogroups. DA-INT haplotypes are spread in a very diffuse way and are very rare. Namely, two haplotypes of this haplogroup were recorded in the Southern Serbia, while two others have been reported from a tributary to the Upper Danube in Austria and the eastern part of the Black Sea basin in Turkey (Veličković et al. 2023).

Salmo dentex (Heckel, 1852). The name dentex trout comes from the local vernacular names "zubatak" (on the Neretva), "zubara" (on the Krka), and "zubača" (on the Cetina), all meaning something like "toothed trout" (Ocvirk and Voljč 1981). However, there is no clear definition for this trout, except that it has a large head and a wide-split mouth with large teeth and feed primarily on fish. The questionable species cannot always be assigned to the brown trout either, since it is, for example, a marble trout in the lower Neretva (more on this below, 2.3.).

Fig. 12 Marble trout (*Salmo marmoratus*), ca. 29 cm TL; Cijevna River, Lake Skadar drainage, Montenegro. Photo by Johannes Schöffmann

2.2 *Marble Trout (*Salmo marmoratus *Cuvier, 1829)*

Differs from the other trout in its distinctive marbled markings on the head, back, and sides (Fig. 11). Its capacity to large growth is also remarkable; up to 140 cm in length and up to 30 kg in weight (Marić 1995; Povž et al. 1996). There are two population groups of marble trout, which differ in particular by their mitochondrial genome: The northern populations inhabit the catchment areas of the rivers that flow into the northern Adriatic Sea, from the Po River drainage in Italy in the west to the Slovenian rivers Soča and Rižana in the east, and exhibit the *marmoratus* mtDNA haplogroup (MA haplotypes). The southern populations are found in the Neretva River drainage in Bosnia-Herzegovina and the river system of Lake Skadar in Montenegro (Fig. 12), and the Drin River system in Serbia, Albania, and North Macedonia; only AD haplotypes have been detected in these populations of the southern Adriatic basin (Snoj et al. 2010). Nevertheless, phylogenetic reconstruction based on nDNA single nucleotide polymorphism analysis revealed that northern and southern marble trout are conspecific and indicate that the marble trout represents a monophyletic unit of the same rank as brown trout (Pustovrh et al. 2011, 2012, 2014).

2.3 *Dentex Trout (*Salmo dentex *(Heckel, 1852))*

Dentex trout was first described from Krka, Cetina, and Neretva rivers in the region of modern-day Croatia and Bosnia-Herzegovina. Later the range was extended to other areas in the Western Balkans, such as the Morača-Lake Skadar system in Montenegro or the Vjosa (Aoos) River drainage in Greece. Dentex trout are always large growing (>60 cm TL), predatory fish and are usually found in big rivers, often in connection with a lacustrine habitat. However, their taxonomic status is questionable because genetic studies have shown that the dentex trout of the lower course of the Neretva is actually a marble trout, while the dentex trout in the Lake Skadar system is closely related to the respective brown trout populations (Snoj et al. 2010).

*2.4 Softmouth Trout (*Salmo obtusirostris *(Heckel, 1852))*

Softmouth trout constitutes a well-defined clade within the genus *Salmo* and was first described by Heckel in 1852 (Snoj et al. 2002). The species also differs significantly from other trout in terms of morphological characteristics, in particular the inferior mouth with fleshy lips and a short but deep maxilla that only reaches below the middle of the eye. Corresponding to the four river systems in which the softmouth trout occurs, four subspecies have been described: *Salmo obtusirostris oxyrhynchus* (Steindachner, 1882) from the Neretva drainage; *Salmo obtusirostris krkensis* (Karaman, 1927) from the upper Krka River in Croatia; *Salmo obtusirostris salonitanus* (Karaman, 1927) from the Jadro near Split in Croatia (Fig. 13); *Salmo obtusirostris zetensis* (Hadžišče, 1960) from the lower Zeta River in the Morača-Lake Skadar system in Montenegro. The softmouth trout from the Vrljika River in Croatia (Fig. 14), which drains through an underground connection to the Neretva drainage, could be considered another subspecies (Snoj et al. 2008).

*2.5 Belvica Trout (*Salmo ohridanus *Steindachner, 1892)*

Belvica trout represents another well-defined clade of the genus *Salmo*, which has evolved separately approximately four million years ago (Sušnik et al. 2006; Veličković et al. 2023). The relatively small (<35 cm TL) and slender, small-mouthed, silvery fish is endemic to Lake Ohrid (Fig. 15), where it lives outside the spawning season at depths of 25 to 60 m and feeds on zooplankton and benthic invertebrates. Spawning takes place from the middle of December until the

Fig. 13 Softmouth trout (*Salmo obtusirostris salonitanus*), ca. 22 cm TL; Žrnovnica River, Croatia; translocated population. Photo by Johannes Schöffmann

Fig. 14 Softmouth trout (*Salmo obtusirostris* ssp.), ca. 33 cm TL; Vrljika River, Trebižat/Neretva drainage, Croatia. Photo by Johannes Schöffmann

beginning of February, close to the shoreline in 8–9 m of water (Rakaj and Flloko 1995). The belvica occurs in sympatry with the Ohrid trout, which was divided into four subspecies, later treated as separate species (see above, 2.1).

2.6 *Huchen (*Hucho hucho *(Linnaeus, 1758))*

Like the *Salmo* genus, the *Hucho* genus is a member of the subfamily Salmoninae. The huchen, also known as Danube salmon (Fig. 16), naturally only occurs in the Danube basin. Its distribution in the Western Balkans extends across the drainages of the Sava, Drava, and Great Morava rivers in Slovenia, Croatia, Bosnia-Herzegovina, Montenegro, and Serbia. The most important self-sustaining huchen populations in this region are in the upper Sava drainage, in the upper course of the Kolpa/Kupa, in the drainages of the Una, Vrbas, and Bosna rivers, in the Drina with its tributaries Tara and Lim, as well as in the Ibar River of the Great Morava drainage (Freyhof et al. 2015). Lake Plav and its outflow, the Lim River, constitute the southernmost

Fig. 15 Belvica trout (*Salmo ohridanus*), ca. 22 cm TL; Lake Ohrid, Albania. Photo by Spase Shumka

Fig. 16 Huchen (*Hucho hucho*); Sava Bohinjka, upper Sava drainage, Slovenia. Underwater photo by Benedikt Schöffmann

distribution limit of the species. The huchen is the largest representative of Europe's salmonids. It can reach a total length of up to 183 cm and a weight of up to 60 kg (Holčík et al. 1988); only its closest relative in Siberia, the taimen (*Hucho taimen*), is larger. The largest records from the Balkans include a 58 kg (128 lb) specimen caught in January 1938 by the Bosnian railway worker and angler Halil Sofradžija on the Drina River near the town of Ustikolina in Bosnia-Herzegovina. This case was recorded on photograph and a reportage was later published in the Serbian daily newspaper Politika on February 23, 1938. Dževat Šarkinović, angler from Plav in Montenegro, caught a huchen weighting 41.3 kg (91 lb), 145 cm (57 in) long, in late spring of 1985 in Lake Plav (Fishing Worldrecords 2018). Genetic investigations indicated large-scale geographic structure between the north-western (Austria and Slovenia) distribution area and eastern (Slovakia and Ukraine) or southern (Bosnia-Herzegovina, Montenegro and Serbia) sample sites (Weiss et al. 2011; Marić et al. 2014).

2.7 European Grayling (*Thymallus thymallus,* Linnaeus, 1758)

The genus *Thymallus* belongs to the subfamily Thymallinae. The European grayling (Fig. 17) is the most widespread of the three accepted grayling species native to Europe. Its range extends from the Barents Sea basin west of the Urals, throughout the Atlantic basin to Great Britain and France (except the Loire drainage), the Rhône drainage, the Black Sea basin, and the northern Caspian basin (Volga and Ural river drainages) (Kottelat and Freyhof 2007; Persat et al. 2019). The distribution of the

Fig. 17 European grayling (*Thymallus thymallus*), ca. 25 cm TL; Rzav River, West Morava drainage, Serbia. Photo by Saša Marić

European grayling in the Balkans largely corresponds to that of the huchen. In the Balkans, the European grayling was introduced outside of its natural range, namely in the upper reaches of the Neretva and in the Morača, a tributary of Lake Skadar, as well as in the Soča drainage. The grayling in the Danube basin of the Western Balkans form their own phylogenetic clade (Balkan clade) within the European populations and can be divided into two groups: a northern group with haplotypes found in the Slovenian part of the Danube drainage, and a southern group, consisting of Bosnia-Herzegovina, Montenegro, and Serbia (Marić et al. 2011, 2012).

2.8 Adriatic Grayling (*Thymallus aeliani,* Valenciennes, 1848)

Bianco (2014) reactivated the name *T. aeliani* for Adriatic grayling, while Bravničar et al. (2020) genetically confirmed them as a distinct species. It occurs in the northern Adriatic basin of Italy and Slovenia. Most of the original populations have long since been strongly mixed by stocking grayling from other regions; that also applies to the population of the Soča River drainage in Slovenia (Sušnik et al. 2004) (Fig. 18). The only surviving genetically pure population of Adriatic grayling is found in the Sesia River in Italy (Bravničar 2021).

3 Current State and Threats

The Western Balkans continue to have to most free-flowing and largely intact rivers in Europe. Unlike many other parts of Europe, most of the decline in salmonids in the Western Balkans is due to overfishing rather than habitat destruction. However, this situation appears to be taking a turn for the worse. The increasing obstruction of

Fig. 18 Adriatic grayling (*Thymallus aeliani*), ca. 35 cm TL; Tolminka River, Soča drainage, Slovenia. Photo by Saša Marić

the rivers by dams is a serious problem. According to the NGO Riverwatch, in 2020, 1480 hydropower plants were operating in the Balkans, more than double what existed in 2015, and 108 plants were under construction. An additional 3431 hydro projects are planned for construction across the region, 92% of which are small (see Figs. 19 and 20). This development will undoubtedly have a direct impact on salmonid populations and other fish communities due to loss of habitat, habitat fragmentation, and disruption of the hydrological regime (Weiss et al. 2018). A

Fig. 19 Garvanica water intakes of hydroelectric power plant Džep. Photo by Ninoslav Kostić

Fig. 20 Hydroelectric power plant (power plant and tailrace) on the Ljutina River, Serbia. Photo by Đorđe Gajić

prime example of changes in fish assemblages due to such anthropogenic influences is the Sava River, which until the 1950s was a typical trout zone from the source to the mouth of the Krka River, i.e. about 200 km. This has been changed during the second half of the twentieth century and the trout zone only remained on the first 100 km of the upper reaches of Sava River (Piria et al. 2019).

The discharge of water from the headwaters for the operation of hydroelectric power plants often threatens the habitat of isolated residual populations of already endangered brown trout species such as *Salmo macedonicus* and *S. pelagonicus* in the Vardar River drainage. This applies to the populations of the Došnica and Konjarska Reka, respectively, in North Macedonia, which have so far been spared from artificial stocking and preserved in their pure form up to now. The status of the last remaining pure population of *S. pelagonicus* in the Vardar drainage of Greece is unknown since the construction of the Papadia dam in the source region of the Skopos River (also known as Voras).

Another significant threat to aquatic ecosystems and habitats comes from agriculture in relation to water withdrawal and input of nutrients and fine sediments. In densely populated areas or regions developed for tourism, there is also wastewater from households and industry, which impairs the water quality through eutrophication or toxic substances. It is to be feared that all of these impacts as a result of anthropogenic activities will be exacerbated by ongoing climate change (Stefanidis et al. 2018). In addition, the impacts of climate change will most likely result in a significant alteration in water demand for irrigation and drinking water supply during the summer months with low discharge, which will reduce the availability of habitats

Fig. 21 Heavily polluted stream in southern Kosovo (This designation is without prejudice to position on status and is in the line with United Nations Security Council Resolution 1244 and the International Court of Justice Opinion on the Kosovo declaration of independence); a tributary of the Lepenec River, Vardar drainage. Photo by Johannes Schöffmann

for various aquatic species in mountain rivers that are particularly sensitive to climatic fluctuations (Papadaki et al. 2016). Moreover, in many places the rivers are still being misused for waste disposal. In addition to the aesthetic misery, plastic and chemical residues undoubtedly have a dangerous impact on ecosystems (see Figs. 21 and 22).

In order to counteract the fishing pressure and to strengthen the dwindling populations, restocking measures have been carried out for many decades. Since salmonids from other river drainages or even big catchment areas (e.g. Atlantic basin) were usually used for this purpose, the genetic integrity of the populations was often severely affected, in some cases completely destroyed. In some places, hybridization between two species occurred, such as in the Soča River between native marble trout and non-native brown trout (Sušnik Bajec et al. 2015), or the innate reproductive isolation mechanisms between sympatric species were suppressed, e.g. in the Neretva between local brown trout, marble trout, and softmouth trout (Razpet et al. 2007). It should be noted that a certain degree of interbreeding between coexisting native trout has always occurred, but the frequency of hybridization has not been high enough to break down the species barrier. However, any introduction of non-native counterparts may create a hybridization bridge between native trout resulting in overall panmixia and their genetic breakdown (Schöffmann et al. 2019).

The upper and the lower courses of the Neretva are separated by four large dams and reservoirs for electricity generation. On the upper reaches of the Neretva, a small hatchery run by the municipality of Konjic with the help of the University of Sarajevo is trying to breed authentic fish (local brown trout, marble and softmouth trout) for stock replenishment. However, there is lack of financial resources for

Fig. 22 Heavily polluted stream in Bosnia and Herzegovina; Bregava River, Neretva drainage. Photo by Saša Marić

efficient utilization, thus, not all three species can be reproduced regularly every year. In addition, some of the marble trout spawners used for breeding show typical hybridization characteristics, such as round spots on the sides or a lack of marbling pattern on the head and back. The marble trout stocks in the upper Neretva have hardly recovered despite a few occasional stocking measures, while the stocks of native brown trout and softmouth trout are in good condition, not least thanks to restrictive fishing regulations, e.g. catch and release on a roughly 3 km stretch of the upper reaches (Schöffmann 2018). Unfortunately, the influence of non-native trout is still clearly present. During a flood in November 2021, for example, several thousand non-native brown trout from a fish farm on the upper reaches accidentally got into the river. It also cannot be ruled out that the cheaper, non-native brown trout will still be used for stocking in certain river sections. For some time now, a large fish farm near the town of Konjic has been raising significant amounts of native brown trout for stocking purposes. Rearing this fry is more difficult than with domesticated brown trout. For example, the artificial light in the hall with the breeding tanks must be dimmed during feeding, otherwise the fish will not come to the surface of the water to eat (Safet Nezirević, personal communication). Stocking is carried out by the local fishing association, the costs of which are to be borne by the power plant operator according to the law. The latter fish farm is also currently attempting to establish a breeding stock of softmouth trout.

But there is also a special success story to report and an example where the interbreeding of species could be stopped and reversed. In the Soča River drainage, only a few small, genetically pure marble trout populations remained above natural

Fig. 23 Heavy growth of algae in the middle reaches of the Idrijca in summer 2022. Photo by Johannes Schöffmann

barriers. Spawners from such streams, especially the Zadlaščica, have been used for captive breeding and reintroductions programs. Later, suitable spawners from the so-called hybrid zones were also selected with the help of genetic analyses. Thanks to stocking measures with bred marble trout and the ban on stocking with brown trout, the situation has improved throughout the catchment area of the upper and middle Soča. In the hybrid zones, were hybrids and brown trout still dominated until the end of the twentieth century, mainly marble trout are found again today (Sušnik Bajec et al. 2015; Schöffmann 2021).

The effects of advancing climate change were particularly noticeable in the summer of 2022. Due to persistent droughts and high temperatures, not only the rivers of northern Italy had extremely low water levels, but also the Soča River drainage to the east. In the middle and lower section of the Idrijca, one of the largest tributaries of the Soča and an important salmonid area, the water temperatures rose to around 20 °C. Dense algae growth covered the river bottom, especially in the calmer parts and along the banks (Fig. 23). Most of the last genetically pure marble trout populations are found in some headwaters in Idrijca drainage. This is also the case in the Trebuščica tributary, where the water temperatures in the lower reaches rarely rises above 16 °C, but in July 2022 it was around 18 °C. The growth of algae is also considerable here and the population of cyprinids (e.g. *Barbus* sp. and *Phoxinus lumaireul*) has increased noticeably (Figs. 24 and 25). Moreover, human activities such as road construction can lead to increased nutrient input into the rivers, which also promotes algae growth and thus amplifies the negative climate-related influences. If such situations occur more frequently in the coming years, the composition of the fish fauna could change.

Fig. 24 On the lower reaches of the Trebuščica in July 2022. Photo by Johannes Schöffmann

Fig. 25 On the lower reaches of the Trebuščica in July 2017 (the same spot as in the Fig. 24, with the water level barely higher). Photo by Johannes Schöffmann

Excessive und uncontrolled fishing, often using traps (see Figs. 26 and 27), has drastically depleted or even wiped out native trout stocks in many regions of the Western Balkans over the last century. In particular, most populations (subspecies) of the softmouth trout are severely affected. The subspecies of the Krka River, *S. obtusirostris krkensis*, has in all probability already disappeared. The reasons for this are not entirely clear, as the habitat still appears to be intact. It is likely that

Fig. 26 Fish trap in a tributary of the lower Zrmanja River, Croatia, evidence of illegal fishing. Photo by Johannes Schöffmann

heavy fishing pressure, especially during the Balkan war and shortly thereafter, and the introduction of rainbow trout led to its extinction (Schöffmann 2021). A remnant population was found on an approximately 300 m long section of the upper reaches of the Krka, where a few specimens were still sighted earlier this century, but no specimens were present as of 2016. At this location there is also a large fish farm. While snorkelling in summer 2021, numerous dead rainbow trout were seen at the outflow of this facilities, apparently dying from disease (see Fig. 28). It can therefore be assumed that infectious diseases could have led to the final disappearance of the softmouth trout (Author's own observations).

The occurrences of the subspecies of the Jadro, *S. o. salonitanus*[1], and the Vrljika, *S. o.* ssp., are limited to short sections in the headwaters of the rivers and therefore particularly vulnerable. It seems, however, that these populations are still in a stable state thanks to protective measures such as a ban on fishing in the source areas and no-kill fly fishing in the lower sections. A breeding program was started on the Vrljika a couple of years ago, in which the hatched juveniles are raised to a length of 4–5 cm and then released into the river. Until recently, illegal fishing, especially with traps, was still common here. Efforts are also being made to remove the invasive

[1] As part of a new evaluation of Heckel's type material, a lectotype for the taxon *Salmo obtusirostris* was designated. Since this lectotype originates from the Jadro River, the Jadro softmouth trout becomes the nomotypical subspecies according to the rules of the International Commission on Zoological Nomenclature (Article 47.1): *Salmo obtusirostris obtusirostris* (Heckel 1852). The previously used name *S. o. salonitanus* (or *salonitana*) is therefore invalid (Hochleithner 2021). However, there are controversial opinions about the necessity of this name change.

Fig. 27 Fish trap with three live softmouth trout found in the Vrljika River by Johannes Schöffmann in 2004. Photo by Aleš Snoj

pike (*Esox lucius*) (Dalmatinski portal 2020). In the Neretva River, where a series of four large hydroelectric dams have radically changed the habitats between the upper and lower reaches, softmouth trout occur in significant numbers only in two zones of the watercourse, i.e. in the upper Neretva upstream from the town of Konjic and in the lower Neretva downstream from the town of Mostar, including the left side tributaries Buna and Bunica. In the other tributaries of the Neretva, softmouth trout have already become scarce or have disappeared altogether (Glamuzina et al. 2018). In the main course of the lower Neretva, the population is restricted to the river section between Mostar and Čapljina. However, it remains questionable whether this

Fig. 28 Trout carcasses in the upper Krka River near the outlet of a fish farm. Photo by Johannes Schöffmann

habitat is sufficient to sustain the survival of the species, particularly during the years of water flow changes and observed climatic and precipitation changes. The stocks in the Buna and its tributary Bunica are threatened by proposed dams designed to divert water to another watershed, the Trebišnjica River system in the southeast of Bosnia-Herzegovina (Brljević 2014). The purpose of the project was to maintain the water supply during dry summer periods and also to use it to generate electricity. The Bregava is another left tributary of the lower Neretva which was included in this so-called Upper Horizons project, which aims to divert about 30% of the three river's water volume from the source areas. These areas are located in karst fields at higher elevations where surface water enters the subsoil through natural openings (ponors) and flows to the re-emerge from the underground, which is the actual origin of the river (Milanović and Kukurić 2013; Čokorilo Ilić et al. 2016). Since the construction of a power plant in 2016, a considerable part of the water has been diverted from the headwaters of the Bregava through an open canal and returned about 2 km further downstream. The remaining stock of softmouth trout, which is limited to the upper reaches, is threatened not only by high summer water temperatures and the resulting deterioration in habitat, reinforced by the partial river diversion, but also by non-native brown trout (Author's own observations). The

Fig. 29 Consequence of drought in the source region of the Bregava River, Bosnia and Herzegovina. Photo by Saša Marić

Fig. 30 Consequence of drought is reflected on strong growth of algae and the dominance of cyprinids in the upper reaches of the Bregava River, Bosnia and Herzegovina. Photo by Johannes Schöffmann

change of the ecosystem in the Bregava as a consequence of drought can be seen in the source region (Fig. 29) and is also reflected in the strong growth of algae and the dominance of cyprinids in the upper reaches (Fig. 30).

The above-mentioned Vrljika drains underground to the Tihaljina, which further downstream is called Mlade and finally joins the lower Neretva from the right-hand side as Trebižat. In addition to the softmouth trout population in the Vrljika, there are other populations in this catchment that are isolated from each other and from the Neretva population by high waterfalls and occur at cold springs of the main river and in two tributaries, Vrioštica and Studenčica. The softmouth trout from these waters, like those from the Vrljika, have a somewhat blunter snout than the softmouth trout in the rest of the Neretva catchment, but differ significantly in their spotting patterns: the fish from the Vrljika only have red spots along the flanks and a few small dark ones on the head, those of the Trebižat and its tributaries have, in addition to the red spots distributed over the sides, numerous black spots, especially on the front half of the body. While little is known about their current status, these small populations are poorly protected. Brown trout stocking is detectable at least in the Studenčica, where a small proportion of a haplotype from the non-native Danubian brown trout strain was found (unpublished data). There are no genetic studies on the Vrioštica population yet. Some rainbow trout were observed in all aforementioned river sections except Vrljika (Author's own observations). Besides intensive poaching, the introduced pike also pose a threat to the native fish fauna in the Trebižat (O. Anapović and P. Spasojević, personal communication). The greatest threat to the softmouth trout and other native trout of the Neretva, however, is posed by the planned large hydropower plants in its upper reaches, warn NGOs RiverWatch and EuroNatur. In the Red Book of freshwater fish in Croatia (2006), the softmouth trout from Krka and Neretva were classified as critical endangered (CR) and the Jadro softmouth trout as vulnerable (VU), but whether this is still the case remains to be clarified. The Zeta softmouth trout, *S. o. zetensis*, is only found in small numbers in three limited localities, one short, water-rich tributary of the upper section of the lower Zeta and two in the middle section below a dam. The species was once abundant in the lower Zeta but has been drastically depleted since the late last century due to overfishing and poaching, particularly at their spawning grounds, including the use of electrofishing and explosives (Mrdak et al. 2012; Schöffmann 2021).

The consequences of overexploitation of the other salmonid stocks in the Zeta River are also evident. In 2004, for example, several brown and marble trout up to a size of around 30 cm could still be observed while snorkelling on a roughly one-kilometre stretch of river upstream from the village of Tunjevo, whereas in 2019 only cyprinids and three-spined sticklebacks (*Gasterosteus aculeatus*) were sighted on the same stretch. Most alarmingly, not even juvenile trout were present. Similar observations were made in the Cijevna River, another tributary of the Morača-Skadar Lake system, which used to have good stocks of brown and marble trout. The upper reaches of the Cijevna are in Albania, where the river is called the Cem. Here, long stretches of the river were badly affected by the expansion of the road. However, uncontrolled fishing has long contributed to the depletion of trout stocks (Author's own observations). Moreover, the Albanian government is planning to build several hydropower plants on the Cem and its main tributary (Weiss et al. 2018), which would damage the ecosystem and probably irretrievably destroy the natural habitat.

Some of the southernmost populations of marble trout have also apparently fallen victim to overharvesting. No specimens have recently been found in the lower Radika in North Macedonia, and the two tributaries of the Drin in Albania, Shala and Valbona. A dam is also being built on the lower reaches of the Valbona, where some specimens were still observed in the 1990s (Schöffmann et al. 2019). At that time, fishing with explosives was still common (Schöffmann 1994). Not only marble trout, but also the local brown trout stocks are endangered in Albania. In addition to the uncontrolled fishing with rods and cast nets, the harpoons, which have been used more and more frequently for underwater hunting, have drastically reduced the stocks.

Highly valued as food fish, Lake Ohrid's two salmonid species, belvica (*S. ohridanus*) and Ohrid trout (*S. letnica*) [2], have long been commercially fished. Demand has increased rapidly in recent years due to increasing tourism. The consequences of the fishing pressure on trout stocks are obvious. Compared to the 1990s, when hardly any tourists travelled the Albanian lake's coast, the Ohrid trout landed by fishermen were on average 30–50 cm long, today they are barely 30 cm long, mostly much less, i.e., not even sexually mature. The same applies to the belvica, where the maximum catch sizes are 20–25 cm (Schöffmann 2019). The legal minimum catch sizes for Ohrid trout and belvica are 32 cm and 30 cm, respectively, in Albania and 35 cm and 22 cm, respectively, in North Macedonia. There are fishing ban seasons in Albania from first December to 28th February for Ohrid trout and from first November to 31st January for belvica, in North Macedonia from first December to 20th March for both species (Spirkovski et al. 2017). These differences in protection measures between the two countries bordering the lake make it more difficult to control and prevent illegal fishing. To make matters worse, there has been no concession company in North Macedonia in recent years that issues fishing licences and monitors compliance with fishing rules. Due to this lack of management, there is currently a complete ban on fishing for salmonids in this area, but this does not prevent illegal fishing (Vasil Kostov, personal communication). Both the Ohrid trout and the belvica are artificially propagated for stocking purposes. Not only were the above-mentioned intra-lacustric forms (subspecies or species) of Ohrid trout mixed, but there were also experimental crossbreeds between the two "good" species, male *S. letnica* and female *S. ohridanus* (Rakaj and Flloko 1995). A specimen that showed intermedial external characteristics of both species, e.g. short maxillary and reduced spotting pattern like belvica and a deep body like the Ohrid trout, was caught by Albanian fishermen in 1993 (Schöffmann 2019). In the course of a genetic study, one phenotypically characterized *S. ohridanus* specimen was found that carried a typical *S. letnica* haplotype, as well as a "mixture" of

---

[2] The four ecological trout forms or species of Lake Ohrid (*S. aphelios*, *S. balcanicus*, *S. letnica*, *S. lumi*) are often assigned to one species, *S. letnica*, simply referred to as the Ohrid trout.

Fig. 31 Ohrid trout (*Salmo letnica*), ca. 65 and 55 cm; Lisina Lake, Serbia; translocated population. Photo by Saša Marić

taxon-specific microsatellite alleles. Whether or not this evidence of hybridization stems from natural gene flow, or escapees from a cross-breed experiment is, however, impossible to determine (Sušnik et al. 2006). Rainbow trout, originating from a fish farm on the Albanian shore, were first recorded in Lake Ohrid in 1974 and specimens were regularly found in the fish catches until 1994. Although the facility was later converted into a breeding and rearing station for Ohrid trout, rainbow trout still occasionally appear in the fish catches today (Spirkovski et al. 2017). As potential food competitors, the allochthonous trout pose an additional threat to native salmonids.

In the second half of the last century, Ohrid trout have been introduced into two water reservoirs on the Vlasina Plateau in Serbia, Lakes Vlasina and Lisina, belonging to the Danubian and Aegean drainage basins, respectively. The Ohrid trout were able to establish a self-sustaining population in Lisina Lake (Fig. 31). Moreover, brown trout from the Struma drainage (Aegean basin) were translocated to the adjacent upper Jerma River (Danubian basin), where they replaced completely the local brown trout population (Marić et al. 2022). Another example of translocation across the watershed took place in North Macedonia and Serbia, where brown trout originating from the Radika River (Adriatic basin) were used for large-scale stocking in the Vardar River drainage (Aegean basin) (Marić et al. 2017).

The spread of invasive exotic fish species has changed the composition of fish communities in many bodies of water. The pumpkinseed (*Lepomis gibbosus*), which was introduced to Europe from North America in the late 1800s (Maes 1898), has also become well established in suitable places in the Balkans. Today, pumpkinseeds are dominant in the lower Zrmanja, below the Berberov Buk waterfall, where local brown trout can be found alongside with various species of cyprinids, freshwater

Fig. 32 Pumpkinseed (*Lepomis gibbosus*) are in high density in the lower reaches of Zrmanja. Underwater photo by Benedikt Schöffmann

Fig. 33 Juvenile brown trout (right) in a school of pumpkinseed; lower Zrmanja. Photo by Johannes Schöffmann

blennies (*Salaria fluviatilis*), and black-spot gobies (*Pomatoschistus canestrinii*). Juvenile trout often swim between schools of pumpkinseed in the shore areas (Figs. 32 and 33; Author's personal observation). Pumpkinseed can contribute to a considerable reduction in invertebrates and thus compete with trout and other fish for food. They also have a potential impact on salmonid populations by preying on their roe and brood (Jordan et al. 2009). On the other hand, studies carried out in small rivers in southern England hardly show any influence of the pumpkinseed on the brown trout stocks. However, it is recognized that the situation may change as a result of climate warming, which is likely to favour reproduction of the alien species (Gordon et al. 2010). This already applies to the lower Zrmanja, where the water temperatures at the beginning of June 2022 were just under 20 °C, at least in the shallow water areas preferred by young trout and pumpkinseed (Author's own observation).

There is a serious risk to wild fish stocks associated with transmission of parasites by exotic fish used in aquaculture, as evolutionary the native host and exotic parasite have not had time to develop an equilibrium relationship. Many diseases of salmonids that infect hatchery-reared fish and also occur in wild populations have been imported. For example, rainbow trout from western North America have carried furunculosis to Europe (Elvira 2001). A diverse parasitofauna of rainbow trout has been detected in the surface waters in the Serbian part of the Danube basin (Djikanović et al. 2018). The outbreak of diseases can be promoted by the change in the flow regime and the deterioration of water quality as a result of the construction of dams, as evidenced by the appearance of individuals of softmouth trout with fungal disease in the lower Neretva (Glamuzina et al. 2018).

The huchen (*Hucho hucho*) once inhabited a significant portion of the Danube drainage basin stretching to the Iron Gate and the majority of large and medium-sized tributaries. Typical huchen habitats are montane and sub-montane reaches of large streams and swift rivers with gravel beds, well oxygenated, fast-flowing water and temperatures rarely exceeding 15 °C. Adult fish prefer deep pools and the spawning sites are on very clean gravel in fast flowing water, often in small tributaries. Now the huchen is a threatened species throughout its range of occurrence, and, according to IUCN criteria, it is classified as endangered (EN). Habitat degradation is the most serious negative factor impacting huchen populations. Dam construction, river regulation, and water pollution have led to loss or degradation of spawning sites. Since the late nineteenth century, huchen have been eliminated from approximately 70–90% of their native range. However, decline in the Balkan region has been moderate by comparison (ca. 35%, or about 1000 km), with the majority of the species' remaining intact habitat found in this region. A total of 1822 river km in the Balkan region has been identified as carrying self-sustaining populations of huchen. These are 65% of the world's functional huchen rivers (Freyhof et al. 2015). In the Balkans, the most important river in terms of habitat length is the Drina together with its major tributaries the Lim and Tara, totalling 30% (553 km) of the Balkan huchen distribution. The Zapadna Morava River (West Morava) and its tributaries was once a large huchen habitat, but almost all populations have been lost. The river system is polluted through most of its length, as well as the Ibar River, its largest tributary. Actually, in the entire Zapadna Morava drainage there is just one small huchen population in the upper part of the Ibar River, which is threatened with extinction due to a planned hydropower dam. Overfishing and poaching are also a considerable problem in some regions (especially in the Drina). However, the major threat to these populations is a massive hydropower development plan. A total of 93 dam projects were identified directly in river reaches supporting huchen and a large number of additional projects are located in tributaries or headwater reaches upstream of huchen habitat that will invariably degrade environment conditions downstream. If these plans are carried out, it can be predicted that at least 60–70%

of the Balkan population and about 35–40% of the global populations of huchen would be lost (Freyhof et al. 2015).

As the stretches of river suitable for huchen become shorter, the risk of genetic impoverishment and inbreeding also increases. As a large apex predator, healthy huchen populations need considerable space, with only 10 adult individuals typically in 1 km of river for a medium-sized river (Ratschan 2014). Just 10 of the listed river stretches in the Balkans harbouring permanently huchen are longer than 50 km and could be expected to hold more than 500 adults. A rule-of-thumb for avoiding long-term inbreeding is a minimum of 500 breeders but these numbers have been called into question and new studies consider even 1000 breeders to be needed for the long-term survival and adaptive capacity of species (Frankham et al. 2014).

Attempts have long been made to strengthen huchen populations through artificial breeding and stocking. Huchen have been bred in Slovenia for 100 years. Intensive reproduction began in the 1980s and is now practiced in larger aquaculture facilities at about a dozen sites (Sušnik Bajec et al. 2016). Since adult huchen are not kept in any fish farms in Slovenia, the spawners are always caught in the wild. Supportive stocking measures are carried out mainly in the Sava and its upper tributary Sava Bohinjka, while the tributaries to the middle and lower reaches such as Ljubljanica, Krka, and Kolpa are managed less intensively and the populations are largely dependent on natural spawning. Unlike in previous decades, efforts are now being made to ensure that the three, historically isolated, river system-bound populations (from the Sava, Kolpa and Mura-Drava) are managed separately to preserve local genetic pools and possible adaptive differentiation between populations. The longest barrier-free river section (except a semi-permeable barrier) inhabited by a self-sustaining huchen population in Slovenia is the about 50 km long stretch of the middle Sava from Medvode to Litija. The proposed construction of multiple hydropower plants on this section would split the huchen population into smaller, mutually isolated subunits, leading to increased inbreeding, accelerated genetic drift, a reduction in genetic diversity, and an inability of the compromised population to cope with an ecologically changing environment (Snoj et al. 2022). Two successful artificial reproduction projects have also been conducted in Serbia in the past twenty years: the first was implemented in 2000 with the release of approximately 10,000 fish in the Drina, Vapa, Uvac, Lim, Moravica, and Ibar rivers, and the second project, which began in 2011, is known as the "Artificial reproduction of huchen to supplement the natural population in the Drina River". Under the auspices of this program, 33,300 one (0+) year old and 7200 two (1+) years old huchen were released into the Drina River during 2011 and 2012 (Veličković et al. 2023). Currently in Serbia there is also one huchen brood stock of about 20 individuals originating from the Lim River system.

Breeding and restocking measures, the effectiveness of which is often overestimated, can also have harmful impacts, e.g. the transmission of unsuitable genes or the introduction of pathogens (see also Ihuţ et al. 2014). Restoring self-sustaining populations should therefore always be a top priority.

Declines in European grayling (*Thymallus thymallus*) populations due to water pollution, habitat destruction, river engineering, predation from piscivorous birds, and overfishing are evident throughout their range, including the Balkans. Thus, the populations in the Serbian rivers Uvac, Jadar, and Studenica went extinct, while four new populations were established: in the Rzav (through stocking with fish originating from an unknown source population from Slovenia), Moravica (stocking with fish originating from Studencica, Drina and Una rivers), Đetinja (stocking with fish originating from Drina), and Gradac (stocking with fish originating from an unknown source population from Slovenia, and from Drina and Una rivers). Population decline leads to an increase in management activities that involve rearing and stocking grayling, which may cause a change in genetic architecture and extinction of natural populations (Marić et al. 2011). In the Red Book of freshwater fish in Croatia (2006) the European grayling is listed as vulnerable (VU).

The Adriatic grayling in the Soča River drainage is much more endangered than initially anticipated, as the population is in sharp decline. Furthermore, the Soča population reflects a much higher genetic proportion of the non-native Sava populations than originally believed, and only 50–60% of the original Adriatic gene pool persisted (Sušnik et al. 2004). There are considerations to restore a largely Adriatic grayling population in the Soča drainage by translocating grayling from the Sesia River in Northern Italy, the only remaining pure population of *T. aeliani* (Bravničar 2021). Studies on grayling populations from across Northern Italy have shown that despite decades of stocking with grayling from the Atlantic and Danube catchments, an unexpectedly high proportion of the indigenous genetic signature has been preserved in several Adriatic watercourses (Meraner et al. 2014).

4 Recommendations for Sustainable Use of Salmonid Populations

To ensure the conservation and survival of the native salmonid stocks on the Western Balkans, measures are needed to protect their habitats, abundance, population structure, and genetic diversity. In order to avoid further loss of biodiversity, special attention should be paid to water bodies where endangered species or populations occur. This also means, for example, that there should be no further hydropower development including micro-hydropower in rivers holding self-sustaining huchen populations including spawning streams. The feasibility of restoration measures in habitats formerly inhabited by huchen should be examined. Habitat conservation not only protects rare salmonids, but also other species of fish, invertebrates, etc. Another important factor is the conservation of the fish stocks through controlled fishing and the curbing of poaching. The establishment of sanctuary sectors with a total ban on fishing and/or catch and release fishing sectors for endangered or critically endangered populations is highly recommended. This is especially true for most softmouth trout populations. Establishing an additional protected population in a previously fishless stream can serve as a safeguard against the extinction of a particular population (subspecies, species). Such an attempt has already been carried out successfully. In the middle of the twentieth century, some specimens of the softmouth trout from the Jadro River in Split were translocated into the Žrnovnica, a small stream only a few kilometres south of Split where salmonids were previously absent. Softmouth trout were able to establish themselves due to favourable environmental conditions (Treer et al. 2003).

Stocking measures have too often been considered a viable means of compensating for the declining fish populations without paying attention to the negative effects. However, supporting stocking should only be considered if the self-reproduction of the population remains too low despite good habitat (or habitat improvement) and catch restrictions (or ban of fishing). In order to preserve the genetic integrity of salmonid stocks, restocking with fish of non-local origin should be avoided at all costs. Only spawners from the river system to be stocked should be used for breeding. Ideally, sufficient individuals should be caught each year to maintain genetic diversity and prevent inbreeding. As soon as the population has recovered, the stocking should be discontinued.

References

Apostolidis AP, Madeira MJ, Hansen M, Machordom A (2008) Genetic structure and demographic history of brown trout (*Salmo trutta*) populations from the southern Balkans. Freshw Biol 53: 1555–1566

Bernatchez L (2001) The evolutionary history of brown trout (*Salmo trutta* L.) inferred from phylogeographic, nested clade, and mismatch analyses of mitochondrial DNA variation. Evolution 55:351–379

Bernatchez L, Guyomard R, Bonhomme F (1992) DNA sequence variation of the mitochondrial control region among geographically and morphologically remote European brown trout *Salmo trutta* populations. Mol Ecol 1:161–173

Berrebi P, Tougard C, Dubois S, Shao Z, Koutseri I, Petkovski S, Crivelli AJ (2013) Genetic diversity and conservation of the Prespa trout in the Balkans. Int J Mol Sci 14:23454–23470

Bianco PG (2014) An update on the status of native and exotic freshwater fishes of Italy. J Appl Ichthyol 30:62–77

Bravničar J (2021) Population genomics and conservation of Adriatic grayling *Thymallus aeliani* (Valenciennes, 1848) in Soča River. PhD Thesis, University Ljubljana, Slovenia

Bravničar J, Palandačić A, Sušnik Bajec S, Snoj A (2020) Neotype designation for *Thymallus aeliani* Valenciennes, 1848 from a museum topotype specimen and its affiliation with Adriatic grayling on the basis of mitochondrial DNA. ZooKeys 999:165–178

Brljević D (2014) Education of the public on sustainable water use and the protection of endemic fish in the Neretva River valley (CEPF final project completion report). Croatian Ecological Society "Buna"

Čokorilo Ilić M, Stevanović Z, Ristić Vakanjac V (2016) Environmental aspects and potential impacts of proposed water transfer scheme on sustainable water management in eastern Herzegovina. Environ Earth Sci 75:521

Dalmatinski portal (2020). https://dalmtinskiportal.hr/zivot/zasticena-pastrva-vracena-u-rijeku-vrljiku/67203

Delling B (2003) Species diversity and phylogeny of Salmo with emphasis on southern trouts (Teleostei, Salmonidae). Doctoral thesis, Stockholm University, Stockholm

Djikanović V, Simonović P, Cakić P, Nikolić V (2018) Parasitofauna of allochthonous fish species in the open waters of the Danube River basin (Serbian part) – impact on the native fish fauna. Appl Ecol Environ Res 16:6129–6142

Elvira B (2001) Identification of non-native freshwater fishes established in Europe and assessment of their potential threats to the biological diversity. Document of the 21st Meeting of the Standing Committee of the Convention of the Conservation of European Wildlife and Natural Habitats. Strasbourg, 1–35 pp.

Fishing Worldrecords. www.fishing-worldrecords.com. Archived from the original on 14 September 2018

Frankham R, Bradshaw CJA, Brook BW (2014) Genetics in conservation management: revised recommendations for the 50/500 rules, red list criteria and population viability analyses. Biol Conserv 170:56–63

Freyhof J, Weiss S, Adrović A, Ćaleta M, Duplić A, Hrašovec B, Kalamujić B, Marčić Z, Milošević D, Mrakovčić M, Mrdak D, Piria M, Simonović P, Šljuka S, Tomljanović T, Zabric D (2015) The Huchen *Hucho hucho* in the Balkan region: distribution and future impacts by hydropower development. RiverWatch & EuroNatur, 30 p.

Glamuzina B, Stanić-Koštroman S, Matić-Skoko S, Glamuzina L, Muhamedagić S, Rozić I, Weiss S, Pavličević J (2018) Recent status and life history traits of endangered soft-mouth trout, *Salmo obtusirostris* in the river Neretva catchment (Bosnia and Herzegovina) as a consequence of river alteration. J Appl Ichthyol 34:1160–1168

Gordon HC, Stakėnas S, Cucherousset J (2010) Aliens versus the natives: interactions between introduced pumpkinseed and indigenous brown trout in small streams of southern England. Am Fish Soc Symp 73:347–370

Heckel JJ (1852) Beiträge zu den Gattungen *Salmo, Fario, Salar, Coregonus, Chondrostoma* und *Telestes.* Sitzungsberichte der Mathematisch-Naturwissenschaftlichen Classe der Kaiserlichen Akademie der Wissenschaften Wien, 8: 347–391, pls. 7–14

Hochleithner M (2021) Das Typusmaterial der Weichmaulforelle, *Salmo obtusirostris* (Heckel, 1852). Österreichs Fischerei 74:95–106

Holčík J, Hensel K, Nieslanik J, Skácel J (1988) The Eurasian huchen, *Hucho hucho*, largest salmon of the world. Dr. W. Junk Publisher, Dordrecht-Boston-Lancaster, p 239

Ihuț A, Zitek A, Weiss S, Ratschan C, Holzer G, Kaufmann T, Cocan D, Constantinescu R, Mireşan V (2014) Danube salmon (*Hucho hucho*) in central and South Eastern Europe: a review for the development of an international program for the rehabilitation and conservation of Danube salmon populations. Bulletin UASVM Animal Science and Biotechnologie 71:86–101

Janković D (1963) Trout from Lake Plavsko. Arhiv Bioloških Nauka 15:29–40

Jordan C, Backe N, Wright MC, Tovey CP (2009) Biological synopsis of pumpkinseed (*Lepomus gibbosus*). Can Manuscr Rep Fish Aquat Sci 2886:1–16

Karaman S (1924) Pisces Macedoniae. Hvratska Štamparija, Split, Croatia, p 90

Karaman S (1927) Salmonidi Balkana [Les Salmonidés des Balkans]. Bulletin de la Societé Scientifique de Skoplje 2:253–268

Karaman S (1932) Novi prilozi poznavanju naših Salmonida. Ribarski List, Sarajevo 7(9–10):1–3

Karaman S (1933) Prilozi ihtiologiji Crne Gore. Ribarski List, Sarajevo 8:104–106

Karaman S (1938) Beitrag zur Kenntnis der Süsswasserfische Jugoslawiens. Bulletin de la Societé Scientifique de Skoplje 18:131–139

Kohout J, Šedivá A, Apostolou A, Stefanov T, Marić S, Gaffaroğlu M, Šlechta V (2013) Genetic diversity and phylogenetic origin of brown trout *Salmo trutta* populations in eastern Balkans. Biologia 68:1229–1237

Kottelat M (1997) European freshwater fishes. Biologia 52/Suppl. 5:1–271

Kottelat M, Freyhof J (2007) Handbook of European freshwater fishes. Kottelat, Cornol, Switzerland and Freyhof, Berlin, Germany, p 646

Maes L (1898) Notes sur la pêche fluviale et maritime en Belgique. Administration des Eaux et Forêts, Imprimerie Scientifique, Bruxelles, p 295

Marić D (1995) Endemic fish species of Montenegro. Biol Conserv 72:187–194

Marić S, Nikolić V, Paunović M, Simonović P (2009) Istraživanje životnog ciklusa i populacionih karakteristika mladice u Srbiji. Univerzitet u Beogradu – Biološki fakultet i Ministarstvo životne sredine i prostornog planiranja, Beograd, p 79
Marić S, Simonivić P, Razpet A (2010) Genetic characterization of broodstock brown trout from bled fish-farm, Slovenia. Period Biol 112:145–148
Marić S, Razpet A, Nikolić V, Simonović P (2011) Genetic differentiation of European grayling (*Thymallus thymallus*) populations in Serbia, based on mitochondrial and nuclear DNA analyses. Genet Sel Evol 43:2
Marić S, Kalamujić B, Snoj A, Rezpet A, Lukić-Bilela L, Pojskić N, Sušnik BS (2012) Genetic variation of European grayling (*Thymallus thymallus*) populations in the Western Balkans. Hydrobiologia 691:225–237
Marić S, Razpet A, Nikolić V, Snoj A, Simonović P (2014) Analysis of genetic structure of huchen (*Hucho hucho*) in Serbia inferred from mitochondrial and nuclear DNA. Acta Vet (Beograd) 64: 236–244
Marić S, Sušnik Bajec S, Schöffmann J, Kostov V, Snoj A (2017) Phylogeography of stream-dwelling trout in the Republic of Macedonia and a molecular genetic basis for revision of the taxonomy proposed by S. Karaman Hydrobiologia 785:249–260
Marić S, Stanković D, Sušnik Bajec S, Vukić J, Šanda R, Stefanov T, Nikolić D, Snoj A (2022) Perils of brown trout (*Salmo* spp.) mitigation-driven translocations: a case study from the Vlasina plateau. Southeast Serbia Biological Invasions 24:999–1016
Meraner A, Cornetti L, Gandolfi A (2014) Defining conservation units in a stocking-induced genetic melting pot: unravelling native and multiple exotic genetic imprints of recent and historical secondary contact in Adriatic grayling. Ecol Evol 4:1313–1327
Milanović P, Kukrić N (2013) Hydro-power and groundwater in karst, (prepared for the world water development report). International Groundwater Resources Assessment Centre – GRAC, 6 p. https://www.un-igrac.org/resource/hydropower-and-groundwater-karst
Mrdak D, Nikolić V, Tošić A, Simonović P (2012) Molecular and ecological features of the soft-muzzled trout *Salmo obtusirostris* (Heckel, 1852) in the Zeta River, Montenegro. Biologia 67: 222–233
Ocvirk J, Voljč B (1981) Die endemischen Süßwasserfische der dalmatinischen Adriazuflüsse: Die Zahnforelle. Fliegenfischer 45:28–29
Papadaki C, Soulis K, Muños-Mas R, Martinez-Capel F, Zogaris S, Ntoanidis L, Dimitriou E (2016) Potential impacts of climate change on flow regime and fish habitat in mountain rivers of the South-Western Balkans. Sci Total Environ 540:418–428
Persat H, Weiss S, Froufe E, Secci-Petretto G, Denys GPJ (2019) A third European species of grayling (Actinopterygii, Salmonidae), endemic to the Loire River basin (France), *Thymallus ligericus* n. sp. Cybium 43:233–238
Piria M, Simonović P, Zanella D, Ćaleta M, Šprem N, Paunović M, Tomljanović T, Gavrilović A, Pecina M, Špelić I, Maulić D, Rezić A, Aničić I, Safner R, Treer T (2019) Long-term analysis of fish assemblage structure in the middle section of the Sava River – the impact of pollution, flood protection and dam construction. Sci Total Environ 651:143–153
Poljakov GD, Filipi N, Basho K, Hysenaj A (1958) Peshqit e Shqiperise. Universiteti Shtetëror i Tiranës, Tirana, p 286
Povž M, Jesenšek D, Berrebi P, Crivelli AJ (1996) The marble trout *Salmo trutta marmoratus*, Cuvier 1917 in the Soča River basin, Slovenia, Tour du Valat, Arles, 65 p
Pustovrh G, Sušnik Bajec S, Snoj A (2011) Evolutionary relationship between marble trout of the northern and the southern Adriatic basin. Mol Phylogenet Evol 59:761–766
Pustovrh G, Sušnik Bajec S, Snoj A (2012) A set of SNPs for *Salmo trutta* and its application in supplementary breeding programs. Aquaculture 370–371:102–108
Pustovrh G, Snoj A, Sušnik Bajec S (2014) Molecular phylogeny of *Salmo* of the western Balkans, based upon multiple nuclear loci. Genetic Selection Evolution 46:7
Rakaj N, Flloko A (1995) Conservation status of freshwater fish in Albania. Biol Conserv 72:195–199

Ratschan C (2014) Aspekte zur Gefährdung und zum Schutz des Huchens in Österreich. Denisia, 33. Neue Serie 163:443–462

Razpet A, Sušnik S, Jug T, Snoj A (2007) Genetic variation among trout in the river Neretva basin, Bosnia and Herzegovina. J Fish Biol 70:94–110

Schöffmann J (1994) Zur gegenwärtigen Situation der Marmorierten Forelle (*Salmo marmoratus* Cuvier, 1817) in Albanien, ihrem südlichsten Verbreitungsraum. – on the situation of *Salmo marmoratus* in Albania, the southernmost area of distribution of this species. Österreichs Fischerei 47:132–136

Schöffmann J (2016) Die Forellen (*Salmo* spp.) im Einzugsgebiet des Vardar und des unmittelbar angrenzenden Adriabeckens. - Trouts of the Vardar River basin and the immediately adjacent Adriatic Sea drainage basin. Österreichs Fischerei 69:277–285. + 70 (2017) 79

Schöffmann J (2018) Salmoniden im Einzugsgebiet der Neretva. Österreichs Fischerei 71:264–271

Schöffmann J (2019) Zur Situation der heimischen Forellen (*Salmo* spp.) Albaniens. - on the situation of the native trout (*Salmo* spp.) in Albania. Österreichs Fischerei 72:108–115

Schöffmann J (2021) Trout and salmon of the genus *Salmo*. American Fishery Society, Bethesda, Maryland, p 303

Schöffmann J, Marić S, Snoj A (2019) Trout of Southeast Europe, western and Central Asia. In: Kershner JL, Williams JE, Gresswell RE, Lobón-Cerviá J (eds) Trout and char of the world. American Fishery Society, Bethesda, Maryland, pp 411–456

Simonović P, Marić S, Nikolić V (2007) Trout *Salmo* spp. complex in Serbia and adjacent regions of western Balkans: reconstruction of evolutionary history from external morphology. J Fish Biol 70/Suppl. C:359–380

Snoj A, Melkič E, Sušnik S, Muhamedagić S, Dovč P (2002) DNA phylogeny supports revised classification of *Salmothymus obtusirostris*. Biol J Linn Soc 77:399–411

Snoj A, Bogut I, Sušnik S (2008) Evidence of genetically distinct population of Vrljika softmouth trout *Salmo obtusirostris* Heckel evolved by vicariance. J Fish Biol 72:1945–1959

Snoj A, Marić S, Berrebi P, Crivelli AJ, Shumka S, Sušnik S (2009) Genetic architecture of trout from Albania as revealed by mtDNA control region variation. Genet Sel Evol 41:22

Snoj A, Glamuzina B, Razpet A, Zablocki J, Bogut I, Lerceteau-Köhler E, Pojskić N, Sušnik S (2010) Resolving taxonomic uncertainties using molecular systematics: *Salmo dentex* and the Balkan trout community. Hydrobiologia 651:199–212

Snoj A, Bravničar J, Zabric D, Sušnik Bajec S (2022) Conservation genetics study of huchen in Slovenia recommends river system-based management and indicates self-sustainability of the middle Sava population. Aquat Conserv Mar Freshwat Ecosyst 32:1171–1183

Spirkovski Z, Palluqi A, Flloko A, Miraku T, Kapedani E, Ilik-Boeva D, Talevski T, Trajcevski B, Ritterbusch D, Brämick U, Pietrock M, Peveling R (2017) Fish and fisheries Lake Ohrid – implementing the EU water framework directive in South-Eastern Europe. Deutsche Gesellschaft für Internationale Zusammenarbeit (GIZ) GmbH, Bonn and Eschborn, Germany, Pegi Sh. p.k. Book Publisher, Tirana, p 99

Stefanidis K, Panagopoulos Y, Mimikou M (2018) Response of a multi-stressed Mediterranean river to future climate and socio-economic scenarios. Sci Total Environ 627:756–769

Sušnik Bajec S, Pustovrh G, Jesenšek D, Snoj A (2015) Population genetic SNP analysis of marble and brown trout in a hybridization zone of the Adriatic watershed in Slovenia. Biol Conserv 184:239–250

Sušnik Bajec S, Zabric D, Šantl S, Podgornik S, Jenič A, Čart M, Bric B, Pliberšek K, Javornik L, Snoj A (2016) Elaboration of expertise for huchen habitat and population conservation in the middle Sava river: scientific raport, Ljubljana

Sušnik S, Berrebi P, Dovč P, Hansen MM, Snoj A (2004) Genetic introgression between wild and stocked salmonids and the prospects of using molecular markers in population rehabilitation: the case of the Adriatic grayling (*Thymallus thymallus* L., 1785). Heredity 93:273–282

Sušnik S, Knizhin I, Snoj A, Weiss S (2006) Genetic and morphological characterization of a Lake Ohrid endemic, *Salmo (Acantholingua) ohridanus* with a comparison to sympatric *Salmo trutta*. J Fish Biol 68/Suppl. A:2–23

Taler Z (1950) Visovačka jezerska pastrva (Salmo visovacensis n. sp.). Glasnik Bioloske Sekcije, Hrvatsko Prirodoslovno Društvo, Ser 2B 2–3:118–158

Treer T, Aničić I, Safner R, Odak T, Piria M (2003) Note on growth and condition of endemic soft-muzzled trout *Salmothymus obtusirostris* translocated into a Dalmatian river. Biologia 58:999–1001

Turin P, Zanetti M, Biló MF (2006) Distribuzione e stato delle popolazioni di trota marmorata nelle acque del bacino dell'Alto Adriatico. Biologia Ambientale 20:39–44

Veličković T, Snoj A, Simić V, Šanda R, Vukić J, Barcytė D, Stanković D, Marić S (2023) A new perspective on the molecular dating of the brown trout complex with an extended phylogeographic information on the species in Serbia. Contrib Zool 1:1–28. https://doi.org/10.1163/18759866-bja10046

Weiss S, Marić S, Snoj A (2011) Regional structure despite limited mtDNA sequence diversity found in the endangered huchen, *Hucho hucho* (Linnaeus, 1758). Hydrobiologia 658:103–110

Weiss S, Apostolou A, Đug S, Marčić Z, Mušović M, Oikonomou A, Shumka S, Škrijelj R, Simonović P, Vesnić A, Zabric D (2018) Endangered fish species in Balkan rivers: their distributions and threats from hydropower development. RiverWatch & EuroNatur, p 162

Past and Future of Sturgeon Species (Acipenseridae) in Western Balkans: Case for Permanent Conservation or Sustainable Management

Gorčin Cvijanović, Vesna Đikanović, Laslo Galambos, Polona Pengal, and Marija Smederevac-Lalić

Abstract Sturgeon species in Western Balkans are mainly occupying the Danube River basin, which represents one of the centers for sturgeon diversity. Although there are numerous negative anthropogenic impacts (habitat fragmentation, pollution, fishery pressure) on sturgeons, constructions of hydropower dams in the Iron Gate area severely decreased their natural spawning habitats and only sterlets have a viable population upstream from Iron Gate 2 dam. Review of the status and population trends based on available statistics, pollution impact, composition and change in their diet, the impact of invasive species, as well as restocking and ex situ conservation programs will be presented. Assessment of legislation and sturgeon protection projects and their impact on wild sturgeon populations will be discussed.

Keywords *Acipenser* sp. · *Huso huso* · Parasites · Aquaculture · Endangered species · Protection

1 Introduction

With fossils dating from about 300 million years ago, sturgeons can be considered "living fossils," undergoing little morphological change and with many biological features of ancient fishes (Carmona et al. 2009). Essentially, they maintained their

G. Cvijanović (✉) · M. Smederevac-Lalić
Institute for Multidisciplinary Research, University of Belgrade, Belgrade, Serbia
e-mail: mitrandir@imsi.rs

V. Đikanović
Institute for Biological Research "Siniša Stanković", University of Belgrade, Belgrade, Serbia

L. Galambos
Institute for Nature Conservation of Vojvodina Province, Novi Sad, Serbia

P. Pengal
REVIVO, Institute for Ichthyological and Ecological Research, PE, Ljubljana, Slovenia

V. Simić et al. (eds.), *Ecological Sustainability of Fish Resources of Inland Waters of the Western Balkans*, Fish & Fisheries Series 43,
https://doi.org/10.1007/978-3-031-36926-1_10

present traits (homing behavior, longevity, late maturation, and low fecundity) for more than 200 million years, establishing regular and abundant populations through the Northern Hemisphere. They survived glaciations and mass extinction events, while adapting to different living conditions from cold environment of Siberia and North America to warm Mediterranean waters (Hernando et al. 2009; Jarić et al. 2011a). Since all sturgeons are migratory (anadromous or potamodromous) and spawn in freshwater, they are particularly vulnerable and extremely susceptible to anthropogenic impact, such as fisheries, habitat pollution, and alteration of their migratory routes (Freidrich et al. 2018). This was the main reason for development of numerous strategies for sturgeons conservation in wild and for their aquaculture, not only for market demand (caviar and meat), but also to raise adequate specimens for conservation and repopulation of their natural habitats. Recovery of the wild sturgeons requires a sound knowledge of their taxonomy, biology, and biogeographical distribution.

Sturgeons can be easily identified by elongated snout with the four barbels in front of the ventrally located mouth, five rows of bony scutes on a body, a single dorsal fin situated near a heterocercal tail, no branchiostegal rays and a largely cartilaginous endoskeleton (Helfman et al. 2009). They feed (mostly) on benthic invertebrates, although larger individuals can feed on fishes due to specific morphological jaw suspension. With mobile hyoid arch and the independence of the jaws from neurocranium, sturgeons are able quickly to protrude jaws from the neurocranium (Carroll and Wainwright 2003) and efficiently feed on its pray. They do not rely on vision in pray detection, but rather on touch, chemoreception, and possibly electrolocation through rostral ampullary organs (Helfman et al. 2009). With sturgeons being particularly long-lived fishes, and having rare natural predators beyond the juvenile stage, their natural mortality rates of adults were historically low. With their sexual maturity attaining slowly, and their high fecundity (ovaries may account for 25% of female body mass) and long period of spawning intervals for females (every 3–5 years or longer), sturgeons are susceptible to overexploitation by humans. All 27 sturgeons species, belonging to the family Acipenseridae and grouped in four genera (*Acipenser*, *Huso*, *Scaphirhynchus*, and *Pseudoscaphirhyncus*), are listed on the IUCN Red List of Threatened Species. With 85% of sturgeon at risk of extinction, this group of fishes is regarded as most threatened animals group in the world.

With many factors influencing the decline of sturgeons populations, such as unsustainable exploitation, blocking of migration routes, destruction of habitats, water pollution, loss of genetic integrity of native species, and change in hydrological regime, question should be addressed about measures which supposed to stop rapid disappearance of this group of fishes. Due to demand for their caviar and meat, fishing pressure was affecting populations for over the 150 years. Ristić (1963a) reported significant decrease of sturgeons catch in Low Danube Section even before construction of Iron Gate dams. Although legal, fishing was unsustainable. Mrdak (2009) reported dramatic decrease of sturgeon catch in Skadar Lake drainage, from 1947–1966 period (two tones) to 1967–1976 (no catch), due to construction of stationary fish weir. This was governmentally constructed and controlled fish weir (from 1960 until the late 1980s), in order to harvest fish migrating from Adriatic Sea to Lake Skadar, via Bojana/Buna River (Mrdak 2009). With governments passing legislations on protection and implementing complete fish ban on sturgeons catch, the illegal catch (and caviar trade) is still occurring and continuously providing negative effect to any restoration actions. According to Jahrl et al. (2021), 1/3 of sturgeon caviar and meat products in Low Danube countries were sold illegally. Additionally, 19% of all samples came from wild sturgeon, which makes their origin and trade illegal. Habitat alteration (and subsequent loss) is due to construction of hydropower infrastructure, and enforcement of measures for navigation and flood protection. On the Volga River, after the construction of dams, sturgeons lose 80–85% of history spawning grounds, with beluga spawning grounds reduced almost 100% (Secor et al. 2000). Dams construction in Iron Gate area of the Danube River influenced variation in composition of bottom fauna (Janković et al. 1994; Djikanovic et al. 2015). Even in situation where there was no dam obstruction (e.g., the Ural River), sturgeons lost 50% of spawning grounds due to pollution and sedimentation (Secor et al. 2000). According to Freidrich et al. (2018), loss of genetic diversity is caused either by careless selection of breeders in stocking or ex situ conservation programs or by demographic bottleneck in natural populations. Danger of misidentification was exposed in 2016, when state authority approved stocking of Mura and Sava River with sterlet (15 and 93 specimens were released, respectfully), without previous survey, subsequently only to discovery that specimens belong to nonindigenous Siberian sturgeon (Govedič and Friedrich 2018). Once exotic sturgeon species, genotypes, or hybrids are introduced (deliberately or unintentionally), it can lead to loss of genetic characteristics of wild species, compromises its adaptation to a given set of habitat features and diminishes its population fitness by outbreeding depression (Freidrich et al. 2018).

Although the term "Western Balkans" is political, mostly used to describe area of former SFR Yugoslavia and Republic of Albania, hydrologically it represents area of Balkan peninsula (mainly) related to Black and Adriatic Sea drainage. In this contest, historically area of Western Balkan was inhabited by seven sturgeon species, sterlet (*Acipenser ruthenus*), Russian sturgeon (*Acipenser gueldenstaedtii*), Adriatic sturgeon (*Acipenser naccarii*), ship sturgeon (*Acipenser nudiventris*), stellate sturgeon (*Acipenser stellatus*), European Atlantic sturgeon (*Acipenser sturio*), and Beluga (*Huso huso*). With the Danube River Basin being biggest contributor to the Black Sea drainage in the Balkans, most of the sturgeons were inhabiting this river and its tributaries, with only Adriatic sturgeon and European Atlantic sturgeon entering rivers of Adriatic Sea drainage. However, after construction of Iron Gates (IG) Dam 1 and 2 (in 1970 and 1984, respectively) on the Danube River, sturgeons spawning migratory routes were blocked. Although the Danube River Basin experienced rapid decrease in sturgeons catch by mid-nineteenth century (Lenhardt et al. 2008), significant decline of sturgeons number was after constructions of IG dams and increase of intensive fishing pressure. Adriatic sturgeon and European Atlantic sturgeon experienced similar trend, with increased fishing pressure, pollution, and (in some cases) regulation/alteration to Adriatic Sea drainage rivers. Rare sturgeons research through the region of West Balkans are not sufficient to give better insight into real condition of wild populations.

2 Western Balkans Sturgeons Species

One of the most distinguishing differences between genus Huso and Acipenser is that former genus has a large mouth shaped as a crescent moon, occupying almost the entire width of the lower surface of the head, while latter genus have a suitably large transversal mouth, occupying only part of the lower surface of the head (Maximov et al. 2014).

2.1 Acipenser Ruthenus Linnaeus, 1758 (eng. sterlet, rus. стерлядь, srp. ке□ига, hrv. kečiga, slv. kečiga)

2.1.1 Distribution (Past and Present) in Europe and Western Balkans

In Europe occupy rivers in the drainages of Black, Caspean, Azov, Kara, and White Seas, while in Siberia occupy rivers in drainages of Ob and Yenisei River (Sokolov and Berdicheskii 1989; Freidrich et al. 2018). In the Danube and its tributaries: Sava, Alt, Tisza, Drava, Mura, Vah, Morava, Inn, Isar; Dnjeper, Dniester, Don, Kuban, and their main tributaries in the Black Sea range (Freidrich et al. 2018). In the Danube, regularly occurred up to Vienna, frequently to Linz, Passau, Regensburg, and even Ulm, with large autochthonous population between Passau and Regensburg (Hensel and Holčik 1997). In Western Balakans, abundant populations through the range, in the Danube, Sava, Kupa, Drava, and Mura River (Hensel and Holčik 1997; Govedič and Friedrich 2018).

Presently (Fig. 1), one small isolated population with limited reproduction occurs in the fragmented section of the Upper Danube (Austria), with populations in Middle Danube experiencing recovery in Slovakia and Hungary (Freidrich 2018; Freidrich et al. 2018). In Western Balkans, most abundant populations are near Smederevo, Belgrade, and Zemun, with viable population in Tisza, while populations in upper section of Danube in Serbia experienced severe decline (Cvijanović et al. 2017). Present in Sava, Drava, and Mura River in Croatia (Mrakovčić et al. 2006), while sporadic occurrence in Drava River (Slovenia) could be attributed to downstream migrations from Austria and its reintroduction program (Govedič and Friedrich 2018).

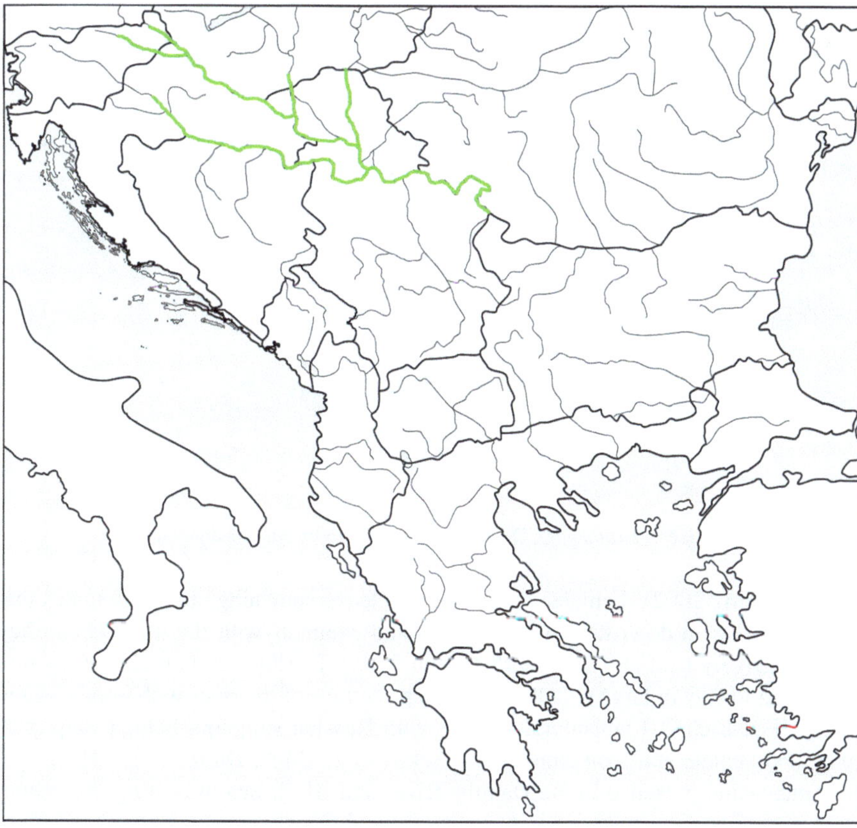

Fig. 1 Present distribution of *Acipenser ruthenus* in Western Balkans

2.1.2 Species Description, Biology, and Habitat

In dorsal row 12–17 scutes, with first dorsal scute not fused with head. Number of lateral scutes is 56–71, while there are 10–19 scutes in ventral row. Lower lip interrupted in middle, with four long fimbriate barbels above upper lip (Kottelat and Freyhof 2007) (Fig. 2). Maximal length and weight are 125 cm and 16 kg, respectively (Berg 1962; Birstein 1993; Freidrich et al. 2018). Maximum age is over 25 years (in aquaculture is even 33 years; Shljahtin 2006). Maturation is between 3–5 and 4–8 years for males and females, respectively (Reinartz 2002; Freidrich et al. 2018). Potamodromous (anadromous populations extirpated), occupying large rivers, usually in deep waters and with the fast current (Birstein 1993; Kottelat and Freyhof 2007). Spawns in April–June, on gravel and with strong current, where males dominate with 60–70% in number (Sokolov and Berdicheskii 1989). In the Danube River fecundity of the specimen (with body weight of 1 kg and age of 6+) is about 35,000 eggs (Ristić 1963c). Incubation lasts 4–5 days, after which larva drift in area with slow current and sand bottom (Sokolov and Berdicheskii 1989; Birstein

Fig. 2 Sterlets (*Acipenser ruthenus*), Danube, Serbia (Photo by M. Smederevac-Lalić)

1993; Reinartz 2002). Females are the first to leave spawning site, with males that follow, and swim downstream to sandbars and channels with the mud where they feed intensively (Sokolov and Berdicheskii 1989). During the winter, sterlets generally tend to stay in large shoals within deep sections of the river with good oxygen supply (Friedrich 2012). Natural hybrids with Russian sturgeon, beluga, ship sturgeon, and stellate sturgeon can occur (Tsekov et al. 2008; Dudu et al. 2011), with their increasing presence in the Danube River and Black Sea indicating decreasing trend in sturgeon stocks.

2.1.3 Conservation Status (IUCN Red List)

Endangered (EN), A2cde

Although local populations in the Danube River are still surviving, it is estimated that the native populations have declined by 60–70% over the last three generations (Gessner et al. 2022a).

2.2 *Acipenser Gueldenstaedtii Brandt & Ratzeburg, 1833 (Eng. Russian Sturgeon, Rus. русский осётр, Srp. руска јесетра, Hrv. Dunavska Jesetra, Slv. kašikar/Ruski Jester)*

2.2.1 Distribution (Past and Present) in Europe and Western Balkans

Used to occur in the Black Sea and its tributaries: the Danube (and its tributaries Olt, Sava, Tisza, Drava, Mura, Morava), Dnjester, Dnjeper, Don, and Kuban rivers (Freidrich et al. 2018). Prior to construction of Iron Gate dams on the Danube River, migrated upstream as far as Bratislava, sometimes Vienna and Regensburg (Birstein et al. 1997). In Western Balkans (Fig. 3), used to occur in the Croatian section of Drava and Sava rivers, as well as the Slovenian section of the Drava, Mura, and Sava rivers (Govedič and Friedrich 2018; Ćaleta et al. 2019). Ristić

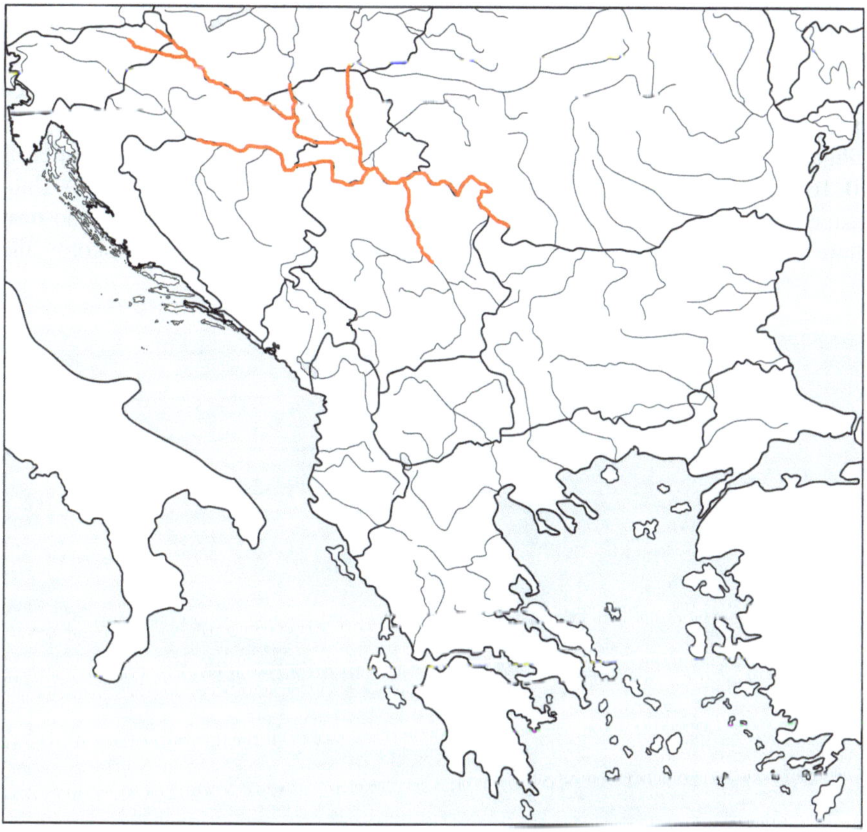

Fig. 3 Past distribution of *Acipenser gueldenstaedtii* in Western Balkans

(1967a, b) reported Russian sturgeon spawning grounds in the Serbian section of the Danube River at 892, 903, 912, 929, and 944 rkm.

Although catch statistic of Russian sturgeon from Serbia section of Danube River suggested extinction risk by mid-twenty-first century (Lenhardt et al. 2006a), sporadic findings in this section, as well as trace element assessment of juveniles' pectoral fin rays, suggest existence of special resident (potamodromous) form (Hensel and Holčik 1997; Jarić et al. 2011b). However, sporadic findings of Russian sturgeon could also be attributed to escape or released specimens from aquaculture, so the only section of the Danube River drainage in Western Balkans to have wild specimens is downstream form Iron Gate 2 Dam.

2.2.2 Species Description, Biology, and Habitat

Number of dorsal scutes is 8–19, while number of lateral and ventral scutes is 24–50 and 6–13, respectively (Kottelat and Freyhof 2007). Lower lip interrupted in middle, with the base of not fimbriate barbels closer to tip of snout than the mouth. Maximal length and weight are 236 cm and 115 kg, respectively (Birstein 1993; Kottelat and Freyhof 2007) (Figs. 4 and 5) Average body length and weight of males are 129.5 cm and 12.7 kg (respectively), while average body length and weight of females are 145.8 cm and 22.12 kg (Alavi-Yaganeh and Falahatkar 2013; Freidrich et al. 2018). Maximum age for males is 52 years, while females max age is 62 years (Shljahtin 2006). Maturation for males is 8–13 years, while females maturation is 10–16 years (Shljahtin 2006; Kottelat and Freyhof 2007; Freidrich et al. 2018). Until maturation, they spend time in sea, on mud and sand bottom of estuary and near shore zones, where they feed molluscs, crustaceans, and small fishes (Sokolov and

Fig. 4 Russian sturgeon (*Acipenser gueldenstaedtii*), Danube, Serbia (Photo by Đ. Petrović)

Fig. 5 YOY Russian sturgeon (*Acipenser gueldenstaedtii*), Czech Republic (Photo M. Smederevac-Lalić)

Berdicheskii 1989). Depth range in sea is from 2 up to 100 m (Vescei 2001). In the big rivers they occupy deep waters (2–30 m) with moderate current (Vescei 2001; Birstein et al. 1997), where they spawn on rocky and gravel substrate (Vescei 2001; Bloesch et al. 2006). Spawning is in July, with temperature 17–23 °C (Kottelat and Freyhof 2007). During upstream migrations individuals swim near bottom, while on downstream swim near surface in the fast current (Ristić 1963a).

2.2.3 Conservation Status (IUCN Red List)

Critically endangered (CR), A2bcde

Rare in the Black Sea basin, with estimation of decline by 90% of natural populations over the last three generations, and with the Danube River migrants becoming extirpated within the next 10 years (Gessner et al. 2022b).

2.3 *Acipenser Naccarii Bonaparte, 1836 (Eng. Adriatic Sturgeon, Rus. адриатический осётр, Alb. Blini I Adriatikut, Cnr. Jesetra Tuponoska, Hrv. Jadranska Jesetra, Slv. Jadranski Jester, Srp. јадранска јесетра)*

2.3.1 Distribution (Past and Present) in Europe and Western Balkans

Was endemic in the Northern part of the Adriatic Sea, the Po River (and its tributaries), as well as Adige, Brenta, Bacchiglione, Piave, Livenza, Tagliamento, Sile, Soča, Cetina, Mirna, Raša, Zrmanja, Krka, Neretva, Bojana/Buna, Drin and

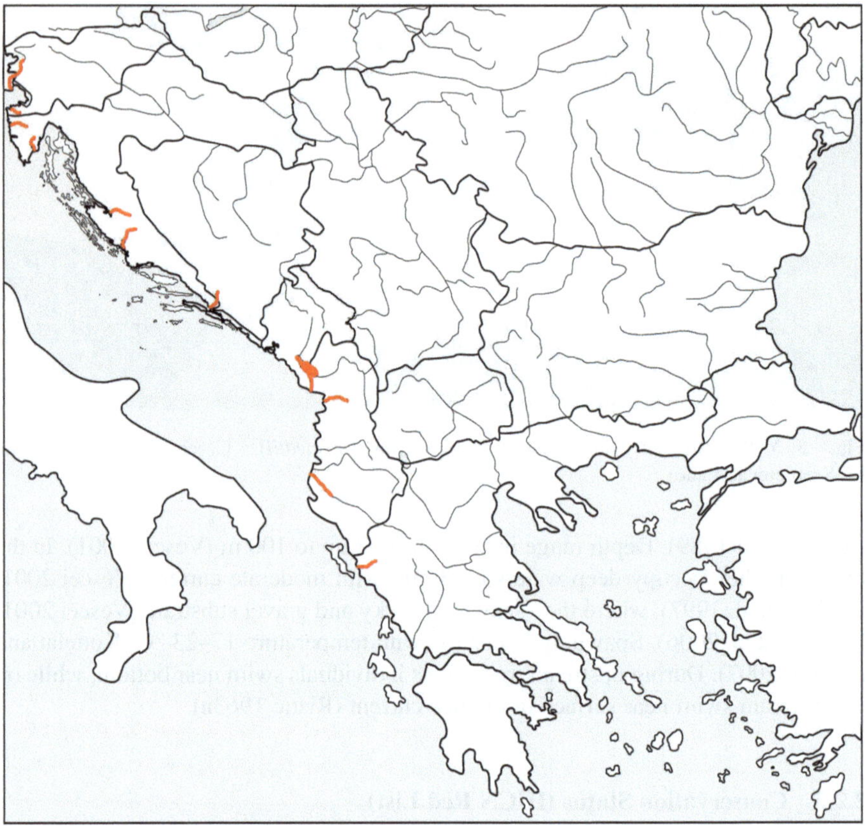

Fig. 6 Past distribution of *Acipenser naccarii* in Western Balkans

Vjosa rivers, as well as Skadar/Skodra Lake on the Montenegro-Albania border (Knežević 1985; Mrakovčić et al. 2006; Marić and Milošević 2011; Noakes and Bouvier 2013; Shumka et al. 2018; Freidrich et al. 2018; Ćaleta et al. 2019). Western Balkans (past) distribution in Fig. 6.

A small stock exists in the Po River, in a large part due to supportive stocking (Bronzi et al. 2022). In Western Balkans, statuses of species in Bojana/Buna and Drin rivers are unclear, although Ludwig and Kirschbaum (1998) reported presence of wild populations in Bojana/Buna basin.

2.3.2 Species Description, Biology, and Habitat

In dorsal row 10–14 scutes, while the number of scutes in lateral and ventral row are 32–42 and 8–11, respectively (Kottelat and Freyhof 2007) (Fig. 7). Barbels are not fimbriate, and their base is closer to tip of snout than to mouth. Snout is short and blunt, while lower lip is interrupted in middle (Kottelat and Freyhof 2007). Between

Fig. 7 Adriatic sturgeon (*Acipenser naccarii*) (Photo by T. Friedrich)

dorsal and lateral scutes, there are many small granuliform/star-shaped bony denticles (Kottelat and Freyhof 2007). Maximal body length and maximal body weight are 200 cm and 90 kg, respectively, while the average length is 100–180 cm (Hernando et al. 2009; Freidrich et al. 2018). Maximal age is over 50 years, while the age of first reproduction of males and females is 6–8 years and 8–13 years, respectively (Kottelat and Freyhof 2007; Freidrich et al. 2018). Anadromous species, but in sea remain in costal/estuarian waters (with depts at 10–40 m), while in freshwaters occupy large deep rivers (Bronzi et al. 2022). Hernando et al. (2009) reported presence of a landlocked population in the Ticino River. Reproduction occurs from May to July, on sandy and muddy substrate, with a rapid water flow (Bronzi et al. 2022). Benthic invertebrates and fish are main diet items (Kottelat and Freyhof 2007).

2.3.3 Conservation Status (IUCN Red List)

Critically Endangered (CR), A2bcde

Although restocking and possibly residual populations in Italy are thought to be slightly recovering, estimations are that natural populations decline by more than 99% over the last three generations, and it is now considered extinct from rivers in Croatia, Bosnia, and Herzegovina, Montenegro and Greece (Bronzi et al. 2022).

2.4 *Acipenser Nudiventris Lovetzky, 1828 (Eng. Ship Sturgeon, Rus. шип, Srp. сим, Hrv. Sim, Slv. Glatki Jeseter)*

2.4.1 Distribution (Past and Present) in Europe and Western Balkans

Common in the Danube River (Fig. 8) and its tributaries Sava, Tisza, Drava, Mura, Vah, Una, Kupa, Lonja, Maros, Morava, Prut, and Siret rivers (Hensel and Holčik 1997; Ćaleta et al. 2019). Historically, also present in Dnjester, Don, Kuban, Enguri, Rioni, and Sakarya rivers (Freidrich et al. 2018). Maximum body length and maximum body weight are 220 cm and 120 kg, respectively, while average length is 120,150 cm (Freidrich et al. 2018).

With only two specimens caught in 2003 and 2010 in the Danube River (Smonović et al. 2005; Guti 2011), and one in the Mura River in 2005 (Guti 2006), ship sturgeon is considered as possibly extinct in the Danube River Basin

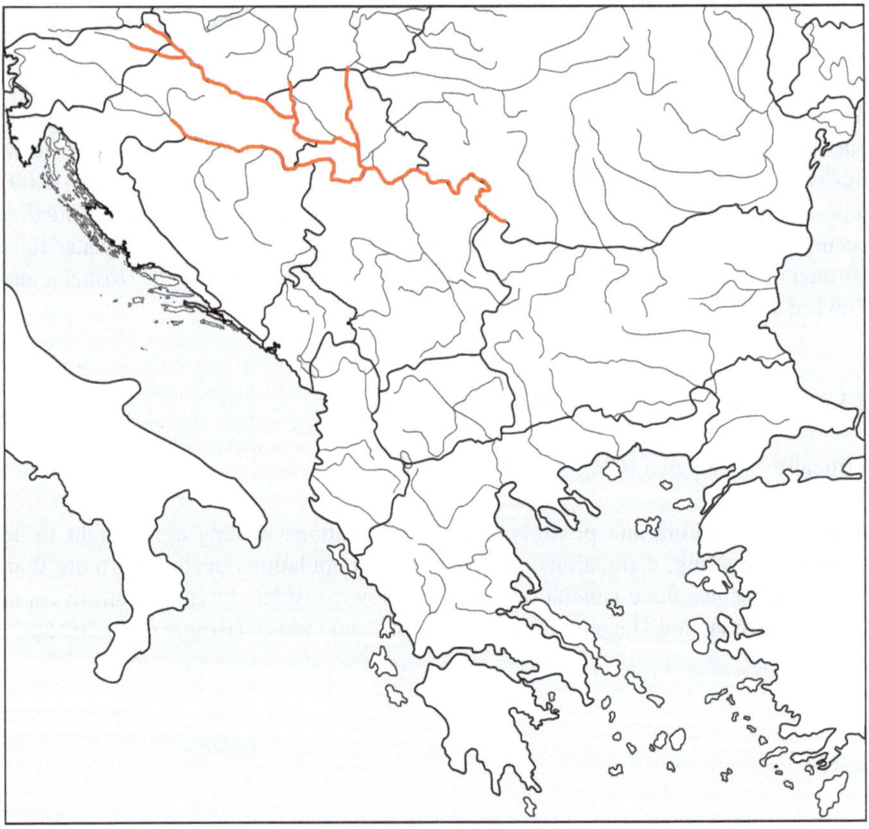

Fig. 8 Past distribution of *Acipenser nudiventris* in Western Balkans

(Jarić et al. 2016; Freidrich et al. 2018). This is in concordance with a recent eDNA study, which didn't provide evidence for this species existence in the Danube River (Freyhof et al. 2022).

2.4.2 Species Description, Biology, and Habitat

Most distinguished characteristic from other sturgeons species in Europe is continuous lower lip, with no interruption in middle. Numbers of dorsal and lateral scutes are 11–17 and 49–74, respectively (Fig. 9). First dorsal scute is larger than the following ones, fused with the head (Kottelat and Freyhof 2007). Maximum body length is 220 cm, while maximal body weight is 120 kg (Freidrich et al. 2018). Average body length is between 120 and 150 cm. Life expectancy is over 30 years, with maximal recorded age of 36 years (Sokolov and Vasilev 1989; Freidrich et al. 2018). Forms both anadromous and resident populations, with only resident strain occupying the Danube River Basin (Hensel and Holčik 1997). Males spawn first time at 6–12 years, while females spawn at 12–18 years (Kottelat and Freyhof 2007; Freidrich et al. 2018). Spawning is from March to May in rivers main flow (fast current and water temperature 10–15 °C), with rocky and gravel bottom (Sokolov and Vasilev 1989; Reinartz 2002; Kottelat and Freyhof 2007). Males and females spawn every 1–2 and 2–3 years, respectively. Ship sturgeon spends the winter in deep areas with clay bottom, without feeding (Friedrich 2012). Feeds on a wide variety of molluscs, crustaceans, and benthic fishes (Kottelat and Freyhof 2007).

Fig. 9 Ship sturgeon (*Acipenser nudiventris*), Danube, Serbia (Photo by P. Simonović)

2.4.3 Conservation Status (IUCN Red List)

Critically Endangered (CR), A2cde

Although there are no recordings in the Danube River over the 10 years, it is suspected that this species has undergone a population decline of more than 99% in the past three generations (Freyhof et al. 2022).

2.5 Acipenser stellatus Pallas, 1771 (Eng. Stellate/Starry Sturgeon, Rus. севрюга, Alb. Blini Turigjate, Cnr. Pastruga, Hrv. Pastruga, Slv. Pastruga, Srp. паструга)

2.5.1 Distribution (Past and Present) in Europe and Western Balkans

In the Black Sea and its drainage, with the Danube River (and its tributaries Sava, Tisza, Drava, Mura, Jiu, Olt, Siret, and Prut rivers), Dnjeper, Dniester, Don, Kuban, Enguri, Rioni, Tskhenistskali, Coruh, Yesilirmak, Kizilirmak rivers, as well as in the Aegean Sea and its tributaries Struma/Strymonas and Evros rivers (Paschos et al. 2008; Freidrich et al. 2018). Western Balkans (past) distribution is in Fig. 10.

It is considered extinct in the Aegan Sea and upstream from Iron Gate dams in the Danube River (Paschos et al. 2008; Freidrich et al. 2018; Ćaleta et al. 2019). Small wild stock is restricted to the coastal waters of Romanian Black Sea section, with some indication of successful spawning in the Danube River (Holostenco et al. 2013).

2.5.2 Species Description, Biology, and Habitat

In dorsal row, there is between 9 and 16 scutes, while lateral and ventral row have 26–43 and 9–14, respectively (Kottelat and Freyhof 2007) (Figs. 11 and 12). Base of barbels closer to mouth than to tip of snout, with lower lip interrupted in the middle (Kottelat and Freyhof 2007). Maximal body length and weight are 290 cm and 80 kg, respectively (Freidrich et al. 2018). Average body length is between 120 and 180 cm, and maximal age is 35 years (Freidrich et al. 2018). Anadromous species, with spawning migration occurring from April to September. Some studies show that stellate sturgeon specimens feed at the river mouth until they reach 3–4 years old (Holostenco et al. 2013). Males reproduce for the first time at 6–12 years, while females reproduce at 8–14 years, with males and females reproducing every 2–3 and 3–4 years, respectively (Kottelat and Freyhof 2007; Freidrich et al. 2018). Females remain at spawning sites for 10–12 days, while males remain around six weeks (Kottelat and Freyhof 2007). Larvae lift in the water column and drift with current (Holostenco et al. 2013). Stomach content of juveniles, feeding in the river mouth

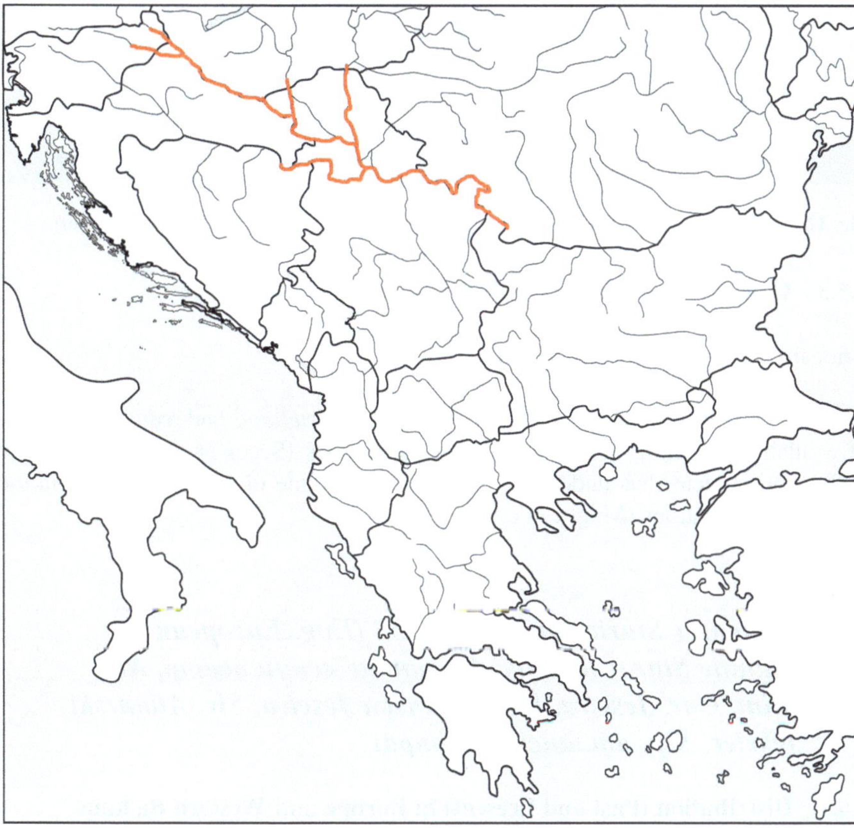

Fig. 10 Past distribution of Acipenser stellatus in Western Balkans

Fig. 11 Stellate sturgeon (*Acipenser stellatus*), Danube, Romania (Photo by S. Hont)

area of the Black Sea, contains little crustaceans and worms (polychaetes), while at sea it feeds on crustaceans, molluscs, and benthic and pelagic fishes (Kottelat and Freyhof 2007, Holostenco et al. 2013).

Fig. 12 YOY stellate sturgeon (*Acipenser stellatus*), Danube, Romania (Photo by S. Hont)

2.5.3 Conservation Status (IUCN Red List)

Critically Endangered (CR), A2bcde

After the construction of dam on the Volga River, *A. stellatus* had reduction by 40% of available spawning grounds in the upper reaches (Secor et al. 2000; Maltsev 2009). This species has undergone a population decline of more than 95%, in the past three generations (Mugue et al. 2022).

2.6 Acipenser Sturio Linnaeus, 1758 (Eng. European Atlantic Sturgeon, Rus. Атланти□еский осётр, Alb. Blini, Cnr. Jesetra, Hrv. Atlanska Jesetra, Slv. Atlanrski Jeseter, Srp. атланска јесетра)

2.6.1 Distribution (Past and Present) in Europe and Western Balkans

In the North Sea, with tributaries Eider, Elbe, Weser, Ems, Rhine, Maas, Scheldt, Thames, Trent, Severn, Seine; in the Atlantic Coast, with tributaries Loire, Gironde-Garonne-Dordogne, Adour, Douro, Guadiana, Guadalquivir; in the Mediterranean Sea, with its tributaries Ebro, Rhone, Saone, Tiber; in the Adriatic Sea, with tributaries Po, Adige, Isonzo, Soča, Mirna, Raša, Zrmanja, Krka, Cetina,Nereteva, Drin, Buna, Pinio, as well as Skadar/Skodra Lake; the Aegan Sea with tributaries Struma, Meric, Evros, Nestos, Pinos; in the Black Sea with tributaries Danube, Rioni, Ingouri, Kizilirmak, Sakarya (Knežević 1985; Sokolov and Berdicheskii 1989; Mrakovčić et al. 2006; Paschos et al. 2008; Marić and Milošević 2011; Freidrich et al. 2018; Ćaleta et al. 2019). Western Balkans (past) distribution in Fig. 13.

Present in the Gironde-Dordogne-Garonne Basin, with its marine distribution area extending from the Bay of Biscay to the North Sea (Freidrich et al. 2018). Talevski et al. (2009) suggest that *A. sturio* shouldn't be considered as the inhabitants of the Skadar/Shkodra Lake. Reports for the Ombla River near Dubrovnik (Croatia) are not confirmed (Ćaleta et al. 2019). Its presence in the Iron Gate area on the Danube River is highly unlikely, since last few specimens were caught during 1948–1954 period (Lenhardt et al. 2006a).

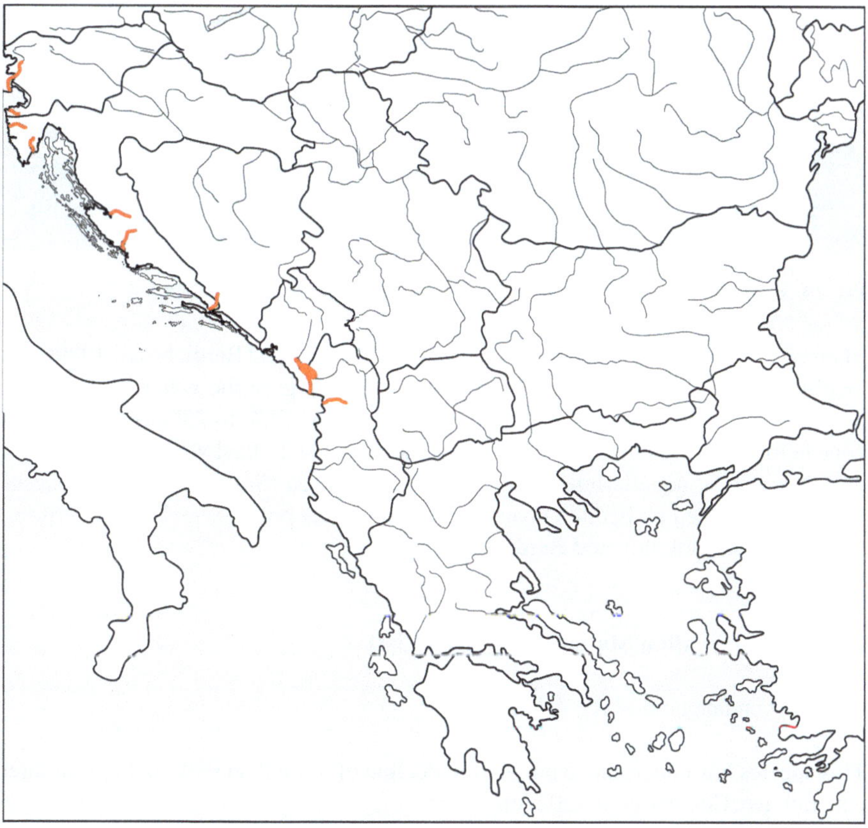

Fig. 13 Past distribution of Acipenser sturio in Western Balkans

2.6.2 Species Description, Biology, and Habitat

Number of dorsal scutes is 9–16, with first dorsal scute fused with head. Number of lateral and ventral scutes are 24–40 and 9–14, respectively (Kottelat and Freyhof 2007) (Fig. 14). There are many rhombic denticles between dorsal and lateral scutes (Sokolov and Berdicheskii 1989). Barbels are not fimbriate, and their base is midway between mouth and tip of snout (or closer to mouth). Lower lip interrupted in middle. Maximal length and weight are 600 cm and 1000 kg, respectively (Sokolov and Berdicheskii 1989; Kottelat and Freyhof 2007). Average body length is between 150 and 300 cm, and maximal age is over 60 years (Freidrich et al. 2018). Anadromous species, with spawning migration occurring from March to August. They migrate in the main river channel at depths of 2–8 m and temperature 12–17.5 °C (Sokolov and Berdicheskii 1989), with males entering rivers two three weeks sooner than females. Males reproduce for the first time at 10–12 years, while females reproduce at 13 16 years, with males and females reproducing every year and every second year, respectively (Kottelat and Freyhof 2007; Freidrich et al. 2018).

Fig. 14 European Atlantic sturgeon (*Acipenser sturio*) (Photo by I. R. Le Barh)

Their eggs are sticky and adhere to the bottom (Sokolov and Berdicheskii 1989), and incubation period varies from 3 to 14 days (depending on the water temperature). Juveniles spend their first years in the brackish waters (5‰ to 25‰) of the estuary zone before moving out to sea (Holostenco et al. 2013). In freshwater, juveniles feed on the larvae of aquatic insects, worms, and Mollusca (Sokolov and Berdicheskii 1989). Adults feed on benthic invertebrates (mollusca, polychaete, isopod, shrimps) and small fish (Sokolov and Berdicheskii 1989).

2.6.3 Conservation Status (IUCN Red List)

Critically endangered (CR), A2cde

This species has undergone a population decline of more than 99%, in the past three generations (Gessner et al., 2022c).

2.7 Huso Huso Linnaeus, 1758 (Eng. Beluga, Rus. белуга, Alb. Blini Turishkurte, Cnr. Moruna, Hrv. Moruna, Slv. Beluga, Srp. моруна)

2.7.1 Distribution (Past and Present) in Europe and Western Balkans

The Black Sea, with the Danube River and its tributaries (Sava, Tisza, Drava, Mura, Jui, Morava, Olt), Dnjepr, Dniester, Don, Kuban, Enguri, Rioni, Coruh, Yesilirmak, Kizilirmak and Sakarya rivers, and the Adriatic Sea and the Po River (Freidrich et al. 2018). Western Balkans (past) distribution in Fig. 15.

In the 1970s, species has been extirpated from the Adriatic Sea (Gessner et al. 2010).

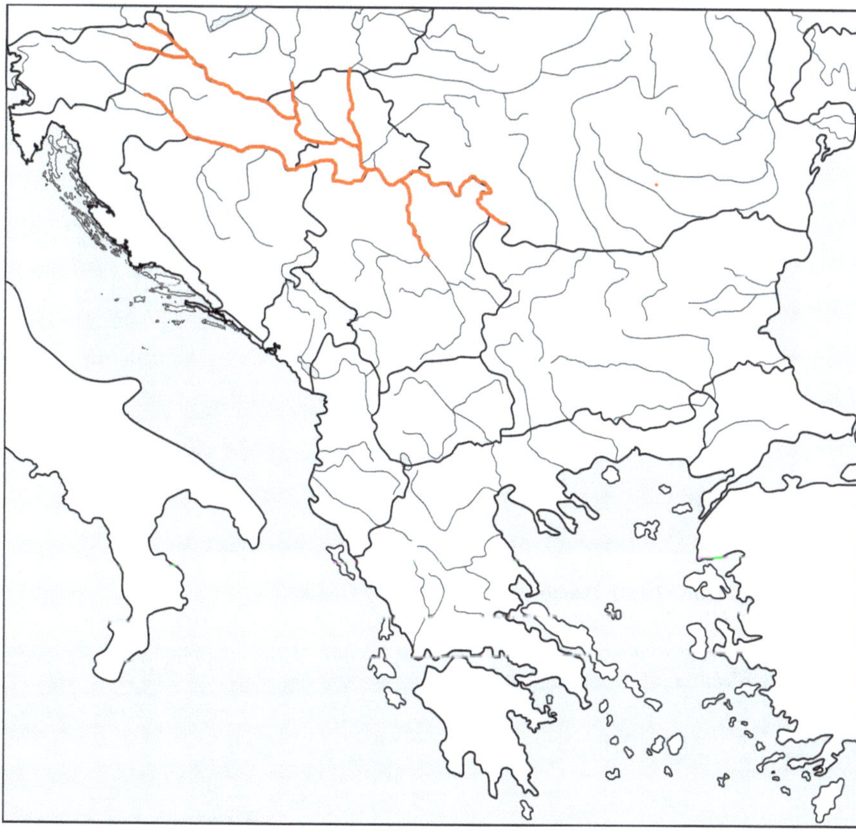

Fig. 15 Past distribution of Huso huso in Western Balkans

2.7.2 Species Description, Biology, and Habitat

Number of scutes in dorsal, lateral, and ventral rows are 9–17, 37–53 and 7–14, respectively (Kottelat and Freyhof 2007) (Figs. 16 and 17). In mature specimen scutes are covered with a soft skin (Sokolov and Berdicheskii 1989). Crescent-shaped large mouth, with lower lip interrupted in middle. Maximal length and weight are 800 cm and 2000 kg, respectively, while maximum age is over 100 years (Freidrich et al. 2018). Anadromous species, and during its stay in marine environment it occupies pelagic zone, while descending as deep as 160 m, or even 180 m (in the Black Sea, Sokolov and Berdicheskii 1989). Its vertical distribution is varying due to availability of its food organisms. The Danube River migration starts in January (with early spring), or in the second half of March (late spring) at water temperature of 4–5 °C (Maximov et al. 2014). Maturation of males and females is 10–16 years and 14–20 years, respectively (Freidrich et al. 2018). During the mid-twentieth century, 24 years female had body length, weight and fecundity of 264 cm, 148 kg, and 594,000 eggs, respectively (Ristić 1963a). Danube specimens

Fig. 16 Beluga (*Huso huso*), Danube, Serbia. (Photo by G. Žikić)

Fig. 17 YOY beluga (*Huso huso*), Danube, Romania. (Photo by S. Hont)

with Spawns every 3–4 years, during April–June, in strong-current habitats of large and deep rivers, with stone or gravel bottom (Kottelat and Freyhof 2007). They share the same spawning ground, in the same period of time, with sterlet (Holostenco et al. 2013). Larvae are pelagic for 7–8 days, and drift with current, while YOY migrates downstream in crowds during their first summer (Kottelat and Freyhof 2007). It is solitary species, but it agglomerates in larger groups during the winter (Maximov et al. 2014). Start feeding on vertebrates at a very early age, consuming shad (*Alosa* spp.), roach (*Rutilus rutilus*) and other fishes, with larger individuals consume aquatic birds and even baby seals (Sokolov and Berdicheskii 1989).

2.7.3 Conservation Status (IUCN Red List)

Critically Endangered (CR), A2bcd

This species has undergone a population decline of more than 90%, in the past three generations (Gessner et al. 2010).

3 Sturgeons Research (and Conservation Assessment) in Western Balkans

Sturgeons in Western Balkans region were group of interest for scientists over the 100 years, with most authors reporting them as part of fish communities in Albania (Poljakov et al. 1958), Bosnia and Hercegovina (Ćurčić 1910), Croatia (Trojanović 1934; Vuković and Ivanović 1971), Montenegro (Knežević 1985), Slovenia (Munda 1926). However, more thorough research started in the second half of the twentieth century, with the work of Janković (1958) and Ristić (1959, 1963a, 1963b, 1963c, 1967a, 1967b, 1967c, 1967d, 1969a, 1969b, 1970, 1971).

Although Taler (1948, 51) did report catch of two beluga from 1948, from the Sava River (near Lonje) and the Danube River (near Kladovo), with body weight 70 kg and 187 kg, respectively, along with report that the sturgeon catch in 1938 was 44,940 kg, most of information were description of beluga, ship, sterlet, Russian, Atlantic, stellate, Adriatic sturgeon distribution.

Perhaps first significant and detailed study of sturgeons in Western Balkans was survey on sterlet biology and ecology (Janković 1958), between 1951 and 1957. On the Danube River section, between 1215 rkm and 991 rkm, sterlets were analyzed for length-weight relationship, morphology, fecundity, diet, and age structure. Some conclusions were that sterlet catch mostly comprised of 3+ age specimens, with significant portion of 1+ and 2+ specimens, which suggests unsustainable spawning. Additionally, diet composition revealed that Trichoptera, Chironomidae, and Amphipoda were present through the year. Ristić (1959) was monitoring sterlet migrations (along with carp, catfish pikeperch, and bream) in the Danube, Sava, and Tisa Rivers, during 1952–1955, and reported both upstream and downstream migrations, with specimens almost reaching the Danube River Delta (in winter of 1955). Significant research was from 1951 to 1957 (Ristić, 1963a,b,c) when 6597 specimens of Russian sturgeon, beluga, stellate sturgeon, and sterlet were analyzed in the Danube River section from Sip (944 rkm) to Grabovica (886 rkm). List of islands and sand surges, along with table with fishermen data (number of fishermen on each sector, nets, hooks, type of gear), was provided. Some of the findings included males to female ratio for beluga (17.6:1), stellate (1.1:1), Russian sturgeon (2.6:1), based on which author concluded that spawning success would have negative trend. Provides description of beluga and Russian sturgeon, with the biggest specimens being 264 cm/148 kg (April 1955, near Sip) and 215 cm/47 kg (1951, near Sip), respectively. Data about spring and winter forms were provided, as well as average Russian sturgeon in Iron Gate section of the Danube River (20 kg, 17–18+, 125–135,000 eggs). Decline of catch was supported with data from the 1900s; Romania 350–400,000 kg, Bulgaria 12–150,000 kg, Serbia 120,000 kg annually, while in the 1960s all three countries had 120,000 kg annually. Ristić (1963b) also provided assessment of Atlantic sturgeon and ship sturgeon decline and gave insight about ship sturgeon protection (law in 1937), which was not successful. Data about stellate sturgeon and its catch in this section of the Danube River, from the 1890s and 1900s (average specimens of 180–200 cm, 80 kg, and fecundity for 15+, 13 kg, 145–160,000 eggs) was reported. Ristić (1963c) also provided description of sterlet and its two forms (long and short snout), biggest caught specimen (body length and weight 100 cm and 9 kg, respectively). It also provided data on correlation between water level fluctuation and catch of beluga, overview of age classes and spawning grounds (near Sip, Mala and Velika Vrbica, Milutinovac, Ljubičevac) in this section of the Danube River. Another detail and comprehensive survey was from 1948 to 1954 (Ristić 1967a, b, c, d) when 3151 specimen of Russian sturgeon (2269 males, 882 females) were assessed on the Danube River section between Sip (rkm 944) and Grabovica (rkm 886). Hydrological data (annual water flow of 5600 m^3/s) was available, with data of bottom substrate shift from rock/gravel (Sip), gravel/send (Davidovac-Kostel section) to gravel/sand/mud (Grabar), also available. Data of males to female ratio for spring and winter forms showed 4:1 and 1.9:1, respectively. Reports of Russian sturgeon catch were available, with highest being in Korbovo (rkm 912), then Sip, Ljubičevac, Davidovac-Kladušnica-Kladovo section, then Vajuga, and Kostel. Observation was that increase of velocity (m/s) and water level was good for spawning. Reports of spawning places for winter forms were at Sip, Ljubičevac, and Vajuga (with water temperature 6–11 °C), while spawning places for spring forms were Korbovo, Kostel, Davidovac, Kladušnica, and Kladovo (12–14 °C). Additionally, wintering places were reported at Sip, Vajuga, Ljubičevac, with possibly at Veliki and Mali Kazan, Golubinje, Gospojin Vir, Brnjica. Data was provided for age structure of winter forms (8–14+ in males and 10–23+ in females) and spring forms (7+ in males and 10–12+ in females). Conclusion was that the populations are decreasing. Following years, Ristić (1969a, b, 1970, 1971) provided comprehensive data for sterlet. From 1949 to 1967 (Ristić 1969a), 5137 sterlets were analyzed in the Danube, Sava, and Tisza Rivers. Ratio of specimens younger than 4+ in total catch was 94–99% in upper section, 50–53% in mid-section, with 79% in lower section of the Danube River, while in the Sava River it was 92% and in the Tisza River 93%. This survey provided basis for protection by defining protective measures (47 cm for sterlet total length, age of 4+, mash size for nets >45 mm). In a survey (Ristić 1969b) at the Danube River between 1142 and 1144 rkm, 1975 specimens were collected, along with geological and hydrological data. Age structure of population was assessed, with 0+/1+ (41%), 2+/3+ (42%) and 3+/and older (17%). Recaptured specimens (1.8% of all marked specimens) couldn't provide some reliable conclusions. Some conservation measures were recommended, such as ban on sterlet rings and nets with mesh size smaller than 50 mm, with also suggestion to put the fishing ban for 60 days (between April–May). Also, Ristić

(1969b) reported spawning places in the Danube River, at 1312 rkm, 956 rkm, 944 rkm, in the Sava at 149 rkm, and in the Tisza at 134 rkm. In 1952–1962 survey (Ristić 1970) on the Danube River, marked and recaptured sterlet showed that they migrated mainly downstream, with distances from 210 km to 322 km, and even 524 km. Analysis of 7313 sterlets in the Danube, Sava, and Tisza Rivers (Ristić 1971), during 1948–1969, provided data on two morphs, short and long snout. With the age of specimens between 0+ and 19+, of total number of samples, specimens with long snout were prevailing (67%), with 33% of short snout. Observation of spawning dynamic shows that short snout specimens and long snout are successive (long snout starlet arrives after short snout).

During the 1980s and 1990s, sporadic reports of sturgeons were part of fish community surveys in Albania (Rakaj 1995) and Montenegro (Knežević 1985; Marić, 1995), while reports in Slovenia focused on sterlet (Pintarič 1982; Gregori 1988; Jeremko 1998). Despite construction of Iron Gate dams, research in Serbia (Stamenković 1991; Janković 1993) was still focusing on that section of the Danube River, with reports of decrease in numbers of localities for fishing sturgeons and mass upstream migrations of sterlets after construction of the dams.

At the beginning of the twenty-first century, number of publications concerning sturgeons and their protection in Western Balkans rapidly increased (Lenhardt et al. 2004, 2005a, 2006a, 2009, 2012, 2014; Jarić et al. 2009a, b, 2010, 2011a, b, c, 2014, 2015; Smederevac-Lalić et al. 2011; Cvijanović et al. 2015, 2017). Although beluga, Russian sturgeon, Atlantic sturgeon, and ship sturgeon are investigated (Lenhardt et al. 2005a, 2006a; Jarić et al. 2009b, 2011a), main focus was on sterlet (Lenhardt et al. 2004, 2009, 2012; Ognjanović et al. 2008, Jarić et al. 2011c; Cvijanović et al. 2015, 2017), with nonindigenous paddlefish (*Polyodon spathula*) and Siberian sturgeon (*Acipenser baerii*) also subject of interest (Lenhardt et al. 2006b; Jarić et al. 2019; Govedič and Friedrich 2018). Sterlet survey in the Danube River, between 2002–2003 (Lenhardt et al. 2004) showed shift of length-frequency distribution toward lower length classes, as well as decrease in sterlet catch. In spring and fall of 2001, at the Danube River downstream from Iron Gate 2 dam, females of beluga, Russian sturgeon, and sterlet were used for comparative study (Lenhardt et al. 2005a) of the biometry, gravimetry, and protein contents of the oocytes. It showed that beluga oocyte diameter, volume, wet and dry mass were the largest followed by Russian sturgeon, then starlet. Oocyte water and relative protein content of all three sturgeons were very similar, with the water content results at the lower range. Based on the beluga and Russian sturgeon catch data (period 1960–1997) time-dependent model was deployed (Lenhardt et al. 2006a), and it predicts extinction of Russian sturgeon around the middle of century, with beluga being extinct at the middle of the millennium. Existence of two morphs in sterlet, short and long snout, was discussed in Ognjanović et al. (2008). Lenhardt et al. (2008) provide data on sturgeon catch decline in the Danube River, as well as review of scientific research and activities related to their protection. Change in Fulton's condition factor in juvenile sterlets during the year was reported by Lenhardt et al. (2009). Additionally, difference in hepatosomatic index, which is due to difference in sterlet diet in nature through the year, was also reported, as well as prevalence and intensity of parasite *Skrjabinopsolus semiarmatus*. Jarić et al. (2009a) assessed different population viability analysis (PVA) methods and models developed for sturgeon species and suggested that all sturgeon PVAs generally produced the conclusion that slow physical growth and discontinuous spawning of sturgeons result in low intrinsic population growth rates, slow recovery from exploitation, and high sensitivity to harvest. Additionally, they suggested that the minimum length limit should be at least several cm longer than the length at which individuals reach maturity in order to enable at least one or two spawning events to occur prior to harvest. Also, authors concluded that stocking can represent only a temporary measure, and restoring the spawning habitat remains the only long-term solution for maintaining viable populations. Another analysis (Jarić et al. 2009b) deployed five different statistical methods, based on sighting data, for assessment of sturgeons extinction probability. All methods provide significant probability that the Atlantic sturgeon is extinct (from '66 and '70), while it suggests that ship sturgeon is still present, but it will be extinct in the next couple of decades. Another PVA analysis (Jarić et al. 2010) revealed a large sensitivity of the Danube sturgeon populations to changes in the natural mortality, fecundity, age at maturity, and spawning frequency. It showed that species with the lowest proportion of the adults in a population, such as the ship sturgeon and the Atlantic sturgeon, were also much more vulnerable than the other species at the same population size. Sterlet has higher adult mortality than the other sturgeon species, which probably reduces the positive effects of the stocking with adults. Analysis (Jarić et al. 2011b) of Sr:Ca ratio in the fin ray of the juvenile Russian sturgeon caught below the Djerdap II dam indicates that this individual has never left the freshwater. This finding could be explained either by the theory of existence of a special resident, potamodromous form of the Russian sturgeon in the Danube or as a specimen escaped or released from some aquaculture facility. Accumulation of heavy metals in muscle, liver, and gills of sterlet was assessed in Jarić et al. (2011c), and liver appeared to be the main heavy metal storage tissue, while the lowest levels of analyzed metals were in muscle. Status and trends in management of sturgeon species and the development of its aquaculture (in Romania, Bulgaria, Serbia, Ukraine, and Moldova) was assessed in Smederevac-Lalić et al. (2011). Statistical analysis (Lenhardt et al. 2012) of 15 morphological traits of wild and reared sterlets showed significant differentiation in 11 traits, with reared specimens having significantly shorter pectoral fins and stockier body. Lenhardt et al. (2014) gave insight to impact of pollution on sterlet, genetic analysis of the Danube sturgeons, sturgeon catch and protection in Serbia. Jarić et al. (2014) assessed trends in modeling over time within sturgeon research and show that population models mainly focused upon the impact of fisheries, dams, stocking, genetics, and sensitivity analysis, while habitat suitability models, hydrodynamic models, bioenergetic models, and general statistical models focus on species distribution, survival, growth, and spawning migrations. A simple life-table metamodel was introduced by Jarić et al. (2015), which facilitates population and fishery assessments across entire groups of fish. The method also allows utilizing life history data from well-studied species to infer fishing mortality thresholds for other, poorly studied species within the same group. Input parameters required to apply the presented approach include female fecundity, age at maturity, longevity, adult, subadult and YOY survival, and spawning frequency. During 2007–2009, wild starlet populations, from Middle and Lower Danube and Lower Tisza rivers, were investigated (Cvijanović et al. 2015; Cvijanović et al. 2017) for genetic diversity with the use of mtDNA and microsatellite loci. These analyses showed variation in the detection of a genetic bottleneck, as well as revealing different effective population size.

Recently, new data on distribution and assessment of sturgeons in Western Balkans were available in Albania (Shumka et al. 2018), Croatia (Ćaleta et al. 2019), and Slovenia (Govedič and Friedrich, 2018). Also, ongoing research of sterlet in the Danube River (project WePass 2), provides some insight into migration patterns. Sterlet specimen caught and tagged in October 2021, near Ljubičevac (890 rkm), was detected some 14 days later downstream, at the 600 rkm (Cvijanović et al. unpublished data). Data suggest that tagged sterlet migrated downstream through turbines, which is in concordance with conclusion of Honsig-Erlenburg and Friedl (1999). However, sterlet specimens caught near Ljubičevac (890 rkm) in Jun 2022, tagged and released downstream at Prahovo (860 rkm), were migrating upstream and reached Iron Gate 2 Dam (Cvijanović et al. unpublished data).

4 Parasites and Food Composition

Sturgeons are unique among fish, in possessing a remarkably diverse assemblage of host-specific parasites (Skryabina 1974; Choudhury and Dick 2001). Information on the sturgeon parasites composition and distribution are summarized (see Table 1 from Choudhury and Dick 2001). The parasites of acipenserids (sturgeons) exhibit no apparent specificity for either of the two major lineages of sturgeons, Huso or Acipenser L., where their species coexist (Ponto-Caspian region and Amur River drainage). Only two parasites, *P. hydriforme*, considered by Raikova (1994) to be an “ancient parasite,” and the intestinal fluke, *Crepidostomum auriculatum* Wedl, occur in every major drainage basin where sturgeons occur (Choudhury and Dick 2001). The absence, in Europe, of *C. auriculatum* in acipenserids other than the sterlet (*A. ruthenus* L.) is noteworthy and, lacking satisfactory contemporary ecological reasons, requires a historical explanation. Its presence in sterlet in the Ponto-Caspian drainage (Skryabina 1974; also lists a historical record in the Rhine) may be the result of a single dispersal event as recent as the Pleistocene. The highest number of endemic parasites is found in sturgeons of the Ponto-Caspian basins (seven parasite species).

The parasite fauna supports the hypothesis of a former continuous distribution of Nearctic and eastern Palaearctic (Siberian) sturgeons that was subsequently fragmented by geological and climatic events, resulting in major displacement of ancestral sturgeon populations in North America. The parasites also reflect a historical connection with marine and brackish water environments that pre-dates the diversification of sturgeons, indicating that ancestral acipenserids were diadromous (Choudhury and Dick 2001). The parasites of sturgeons have been studied by several authors (Dubinin 1952; Dogiel and Bykhovskiy 1939; Shulman 1954; Nechaeva 1964; Skryabina 1974).

Sterlet presents the only sturgeon species which is a still numerous in the Serbian part of the River Danube. The sterlet collected along the Danube's course through the Belgrade Region were examined and 13 helminth species have been recorded: four species of Trematoda (*Skrjabinopsolus semiarmatus*-dominant species, *Sanguinicola inermis*, *Posthodiplostomum cuticola* and *Azygia lucii*), one species of Cestoda (*Amphilina foliacea*), four species of Nematoda (*Contracaecum bidentatum*, *Contracaecum sinipercae*, *Cystoopsis acipenseris,* and *Capillospiura ovotrichuria*) and four species of Acanthocephala (*Acanthocephalus anguillae*, *Acanthocephalus lucii*, *Pomphorhynchus laevis* and *Pomphorhynchus bosniacus*). Three of these mentioned species are specific for representatives of Acipenseridae—*Skrjabinopsolus semiarmatus*, *Contracaecum bidentatum,* and *Amphilina foliacea* after Moravec et al. (1989), also according to the opinion of Kakacheva-Avramova (1983), who listed the above-mentioned species and the species *Cystoopsis acipenseris*, while many other parasite species of fish host other types of parasites.

Relevant researches on the parasite fauna of sterlet in the Danube River are very rare (Moravec et al. 1989). In Serbia, only a few data are able to cited here: Jankovic (1957); Cakić (1986a, b). Janković (1957) reported the following endo-parasitic species—*Amphilina foliacea*, *Ascaris* sp., *Echinorhynchus gadi*, and undetermined representatives from Class Trematoda, Subclass Digenea. After almost 30 years, by examination of Danubian starlet helminths, the presence of three helminth parasitic species—*Skrjabinopsolus semiarmatus*, *Contracaecum bidentatum,* and *Acanthocephalus anguillae*—has been published (Cakić et al. 2008).

The main food of the sterlet is various benthic organisms, the most important of which are insect larvae, primarily Chironomidae, Trichoptera, Ephemeroptera, and Simuliidae. It also feeds on larvae of Plecoptera and Heleidae, small molluscs of the genera Sphaerium, Pisidium, and Viviparus, representatives of Oligochaeta, Polychaeta, Hirudinea, and other benthic invertebrates. During the spawning period of other fish species, a significant part of their diet is their eggs.

After study of Ristić (1970), there were only sporadic research activities, mostly focused on sterlet food composition, growth, and length-weight relationship (Janković et al. 1994). More detailed analyses were originated at the beginning of the twenty-first century. In 2002 and 2003, research conducted on specimens collected from the Danube near Belgrade involved analysis of morphological variability, sterlet growth, and its parasites (Kolarević 2004; Lenhardt et al. 2004, 2009).

The diet of sterlet mostly comprised bottom fauna taxa. Sterlet in the Danube feed mainly on larvae of Trichoptera, Chironomidae (Diptera), and Gammaridae (Amphipoda) throughout the year and represented the main part of its diet (Janković 1958). In a study conducted in 1986, Trichoptera and Chironomidae were again found to be the main source of its food in the Danube River, while its main food in the Danube section between the two dams in the Djerdap gorge were Hirudinea (91%), which is probably due to the shift in benthic fauna composition caused by the impoundment by the dam (Janković et al. 1994). The most recent studies of sterlet nutrition, conducted by Lapkina et al. (2005), have shown that Chironomidae were dominant sterlet food in July, while the leeches were dominant in August and September (representatives of six genera and three families). Leeches represented 100% of the sterlet diet during September, which has led to an increased sterlet growth (2.2 g/day), in comparison with its growth in July (1.8 g/day; Lapkina et al. 2005).

Analysis of data for sterlet's diet on nine locations along the Danube River in Serbia showed that a total of 12 groups of food of invertebrate fauna have been identified. Trichoptera (3.16–54.43%), Chironomidae (4.32–95.77%), and Gammaridae (0.51–98.61%) comprised the main part of sterlet diet, especially in sector 1189–1043 rkm, with a reduction in food composition variability close to the Djerdap I dam and in the reservoir between the two dams. The other components of the diet were Corophium sp., Asselus sp., Mollusca, Oligochaeta, Annelida, Insecta, Nematoda, and Hirudinea. The organisms typical of lithorheophilic and psammorheophilic biocoenoses play a considerable role in the diet of the sterlet. Composition of bottom fauna as food items varies due to changes in environmental conditions induced mainly by the construction of two dams: Djerdap I (943 river km, 1970) and Djerdap II (863 river km, 1984) (Djikanovic et al. 2015).

Comparing seasonal aspects of sterlet feeding in the River Danube shows that the same bottom fauna groups participate in its diet throughout the year with some differences in prey proportions according to season. Stomach contents depend on the bottom fauna composition of each river sector as well as of sampling season. In May, Trichoptera, Amphipoda, and Hirudinea were most abundant, whereas in September and October, representatives of Chironomidae, Oligochaeta, Gammaridae, and Corophium sp. were mainly present in the diet (Janković 1958; Janković et al. 1994; Strelnikova 2012). Upstream of the Belgrade section, Chironomidae and Trichoptera have great importance in the sterlet diet. In the Belgrade section Amphipoda and Hirudinea are also important, whereas Oligochaeta, Gammaridae, and Chironomidae are also important in the downstream sector (Janković 1958; Janković et al. 1994; Strelnikova 2012).

Analysis of data on the structure of the macrozoobenthos community before construction of the Iron Gate Dam on the River Danube and after its construction indicates that certain global changes have occurred in the structure of this community. Among these changes are decreases in the abundance of populations of amphipods, some species of gastropods and bivalves, and increases in the abundance of pelophilous and phytophilous forms of the group Oligochaeta and pulmonate gastropods (Simić and Simić 2004). The abundance of amphipod crustaceans, especially species of Ponto Caspian origin, has declined in the Iron Gate region. Representatives of Gammaridae and Corophium sp. were encountered in later years (Simić and Simić 2004). It can be seen from the results that Ponto-Caspian species such as Corophium sp. is more sensitive than Gammaridae (Simić and Simić 2004). During research conducted under the international project "Joint Danube Survey 1 and 2," Mollusca dominated both in terms of the number of species and their relative abundance, followed by representatives of Chironomidae and Oligochaeta (Paunović et al. 2010).

Future effects of climate change on benthic macroinvertebrates in the Danube River are expected. Pressures on the Danubian aquatic biodiversity, particularly on the aquatic macrozoobenthos community, as a food resources, come from the invasion of non-native species, as well as from land use change, water stress, pollution, habitat fragmentation, flow regulations (DRB climate change adaptation, ICPDR, 2017–2018). Impacts on macrozoobenthos are: (1) Changes in timing and magnitude of low flow and flood events; (2) Changes in the hydrological regime may create new conditions to which existing hydrobionts can be poorly adapted; (3) increase of water temperature during summer months can lead to reduced oxygen and CO2 solubility pressuring macrozoobenthos community structure and composition (taxa correlated to high temperature, including some dipterans and oligochaetes, might benefit from global warming and increase their abundance and range of distribution).

Differences in invertebrate community composition during JDS expedition are recorded. Compared to the results from JDS3 (2013), a similar diversity pattern occurred, however, the number of taxa of Gastropoda groups found during JDS4 (2019) has doubled. On the other hand, several Ponto-Caspian species native to the Lower Danube River stretch found during JDS3 were now seen to be missing. In addition, species from genus Pisidium sp. are completely missing in the taxa lists from the middle and lower reaches. Changes in the macrozoobenthos community can greatly affect the starlet population as well as their food source, and perhaps influence their feeding patterns. Also, changes in macrozoobenthos could effect on the presence of some helminths species in the starlet intestine, their prevalence, and intensity of infestation.

5 Review of Legislation Measures in Force

On the European continent the Danube and Rioni River remained refuges of critical importance for the last reproducing wild sturgeon populations. Recent assessments of the sturgeon stocks are based on a very scattered and patchy set of incoherent information. Danube Basin countries and the Black Sea range states of Turkey and Georgia adopted the Pan European Action Plan for Sturgeons and agreed to fully implement the comprehensive set of measures (Recommendation No. 199 (2018) of the Bern Convention).

The conservation of migratory fish species, particularly sturgeon species, protection of their habitats, and ecological corridors are covered in most of the countries in their conservation, biodiversity policy or river management, particularly with regard to the implementation of WFD requirements. But, plans on the national or regional level often do not explicitly target particularly sturgeon species. This reflects the often-general nature of national conservation and biodiversity policy plans rather than a certain measure for protection of species and habitats.

Sturgeons are in general covered by various national legal instruments, fishing regulations, bans and restrictions, but protection of these species needs a holistic approach connecting international waters, coastal areas and river systems, including a well-coordinated national and international legislation (Bern Convention, the Bonn Convention on Migratory Species, CITES, Habitats Directive, Water Framework Directive or the Marine Strategy Framework Directive). In 2019 representatives from the Danube Basin and the Black Sea issued the Galati Declaration on Sturgeon Conservation in the Danube Basin and Black Sea. That was the first ever joint discussion of 120 delegates of fisheries, water and environmental administration, scientific community and civil society from Danube and Black Sea states about necessary conservation measures, and urgent need for population monitoring as a prerequisite to conservation measures (DSTF). All of these policies are the first step in structuring and implementing successful conservation programs worldwide and providing the necessary tools for effective international control. Government decisions and policies should be based on the results of scientific investigations. A common proactive approach of monitoring and regulating of endangered sturgeon species will lead to the recovery of the populations (Bănăduc et al. 2016).

The Bern Convention is focused on European nature protection, for both members and some non-member countries of Europe. This agreement is relevant to the protection of listed species in the Balkans. It includes annexes of species, with varying levels of protection, listed under Appendix II (Strictly Protected Fauna) or Appendix III (Protected Fauna) of the Bern Convention. Citizens may issue complaints concerning the lack of implementation.

The European Habitats Directive (92/43/EEC) is the backbone of legally binding European species protection and conservation. The relevant appendices are Appendix II (species of community interest, for which protection areas must be assigned), Appendix IV (strictly protected species), and Appendix V (species whose exploitation is compatible with a favorable conservation status) (Weiss et al. 2018).

Action Plan for the conservation of Sturgeons in the Danube River Basin was submitted and approved in the Bern Convention and was adopted by all Danubian countries. Sturgeon Action Plan is an important document and it was expected to be legally binding instrument (Bănăduc et al. 2016). But was never truly implemented despite being supported by a wide range of protective international legislation. In 2012 the Danube Sturgeon Task Force (DSTF) was established as a network of dedicated volunteers from scientific, governmental, and nongovernmental organizations. The Sturgeon Action Plan was updated and streamlined into the program "Sturgeon 2020" (Freidrich 2018).

The list of the protecting international legislation that imply on the sturgeon fish species and that are adopted by the Western Balkan countries: CITES (Convention on International Trade in Endangered Species of Wild Fauna and Flora), Bonn (Convention on the Conservation of Migratory Species of Wild Animals), Bern (Bern Convention on the Conservation of European Wildlife and Natural Habitats), CBD (Convention on Biological Diversity), Ramsar (Convention on Wetlands) (Freidrich 2018).

The CITES was probably the most significant international act protecting sturgeons. Listing of the sturgeon species on Annex I and II which defines whether or not there is a possibility of placing it on the market with special permits, system of export quotas, an universal labeling system for the identification of caviar, international cooperation and harmonization of management and conservation measures, joint stock assessment and enhancement, and an improved monitoring and reporting as well as sustainable management measures put in force such as moratorium indicated that only the Convention on International Trade in Endangered Species of Wild Fauna and Flora (CITES) had and real effects in the implementation process (Jarić et al. 2017).

The moratorium was introduced by national governments in 2006 (under the CITES guidelines) for a period of ten years, in response to the rapid decline of migratory sturgeon species in Europe. That was drastic policy intervention that affected local fisheries in the Danube River basin. The ban was extended 2016 for another at least five years. The ban covers five species of sturgeon from the Danube River basin—Russian sturgeon, sterlet, stellate sturgeon, beluga, and ship sturgeon. Although this fishing ban is considered an important step toward sturgeon conservation, it reduces the ability to estimate sturgeon stocks. However, recent findings confirm that the ban does not prevent poaching (Teodorescu and van den Kommer, 2020).

5.1 Serbia

The management of the sturgeon species in Serbia is addressed by the following national legislations: Law on Protection and sustainable use of the fish resources (Official Gazette of the RS, no. 128/2014 and 95/2018—other laws) and a set of by-laws: Order of measures for the preservation and protection of the fish stock (Official Gazette of the RS, no. 56/ 2015 and 94/2018), The Decree of fishing tools, fishing methodology and facilities for commercial fishing, as well as on the method, tools, equipment, and means used for recreational fishing (Official Gazette of the RS, no. 9/2017 and 34/2018). Law on Nature Protection (Official Gazette of RS, No. 36/2009, 88/2010, 91/2010–amended, 14/2016 and 95/2018–other laws), Rulebook on the proclamation and protection of strictly protected and protected wild species of plants, animals, and mushrooms (Official Gazette of RS, No. 5/2010, 47/2011, 32/2016 and 98/2016). For the strictly protected wild fish species (such as sturgeons), prohibition of their use, destruction and all actions that may threaten wild species and their habitats is defined. Also, measures and actions to manage their population is proscribed. In Serbia, all 5 sturgeon species are protected by the ban: Russian sturgeon, sterlet, stellate sturgeon, beluga, and ship sturgeon. Fishing on sturgeons is prohibited, except for scientific research purposes, and the most important measures are aimed at preserving their habitats in relation to the space they inhabit and the quality of the habitat itself.

5.2 Slovenia

Sterlet was one species of sturgeon included in the Slovenian freshwater fish species list because it was caught during the 80s and 90s of the last century. Since 1992, it has been listed on the Red List of freshwater fish of Slovenia, firstly assessed as "extinct" and fully protected (Ur. l. RS 1993). Then the status was changed to "rare" in 2003 (Ur. l. RS 2002) and remained protected (Ur. l. RS 2004). Other Danube sturgeons are not specially assessed by national legislation, except European sturgeon that is included in the list of marine fish as endangered (Ur. l. RS 2002) (Govedič and Friedrich 2018).

5.3 Croatia

The only sturgeon species listed in Croatian regulation regarding closed season and minimal catch size is sterlet. The moratorium on sturgeon fishing in Croatia was not established due to strong resistance from commercial fishermen, so the catch quota was reduced (Piria, personal communication). Catch quota according to new Freshwater Fisheries Act from 2019, for sterlet catch from the Sava is 100 kg and from the Danube fishing zone is 500 kg. Closed seasons last from 1 March–31 May and minimum catch size is 40 cm for the only sturgeon species listed in Regulation in force from 2005 on the protection of fish in freshwater fisheries. Fish species that are strictly protected by the Nature Protection Act under threatened category RE are Russian sturgeon, stellate sturgeon, Atlantic sturgeon, beluga, and ship sturgeon while Adriatic sturgeon is evaluated as CR (EN).

5.4 Bosnia and Herzegovina

Fishing in Bosnia and Herzegovina is regulated by the following laws and by-laws depending on the entities:

Laws regarding the protection and preservation of wildlife and fish fauna:

Law on Environmental Protection (Official Gazette of FBiH no. 33, 2003).

Order on new measures for research or conservation to prevent a significant negative impact on species by deliberate capture or killing of species (Official Gazette of FBiH no. 65, 2006).

Law on Environmental Protection (Official Gazette of RS no. 113, 2008).

Order on the establishment of a monitoring system for the deliberate keeping and killing of protected animals (Official Gazette of RS no. 113, 2008).

Law on Environmental Protection (Official Gazette BD no. 24, 2004).

Fisheries and Aquaculture Laws:

Law on Freshwater Fisheries (Official Gazette of FBiH no. 64, 2004).
Law on Freshwater Fisheries (Official Gazette BD no. 35, 2005).
Law on Fisheries (Official Gazette of RS no. 72, 2012).
Sturgeon species are protected by the closed season and minimal catch size. Sterlet is protected from 1 March to 31 May, minimal catch size 40 cm, while sturgeon closed season is in April and minimal catch size 50 cm.

5.5 *Montenegro*

Fishing in Montenegro is regulated, by several documents. In case of freshwater fishing, currently there are two basic documents: Freshwater Fishing and Aquaculture Act, from 2018 (Zakon o slatkovodnom ribarstvu i akvakulturi, broj: 01–290/2, od 16.03.2018. god.) and Order on Prohibitions and Restrictions on Fishing and Protective Measures of Fish Fund, dated May 27, 2016 (Naredba o ribolovnim zabranama, ograničenjima i mjerama za zaštitu ribljeg fonda," broj 33/2016, od 27.5.2016. god.). There are no explicit legislation measures regarding sturgeons species.

5.6 *Macedonia*

Fishing in Macedonia is regulated by the Fisheries Law, which deals with commercial fishing, sport fishing, and fish production. There are no explicit legislation measures regarding sturgeons species.

5.7 *Albania*

This Law on Fisheries (No. 80/2017, 64/2012) regulates all fishery activities and their management and aims at ensuring the protection of the marine life and internal waters. Article 37 prohibits the fishing of sturgeon—Atlantic sturgeon, in particular catching, keeping or transiting on board, intentional landing and marketing of the following marine organisms, including for human consumption, is prohibited at any time, in any area, and using any type of equipment or gear (https://www.faoadriamed.org/html/legislation/LegALBComp.html).

Major deficits in the management and policy relate to their implementation. Different legal, strategic, and management documents must be better harmonized. Data related to sturgeons monitoring are lacking in practice. This is the case even if such assessments are required by nature and biodiversity protection plans or river basin management. Example in Serbia: The call for sterlet monitoring was announced but never finished tender procedure and put in motion. Development of specific regional, national, and international Action Plans for the Danube sturgeons are efforts of the experts to help policy and management to respect minimum ecological requirements as well as to identify and promote compromise solutions for their protection (Measures project).

6 Aquaculture

Aquaculture plays a very important role in the reproduction of endangered fish species, as well as the breeding of their fry for stocking purposes, especially for endangered species such as sturgeons in the waters where their numbers have declined, their introduction into waters from which they have disappeared (reintroduction) or for further breeding in ex situ conditions (Marković et al. 2016). The basic idea of ex situ is to establish closed life cycle units, which should serve as living gene banks and at a later stage releasing them into the nature should stabilize and strengthen wild populations (Freidrich et al. 2018). Numerous attempts to stabilize populations by hatchery and release programs have been made in the Danube over the past few decades. Sterlet was example of such efforts (Lenhardt et al. 2020). In order to prevent the occurrence of inbreeding and outbreeding, genetic assessments of the suitability of specimens intended for stocking should be recognized as a priority, therefore ex situ measures cannot be sustained without in situ measures and must always be linked to the natural population in order to maintain the natural gene pool and it is still just solutions until in situ habitats and populations are re-established (Lenhardt et al. 2020; Freidrich et al. 2018). Hatchery should provide semi-natural conditions to allow the sturgeon to adapt in the wild and exhibit home behavior based on conditions such as water chemistry, diet, flow rates, temperatures, etc. (Freidrich et al. 2018).

6.1 Serbia

According to the Law on Protection and Sustainable Use of Fish Stocks (Official Gazette of RS, No. 128/2014), stocking of a fishing area may be conducted only with autochthonous specimens reared by the organization with special competences, as defined by special regulations, while the Public Procurement Law (Official Gazette of RS, No. 124/2012) established that all fishing area stakeholders are obliged to invite for public procurements. Complicated public procurement procedures make such companies discouraged to produce specific stocking material. In addition, there are also no governmental subventions for these purposes, and when faced with the limited and unreliable market, hatcheries become reluctant to invest in certain species. Additional problem is inadequate knowledge of broodstock origin and the absence of broodfish tagging. This can lead to outbreeding errors and consequently to serious negative effects in wild populations. Potential and planned stocking programs in Serbia are therefore postponed (Lenhardt et al. 2020).

There was one facility, located in Kusjak (Fenix) near the Iron Gate II dam, where beluga and Russian sturgeon were kept and attempted to hatch.

Aquaculture facilities for sturgeons in Serbia are situated in the Vojvodina autonomous province, near Bačko Petrovo Selo (PR STR TISA), Temerin (Mehanika Group), Plandište (Fish cooperative MALGURA), Vršac (two companies, Aquacultura and Fish Farm) and Ledinci (Ledico Food). All six facilities are raising sterlet, with five facilities also raising Siberian sturgeon, two are raising Russian sturgeon and one facility is raising hybrid (Acipenser baerii x Acipenser ruthenuss). All facilities are private property. Unfortunately, they were not established as an ex situ facility, but rather for commercial purposes. Also, these were not registered objects by the authorized Ministry for Environment.

6.2 Slovenia

Slovenian hatcheries breed at least three sturgeon species: indigenous sterlet, nonindigenous Siberian sturgeon, and paddlefish. According to a recent study (Govedič and Friedrich 2018) sterlets have been reared in Požeg water reservoir, also in Turn ponds (Turnovi ribniki) in the area of Rače. Paddlefish were cultivated in a facility close to Rogaška Slatina. Russian sturgeon was reared in Požeg water reservoir. Siberian sturgeon is cultivated in an aquaculture facility at Dvor in the Dolenjska region along the Krka River.

6.3 Croatia

According to Freshwater aquaculture report for EU Commission (EUMOFA 2021), aquaculture of sturgeon species in Croatia, or the production of caviar, was not reported for period 2009–2019.

7 Conclusions

According to Freidrich et al. (2018), conservation status of all sturgeon species in Europe has become critical without signs of recovery. Bearing this in mind, question about lack of previous action success should be made.

Sturgeons are fish group with characteristic traits, such as longevity, late maturation, spawning in freshwater (whether they are anadromous or potamodromous), and homing behavior, all of which require complex approach for their conservation. Since they move through different habitats for spawning, feeding, and wintering, they often undertake long migrations and cross borders. One of issues in sturgeon protection programs is lack of integrated cross-border management. Also, since sturgeons life cycle is long and fish mature late, only long-term protection/recovery programs over several decades could have chance for success. Without political will, lack of coordination, and clear responsibility, as well as lack of resources, successfully implementing any restoration program is futile. Although Action plan for sturgeon protection was developed in 2005. (Lenhardt et al. 2005b), its ratification and implementation are lacking.

Although the research of sturgeons is ongoing over a century, there is need for data update. Based on available data, during the mid-twentieth century, research of sturgeons in the Western Balkans was oriented to life history, biology, and ecology (Janković 1958; Ristić 1963a, b, c, 1967a, b, c, d, 1969a, b). Due to lack of different sturgeons species, as a consequence of Iron Gate dams construction, as well as lack of funding, later research was focused on sterlet morphometry, pollution assessment, and genetic (Jarić et al. 2011c; Lenhardt et al. 2012, 2014; Cvijanović et al. 2015, 2017), i.e., more laboratory-oriented research. In the last decade, development of population models and extinction models (Jarić et al. 2009a, 2010, 2014, 2016; Simić et al. 2014) provided tools for decision makers to adapt conservation measures. However, since the sturgeon environment is changing, and there is uncertainty about their response to these changes, more recent data should be available in order to have precise population assessment/simulation. Even knowledge about hybridization and identification could help avoid misidentification of species in wild, such as 2009. Sora and Mura rivers substitution of Siberian sturgeon for sterlet (Govedič and Friedrich 2018).

In Russia (around 1850), attempts for sturgeons artificial breeding were made, in order to maintain, restock, and enhance the wild populations (Carmona et al. 2009). With over a century of ex situ conservation efforts, this measure is still unavoidable action for sturgeon restoration, especially in Western Balkans. Although ongoing development of hatchery facility in Slovenia (as a part of LIFEBoat4sturgeons) should provide sterlets for restocking of Mura River, there is more need for development of ex situ sturgeon hatcheries. Most of the aquaculture facilities are commercially oriented, so the local government and authorities could encourage and help implement more ex situ conservation programs.

One of the things that are missing in all the countries of Western Balkans, due to lack of political commitment and funding, are monitoring programs. Apart from sporadic projects (i.e., MEASURES), there is no assessment of natural populations, distribution of endangered species, and evaluation of efficiency of implemented conservation measures.

Balkan region is the target of one of the most ambitious hydropower expansions plans in the world, with over the 2700 projects planned (Weiss et al. 2018). Despite the fact that the Balkan region has some of Europe's most pristine rivers and is a global hotspot for biodiversity, over one third of hydropower projects are located in protected areas (Weiss et al. 2018). If project of dams construction will materialize, it would have irreversible effects on water flow, as well as access to sturgeons spawning grounds. With lack of political will to deal with environmental protection issues in Western Balknas, long history of unsustainable sturgeon fishing and poaching, as well as increase in human population, urbanization and water pollution, only course of action is permanent conservation. In order to restore wild sturgeons in waters of Western Balkans, joint effort of all region countries is necessity. Otherwise, future generations wouldn't be able to know and admire these incredible creatures and will see their local common names as something of the past.

Acknowledgments We would like to thank Dr Davor Zanella for his help in sharing the information about sturgeons in Croatia. We would also like to thank to Dr Mirjana Lenhardt, Dr Jelena Kolarević, and Dr Ivan Jarić for sharing enthusiasm for sturgeons, as well as work space over the years.

References

Alavi-Yaganeh MS, Falahatkar B (2013) Some biological aspects of Russian sturgeon broodstocks, *Acipenser gueldenstaedtii* Brandt & Ratzeburg, 1833 in southeastern Caspian Sea (case study; Torkaman & Mian-Ghale Fishing Stations). J Fish Sci Technol 2(3):89–93

Bănăduc D, Rey S, Trichkova T, Lenhardt M, Curtean-Bănăduc A (2016) The Lower Danube River–Danube Delta–North West Black Sea: A pivotal area of major interest for the past, present and future of its fish fauna — A short review. Sci Total Environ 545–546:137–151

Berg LS (1962) Freshwater fishes of the U.S.S.R. and adjacent countries (4th ed., Vol. 1). Israel program for scientific translations. (Original work published 1948)

Birstein VJ (1993) Sturgeons and paddlefishes: threatened fishes in need of conservation. Conserv Biol 7(4):773–787

Birstein VJ, Waldman JR, Bemis WE (eds) (1997) Sturgeon biodiversity and conservation. Kluwer Academic Publishers

Bloesch J, Jones T, Reinartz R, Striebel B (Eds.). (2006) Action plan for the conservation of sturgeons (Acipenseridae) in the Danube River basin (Nature and Environment, 144). Council of Europe Publishing

Bronzi P, Congiu L, Freyhof J (2022) *Acipenser naccarii*. The IUCN red list of threatened species 2022: e.T224A135061929

Cakić P (1986a) A contribution to the fauna of fish ecto - and endoparasites in the channels of Pančevački rit. University of Belgrade, Center for Multidisciplinary Sciences, Belgrade, p 133. (in Serbian)

Cakić P (1986b) *Azygia lucii* (Müller, 1776) Luhe, 1909 new parasitic Trematoda foe sterlet (*Acipenser ruthenus* L.). Arch Biol Sci Belgrade 38(1-4):7–8

Cakić PD, Đikanović VĐ, Kulišić ZB, Paunović MM, Jakovčev-Todorović DG, Milošević SM (2008) The fauna of Endoparasites in *Acipenser ruthenus* Linnaeus 1758 from the Serbian part of the Danube river. Arch Biol Sci 60(1):103–107

Ćaleta M, Maričić Z, Buj I, Zanella D, Mustafić P, Duplić A, Horvatić S (2019) A review of extant Croatian freshwater fish and lampreys: annotated list and distribution. Croat J Fish 77:136–323

Carmona R, Domezain A, García-Gallego M, Hernando JA (2009) Biology, conservation and sustainable development of sturgeons – Fish and fisheries series, vol 29. Springer, The Netherlands, p 467

Carroll AM, Wainwright PC (2003) Functional morphology of prey capture in the sturgeon, *Scaphirhynchus albus*. J Morphol 256:270–284

Choudhury A, Dick TA (2001) Sturgeons (Chondrostei: Acipenseridae) and their metazoan parasites: patterns and processes in historical biogeography. J Biogeogr 28:1411–1439

Ćurčić V (1910) National fisheries of Bosnia and Herzegovina. Glasnik Zemaljskog muzeja u Bosni i Hercegovini 2-3:379–488. (in Bosnian)

Cvijanović G, Adnađević T, Lenhardt M, Marić S (2015) New data on sterlet (*Acipenser ruthenus* L.) genetic diversity in the middle and lower Danube sections, based on mitochondrial DNA analyses. Genetica 47(3):1051–1062

Cvijanović G, Adnađević T, Jarić I, Lenhardt M, Marić S (2017) Genetic analysis of sterlet (*Acipenser ruthenus* L.) populations in the middle and lower Danube sections. North-Western J Zool 13(1):34–43

Djikanovic VDJ, Skoric SB, Lenhardt MB, Smederevac-Lalic MM, Visnjic-Jeftic ZV, Spasic SZ, Mickovic BM (2015) Review of sterlet (*Acipenser ruthenus* L. 1758) (Actinopterygii: Acipenseridae) feeding habits in the River Danube, 1694-852 river km. J Nat Hist 49(5–8): 411–417

Dogiel VA, Bykhovskiy BE, (1939) Parasity ryb Kaspiiskogo morya. Trudy po kompleksnomu izucheniyu

Dubinin VB (1952) Parazitofauna molodi Noei osetrovikh ryb Nizhnei Volgi. [In Russian], Uchen. Zapiski Leningradsk, Gos. Univ., 141, seryia Biologich. Nauk 28:238–251

Dudu A, Suciu R, Paraschiv M, Georgescu SE, Costache M, Berrebi P (2011) Nuclear markers of Danube sturgeons hybridization. Int J Mol Sci 12(10):6796–6809

EUMOFA (2021) Freshwater aquaculture in the EU. Publications office of the European Union

Freidrich T (2018) Danube Sturgeons: Past and Future. In: Schmutz S, Sendzimir J (eds) Riverine ecosystem management. Aquatic ecology series, vol 8, pp 507–518

Freidrich T, Gessner J, Reinartz R, Striebel B (2018) Pan-European action plan for sturgeons. Convention on the Conservation of European Wildlife and Natural Habitats, Strasbourg, France, p 85

Freyhof J, Chebanov M, Pourkazemi M (2022) *Acipenser nudiventris*. The IUCN red list of threatened species 2022: e.T225A135062224

Friedrich T (2012) Historical distribution, current situation and future potential of sturgeons in Austrian Rivers. Institute of Hydrobiology and Aquatic Ecosystem Management, p 84

Gessner J, Chebanov J, Freyhof J (2010) *Huso huso*. The IUCN red list of threatened species 2022: e.T10269A3187455

Gessner J, Freyhof J, Kottelat M, Friedrich T (2022a) *Acipenser ruthenus*. The IUCN red list of threatened species 2022: e.T227A135062526

Gessner J, Freyhof J, Kottelat M (2022b) *Acipenser gueldenstaedtii*. The IUCN red list of threatened species 2022: e.T232A135063140

Gessner J, Williot P, Rochard E, Freyhof J, Kottelat M (2022c) *Acipenser sturion*. The IUCN red list of threatened species 2022: e.T230A137215851

Govedič M, Friedrich T (2018) First review of recent record of sturgeons and paddlefishes (Acipenseriformes) in the Danube River basin in Slovenia. Natura Sloveniae 20(2):5–16

Gregori J (1988) Sterlet in Črnečah. Fisherman newsletter 1–2:27

Guti G (2006) Past and present status of sturgeons in Hungary. IAD Conference 2006

Guti G (2011) Some results of ichthyological research in the Danube research Institute of the Hungarian Academy of sciences. Presentation at the University of Natural Resources and Life Sciences, Vienna, 25.05.2011

Helfman GS, Colette BB, Facey DE, Bowen BW (2009) The diversity of fishes: biology, evolution, and ecology. John Wiley & Sons, Chichester

Hensel K, Holčík J (1997) Past and current status of sturgeons in the upper and middle Danube River. Environ Biol Fish 48(1–4):184–200

Hernando JA, Domezain A, Zabala C, Cabrera R, Domezain J, Soriguer MC (2009) The regression of sturgeons in southern Europe. In: Biology, conservation and sustainable development of sturgeons – fish and fisheries series, vol 29. Springer, The Netherlands, pp 3–23

Holostenco D, Onără DF, Suciu R, Hont S, Paraschiv M (2013) Distribution and genetic diversity of sturgeons feeding in the marine area of the Danube Delta biosphere reserve. Sci Ann Danube Delta Inst 19:25–34

Honsig-Erlenburg W, Friedl M (1999) Zum Vorkommen des Sterlets (*A.ruthenus* L.) in Kärnten. Österreichs Fischerei 52:129–133

ICPDR – IKSD, Report, Danube River Basin Climate Change Adaptation: "Integrating and editing new scientific results in climate change research and the resulting impacts on water availability to revise the existing adaptation strategies in the Danube River basin". 2017–2018

Jahrl J, Boner M, Striebel B, Ludwig A (2021) evidence for trafficking of critically endangered sturgeon in the lower Danube region, final report 2021, WWF Austria & WWF CEE, 2021

Janković D (1957) Paraziti ograna za varenje i trbušne duplje dunavske kečige. Ribarstvo Jugoslavije 12(6):126. Zagreb. (in Serbian)

Janković D (1958) Ecological research on Danubian sterlet (*Acipenser ruthenus* L.). Institute Biologique Beograd, Monographies 2:1–131. (in Serbian)

Janković D (1993) Populations of Acipenseridae prior and after the construction of the HEPS Djerdap I and II. Ichthyologia 25(1):29–34

Janković D, Krpo J, Lenhardt M, Hegediš A (1994) Ecological characteristics of the Danubian fish species. In: Jankovic D, and Jovicic M, editors. The Danube in Yugoslavia. Institute for biological research "Siniša Stanković", Belgrade; Institute for development of water resources "Jaroslav Černi", Belgrade; Commission of the European communities Brussels, Belgium; Ministry of science and technology of Republic of Serbia, Belgrade, pp 159–173

Jarić I, Lenhardt M, Cvijanović G, Ebenhard T (2009a) Population viability analysis and potential of its application to Danube sturgeons. Arch Biol Sci Belgrade 61(1):123–128

Jarić I, Lenhardt M, Cvijanović G, Ebenhard T (2009b) *Acipenser sturio* and *Acipenser nudiventris* in the Danube – extant or extinct? J Appl Ichthyol 25:137–141

Jarić I, Ebenhard T, Lenhardt M (2010) Population viability analysis of the Danube sturgeon population in a vortex simulation model. Rev Fish Biol Fish 20:219–237

Jarić I, Ebenhard T, Lenhardt M (2011a) Population viability analysis of the Danube sturgeon population in a vortex simulation model. Rev Fish Biol Fish 20:219–237

Jarić I, Lenhardt M, Pallon J, Elfman M, Kalauzi A, Suciu R, Cvijanović G, Ebenhard T (2011b) Insight into Danube sturgeon life history: trace element assessment in pectoral fin rays. Environ Biol Fish 90:171–181

Jarić I, Višnjić-Jeftić Ž, Cvijanović G, Gačić Z, Jovanović LJ, Skorić S, Lenhardt M (2011c) Determination of different heavy metal and trace element accumulation in liver, gills, intestine and muscle of sterlet (*Acipenser ruthenus*) from the Danube River in Serbia by ICP-OES. Microchem J 98:77–81

Jarić I, Gessner J, Acolas M-L, Lambert P, Rochard E (2014) Modelling attempts utilized in sturgeon research: a review of the state-of-the art. J Appl Ichthyol 30:1379–1386

Jarić I, Gessner J, Lenhardt M (2015) A life-table metamodel to support the management of data deficient species, exemplified in sturgeons and shads. Environ Biol Fish 98:2337–2352

Jarić I, Gessner J, Solow AR (2016) Inferring functional extinction based on sighting records. Biol Conserv 199:84–87

Jarić I, Riepe C, Gessner J (2017) Sturgeon and paddlefish life history and management: experts' knowledge and beliefs. J Appl Ichthyol 34(2):244–257

Jarić I, Bronzi P, Cvijanović G, Lenhardt M, Smederevac-Lalić M, Gessner J (2019) Paddlefish (Polyodon spathula) in Europe: an aquaculture species and a potential invader. J Appl Ichthyol 35(1):267–274. https://doi.org/10.1111/jai.13672

Jeremko B (1998) Sterlet from channel of Drava. Fisherman newsletter 5:120

Kakacheva-Avramova D (1983) Helminths of freshwater fishes in Bulgaria. Bulg Acad Sci Sofia:261

Knežević B (1985) Fishes of Lake Skadar. General introduction. Survey on previous and recent investigations. In: Karaman GS, Beeton AM (eds) The biota and limnology of Lake Skadar. Univeryitet "Veljko Vlahović", Institut ya Biološka i medicinska istraživanja u SRCG, Biološki zavod. Smithsonian Institution, Center for Great Lakes Studies Univ. Wisconsin – Titograd, Washington DC, Milwaukee, pp 311–316

Kolarević J (2004) Population - ecological study of sterlet (*Acipenser ruthenus* L.) in waters of Danube on territory of Belgrade. MSc thesis, University of Belgrade, Serbia

Kottelat M, Freyhof J (2007) Handbook of European freshwater fishes. Kottelat, Cornol, Switzerland and Freyhof, Berlin, Germany

Lapkina LN, Strelnikova AP, Chuiko GM, Lenhardt M, Kolarevic J (2005) International conference "structual-functional characteristics of the biosystem of the North". Leeches as the object of fish feeding. In: Smirnova SL (ed) September 27. Petrozavodski State University, Russia, pp 214-217

Lenhardt M, Cakić P, Kolarević J, Mićković B, Nikčević M (2004) Changes in sterlet (*Acipenser ruthenus* L.) catch and length frequency distribution in the Serbian part of the Danube River during the twentieth century. Ecohydrol Hydrobiol 4(2):193–197

Lenhardt M, Finn RN, Cakić P, Kolarević J, Krpo-Ćetković J, Radović I, Fyhn HJ (2005a) Analysis of the post-vitellogenic oocytes of three species of Danubian Acipenseridae. Belg J Zool 135(2): 205–207

Lenhardt M, Hegediš A, Jarić I (2005b) Action plan for sturgeon species management in fishery waters of republic Serbia. - Institute for Biological Research "Siniša Stanković", Belgrade; Developed for Ministry of Science and Environmental Protection of Republic Serbia, pp 1–21

Lenhardt M, Jarić I, Kalauzi A, Cvijanović G (2006a) Assessment of extinction risk and reasons for decline in sturgeon. Biodivers Conserv 15(6):1967–1976

Lenhardt M, Hegediš A, Mićković B, Višnjić Jeftić Ž, Smederevac M, Jarić I, Cvijanović G, Gačić Z (2006b) The first record of north American paddlefish (*Polyodon spathula* Walbaum, 1792) in Serbian part of the Danube River. Arch Biol Sci 58(3):27P–28P

Lenhardt M, Jarić I, Cvijanović G, Smederevac-Lalić M (2008) The key threats to sturgeons and measures for their protection in the lower Danube region. In: Lagutov V (ed) Rescue of Sturgeon Species in the Ural River basin. Springer, Dordrecht, pp 87–96

Lenhardt M, Jarić I, Cakić P, Cvijanović G, Gačić Z, Kolarević J (2009) Seasonal changes in condition, hepatosomatic index and parasitism in sterlet (*Acipenser ruthenus* L.). Turk J Vet Anim Sci 33(3):209–214

Lenhardt M, Jarić I, Cvijanović G, Kolarević J, Gačić Z, Smederevac-Lalić M, Višnjić-Jeftić Ž (2012) Comparison of morphological characters between wild and cultured sterlet (*Acipenser ruthenus* L.). Slov Vet Res 49:177–184

Lenhardt M, Smederevac-Lalić M, Djikanović V, Cvijanović G, Vuković-Gačić B, Gačić Z, Jarić I (2014) Biomonitoring and genetic analysis of strugeons in Serbia: a contribution to their conservation. Acta Zoologica Bulgarica 7:69–73

Lenhardt M, Smederevac-Lalić M, Hegediś A, Skorić S, Cvijanović G, Višnjic-Jeftić Ž, Djikanović V, Jovičić K, Jaćimović M, Jarić I (2020) Human impacts on fish Fauna in the Danube River in Serbia: current status and ecological implications. In: Bănăduc D, Curtean-Bănăduc A, Pedrotti F, Cianfaglione K, Akeroyd J (eds) Human Impact on Danube Watershed Biodiversity in the XXI Century. Geobotany studies. Springer, Cham

Ludwig A, Kirschbaum F (1998) Comparison of mitochondrial DNA sequences between the European and the Adriatic sturgeon. J Fish Biol 52:1289–1291

Maltsev SA (2009) Conservation of the sturgeon fish in lower Volga. In: Biology, conservation and sustainable development of sturgeons – fish and fisheries series, vol 29. Springer, The Netherlands, pp 265–273

Marić D (1995) Endemic fish secies of Montenegro. Biol Conserv 72:187–194

Marić D, Milošević D (2011) Catalog of freshwater fishes (Osteichthyes) of Montenegro. Montenegrin Academy of Science and Arts, Podgorica 5(4):114

Marković Z, Stanković M, Rašković B, Sekulić N, Poleksić V (2016) Aquaculture in service of edangered fish species protection in Serbia. Ecological and economic significance of fauna of Serbia. Scientific meetings of Sebian Academy of Science and Arts 171(12):173–197

Maximov V, Tiganov G, Paraschiv M, Nenciu MI, Zaharia T (2014) Preliminary data on the monitoring of sturgeon species in Romania marine waters. J Environ Prot Ecol 15(3):933–943

Moravec F, Holcik J, Meszaros J (1989) Notes on the helminthfauna of the sterlet (*Acipenser ruthenus* L.) in Slovakia. Biologia (Bratislava) 44(2):151–155

Mrakovčić M, Brigić A, Buj I, Ćaleta M, Mustafić P, Zanella D (2006) Red book of freshwater fish of Croatia. Ministry of Culture, State Institute for Nature Protection, Republic of Croatia

Mrdak D (2009) Environmental risk assessment of the Morača dams: fish fauna of Morača river canyon and Skadar Lake. Project report, WWF MedPO and Green Home, p 76

Mugue N, Friedrich T, Chebanov M, Ruban G (2022) *Acipenser stellatus*. The IUCN red list of threatened species 2022: e.T229A135062806

Munda A (1926) Ribe v slovenskih vodah. Ljubljana, Slovensko ribarsko društvo v Ljublani, 63 str. (in Slovenian)

Nechaeva NL (1964) Parazitofauna molodi osetrovykh ryb Kaspiisko–Kurinskogo rajona. [In Russian] Trudy VNIRO 54:223–238

Noakes DL, Bouvier LD (2013) Threatened fishes of the world: the end of a series. Environ Biol Fish 96:1135–1149

Ognjanović D, Nikolić V, Simonović P (2008) Morphometrics of two morphs of sterlet, *Acipenser ruthenus* L., in the middle course of the Danube River (Serbia). J Appl Ichthyol 24:126–130

Paschos I, Perdikaris C, Gouva E, Nathanailides C (2008) Sturgeons in Greece: a review. J Appl Ichthyol 24:131–137

Paunovic M, Csányi B, Simic V, Djikanovic V, Petrovic A, Miljanovic B, Atanackovic A (2010) Chapter 11: community structure of the aquatic macroinvertebrates of the Danube River and its main tributaries in Serbia. In: Paunovic M, Simonovic P, Simic V, Simic S (eds) The Danube in Serbia – the results of National Program of the second joint Danube survey 2007. Directorate for Water Management, Belgrade (Serbia), pp 183–205

Pintarič R (1982) Unusual news: sterlet again in Slovenia! Fisherman newsletter 12:320

Poljakov GD, Filipi N, Basho K, Hysenaj A (1958) Fishes of Alabania. Tirana 187. (in Albanian)

Raikova EV (1994) Life cycle, cytology, and morphology of *Polypodium hydriforme*, a coelenterate parasite of the eggs of acipenseriform fishes. J Parasitol 80:1–22

Rakaj N (1995) Ichthyofauna of Albania. University of Tirana, Tirana, p 700. (in Albanian)

Reinartz R (2002) Sturgeons in the Danube River. Biology, status, conservation. International Association for Danube Research (IAD)

Ristić MĐ (1959) Monitoring of migrations of economicaly important fish species in the Danube and its tributaries. Fisheries of Yugoslavia 14(2):25–30

Ristić MĐ (1963a) Fishery biological studie of Acipenseridae in Jugoslav section of lower part of Danube. Fisheries of Yugoslavia 18(2):48–58

Ristić MĐ (1963b) Fishery biological studie of Acipenseridae in Jugoslav section of lower part of Danube. Fisheries of Yugoslavia 18(3):73–80

Ristić MĐ (1963c) Fishery biological studie of Acipenseridae in Jugoslav section of lower part of Danube. Fisheries of Yugoslavia 18(4):103–111

Ristić MĐ (1967a) Biological differentiation of spring and winter forms of *Acipenser güeldenstädti* Brandt in Jugoslav section of lower Danube River. Fisheries of Yugoslavia 21(2):25–37

Ristić MĐ (1967b) Biological differentiation of spring and winter forms of *Acipenser güeldenstädti* Brandt in Jugoslav section of lower Danube River. Fisheries of Yugoslavia 21(3):53–64

Ristić MĐ (1967c) Biological differentiation of spring and winter forms of *Acipenser güeldenstädti* Brandt in Jugoslav section of lower Danube River. Fisheries of Yugoslavia 21(4):85–95

Ristić MĐ (1967d) Biological differentiation of spring and winter forms of *Acipenser güeldenstädti* Brandt in Jugoslav section of lower Danube River. Fisheries of Yugoslavia 21(5):113–122

Ristić MĐ (1969a) Characteristic of sterlet (*Acipenser ruthenus* L.) growth in Danube, Sava and Tisza Rivers. Fisheries of Yugoslavia 24(2):21–31

Ristić MĐ (1969b) Determining spawning grounds, migration patterns and age structure of sterlet (*Acipenser ruthenus* L.) in fishing area of the Danube River – "Vodice" km 1142–1144. Fisheries of Yugoslavia 24(4):69–77

Ristić MĐ (1970) Migrations of fishes in the Danube River and its tributaries, and its effect on condition and dynamic of economically important fish species, as well as fishing. Fisheries of Yugoslavia 25(1):1–15

Ristić MĐ (1971) Contribution to clarification of existence of short snout and long snout morphs in sterlet (*Acipenser ruthenus* L. *Morpha kamensis* et typica-Lovetzky 1834, Berg L.S. 1932–1934) as a consequence of forms in the Danube, Sava and Tisza River. Fisheries of Yugoslavia 26(2):25–28

Secor DH, Arefjev V, Nikolaev A, Sharov A (2000) Restoration of sturgeons: lessons from Caspian Sea sturgeon researching program. Fish Fish 1:215–230

Shljahtin GB (ed) (2006) Red book of Saratov region: mushrooms, lichens, plants, animals. Publishing Chamber of Commerce and Industry of the Saratov region

Shulman SS (1954) Obzor fauny parasitov osetrovikh ryb SSSR [in Russian]. Trudy Leningradskogo obshchestva estetstvoispttatelei 72:190–254

Shumka S, Meulenbroek P, Schiemer F, Šanda R (2018) Fishes of the river Vjosa – an annotated checklist. Acta ZooBot Austria 155:163–176

Simić V, Simić S (2004) Macroinvertebrates and fishes in the part of the Danube flowing through the iron gate National Park and possibilities of their protection under in situ and ex situ conditions. Arch Biol Sci 56:53–57

Simić VM, Simić SB, Piperac MS, Petrović A, Milošević D (2014) Commercial fish species of inland waters: a model for sustainability assessment and management. Sci Total Environ 497: 642–650

Simonović P, Budakov LJ, Nikolić V, Marić S (2005) Recent record of the ship sturgeon *Acipenser nudiventris* in the middle Danube (Serbia). Biologia Bratislava 60(2):1–2

Skryabina ES (1974) Gel'minty osetrovykh ryb (Acipenseridae, Bonaparte, 1831). Izdatel'stvo Nauka, Moscow. [in Russian with English abstract]

Smederevac-Lalić M, Jarić I, Višnjić-Jeftić Ž, Skorić S, Cvijanović G, Gačić Z, Lenhardt M (2011) Management approaches and aquaculture of sturgeons in the lower Danube region countries. J Appl Ichthyol 27(3):94–100

Sokolov LI, Berdicheskii LS (1989) Acipensiridae. In: Holčik J (ed) The freshwater fishes of Europe. Vol. 1., Part II. General introduction to fishes Acipenseriformes, pp 150–153. AULA-Verlag

Sokolov LI, Vasilev VP (1989) *Acipenser nudiventris* Lovetsky, 1828. In: Holčík J (ed) The freshwater fishes of Europe. Vol. 1, Part II. General introduction to fishes, Acipenseriformes. AULA-Verlag, pp 206–226

Stamenković S (1991) Evolution des captures de trois especes d'esturgeons *Huso huso*, *Acipenser gueldenstaedti*, et *Acipenser stellatus* dans la partie Yougoslave du Danube. In: Williot, P. (ed.): Acipenser. Cemagref Publ. [In French]

Strelnikova AP (2012) Feeding of juvenile sterlet (*Acipenser ruthenus*, Acipenseridae) in the Danube River midstream. J Ichthyol 52:85–90

Taler Z (1948) Sterlet, sturgeon and its relatives. Fisheries of Yugoslavija 3(7–8):57

Talevski T, Milošević D, Marić D, Petrović D, Talevska M, Talevska A (2009) Anthropogenic influence on Biodiversety of Ichthyofauna and Macrophyte vegetation from Lake Ohrid and Lake Skadar. J Int Environ Appl Sci 4(3):317–324

Teodorescu D, Van den Kommer M (2020) Economic decline, fishing bans, and obstructive politics: is there a future for small scale fisheries in Romania's Danube Delta? In: Pascual-Fernández J, Pita C, Bavinck M (eds) Small scale fisheries in Europe: status, resilience and governance. MARE publication series, vol 23. Springer, Cham

Trojanović S (1934) On freshwater fish. Priroda 24(1):13–19. (in Croatian)

Tsekov A, Ivanova P, Angelov M, Atanasova S, Bloesch J (2008) Natural sturgeon hybrids along Bulgarian Black Sea coast and in Danube River. Acta Zoologica Bulgarica 60:311–316

Vescei P (2001) Threatened fishes of the world: *Acipenser gueldenstaedtii* Brandt & Ratzenburg, 1833 (Acipenseridae). Environ Biol Fish 60:362

Vuković T, Ivanović B (1971) Freshwater fishes of Yugoslavia. Zemaljski Muzej, Sarajevo. BIH

Weiss S, Apostolou A, Đug S, Marčić Z, Mušović M, Oikonomou A, Shumka S, Škrijelj R, Simonović P, Vesnić A, Zabric D (2018) Endangered fish species in Balkan Rivers: their distributions and threats from hydropower development. Riverwatch & EuroNatur, 162 p

Part III
Effects of Stressors on Inland Water Ecosystems and Fish Resources

Summary

In the current era, which many refer to as the Anthropocene due to the complete dominance of humans, the number of factors that have a stressful effect on ecosystems has increased many times over. The prevailing view is that inland water ecosystems are more vulnerable and are losing their stability and functionality faster than other ecosystems. The chapters in this part address the effects of eutrophication, toxic substances, and allochthonous species on the stability of fish communities in inland waters. Stressors reduce the viability of fish communities and weaken the overall ecological services of inland waters.

Eutrophication of Fishing Waters and the Influence of Cyanobacterial Occurrence and Blooming on Fish Resources: Case Studies in Serbia

Snežana Simić, Nevena Đorđević, Nada Tokodi, Damjana Drobac Backović, and Zoran Marinović

Abstract The blooming of cyanobacteria is followed by the change in the ecological factors of the water ecosystem, which can result in changes in biocenosis, particularly in fish communities, including changes in species composition, abundance, and biomass. Some blooming cyanobacteria can produce harmful metabolites, e.g. cyanotoxins, that can have an adverse effect on the fish health and production. In this chapter, the relationship between eutrophication, cyanobacterial blooming as well as the fish communities in Serbian fishing waters will be described. Based on the presence and abundance of cyanobacteria, the degree of health risk due to the usage of these waters was estimated according to the World Health Organization recommendation. The correlation between the presence and abundance of cyanobacteria, cyanotoxin concentrations, and the recorded fish mortality cases has been particularly analyzed. Mass fish mortality has been recorded in some cases, and accumulated toxins negatively affected the quality and health safety of fish used in the human diet. Therefore, fishing should be prohibited during blooming conditions, especially when the presence of cyanotoxins in the water is detected. The negative consequences of cyanobacterial blooms are multiple; they affect the environment, the human health, and have an economic impact on local and broader society.

Keywords Cyanobacteria · Fishery · Health risk · Trophic state · Reservoirs

S. Simić (✉) · N. Đorđević
Department of Biology and Ecology, Faculty of Science, University of Kragujevac, Kragujevac, Serbia
e-mail: snezana.simic@pmf.kg.ac.rs

N. Tokodi · D. Drobac Backović
Department of Biology and Ecology, Faculty of Sciences, University of Novi Sad, Novi Sad, Serbia

Z. Marinović
Department of Aquaculture, Hungarian University of Agriculture and Life Sciences, Gödöllő, Hungary

V. Simić et al. (eds.), *Ecological Sustainability of Fish Resources of Inland Waters of the Western Balkans*, Fish & Fisheries Series 43,
https://doi.org/10.1007/978-3-031-36926-1_11

1 Eutrophication, Cyanobacterial Blooming, Cyanotoxins, and Adverse Effect on the Fisheries

In the conditions of climate change and increasingly intense anthropogenic influence, water quality decline and eutrophication of many freshwater ecosystems have been observed in the last few decades. The term eutrophication has been defined many times by several authors. Smith and Schindler (2009) defined eutrophication as "excessive plant growth resulting from nutrient enrichment by human activity."

Eutrophication is mainly caused by the excessive anthropogenic release of nutrients (phosphorus and nitrogen) (Smith 2003; Diaz and Rosenberg 2008; Simić et al. 2017). Excess phosphorus inputs to lakes usually come from sewage, industrial discharges, and runoff from agricultural land, construction sites, and urban areas. Over time, many countries have regulated point sources of nutrients, such as municipal and industrial discharges. Diffuse sources of nutrients, such as runoff from agricultural or urban lands, have replaced point sources as the main driver of eutrophication in many regions (Peters 1986). An important driver of diffuse nutrient input is an excessive application of fertilizer or manure, which causes phosphorus to accumulate in soils (Sakan et al. 2007) and phosphorus-rich soils are washed into lakes. Similar situation occurs in Serbia (Svirčev et al. 2008; Karadžić et al. 2010; SEPA 2013; Simić et al. 2017). In water bodies which are intensively used for fishing, an increased intake of phosphorus can also occur as large amounts of a loose mixture of diverse foodstuffs gets introduced to a specific location as hookbait (Niesar et al. 2004). Eutrophication could have multiple significant consequences for freshwater fisheries (Winfield 2016; Bănăduc et al. 2022). Consequences for fisheries may initially be positive through increased fish production, but in most systems, they quickly become negative through ecological mechanisms including deoxygenation, sedimentation, and shifts in competitive relationships within the fish community as well as changes in qualitative composition and biomass of the fish community (Winfield 2016; Simić et al. 2022).

The negative consequences of highly eutrophic waters are especially pronounced due to the excessive development of phytoplankton, which can lead to blooms of Cyanobacteria, Dinophyta, Bacillariophyta, and Euglenophyta. Blooms can be dangerous if associated with negative consequences in the aquatic ecosystem, such as increased turbidity, mortality of aquatic organisms due to oxygen depletion by the microbial degradation of senescent blooms, reduction of submerged aquatic vegetation, disruption of ecosystem stability, negative impact on food chains, production of various taste and odor compounds, and highly active toxic substances (Carmichael and Falconer 1993; Scheffer et al. 1993; Chorus and Bartram 1999). This results in disturbances in the water supply, making the water unusable for irrigation, sports, and recreational activities. Finally, blooms could have a direct and/or indirect negative impact on human health (Đorđević and Simić 2014; Watson et al. 2015; Simić et al. 2017; Svirčev et al. 2017a; Tokodi et al. 2020). Specifically, cyanobacterial harmful algal blooms or CyanoHABs threaten the ecological integrity and sustainability of aquatic ecosystems that are depended upon for drinking water, irrigation, fishing, and recreation (Paerl and Otten 2013).

1.1 Cyanobacteria and Cyanotoxins

Cyanobacteria are ubiquitous photosynthetic prokaryotes able to tolerate a wide range of environmental conditions, giving them a competitive advantage in these ecosystems. Aquatic ecosystems across the globe are changing due to both natural and anthropogenic effects, and cyanobacteria seem to be thriving, mostly in the form of extensive proliferation. This accelerated rate of cyanobacterial reproduction in aquatic ecosystems that results in a drastic increase in total biomass is usually defined as cyanobacterial blooming (Fleming et al. 2002). Blooms are easily noticeable and often manifested by a change of the watercolor, turbidity, and reduction in transparency. On the water surface, they can also be presented as visible accumulations in the form of coating, surface scum, or lumpy thick formations. In recent years, the general consensus is that anthropogenic nutrient over-enrichment (namely phosphorous, and nitrogen) of freshwater and marine ecosystems has caused and promoted bloom expansion and persistence (Huisman et al. 2005; Paerl and Paul 2011). Furthermore, climate change provides an additional catalyst for their expansion as rising global temperatures and changing precipitation patterns both stimulate cyanobacterial blooms (Jöhnk et al. 2008; Paul 2008; Paerl and Huisman 2009).

Cyanobacteria can produce many different biologically active substances, as a result of their physiological and ecological diversity, exceptional adaptability, and rich genetic potential. Some of them can include highly valuable metabolites, and biomodulators, but also products of cyanobacteria that can cause disease or have a lethal effect (cyanotoxins) (Singh et al. 2006). It is important to highlight that there is a significant difference between non-toxic, potentially toxic, and toxic cyanobacteria. Furthermore, since cyanobacterial toxin production is a strictly strain-specific property that is genetically determined (Vezie et al. 2002), all strains possessing genes encoding the synthesis of certain toxins are considered potentially toxic (Zurawell et al. 2005; Burkholder 2009).

Most species do not synthesize cyanotoxins during the first stages of cell growth, but only after entering the secondary metabolism, which is thought to be conditioned by stress factors. Therefore, it is generally accepted that cyanotoxins are a product of secondary metabolism of cyanobacteria (Carmichael 1992; Kurmayer and Christiansen 2009). The possible evolutionary advantages of toxin production can be classified into two strategies: competitive advantage and physiological assistance. The first strategy explains how the products of cyanobacteria represent a defense mechanism, in response to a feeding pressure and/or competition for resources. The second strategy considers the contribution of secondary metabolites in cell physiology through benefits in achieving homeostasis, photosynthetic efficiency, and enhanced growth rate (Holland and Kinnear 2013).

Cyanobacterial toxins are a biochemically and functionally diverse group of physiologically active compounds and can be divided into several categories. Depending on their effect on the target organ, they can conditionally be divided into four groups (Carmichael 1992; Codd et al. 2005): hepatotoxins (microcystins (MCs), nodularins (NOD)); neurotoxins (anatoxins (ATX) and saxitoxins (STX)); dermatotoxins (aplysiatoxin, debromoaplysiatoxin, and lyngbyatoxin); and cytotoxins (cylindrospermopsin (CYN) and lipopolysaccharides (LPS)). On the other hand, based on their chemical structure, cyanotoxins can be divided into three groups (Codd et al. 2005): cyclic peptides (MCs and NOD); alkaloids (ATX, STX, CYN, aplysiatoxin, and lyngbyatoxin); and LPS (lipopolysaccharide endotoxins). It should be emphasized that there is a health risk due to exposure to cyanotoxins, especially the most common ones called microcystins (MCs).

The establishment of a tolerable daily intake (TDI) of 0.04 μg/kg body weight per day, as well as the recommended limit value of MC-LR in drinking water of 1 μg/L (WHO 1998), aims to limit chronic exposure to cyanotoxins and potential consequences for human health. Therefore, the formation of legislation with guideline values for different human exposure pathways to cyanotoxins should be done in order to prevent potential health risks. In many countries, measures to protect public health from cyanotoxins have been implemented by applying the guideline values proposed by the WHO. The recommended values do not represent international standards, but some countries adhere to them, and some have set their own national standards based on their national, economic, and environmental factors (Svirčev et al. 2011). However, very few countries have set guidelines or acceptable levels for cyanotoxins in food, and they are usually only set up for fish and shellfish used for human consumption. The only exception concerns STX in bivalve shellfish since this issue is already covered by regulations in Europe, USA, Canada, Australia, and New Zealand for marine biotoxins (STX can be produced both by marine dinoflagellates and by freshwater cyanobacteria). In the USA, some States have health advisories in case of fish contamination by MCs, and sometimes also by ATX-a, CYN, and STX. Victoria State (Australia) defined acceptable levels of cyanobacterial toxins in seafood (fish, prawns, and mussels) harvested from the lakes. The identified toxins for the risk assessment were MCs, NOD, STX, and CYN (Mulvenna et al. 2012; Testai et al. 2016a). Another problem is that values vary considerably between different countries: for MCs, five authorities give values ranging from 5.6 μg/kg for fish (France) to 51 μg/kg for molluscs (Australia); for CYN, three authorities give values from 18 μg/kg (two states in Australia) to 70 μg/kg (California, USA); for STX in fish, prawns, and shellfish, the value of 800 μg/kg is the same in these two Australian states and in Canada, while for ATX only California, USA, gives a value set at 5000 μg/kg for fish (Testai et al. 2016b). Choices of control measures can be optimized depending on the conditions in the specific aquatic setting that lead to blooms, the local patterns of consumption potentially leading to exposure, and the available institutional capacity for operational monitoring (Testai et al. 2016b). Regrettably, there are no guideline values for cyanotoxins in drinking water supply, or food when it comes to the Republic of Serbia. In order to prevent any potential risks, guideline values should be implemented into national regulations as soon as possible, but at the same time, to control eutrophication, and subsequently reduce cyanobacterial growth and exposure to cyanotoxins (Drobac et al. 2013; Đorđević 2021).

1.2 Adverse Effects on the Fish Stocks and Health Hazards after Cyanotoxin Exposure

Eutrophication, along with cyanobacterial blooms, can have negative impacts on the overall structure and function of an ecosystem. The nutrient concentration in a waterbody can be affected directly by nutrient loading and indirectly by the presence of fish and fisheries. The changes in the fish community structure of Serbian reservoirs in recent decades have been attributed to both the alteration of nutrient loading and eutrophication pressure, as well as changes in the local climate (Đorđević et al. 2015; SEPA 2013, 2019).

Consumption of fish from waters affected by cyanobacterial blooms represents an important route of human exposure to cyanotoxins (Ibelings and Chorus 2007; Wilson et al. 2008; Peng et al. 2010; Poste et al. 2011). Researchers from China indicated that aquatic products from three major lakes were unsafe for human consumption after exposure to MCs, which made estimated daily intake values 1.5 to 148 times higher than the TDI (Peng et al. 2010). Similarly, TDI was exceeded for several fish products originating from, e.g., Greek lakes (Papadimitriou et al. 2010), Funil and Furnas reservoirs in Brazil (Deblois et al. 2008), and recently from Pike River Ecological Reserve in the USA (Skafi et al. 2021), indicating that the need for legislation for this type of human exposure to cyanotoxins is even more relevant.

Besides the aforementioned exposure route, cyanotoxin exposure could also occur via ingestion of contaminated drinking water, dermal contact, and inhalation during recreational activities (including fishing), consumption of plants contaminated during irrigation, consumption of cyanobacterial dietary supplements and through hemodialysis (Drobac et al. 2013). Extensive research in Serbia demonstrated a connection between the high incidence of primary liver cancer in the three critical districts with the appearance of water blooms in the reservoirs used for the drinking water supply of these districts. Simultaneously, there was no correlation with the main risk factors—incidence of chronic viral hepatitis (HBV and HCV) and liver cirrhosis mortality in Central Serbia from 1999 to 2008 based on the epidemiological data (Svirčev et al. 2013). Moreover, further research during the same time period revealed that the same critical districts had significantly higher incidences of nine more cancers (brain, heart, mediastinum, and pleura, ovary, testicular, gastric, colorectal, retroperitoneum and peritoneum, leukemia, malignant melanoma of skin), indicating that cyanobacterial toxic metabolites could act synergistically with other risk factors causing the higher incidence of the examined cancers (Svirčev et al. 2014a). Globally, several studies have found a positive correlation between toxigenic cyanobacteria/cyanotoxins and adverse health effects—with examples from Australia, China, Serbia, Sweden, UK, Portugal, Brazil, USA, and Canada. A review of 42 studies provided evidence in favor of cyanotoxin involvement in human health disorders. Although epidemiological studies cannot prove the cause of diseases related to toxic cyanobacteria, they can greatly increase the understanding of the hazards and risks involved and potentially aid the design and implementation of effective risk management policies (Svirčev et al. 2017a).

2 The Serbian Experience

Cyanobacterial blooms have been reported in all parts of the world. It is possible that the oldest information about the occurrence of cyanobacteria and algae blooming comes from the period of the Han dynasty in China 1000 years ago, when it was recorded that the river water was of a distinctly dark green color (Chorus and Bartram 1999). Nowadays, the consensus among experts is that the frequency of harmful cyanobacterial blooms has been increasing worldwide over the past few decades. This trend is of great concern, as it may have negative effects on the biodiversity and functioning of aquatic food webs and also threatens the use of affected waters for drinking, bathing, fishing, and other recreational uses (Huisman et al. 2018).

Global geographical and historical overview of cyanotoxin distribution and cyanobacterial poisonings that has been performed on 468 selected articles showed that there were 1118 recorded identifications of major cyanotoxins in 869 freshwater ecosystems from 66 countries throughout the world until 2018 (Svirčev et al. 2019). MCs were the most often recorded cyanotoxins worldwide (63%), then CYN (10%), ATX (9%), and STX (8%), while NOD was the most rarely recorded cyanotoxin (2%), and in some reports, cyanotoxins were not analyzed or specified (9%). Furthermore, there were 183 recorded cyanotoxin poisonings of humans and/or animals, while the true global extent of cyanotoxins and associated poisonings is likely to be greater than found in the available literature (Svirčev et al. 2019).

Research on the occurrence of toxic cyanobacteria in many aquatic ecosystems in the Republic of Serbia has been published in numerous publications. One of the earliest data on algal research in Serbia was published almost 140 years ago in the work "Fragmenta Phycologiae Bosniaco-Serbicae" (Scharschmidt 1883), with a list of 46 algal species. A more detailed approach to the research of aquatic ecosystems in the Republic of Serbia began to develop after World War II (Blaženčić 1986). Research was not systematic, so it is likely that number of present and blooming species of cyanobacterial was grossly underestimated. An extensive review of publications until 2012 showed that cyanobacterial blooms were found in almost 80% of the investigated aquatic ecosystems, which included 65 aquatic ecosystems. More than 70 species, including blooms of 24 species from 13 genera, were recorded, and *Microcystis aeruginosa* and *Aphanizomenon flos-aquae* were found to be the most numerous species in the vast majority of blooming events. Consequently, a very clear indication of the omnipresence of cyanobacterial blooms in different water bodies throughout the Republic of Serbia was noted (Svirčev et al. 2014b; Svirčev et al. 2017b).

Although the fishponds were not the focus of most of the aforementioned research, they are commonly affected by blooms. In the period from 1949 until 2011, research was conducted in 8 fishponds in Serbia, with *Microcystis aeruginosa* found most frequently blooming, while invasive *Raphidiopsis raciborskii* was also noted in Kapetanski rit fishponds (Milovanović and Živković 1953, 1959; Milovanović 1963; Ristić et al. 1979; Ćirić et al. 2010). Furthermore, cyanotoxins were detected in water, fish liver and muscle, but also in the sediment and aquatic plants (World Bank Report, DM 4307 2011; Tokodi et al. 2013; Drobac et al. 2016). Additionally, histological alterations in the fish tissues of the liver, kidneys, gills, intestine, and muscle were also observed in blooming fishponds (Drobac et al. 2016), suggesting that cyanobacteria and their toxins in fishponds could be hazardous not only to fish quality, but ultimately to the economy, human health, and the environment as well. Even though cyanobacterial investigation in Serbian fishponds commenced during the 1950s, in recent decades an increase in scientific interest regarding cyanotoxin presence in water and fish tissue is arising, mostly due to consequential negative health and economic impacts of this phenomenon (Drobac 2018).

One of the most comprehensive research was conducted on 13 undisclosed fishponds in Vojvodina, where cyanobacterial blooming was inspected from May to September of 2011 (Drobac 2015; Drobac et al. 2016). By measuring Chlorophyll *a* (Chl *a*) concentrations, it was determined that water ranged from meso-eutrophic (nine fishponds) to eutrophic (two fishponds) and polytrophic (two fishponds). Such high pigment concentrations may indicate the presence of a large number of cyanobacteria, which was confirmed by microscopic phytoplankton analyses. In all investigated fishponds, the number of cyanobacterial cells was around or over 10,000 cells/mL, delineating cyanobacterial blooming in all fishponds. The highest number of cyanobacterial cells recorded was 87,979,440 cells/mL, with the dominant species being *Microcystis aeruginosa* (78,205,440 cells/ml). This species was the most abundant in the majority of fishponds, alongside *Pseudanabaena limnetica* and *Planktothrix agardhii*, while more than 20 cyanobacterial species have been recorded, including highly invasive *Raphidiopsis raciborskii* in one of the fishponds (Drobac 2015; Drobac et al. 2016).

More than a dozen genera present are potential cyanotoxin producers, hence cyanotoxins were anticipated and affirmed in water samples via ELISA tests. MCs/NOD (most probably MCs based on the genera present) were detected in 85% of fishponds, while low levels of STX were noted in 38% of investigated fishponds. Concomitantly, fishponds with higher concentrations of cyanotoxins also contained the utmost cyanobacterial abundance and the highest Chl *a* concentration (Drobac 2015; Drobac et al. 2016). Additionally, the toxicity of water sampled during the fall of the same year was verified by *Artemia salina* toxicity test (Tokodi et al. 2013). Moreover, the presence of MCs was validated with in vitro protein phosphatase 1 inhibition assay (PPI) in the same water samples (Tokodi et al. 2014).

Considering that the fish intended for human consumption are cultivated in these fishponds, cyanotoxin contamination of fish meat was also analyzed. Liquid chromatography with tandem mass spectrometry (LC-MS/MS) showed that 40% of tested samples of common carp (*Cyprinus carpio*) muscle tissue contained circa 60 μg MC-RR/kg dry weight (Drobac 2015; Drobac et al. 2016). MC-RR is considered to be about ten times less toxic MCs variant than MC-LR (Sivonen and Jones 1999), accordingly, since some authorities defined guideline value for MC-LR in fish to be around 5 μg/kg (Chorus and Welker 2021), in this regard value for MC-RR in fish would comparatively be 50 μg/kg. Thus, the obtained toxin concentrations in fish muscle from blooming fishponds are close to the suggested limits (Drobac et al. 2016; Drobac 2018). However, with respect to current analytical shortcomings obtained values could only be considered as an estimation of the true cyanotoxin content in complex matrices in which they are bound to proteins, ergo are not detectable (Testai et al. 2016a). Furthermore, literature data do not indicate a high risk of short-term exposure to cyanotoxins through fish muscle tissue, however, the risk is greater if viscera is being consumed (Chorus and Welker 2021).

Cyanobacteria and cyanotoxins can also affect fish tissue, for instance, specimens caught from blooming fishponds in Vojvodina displayed severe histopathological alterations in several organs, such as the liver, kidney, intestines, gills, and muscle. Interestingly, muscle tissue exhibited severe degeneration, which is rarely described (Drobac et al. 2016). Other similar histopathological changes were observed in fish exposed to cyanobacteria and their toxins in other blooming waters in Serbia (Tokodi et al. 2018, 2020) and neighboring country Hungary (Drobac Backović et al. 2021; Marinović et al. 2021).

Research has been realized on several other water bodies in Serbia containing fish that can be used for human consumption, including the Aleksandrovac reservoir (Đorđević et al. 2015, 2017; Svirčev et al. 2016), Vrutci reservoir (Svirčev et al. 2017c), and Ludaš Lake (Tokodi et al. 2018, 2020) among others. In all of the investigated water bodies, cyanobacteria bloomed and produced toxic metabolites. As a result, cyanotoxins have been detected in sampled fish tissues (Tokodi et al. 2018), even in the muscles of commercial fish (Svirčev et al. 2017c). Furthermore, histopathological changes, as well as DNA damage, were observed in various organs of the exposed fish (Tokodi et al. 2018, 2020). In addition to histopathological changes in the liver, biochemical alterations of antioxidant biomarkers were revealed in livers (Perendija et al. 2011) and even more so in gills (Gavrilović et al. 2014) of fish exposed to *Aphanizomenon flos-aquae* bloom in the Gruža Reservoir. All of these aforementioned disorders indicate how deleterious eutrophication and cyanobacterial blooming in fishing waters can be on fish resources.

3 Eutrophication in Selected Fishing Waters of Serbia, and the Potential Toxicity of Cyanobacterial Blooms and Cyanotoxins

Fishery-significant reservoirs, also known as fishery reservoirs or fishponds, are bodies of water specifically designed and managed for the purpose of fish farming and/or recreational fishing. These reservoirs are often artificially created by damming a river or stream, and they can vary greatly in size and depth. Fishery-significant reservoirs are important because they provide a controlled environment for fish farming and help to sustain fish populations for recreational fishing. They also serve as a source of food and income for local communities that rely on fishing as a source of livelihood. However, they can also have negative impacts on the environment, such as altering natural water flows and causing changes in the composition of aquatic plant and animal communities. It is important to carefully manage fishery-significant reservoirs to ensure their sustainability and minimize negative impacts on the environment (Winfield 2016). For the purpose of writing this chapter, we have gathered research results from the past decade (2012–2022). In addition, we have chosen 11 fishery-significant reservoirs with different trophy levels to be included in the chapter. The locations of the eleven reservoirs that belong to different geographical regions and different fishing areas (Fig. 1), including data about length, width, and max. depth, as well as the purpose of the reservoir are shown in Fig. 1 and Table 1.

The trophic state of the reservoirs was determined by Carlson's Trophic State Index—TSI (Carlson 1977), which has been adopted by the US EPA nutrient criteria recommendations for lakes and reservoirs (US-EPA 2000). Data on transparency, total phosphorus, and Chl *a* concentration were taken from the report of the Serbian Environmental Protection Agency (SEPA 2012, 2013, 2014, 2015, 2017a, b, 2018, 2019, 2020). According to Carlson (1977) lake classification regarding trophic condition range, several categories of TSI values are possible, and results for eleven selected reservoirs in Serbia are shown in Table 2.

The Vlasina reservoir is oligotrophic, while the Potpeć and Šumarice reservoirs are mesotrophic. The increased level of eutrophication was observed in 5 reservoirs for water supply (Barje, Garaši, Vrutci, Bovan, Gruža). The hypereutrophic conditions in three reservoirs intended only for sports and recreation prevailed (Aleksandrovac, Palić, Ludaš) (Table 2).

According to the literature, the assessment of the level of hazard and health risk was based on the abundance of cyanobacteria and the WHO's recommendations (Chorus and Bartram 1999). This approach is commonly used in Serbia for recreational water (Table 3). According to the WHO guidelines, the concentration of MCs in drinking water should not exceed 1 μg/L. The guidelines for the abundance of cyanobacteria cells in water are based on legislation in the Czech Republic. Therefore, the limit between low and moderate risk is set at 2000 cells/mL (Chorus 2012) (Table 3).

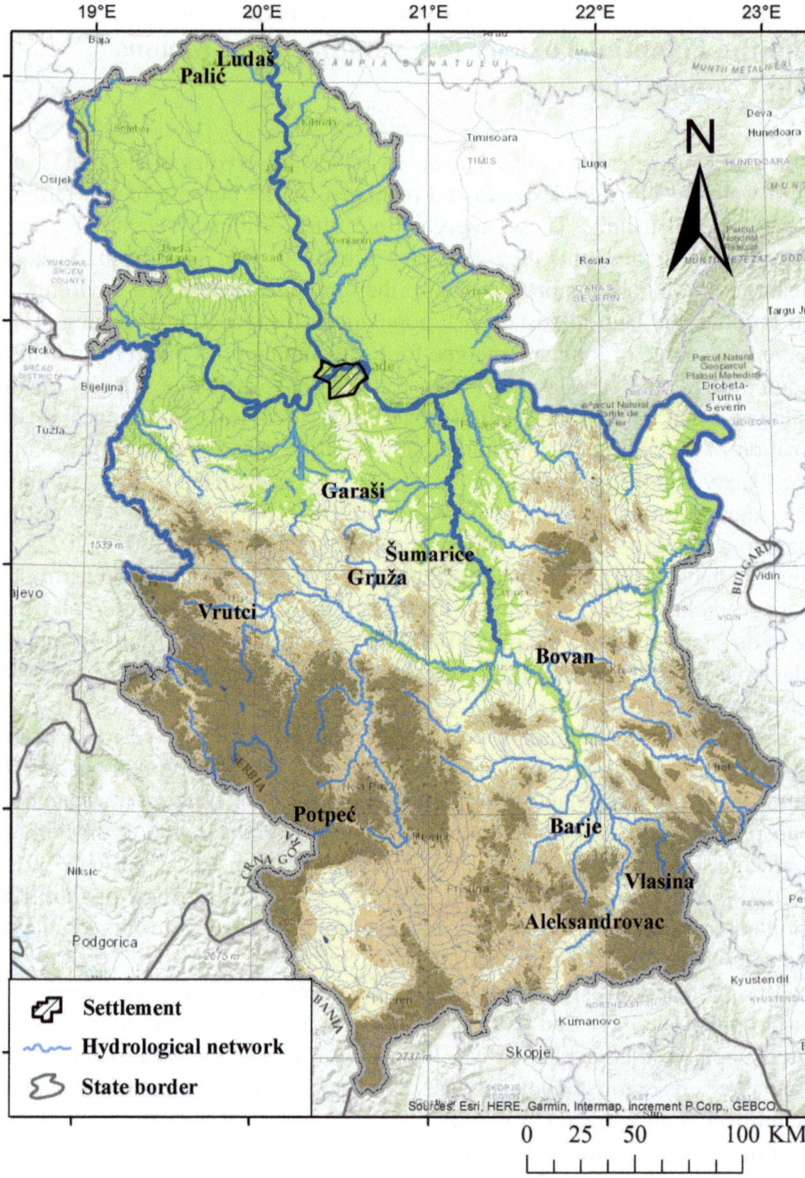

Fig. 1 Locations of the selected reservoirs in Serbia. Modified and cited from Đorđević (2021)

Low level of risk includes short-term adverse health outcomes, most often in the form of skin irritations and gastrointestinal illnesses. Moderate risk level also implies short-term health consequences, but also potential long-term illnesses, while the high risk level, in addition to short-term and long-term illnesses, also includes the potential for acute poisoning even with a fatal outcome (WHO 2003).

Table 1 Basic morphological data of the selected reservoirs

Reservoir	Geographic coordinates	Surface (ha)	Length (km)	Width (km)	Max. depth (m)	Purpose
Vlasina	42° 42′ 0″ N 22° 20′ 0″ E	1500	9	3.5	25–35	WS/SR/HP/FC
Potpeć	43° 30′ 31″ N 19° 35′ 42″ E	700	20	-	40	WS/SR/HP/FC
Šumarice	44° 1′ 48.75″ N 20° 52′ 28.17″ E	22	1.5	0.175	15	SR/FC
Barje	42° 48′ 54.2″ N 21° 48′ 52.66″ E	130	7.3	0.3	25	WS/SR/FC
Garaši	44° 17′ 17.81″ N 20° 28′ 44.65″ E	65	1.8	0.07-0.420	20	WS/SR/FC
Vrutci	43°50′34″N 19°41′36″E	192	8	-	60	WS/SR/HP/FC
Bovan	43° 38′ 45.79″ N 21° 42′ 27.91″ E	400	8	0.5	50	WS/SR/FC
Gruža	43° 55′ 18.84″ N 20° 41′ 20.04″ E	900	10	0.3–2.8	35	WS/SR/FC
Aleksandrovac	42° 29′ 22.35″ N 21° 53′ 57.61″ E	12	0.5	0.25	2.5–4	SR/I
Lake Palić	46° 4′ 30″ N 19° 45′ 30″ E	560	8	1	2.5–3	SR
Ludaš Lake	46° 5′ 58″ N 19° 49′ 27″ E	328	4.5	0.01	1–2	SR

WS water supply, *SR* sports and recreational, *I* irrigation, *HP* hydro power, *FC* Flood control, – no data

The literature data summarized in previously published studies and reports (Karadžić et al. 2010; Sedmak and Svirčev 2011; SEPA 2012, 2013; Đorđević and Simić 2014; Kostić et al. 2014, 2015, 2016; Simić et al. 2014, 2017, 2018, 2021; SEPA 2014, 2015, 2017a, b, 2018, 2019, 2020; Svirčev et al. 2017b, c; Blagojević Ponjavić et al. 2019, Trbojević et al. 2019; Čađo et al. 2020; Jovanović 2020; Blagojević 2021; Đorđević 2021) were used to create a list of cyanobacteria, the occurrence of blooming species, the presence of invasive taxons, as well as the phytoplankton and cyanobacteria abundance in waterbodies that are of interest for water supply and/or recreation (fishing waters) (lake and reservoirs) (Table 4).

Data on ichthyofauna for the past decade, from 2012 to 2022, were collected as part of the Fishing Water Management Program (Simić et al. 2001–2021) in accordance with the Law on the Protection and Sustainable Use of Fish Fund (Official Gazette of RS, 128/2014). This includes qualitative analysis, total fish biomass, and biomass by families (kg/ha) for the Vlasina, Potpeć, Šumarice, Barje, Garaši, Vrutci, Bovan, Gruža, and Aleksandrovac reservoirs (Table 5; Fig. 2). A qualitative and quantitative analysis (percentage composition of fish families in total biomass (%)) of the ichthyofauna in the fishing area of Palić and Ludaš Lake was conducted using data collected from the "Report on the performed monitoring in the

Table 2 The trophic state of the Serbian reservoirs according to Carlson's Trophic State Index—TSI (Chl *a*). The trophic parameters of reservoirs (transparency, total phosphor, Chl *a*)

Reservoir	Transparency (m)	Total P (mg/l)	TSI (TP)	Chl *a* (μg/L)	TSI (Chl *a*)	Categories according to TSI (Chl *a*)[a]
Vlasina	1.6–3	0.03	52.73	1.08	31.32	O
Potpeć	1–4	0.03	53.22	4.70	45.75	M
Šumarice	1–3	0.04	47.37	6.32	48.65	M
Barje	1.3–3.2	0.05	61.46	8.42	51.47	ME
Garaši	1.1–1.6	0.12	72.98	16.73	58.21	ME
Vrutci	0.7–2	0.07	64.59	26.19	62.60	E
Bovan	0.9–1	0.12	73.69	33.10	64.90	E
Gruža	0.25–2	0.12	73.46	37.47	66.12	E
Aleksandrovac	0.15–1.4	0.35	88.79	142.00	79.19	H
Lake Palić	–	0.36	88.91	525.00	92.02	H
Ludaš Lake	–	0.29	85.95	463.00	90.78	H

[a]*O* oligotrophic, *M* mesotrophic, *ME* mildly eutrophic, *E* eutrophic, *H* hypereutrophic, – no data

Table 3 Risk levels related to the abundance of cyanobacteria in drinking and recreational water (Chorus 2012)

Level of risk	Drinking water (raw water)[a]	Recreational water[b]
Low level of risk	<2000 cells/mL	<20,000 cells/mL
Moderate level of risk	2000–100,000 cells/mL	20,000–100,000 cells/mL
High level of risk	>100,000 cells/mL	>100,000 cells/mL

[a]According to the legislation of the Czech Republic
[b]Based on WHO recommendations (WHO 2003)

fishing area of the Special Nature Reserve "Ludaš Lake" and Nature Park "Palić"" (Miljanović et al. 2016) (Table 5; Fig. 3). In Figs. 2 and 3, the biomass of fish is plotted against the trophic status of the reservoirs.

The data on the presence of cyanotoxins and the impact on fish were also collected, due to the health effects that cyanobacteria and their toxins have, both on fish and potentially on humans which use cyanotoxin-infected fish in their diet, especially from waters used for fishing purposes (Drobac et al. 2013; Drobac 2015; Đorđević et al. 2015, 2017; Tokodi et al. 2018, 2020).

3.1 Reservoir Vlasina

Vlasina is a semi-artificial reservoir in Southeast Serbia (Table 1) that is the highest and largest reservoir in Serbia. It was created in 1947–51 when the peat bog Vlasinsko blato (Vlasina mud) was closed off by a dam and submerged by the waters of incoming rivers, chiefly the Vlasina River. According to WHO recommendations, the oligotrophic Vlasina reservoir has very low levels of phytoplankton

Table 4 The list of blooming and invasive cyanobacteria taxa, phytoplankton, and cyanobacteria abundance in Serbian reservoirs. The level of health risk is based on WHO guidelines (WHO 2003; Chorus 2012)

Reservoir	Blooming Cyanobacteria taxa* Invasive Cyanobacteria taxa**	Abundance of Cyanobacteria (cells/mL)	Abundance of phytoplankton (cells/mL)	Level of health risk WHO water supply reservoir (Chorus 2012)	Level of health risk WHO reservoirs for sports and recreational purposes (WHO 2003)
Vlasina	–	87	1112	Low	Low
Potpeć	*Planktotarix rubescens* (De Candolle ex Gomont) Anagnostidis & Komárek**	124	1253	Low	Low
Šumarice	*Aphanizomenon flos-aquae* Ralfs ex Bornet & Flahault* *Carysosporum bergii* (Ostenfeld) Zapomelova et al.* *Cuspidothrix issatschenkoi* (Usachev) Rajaniemi et al. ** *Microcystis aeruginosa* Kütz* *Sphaerospermopsis aphanizomenoides* (Forti) Zapomelová et al. ** *Raphidiopsis raciborskii* (Woloszynska) Aguilera et al. **	29,568	49,280	/	Moderate
Barje	*Dolichospermum planctonicum* (Brunnthaler) Wacklin, L. Hoffmann & Komárek *	2096	13,100	Moderate	Low
Garaši	*Planktothrix agardhii* (Gomont) Anagnostidis & Komárek *	94,265	105,833	Moderate	Moderate
Vrutci	*Planktothrix rubescens* (De Candolle ex Gomont) Anagnostidis & Komárek *	76,223	117,810	Moderate	Moderate
Bovan	–	14,159	113,275	Moderate	Low
Gruža		90,594	183,501	Moderate	Moderate

(continued)

Table 4 (continued)

Reservoir	Blooming Cyanobacteria taxa* Invasive Cyanobacteria taxa**	Abundance of Cyanobacteria (cells/mL)	Abundance of phytoplankton (cells/mL)	Level of health risk WHO water supply reservoir (Chorus 2012)	Level of health risk WHO reservoirs for sports and recreational purposes (WHO 2003)
	Aphanizomenon flos-aquae Ralfs ex Bornet & Flahault* *Microcystis aeruginosa* (Kützing) Kützing * *Cuspidothrix issatschenkoi* (Usachev) Rajaniemi et al.** *Planktothrix rubescens* (De Candolle ex Gomont) Anagnostidis & Komárek**				
Aleksandrovac	*Chrysosporum bergii* (Ostenfeld) Zapomelova et al.** *Planktothrix* sp.K. Anagnostidis & J. Komárek *Raphidiopsis raciborskii* (Woloszynska) Aguilera et al. */**	10,283,158	12,097,833	/	High
Lake Palić	*Microcystis aeruginosa* (Kützing) Kützing* *Raphidiopsis raciborskii* (Woloszynska) Aguilera et al.** *Sphaerospermopsis aphanizomenoides* (Forti) Zapomelová et al. **	211,400,000	248,400,000	/	High

Ludaš Lake	*Microcystis aeruginosa* (Kützing) Kützing * *M. wesenbergii* (Komárek) Komárek * *Planktothrix agardhii* (Gomont) Anagnostidis & Komárek *	17,204,000	25,300,000	/	High

* Blooming species; ** Invasive species; – Blooming and invasive cyanobacterial taxa were not recorded; / it is not used for the water supply

Table 5 The fish species important for fisheries in selected reservoirs in Serbia

Reservoir / Fish species	Vlasina	Potpeć	Šumarice	Barje	Garaši	Vrutci	Bovan	Gruža	Aleksandrovac	Lake Palić	Ludaš Lake
Salmonidae											
Salmo trutta (Linnaeus, 1785)	+	+	–	–	–	–	–	–	–	–	–
Hucho hucho (Linnaeus, 1758)	–	+	–	–	–	–	–	–	–	–	–
Thymallus thymallus (Linnaeus, 1758)	–	+	–	–	–	–	–	–	–	–	–
Percidae											
Micropterus salmoides (Lacepède, 1802)	–	–	+	–	–	–	–	–	–	–	–
Perca fluviatilis (Linnaeus, 1758)	+	–	+	+	+	+	+	+	–	+	–
Sander lucioperca (Linnaeus, 1758)	–	–	+	+	+	–	+	+	–	–	+
Centrarchidae											
Lepomis gibbosus (Linnaeus, 1758)	+	+	+	–	–	–	+	+	–	–	+
Esocidae											
Esox lucius (Linnaeus, 1758)	+	–	+	–	–	–	+	–	–	–	–
Cyprinidae											
Abramis brama (Linnaeus, 1758)	+	–	+	+	+	+	+	+	+	–	–
Alburnoides bipunctatus (Bloch, 1782)	–	–	–	–	–	–	+	–	–	–	–
Alburnus alburnus (Linnaeus, 1758)	+	+	+	+	+	+	+	+	+	–	–
Barbus balcanicus (Kotlík, Tsigenopoulos, Ráb & Berrebi, 2002)	–	+	–	–	–	–	–	–	–	–	–
Barbus barbus (Linnaeus, 1758)	–	+	–	–	–	–	–	–	–	–	–
Blicca bjoerkna (Linnaeus, 1758)	–	–	–	–	+	–	+	–	–	–	–
Carassius carassius (Linnaeus, 1758)	+	–	–	–	–	–	–	–	–	–	–
Carassius gibelio (Bloch, 1782)	+	+	+	–	–	+	+	+	+	+	+
Chondrostoma nasus (Linnaeus, 1758)	–	+	+	–	–	+	–	–	–	–	–
Ctenopharyngodon idella (Valenciennes, 1844)	–	–	–	–	–	–	+	–	+	–	–

Cyprinus carpio (Linnaeus, 1758)	–	+	+	–	–	+	+	+	+	+	–
Gobio obtusirostris (Valenciennes, 1842)	–	+	–	–	–	–	–	–	–	–	–
Hypophthalmichthys molitrix (Valenciennes, 1844)	–	–	+	–	–	+	+	–	–	–	+
Hypophthalmichthys nobilis (Richardson, 1845)	–	–	–	–	–	+	+	–	+	–	–
Hypophthalmichthys sp.	+	–	–	–	–	–	–	–	–	–	–
Leucaspius delineatus (Heckel, 1843)	–	–	–	–	–	–	+	–	–	–	–
Leuciscus aspius (Linnaeus, 1758)	–	–	–	–	–	+	–	–	–	–	–
Pseudorasbora parva (Temminck & Schlegel, 1846)	–	+	–	–	–	–	–	–	–	–	+
Rhodeus sericeus amarus (Bloch, 1782)	–	–	+	–	–	–	–	–	–	–	–
Rutilus rutilus (Linnaeus, 1758)	+	–	+	+	+	+	+	+	–	–	–
Scardinius erythrophthalmus (Linnaeus, 1758)	–	–	+	–	–	–	+	–	–	–	–
Squalius cephalus (Linnaeus, 1758)	+	+	–	+	–	+	–	–	–	–	–
Vimba vimba (Linnaeus, 1758)	–	–	–	–	–	–	+	–	–	–	–
Ictaluridae											
Ameiurus nebulosus (Lesueur, 1819)	+	–	+	+	–	–	–	+	–	–	–
Siluridae											
Silurus glanis (Linnaeus, 1758)	+	–	+	–	+	+	+	+	+	–	–
Σ	13	13	16	7	7	12	16	10	7	3	5

+ present species; – not detected

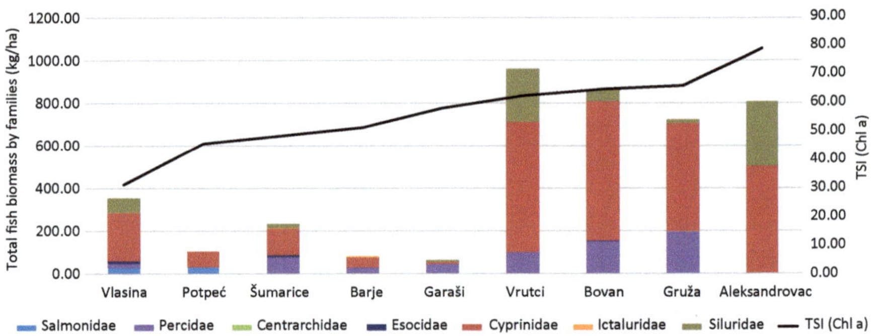

Fig. 2 The relationship between fish biomass (total fish biomass and biomass by families) (kg/ha) and trophic state index TSI (Chl *a*) in Serbian reservoirs. The primary y-axis shows fish biomass, while the secondary y-axis displays TSI (Chl *a*)

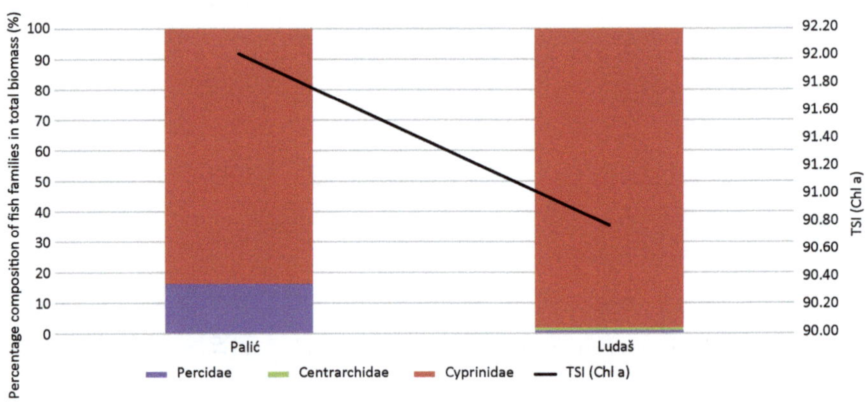

Fig. 3 The relationship between the percentage composition of fish families in total biomass (%) and trophic state index TSI (Chl *a*) in Lake Palić and Ludaš. The primary y-axis shows the percentage composition of fish families in total biomass, while the secondary y-axis displays TSI (Chl *a*)

and cyanobacteria abundance (Chorus and Bartran 1999). In relation to the abundance of cyanobacteria, this fishing water is classified as having a low probability of adverse health effects. Cyanobacteria are present in this reservoir, but the abundance (87 cell/mL) does not pose any potential danger (Table 4). Since the formation of the reservoir, the trophic state has not changed (Cvijan and Laušević 1991). In this reservoir, the 13 species important for fisheries from six families were dominant and frequently present. The overall fish biomass was measured at 352 kg/ha. The Cyprinidae family made up the largest portion of the total fish biomass at 64%, followed by the Siluridae family at 18%, and the Salmonidae family at 9% (Fig. 2; Table 5). The Vlasina reservoir is a popular destination for recreational fishing, and the low level of eutrophication of the reservoir has a positive impact on the health of fish populations and the overall productivity of the fishery.

3.2 Reservoir Potpeć

Potpeć is a reservoir on the Lim River, in West Serbia (Table 1). It was created for hydroelectric power purposes by a concrete gravity dam. This mesotrophic reservoir is characterized by a very low abundance of phytoplankton and cyanobacteria (124 cells/mL) (Table 4). Accordingly, a low level of health risk was established in this reservoir (Table 4). No cyanobacterial blooms were recorded, but the occurrence of invasive species *Planktothrix rubescens* was an important find. The 13 species important for fisheries from three families were dominant and frequently present. The average biomass was 103 kg/ha, with the Cyprinidae family comprising the majority at 68%, and the Salmonidae family making up 31% (Fig. 2; Table 5). The Potpeć reservoir is used for recreational fishing, and it is crucial to maintain a low level of trophy in order to support healthy fish populations and a productive fishery.

3.3 Reservoir Šumarice

The Šumarice reservoir is located in Central Serbia (Table 1) and is used exclusively for sports and recreational activities. It has a mesotrophic character with an average cyanobacterial count of 29,568 cells/mL, thus making it an accumulation with a medium level of health risk (Table 4). Some of the potentially toxic and invasive cyanobacteria (Kaštovský et al. 2010; Kokocinski et al. 2017), such as *Aphanizomenon flos-aquae*, *Chrysosporum bergii*, *Cuspidothrix issatschenkoi, Microcystis aeruginosa*, *Sphaerospermopsis aphanizomenoides*, *Raphidiopsis raciborskii* were recorded in this reservoir (Table 4).

Cyanobacteria were present in the reservoir, but were not blooming until 2014 (Simić et al. 2017; Đorđević 2021). The condition in the reservoir changed after an accidental inflow of sewage into the reservoir after extreme rainfall in May 2014 (Simić et al. 2017; Đorđević 2021). After extreme precipitation, the sewage collector bursted, leading to an episodic increase in the nutrient concentration of the reservoir, and thus it changed the composition of the cyanobacterial community (predominantly from August to November 2014). The suppression of all other species led to the domination of *Aphanizomenon flos-aquae*, thus commencing the first cyanobacterial blooming in the Šumarice reservoir (Fig. 4). The abundance of *A. flos-aquae* during the blooming event (4,010,000 cells/mL) was not in accordance with the required values of these parameters for the correct management of recreational water (Chorus and Bartram 1999). According to the WHO (Chorus and Bartram 1999), the detected abundance and blooming indicates a major health threat, and for that reason, the use of the Šumarice reservoir was prohibited by the Institute for Health Protection (Simić et al. 2017; Đorđević 2021). Despite the hypereutrophic conditions, the cyanobacterial bloom had no significant impact on the fish community or caused accidental fish kills (Simić et al. 2017; Đorđević 2021). Because of the

Fig. 4 The mesotrophic reservoir Šumarice (left); The blooming of *Aphanizomenon flos-aquae* in Šumarice reservoir (2014) (right). Photo by Snežana Simić

sewage water injection in the Šumarice reservoir, the change from mesotrophic to eutrophic condition is still present. During the research carried out in July 2022, qualitative analysis of the phytoplankton of the Šumarice reservoir showed that 16 taxa of Cyanobacteria were recorded, the dominant species in the phytoplankton were cyanobacteria *Chrysosporum bergii* and *A. flos-aquae* (Simić and Đorđević 2022). The presence and increase in the abundance of Cyanobacteria, especially the presence of the invasive taxon *Raphidiopsis raciborskii*, requires special attention, because of *R. raciborskii* is potentially toxic, and these taxa was also associated with the occurrence of fish mortality in some Serbian reservoirs (Đorđević et al. 2015). All cyanobacteria which were recorded in the Šumarice reservoir are potentially toxic and can produce MCs, ATX, STX, and CYN.

In the Šumarice reservoir, 16 species important for fisheries from six families were dominant and frequently present. The overall fish biomass was recorded at 232 kg/ha. The Cyprinidae family made up the largest proportion of the average total fish biomass, with 53%. The Percidae and Siluridae families followed, making up 34% and 7%, respectively (Table 5; Fig. 2). In this reservoir, no accidental fish mortality associated with cyanobacterial blooming, was recorded.

3.4 Reservoir Barje

The artificial reservoir Barje is located in southern Serbia (Table 1). It was created by the construction of a high dam in the gorge of the river Veternica for water supply, irrigation, flood protection, and recreative fishing. This mildly eutrophic reservoir with cyanobacteria abundance of 2096 cells/mL is characterized as a medium health risk from the aspect of health and use of this water for drinking, and low health risk from the aspect of sports and recreational activities (Chorus and Bartram 1999)

(Table 4). An earlier study of this reservoir indicated that the water was mesotrophic, that cyanobacteria occurred in phytoplankton in low abundance (Simić 2004; SEPA 2018). The blooming of cyanobacteria, namely *Dolichospermum planctonicum*, was rarely recorded. In recent years, there has been a deterioration in the condition of this reservoir and a transition from a mesotrophic to a eutrophic state occurred. In the summer of 2020, the blooming of *Aphanizomenon flos-aquae* species was recorded (unpublished data). Blooming was a consequence of the inflow of a large number of polluting substances via the Veternica River which reaches the reservoir, especially after heavy rainfall. In this reservoir, the seven species important for fisheries from three families were dominant and frequently present. The majority of the total fish biomass, which was 81 kg/ha, was made up of fish from the Cyprinidae family (57%). The Percidae and Ictaluridae families contributed 39% and 4%, respectively (Table 5; Fig. 2).

3.5 Reservoir Garaši

The Garaši reservoir was formed at the mouth of two small rivers, Velika Bukulja and Mala Bukulja, in central Serbia for water supply (Table 1). This mildly eutrophic reservoir with a cyanobacterial abundance of 94,265 cells/mL presents a medium health risk according to WHO recommendations (Chorus and Bartram 1999) (Table 4). In previous studies based on the calculated trophic index (TSI), the Garaši reservoir could be considered as mesotrophic. Dominating cyanobacteria were: *Planktothrix agardhii*, *Aphanizomenon flos-aquae*, *Microcystis aeruginosa*, and *Anabaena affinis* (Karadžić et al. 2010). The trophic status of this reservoir has been deteriorating in recent years (SEPA 2017a, 2021).

In this mildly eutrophic reservoir, seven species important for fisheries from three families were dominant and frequently present. The overall average fish biomass was recorded at 61 kg/ha, with the Percidae family making up the largest portion at 75%. Cyprinidae and Siluridae contributed 18% and 7%, respectively (Table 5; Fig. 2). There have been no recorded instances of accidental fish mortality associated with cyanobacterial blooms in the Garaši reservoir.

3.6 Reservoir Vrutci

The Vrutci reservoir was created by damming the Đetinja River in Western Serbia (Table 1). It is a temperate, mid-altitude reservoir. It is primarily used for drinking water supply of the city of Užice, its water is also used by a small hydropower plant, as well as for flood control and sediment retention. In addition, Vrutci reservoir is a destination for recreation and fishing, even restocking has been done repeatedly in order to manage the fish stocks. The eutrophic reservoir Vrutci is characterized by high primary production of phytoplankton (76,223 cells/mL) and high production of fish biomass (959.34 kg/ha) (Table 4; Fig. 2). In relation to WHO recommendations during 2018 (SEPA 2019), Vrutci reservoir with a cyanobacteria abundance of 76,223 cells/mL presented a medium health risk both from the aspect of drinking water use and from the aspect of water use for sports and recreational activities (Chorus and Bartram 1999). The overall fish biomass was recorded at 959 kg/ha. The Cyprinidae family made up the largest percentage of the average total fish biomass at 64%, followed by Siluridae at 25% and Percidae at 11% (Table 5; Fig. 2). All of these activities contributed to water quality deterioration and eutrophication (Đenić 2014).

As the reservoir is a sight of intense anthropogenic activity, inevitably, in December 2013, a cyanobacterial blooming occurred in form of purple-red floating mats and water color become red (Fig. 5; Kostić et al. 2014, 2015, 2016; Drobac 2015; Svirčev et al. 2017c; Blagojević et al. 2019, 2021). Fish mortality and cyanotoxin accumulation in the fish tissue also occurred during notable blooming event in the Vrutci reservoir (Drobac 2015; Svirčev et al. 2017c).

Institute of Public Health of Serbia "Dr Milan Jovanović Batut" determined that the blooming cyanobacteria was *Planktothrix rubescens*, and that its cells were present in the treated water (10,000 cells/L) and even in the piped-water network (1000 cells/L), where MC-LR was also detected, but in concentration below the guideline value (1 μg/L) proposed by the WHO (Institute of Public Health Serbia "Dr Milan Jovanović Batut" 2014). Nonetheless, the use of tap water for drinking and cooking purposes was prohibited. Ensuing investigation by the Agency for Environmental Protection showed that the highest number of *P. rubescens* cells was recorded at the entrance of the lake at 0.5 m depth (107,900 cells/mL), and near the dam at 10 m depth (98,936 cells/mL) (The Agency for Environmental protection 2013). This is consistent with biology of the species which typically occupies the niche in metalimnion (Drobac 2015, 2018).

Fig. 5 The blooming of *Planktothrix rubescens* in Vrutci reservoir. Photo by Snežana Simić

This distribution influenced the toxicity of the water, which was recorded with the *Artemia salina* toxicity test at greater depths, especially at a depth of 14 meters in metalimnion, while toxicity was not observed in shallower water, as well as in the tap water (Drobac 2015). In order to determine the cyanotoxins responsible for the toxicity, LC-MS/MS analyses were performed at Anahem laboratory in Belgrade, Serbia, Demokritos in Athens, Greece, and Åbo Akademi University in Turku, Finland. Several MC forms were found in biomass (dmMC-RR 11.694 μg/L, dmMC-LR 3.690 μg/L, MC-LR 53 μg/L, and MC-YR 20 μg/L January 2014) and water (dmMC-RR 2.1 μg/L April 2014) from Lake Vrutci, as well as in tap water (dmMC-RR 5.7 μg/L January 2014) (Svirčev et al. 2017c). Albeit, high concentrations of dmMC-RR have been detected in tap water, it is about four to five times less toxic form than the referenced MC-LR (Sivonen and Jones 1999).

Less toxic forms of MC were also found in fish tissue sampled in April 2014. Although the bloom was observed in December of 2013, the muscles of fish sampled in September/October of the same year already contained toxins, indicating the earlier presence of cyanobacteria in Vrutci reservoir (Drobac 2015; Svirčev et al. 2017c). Lower cyanotoxin concentrations obtained in the autumn samples may be due to the analysis of solely muscles, while the complete fish (muscles and viscera) sampled in April 2014 was analyzed for the presence of cyanotoxins. This confirms that if the entrails are removed, exposure to cyanotoxins could be reduced.

In the blooming period, slight fish mortality was also observed. Eyewitnesses described that the fish were swimming disoriented in circles, and dead fish had distinctly red-colored gills (Svirčev et al. 2017c). In order to assess human exposure to cyanotoxins, a questionnaire was conducted and one of the questions was whether the respondents noticed dead animals in and around the reservoir. The positive answer gave 13% of the respondents, and of those more than half (55%) declared that they saw dead fish in the Vrutci reservoir. One of the respondents stated that the aquarium fish died in the tap water during the blooming period. The Questionnaire also revealed that 32% of the respondents consumed fish from the reservoir in the last three years, and some of them (17 respondents) continued even during blooming. Therefore, in addition to other ways of exposure (e.g., tap water, reservoir water for recreation), the citizens of Užice could have been exposed to cyanotoxins during fishing and consumption of contaminated fish (Drobac 2015).

As a result of the exposure, some health disorders reported by the respondents can potentially be linked to cyanobacterial blooming, especially the most common gastrointestinal problems, followed by skin and eye irritation. Statistical analyses of epidemiological data from the Institute of Public Health of Užice, affirmed that the number of people suffering from diseases of digestive system was significantly increased during the period from 2012 to 2015, compared to the previous period of four years. Skin diseases other than infections of skin and subcutaneous tissue were also significantly increased in that period (Svirčev et al. 2017c). It is likely that cyanobacteria and their toxins can have a detrimental effect on human health directly and/or indirectly through consumption of fish and other aquatic organisms that have accumulated cyanotoxins.

3.7 Bovan Reservoir

The reservoir Bovan was created between 1978 and 1984 by building a dam in the middle of the Sokobanjska Moravica River in Southeast Serbia (Table 1). The main purpose of constructing the reservoir was to control the flow of the Morava River and protect the Djerdap I reservoir. The reservoir Bovan serves as a source of irrigation and electricity, as well as a way to collect sediment from the river. It is also an important source of drinking water for the municipality of Aleksinac. The reservoir is a popular tourist destination, known for its recreational fishing, sailing, and bathing areas (Zlatković et al. 2022). The water is eutrophic with a cyanobacteria abundance of 14,159 cells/mL (Table 4), and according to WHO recommendations during 2014 (SEPA 2015), the lake was of medium health risk from the aspect of drinking water use, and low health risk from the aspect of sports and recreation (Chorus and Bartram 1999). No blooms of cyanobacteria were recorded in the reservoir. During the research conducted in 2019, three potentially harmful cyanobacteria were detected in reservoir Bovan: *Pseudanabaena catenata*, *Cylindrospermum stagnale*, and *Aphanocapsa holsatica* (Zlatković et al. 2022).

In the Bovan reservoir in fish biomass, the 16 species important for fisheries from five families were dominant. The overall fish biomass was recorded at 862 kg/ha. The Cyprinidae family made up the largest proportion of the average total fish biomass at 76%, followed by the Percidae family at 18% and the Siluridae family at 6% (Table 5; Fig. 2). Accidental fish mortality due to cyanobacterial blooms have not been recorded in this reservoir, but the presence of potentially toxic cyanobacteria and a high level of trophy may lead to blooms in the future. Therefore, it is important to carefully monitor the water quality of the multifunctional reservoir Bovan, which serves both as a recreational spot for activities such as fishing and as a source of drinking water.

3.8 Gruža Reservoir

The Gruža Reservoir was created by damming the middle part of the Gruža River for the following purposes: providing a water supply for the City of Kragujevac (Central Serbia) (Table 1), the surrounding villages and the industry; as a protection against flooding, for retaining the sediment and to increase low flows in the downstream section of the River Gruža during drought events. This eutrophic reservoir with a cyanobacterial abundance of 90,594 cells/mL presents a medium health risk according to WHO recommendations (Chorus and Bartram 1999) (Table 4). The blooming of the potentially toxic species *Aphanizomenon flos-aquae* and *Microcystis aeruginosa* is often present in the summer months, in the shallower parts of reservoir (Fig. 6; Sedmak and Svirčev 2011; Jovanović 2020). The trophic status of the reservoir is mesotrophic with a tendency toward eutrophic status since its formation. The morphometric characteristics of the reservoir, including its small surface-to-depth ratio and the inflow of a large amount of N and P compounds from the nearby agricultural land, contribute to the favorable conditions for the growth of cyanobacteria (Ranković and Simić 2005).

From the fish biomass perspective, 10 species important for fisheries from five families were dominant. The overall fish biomass was recorded at 721 kg/ha. The Cyprinidae family made up the majority of the average total fish biomass at 70%, followed by Percidae at 27%, while the share of Ictaluridae was significantly lower (Table 5; Fig. 2). There have been no recorded instances of fish mortality due to cyanobacterial blooms in this reservoir. However, the presence of cyanobacterial blooms has been shown to affect the biochemical and physiological processes of fish in this reservoir (Perendija et al. 2011). A study of the biochemical and ultrastructural changes in the liver of the European perch (*Perca fluviatilis*), in response to *Aphanizomenon flos-aquae* bloom in the Gruža Reservoir, showed the influences of an *A. flos-aquae* bloom on antioxidant biomarkers and on histopathological alterations in the liver of the *P. fluviatilis* (Perendija et al. 2011).

3.9 Reservoir Aleksandrovac

The Aleksandrovac reservoir is a shallow artificial lake located in the southernmost part of Serbia (Table 1). It was created in 1963–1964 and initially used for irrigation and later for sports and recreational purposes and, above all, for fishing. This hypereutrophic reservoir with a cyanobacteria abundance of 10,283,158 cells/mL is a lake of high health risk in relation to WHO recommendations (Table 4; Đorđević et al. 2015; Đorđević 2021; Chorus and Bartram 1999). In the Aleksandrovac reservoir in fish biomass, the seven species important for fisheries from two families were dominant. The fish biomass (804 kg/ha) was primarily composed of fish from

Fig. 6 The eutrophic reservoir Gruža; The blooming of *Aphanizomenon flos-aquae* in Gruža reservoir. Photo by Nevena Đorđević

the Cyprinidae family (63%) and Siluridae family (37%), with *Carassius carassius* representing the majority (70%) of the Cyprinidae family biomass (Table 5; Fig. 2).

Aleksandrovac reservoir was formed on salines, and as alkaline salt microcumulation with extreme ecological conditions, it is suitable for the development of alien and potentially invasive species. The appearance and specific blooming of species common for tropical regions in the Aleksandrovac reservoir demonstrate the potential hazard to the environment (including fish mortality). This is validated by its history of eutrophication, cyanobacterial blooming, and prior fish mortalities (Đorđević and Simić 2014; Đorđević 2021). Efforts for restoration of the lake were futile (2009–2010), and in December 2012, the last fish mortality was recorded, when over 1.7 tons of fish have been killed again (Drobac 2015, 2018; Đorđević et al. 2015; Svirčev et al. 2016; Đorđević 2021) (Fig. 7). This event attracted media and scientific attention, and the search for the culprit initiated.

Investigation eliminated bacterial infections, heavy metal poisoning, and wastewaters from industrial pollutants (Đorđević and Simić 2014; Svirčev et al. 2016). The highest probability was that the fish mortality was a result of oxygen depletion in the water, because right before the fish kill, ice cover formed on the lake surface, due to a sudden temperature drop below zero degrees. However, it was stated that the ice on the lake had been broken up, furthermore, a few days after the fish kill concentration of dissolved oxygen was measured to be 7.8 mg/L O_2 (Svirčev et al. 2016), revealing that the oxygen level would not threaten fish survival (Kramer 1987).

Other physicochemical and microbiological analyses conducted one month before the fish kill by the Institute for Water Resources "Jaroslav Černi" as a part of the monitoring of the general condition of the lake, were all within the normal limits. Nonetheless, several parameters were ambiguous. Increased values of Chl *a* (214.2 mg/m^3) and Phaeophytin (53.7 mg/m^3) pointed to the eu-polytrophic and polytrophic state, as well as potential cyanobacterial blooming. This would explain the yellow-brown color of the water at the time of sampling (Svirčev et al. 2016).

Microscopic examination revealed that the blooming cyanobacterial species was *Raphidiopsis raciborskii* (Đorđević et al. 2015; Svirčev et al. 2016), which was

Fig. 7 The fish mortality in the Aleksandrovac reservoir (2012). Photo by Nevena Đorđević

Fig. 8 The hypereutrophic reservoir Aleksandrovac and the blooming of *Raphidiopsis raciborskii*. Photo by Nevena Đorđević

noted (low abundance 37 trichomes/mL) for the first time in Aleksandovac reservoir two years prior to a blooming event (Simić et al. 2011; Đorđević and Simić 2014) (Fig. 8). Over a two-year period, there was a significant increase in the abundance of this species, marked by intense blooming and the highest recorded abundance in October (2.38×10^6 trichomes/ml), observed two months before the mass fish mortality (Đorđević et al. 2015). This originally tropical and subtropical species is not only highly invasive (Sinha et al. 2012), but also potentially toxic, as it can produce several cyanotoxins, including CYN and STX (Đorđević et al. 2015; Meriluoto et al. 2017; Vico et al. 2020). Continuous monitoring of the reservoir from 2012 to 2014 revealed the first detection of CYN, with concentrations ranging from 1.91 to 24.28 μg/L. (Đorđević et al. 2015; Đorđević 2021).

Artemia salina toxicity test substantiated intracellular toxicity of *Raphidiopsis raciborskii* biomass. Accordingly, a realistic scenario would be that the toxic intracellular content was released in the water as a result of cyanobacterial cell lysis, caused by freezing and thawing due to temperature fluctuations (Drobac 2015, Svirčev et al. 2016). The decay of cyanobacterial biomass is reflected in a reduced number of *R. raciborskii* filaments (from 2,069,105 filaments/mL to 1,249,103 filaments/mL), Chl *a* level (from 214.2 mg/m^3 to 17.2 mg/m^3), Phaeophytin (from 53.7 mg/m^3 to 5.86 mg/m^3), total organic carbon (from 34.38 mg/L to 22.36 mg/L), an increase of ammonium ion (from 0.08 mg N/L to 0.9 mg N/L) in samples taken in November and December, respectively (Svirčev et al. 2016).

Based on the aforementioned, it can be concluded that the toxic content from the cyanobacterial cells led to fish poisoning, but what has not yet been fully clarified is which cyanotoxins are responsible for the poisoning. Potential agents could be CYN (24.28 μg/L of CYN was detected in November) (Đorđević et al. 2015; Đorđević 2021) or yet uncharacterized cyanotoxin(s) from this cyanobacterium (Svirčev et al. 2016), as it was the case in Hungary, where an unidentified neurotoxic agent(s) from *Raphidiopsis raciborskii* was also connected to fish mortality (Vehovszky et al. 2015). In addition, the methanol extract from the same natural population of

R. raciborskii collected at Aleksandrovac reservoir showed significant genotoxicity even in low concentrations and increased oxidative stress in rat liver after a single exposure during in vivo laboratory studies (Đorđević et al. 2017; Đorđević 2021). Regardless of the exact toxic agent, gathered data originating from this case study imply how cyanobacterial blooming can adversely influence fish resources. Mainly, cyanobacteria can cause fish mortality by altering pH value (photosynthesis decreases dissolved CO_2), reducing oxygen level (decay and microbial decomposition of cyanobacterial biomass), increasing ammonia (decay of cyanobacterial biomass) and stress, as well as by synthesizing cyanotoxins (Bury et al. 1995, 1996a).

3.10 Lake Palić

The Palić Lake is a shallow Pannonian Lake that was created about a million years ago when the wind blew and created pits separated by dunes. It is mostly recharged by atmospheric precipitation and is the largest natural lake in Serbia (Table 1). This hypereutrophic lake has high levels of cyanobacteria (cyanobacteria abundance of 211,400,000 cells/mL), which can be harmful to humans according to WHO recommendations (Chorus and Bartram 1999). Blooming of the potentially toxic species *Microcystis aeruginosa* was recorded in the lake, as well as the appearance of the invasive, potentially toxic species *Raphidiopsis raciborskii* and *Sphaerospermopsis aphanizomenoides* (Table 4). In Lake Palić, three species important for fisheries from two families were dominant in the fish biomass. In the fish biomass, the Cyprinidae and Percidae families were the most dominant, making up 84% and 16%, respectively (Table 5; Fig. 3).

The lake itself was in a very bad condition since the late sixties of the last century. The lake has recreational purposes, but it is also a collector for treated municipal wastewater coming from the lagoons for active sludge water treatment (Raičević et al. 2011). The sewage discharges from rapidly developing towns in the watershed and the growing use of fertilizers in agriculture increased the nutrient load to the lake in the last decades. A steady increase in phosphorus loading is the most important factor of lake eutrophication, which leaves negative consequences for the fish population. Accidental fish mortality, which was associated with the occurrence of cyanobacterial blooms in this reservoir, was repeatedly recorded.

3.11 Ludaš Lake

Lake Ludaš is the second largest aeolian lake in Serbia. It is located in the northern part of Serbia between the Danube and Tisa rivers, and close to the Hungarian border (Table 1). It was formed approximately one million years ago on the sandy terrain of the Bačka loess plateau (Rudić et al. 2015). The lake is a wetland area of international significance, a Special Nature Reserve, a Ramsar site, and an Important Bird and Biodiversity Area (IBA area). The lake is hypereutrophic, with a cyanobacteria abundance of 17,204,000 cells/mL, and it also presents high health risks according to WHO recommendations (Chorus and Bartram 1999). Blooming of the potentially toxic species *Microcystis aeruginosa* and *M. wesenbergeii*, as well as the massive appearance of the also potentially toxic species *Planktothrix agardhii* was recorded in the lake (Table 4).

In the fish biomass, the five species important for fisheries from three families were observed. The Cyprinidae family made up 98% of the total fish biomass (Table 5; Fig. 3). Accidental fish mortality associated with the occurrence of cyanobacterial blooms in this reservoir was recorded.

Anthropogenic activities have led to the deterioration of water quality and cyanobacterial proliferation (Ćirić et al. 2020; Marinović 2021). As a result, during the summer of 2006, 603 μg/L MC-LR have been measured, making it one of the highest concentrations ever documented in Serbia (Simeunović 2009). Although measures have been sought to solve this tenacious cyanobacterial problem (Tokodi et al. 2018; Dziga et al. 2019), perpetual blooming persists for more than half a century. Accumulation of cyanotoxins and histopathological changes in fish tissue were monitored in Lake Ludaš through the years (Tokodi 2016; Tokodi et al. 2018, 2020; Marinović 2021).

During the investigation in 2011 and 2012, cyanobacteria were still blooming, and the most abundant species were *Limnothrix redekei*, *Pseudanabaena limnetica*, *Planktothrix agardhii*, and *Microcystis* sp. Invasive species *Raphidiopsis raciborskii* and *Sphaerospermopsis aphanizomenoides* were also present (Tokodi et al. 2018). Toxicity of water and biomass was established with the *Artemia salina* toxicity test, ELISA tests indicated the presence of MC/NOD and STX, and high MC-LR concentration was detected with PPI assay (up to 828 μg equiv./L MC-LR). Furthermore, MC-RR was found in the rhizome of aquatic plants (*Phragmites australis* 0.22 ng/g dw, *Typha latifolia* 0.05 ng/g dw, *Nymphaea elegans* 0.02 ng/g dw) from Ludaš Lake. Fish from the lake (*Carassius gibelio*) also accumulated cyanotoxins in their tissues, MC-LR was found in the intestines (0.27 ng/g dw) and kidneys (0.07 ng/g dw), while MC-RR was found in fish gills (0.25 ng/g dw), gonads (0.06 ng/g dw), and muscle (0.06 ng/g dw). In the liver, kidney, intestines, and gills of fish histopathological changes were noticed. DNA damage of the liver, gills, and blood of *Prussian carp* was also detected with the comet assay during the investigation in 2012 (Tokodi 2016; Tokodi et al. 2018). Further research conducted in 2018 showed the presence of MC and STX coding genes in biomass samples from Lake Ludaš. Thus, as expected, several variants of MCs were detected in the water (MC-LR 0.1 μg/L, MC-dmLR 0.023 μg/L, MC-RR 0.006 μg/L, MC-dmRR 0.003 μg/L, MC-LF present), however, this time, they were not found in the fish tissue. Still, histopathological changes were observed in liver, kidney, and gills of Prussian carp (*C. gibelio*) from Lake Ludaš (Tokodi et al. 2020; Marinović 2021). Prussian carp is a frequent catch of local fishermen who often come to one part of the lake for fishing, meanwhile, they can be exposed to cyanotoxins through inhalation, direct dermal contact and finally through contaminated food. In like manner, the birds, for which this Lake is well-known for, are also jeopardized (Tokodi et al. 2020).

Moreover, the entire biota is endangered by the ongoing cyanobacterial blooming. It was demonstrated that prolonged exposure to MCs from cyanobacterial bloom altered the oxidative status of the liver of *Pelophylax* kl. *esculentus* frogs in Lake Ludaš. The defense mechanisms were not able to protect cells and oxidative damage to proteins and lipids in the frog livers resulted (Gavrilović et al. 2021). Furthermore, blooming conditions and cyanobacterial toxins in Lake Ludaš disrupted the prooxidant/antioxidant balance in frog tissues, also higher sensitivity with lower antioxidant defense system capability in the skin, compared to muscle, of frogs living in such conditions was observed (Gavrilović et al. 2020).

Conspicuously, epidemiological research conducted in Vojvodina, Serbia indicated that incidence of melanoma was higher in municipalities with blooming water bodies used for fishing, recreation, and irrigation; however, results could be considered as inconclusive, due to the restrictions in the cancer database. Concurrently, literature data support the hypothesis that cyanobacteria could be a potential risk cofactor, which together with ultraviolet radiation may increase the incidence of melanoma in people exposed during recreational and occupational activities alongside blooming waters (Drobac Backović et al. 2020). It is indicative that cyanobacteria and their toxic metabolites have adverse effects on a wide spectrum of organisms, viewed from an ecological, health, and economic standpoint, thus they are deserving of dedicated professional and scientific attention.

4 Effects of Cyanotoxins on Fish

Previous parts of this chapter have extensively described the occurrence and potential toxicity of cyanoHABs and cyanotoxins, and have detailed the cyanoHAB events in Serbia, but also summarized the studies which assessed the effects of cyanobacterial blooms on fish in blooming waterbodies of Serbia. In this subsection, we will discuss the impact of these blooming events on fish. Firstly, we will outline the exposure routes of fish to cyanotoxins, then we will detail the absorption and distribution mechanisms, accumulation patterns, and lastly briefly outline toxic effects exerted at various levels of organization.

4.1 Exposure Routes and Absorption

As fish are important members of freshwater ecosystems, they are prone to direct exposure to cyanotoxins. The main exposure routes are through direct contact with waterborne toxins, or through their diet (Svirčev et al. 2015). Consequently, the organs which are in the first line of cyanotoxin action are the skin, gills, and intestines. Additional exposure routes are through intraperitoneal injection or gavage, but those are reserved for a laboratory setting, and therefore will be omitted in this review.

The absorption of a given cyanotoxin by the gills and/or intestines (but also the cellular uptake) is highly dependent on the structure of the cyanotoxin, and consequently on the type of transport that is necessary to facilitate its transfer through the plasma membrane. Most cyanotoxins are hydrophilic, and therefore necessitate active transport for their absorption/uptake (Buratti et al. 2017). The most common example is the uptake of MCs and NOD by the organic anion transport system (OATP) (Campos and Vasconcelos 2010; Steiner et al. 2016; Buratti et al. 2017; Chen et al. 2021). On the other hand, small molecules such as STX and gonyautoxins can permeate the intestines through a paracellular pathway, and therefore do not need an active transport mechanism (Torres et al. 2007).

The tissue distribution of cyanotoxins also depends on the expression of the particular transport mechanisms in given cells. In zebrafish, MCs are transported by Oatp1d1, Oatp1f2, and Oatp1f4 (Steiner et al. 2016). However, the tissue distribution of these transporters is different with the highest expression being registered in the liver, followed by the brain, intestines, and gills (Popovic et al. 2010; Steiner et al. 2016). This in turn can explain why MCs are considered hepatotoxins, but why they also cause adverse effects in other organs. Furthermore, the affinity of different MC congeners varies across different Oatp isoforms, hence the adverse effects of blooming events containing various MC congeners on different organs might vary. Similar can be presumed to occur with other cyanotoxins as well.

4.2 Accumulation and Biotransformation

Upon entering the organism/cell, cyanotoxins either exert their toxic effect, or accumulate within the body. Accumulation of cyanotoxins in aquatic organisms is well-known and well reported. Invertebrates (especially zooplanktonic grazers) can accumulate large quantities of cyanotoxins and are reported to show potential for biomagnification (Watanabe et al. 1992; Ferrão-Filho and Kozlowsky-Suzuki 2011; Pham and Utsumi 2018). Fish can also accumulate cyanotoxins, however, both biomagnification and biodilution have been reported, and the outcome mainly depends on the diet. Zooplanktivorous fish showed a tendency toward biomagnification, while other fish showed a tendency toward biodilution (Kozlowsky-Suzuki et al. 2012). Interestingly, phytoplanktivorous fish showed the lowest tendency toward cyanotoxin accumulation, a result later corroborated by Flores et al. (2018). The exact reason causing this distinction is unknown, but it is presumed that phytoplanktivorous fish have better developed detoxication mechanisms, hence do not accumulate large quantities of toxic compounds.

To defend against the toxic action of various compounds, organisms have developed specific pathways through which they can excrete the compound in its original form, or to biotransform it to cause less adverse effects and facilitate its excretion (van der Oost et al. 2003; Dekant 2009). Biotransformation is catalyzed by the enzymes of biotransformation and is commonly divided into two steps: phase I and phase II (Beiras 2018; Gerba 2019). MCs and NOD are known to conjugate to glutathione or cysteine during the phase II of biotransformation which effectively facilitates their excretion through urine or feces (Buratti et al. 2017).

4.3 Toxic Effects of Cyanotoxins in Fish

As mentioned previously, cyanotoxins can exert various adverse effects on the organism after absorption. The intensity of the adverse effect depends on the toxic action of the cyanotoxin, on its concentration, but also of the compound mix which can be observed within the bloom. MCs, NOD, CYN, ATX, STX are amongst the most studied cyanotoxins, and the mechanisms of their actions are well described (Campos and Vasconcelos 2010; Buratti et al. 2017; Du et al. 2019; Metcalf and Souza 2019; Christensen and Khan 2020).

The toxic action of most cyanotoxins can be observed at different levels of organization. For instance, at the molecular level, MCs and NOD covalently bind to protein phosphatases thus inducing hyperphosphorylation of various proteins (Campos and Vasconcelos 2010; Buratti et al. 2017; Banerjee et al. 2021). Hyperphosphorylation of actin filaments can be morphologically observed as alterations in the cytoskeleton or loss of cell shape frequently observed in hepatocytes. Cyanotoxins can also induce an increased production of reactive oxygen species (ROS) thus causing oxidative stress, DNA fragmentation, and apoptosis. Some cyanotoxins can even directly cause transcription alterations thus inducing immunotoxicity, tumorigenicity, or reproductive toxicity. Some of these pathways are summarized in Fig. 9.

4.4 Case Studies of Fish Intoxications in the Western Balkans

Cyanobacterial blooming and the adverse effects they can cause have been extensively studied in the waterbodies of Serbia over the past decades. Amongst Serbian waterbodies, lakes Ludaš, Aleksandrovac, Gruža, and Vrutci have been studied the most. As previously mentioned, cyanobacterial blooms were extensive with cell numbers reaching even up to tens of millions per milliliter. Cyanotoxins (mostly MCs) have also been measured in concentrations higher than the ones recommended by the WHO as safe. Therefore, the effects of such blooming events on fish were expected.

The most frequently detected cyanotoxins in the Serbian waterbodies were MCs. As they are considered hepatotoxins, liver damage in fish living in these blooming ecosystems was expected. Compared to control individuals reared in control recirculating conditions (Fig. 10a–c), fish caught from Lake Ludaš (Prussian carp *Carassius gibelio*) displayed histopathological alterations such as the loss of the

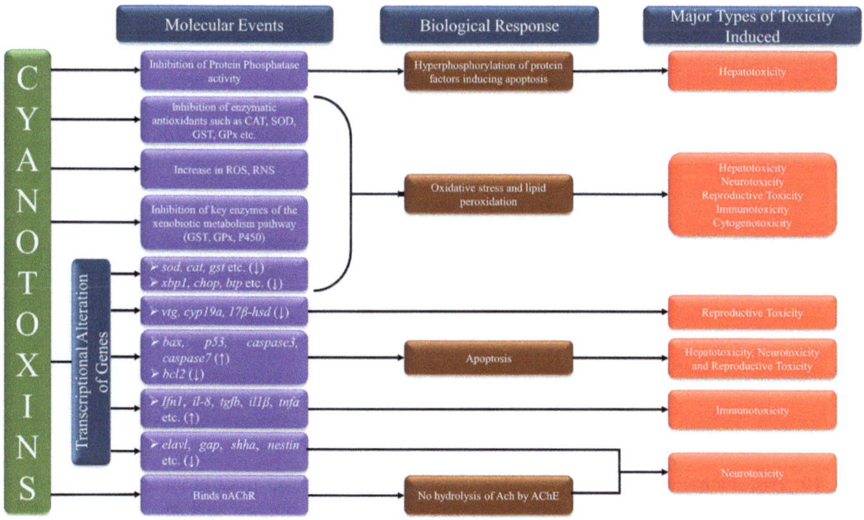

Fig. 9 Summary of different molecular mechanisms responsible for various cyanotoxin-induced toxicities in fish. Cited from Banerjee et al. 2021 with permission from the publisher

typical cord-like parenchymal structure, dissociation of cells, loss of polygonal shape and rounding of cells, vacuolizations, chromatin condensation, and karyopyknosis (Fig. 10d, e) (Tokodi et al. 2018, 2020).

Histopathological lesions were similar throughout the investigated years, and qualitatively, there were no significant differences between seasons. However, when a semi-quantitative analysis was employed to compare hepatic lesions between March and September 2018, the liver index quantifying total hepatic lesions was significantly higher in fish caught during September compared to those caught during March (Marinović 2021). These results indicated that the tissue damage was higher after the peak of the bloom, and as the bloom started to collapse. Hepatic damage in fish (common carp *Cyprinus carpio*) sampled from blooming fishponds was even more severe and in addition to the mentioned lesions, it included necrotic fields and lysis of nuclear and cell membranes (Fig. 10f, g) (Drobac et al. 2016). Necrosis and apoptosis were also apparent in the livers of European perch *Perca fluviatilis* caught from the blooming Gruža Reservoir (Perendija et al. 2011).

Histopathological alterations observed in fish caught from Lake Ludaš and fishponds are commonly reported in fish exposed to cyanobacterial blooms or purified cyanotoxins (Carbis et al. 1996, 1997; Kotak et al. 1996; Fischer and Dietrich 2000; Ernst et al. 2006; Li et al. 2007; Atencio et al. 2008a, b; Jiang et al. 2011; Svirčev et al. 2015; Drobac Backović et al. 2021; Marinović et al. 2021). Similar alterations were detected in livers of mammals exposed to cyanotoxins (Hooser et al. 1990; Gupta et al. 2003; Sedan et al. 2015). However, it is not possible to completely ascertain which alterations occurred due to direct cyanotoxin action, and which represent a more generalized response to different stressors. For example,

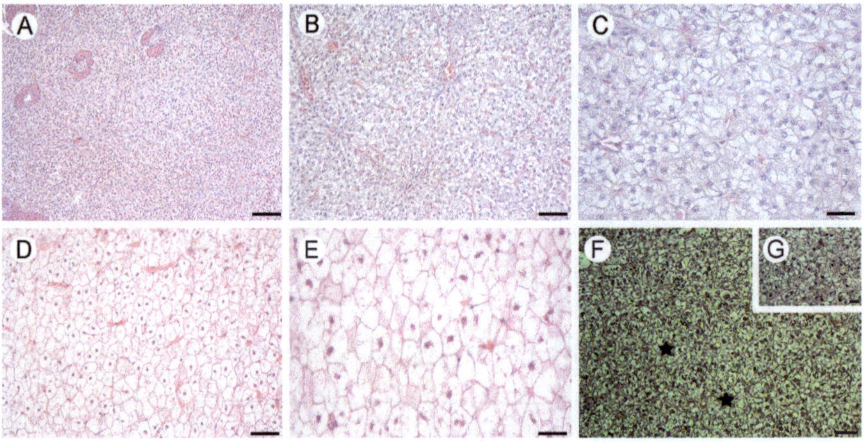

Fig. 10 Histopathological alterations observed in the livers of fish from blooming ecosystems in Serbia. (**a–c**) Livers of control individuals reared in a recirculating system of the Hungarian University of Agriculture and Life Sciences displayed normal hepatic structure with the typical cord-like parenchymal structure, polygonal hepatocytes, and large nuclei with prominent nucleoli. (**d**, **e**) Livers of fish caught from Lake Ludaš displayed the loss of the cord-like structure, intense fatty accumulation and vacuolization, hypertrophy, and karyopyknosis. (**f**, **g**) Livers of fish caught in blooming fishponds additionally showed necrotic fields (asterisk), karyopyknosis, and karyolysis. H&E. Scale bars: (**a**) –100 μm; (**b**, **d**, **f**) –50 μm; (**c**, **e**) –25 μm; (**g**) –20 μm. Cited from Drobac et al. (2016) and Marinović (2021) with permission from the publisher

hepatocyte vacuolization is observed after exposure of fish to various xenobiotics such as metals or PAHs (Wolf and Wolfe 2005; Nikolić et al. 2020) but can also be a result of imbalances in the diet (Carvalho et al. 2021; Crowe et al. 2021; Zhong et al. 2022). On the other hand, as MCs are known to inhibit protein phosphatases (PP1 and PP2A) and cause hyperphosphorylation of various proteins (including actin filaments), alterations such as cytoskeletal damages, rounding of cells, and loss of the parenchymal structure might be the result of direct MC action.

When comparing to control kidneys (Fig. 11a–c), kidneys of fish caught from Lake Ludaš displayed intense degeneration of both tubules and glomeruli (Tokodi et al. 2018, 2020). The most intense tubular alteration was the vacuolization of tubular epithelial cells which in some cases led to complete detachment from the basal membrane (Fig. 11d, e). Many of the tubular epithelial cells displayed pyknotic nuclei indicative of necrosis, and in many cases such cells showed signs of membrane lysis with no discernible boundary between them. Glomerulopathy was mainly expressed through dilatation of the Bowman's capsule and glomerular atrophy. Necrosis and hyalinization were also observed in the interstitium, accompanied by macrophage infiltrations. The semi-quantitative analysis conducted in 2018 displayed that there were no significant differences in the kidney index between the fish sampled in March and September, however, the tubular alterations were more intense during March (Marinović 2021). Similar alterations were observed in

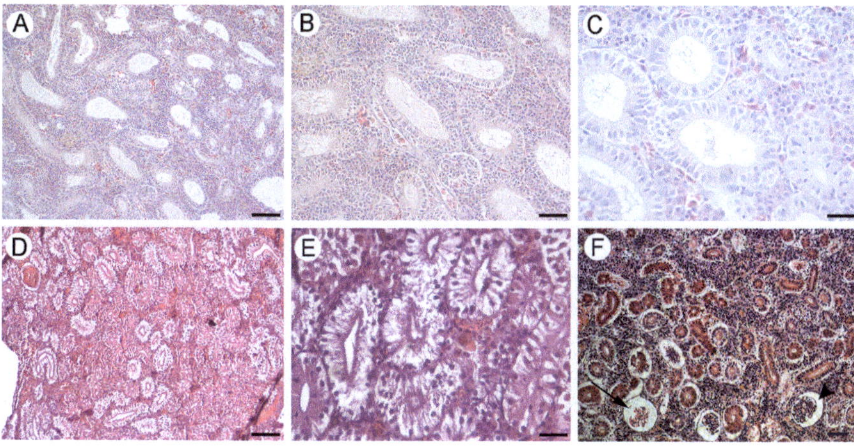

Fig. 11 Histopathological alterations observed in kidneys of fish from blooming ecosystems in Serbia. (**a–c**) Kidneys of control individuals reared in a recirculating system of the Hungarian University of Agriculture and Life Sciences demonstrated normal renal structure. (**d**, **e**) Kidneys of fish caught from Lake Ludaš displayed tubular alterations in the form of intense vacuolization, detachment of epithelial cells from the basal membrane, pyknotic nuclei. (**f**) Glomerulopathy characterized by glomerular necrosis (arrow) and dilatation of the Bowman's capsule (arrowhead) in fish caught from blooming fishponds. H&E. Scale bars: (**a**, **d**) –100 μm; (**b**, **f**) –50 μm; (**c**, **e**) – 25 μm. Cited from Drobac et al. (2016) and Marinović (2021) with permission from the publisher

fish sampled from the blooming fishponds (Fig. 11f), however, they were more severe (Drobac et al. 2016).

Since kidneys receive a large portion the post-branchial blood (Svirčev et al. 2015) and express Oatps (Popovic et al. 2010; Steiner et al. 2016), MCs can be easily absorbed and exert toxic action in the kidney cells. Fischer and Dietrich (2000) showed that common carp kidneys indeed uptake MCs, and the pathologies are exerted as rapidly as in the liver. Alterations observed in fish caught from Lake Ludaš and fishponds are commonly reported in fish exposed to cyanobacterial blooms or purified cyanotoxins (Carbis et al. 1996; Fischer and Dietrich 2000; Li et al. 2008; Mitsoura et al. 2013; Drobac Backović et al. 2021; Marinović et al. 2021). Most of the alterations are reversible, however, necrosis and hyalinization are irreversible alterations and can thus lead to permanent difficulties in organ function.

Compared to control samples (Fig. 12a), histopathological alterations in the gills of fish sampled from Lake Ludaš were apparent (Tokodi et al. 2018, 2020). They mostly included hyperplasia of the interlamellar epithelium with an apparent proliferation of epithelial, but also chloride and mucous cells (Fig. 12b–g). In several cases, this proliferation led to the complete fusion of secondary lamellae (Fig. 12c). Hypertrophy of epithelial and mucous cells, epithelial lifting and clubbing were also observed. Telangiectasia and hemorrhage were most prominent circulatory disturbances. Seasonal differences were not apparent (Marinović 2021). Similar alterations (Fig. 12e-g) were observed in fish sampled from fishponds (Drobac et al. 2016).

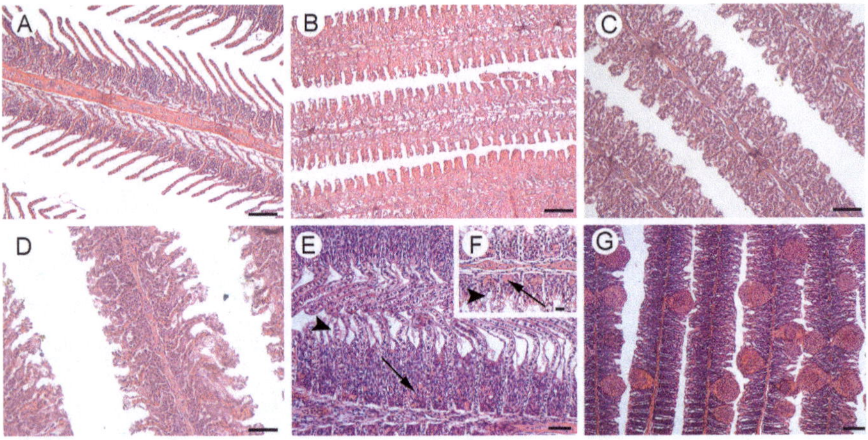

Fig. 12 Histopathological alterations observed in gills of fish from blooming ecosystems in Serbia. (**a**) Gills of control individuals reared in a recirculating system of the Hungarian University of Agriculture and Life Sciences displayed normal structure. (**b–d**) Gills of fish caught from Lake Ludaš showed hyperplasia of the interlamellar epithelium, epithelial hypertrophy, and clubbing. (**e–g**) Gills of fish caught from blooming fishponds displayed epithelial lifting (**e**, arrowhead) and proliferation of chloride (**e**, arrow; **f**, arrow) and mucous (**f**, arrowhead) cells, and telangiectasia (**g**). **H&E**. Scale bars: **a**, **b**, **e** –100 μm; **c**, **d**, **e** – 50 μm; **f** – 20 μm. Cited from Drobac et al. (2016) and Marinović (2021) with permission from the publisher

As previously mentioned, since gills are in direct contact with water, they are also in direct contact with the waterborne cyanotoxins released during cyanoHABs. They are also very sensitive to waterborne pollutants as they make up to 50% of the total surface of the fish (Lujić et al. 2015). Many alterations observed in our studies were also reported in other studies describing the effects of cyanoHABs on fish (Carbis et al. 1997; Svirčev et al. 2015; Drobac Backović et al. 2021; Marinović et al. 2021). Alterations such as epithelial hyperplasia, hypertrophy, and lifting are primarily considered defense mechanisms, and are observed during exposure of fish to a vast array of xenobiotics (Mallatt 1985). Hyperplasia of the interlamellar epithelium reduces the respiratory surface, while epithelial hypertrophy and lifting increase the distance between the water and blood, thus increasing the diffusion distance (Mallatt 1985; Agamy 2013; Lujić et al. 2015). However, alterations in the chloride cells might be the result of direct MC action as it is known they can lead to disbalances in concentrations of Na^+ and K^+ ions, as well as to problems in regulation of many other ions (Gaete et al. 1994; Bury et al. 1996b).

Intestines are another organ that is in the direct contact with cyanotoxins. However, the results are not as uniform as was the case of the previous three organs. Histopathological alterations were observed during the sampling at Lake Ludaš during 2011/2012 (Tokodi et al. 2018). They were mostly in the form of edematous alterations in the lamina propria with consequent dilation and hypertrophy of goblet cells. Necrosis and desquamation of enterocytes at the apical parts of the villi were indicative of necrotic enteritis. However, during the sampling of 2018, no intestinal alterations were observed (Tokodi et al. 2020). Alterations similar to those observed during 2011/2012 were also observed in the intestines of fish sampled from the blooming fishponds (Drobac et al. 2016).

Alterations in spleen, gonads, and muscles were not observed in any of the fish sampled in Lake Ludaš or fishponds (Drobac et al. 2016; Tokodi et al. 2018, 2020). The only exception are the muscles of fish sampled from the blooming fishponds which displayed separation of single muscle fibers, lifting of the muscle fascicles from the perimysium, necrosis of single muscle fibers, and fatty acid inclusion (Drobac et al. 2016).

In addition to the assessment of morphological alterations caused by cyanobacterial blooming, several studies have detailed the effects of cyanobacterial blooms on the molecular level. DNA damage was quantified in the blood, livers, and gills of fish caught from Lake Ludaš during 2012 by comet assay (Tokodi et al. 2018). Liver and gill cells sustained more DNA breaks compared to blood cells. DNA damage in different fish and mammalian cells after exposure to cyanobacterial extracts or purified toxins has been vastly reported (Gaudin et al. 2008; da Silva et al. 2014; Banerjee et al. 2021; Bittner et al. 2021; Rabelo et al. 2021). It is presumed that cyanotoxins can either directly induce DNA breaks, or more likely, indirectly through the enhanced production of ROS and consequent oxidative stress. Indeed, the induction of oxidative stress in cells and fish exposed to cyanobacterial blooms or purified cyanotoxins has also been well documented. European perch (Perendija et al. 2011), as well as three cyprinid fish species (roach *Rutilus rutilus*, white bream *Blicca bjoerkna*, and Prussian carp) (Gavrilović et al. 2014) sampled from the Gruža Reservoir during cyanobacterial blooming, exhibited signs of oxidative stress. The levels of catalase (CAT), glutathione peroxidase (GPx), and -SH groups in the livers of European perch significantly diminished during the blooming, while no differences were observed in the levels of total superoxide dismutase (SOD), glutathione-S-transferase (GST), and glutathione reductase (GR). In the three mentioned cyprinid species, the levels of liver MnSOD were uniformly higher and the levels of CAT and GPx were uniformly lower, while GR and GST levels were variable depending on the species. In gills, on the other hand, SOD, GR, and GST levels were uniformly lower, while the levels of CAT and GPx were variable depending on the species. These results indicate the presence of oxidative stress in the tissues of fish from blooming waterbodies which agrees with several other studies (Jos et al. 2005; Prieto et al. 2006; Chen et al. 2012; Garcia and Martinez 2012; Hou et al. 2015; Marinović 2021).

To summarize, fish living in the blooming ecosystems of Serbia have displayed alterations at various levels of the organization. Most studied were the histopathological effects. Livers, kidneys, and gills of fish from Lake Ludaš and blooming fishponds displayed the most adverse alterations which were in line with observations of other authors (reviewed in Svirčev et al. 2015). However, histopathological alterations, as well as various molecular markers, are unspecific and can be the result of exposure to various xenobiotics. The most common example is the vacuolization of hepatocytes which can occur as a result of the action of various pollutants (Wolf and Wolfe 2005; Nikolić et al. 2020), but also due to food imbalances (Carvalho et al. 2021; Crowe et al. 2021; Zhong et al. 2022). Similarly, hyperplasia of the interlamellar epithelium and epithelial hypertrophy and lifting are common gill alterations observed in many different studies as they are the primary defense mechanisms against the toxin action (Mallatt 1985; Lujić et al. 2015;). Therefore, to be certain that a particularly toxic substance is causing the observed effect, extrapolation of laboratory results of exposure to purified toxins is needed. In the case of MCs, many laboratory studies have corroborated results observed in fish caught from blooming ecosystems (reviewed in Svirčev et al. 2015). Furthermore, studies of the exact molecular mechanisms of toxin action during laboratory in vivo or in vitro studies are of crucial importance for understanding toxic action at the higher levels of biological organization as well.

Most of the histopathological alterations of fish from the blooming ecosystems of Serbia were regressive, thus implying the tissue is expected to recover after the toxic substance is removed. This is particularly the case for structural changes observed in the liver and kidneys or the defensive mechanisms of gills. Given that these alterations were observed in fish throughout the year indicates that the irritants causing these alterations are continuously present within the ecosystems. Indeed, the blooming was continuously observed in Lake Ludaš throughout the year with increases and decreases in cell numbers, therefore we can assume that this phenomenon is responsible for the manifested histopathological alterations.

Even though cyanotoxin production is the most dangerous component of cyanoHABs, it is not the only stress factor for fish. Apart from cyanotoxin production, cyanoHABs are usually responsible for physiochemical changes in water characteristics such as low oxygen concentrations and high pH and ammonia content which can also exert negative effects on fish (Burdick et al. 2020b). During hypoxia, fish accelerate the ventilation rate (Heath 1995) and gill filaments rearrange themselves so that the interlamellar cell mass regresses leading to a larger respiratory surface (Nilsson 2007). This may lead to an increased absorption of waterborne xenobiotics. On the other hand, hyperplasia is the most common defensive mechanism observed in fish gills during cyanobacterial blooms, however, this alteration significantly restricts the respiratory surface. This in turn leaves fish vulnerable to acute hypoxia which is a frequent occurrence during the collapse of cyanobacterial blooms. Sudden and acute hypoxia coupled with cyanotoxin action may be responsible for many mass fish deaths which are known to occur during cyanoHABs (Burdick et al. 2020a). Apart from hypoxia, ammonia and high pH can have a synergistic effect as the concentration of the toxic non-ionized ammonia form is greater at high pH (Heath 1995). This form can easily defuse through the gills and increase its blood concentration which can become toxic (Burdick et al. 2020b). Therefore, cyanoHABs can have a much wider negative effect on fish, and even though the cyanotoxin action is considered to mostly be subacute, the synergistic effects of various stress factors caused by cyanoHABs can lead to acute action and even mass fish mortalities.

5 Conclusion

Due to climate change and increased human influence, there has been a significant shift in the trophic status of the fishing waters of Serbia. This has affected the health of fish, the composition of fish communities, and the total biomass of species important for fisheries. In the oligotrophic reservoir Vlasina, which has a high diversity and biomass of fish, a positive impact of the trophic status on the fish population was observed. The biomass and abundance of the fish community in selected reservoirs in Serbia generally remain stable at mesotrophic (Potpeć, Šumarice) and mildly eutrophic levels (Barje, Garaši) and increase sharply to maximum values in moderately and highly eutrophic reservoirs. In eutrophic and hypereutrophic reservoirs, cyanobacterial blooms and the presence of toxic cyanobacteria are common occurrence. The results showed that the Vrutci, Gruža, and Bovan reservoirs were eutrophic on average, and their total fish biomass was higher than that in mesotrophic and oligotrophic reservoirs. These reservoirs also had frequent blooms of potentially toxic cyanobacteria such as *Planktothrix rubescens* (Vrutci), *Aphanizomenon flos-aquae*, and *Microcystis aeruginosa* (Gruža). According to WHO recommendations and the abundance of cyanobacteria, Gruža and Vrutci are classified as waters with a moderate probability of adverse health effects. Certain biochemical and histopathological changes were observed in the tissues of fish from these reservoirs during blooming conditions.

In hypereutrophic reservoirs, the diversity of fish species decreases significantly, and the biomass of fish reaches maximum values (Aleksandrovac, Lake Palić, Ludaš Lake). The hypereutrophic reservoirs Aleksandrovac, Lake Palić, and Ludaš Lake are classified as waters with a high risk of health effects according to WHO recommendations. Fish in these waters with massive cyanobacteria development and their toxins have also manifested numerous and various biochemical and histopathological changes in their tissues, as well as accumulated cyanotoxins, even in muscles intended for human consumption, thereby potentially endangering the health of consumers. In these conditions, fish mortality often occurs. The occurrence of the invasive tropical species *Raphidiopsis raciborskii* in Aleksandrovac, Palić, and Šumarice is particularly significant. *R. raciborskii* is capable of producing a very potent toxin CYN, and in the case of a massive occurrence, it could lead to fish mortality.

When preparing the Fishing Water Management Program and anticipating future activities, it is essential to take into account the monitoring results related to the presence and abundance of cyanobacteria, due to their direct negative effects in the form of cyanotoxins and the risks to fish health and quality, which can also indirectly impact human health. Training of fishery water managers and users (primarily fishermen) is necessary, with the goal of identifying potential risks and implementing measures for continuous monitoring, including temporary or permanent water use prohibition if results indicate a high risk of adverse health effects.

Acknowledgments The investigation was supported by Grant (Agreement No. 451-03-47/2023-01/200122 and No. 451-03-47/2023-01/200125) funded by the Serbian Ministry of Science, Technological Development and Innovation. The authors would like to express their gratitude to Prof. Zorica Svirčev and Jussi Meriluoto, PhD for their guidance and support during the research.

References

Agamy E (2013) Impact of laboratory exposure to light Arabian crude oil, dispersed oil and dispersant on the gills of the juvenile brown spotted grouper (*Epinephelus chlorostigma*): a histopathological study. Mar Environ Res 86:46–55

Atencio L, Moreno I, Jos A, Pichardo S, Moyano R, Blanco A, Cameán AM (2008a) Dose-dependent antioxidant responses and pathological changes in tenca (*Tinca tinca*) after acute oral exposure to *Microcystis* under laboratory conditions. Toxicon 52:1–12

Atencio L, Moreno I, Prieto AI, Moyano R, Molina AM, Cameán AM (2008b) Acute effects of microcystins MC-LR and MC-RR on acid and alkaline phosphatase activities and pathological changes in intraperitoneally exposed tilapia fish (*Oreochromis* sp.). Toxicol Pathol 36:449–458

Bănăduc D, Simić V, Cianfaglione K, Barinova S, Afanasyev S, Öktener A, McCall G, Simić S, Curtean-Bănăduc A (2022) Freshwater as a sustainable resource and generator of secondary resources in the 21st century: stressors, threats, risks, management and protection strategies, and conservation approaches. Int J Environ Res Public Health 19(24):16570. https://doi.org/10.3390/ijerph192416570

Banerjee S, Maity S, Guchhait R, Chatterjee A, Biswas C, Adhikari M, Pramanick K (2021) Toxic effects of cyanotoxins in teleost fish: a comprehensive review. Aquat Toxicol 240:105971

Beiras R (2018) Biotransformation. In: Beiras R (ed) Marine pollution: sources, fate and effects of pollutants in coastal ecosystems. Elsevier, Amsterdam, pp 205–214

Bittner M, Štern A, Smutná M, Hilscherová K, Žegura B (2021) Cytotoxic and genotoxic effects of cyanobacterial and algal extracts - microcystin and retinoic acid content. Toxins 13:107

Blagojević Ponjavić A (2021) Phytoplankton dynamics and the phenomenon of bloom of cyanobacterium *Planktothrix rubescens* in the Vrutci reservoir. Dissertation, University of Belgrade

Blagojević Ponjavić A, Kostić D, Marjanović P, Trbojević I, Popović S, Predojević D, Subakov Simić G (2019) Bloom of the potentially toxic cyanobacterium *P. rubescens*: seasonal distribution and possible drivers of its proliferation in the Vrutci reservoir (Serbia). Oceanol Hydrobiol Stud 48(4):316–327

Blaženčić J (1986) Review of development of algology in Serbia from 1883. To 1983. U: Glasnik instituta za botaniku i botaničke bašte Univerziteta u Beogradu 20:99–108

Buratti FM, Manganelli M, Vichi S, Stefanelli M, Scardala S, Testai E, Funari E (2017) Cyanotoxins: producing organisms, occurrence, toxicity, mechanism of action and human health toxicological risk evaluation. Arch Toxicol 91:1049–1130

Burdick SM, Hereford DM, Conway CM, Banet NV, Powers R, Martin BA, Elliott DG (2020a) Mortality of endangered juvenile lost river suckers associated with cyanobacteria blooms in mesocosms in upper Klamath Lake, Oregon. Trans Am Fish Soc 149:245–265

Burdick SM, Hewitt DA, Martin BA, Schenk L, Rounds SA (2020b) Effects of harmful algal blooms and associated water-quality on endangered lost river and shortnose suckers. Harmful Algae 97:101847

Burkholder JM (2009) Phytoplankton and episodic suspended sediment loading: phosphate partitioning and mechanisms for survival. Limnol Oceanog 37:974–988

Bury NR, Eddy FB, Codd GA (1995) The effects of the cyanobacterium *Microcystis aeruginosa*, the cyanobacterial hepatotoxin microcystin-LR and ammonia on growth rate and ionic regulation of brown trout. J Fish Biol 46(6):1042–1054

Bury NR, Eddy FB, Codd GA (1996a) Stress responses of brown trout, *Salmo trutta* L. to the cyanobacterium *Microcystis aeruginosa*. Environ Toxic Water 11:187–194

Bury N, Flik G, Eddy F, Codd G (1996b) The effects of cyanobacteria and the cyanobacterial toxin microcystin-LR on Ca2+ transport and Na+/K+-ATPase in tilapia gills. J Exp Biol 199:1319–1326

Čađo S, Novaković B, Đurković A, Denić Lj, Dopuđa Glišić T, Stojanović Z, Veljković N, Domanović M, Žarić D (2020) Ekološki potencijal akumulacija za vodosnabdevanje u Srbiji. Voda i sanitarna tehnika 3-4:19–40

Campos A, Vasconcelos V (2010) Molecular mechanisms of microcystin toxicity in animal cells. Int J Mol Sci 11:268–287

Carbis CR, Rawlin GT, Mitchell GF, Anderson JW, McCauley I (1996) The histopathology of carp, *Cyprinus carpio* L., exposed to microcystins by gavage, immersion, and intraperitoneal administration. J Fish Dis 19:199–207

Carbis CR, Tawlin GT, Grant P, Mitchell GF, Anderson JW, McCauley I (1997) A study of feral carp, *Cyprinus carpio* L., exposed to *Microcystis aeruginosa* at lake Mokoan, Australia, and possible implications for fish health. J Fish Dis 20:81–91

Carlson RE (1977) A trophic state index for lakes. Limnol Oceanogr 22:361–368

Carmichael WW (1992) Cyanobacteria secondary metabolites – the cyanotoxins. J App Bacteriol 72(6):445–459

Carmichael WW, Falconer IR (1993) Algal toxins in seafood and drinking water. Academic Press, London

Carvalho M, Montero D, Torrecillas S, Castro P, Zamorano MJ, Izquierdo M (2021) Hepatic biochemical, morphological and molecular effects of feeding microalgae and poultry oils to gilthead sea bream (*Sparus aurata*). Aquaculture 532:736073

Chen Y, Zeng SF, Cao YF (2012) Oxidative stress response in zebrafish (*Danio rerio*) gill experimentally exposed to subchronic microcystin-LR. Environ Monit Assess 184:6775–6787

Chen G, Wang L, Wang M, Hu T (2021) Comprehensive insights into the occurrence and toxicological issues of nodularins. Mar Pollut Bull 162:111884

Chorus I (2012) Current approaches to cyanotoxin risk assessment, risk management and regulations in different countries. Federal Environmental Agency (Umweltbundesamt – UBA), Dessau-Roßlau, Germany, p 151

Chorus I, Bartram J (1999) Toxic cyanobacteria in water. E&F Spon, London

Chorus I, Welker M (2021) Toxic cyanobacteria in water, 2nd edn. CRC Press, Boca Raton, FL. on behalf of the World Health Organization, Geneva, CH

Christensen VG, Khan E (2020) Freshwater neurotoxins and concerns for human, animal, and ecosystem health: a review of anatoxin-a and saxitoxin. Sci Total Environ 736:139515

Ćirić M, Marković Z, Dulić Z, Subakov-Simić G (2010) First report of cyanobacterium *Cylindrospermopsis raciborskii* from carp ponds in Serbia. In: The 8th international conference on toxic cyanobacteria (ICTC8), Istanbul, Turkey

Ćirić M, Gavrilović B, Dojčinović B, Reh SČ, Zhou Y, Song C, Cao X (2020) Past studies and potential measures for rehabilitation of the shallow Lake (Lake Ludaš). J Geogr Inst "Jovan Cvijić" 70:71–80. https://doi.org/10.2298/IJGI2001071C

Codd GA, Lindsay J, Young LF, Morrison LF, Metcalf JS (2005) Harmful cyanobacteria. In: Huisman J, Matthijs HCP, Visser PM (eds) Harmful cyanobacteria. Springer, Heidelberg, pp 1–23

Crowe BH, Harris JO, Bansemer MS, Stone DAJ (2021) Restricted feeding and dietary energy levels affect liver structure in cultured yellowtail kingfish (*Seriola lalandi*, Valenciennes) at summer water temperatures. Aquac Res 52:6074–6086

Da Silva CA, De Morais ECP, Costa MDM, Ribas JLC, Guiloski IC, Ramsdorf WA, Zanata SM, Cestari MM, Oliveira Ribeiro CA, Magalhães VF, Trudeau VL, Silva De Assis HC (2014) Saxitoxins induce cytotoxicity, genotoxicity and oxidative stress in teleost neurons in vitro. Toxicon 86:8–15

Deblois CP, Aranda-Rodriguez R, Giani A, Bird DF (2008) Microcystin accumulation in liver and muscle of tilapia in two large Brazilian hydroelectric reservoirs. Toxicon 51:435–448. https://doi.org/10.1016/j.toxicon.2007.10.017

Dekant W (2009) The role of biotransformation and bioactivation in toxicity. In: Luch A (ed) Molecular, clinical and environmental toxicology. Birkhäuser, Basel, pp 57–86

Đenić D (2014) Cyanobacterial blooms in Vrutci reservoir. Master's thesis, University of Novi Sad

Diaz RJ, Rosenberg R (2008) Spreading dead zones and consequences for marine ecosystems. Science 321:926–929

Đorđević N (2021) The effect of changes of ecological factors on composition and dynamics of phytoplankton cyanobacteria (Cyanobacteria) in small reservoirs. Dissertation, University of Kragujevac

Đorđević N, Simić S (2014) Cyanobacterial blooms in oligosaline and alkaline microaccumulation before and after rehabilitation. Pol J Environ Stud 23(6):1975–1982

Đorđević N, Simić S, Ćirić A (2015) First identification of the cylindrospermopsin (CYN) - producing cyanobacterium *Cylindrospermopsis raciborskii* (Woloszyńska) Seenayya & Subba Raju in Serbia. Fresenius Environ Bull 24(11a):3736–3742

Đorđević NB, Matić SL, Simić SB, Stanić SM, Mihailović VB, Stanković NM, Stanković VD, Ćirić AR (2017) Impact of the toxicity of *Cylindrospermopsis raciborskii* (Woloszynska) Seenayya & Subba Raju on laboratory rats in vivo. Environ Sci Pollut R 24(16): 14259–14272. https://doi.org/10.1007/s11356-017-8940-6

Drobac D (2015) Human exposure to cyanotoxins and their health effects. Dissertation, University of Novi Sad

Drobac D (2018) Cyanotoxin exposure and human health. Andrejević Endowment, Belgrade, p 93

Drobac Backović D, Tokodi N, Nikolin B, Svirčev Z (2020) Cyanobacteria – insidious foe of the skin? J Water Health 18(3):314–330. https://doi.org/10.2166/wh.2020.001

Drobac Backović D, Tokodi N, Marinović Z, Lujić J, Dulić T, Simić SB, Đordević NB, Kitanović N, Šćekić I, Urbányi B, Meriluoto J, Svirčev Z (2021) Cyanobacteria, cyanotoxins, and their histopathological effects on fish tissues in Fehérvárcsurgó reservoir. Hungary Environ Monit Assess 193:554. https://doi.org/10.1007/s10661-021-09324-3

Drobac D, Tokodi N, Simeunović J, Baltić V, Stanić D, Svirčev Z (2013) Human exposure to cyanotoxins and their health effects. Arh Hig Rada Toksikol 64(2):305–316

Drobac D, Tokodi N, Lujić J, Marinović Z, Subakov-Simić G, Dulić T, Važić T, Nybom S, Meriluoto J, Codd GA, Svirčev Z (2016) Cyanobacteria and cyanotoxins in fishponds and their effects on fish tissue. Harmful Algae 55:66–76. https://doi.org/10.1016/j.hal.2016.02.007

Du X, Liu H, Yuan L, Wang Y, Ma Y, Wang R, Chen X, Losiewicz MD, Guo H, Zhang H (2019) The diversity of cyanobacterial toxins on structural characterization, distribution and identification: a systematic review. Toxins 11:530

Dziga D, Tokodi N, Backović Drobac D, Kokociński M, Antosiak A, Puchalski J, Strzałka W, Madej M, Meriluoto J, Svirčev Z (2019) The effect of a combined hydrogen peroxide-MlrA treatment on the phytoplankton community and microcystin concentrations in a mesocosm experiment in Lake Ludoš. Toxins 11(12):725. https://doi.org/10.3390/toxins11120725

Ernst B, Hoeger SJ, O'Brien E, Dietrich DR (2006) Oral toxicity of the microcystin-containing cyanobacterium *Planktothrix rubescens* in European whitefish (*Coregonus lavaretus*). Aquat Toxicol 79:31–40

Ferrão-Filho ADS, Kozlowsky-Suzuki B (2011) Cyanotoxins: bioaccumulation and effects on aquatic animals. Mar Drugs 9:2729–2772

Fischer WJ, Dietrich DR (2000) Pathological and biochemical characterization of microcystin-induced hepatopancreas and kidney damage in carp (*Cyprinus carpio*). Toxicol Appl Pharmacol 164:73–81

Fleming L, Rivero C, Burns J, Williams C, Beana J, Shea K, Stinn J (2002) Bluegreen algal (cyanobacterial) toxins, surface drinking water, and liver cancer in Florida. Harmful Algae 1: 157–168

Flores NM, Miller TR, Stockwell JD (2018) A global analysis of the relationship between concentrations of microcystins in water and fish. Front Mar Sci 5:30

Gaete V, Canelo E, Lagos N, Zambrano F (1994) Inhibitory effects of *Microcystis aeruginosa* toxin on ion pumps of the gill of freshwater fish. Toxicon 32:121–127

Garcia CZ, Martinez CBR (2012) Biochemical and genetic alterations in the freshwater neotropical fish Prochilodus lineatus after acute exposure to *Microcystis aeruginosa*. Neotrop Ichthyol 10: 613–622

Gaudin J, Huet S, Jarry G, Fessard V (2008) In vivo DNA damage induced by the cyanotoxin microcystin-LR: comparison of intra-peritoneal and oral administrations by use of the comet assay. Mutat Res Toxicol Environ Mutagen 652:65–71

Gavrilović B, Despotović S, Gavrić J, Borković-Mitić S, Ognjanović B, Pavlović S, Saičić Z (2014) Changes in antioxidant enzyme activities in the livers and gills of three cyprinids after exposure to a cyanobacterial bloom in the Gruža reservoir. Serbia Ecol Indic 38:41–148. https://doi.org/10.1016/j.ecolind.2013.10.016

Gavrilović BR, Prokić MD, Petrović TG, Despotović SG, Radovanović TB, Krizmanić II, Ćirić MD, Gavrić JP (2020) Biochemical parameters in skin and muscle of *Pelophylax* kl. *esculentus* frogs: influence of a cyanobacterial bloom *in situ*. Aquat Toxicol 220:105399

Gavrilović BR, Petrović TG, Radovanović TB, Despotović SG, Gavrić JP, Krizmanić II, Ćirić MD, Prokić MD (2021) Hepatic oxidative stress and neurotoxicity in *Pelophylax* kl. *esculentus* frogs: influence of long-term exposure to a cyanobacterial bloom. Sci Total Environ 750:141569. https://doi.org/10.1016/j.scitotenv.2020.141569

Gerba CP (2019) Environmental toxicology. In: Brusseau ML, Pepper IL, Gerba CP (eds) Environmental and pollution science. Academic Press, Elsevier, Cambridge, pp 511–540

Gupta N, Pant SC, Vijayaraghavan R, Rao PVL (2003) Comparative toxicity evaluation of cyanobacterial cyclic peptide toxin microcystin variants (LR, RR, YR) in mice. Toxicology 188:285–296

Heath AG (1995) Water pollution and fish physiology, 2nd edn. CRC Press

Holland A, Kinnear S (2013) Interpreting the possible ecological role(s) of cyanotoxins: compounds for competitive advantage and/or physiological aide? Mar Drugs 11(7):2239–2258

Hooser SB, Beasley VR, Basgall EJ, Carmichael WW, Haschek WM (1990) Microcystin-LR-induced ultrastructural changes in rats. Vet Pathol 27:9–15

Hou J, Li L, Xue T, Long M, Su Y, Wu N (2015) Hepatic positive and negative antioxidant responses in zebrafish after intraperitoneal administration of toxic microcystin-LR. Chemosphere 120:729–736

Huisman JM, Matthijs HCP, Visser PM (2005) Harmful cyanobacteria. Springer aquatic ecology series 3. Springer, Dordrecht

Huisman J, Codd GA, Paerl HW et al (2018) Cyanobacterial blooms. Nat Rev Microbiol 16:471–483. https://doi.org/10.1038/s41579-018-0040-1

Ibelings BW, Chorus I (2007) Accumulation of cyanobacterial toxins in freshwater "seafood" and its consequences for public health: a review. Environ Pollut 150:177–192. https://doi.org/10.1016/j.envpol.2007.04.012

Institute of Public Health Serbia "Dr Milan Jovanović Batut" (2014) Report on the request for access to information of public importance No.8301/1, Belgrade. Reported on the internet by the Association of citizens „U ime naroda: http://uimenaroda.rs/predmeti/li%C4%8Dni-stav/190-u%C5%BElce-vodosnabdevanje-slika-srbije.html. Accessed 21 January 2016

Jiang J, Gu X, Song R, Zhang Q, Geng J, Wang X, Yang L (2011) Time-dependent oxidative stress and histopathological changes in *Cyprinus carpio* L. exposed to microcystin-LR. Ecotoxicology 20:1000–1009

Jöhnk KD, Huisman J, Sharples J et al (2008) Summer heatwaves promote blooms of harmful cyanobacteria. Glob Change Biol 14:495–512

Jos Á, Pichardo S, Prieto AI, Repetto G, Vázquez CM, Moreno I, Cameán AM (2005) Toxic cyanobacterial cells containing microcystins induce oxidative stress in exposed tilapia fish (*Oreochromis* sp.) under laboratory conditions. Aquat Toxicol 72:261–271

Jovanović J (2020) The distribution of cyanobacteria in surface waters in relation to water supply and recreation in Serbia. Dissertation, University of Belgrade

Karadžić V, Subakov Simić G, Krizmanić J, Natić D (2010) Phytoplankton and eutrophication development in the water supply reservoirs Garaši and Bukulja (Serbia). Desalination 255(1–3): 91–96

Kaštovský J, Hauer T, Mareš J, Krautová M, Bešta T, Komárek J, Desortová B, Heteša J, Hindáková A, Houk V, Janeček E, Kopp R, Marvan P, Pumann P, Skácelová O, Janeček E (2010) A review of the alien and expansive species of freshwater cyanobacteria and algae in The Czech Republic. Biol Invasions 12(10):3599–3625

Kokociński M, Akçaalan R, Salmaso N, Stoyneva-Gärtner MP, Sukenik A (2017) Expansion of alien and invasive cyanobacteria. In: Meriluoto J, Spoof L, Codd GA (eds) Handbook on cyanobacterial monitoring and cyanotoxin analysis. Wiley, Chichester, pp 28–39

Kostić D, Blagojević A, Subakov Simić G, Ivetić M, Naunović Z (2014) Jesenje cvetanje potencijalno tokične cijanobakterije *Plantothrix rubescens* u višenamenskoj akumulaciji Vrutci. Voda i sanitarna tehnika 44(1):5–12

Kostić D, Blagojević A, Subakov Simić G, Predojević D, Naunović Z, Jaćimović N, Grašić S (2015) "Blooming" reservoir response to a high inflow event - case study: the Vrutci reservoir (Western Serbia). In: 7th eastern European Young water professionals conference, Belgrade, pp 97–104

Kostić D, Marjanović P, Marjanović M, Blagojević A, Trbojević I, Predojević D, Subakov Simić G, Vulić D, Obradović V, Naunović Z (2016) Drivers of phytoplankton blooms in the Vrutci reservoir during 2014–2015 and implications for water supply and management. Water Res Manag 6(3):3–12

Kotak BG, Semalulu S, Fritz DL, Prepas EE, Hrudey SE, Coppock RW (1996) Hepatic and renal pathology of intraperitoneally administered microcystin-LR in rainbow trout (*Oncorhynchus mykiss*). Toxicon 34:517–525

Kozlowsky-Suzuki B, Wilson AE, da Silva F-FA (2012) Biomagnification or biodilution of microcystins in aquatic foodwebs? Meta-analyses of laboratory and field studies. Harmful Algae 18:47–55

Kramer DL (1987) Dissolved oxygen and fish behavior. Environ Biol Fish 18:81–92

Kurmayer R, Christiansen G (2009) The genetic basis of toxin production in cyanobacteria. Fr Rev 2:31–50

Li L, Xie P, Chen J (2007) Biochemical and ultrastructural changes of the liver and kidney of the phytoplanktivorous silver carp feeding naturally on toxic *Microcystis* blooms in Taihu Lake, China. Toxicon 49:1042–1053

Li L, Xie P, Guo L, Ke Z, Zhou Q, Liu Y, Qiu T (2008) Field and laboratory studies on pathological and biochemical characterization of microcystin-induced liver and kidney damage in the phytoplanktivorous bighead carp. Sci World J 8:121–137

Lujić J, Matavulj M, Poleksić V, Rašković B, Marinović Z, Kostić D, Miljanović B (2015) Gill reaction to pollutants from the Tamiš River in three freshwater fish species, *Esox lucius* L. 1758, *Sander lucioperca* (L. 1758) and *Silurus glanis* L. 1758: a comparative study. Anat Histol Embryol 44:128–137

Mallatt J (1985) Fish gill structural changes induced by toxicants and other irritants: a statistical review. Can J Fish Aquat Sci 42:630–648

Marinović Z (2021) Cyanobacterial blooming, presence of cyanotoxins and their effects on fish of several select freshwater ecosystems of Serbia and Hungary. Dissertation, University of Novi Sad

Marinović Z, Tokodi N, Backović DD, Šćekić I, Kitanović N, Simić SB, Đorđević NB, Ferincz Á, Staszny Á, Dulić T, Meriluoto J, Urbányi B, Lujić J, Svirčev Z (2021) Does the Kis-Balaton water protection system (KBWPS) effectively safeguard Lake Balaton from toxic cyanobacterial blooms? Microorganisms 9:960. https://doi.org/10.3390/microorganisms9050960

Meriluoto J, Blaha L, Bojadžija G, Bormans M, Brient L, Cood GA, Drobac D, Faassen E, Fastner J, Hiskia A, Ibelings B, Kaloudis T, Kokocinski M, Kurmayer R, Quesada A, Pantelić D, Salmaso N, Tokodi N, Triantis T, Visser P, Svirčev Z (2017) Toxic cyanobacteria and cyanotoxins in European waters - recent progress achieved through the CYANOCOST action and challenges for further research. AIOL 8(1):161–178

Metcalf JS, Souza NR (2019) Cyanobacteria and their toxins. In: Separation science and technology. Elsevier Inc, Amsterdam, pp 125–148

Miljanović B, Jurca T, Pogrmić S, Bajić A, Živković M, Šipoš Š, Pankov N (2016) Report on the performed monitoring in the fishing area of the special nature reserve "Ludaš Lake" and Nature Park "Palić". Faculty of Sciences, University of Novi Sad

Milovanović D (1963) Phytoplankton and primary production in fish ponds „Koluta". Arch Biol Sci 6:3–16

Milovanović D, Živković A (1953) Plankton production investigations in Ečka fish ponds. Proceedings S.A.N. XXIX 197–264

Milovanović D, Živković A (1959) Phytoplankton production in Živača fishpond (II contribution to limnology of lentic waters in Panonia valley). Arch Biol Sci 2:1–7

Mitsoura A, Kagalou I, Papaioannou N, Berillis P, Mente E, Papadimitriou T (2013) The presence of microcystins in fish *Cyprinus carpio* tissues: a histopathological study. Int Aquat Res 5:8

Mulvenna V, Dale K, Priestly B, Mueller U, Humpage A, Shaw G, Allinson G, Falconer I (2012) Health risk assessment for cyanobacterial toxins in seafood. Int J Environ Res Pub He 9:807–820

Niesar M, Arlinghaus R, Rennert B, Mehner T (2004) Coupling insights from a carp, *Cyprinus carpio*, angler survey with feeding experiments to evaluate composition, quality and phosphorus input of groundbait in coarse fishing. Fish Manag Ecol 11:225–235

Nikolić D, Skorić S, Rašković B, Lenhardt M, Krpo-Ćetković J (2020) Impact of reservoir properties on elemental accumulation and histopathology of European perch (*Perca fluviatilis*). Chemosphere 244:125503

Nilsson GE (2007) Gill remodeling in fish - a new fashion or an ancient secret? J Exp Biol 210: 2403–2409

Official Gazette of the Republic of Serbia 128/2014. Law on protection and sustainable use of fish stock

Paerl HW, Huisman J (2009) Climate change: a catalyst for global expansion of harmful cyanobacterial blooms. Environ Microbiol Rep 1(1):27–37

Paerl HW, Otten TG (2013) Harmful cyanobacterial blooms: causes, consequences, and controls. Microb Ecol 65:995–1010. https://doi.org/10.1007/s00248-012-0159-y

Paerl HW, Paul V (2011) Climate change: links to global expansion of harmful cyanobacteria. Water Res 46:1349–1363

Papadimitriou T, Kagalou I, Bacopoulos V, Leonardos ID (2010) Accumulation of microcystins in water and fish tissues: an estimation of risks associated with microcystins in most of the Greek Lakes. Environ Toxicol 25:418–427. https://doi.org/10.1002/tox.20513

Paul VJ (2008) Global warming and cyanobacterial harmful algal blooms. In: Hudnell HK (ed) Cyanobacterial harmful algal blooms: state of the science and research needs, Advances in experimental medicine and biology, vol 619. Springer, Heidelberg, pp 239–257

Peng L, Liu Y, Chen W, Liu L, Kent M, Song L (2010) Health risks associated with consumption of microcystin contaminated fish and shellfish in three Chinese lakes: significance for freshwater aquacultures. Ecotoxicol Environ Saf 73:1804–1811. https://doi.org/10.1016/j.ecoenv.2010.07.043

Perendija B, Despotović S, Radovanović T, Gavrić J, Mitić Borković S, Pavlović SZ, Ognjanović B, Simić S, Pajović S, Saičić Z (2011) Biochemical and ultrastructural changes in the liver of European perch (*Perca fluviatilis* L.) in response to cyanobacterial bloom in the Gruža reservoir. Arch Biol Sci 63(4):979–989

Peters R (1986) The role of prediction in limnology. Limnol Oceanogr 31:1143–1159

Pham T-L, Utsumi M (2018) An overview of the accumulation of microcystins in aquatic ecosystems. J Environ Manag 213:520–529

Popovic M, Zaja R, Smital T (2010) Organic anion transporting polypeptides (OATP) in zebrafish (*Danio rerio*): phylogenetic analysis and tissue distribution. Comp Biochem Physiol - A Mol Integr Physiol 155:327–335

Poste AE, Hecky RE, Guildford SJ (2011) Evaluating microcystin exposure risk through fish consumption. Environ Sci Technol 45:5806–5811. https://doi.org/10.1021/es200285c

Prieto AI, Jos Á, Pichardo S, Moreno I, Cameán AM (2006) Differential oxidative stress responses to microcystins LR and RR in intraperitoneally exposed tilapia fish (*Oreochromis* sp.). Aquat Toxicol 77:314–321

Rabelo JCS, Hanusch AL, de Jesus LWO, Mesquita LA, Franco FC, Silva RA, Sabóia-Morais SMT (2021) DNA damage induced by cylindrospermopsin on different tissues of the biomonitor fish *Poecilia reticulata*. Environ Toxicol 36:1125–1134

Raičević V, Božić M, Rudić Ž, Lalević B, Kiković D (2011) The evolution of the eutrophication of the Palić Lake (Serbia). Afr J Biotechnol 10:1736–1744

Ranković B, Simić S (2005) Phytoplankton of Gruža reservoir. In: Čomić Lj, Ostojić A (eds) Gruža Reservoir-Monography. Faculty of Natural Sciences, Kragujevac, pp 65-78

Ristić O, Gajin S, Gantar M, Matavulj M (1979) Microbiological studies of some fish ponds in Vojvodina. In: Proceedings of the II Congress of Ecologists in Yugoslavia. Zagreb, 1923–1935

Rudić Ž, Vujović B, Božić M, Raičević V (2015) Lake Ludaš special nature reserve – bacteriological point of view, VII international conference “Water & Fish” – abstract book, University of Belgrade, Faculty of Agriculture, Zemun-Belgrade, Serbia

Sakan S, Gržetić I, Đorđević D (2007) Distribution and fractionation of heavy metals in the Tisza (Tisza) river sediments. Environ Sci Pollut Res Int 14(4):229–236

Scharschmidt J (1883) Fragmenta phycologiae Bosniaco-Serbicae. Magyar novenytani lapok. Budapest 7(75):33–39

Scheffer M, Hosper SH, Meijer ML, Moss B, Jeppesen E (1993) Alternative equilibria in shallow lakes. Trends Ecol Evol 8:275–279

Sedan D, Laguens M, Copparoni G, Aranda JO, Giannuzzi L, Marra CA, Andrinolo D (2015) Hepatic and intestine alterations in mice after prolonged exposure to low oral doses of microcystin-LR. Toxicon 104:26–33

Sedmak B, Svirčev Z (2011) Cijanobakterije i njihovi toksini-ekološki i toksikološki rizici i cvetanje cijanobakterija u Srbiji. Visoka šola za varstvo okolja, Velenje, Slovenija

Simeunović J (2009) Ecophysiological characteristics of potentially toxic and toxic aquatic strains of cyanobacteria in the Vojvodina area. Dissertation, University of Novi Sad

Simić S (2004) Changes in structure of the phytoplankton and processes of eutrophication in the Barje reservoir (Serbia). Krag J Sci 26:53–64

Simić S, Đorđević N (2022) Report on the biological analysis of the phytoplankton accumulation Šumarice and assessment of the ecological potential of the accumulation. NGO Ekomar

Simić V, Simić S, Petrović A, Veličković T et al (2001–2021) Fisheries management plans for the protection and sustainable use of fish stocks (2001–2021). Faculty of Science, University of Kragujevac

Simić S, Mišćević M, Đorđević N, Popović N (2011) Cyanobacteria in Aleksandrovac Lake- before and after revitalisation. In: abstracts of the 16th Academy of Studenica, cyanobacteria and human health, Novi Sad. Serbia, 1–3 July 2011, p 42

Simić S, Komárek J, Đorđević N (2014) The confirmation of the genus *Glaucospira* (*Cyanobacteria*) and the occurrence of *Glaucospira laxissima* (G. S West) comb nova in Serbia. Cryptogamie algol 35(3):259–267

Simić SB, Đorđević NB, Dj M (2017) The relationship between the dominance of cyanobacteria species and environmental variables in different seasons and extreme precipitation. Fundam Appl Limnol 190(1):1–11

Simić S, Đorđević N, Mitrović A (2018) Preliminary assessment of the degree of vulnerability and health risk in some fishing waters based on cyanobacteria in 2017. In 8th International Conference "Water & Fish" Faculty of Agriculture, Belgrade-Zemun, Serbia, p 394

Simić V, Simić S, Petrović A, Đorđević N, Simović P (2021) Report on the biological analysis of the Barje accumulation and assessment of the potential of the accumulation. Faculty of Science, University of Kragujevac, Serbia

Simić V, Bănăduc D, Curtean Bănăduc A, Petrović A, Veličković T, Stojković Piperac M, Simić S (2022) Assessment of the ecological sustainability of river basins based on the modified the ESHIPPOfish model on the example of the Velika Morava basin (Serbia, Central Balkans). Front Environ Sci 10:952692. https://doi.org/10.3389/fenvs.2022.952692

Singh S, Kate BN, Banerjee US (2006) Bioactive compounds from cyanobacteria and microalgae: an overview. Crit Rev Biotechnol 25(3):73–95

Sinha R, Pearson L, Davis T, Burford M, Orr P, Neilan B (2012) Increased incidence of *Cylindrospermopsis raciborskii* in temperate zones – is climate change responsible? Water Res 46(5):1408–1419. https://doi.org/10.1016/j.watres.2011.12.019

Sivonen K, Jones G (1999) Cyanobacterial toxins. In: Chorus I, Bartram J (eds) Toxic cyanobacteria in water: a guide to public health significance, monitoring and management. Spon E&FN, London, pp 41–111

Skafi M, Vo Duy S, Munoz G, Dinh QT, Simon DF, Juneau P, Sauvé S (2021) Occurrence of microcystins, anabaenopeptins and other cyanotoxins in fish from a freshwater wildlife reserve impacted by harmful cyanobacterial blooms. Toxicon 194:44–52. https://doi.org/10.1016/j.toxicon.2021.02.004

Smith VH (2003) Eutrophication of freshwater and coastal marine ecosystems a global problem. Environ Sci Pollut Res 10:126–139. https://doi.org/10.1065/espr2002.12.142

Smith VH, Schindler DW (2009) Eutrophication science: where do we go from here? Trends Ecol Evol 24:201–207

Steiner K, Zimmermann L, Hagenbuch B, Dietrich D (2016) Zebrafish Oatp-mediated transport of microcystin congeners. Arch Toxicol 90:1129–1139

Svirčev Z, Krstić S, Marković SB, Plavša J, Lazić L (2008) Methods for management of eutrophication in freshwater ecosystems in Vojvodina. Geogr Pannonica 12(1):4–11

Svirčev Z, Baltić V, Simeunović J, Tokodi N, Drobac D (2011) Cyanotoxin legislation. 16th academy of Studenica: cyanobacteria and human health, 1–3 Jul, Novi Sad, Srbija

Svirčev Z, Drobac D, Tokodi N, Vidović M, Simeunović J, Miladinov-Mikov M, Baltić V (2013) Epidemiology of primary liver cancer in Serbia and possible connection with cyanobacterial blooms. J Environ Sci Health C Environ Carcinog Ecotoxicol Rev 31(3):181–200

Svirčev Z, Drobac D, Tokodi N, Lužanin Z, Munjas AM, Nikolin B, Vuleta D, Meriluoto J (2014a) Epidemiology of cancers in Serbia and possible connection with cyanobacterial blooms. J Environ Sci Health C Environ Carcinog Ecotoxicol Rev 32(4):319–337

Svirčev Z, Tokodi N, Drobac D, Codd GA (2014b) Cyanobacteria in aquatic ecosystems in Serbia: effects on water quality, human health and biodiversity. Syst Biodivers 12(3):261–270

Svirčev Z, Lujić J, Marinović Z, Drobac D, Tokodi N, Stojiljković B, Meriluoto J (2015) Toxicopathology induced by microcystins and nodularin: a histopathological review. J Environ Sci Heal Part C 33:125–167

Svirčev Z, Obradović V, Codd GA, Marjanović P, Spoof L, Drobac D, Tokodi N, Petković A, Nenin T, Simeunović J, Važić T, Meriluoto J (2016) Massive fish mortality and *Cylindrospermopsis raciborskii* bloom in Aleksandrovac Lake. Ecotoxicology 25(7): 1353–1363

Svirčev Z, Drobac D, Tokodi N, Mijović B, Codd GA, Meriluoto J (2017a) Toxicology of microcystins with reference to cases of human intoxications and epidemiological investigations of exposures to cyanobacteria and cyanotoxins. Archives of Toxicology 91(2):621-650. ISSN: 03405761, https://doi.org/10.1007/s00204-016-1921-6

Svirčev Z, Tokodi N, Drobac D (2017b) Review of 130 years of research on cyanobacteria in aquatic ecosystems in Serbia presented in a Serbian cyanobacterial database. Adv Oceanogr Limnol 8(1):153–160. https://doi.org/10.4081/aiol.2017.6360

Svirčev Z, Drobac D, Tokodi N, Đenić D, Simeunović J, Hiskia A, Kaloudis T, Mijović B, Šušak S, Protić M, Vidović M, Onjia A, Nybom S, Važić T, Palanački Malešević T, Dulić T, Pantelić D, Vukašinović M, Meriluoto J (2017c) Lessons from the Užice case: how to complement analytical data. In: Meriluoto J, Spoof L, Codd GA (eds) Handbook of cyanobacterial monitoring and cyanotoxin analysis. Wiley, Chichester, pp 298–308

Svirčev Z, Lalić D, Bojadžija Savić G, Tokodi N, Drobac Backović D, Chen L, Meriluoto J, Codd GA (2019) Global geographical and historical overview of cyanotoxin distribution and cyanobacterial poisonings. Arch Toxicol 93(9):2429–2481. https://doi.org/10.1007/s00204-019-02524-4

Testai E, Buratti F, Funari E et al. (2016a) Review and analysis of occurrence, exposure and toxicity of cyanobacteria toxins in food. EFSA supporting publication 998E p 309. https://doi.org/10.2903/sp.efsa.2016.EN-99

Testai E, Buratti FM, Funari E, Manganelli M, Vichi S, Arnich N et al (2016b) Review and analysis of occurrence, exposure and toxicity of cyanobacteria toxins in food. EFSA Supporting Publications: EN-998, Parma, p 309

The Agency for Environmental Protection (Agencija za zaštitu životne sredine) (2013) http://www.sepa.gov.rs/index.php?menu=504000000&id=20013&akcija=showArhiva. Accessed 21 January 2016

The Environmental Protection Agency - SEPA (2012) The results of surface and underground water quality testing for 2011, The Ministry of Energy, Development and Environmental Protection of the Republic of Serbia p 638. Accessed 15 July 2022

The Environmental Protection Agency - SEPA (2013) The Status of lakes and reservoirs in 2012, The Ministry of Energy, Development and Environmental Protection of the Republic of Serbia p 43. Accessed 15 July 2022

The Environmental Protection Agency - SEPA (2014) The results of surface and underground water quality testing for 2013, Ministry of Agriculture and Environmental Protection of the Republic of Serbia p 408. Accessed 15 July 2022

The Environmental Protection Agency - SEPA (2015) The results of surface and underground water quality testing for 2014, Ministry of Agriculture and Environmental Protection of the Republic of Serbia p 474. Accessed 15 July 2022

The Environmental Protection Agency - SEPA (2017a) The results of surface and underground water quality testing for 2015, Ministry of Agriculture and Environmental Protection of the Republic of Serbia p 408. Accessed 15 July 2022

The Environmental Protection Agency - SEPA (2017b) The results of surface and underground water quality testing for 2016, Ministry of Environmental Protection of the Republic of Serbia p 370. Accessed 15 July 2022

The Environmental Protection Agency - SEPA (2018) The results of surface and underground water quality testing for 2017, Ministry of Environmental Protection of the Republic of Serbia p 356. Accessed 15 July 2022

The Environmental Protection Agency - SEPA (2019) The results of surface and underground water quality testing for 2018, Ministry of Environmental Protection of the Republic of Serbia p 442. Accessed 15 July 2022

The Environmental Protection Agency - SEPA (2020) The results of surface and underground water quality testing for 2019, Ministry of Environmental Protection of the Republic of Serbia p 390. Accessed 15 July 2022

The Environmental Protection Agency - SEPA (2021) The results of surface and underground water quality testing for 2020, Ministry of Environmental Protection of the Republic of Serbia p 317. Accessed 15 July 2022

Tokodi N (2016) Toxic cyanobacteria from the territory of the Republic of Serbia. Dissertation, University of Novi Sad

Tokodi N, Drobac D, Simeunović J, Svirčev Z (2013) Assessment of acute cyanotoxicity using *Artemia salina* bioassay in water samples from fishponds. Paper presented at the 17th international eco-conference, 10th environmental protection of urban and suburban settlements, Novi Sad, Serbia, 25–28 September 2013

Tokodi N, Drobac D, Simeunović J, Svirčev Z (2014) Microcystin concentrations in fishpond waters. J Nat Sci Matica Srpska, Novi Sad 127:35–42

Tokodi N, Drobac D, Meriluoto J, Lujić J, Marinović Z, Važić T, Nybom S, Simeunović J, Dulić T, Lazić G, Petrović T, Vuković-Gačić B, Sunjog K, Kolarević S, Kračun-Kolarević M, Subakov-Simić G, Miljanović B, Codd GA, Svirčev Z (2018) Cyanobacterial effects in Lake Ludoš, Serbia - is preservation of a degraded aquatic ecosystem justified? Sci Total Environ 635:1047–1062. https://doi.org/10.1016/j.scitotenv.2018.04.177

Tokodi N, Drobac Backović D, Lujić J, Šćekić I, Simić S, Đorđević N, Dulić T, Miljanović B, Kitanović N, Marinović Z, Savela H, Meriluoto J, Svirčev Z (2020) Protected freshwater ecosystem with incessant cyanobacterial blooming awaiting a resolution. Water 12(1):129. https://doi.org/10.3390/w12010129

Torres R, Pizarro L, Csendes A, García C, Lagos N (2007) GTX 2/3 epimers permeate the intestine through a paracellular pathway. J Toxicol Sci 32:241–248

Trbojević I, Blagojević A, Kostić D, Marjanović P, Krizmanić J, Popović S, Subakov Simić G (2019) Periphyton development during summer stratification in the presence of a metalimnetic bloom of *Planktothrix rubescens*. Limnologica 78:125709

US-EPA (2000) Ambient Water Quality Criteria Recommendations Lakes and Reservoirs in Nu-trient Ecoregion, XI Office of Water, EPA 822-B-00-012

van der Oost R, Beyer J, Vermeulen NPE (2003) Fish bioaccumulation and biomarkers in environmental risk assessment: a review. Environ Toxicol Pharmacol 13:57–149

Vehovszky Á, Kovács WA, Farkas A, Györi J, Szabó H, Vasas G (2015) Pharmacological studies confirm neurotoxic metabolite(s) produced by the bloom-forming *Cylindrospermopsis raciborskii* in Hungary. Environ Toxicol 30(5):501–512

Vezie C, Rapala J, Vaitomaa J, Seitsonen J, Sivonen K (2002) Effect of nitrogen and phosphorus on growth of toxic and nontoxic *Microcystis* strains and on intracellular microcystin concentration. Microb Ecol 43:443–454

Vico P, Bonilla S, Cremella B, Aubriot L, Iriarte A, Piccini C (2020) Biogeography of the cyanobacterium *Raphidiopsis* (*Cylindrospermopsis*) *raciborskii*: integrating genomics, phylogenetic and toxicity data. Mol Phylogenet Evol 148:106824

Watanabe MM, Kaya K, Takamura N (1992) Fate of the toxic cyclic heptapeptides, the microcystins, from blooms of *Microcystis* (cyanobacteria) in a hypertrophic lake. J Phycol 28: 761–767

Watson SB, Whitton BA, Higgins SN, Paerl HW, Brooks BW, Wehr JD (2015) Harmful Algal Blooms. In: Wehr JD, Sheath RG, Kociolek JP (eds) Freshwater Algae of North America: ecology and classification, 2nd edn. Academic Press (imprint of Elsevier), Cambridge, MA, pp 873–920

WHO (World Health Organization) (1998) Guidelines for drinking- water quality. 2nd ed. Addendum to Vol. 2. WHO, Geneva

Wilson AE, Gossiaux DC, Höök TO, Berry JP, Landrum PF, Dyble J et al (2008) Evaluation of the human health threat associated with the hepatotoxin microcystin in the muscle and liver tissues of yellow perch (*Perca flavescens*). Can J Fish Aquat Sci 65:1487–1497. https://doi.org/10.1139/F08-067

Winfield IJ (2016) Eutrophication and freshwater fisheries. In: Craig JF (ed) Freshwater fisheries ecology. Wiley Blackwell, Oxford, pp 781–795

Wolf JC, Wolfe MJ (2005) A brief overview of nonneoplastic hepatic toxicity in fish. Toxicol Pathol 33:75–85

World Bank Report DM 4307 (2011) Introducing Daphnia grazing to control global warming associated cyanobacterial toxic blooms in fishing pond. In: World Health Organization 1998.

Guidelines for Drinking-water Quality, 2nd edition. Addendum to Vol. 2. WHO, Geneva, Switzerland

Zhong L, Liu H, Zhang H, Zhang W, Li M, Huang Y, Yao J, Huang X, Geng Y, Chen D, Ouyang P, Yang S, Luo W, Yin L (2022) High starch in diet leads to disruption of hepatic glycogen metabolism and liver fibrosis in largemouth bass (*Micropterus salmoides*), which is mediated by the PI3K/Akt signaling pathway. Front Physiol 0:1023

Zlatković S, Medić O, Predojević D, Nikolić I, Subakov-Simić G, Onjia A, Berić T, Stanković S (2022) Spatio-temporal dynamics in Physico-chemical properties, phytoplankton and bacterial diversity as an indication of the Bovan reservoir water quality. Water 14(3):391. https://doi.org/10.3390/w14030391

Zurawell RW, Chen H, Burke JM, Prepas EE (2005) Hepatotoxic cyanobacteria: a review of the biological importance of microcystins in freshwater environments. J Toxicol Environ Health B Crit Rev 8(1):1–37

Pollution of Fishing Waters of the Western Balkan: Potentially Toxic Elements and Their Impact on the Ecological Sustainability of Fish Resources

Aleksandra Milošković and Nataša Kojadinović

Abstract The increase in the world's population increases the levels of pollutants in the aquatic environment. Among different xenobiotics, the concentration levels of potentially toxic elements are very important due to their negative effects on living organisms, including humans. Western Balkan mainly includes underdeveloped countries with the unresolved issue of wastewater and consequently a large number of pollutants in fishing water. All commercial fishing waters (except the Tisza River) are transboundary and there are no borders for pollutants in water, so we can observe Western Balkan as a single territory regarding aquatic pollution. The aim of this chapter is to present a comprehensive traditional literature overview of studies related to the contamination of fishing waters (used for commercial fishing) of the Western Balkans, with special reference to pollution with potentially toxic elements (PTEs) and their impact on the ecological sustainability of fish resources. This review has shown that research regarding this issue has been very limited over the last two decades. The most studied fish species were from the Danube and Sava rivers, and they were also the most polluted with PTEs. Elevated concentrations of Fe, Zn, Hg, Cd, As, and Ni were found in fish species in the Danube River, while elevated concentrations of Hg were recorded in fish species sampled from the Sava River, downstream from Zagreb. Transboundary lakes used for commercial fishing have been very little investigated and there are only a few studies on PTEs in fish. However, elevated concentrations of Cd and Pb were reported in fish from Lake Prespa.

A. Milošković (✉)
Institute for Information Technologies Kragujevac, University of Kragujevac, Kragujevac, Serbia
e-mail: aleksandra@uni.kg.ac.rs

N. Kojadinović
Institute of Biology and Ecology, Faculty of Science, University of Kragujevac, Kragujevac, Serbia
e-mail: natasa.kojadinovic@pmf.kg.ac.rs

V. Simić et al. (eds.), *Ecological Sustainability of Fish Resources of Inland Waters of the Western Balkans*, Fish & Fisheries Series 43,
https://doi.org/10.1007/978-3-031-36926-1_12

Keywords Commercial fishing · Transboundary waters · Danube River · Sava River

1 Introduction

The sustainability of aquatic ecosystems implies that healthy ecosystems produce high social and economic benefits and remain ecologically sustainable at the same time (Arlinghaus et al. 2002). The same authors emphasize that strong sustainability in fisheries includes environmental quality, biological integrity, ecosystem health, and biodiversity. Healthy aquatic ecosystems (habitat quantity and quality) support fish populations and productivity (Lapointe et al. 2014), yet these ecosystems are among the most globally imperiled (Dudgeon et al. 2006).

The Balkan Peninsula, hosting a rich freshwater ichthyofauna, is considered one of the world's biodiversity hotspots (Peter 2006). In the Western Balkan countries inland fisheries include subsistence fisheries (destined for direct household consumption), recreational fisheries, and commercial fisheries (Funge-Smith and Bennett 2019). In the region, many aquatic resources are transboundary, including large lakes Skadar, Ohrid, Prespa, and Dorjan, and large rivers Sava and Danube, and therefore they are equally used for commercial fishery by two or more countries.

Freshwater quality varies significantly across the countries of Western Balkan—Slovenia, Croatia, Serbia, Bosnia and Herzegovina, Montenegro, North Macedonia, and Albania, starting from the pristine mountain streams to the rivers and lakes polluted by urban and industrial wastewater and agricultural run. Urban pollution prevails in all countries, while mining effluents affect mainly Albanian rivers, industrial pollution is important in North Macedonia, and Bosnia and Herzegovina, and agricultural pollution is widespread in Serbia. Also, wastewater treatment in the region is nonexistent or poor. In Serbia, for example, many large industrial facilities discharge their wastewater directly into rivers with minimal or without treatment (Sunjog et al. 2019).

In the Anthropocene, among the primary threats to freshwater ecosystems and fish are anthropogenic activities that lead to water pollution. The most frequent contaminants are in the form of heavy metals (Khoshnood 2016). Heavy metals or nowadays named potentially toxic elements are the most important since they are not naturally degraded over time, remaining in the water or sediment for a long time (Ardeshir et al. 2017), and have a high potential to enter and accumulate in food chains, reaching humans as end consumers.

With the exception of Slovenia and Croatia, which are EU Member States, other Western Balkan countries are in the process of integration (Skoulikidis et al. 2022), with different water management traditions and approaches, reflecting differences in political systems and administrative structures. In most Western Balkan countries water chemical quality monitoring is usually neglected or inaccessible, whereas scientific studies focused on pollution are often written in the native language (Skoulikidis et al. 2022).

This chapter offers a traditional literature overview of studies related to the contamination of fishing waters (used for commercial fishery) of the Western Balkans, with special reference to pollution with potentially toxic elements (PTEs) and their impact on the ecological sustainability of fish resources.

2 Inland Water Resources of the Western Balkans

The Western Balkan countries are Slovenia, Croatia, Bosnia and Herzegovina, Serbia, Montenegro, Albania, and North Macedonia. The countries of the Western Balkans are coastal, apart from Serbia and North Macedonia. The largest part of the Adriatic coast belongs to Croatia (1777 km), followed by Albania (476 km), and Montenegro (293.5 km). On the other hand, Slovenia and Bosnia and Herzegovina have access to the Adriatic Sea in lengths of 43 km and 21.2 km, respectively. Regardless of the length of the Adriatic coast, the economic importance of the countries with Adriatic Sea access is large.

Since seawater is beyond the scope of this book and this chapter is focused on inland waters, below, in subsections, a brief overview of the inland water resources of these countries is given.

2.1 Inland Waters of Slovenia and Pollution Pressures

Slovenia is characterized by a large variety of inland water forms—rivers, streams, natural and artificial lakes, swamps, and marshes (Kolbezen 1998). The total river network combined length of nearly 28,000 km, and the longest river is the Sava, whose Slovenian part is 221 km long. Almost half of the total area of standing water is artificial reservoirs (Hrvatin et al. 2020).

Slovenia is the most industrialized part of the former Yugoslavia with an industrial production of about 35% of the gross domestic product (Zelenakova et al. 2020). Slovenia also has advanced agriculture. This led to the impact on the quality of Slovenian waters in the second half of the twentieth century. At the end of the eighties of the twentieth century, the number and length of excessively polluted parts of watercourses directly exposed to pollution increased. During the 1990s, the situation stabilized and improved thereafter (Bat and Uhan 2004).

2.2 Inland Waters of Croatia and Pollution Pressures

There are two river basins in Croatia—the continental part of the country belongs to the Danube River Basin, and the coastal area is part of the Adriatic Sea Basin. The largest rivers in Croatia are the Danube, Sava, and Drava. In the continental part of

Croatia, there are not many natural lakes, but there are many multi-purpose artificial lakes/reservoirs.

Water quality pressures can be marked as diffuse and point pollution from different sources. The main point pollution sources are sewage systems that discharge sewage water directly into the environment, while the most common diffuse pollution sources are wastewater from settlements without sewage systems, agricultural fields, roads, and farms. Agricultural activities and farming are stressed as the most significant diffuse pollution across the continental part of the country (Tadić et al. 2020).

2.3 *Inland Waters of Bosnia and Herzegovina and Pollution Pressures*

In Bosnia and Herzegovina, two river basins are the Black and Adriatic Sea basins. The largest rivers Sava, Una, and the Drina are at the same time natural borders of the country (more than half the country's borders). There are several natural lakes in Bosnia and Herzegovina and nearly 30 artificial reservoirs, mainly made for hydropower production and flood protection, as well as for the drinking water supply and irrigation (Hadžić and Imamović 2020).

Pollution is highest in the lower part of the Sava, Bosna, and Vrbas rivers. Before the war, Bosnia and Herzegovina was the industrial heartland of the pre-existing Federal Republic of Yugoslavia and most of the rivers were severely polluted by industrial wastewater discharges (World bank 2003). After the war, existing sewage infrastructure and facilities for the purification of wastewater from industry were significantly damaged. Since the Dayton Agreement, Bosnia and Herzegovina is a decentralized state with two entities (the Federation of Bosnia and Herzegovina and the Republic of Srpska) which politically and economically further complicates both monitoring and water management. The same problem is with transboundary rivers Sava, Una, and the Drina.

2.4 *Inland Waters of Serbia and Pollution Pressures*

The waters of Serbia include rivers, lakes, and reservoirs and gravitate to three seas: the Black Sea, the Adriatic Sea, and the Aegean Sea. The largest river in the territory of Serbia is the Danube River with the significant tributaries Tisa, Sava, and Velika Morava. Velika Morava is also the largest river with the entire river course on the territory of Serbia. The largest natural lakes are situated in the northern province of Vojvodina. A lot of reservoirs are built after World War II for water supply and hydropower production.

In Serbia, according to CEDEF (2015), 5–10% of wastewater is processed. On the other hand, more than 50% of industrial facilities have no water purification systems before discharge into the recipient (Sunjog et al. 2019). Occasional accidental pollution also happens commonly (Sakan et al. 2007; Teodorović 2009).

2.5 *Inland Waters of Montenegro and Pollution Pressures*

Water resources classify Montenegro as a country rich in water. The waters of Montenegro belong to two watersheds, the Black Sea, and the Adriatic Sea watershed. The most important rivers are Tara, Morača, Lim, Piva, and Zeta. Bojana River is ranked third in the Mediterranean by the amount of water that is released into the Adriatic Sea. Skadar Lake is the largest lake in the Balkans, shared between Montenegro and Albania (Pešić et al. 2018).

Besides natural pressures, the quality of freshwater ecosystems in Montenegro is affected by various anthropogenic activities—tourism, traffic, chemical and pharmaceutical industry, mines and quarries, agriculture and intensive livestock, meat and food industry, wood and stone processing, etc. (Kračun-Kolarević et al. 2019). The major point source of pollution in Montenegro is the discharge of municipal and industrial wastewater, without wastewater treatment, directly into the water bodies (Vukašinović-Pešić et al. 2019).

2.6 *Inland Waters of North Macedonia and Pollution Pressures*

The water resources in North Macedonia are unequally distributed, so the western part of the country is richer. The river basin is separated between the Aegean basin, the Adriatic basin, and the Black Sea basin. The most important rivers are Vardar, Crni Drim, Treska, Crna reka, Pčinja, and Bregalnica. The most significant water resources are the tectonic natural lakes Ohrid, Prespa, and Dojran. All three lakes are transboundary. Ohrid Lake lies two third in Macedonia and one-third in Albania. North Macedonia shares Prespa Lake with Albania and Greece, while Dojran Lake is shared only with Greece.

Wastewater treatment is almost non-existing in North Macedonia. As a result, downstream of densely populated areas, as a result of untreated municipal and industrial wastewater discharges, pollution of rivers is high. Agriculture has been reported as a significant polluter of water resources (World Bank 2003). Still, there is no regular monitoring of water quality or industrial discharges.

2.7 Inland Waters of Albania and Pollution Pressures

Albania is rich in water resources like rivers, lakes, and artificial reservoirs. Albania has a dense network of rivers belonging to the Black Sea, the Adriatic Sea, and the Aegean Sea. The largest river is Drim. Other important rivers are Bojana, Erzen, Škumbin, Semani, and Vjosa. The most important water resources are the transboundary lakes Skadar, Ohrid, and Prespa.

The main problem of water pollution is the discharge of urban and industrial wastewater without treatment directly into rivers and lakes. Another problem is leaching wastewaters from solid wastes of urban and industrial solid wastes, often located close to rivers and lakes. Monitoring is irregular and the information is often not reliable or even contradictory (World Bank 2003). The present state of water pollution in Albania is a real risk to the economy and human health, and the implementation of wastewater treatment and waste disposal management should be of the highest priority for the Albanian government (Cullaj et al. 2005).

3 Inland Waters for Commercial Fishing in Countries of Western Balkan

Marine commercial fishing is allowed in all countries with access to the Adriatic Sea (UNEP-MAP-RAC/SPA 2014). On the other hand, in Slovenia inland commercial fishing is prohibited and only sports and recreational fishing are allowed (FAO 2022).

Figure 1 shows inland commercial fishing waters in the countries of the Western Balkans. When it comes to Croatia, Bosnia and Herzegovina, and Serbia commercial fishing is limited to large rivers. In Croatia, commercial inland fishing is allowed only in the Danube River and lower parts of the Sava River. The commercial inland fishery has low economic value and a small impact on the economy in Croatia, so it will gradually be replaced by recreational fisheries. When it comes to Bosnia and Herzegovina, commercial inland fishing is allowed on the Sava River. Commercial fishing in Serbia is allowed on the Danube, Sava, and Tisa rivers.

In Montenegro, North Macedonia, and Albania commercial fishing is allowed in large transboundary lakes. In Montenegro commercial fishing is limited to Skadar Lake. In Albania commercial fishing takes place in Skadar, Ohrid, and Prespa lakes. Commercial fishing in North Macedonia is allowed in Ohrid, Prespa, and Dojran lakes.

Fig. 1 Inland commercial fishing waters in the countries of the Western Balkans (Danube, Sava, and Tisza rivers; Skadar, Ohrid, Prespa, and Dojran lakes)

3.1 Danube River: Main Characteristics and Anthropogenic Pressures

Danube River, the second largest river in Europe, just behind the Volga, forms one of the continent's largest and most important water systems. It is a multinational river basin, shared by nineteen countries: Albania, Austria, Bosnia and Herzegovina, Bulgaria, Croatia, Czech Republic, Germany, Hungary, Italy, Macedonia, Moldova, Poland, Romania, Serbia, Montenegro, Slovakia, Slovenia, Switzerland, and Ukraine. Western Balkan countries are in the middle sector of the Danube River, and here the river receives its largest tributaries, substantially increasing its flow—Drava, Tisza, and Sava (Gâştescu 2017). The Danube River basin has high environmental, social, and economic value, including numerous important natural areas, major resources for drinking water supply, agriculture, industry, fishing, tourism and recreation, power generation, and river navigation.

The economic development in the countries of the Danube region brought both improvements in life quality and a threat to the environment and the Danube River itself. Urban settlement development, an increase in industrial activities, and extensive agriculture are potential sources of pollution if not properly managed and can have a negative impact on the water quality and aquatic life. Danube River passes through numerous large cities, including four capitals Vienna, Bratislava, Budapest, and Belgrade, responsible for a large number of hazardous substances from industry and untreated or insufficiently treated wastewater. Industrial development led to chemical stress factors, among which the most important are heavy metals. Danube River is the main waterway in Europe, so transport also represents a significant pressure on the river. Recently, emerging contaminants facing the Danube River are microplastics and pharmaceutical waste (Mănoiu and Crăciun 2021).

Although it flows through many countries, the aquatic habitats of the watershed can be considered as part of a single system so harmful activities in one section always affect the other sections (Galatchi and Tudor 2007). This obliges all countries, whether members of the European Union or associated countries, to make considerable efforts to diminish the pollution of the whole Danube flow. Since 1998, water management in the Danube River basin has been overseen by the International Commission for the Protection of the Danube River (ICPDR). So far, monitoring of the water quality of the Danube River has been conducted through four JDS (Joint Danube Survey) in 2001, 2007, 2013, and 2019 (Simionov et al. 2021).

3.2 Sava River: Main Characteristics and Anthropogenic Pressures

The Sava River springs in Slovenia and flows a distance of 945 km through Croatia, Bosnia and Herzegovina, and Serbia while it flows into the Danube in Belgrade (Zuliani et al. 2019). The Sava River basin covers significant areas of these countries, where it has historically been a vital freshwater resource for the general population, agriculture and industrial development, a hydropower generator, a relevant communication route, and a tourist attraction (International Sava River Basin Commission 2008).

Sava River flows through the capital cities of Slovenia, Croatia, and Serbia (Ljubljana, Zagreb, and Belgrade), with an overall of 8.2 million population in the basin. According to the International Sava River Basin Commission (International Sava River Basin Commission 2016), it is under significant anthropogenic pressure. Komatina and Grošelj (2015) emphasize that the upper reaches of the Sava River are largely influenced by carbonate mineral weathering, the middle by agricultural activity and biological processes related to eutrophication, while the lower reaches are influenced mostly by stressors related to high pollution from industrial processing along with untreated municipal wastewater discharges. Milačič et al. (2010) found that the Sava River was a moderately polluted European river with hotspots of Cr and Ni contamination in areas affected by metal and steel production industries and hotspots of Hg contamination in areas affected by chemical and Chlor-alkali industries. The same authors emphasize that concentrations of PTE increased downstream.

Slovenia and Croatia have reported on the establishment of the monitoring program in line with the principles of the EU Water Frame Directive (WFD 2000/

60/EC) while the other Sava countries Bosnia and Herzegovina and Serbia are in the development phase (International Sava River Basin Commission 2016).

3.3 Tisza River: Main Characteristics and Anthropogenic Pressures

With respect to its length of 966 km, the Tisza River forms the largest tributary of the Danube. The catchment of the Tisza River is shared by five countries: Slovakia, Ukraine, Romania, Hungary, and Serbia. The lower sector of the Tisza River lies in the territory of Serbia. As an international river, the Tisza has great economic importance in this part of Europe.

Anthropogenic influence causes permanent pollution of the Tisza River, whether it originates from communal, industrial, or agricultural activities (Sakan et al. 2009). In the middle course of the Tisza River, in the Pannonian part of the catchment area, agriculture is developed. On the other hand, the main industrial plants are in Romania and Hungary, although they also exist in Ukraine, Slovakia, and Serbia. In Romania, numerous severe industrial accidents happened at the beginning of the twenty-first century—e.g., cyanide spills at Baia Mare and heavy metal contamination at Baia Borsa. Sources of inorganic pollution in Serbia are wastewater from factories and municipal sewage discharges (Sakan et al. 2009). Considering the fact that the Tisza River is part of the Danube basin, international activities on the protection of the Tisza River take place within the framework of the International Commission for the Protection of the Danube River (ICPDR).

3.4 Skadar Lake: Main Characteristics and Anthropogenic Pressures

Skadar Lake is the largest transboundary lake in the Balkans, shared by Montenegro and Albania. Lake represents a hot spot of biodiversity in the Western Balkan (Pešić et al. 2018). Fishes are the most significant natural resource in the lake, in terms of contribution to the local economy. Montenegro and Albania take part in commercial fishing on Skadar Lake. Besides its economic importance for neighboring countries, Skadar Lake is also important from a larger regional perspective, considering its particular hydrological and ecological values, taking into account the connection of the lake with Ohrid and Prespa lakes through the Drim River and with the Adriatic sea through the Bojana River (CETI and APAVA 2007).

The anthropogenic pressure on Lake Skadar is significant. On the Albanian side of the lake, untreated wastewater from Skadar city is discharged directly into the lake. The main diffuse sources are domestic wastewater from households not connected to the sewer system and agriculture (Vukašinović-Pešić et al. 2019). In

Montenegro, the main pressures are wastewater from Podgorica, Nikšić, Cetinje, Danilovgrad, and Rijeka Crnojevića which are discharged via the tributaries of Skadar Lake, Morača, Zeta, and Crnojevića rivers. The main industries are located in the watershed—the steel factory in Nikšić and the aluminum plant in Podgorica (Vukašinović-Pešić et al. 2019).

Currently, there is no permanent monitoring of the lake (Pešić et al. 2018). In order to prevent contamination pressures, an effective bilateral management structure based on relevant Albanian and Montenegrin institutions should be formed (Pešić et al. 2018).

3.5 Ohrid Lake: Main Characteristics and Anthropogenic Pressures

Ohrid Lake is located on the Balkan Peninsula between Albania and North Macedonia. Formed between 3.5 and 4 million years ago, it is considered one of the oldest lakes in the world (Popovska and Bonacci 2007). There are numerous endemic freshwater organisms in Ohrid Lake, for example, ten of the 17 identified fish species are endemic (Watzin et al. 2002). Because of its high biodiversity and endemism, as well as its unique cultural heritage, Lake Ohrid and its surroundings are not only of local, but international significance.

The most significant point sources of pollution in Albania are the discharge of treated wastewater from the Pogradec wastewater treatment plant and the discharge of untreated wastewater from villages and settlements which are not connected to a central sewerage system (GIZ 2017). In Macedonia, the majority (approx. 80%) of the settlements in the watershed of the lake are now connected to sewerage systems and treatment plants (Watzin et al. 2002). However, there are point sources of pollution in the Ohrid Lake basin—wastewaters from sparsely built-up areas and direct industrial discharges. In Albania, the water quality of Lake Ohrid is affected by agriculture and industrial activities, especially in the metallurgical, chemical, and mining industries (coal mines and highly toxic chromium mines due to the high content of heavy metals and nickel availability). In Macedonia, pollution from agriculture is concentrated in the plains north of the lake and along the watersheds of its main tributaries (GIZ 2017).

3.6 Prespa Lake: Main Characteristics and Anthropogenic Pressures

Prespa is a transboundary lake as well. Lake Prespa consists of two interlinked lakes—Macro Prespa and Micro Prespa. Water from Micro Prespa flows into Macro Prespa via a weir, which is used to artificially control its surface water level. The catchment is shared between Macedonia, Albania, and Greece. The water resources of the Prespa are used for different purposes: water supply, industrial use, and irrigation of agricultural regions in all three countries. In addition to agriculture, tourism and fisheries are of economic importance in the lake basin (UNDP 2014).

No significant point sources of pollution exist in the Albanian part of the Prespa Lake basin. On the other hand, in the Macedonian part of Prespa Lake, the wastewater pressure comes from 16,825 inhabitants living in 44 settlements in the basin. Industrial pollution originates from several small to medium-sized plants. Nonpoint pollution sources from agriculture and untreated wastewater from settlements have a less harmful impact on the lake in Albania than in Macedonia. Lastly, illegal landfills and dumps in the villages, especially near the shoreline, are potential sources of diffuse pollution (GIZ 2017).

3.7 Dojran Lake: Main Characteristics and Anthropogenic Pressures

Dojran Lake is located in the southeastern part of the Balkan and it is shared by Macedonia and Greece. Of three natural lakes in North Macedonia, Dojran Lake is the smallest, but not less significant water resource. The watershed of the lake belongs to the river basin of the Vardar which gravitates toward Aegean See. Dojran Lake has a very rich biodiversity. The main economy in the region is tourism and fishery.

There is insufficient scientific baseline data for the condition of Lake Dojran and the clear connection between the contribution of human and natural factors to the current state of the lake (Popovska et al. 2005). The anthropogenic pressure has not changed in the last 20 years, but the significantly reduced water level and the lack of mechanisms of self-purification have led to the hyper-pollution of Lake Dojran. This implies that it is necessary to establish a permanent monitoring system, including the measurement of pesticides and heavy metals, in order to have constant insight into the status of the lake's system and take preventive measures (Popovska and Bonacci 2008).

4 Pollution of Fishing Waters of the Western Balkan: Potentially Toxic Elements in Fish Species

In inland waters, the problem of overfishing is not as great as in marine fisheries, but environmental degradation is the biggest problem facing sustainable development (Arlinghaus et al. 2002). Inland fisheries face numerous threats, including alternative freshwater use (e.g., hydropower, irrigation, and municipal water supply), and high vulnerability to effects of land-based activities such as runoff and pollution (Youn et al. 2014).

Globally, freshwater fish are the most heavily exploited aquatic resource, representing about 20 million tons per year of the world's aquatic food per year (FAO 2008). Fish are also considered to be one of the most sensitive aquatic organisms to toxic substances found in water. Among numerous pollutants, in this chapter, we discuss pollution of commercial fishing waters of Western Balkan with PTEs based on bioaccumulation in fish tissues.

4.1 Methodology

The objective of this chapter is to collect and evaluate all available literature on the PTEs in fish species from commercial fishing waters of Western Balkan countries. The review focuses on studies from Croatia, Bosnia and Herzegovina, Serbia, Montenegro, Albania, and North Macedonia. Slovenia is neglected because inland commercial fishing is not allowed.

Several international databases, including Scopus, PubMed, Web of Science, and Science Direct were searched for publications from 1 January 2000 to 1 July 2022 to retrieve the related articles. We searched for articles using the keywords "heavy metals" OR "metals" OR "elements" OR "trace metals" OR "toxic elements" OR "toxic metal" OR "potentially toxic elements" AND "fish" AND "the name of the ecosystem (Danube River, Sava River, Tisa River, Skadar Lake, Ohrid Lake, Prespa Lake, and Dojran Lake)" AND "the name of the country (Croatia, Bosnia and Herzegovina, Serbia, Montenegro, Albania, and North Macedonia)." After the primary screening, all eligible articles in English were downloaded. Moreover, the reference articles were screened to retrieve more articles.

The inclusion criteria were: (1) full text published in the English language; (2) research performed in Western Balkan countries; (3) papers on fish species from commercial fishing waters; (4) reporting of the concentration of PTEs (mean and/or range). Review articles and books, letters to editors, and conference papers were excluded due to a lack of a peer-review process.

Extracted data from the articles included the name of the ecosystem and the name of the country; examined fish species and fish tissue; a list of assessed PTEs with elements above the maximum permitted concentrations (MPCs) in fish meat for utilization in the human diet reported by authors, and references.

4.2 Potentially Toxic Elements in Fish Species

Guided by the goal of this chapter to identify research studies that contain information about PTEs in fish species in commercial fishing waters in Western Balkan, 35 separate studies were identified. Summary information about these studies is presented in Table 1, including the name of the ecosystem and the name of the country; examined fish species and fish tissue; a list of assessed PTEs with elements above the MPCs in fish meat for utilization in the human diet reported by authors, and references. Among the commercial fishing waters in the Western Balkans, most studies on PTEs in fish tissues have been conducted in the Danube and Sava rivers. When it comes to lakes where commercial fishing is allowed, only individual studies were found.

In the largest number of studies, a total of 16 studies, potentially toxic elements in fish tissues in the flow of the Danube through Serbia were studied, while one study was conducted in the Croatian part of the Danube. Almost all research on PTEs in fish in the Danube River in Serbia was conducted in the vicinity of Belgrade, where the Danube receives all wastewater without any wastewater treatment (Mănoiu and Crăciun 2021). A study by Zrnčić et al. (2013) covered the entire flow of the Danube through Croatia. According to the latter review, the concentration of Hg above MPC in the Croatian part of the Danube River, and concentrations of Hg, Cd, Ni, and As above the MPC in the Serbian part of the Danube River were recorded (Table 1). This is in line with the statement of Pavlović et al. (2016) that several locations along the Danube River are affected by pollution, particularly in industrial and agricultural areas, in the middle and lower part of the Danube. Additionally, Rico et al. (2016) further emphasized that the Danube River is primarily polluted with heavy metals (PTEs). Concentrations of essential elements Fe and Zn above the MPCs were detected in fish species in the Danube as well. Jarić et al. (2011) detected elevated concentrations of Fe in the entire flow of the Danube through Serbia, while in the study of Subotić et al. (2013a), concentrations of Zn above the MPC were detected only in fish species in the flow in the vicinity of Belgrade (Table 1). These data are very important because these trace elements are biologically essential, but in high concentrations, they can cause harmful effects.

The presence of PTEs in fish in the Sava River was investigated in 13 studies. Six studies were based on fish species in the flow through Croatia, five in the Serbian part of the Sava, and two studies covered the entire flow of the Sava River, from the spring to the confluence with the Danube. According to the results of this review, the concentration of Hg above the MPC was recorded in fish species from the Sava River (Table 1). Bošnir et al. (2003) found elevated concentrations of Hg in fish species from the Sava River in the vicinity of Zagreb. In the study by Zuliani et al. (2019), elevated concentrations of Hg were registered in the fish species in the Sava River in the vicinity of Zagreb and downstream. This is in line with Milačič et al. (2010) who emphasized that the Sava River was a moderately polluted river, with an increase in the PTE concentrations downstream, with hotspots of Hg contamination. Komatina and Grošelj (2015) also pointed out that the middle section of the Sava River is under the influence of agricultural activities, while the lower part is mostly affected by industrial pollution and untreated municipal wastewater.

PTEs in fish species from the Serbian part of the Tisa River were investigated only in three studies. The Tisza River has been contaminated in the past by numerous industrial wastes from the Carpathian mountain region in Romania, which has a long tradition of mining, especially Au, Ag, Pb, Zn, Cu, Cd, and Mn (Triebskorn et al.

Table 1 Summary of the studies reporting concentrations of PTEs found in fish species in Western Balkan countries (in chronological order)

Name of ecosystem	Species/tissue	PTEs	Reference
Sava River/Croatia	*Leucius leucius* L., *Rutilus rutilus* L., *Abramis brama* L., *Barbus barbus* L., *Carassius auratus gibelio Bloch, Ictalurus nebulosus Lesneu*	Pb, Cd, As, **Hg**[a]	Bošnir et al. (2003)
Sava River/Croatia	*Leuciscus cephalus* L./gills	Zn, Fe, Cu, Mn, Cd	Dragun et al. (2007)
Sava River/Croatia	*Squalius cephalus* L./gills	Cd, Zn, Cu, Fe, Mn	Dragun et al. (2009)
Sava River/Croatia	*Squalius cephalus* L./liver	Zn, Cu, Fe, Mn, Cd	Podrug et al. (2009)
Danube River/Serbia	*Alosa immaculata*/muscle, liver, gills	Al, **As**, **Cd**, Cu, B, Ba, Fe, Mg, Sr, Zn, Li, Co, Cr, Mn, Mo, Ni, Pb	Visnjic-Jeftic et al. (2010)
Danube River/Serbia	*Acipenser ruthenus*/muscle, gills, liver, intestine	Ag, Al, As, B, Ba, **Cd**, Co, Cr, Cu, **Fe**, Mn, Mo, Ni, Pb, Se, Sr, Zn, Li	Jarić et al. (2011)
Danube River/Serbia	*Barbus barbus*/muscle, liver, gills, gonads	Al, As, B, Ba, Cd, Co, Cr, Cu, Fe, Li, Mn, Mo, Ni, Pb, Sr, Zn	Sunjog et al. (2012)
Danube River/Serbia	*Hypophthalmichthys molitrix, Abramis brama, Blicca bjoerkna, Cyprinus carpio, Silurus glanis*/muscle, gills, liver, gonads	Al, As, B, Ba, Cd, Co, Cr, Cu, Fe, Li, Mn, Mo, Ni, Pb, Se, Sr, Zn	Lenhardt et al. (2012)
Ohrid Lake/Albania	*Salmo letnica, Salmothymus ohridanus*/gill, muscle, liver	Cr, Co, Ni	Malaj et al. (2012)
Sava River/Croatia	*Squalius cephalus* L./gut, gastrointestinal tissue	Zn, Cu, Fe, Mn, Cd	Filipović Marijić and Raspor (2012)
Danube River/Serbia	*Sander lucioperca, Silurus glanis, Cyprinus carpio, Neogobius gymnotrachelus* and *Neogobius melanostomus*/muscle, liver, whole-body composite (in the case of gobies)	As, Cu, Fe, **Hg**, Mn, **Zn**	Subotić et al. (2013a)
Tisza River/Serbia	*Esox lucius, Acipenser ruthenus, Brama brama, Cyprinus carpio*/liver	Al, As, Cd, Co, Cr, Cu, Fe, Hg, Mn, Ni, Pb, Se, Sr, Zn	Štrbac et al. (2013)

(continued)

Table 1 (continued)

Name of ecosystem	Species/tissue	PTEs	Reference
Danube River/Serbia	*Sander lucioperca, Silurus glanis, Lota lota, Cyprinus carpio*/muscle, liver, gills	Al, As, B, Ba, Cd, Co, Cr, Cu, Fe, Hg, Li, Mn, Mo, Ni, Pb, Se, Sr, Zn	Subotić et al. (2013b)
Danube River/Croatia	*Chondrostoma nasus, Ctenopharyngodon idella, Leuciscus idus, Alburnus alburnus, Rutilus rutilus, Carassius gibelio, Cyprinus carpio, Abramis brama, Aspius aspius, Esox lucius, Sander lucioperca, Silurus glanis, Lota lota, Hypophthalmichthys molitrix, Lota lota*/muscle	**Hg**, As, Cd, Pb	Zrnčić et al. (2013)
Danube and Sava rivers/ Serbia	*Barbus barbus, Lota lota, Cyprinus carpio, Abramis brama, Sander lucioperca, Hypophthalmichthys molitrix, Acipenser ruthenus, Silurus glanis, Blicca bjoerkna*/muscle, gills, liver, gonads	Al, As, B, Ba, **Cd**, Co, Cr, Cu, Fe, **Hg**, Li, Mn, Mo, Ni, Pb, Se, Sr, **Zn**	Jovičić et al. (2014)
Sava River/Croatia	*Squalius cephalus* L./gastro-intestinal tissue	Zn, Fe, Mn, Cd, Ag	Filipović Marijić et al. (2014)
Danube River/Serbia	*Silurus glanis*/muscle, gills, spleen, liver, kidneys, intestine, gizzard, heart, brain, gallbladder, swim bladder, vertebra, operculum, gonads	As, Cd, Co, Cr, Cu, Fe, Hg, Mn, Ni, Pb, Se, Zn	Jovičić et al. (2015)
Danube River/Serbia	*Barbus barbus, Acipenser ruthenus*/muscle, gills, liver	As, B, Ba, Cd, Co, Cu, Fe, **Hg**, Li, Mn, Mo, **Ni**, Pb, Se, Sr, Zn	Rašković et al. (2015)
Tisza River/Serbia	*Esox lucius, Acipenser ruthenus, Brama brama, Cyprinus carpio*/liver, gills, gonads, brain	Al, As, B, Cd, Co, Cr, Cu, Fe, Hg, Mn, Ni, Pb, Se, Sr, Zn	Štrbac et al. (2015)
Danube River/Serbia	*Pelecus cultratus, Gymnocephalus cernua, Perca fluviatilis*/muscle, liver, gills, gonads	Al, As, B, Ba, Cd, Co, Cr, Cu, Fe, **Hg**, Li, Mn, Mo, Ni, Pb, Se, Sr, Zn	Subotić et al. (2015)
Danube River/Serbia	*Hypophthalmichthys molitrix, Cyprinus carpio, Silurus glanis*/muscle, digestive tract, liver	Pb, Cd, Hg, As	Milanov et al. (2016)

(continued)

Table 1 (continued)

Name of ecosystem	Species/tissue	PTEs	Reference
Danube River/Serbia	*Hypophthalmichthys molitrix, Cyprinus carpio, Silurus glanis*/muscle, digestive tract, liver	As, Cd, **Hg**, Pb	Ivanović et al. (2016)
Danube River/Serbia	*Barbus barbus*/gills, muscle, intestine, liver	Al, As, B, Ba, Co, Cr, Cu, Hg, Mn, Mo, Ni, Pb, Se, Sr, Zn	Morina et al. (2016)
Sava River/Croatia, Bosnia and Herzegovina, and Serbia	*Alburnus alburnus*/whole-body composite	Al, As, Cd, Co, Cr, Cu, Fe, Hg, Mn, Ni, Pb, Se, Sn, Zn	Kolarevic et al. (2016)
Danube, Sava, and Tisa rivers/Serbia	*Sander lucioperca, Silurus glanis, Abramis brama, Barbus barbus, Chondrostoma nasus, Squalius cephalus*/muscle	Al, As, **Cd**, Co, Cr, Cu, Fe, **Hg**, Mn, Ni, Pb, Zn	Milošković et al. (2016)
Danube River/Serbia	*Carassius gibelio, Barbus barbus, Abramis brama, Cyprinus carpio, Stizostedion lucioperca, Silurus glanis*/muscle	Pb, **Cd**, Hg, Cu, Fe, Zn, As	Jovanović et al. (2017)
Sava River/Serbia	*Abramis brama*/muscle, gills, liver	Al, As, B, Ba, Cd, Co, Cr, Cu, Fe, Li, Mn, Mo, Ni, Pb, Sr, Zn	Kostić et al. (2017)
Sava River/Serbia	*Alburnus alburnus*/whole-body composite	Al, As, Cd, Co, Cr, Cu, Fe, Hg, Mn, Ni, Pb, Se, Sn, Zn	Jovanović et al. (2018)
Skadar Lake/Montenegro	*Scardinus knezevici, Alburnus scoranza, Cyprinus carpio, Rutilus prespensis, Anguilla anguilla, Perca fluviatilis*/muscle	Zn, Mn, Cu, Fe, Cr, Ni, Co, Hg, Pb, Cd, As	Rakočević et al. (2018)
Skadar, Ohrid, Prespa and Dojran lakes/Montenegro, North Macedonia	*Alburnus scoranza* (Ohrid, Skadar), *Alburnus belvica* (Prespa), *Alburnus macedonicus* (Dojran)/whole-body composite	Al, As, **Cd**, Co, Cr, Cu, Fe, Hg, Mn, Ni, **Pb**, Se, Sn, Zn (Prespa Lake)	Milošković et al. (2018)
Sava River/Croatia, Bosnia and Herzegovina, and Serbia	*Squalius cephalus, Rutilus rutilus, Esox lucius, Barbus barbus, Sander lucioperca*/muscle tissue	As, Zn, Cu, Cd, Cr, Pb, Ni, **Hg**	Zuliani et al. (2019)
Danube River/Serbia	*Chondrostoma nasus, Vimba vimba*/muscle, liver	Al, As, B, Ba, Cd, Co, Cr, Cu, Fe, Hg, Li, Mn,	Subotić et al. (2019)

(continued)

Table 1 (continued)

Name of ecosystem	Species/tissue	PTEs	Reference
		Mo, Ni, Pb, Sr, Zn,	
Sava River/Serbia	*Alburnus alburnus*/muscle tissue	Al, As, Cd, Co, Cr, Cu, Fe, Hg, Mn, Ni, Pb, Se, Sn, Zn	Jovanović Marić et al. (2020)
Danube River/Serbia	*Abramis brama*/muscle, gills, liver, gonads	Al, Fe, Cr, As, Sr, Mn, Cd, Zn, Mo, Cu, Li, Ni, B, Co, Pb, Ba	Kostić-Vuković et al. (2021)
Danube River/Serbia	*Vimba vimba*/muscle, liver	Al, As, B, Ba, Ca, Cd, Co, Cr, Cu, Fe, Hg, K, Li, Mg, Mn, Mo, Na, Ni, P, Pb, S, Se, Si, Sr, Zn	Subotić et al. (2021)

[a]Data in bold—PTEs above the MPCs reported by the authors

2008). None of the three studies carried out in the Serbian course of the Tisza River reported elevated concentrations of PTEs in fish (Table 1). However, in a study by Štrbac et al. (2013) elevated concentrations of Pb and Cd were found in the liver of the examined fish, which indicates the need for constant monitoring in the future.

Transboundary lakes used for commercial fishing in Western Balkan—Skadar, Ohrid, Prespa, and Dojran lakes have been very limitedly investigated for potentially toxic elements in fish. The study by Milošković et al. (2018) included all four lakes, while the studies by Malaj et al. (2012) and Rakočević et al. (2018) included Ohrid and Skadar, respectively. According to Malaj et al. (2012), there were no elevated elements in the fishes *Salmo letnica* and *Salmothymus ohridanus* in Ohrid Lake. The first and only study of Skadar Lake by Rakočević et al. (2018) suggested that, according to the concentrations of PTEs in fish muscle, Skadar Lake could be still regarded as an unpolluted ecosystem and the consumption of fish from Skadar Lake not pose a risk to human health. The study by Milošković et al. (2018) confirmed this and highlighted elevated concentrations of Cd and Pb in fish species in Prespa Lake. The presence of toxic Pb in Prespa Lake water may originate from several medium-sized enterprises from industrial branches, including food processing, poultry farming, textiles, metal processing, wood processing, and the chemical industry (Krstic 2012). In the review of metal pollution in the aquatic biota of Lake Skadar, including fish, by Vukašinović-Pešić and Blagojević (2018), an integrated approach combining the analysis of heavy metals in water, sediment, and the aquatic environment was proposed.

5 Conclusions

This chapter provides basic information about commercially important fishing waters in the Western Balkan countries and an overview of the research studies on PTEs in fish species in these waters over the last two decades. Analysis of available literature showed significant pollution by PTEs (Fe, Za, Hg, As, Cd, and Ni) in the fish species in the Danube River. Elevated concentrations of Hg in fish species in the Sava River downstream from Zagreb were recorded. Cd and Pb concentrations above the MPC in fish meat were determined in fish species in Prespa Lake. Given the social issues that have characterized these developing countries throughout history, significant impacts on water resources were expected.

Overall, this review has shown that research regarding this issue has been limited over the last two decades. Considering that almost all waters used for commercial fishing in the Western Balkans are transboundary, cooperation between countries is necessary in terms of monitoring PTE in fish from these waters. In the future, to fill the gap in the fragmented research related to the PTEs of these waters, it is important to make available more data regarding potentially toxic elements in fish.

Acknowledgment The investigation was supported by Grant (Agreement No. 451-03-68/2022-14/200378) funded by the Serbian Ministry of Education, Science and Technological Development.

References

Ardeshir RA, Movahedinia A, Rastgar S (2017) Fish liver biomarkers for heavy metal pollution: a review article. Am J Toxicol 2:1–8

Arlinghaus R, Mehner T, Cowx GI (2002) Reconciling traditional inland fisheries management and sustainability in industrialized countries, with emphasis on Europe. Fish Fish 3:261–316. https://doi.org/10.1046/j.1467-2979.2002.00102x

Bat M, Uhan J (2004) Vode (Waters). In: Lovrenčak F, Pavlovec R (eds) Narava Slovenije. Mladinska knjiga, Ljubljana, pp 103–145

Bošnir J, Puntarć D, Škes I et al (2003) Toxic metals in freshwater fish from the Zagreb area as indicators of environmental pollution. Coll Antropol 27(1):31–39

CEDEF—Central European Development Forum (2015) Usage and treatment of communal and industrial wastewaters in Serbia. http://www.cedeforum.org

CETI, APAWA (2007) The strategic action plan (SAP) for Skadar Lake, Albania & Montenegro. World Bank and Global Environment Facility (GEF), Washington, DC, 82 pp. Retrieved from https://iwlearn.net/resolveuid/b8fab0fe-f54b-4113-814e-850a52eefc04

Cullaj A, Hasko A, Miho A et al (2005) The quality of Albanian natural waters and the human impact. Environ Int 31:133–146. https://doi.org/10.1016/j.envint.2004.06.008

Dragun Z, Raspor B, Podrug M (2007) The influence of the season and the biotic factors on the cytosolic metal concentrations in the gills of the European chub (*Leuciscus cephalus* L.). Chemosphere 69(6):911–919. https://doi.org/10.1016/j.chemosphere.2007.05.069

Dragun Z, Podrug M, Raspor B (2009) Combined Use of Bioindicators and Passive Samplers for the Assessment of River Water Contamination With Metals. Arch Environ Contam Toxicol 57: 211–220. https://doi.org/10.1007/s00244-008-9264-9

Dudgeon D, Arthington AH, Gessner MO et al (2006) Freshwater biodiversity: importance, threats, status and conservation challenges. Biol Rev 81:163–182. https://doi.org/10.1017/S1464793105006950

FAO (2008) The state of world fisheries and aquaculture. FAO, Rome

FAO (2022) Slovenia. Text by Bunič, P. Fisheries and Aquaculture Division (online). Rome (cited Sunday, September 4th, 2022). https://www.fao.org/fishery/en/countrysector/naso_slovenia

Filipović Marijić V, Raspor B (2012) Site-specific gastrointestinal metal variability in relation to the gut content and fish age of indigenous European chub from the Sava River. Water Air Soil Pollut 223:4769–4783. https://doi.org/10.1007/s11270-012-1233-2

Filipović Marijić V, Vardić Smrzlić I, Raspor B (2014) Does fish reproduction and metabolic activity influence metal levels in fish intestinal parasites, acanthocephalans, during fish spawning and post-spawning period? Chemosphere 112:449–455. https://doi.org/10.1016/j.chemosphere.2014.04.086

Funge-Smith S, Bennett A (2019) A fresh look at inland fisheries and their role in food security and livelihoods. Fish Fish 20:1176–1195. https://doi.org/10.1111/faf.12403

Galatchi LD, Tudor M (2007) Europe as a source of pollution-the main factor for the eutrophication of the Danube Delta and Black Sea. In: NATO security through science series C: environmental security, pp 57–63. https://doi.org/10.1007/978-1-4020-5098-5_5

Gâştescu P (2017) The Danube-hydrographic polarisation European axis. State-of-the-art. Riscurişicatastrofe, No. XVI 20(1):9–23. https://doi.org/10.24193/rcj2017_01

GIZ (2017) Initial characterisation of Lakes Prespa, Ohrid and Shkodra/Skadar. Implementing the EU Water Framework Directive in South Eastern Europe. Project: Conservation and sustainable use of biodiversity at Lakes Prespa, Ohrid and Shkodra/Skadar (CSBL). Bonn and Eschborn, Germany. Retrieved from https://www.giz.de/en/downloads/Initial%20Characterisation%20of%20Lakes%20Prespa_Ohrid%20and%20ShkodraSkadar%20-%20EN.pdf

Hadžić E, Imamović A (2020) Water resources in Bosnia and Herzegovina. In: Negm A, Romanescu G, Zelenakova M (eds) Water resources management in Balkan countries. Springer water. Springer, Cham. https://doi.org/10.1007/978-3-030-22468-4_8

Hrvatin M, Komac B, Zorn M (2020) Water resources in Slovenia. In: Negm A, Romanescu G, Zelenakova M (eds) Water resources management in Balkan countries. Springer water. Springer, Cham. https://doi.org/10.1007/978-3-030-22468-4_3

International Sava River Basin Commission (2008) International Sava River Basin Commission website. https://www.savacommission.org/basin_about

International Sava River Basin Commission (2016) 2nd Sava River Basin Analysis. The document is available at the ISRBC website: http://savacommission.org/publication

Ivanović J, Janjić J, Baltić M et al (2016) Metal concentrations in water, sediment and three fish species from the Danube River, Serbia: a cause for environmental concern. Environ Sci Pollut Res Int 23(17):17105–17112. https://doi.org/10.1007/s11356-016-6875-y

Jarić I, Višnjić-Jeftić Ž, Cvijanović G et al (2011) Determination of differential heavy metal and trace element accumulation in liver, gills, intestine and muscle of sterlet (*Acipenser ruthenus*) from the Danube River in Serbia by ICP-OES. Microchem J 98(1):77–81. https://doi.org/10.1016/j.microc.2010.11.008

Jovanović DA, Marković RV, Teodorović VB et al (2017) Determination of heavy metals in muscle tissue of six fish species with different feeding habits from the Danube River, Belgrade-public health and environmental risk assessment. Environ Sci Pollut Res 24(12):11383–11391. https://doi.org/10.1007/s11356-017-8783-1

Jovanović J, Kolarević S, Milošković A et al (2018) Evaluation of genotoxic potential in the Velika Morava River Basin in vitro and in situ. Sci Total Environ 621:1289–1299. https://doi.org/10.1016/j.scitotenv.2017.10.099

Jovanović Marić J, Kračun-Kolarević M, Kolarević S et al (2020) Sensitivity of bleak (*Alburnus alburnus*) in detection of the wastewater related pressure in large lowland rivers. B Environ Contam Tox 105:224–229. https://doi.org/10.1007/s00128-020-02944-4

Jovičić K, Višnjić-Jeftić Ž, Đikanović V et al (2014) Assessment of fish stocks and elemental pollution in the Danube, Sava and Kolubara rivers on the territory of the city of Belgrade, Serbia. Acta Zool Bulg 7:179–184

Jovičić K, Nikolić DM, Višnjić-Jeftić Ž et al (2015) Mapping differential elemental accumulation in fish tissues: assessment of metal and trace element concentrations in wels catfish (*Silurus glanis*) from the Danube River by ICP-MS. Environ Sci Pollut Res Int 22(5):3820–3827. https://doi.org/10.1007/s11356-014-3636-7

Khoshnood Z (2016) Using biomarkers in ecotoxicology: what and why? Focus Sci 2(2):1–2

Kolarevic S, Aborgiba M, Kracun-Kolarevic M et al (2016) Evaluation of genotoxic pressure along the Sava River. PLoS One 11(9). https://doi.org/10.1371/journal.pone.0162450

Kolbezen M (1998) Hidrografija (Hydrology). In: Gams I, Vrišer I (eds) Geografija Slovenije. Slovenska matica, Ljubljana, pp 139–172

Komatina D, Grošelj S (2015) Transboundary water cooperation for sustainable development of the Sava River Basin. In: Milačič R, Ščančar J, Paunović M (eds) The Sava river, the handbook of environmental chemistry, vol 31. Springer, Heidelberg, pp 1–25. https://doi.org/10.1007/978-3-662-44034-6_1

Kostić J, Kolarević S, Kračun-Kolarević M et al (2017) The impact of multiple stressors on the biomarkers response in gills and liver of freshwater breams during different seasons. Sci Total Environ 601-602:1670–1681. https://doi.org/10.1016/j.scitotenv.2017.05.273

Kostić-Vuković J, Kolarević S, Kračun-Kolarević M et al (2021) Temporal variation of biomarkers in common bream *Abramis brama* (L., 1758) exposed to untreated municipal wastewater in the Danube River in Belgrade, Serbia. Environ Monit Assess 193(8):465. https://doi.org/10.1007/s10661-021-09232-6

Kračun-Kolarević M, Kolarević S, Jovanović J et al (2019) Impact of pollution on rivers in Montenegro: ecotoxicological perspective. In: Pešić V, Paunović M, Kostianoy A (eds) The rivers of Montenegro. The handbook of environmental chemistry, vol 93. Springer, Cham. https://doi.org/10.1007/698_2019_425

Krstic S (2012) Environmental changes in lake catchments as a trigger for rapid eutrophication-Prespa Lake case study. In: Studies on environmental and applied geomorphology, pp 63–118. IntechOpen, London. https://doi.org/10.5772/27246

Lapointe NWR, Cooke SJ, Imhof JG et al (2014) Principles for ensuring healthy and productive freshwater ecosystems that support sustainable fisheries. Environ Rev 22:110–134. https://doi.org/10.1139/er-2013-0038

Lenhardt M, Jarić I, Višnjić-Jeftić Ž et al (2012) Concentrations of 17 elements in muscle, gills, liver and gonads of five economically important fish species from the Danube River. Knowl Manag Aquat Ecosyst 407:02. https://doi.org/10.1051/kmae/2012028

Malaj E, Rousseau DPL, Du Laing G et al (2012) Near-shore distribution of heavy metals in the Albanian part of Lake Ohrid. Environ Monit Assess 184:1823–1839. https://doi.org/10.1007/s10661-011-2081-z

Mănoiu VM, Crăciun AI (2021) Danube river water quality trends: A qualitative review based on the open access web of science database. Ecohydrol Hydrobiol 21(4):613–628. https://doi.org/10.1016/j.ecohyd.2021.08.002

Milačič R, Ščančar J, Murko S et al (2010) A complex investigation of the extent of pollution in sediments of the Sava River. Part 1. Selected elements. Environ Monit Assess 163:263–275. https://doi.org/10.1007/s10661-009-0832-x

Milanov R, Krstić M, Marković R et al (2016) Analysis of heavy metals concentration in tissues of three different fish species included in human diet from Danube river. Acta Veterinaria 66(1): 89–102. https://doi.org/10.1515/acve-2016-0007

Milošković A, Dojčinović B, Kovačević S et al (2016) Spatial monitoring of heavy metals in the inland waters of Serbia: a multispecies approach based on commercial fish. Environ Sci Pollut Res Int 23(10):9918–9933. https://doi.org/10.1007/s11356-016-6207-2

Milošković A, Milošević Đ, Radojković N et al (2018) Potentially toxic elements in freshwater (*Alburnus* spp.) and marine (*Sardina pilchardus*) sardines from the Western Balkan Peninsula:

An assessment of human health risk and management. Sci Total Environ 644:899–906. https://doi.org/10.1016/j.scitotenv.2018.07.041

Morina A, Morina F, Djikanović V et al (2016) Common barbel (*Barbus barbus*) as a bioindicator of surface river sediment pollution with Cu and Zn in three rivers of the Danube River Basin in Serbia. Environ Sci Pollut Res Int 23(7):6723–6734. https://doi.org/10.1007/s11356-015-5901-9

Pavlović P, Mitrović M, Đorđević D et al (2016) Assessment of the contamination of riparian soil and vegetation by trace metals—A Danube River case study. Sci Total Environ 540:396–409. https://doi.org/10.1016/j.scitotenv.2015.06.125

Pešić V, Karaman G, Kostianoy A (eds) (2018) Lake Skadar/Shkodra environment. The handbook of environmental chemistry, vol 80. Springer, Cham. 508 pp

Peter F (2006) Biodiversity of European freshwater fish—threats and conservation priorities at the catchment scale. Diploma thesis. Department of Geography, Basel

Podrug M, Raspor B, Erk M et al (2009) Protein and metal concentrations in two fractions of hepatic cytosol of the European chub (*Squalius cephalus* L.). Chemosphere 75(7):843–849. https://doi.org/10.1016/j.chemosphere.2009.01.079

Popovska C, Bonacci O (2007) Basic data on the hydrology of Lakes Ohrid and Prespa. Hydrol Process 21(5):658–664. https://doi.org/10.1002/hyp.6252

Popovska C, Bonacci O (2008) Ecohydrology of Dojran Lake. In: Hlavinek P, Bonacci O, Marsalek J, Mahrikova I (eds) Dangerous pollutants (xenobiotics) in urban water cycle. NATO science for peace and security series. Springer, Dordrecht. https://doi.org/10.1007/978-1-4020-6795-2_14

Popovska C, Gesovska V, Ivanoski D (2005) Ecological and hydrological state of Dojran Lake. Vodoprivreda 216-218:175–180

Rakočević J, Suković D, Marić D (2018) Distribution and relationships of eleven trace elements in muscle of six fish species from Skadar Lake (Montenegro). Turk J Fish Aquat Sci 18:647–657. https://doi.org/10.4194/1303-2712-v18_5_01

Rašković B, Poleksić V, Višnjić-Jeftić Ž et al (2015) Use of histopathology and elemental accumulation in different organs of two benthophagous fish species as indicators of river pollution. Environ Toxicol 30(10):1153–1161. https://doi.org/10.1002/tox.21988

Rico A, Van den Brink PJ, Leitner P et al (2016) Relative influence of chemical and non-chemical stressors on invertebrate communities: a case study in the Danube River. Sci Total Environ 571:1370–1382. https://doi.org/10.1016/j.scitotenv.2016.07.087

Sakan S, Gržetić I, Đorđević D (2007) Distribution and fractionation of heavy metals in the Tisa (Tisza) river sediments. Env Sci Pollut Res 14(4):229–236. https://doi.org/10.1065/espr2006.05.304

Sakan S, Đorđević D, Manojlović D et al (2009) Assessment of heavy metal pollutants accumulation in the Tisza river sediments. J Environ Manag 90(11):3382–3390. https://doi.org/10.1016/j.jenvman.2009.05.013

Simionov IA, Cristea DS, Petrea SM et al (2021) Preliminary investigation of lower Danube pollution caused by potentially toxic metals. Chemosphere 264(1):128496. https://doi.org/10.1016/j.chemosphere.2020.128496

Skoulikidis NT, Zogaris S, Karaouzas J (2022) Rivers of the Balkans. In: Tockner K, Zarfl C, Robinson CT (eds) Rivers of Europe, 2nd edn. Academic, Cambridge, MA; Elsevier, Amsterdam, pp 593–653 https://doi.org/10.1016/B978-0-08-102612-0.00011-0

Štrbac S, Šajnović A, Budakov L et al (2013) Metals in the sediment and liver of four fish species from different trophic levels in Tisza River, Serbia. Chem Ecol 30(2):169–186. https://doi.org/10.1080/02757540.2013.841893

Štrbac S, Kašanin-Grubin M, Jovančićević B et al (2015) Bioaccumulation of heavy metals and microelements in silver bream (*Brama brama* L.), Northern pike (*Esox lucius* L.), sterlet (*Acipenser ruthenus* L.), and common carp (*Cyprinus carpio* L.) from Tisza River, Serbia. J Toxicol Environ Health A 78(11):663–665. https://doi.org/10.1080/15287394.2015.1023406

Subotić S, Višnjić Jeftić Ž, Spasić S et al (2013a) Distribution and accumulation of elements (As, Cu, Fe, Hg, Mn, and Zn) in tissues of fish species from different trophic levels in the Danube River at the confluence with the Sava River (Serbia). Environ Sci Pollut Res Int 20(8): 5309–5317. https://doi.org/10.1007/s11356-013-1522-3

Subotić S, Spasić S, Višnjić-Jeftić Z et al (2013b) Heavy metal and trace element bioaccumulation in target tissues of four edible fish species from the Danube River (Serbia). Ecotoxicol Environ Saf 98:196–202. https://doi.org/10.1016/j.ecoenv.2013.08.020

Subotić S, Višnjić-Jeftić Ž, Spasić S et al (2015) Concentrations of 18 Elements in Muscle, Liver, Gills, and Gonads of Sichel (*Pelecus cultratus*), Ruffe (*Gymnocephalus cernua*), and European Perch (P*erca fluviatilis*) in the Danube River near Belgrade (Serbia). Water Air Soil Pollut 226:287. https://doi.org/10.1007/s11270-015-2544-x

Subotić S, Višnjić-Jeftić Ž, Đikanović V et al (2019) Metal accumulation in muscle and liver of the Common Nase (*Chondrostoma nasus*) and Vimba Bream (*Vimba vimba*) from the Danube River, Serbia: Bioindicative Aspects. Bull Environ Contam Toxicol 103(2):261–266. https://doi.org/10.1007/s00128-019-02657-3

Subotić S, Višnjić-Jeftić Ž, Bojović S et al (2021) Seasonal variations of macro-, micro-, and toxic elements in tissues of vimba bream (*Vimba vimba*) from the Danube River near Belgrade, Serbia. Environ Sci Pollut Res 28:63087–63101. https://doi.org/10.1007/s11356-021-15073-0

Sunjog K, Gačić Z, Kolarević S et al (2012) Heavy metal accumulation and the genotoxicity in barbel (*Barbus barbus*) as indicators of the Danube river pollution. Sci World J. https://doi.org/10.1100/2012/351074

Sunjog K, Kolarević S, Kračun-Kolarević M et al (2019) Seasonal variation in metal concentration in various tissues of the European chub (*Squalius cephalus* L.). Environ Sci Pollut Res 26:9232–9243. https://doi.org/10.1007/s11356-019-04274-3

Tadić L, Šperac M, Karleuša B, Rubinić J (2020) Water quality status of Croatian surface water resources. In: Negm A, Romanescu G, Zelenakova M (eds) Water resources management in Balkan countries. Springer water. Springer, Cham. https://doi.org/10.1007/978-3-030-22468-4_6

Teodorović I (2009) Ecotoxicological research and related legislation in Serbia. Environ Sci Pollut R 16:123–129. https://doi.org/10.1007/s11356-009-0152-2

Triebskorn R, Telcean I, Casper H et al (2008) Monitoring pollution in River Mureş, Romania, part II: metal accumulation and histopathology in fish. Environ Monit Assess 141(1–3): 177–188. https://doi.org/10.1007/s10661-007-9886-9

UNDP (2014) Prespa Lake watershed management plan. United Nations Development Program, Skopje, Macedonia

UNEP-MAP-RAC/SPA (2014) Status and Conservation of Fisheries in the Adriatic Sea. By H. Farrugio & Alen Soldo. Draft internal report for the purposes of the Mediterranean Regional Workshop to Facilitate the Description of Ecologically or Biologically Significant Marine Areas, Malaga, Spain, 7–11 April 2014

Visnjic-Jeftic Z, Jaric I, Jovanović Lj et al (2010) Heavy metal and trace element accumulation in muscle, liver and gills of the Pontic shad (*Alosa immaculata* Bennet 1835) from the Danube River (Serbia). Microchem J 95(2):341–344. https://doi.org/10.1016/j.microc.2010.02.004

Vukašinović-Pešić V, Blagojević N (2018) Metal pollution: evidences from plants, aquatic invertebrates and fish from Lake Skadar. In: Pešić V, Karaman G, Kostianoy A (eds) The Skadar/Shkodra lake environment. The handbook of environmental chemistry, vol 80. Springer, Cham. https://doi.org/10.1007/698_2017_232

Vukašinović-Pešić V, Blagojević N, Savić A et al (2019) The change in the water chemistry of the rivers of Montenegro over a 10-year period. In: Pešić V, Paunović M, Kostianoy A (eds) The rivers of Montenegro. The handbook of environmental chemistry, vol 93. Springer, Cham. https://doi.org/10.1007/698_2019_417

Watzin MC, Puka V, Naumoski T (2002) Lake Ohrid and its watershed, state of the environmental report. Lake Ohrid Conservation Project, Tirana

WFD (2000) Water Framework Directive (WFD), Directive 2000/60/EC of the European Parliament and of the Council establishing a framework for Community action in the field of water policy (OJ L 327, 22.12.2000, p 1)
World Bank (2003) Issues and direction. Vol 2 of water resources management in South Eastern Europe. Notes and facts about the water. International Bank for Reconstruction and Development, Washington, DC. http://web.worldbank.org/archive/website00983A/WEB/PDF/SEE_VOLU.PDF
Youn SJ, Taylor WW, Lynch AJ et al (2014) Inland capture fishery contributions to global food security and threats to their future. Global Food Secur 3:142–148. https://doi.org/10.1016/j.gfs.2014.09.005
Zelenakova M, Negm AM, Minea I (2020) Introduction to "water resources management in Balkan countries". In: Negm A, Romanescu G, Zelenakova M (eds) Water resources management in Balkan countries. Springer water. Springer, Cham. https://doi.org/10.1007/978-3-030-22468-4_1
Zrnčić S, Oraić D, Ćaleta M et al (2013) Biomonitoring of heavy metals in fish from the Danube River. Environ Monit Assess 185(2):1189–1198. https://doi.org/10.1007/s10661-012-2625-x
Zuliani T, Vidmar J, Drinčić A et al (2019) Potentially toxic elements in muscle tissue of different fish species from the Sava River and risk assessment for consumers. Sci Total Environ 650(1): 958–969. https://doi.org/10.1016/j.scitotenv.2018.09.083

Invasive Macroinvertebrate and Fish Species and Their Impact on Fish Resources: A Case Study on Fishing Waters of Serbia

Milena Radenković, Katarina Zorić, and Ana Petrović

Abstract The introduction of species outside their natural range is widely recognized to be one of the main threats to biodiversity and the second leading cause of animal extinctions. Freshwater ecosystems often have high rates of introductions of allochthonous macroinvertebrates and fishes arising from intentional and accidental releases of species. Problems potentially occur when these organisms integrate into the community, interact with native species, and increase competition for resources. It is known that zebra mussel *Dreissena polymorpha*, as well as fish species *Carassius gibelio*, *Ameiurus nebulosus*, *Ameiurus melas*, *Lepomis gibbosus*, and others have a negative impact on the autochthonous fish species of Serbia. Therefore, the aim of the present study is to assess the consequences of introduced macroinvertebrates and fishes, and their impact on fish stock (abundance and biomass) in some Serbian rivers and reservoirs.

Keywords Macroinvertebrate community · Fish community · Biomass · Abundance · Rivers · Reservoirs

Supplementary Information The online version contains supplementary material available at https://doi.org/10.1007/978-3-031-36926-1_13.

M. Radenković (✉) · A. Petrović
Department of Biology and Ecology, Faculty of Science, University of Kragujevac, Kragujevac, Serbia
e-mail: milena.radenkovic@pmf.kg.ac.rs

K. Zorić
Department for Hydroecology and Water Protection, Institute for Biological Research "Siniša Stanković", National Institute of the Republic of Serbia, University of Belgrade, Belgrade, Serbia

V. Simić et al. (eds.), *Ecological Sustainability of Fish Resources of Inland Waters of the Western Balkans*, Fish & Fisheries Series 43,
https://doi.org/10.1007/978-3-031-36926-1_13

1 Introduction

The primary threat to freshwater fish production is loss of habitat. The reasons for the habitat destruction are numerous and diverse, e.g. acidification of watercourses, eutrophication, sedimentation, and chemical pollution from agricultural land use, damming and regulation of rivers particularly for hydroelectric purposes. In addition, it is feared that side effects of the fish farming will affect wild populations negatively due to genetic introgression, ecological competition from escaped hatchery fishes, and the spread of contagious diseases and parasites lethal to fishes (Jonsson and Jonsson 2016). "Biological pollution" (Courtenay 1993), i.e., species introduction into habitats where there were none before, is no less a problem than the ones listed. Many of these introduced species may exhibit invasive characteristics.

Invasive species refer to any live specimen of a species or lower taxon of animals, plants, fungi, or microorganisms whose occurrence in a region outside its natural range has profound impacts on an ecosystem and its services (Magliozzi et al. 2020). Given the long history of biological invasions and their substantial effects, there is clear potential that biological invasions were the result of significant early human impacts on freshwater ecosystems (Dubois et al. 2017). The fish community respond on human pressure, by increasing the number of invasive species (Stamenković et al. 2019). Humans have effectively transported thousands of species around the world, and, with accelerated trade, the rate of introductions has increased over time (Havel et al. 2015).

Ecological impacts of invasive species include the displacement and extinction of native species, alteration of trophic interactions, and disruption of ecosystem functioning. They can also transmit new pathogens and threaten native species' genetic diversity through hybridization (Haubrock et al. 2022). Once established in a new habitat, most invasive species are nearly impossible to eradicate and some of the major pests are extremely expensive to control (Pimentel et al. 2005).

On the territory of Europe there are four invasion corridors, main routes that facilitate entry and dispersal of allochthonous species. The largest is the "northern corridor" linking the Black and Azov seas with the Caspian Sea via the Azov–Caspian waterway, with the Baltic and White seas via the Volga–Baltic waterway, and the White Sea–Baltic Sea waterway. The "central corridor" connects the Black Sea with the Baltic Sea region via Dnieper and the Bug–Prypjat Canal. The "western corridor" links the Mediterranean with the North Sea via the Rhône and Rhine–Rhône Canal. The "southern corridor" via the Danube, the Rhine, and the Main connects the Black Sea with the North Sea (Galil et al. 2007). Serbia, the crossroads of Central and Southeastern Europe, with the Danube and its tributaries is part of the "southern corridor" and is particularly exposed to invasions. A total of 29 macroinvertebrate species (Zorić et al. 2020a) and 25 fish species (Lenhardt et al. 2020) have been introduced so far into the inland waters of Serbia, either intentionally or accidentally.

Aquatic organisms could be characterized as a major component of the assessment processes in aquatic ecosystems, providing a more effective diagnosis of environmental conditions (Van der Schalie et al. 2001; Simić et al. 2014; Milošković et al. 2016). In view of the great biodiversity and vulnerability of freshwater communities to biotic exchange, impacts caused by invasions are certainly a cause of concern. Therefore, this chapter aims to present the relationship of richness, biomass, and diversity between autochthonous (native) and allochthonous (non-native) species, i.e. how biomass and abundance of invasive species affect native species.

2 Methodology

2.1 *Study Sites*

This chapter presents a case study on the fishing waters of Serbia and for this reason, the ecosystems included in this study are the Danube River and Sava River, and the following ten reservoirs: Gruža, Bovan, Vlasina, Gazivode, Šumarice, Vrutci, Potpeć, Barje, Grliško jezero, and Srebrno jezero (Fig. 1).

2.2 *Macroinvertebrates Sampling*

Sampling of benthic macroinvertebrates on the Danube and Sava River was performed using semi-quantitative multi-habitat sampling procedure (MHS). This method is originally recommended for wadeable streams—small and medium-sized rivers (AQEM 2002). For large and very large rivers, such as the Danube and Sava, this procedure is applicable at lower water periods in the littoral zone, up to a maximum water depth of 1.5 m. Although sampling in large and very large rivers is not yet harmonized at the European level (Tubić 2016), MHS is a recommended and well-developed method using as a standardized WFD (The Water Framework Directive) sampling method for the assessment of the ecological status of water bodies.

Before sampling, a detailed assessment of the habitat and proportion of substrate types is mandatory. The sample consists of 20 subsamples collected using standard benthological net with 500 μm mesh size. Subsamples are taken from all types of microhabitats in the research sector (100 m long) according to percentage ratio of each microhabitat, excluding microhabitats which cover less than 5% within the sampled sector. A subsample is a stationary (0.25×0.25 m) sampling accomplished by positioning the net and disturbing the substrate for a distance that equals the square of the frame width upstream of the standard net providing a sampling area of 0.0625 m^2 per subsample.

Sampling of aquatic macroinvertebrate fauna on the Danube River was carried out at eight localities along the main course of the river during July 2019 as part of the JDS 4 expedition.

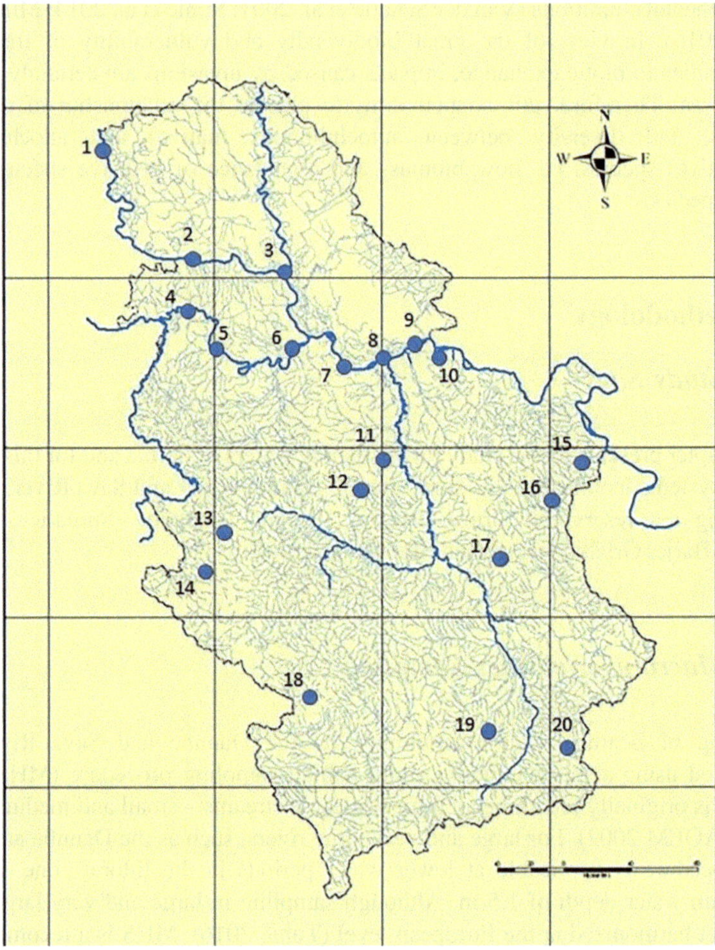

Fig. 1 Map of studied sites; (1) Danube—Bezdan, (2) Danube—Bačka Palanka, (3) Danube—confluence with the Tisza River, (4) Sava—Šabac, (5) Sava—Sremska Mitrovica, (6) Sava—Belgrade, (7) Danube—Pančevo, (8) Danube—confluence with the Velika Morava River, (9) Danube—Banatska Palanka, (10) Srebrno jezero, (11) Šumarice reservoir, (12) Gruža reservoir, (13) Vrutci reservoir, (14) Potpeć reservoir, (15) Danube—upstream from the Timok, (16) Grliško jezero, (17) Bovan reservoir, (18) Gazivode reservoir, (19) Barje reservoir, (20) Vlasina reservoir

The material used for the analysis of the fauna of the Sava River was collected during September 2014 from five localities: Belgrade, Šabac 1, Šabac 2, Sremska Mitrovica 1, and Sremska Mitrovica 2.

At each locality, 20 subsamples were taken, which were pulled into one sample. All samples were sieved through a coarse mesh of 500 μm to remove soft sediment and preserved with ethanol to the final concentration of 70%.

Also, a sampling of aquatic macroinvertebrate fauna was carried out at six reservoirs in Serbia (Bovan, Potpeć, Barje, Gruža, Gliško jezero, and Šumarice reservoirs).

Field research had been done in the 2020 and 2021 years (Barje reservoir, Gruža reservoir, and Grliško jezero). The data for Šumarice, Potpeć, and Bovan reservoirs dates from research conducted in 2015 year.

Aquatic invertebrates were sampled by benthological dredger Ekman type (15 × 15 cm). The material was washed through a sieve with a 210 mm-mesh aperture and preserved on 4% formaldehyde.

In the laboratory, the organisms were sorted, identified, counted, and identified based on guides and keys of Brinkhurst (1971), Ježek (1980), Elliott et al. (1988), Conci and Nielsen (1956), and Pfleger (2000) down to genus or species level. Density and relative abundance were calculated.

2.3 Fish Sampling

Depending on the characteristics of the selected localities, fish sampling was carried out by DC electrofisher "Aquatech" IG 1300 (2.6 kW, 80–470 V) for electrofishing in the littoral zone of reservoirs, and gillnets for rivers and pelagic zone of reservoirs. Stationary and trawling nets of different lengths, widths, and mesh diameters were used depending on the type of waterbody in which fishing is carried out, as well as the fish species that are expected to be caught. After the catch, the abundance of each species was determined, and the length and weight of the specimens were measured.

3 Overview of Invasive Macroinvertebrates and Fishes in Fishing Waters of Serbia

3.1 Composition of the Macroinvertebrate Community of the Investigated Localities

The bottom fauna of the reservoir was composed of 69 invertebrate taxa, belonging to the following faunistic groups: Turbellaria, Limnomedusae, Oligochaeta, Gastropoda, Bivalvia, Amphipoda, Isopoda, Odonata, Trichoptera, Ephemeroptera, Diptera, Megaloptera. The majority of species showed low densities. The dominant taxa were *Limnodrilus hoffmeisteri*, *Chaoborus crystallinus*, *Valvata piscinalis*, and *Dreissena polymorpha* (Supplementary Table 1).

In total, seven allochthonous species were registered: *Branchiura sowerbyi* Beddard, 1892, *Corbicula fluminalis* (O. F. Müller, 1774), *Physella acuta* (Draparnaud, 1805), *Dreissena polymorpha* (Pallas, 1771), *Sinanodonta woodiana* (Lea, 1834), *Craspedacusta sowerbyi* Lankester, 1880, and *Dikerogammarus villosus* (Sowinsky, 1894). In most localities their presence in the community was negligible to moderate, while only in Gliško jezero and Gruža reservoir allochthonous species were important part of the community with percentage participation of 27.86% and 50%, respectively. During the investigation no allochthonous taxa were recorded in Barje and Potpeć reservoirs. On other localities the number of species varied between 1 (Bovan reservoir) and 4 (Gruža reservoir) (Supplementary Table 1).

The Danube and Sava were characterized by greater diversity of macroinvertebrate fauna and allochthonous species as well. In total, 104 taxa were found in the Danube and 155 taxa in the Sava River (Supplementary Tables 2 and 3). The percentage of non-native species on the Danube was below 10% in most sites, while only in the Bezdan locality it was extremely high (74.75%). Allochthonous species were not registered only on locality Timok confluence. The most abundant species were *Corophium curvispinum* (G.O. Sars, 1895), *C. robustum* (G.O. Sars, 1895), *Obesogammarus obesus* (G.O. Sars, 1894), and *Hypania invalida* (Grube, 1860). Unlike the Danube, in the Sava River, the representation of allochthonous fauna along the entire course was uniform (11–13 species per sampling site), except for the Belgrade locality, where only one species was recorded. Percentage participation varied between 0.33 (Belgrade) and 48.25% (Šabac 2). Regarding dominance, species *C. fluminea*, *P. acuta*, and *C. curvispinum* were the most abundant taxa.

3.2 Composition of the Fish Community of the Investigated Localities

The quantitative and qualitative compositions of the fish community were determined after field work.

The presence of allochthonous fish species was recorded at each investigated locality. The Prussian carp *Carassius gibelio* (Bloch, 1782) is present in all localities except Barje reservoir where dominant species are common bream *Abramis brama* (Linnaeus, 1758) and pikeperch *Sander lucioperca* (Linnaeus, 1758). Frequently present, although not in all investigated localities, are also silver carp *Hypophthalmichthys molitrix* (Valenciennes, 1844), bighead carp *Hypophthalmichthys nobilis* (Richardson, 1845), pumpkinseed *Lepomis gibbosus* (Linnaeus, 1758), and *Ameiurus* sp. Pikeperch, common bream, roach *Rutilus rutilus* (Linnaeus, 1758), and Prussian carp dominate the biomass and abundance of the Bovan reservoir (Pavlović et al. 2015; Radenković et al. 2022). The fish community in the Šumarice reservoir consists mainly of rudd *Scardinius erythrophthalmus* (Linnaeus, 1758), roach, pumpkinseed, and *Ameiurus* sp. Although there are also allochthonous species on the Vrutci reservoir, the most abundant species are nase *Chondrostoma nasus* (Linnaeus, 1758), common bream, and perch *Perca fluviatilis* (Linnaeus, 1758). Common bream and Prussian carp stand out in the Gazivode

Fig. 2 Specimens of *Hypophthalmichthys molitrix* caught in the Šumarice reservoir (photo by Vladica Simić)

reservoir (Pavlović et al. 2013). Three species dominate the fish community in the Gruža reservoir and Grliško jezero: pikeperch, Prussian carp, and roach, while in the Vlasina reservoir the dominant species are perch and Prussian carp while in the Srebrno jezero the dominant species are common bream and vimba bream *Vimba vimba* (Linnaeus, 1758). Nase, chub *Squalius cephalus* (Linnaeus, 1758), and Prussian carp are the dominant species in Potpeć reservoir. Along with pikeperch, common barbel *Barbus barbus* (Linnaeus, 1758), and common bream, Prussian carp are the most common species in catches on the Danube River. On the Sava River the most common species caught were common bream, white bream *Blicca bjoerkna* (Linnaeus, 1758), and roach.

From the above, in most of investigated localities Prussian carp stands out as the species that dominates in abundance.

The appearance of large specimens of silver carp (Fig. 2) and bighead carp, as well as grass carp *Ctenopharyngodon idella* (Valenciennes, 1844), is interesting. The presence of all three mentioned species was recorded in the Bovan reservoir during one catch, while bighead carp and silver carp occur in the Gruža, Šumarice, and Vrutci reservoir. All caught specimens are extremely large specimens and therefore have a high contribution in the total biomass (over 50%) in the mentioned catches (Radenković 2019).

Figure 3 shows the percentage ratio of autochthonous and allochthonous fish species throughout their abundance and biomass. In all investigated ecosystems, autochthonous species dominate both in abundance and biomass, with a few

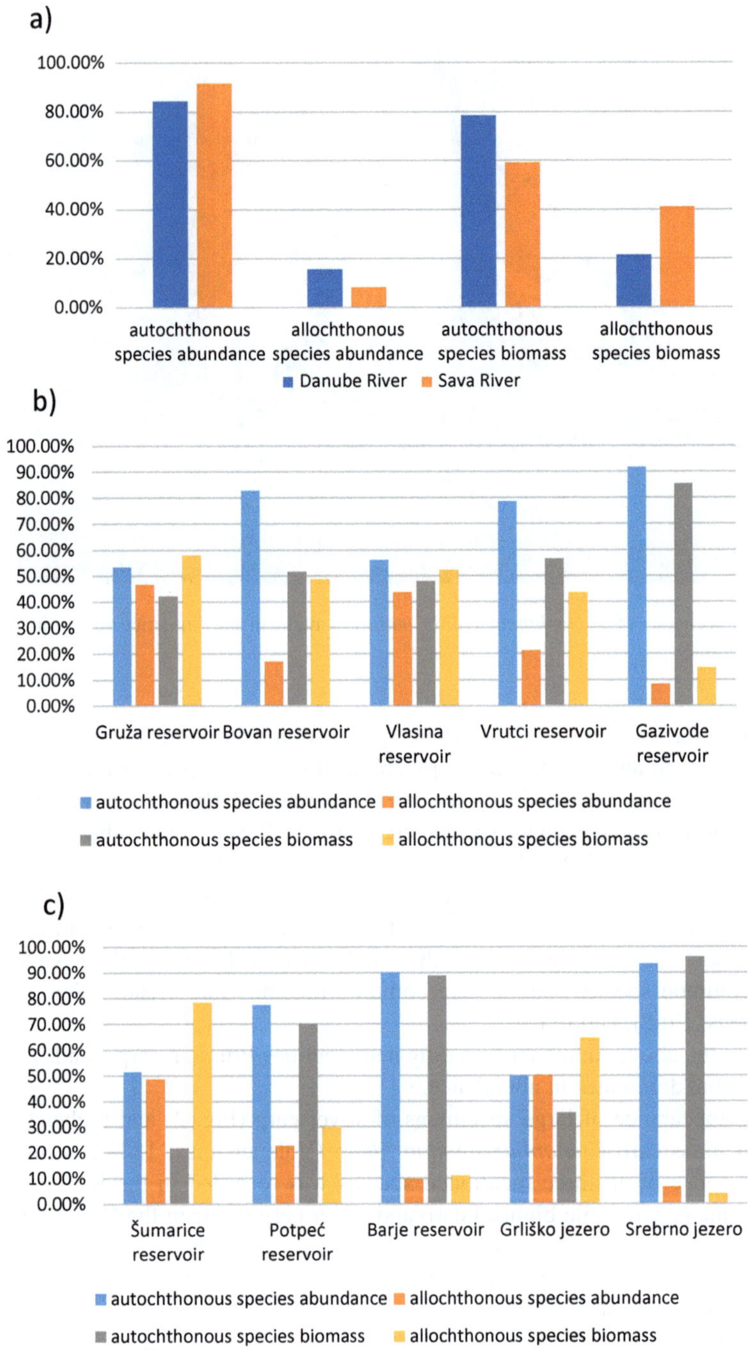

Fig. 3 The percentage ratio of autochthonous and allochthonous fish species throughout their abundance and biomass (**a**) Danube and Sava River, (**b**) Gruža, Bovan, Vlasina, Vrutci, and Gazivode reservoirs, (**c**) Šumarice, Potpeć, Barje, Grliško jezero, and Srebrno jezero reservoirs

exceptions. On the Gruža, Vlasina, and Šumarice reservoirs allochthonous species dominate in biomass (58%, 52%, and 78%, respectively), but not in abundance.

The percentage ratio of allochthonous species in investigated ecosystems is present in Table 1. The most diverse populations of allochthonous species are on the Danube River and Sava River, and on the Gruža, Bovan, Vlasina, Šumarice, and Potpeć reservoirs. On the other hand, only Prussian carp is present as allochthonous species on the Grliško jezero, and on Barje reservoir only *Ameiurus* sp. is present. The Grliško jezero is specific because Prussian carp, although the only allochthonous species caught, occupies 50% of the total abundance and 65% of the total biomass compared to all other caught species. Among allochthonous species in all investigated ecosystems, Prussian carp is the most abundant, except on Šumarice reservoir where the most abundant is pumpkinseed.

4 Role and Impact of Invasive Species in Fishing Waters in Serbia

In European inland waters some communities are now dominated by invasive species, and in some waterbodies these communities have shown very stable composition of dominant species for over a decade (Paunović et al. 2015). While most species fail to thrive or have minor impacts on their new ecosystems the large number of introductions has led to numerous problems (Thomaz et al. 2015).

There are a large number of publications on allochthonous species of river systems in Serbia (Atanacković et al. 2020; Borza et al. 2021; Lucić et al. 2015; Paunović et al. 2005, 2006, 2007, 2008; Pavlović et al. 2006; Raković 2015; Raković et al. 2013; Tomović et al. 2012, 2013; Zorić et al. 2011, 2015, 2020a, b). In total, the allochthonous, introduced aquatic macroinvertebrate fauna of Serbia consists of 29 species distributed within 18 families (Zorić et al. 2020a). On the other hand, the allochthonous fauna of reservoirs is insufficiently treated, so this chapter represents a significant step in that direction.

In reservoirs, macroinvertebrate community tends to be less complex, with reduced diversity and abundance (Ferreira et al. 2009). As expected, the number of allochthonous species in reservoirs is significantly less than in the main course of the Danube and Sava. One of the factors that determine their number in the ecosystem is the distance from the invasive corridor. Based on the overall analyses it has been concluded that allochthonous species are expanding from the Danube as main corridor to other waterbodies in Serbia (Zorić et al. 2020a).

Of all the species present, the snail *P. acuta* is the most widespread allochthonous species, recorded in three reservoirs—Šumarice, Bovan, and Grliško jezero, although without significant participation in the community. The species has already been detected in the Grliško jezero (Novaković 2014), while Šumarice and Bovan reservoirs are new localities where this species is recorded. Bearing in mind that so far this widely adaptable and tolerant species has been registered in eight river basins

Table 1 The percentage ratio of allochthonous species in the investigated ecosystems. The dominant species in certain ecosystems are in bold

	Prussian carp (%)	Silver carp (%)	Bighead carp (%)	Grass carp (%)	Pumpkinseed (%)	*Ameiurus* sp. (%)	*Neogobius* sp. (%)	Largemouth bass (%)	Rainbow trout (%)	Topmouth gudgeon (%)
Danube river	**77.36**	4.53	9.87	0.78	6.17	0.78	0.39	–	–	–
Sava river	**44**	30	–	–	16	8	2	–	–	–
Gruža reservoir	**60**	1	–	–	1	38	–	–	–	–
Bovan reservoir	**89.73**	0.78	2.17	3.12	4.11	1.36	–	–	–	–
Vlasina reservoir	**95.85**	0.41	–	–	0.82	2.90	–	–	–	–
Vrutci reservoir	**93.33**	0.83	5.83	–	–	–	–	–	–	–
Gazivode reservoir	**95.24**	–	–	–	–	4.76	–	–	–	–
Šumarice reservoir	22.82	5.36	–	–	**40.27**	28.86	–	2.68	–	–
Potpeć reservoir	**83.95**	–	–	–	12.34	–	–	–	2.46	1.23
Barje reservoir	–	–	–	–	–	**100**	–	–	–	–
Grliško jezero	**100**	–	–	–	–	–	–	–	–	–
Srebrno jezero	**66.66**	–	33.33	–	–	–	–	–	–	–

and in a total of 77 localities (Zorić et al. 2020a), we assume that the number of findings in reservoirs will also be higher during future research.

All registered species can represent a high-quality source of food for a large number of fish species, and this can be reflected as the possibly positive effect of their introduction of allochthonous species from the perspective of the fish stock.

The potentially negative effects of the presence of allochthonous species of macroinvertebrates on fish communities are known and are reflected primarily in the transfer of parasites that can affect the production and decline of the fish stock, e.g., freshwater myxosporean parasite *Thelohanellus kitauei* is a species causing intestinal giant cystic disease of common carp, which life cycle include *B. sowerbyi* as a definite host (Zhao et al. 2016).

Another possible negative effect can be caused by zebra mussel *D. polymorpha*. The increasing of abundance and formation of the so-called zebra mussel bed can lead to some structural changes of the fish community in favour to benthophagous fish and size deteriorating of other species, e.g. bream (Smirnov et al. 2020). Moreover, Ponto-Caspian gammarids prefer mussel beds as antipredator protection more efficiently than by native gammarid species which represents a positive relationship between invasive Ponto-Caspian species (Kobak et al. 2014).

Also, aggressive behaviour of *D. villosus* expressed not only against other gammarid species competing with them for food resources and space (Dick et al. 2002) but can also negatively affect fish populations through egg and larval predation (Taylor and Dunn 2017).

The aforementioned effects have not been observed in reservoirs of Serbia so far, and they are unlikely considering the low abundance of allochthonous species in the community and the small number of findings.

Certain number of allochthonous fish species have substantial adverse impact on autochthonous ichthyofauna (Lenhardt et al. 2020; Simić et al. 2022). Piria et al. (2017) reported that recent surveys from several Balkan countries have revealed that 15–23% of their fish fauna is alien, with certain catchment such as the Danube River and Pamvotis Lake in Greece having an ichthyofauna comprised of more than 50% and 80% of alien fishes, respectively. Our research showed that in the Danube River, despite the diversity of allochthonous species, their percentage contribution in abundance and biomass is 16% and 22%, respectively. In the Sava River, there is a noticeable increase in the percentage contribution of allochthonous species in biomass in relation to abundance, from 8 to 42%. Actually, during one catch the presence of as many as 14 specimens of silver carp with a total weight of ≈ 100 kg was recorded. Maletin et al. (1994) noted that the contribution of grass carp, silver carp, and bighead carp (the so-called Chinese carps) to the total fish catch from the Danube in Serbia increased steadily during the 1980s and accounted for approximately 10% of total catch at the beginning of the 1990s. The import of bighead, silver and grass carp is a reason for the decrease in tench (*Tinca tinca*) populations due to competition for food (Ćirković et al. 2009). Chinese complex has a positive influence on autochthonous ichthyofauna and the ecosystem in general, contributing to the increase in total ichthyoproduction and maintenance of the biological balance in accordance with the capacity of a given habitat (Maletin et al. 1997). On the other hand, the negative effects of their importation into Serbian waters are illustrated by findings of new parasite in Serbian waters (Lenhardt et al. 2011). The presence of Chinese complex was confirmed in six out of ten reservoirs (Gruža, Bovan, Vlasina, Vrutci, Šumarice, and Srebrno jezero reservoirs), most often silver carp. In all populations of allochthonous species in these reservoirs, the most abundant are in the Srebrno jezero, which represents an oxbow lake along the right Danube bank, namely the bighead carp.

The invasion of Prussian carp coincided with the introduction of the Chinese carps (Piria et al. 2017), and its acclimatization in Serbian waters was rapid, and nowadays, it is found in nearly all waters in Serbia. And in fishing waters covered by this case study, their frequency in the total abundance of allochthonous species ranges from 22% in the Šumarice reservoir to 100% in the Grliško jezero. The Grliško jezero is a good example of the strong invasive potential of this species, where in the total abundance of all species it occupies as much as 50% of the total abundance. Lenhardt et al. (2011) reported that the Prussian carp is an economically important fish species in Serbia, mainly due to its low price. Its negative effect is reflected in a remarkable competition over reproduction and food with other autochthonous species, primarily Crucian Carp *Carassius carassius*, tench, and common carp *Cyprinus carpio* (Maletin et al. 1997; Demeny et al. 2009).

Three North American species, that have adapted well in Serbian waters, are pumpkinseed, black bullhead *Ameiurus melas*, and brown bullhead *Ameiurus nebulosus*. Black bullhead and brown bullhead are difficult to distinguish and for that reason in this chapter we write *Ameiurus* sp. Based on our results, it seems that these three species have adapted better in reservoirs than in rivers. Namely, in the Šumarice reservoir these species have dominant abundance (pumpkinseed 40%, *Ameiurus* sp. 29%) in comparison to other allochthonous. It is important to note that during catch on the Barje reservoir, the only allochthonous species caught was *Ameiurus* sp. Notwithstanding brown bullhead smaller size, the rapid adaptation of the fish has induced its successful reproduction under natural conditions (Sotirov 1968). On the other hand, considering that black bullhead is an undesirable pest, mass removals of this species were organized in some small reservoirs in Serbia for the purpose of removing the species on grounds (Lenhardt et al. 2011). When naturalized allochthonous species become invasive, they pose a serious threat to ecosystem functioning (Mujaju et al. 2021). Pumpkinseed, brown bullhead, and black bullhead have huge negative effects on autochthonous fish fauna by feeding on the bottom fauna, small fishes, fish larvae, and roe (Pujin and Sotirov 1966; Marković and Simović 1997). Lenhardt et al. (2020) state that four species (Prussian carp, pumpkinseed, brown bullhead, and black bullhead) now occupy more than 50% of the territory of Serbia.

Aquatic macroinvertebrates are important components of freshwater ecosystems and significant links in the food chains. Invasive species are well known to restructure freshwater food webs (Vander Zanden et al. 1999), and declining fish production is one of the consequences of changes in food web structure (Stockner et al. 2000). This is evident considering the impact of allochthonous molluscs in reservoirs (e.g. Grliško jezero and Gruža reservoirs) in comparison to other macroinvertebrates

Fig. 4 Colony of *Dreissena polymorpha* on *Sinanodonta woodiana* (photo by Predrag Simović)

groups. The dominant representatives of macroinvertebrate community are *D. polymorpha* and *S. woodiana* in Gruža reservoir, and *D. polymorpha* in Grliško jezero. Consequently, results from this work may indicate that this is the reason for the Prussian carp dominance, because in contrast, in Šumarice reservoir where allochthonous shells did not occupy an ecological niche, dominant allochthonous fish species is pumpkinseed that lives and feeds in littoral (Simonović 2001). Other allochthonous groups of macroinvertebrates (such as some Oligochaeta and Crustacea), although invasive, can represent food for benthivorous species such as common bream, roach, rudd, etc. On the other hand, the shells, especially zebra mussel, which have sharp edges, is unsuitable for ingestion (Pavlović et al. 2015). Most often when we think of invasive macroinvertebrates the first thing that comes to mind is the zebra mussel (Fig. 4). Karatayev et al. (1997) reported that zebra mussel can become enormously abundant and, within a short period of time, can obtain a biomass ten times greater than that of all other native benthic invertebrates.

Conceptually, the issue of allochthonous species is a problem that crosses global boundaries and extends to virtually all waterbodies in the world. Despite our best efforts to manage these species, two things will occur. First, many of these introduced species will not be eradicated from their new locations, and second, new introductions will continue (Cucherousset and Olsen 2011). Since 2000, stocking of Serbian open waters with existing allochthonous species and introduction of new species have been rejected (Lenhardt et al. 2011). The current Law on the Protection and Sustainable Use of Fish Stocks regulates introduction of new species and prohibits the importation of allochthonous fish into natural waters (Anonymous 2018). In practice and in the field, there are problems in the application of existing laws.

Therefore, it is of great importance to involve the public and stakeholders in the process of controlling invasive species. It is necessary to raise the level of awareness about the problem of invasive species through scientific and popular booklets, brochures, leaflets, posters, and tribunes designed for the general public (Zorić et al. 2020a). Integrating results from this chapter with those previously published could be used to draw up a common strategy for managing the fish stock on fishing waters in Serbia.

Acknowledgment The investigation was supported by Grant (Agreement No. 451-03-68/2022-14/200122 and Agreement No. 451-03-68/2022-14/200007) funded by the Serbian Ministry of Science, Technological Development and Innovation. The authors acknowledge the support of the International Commission for the Protection of the Danube River (ICPDR) during Joint Danube Survey 4. The authors would also like to express gratitude to their colleagues from the Department for Hydroecology and Water Protection Institute for Biological Research "Siniša Stanković", University of Belgrade and all participants of the JDS 4 Survey.

References

Anonymous (2018) The law on the protection and sustainable use of fish stocks. Official Gazette of Republic of Serbia 128/2014, 95/2018 (in Serbian)

AQEM (2002) Manual for the application of the AQEM system. A comprehensive method to assess European streams using Benthic macroinvertebrates, developed for the purpose of the water framework directive, version 1; The AQEM Consortium: Duisburg-Essen, Germany, pp 619–630

Atanacković A, Zorić K, Tomović J, Ilić M, Tubić B, Csányi B, Paunović M (2020) Invading the Danube River: range extension of Ponto-Caspian polychaete *Manayunkia caspica* Annenkova, 1929. Acta Zoologica Bulgarica 72(4):619–622

Borza P, Csányi B, Đanić V, Kenderov L, Kladarić L, Lešťáková M, Muc T, Němejcová D, Očadlík M, Paunović M, Rotar B, Szekeres J, Veseli M, Zorić K (2021) Peracarid crustaceans in the River Danube and its tributaries: results of the 4th Joint Danube Survey. BioInvasion Rec 10(3):623–628. https://doi.org/10.3391/bir.2021.10.3.12

Brinkhurst RO (1971) A guide for the identification of British aquatic Oligochaeta. Freshwater biological association, Scientific publication 22. Second edition

Ćirković M, Marković G, Simić V, Maletin S, Milosević N, Momirov D (2009) Reintroduction and repopulation of tench (*Tinca tinca* L.) in fishponds and natural waters (In Serbian with English summary). In: IV international conference "fishery" Belgrade (Serbia), pp 132–137

Conci C, Nielsen C (1956) Odonata. Fauna D'Italia, Vol. I. Sotto gli dell' Accademia Nazionale Iteliana di Entomologia e dell'Unione Zoologica Italiana, Bologna

Courtenay WR (1993) Biological pollution through fish introductions. In: McKnight BN (ed) Biological pollution: the control and impact of invasive species. Indiana Academy of Science, Indianapolis, IN, pp 35–61

Cucherousset J, Olsen JD (2011) Ecological impact of non–native freshwater fishes. Fisheries 36: 215–226

Demeny F, Sipos S, Ittzes I, Szabo Z, Levai P, Bodo I, Urbanyi B, Muller T (2009) Observation of the Crucian carp (*Carassius carassius*) pond culture. In: IV international conference "fishery" Belgrade (Serbia), pp 138–144

Dick JTA, Platvoet D, Kelly DW (2002) Predatory impact of the freshwater invader *Dikerogammarus villosu*s (Crustacea: Amphipoda). Can J Fish Aquat Sci 59:1078–1084. https://doi.org/10.1139/f02-074

Dubois N, Saulnier-Talbot É, Mills K, Gell P, Battarbee R, Bennion H, Chawchai S, Dong X, Francus P, Flower R, Gomes DF, Gregory-Eaves I, Humane S, Kattel G, Jenny JP, Laugdon P, Massaferro J, McGowan S, Mikomägi A, Nguyen MN, Ratnayake AS, Reid M, Rose N, Saros J, Schillereff D, Tolotti M, Valero-Garcés B (2017) First human impacts and responses of aquatic systems: A review of paleolimnological records from around the world. Anthropocene Rev 1–14. https://doi.org/10.1177/2053019617740365

Elliott MJ, Humpesch HU, Macan TT (1988) Larvae of the British Ephemeroptera: a key with ecological notes. Freshw Biol Assoc Sci Pub 49:42–86

Ferreira T, Cortes RV, Morais M (2009) Qualidade ecológica e Gestão Integrada de Albufeiras. INAG, Lisboa, Portugal

Galil BS, Nehring S, Panov VE (2007) Waterways as invasion highways—impact of climate change and globalization. In: Nentwig W (ed) Biological invasions, Ecological studies nr. 193. Springer, Berlin, pp 59–74

Haubrock PJ, Bernery C, Cuthbert RN, Liu C, Kourantidou M, Leroy B, Turbelin AJ, Kramer AM, Verbrugge LNH, Diagne C, Courchamp F, Gozlan RE (2022) Knowledge gaps in economic costs of invasive alien fish worldwide. Sci Total Environ 803:149875. https://doi.org/10.1016/j.scitotenv.2021.149875

Havel JE, Kovalenko KE, Thomaz SM, Amalfitano S, Kats LB (2015) Aquatic invasive species: challenges for the future. Hydrobiologia 750:147–170. https://doi.org/10.1007/s10750-014-2166-0

Ježek J (1980) Ovádovití – Tabanidae. In: Rozkošný R (ed) Klíè vodních larev hmyzu [Identification key to aquatic larvae of insects]. Academia, Praha, pp 415–422. 521 pp (in Czech)

Jonsson B, Jonsson N (2016) Fennoscandian freshwater fishes: diversity, use, threats and management. In: Craig JF (ed) Freshwater fisheries ecology. Wiley, Oxford, pp 101–119

Karatayev AY, Burlakova LE, Padilla DK (1997) The effects of *Dreissena polymorpha* (Pallas) invasion on aquatic communities in Eastern Europe. J Shellfish Res 16(1):187–203

Kobak J, Jermacz Ł, Płąchocki D (2014) Effectiveness of zebra mussels to act as shelters from fish predators differs between native and invasive amphipod prey. Aquat Ecol 48(4):397–408. https://doi.org/10.1007/s10452-014-9492-1

Lenhardt M, Marković G, Hegediš A, Maletin S, Ćirković M, Marković Z (2011) Non–native and translocated fish species in Serbia and their impact on the native ichthyofauna. Rev Fish Biol Fish 21:407–421. https://doi.org/10.1007/s11160-010-9180-8

Lenhardt M, Smederevac-Lalić M, Hegediš A, Skorić S, Cvijanović G, Višnjić-Jeftić Ž, Djikanović V, Jovičić K, Jaćimović M, Jarić I (2020) Human impact on fish fauna in the Danube River in Serbia: Current status and ecological implications. In: Bănăduc D et al (eds) Human impact on Danube watershed biodiversity in the XXI century. Geobotany studies. Springer, Cham, pp 257–280

Lucić A, Paunović M, Tomović J, Kovačević S, Zorić K, Simić V, Atanacković A, Marković V, Kračun-Kolarević M, Hudina S, Lajtner J, Gottstein S, Milošević Đ, Anđus S, Žganec K, Jaklič M, Simčič T, Vilenica M (2015) Aquatic macroinvertebrates of the Sava River. In: Milačić R, Ščančar J, Paunović M (eds) The Sava river. Springer, Heidelberg, pp 335–359

Magliozzi C, Tsiamis K, Vigiak O, Deriu I, Gerasini E, Cardoso AC (2020) Assessing invasive alien species in European catchments: Distribution and impacts. Sci Total Environ 732:138677. https://doi.org/10.1016/j.scitotenv.2020.138677

Maletin S, Djukić N, Stojanović S, Ivanc A, Zderić M, Marić A, Andrić B, Radak L, Miljanović B (1994) Meliorative effect of grass carp (*Ctenopharyngodon idella*) in controlling aquatic macrophytes in the Tisza valley. Tiscia 28:41–45

Maletin S, Djukić N, Miljanović B, Ivanc B (1997) Status of allochthonous ichthyofauna of Pannonian basin in Yugoslavia. Ekologija 32:87–98

Marković G, Simović S (1997) Pumpkinseed (*Lepomis gibbosus* L. 1758)—unwelcome member to the ichthyofauna of the Yugoslav reservoirs (In Serbian with English summary). In: The proceedings of symposium “Ribarstvo Jugoslavije”, pp 123–128

Miloŝković A, Dojčinović B, Kovačević S, Radojković N, Radenković M, Milošević D, Simić V (2016) Spatial monitoring of heavy metals in the inland waters of Serbia: a multispecies approach based on commercial fish. Environ Sci Pollut Res 23:9918–9933. https://doi.org/10.1007/s11356-016-6207-2

Mujaju C, Mudada N, Chikwenhere GP (2021) Invasive alien species in Zimbabwe (Southern Africa). In: Pullaiah T, Ielmini MR (eds) Invasive alien species: Observations and issues from around the world. Chapter 12. Wiley-Blackwell, London, pp 330–361

Novaković B (2014) Distribution of the freshwater snail *Physella acuta* (Dreparnaud, 1805) in the 2009-2013 period. In: Đukić A (ed) Proceedings of the 43rd annual conference "water 2014", Tara, Serbia, June 3–5, 2014. Serbian Water Pollution Control Society, Belgrade, Serbia, pp 77–81

Paunović M, Miljanović B, Simic V, Cakić P, Djikanović V, Jakovcev-Todorović D, Stojanović B, Veljković A (2005) Distribution of non-indigenous tubificid worm *Branchiura sowerbyi* (Beddard, 1892) in Serbia. Biotechnol Biotechnol Equip 3:91–97

Paunović M, Csanyi B, Simić V, Stojanović B, Cakić P (2006) Distribution of *Anodonta (Sinanodonta) woodiana* (Rea, 1834) in inland waters of Serbia. Aquat Invasions 1(3): 154–160. https://doi.org/10.3391/ai.2006.1.3.10

Paunović M, Csanyi B, Knezević S, Simić V, Nenadić D, Jakovcev-Todorović D, Stojanović B, Cakić P (2007) Distribution of Asian clams *Corbicula fluminea* (Muller, 1774) and *C. fluminalis* (Muller, 1774) in Serbia. Aquatic Invasions 2(2):105–112. https://doi.org/10.3391/ai.2007.2.2.3

Paunović M, Borković SS, Pavlović SZ, Saičić ZS, Cakić PD (2008) Results of the 2006 Sava Survey—aquatic macroinvertebrates. Arch Biol Sci 60(2):265–271. https://doi.org/10.2298/ABS0802265P

Paunović M, Csányi B, Simonović P, Zorić K (2015) Invasive alien species in the Danube. In: Liska I (ed) The Danube River Basin. Springer, Berlin

Pavlović SZ, Milošević SM, Borković SB, Simić VM, Paunović MM, Žikić RV, Saičić ZS (2006) A report of *Orconectes (Faxonius) limosus* (Rafinesque, 1817) [Crustacea: Decapoda: Astacidea: Cambaridae: Orconectes: Subgenus Faxonius] in the Serbian part of the River Danube. Biotechnol Biotechnol Equip 1:53–56

Pavlović M, Paunović M, Simić V (2013) Feeding of Eurasian perch (*Perca fluviatilis* L.) in three reservoirs in Serbia. Water Res Manag 3(4):41–46

Pavlović M, Simonović P, Stojković M, Simić V (2015) Analysis of diet of piscivorous fishes in Bovan, Gruža and Šumarice reservoir, Serbia. Iranian J Fish Sci 14(4):908–923

Pfleger V (2000) A field guide in colour to Molluscs. Silverdale Books, Prague. 216 pp

Pimentel D, Zuniga R, Morrison D (2005) Update on the environmental and economic costs associated with alien invasive species in the United States. Ecol Econ 52:273–288

Piria M, Simonović P, Kalogianni E, Vardakas L, Koutsikos N, Zanella D, Ristovska M, Apostolou A, Adrović A, Mrdak D, Tarkan AS, Milošević D, Zanella LN, Bakin R, Ekmekçi FG, Povž M, Korro K, Nikolić V, Škrijelj R, Kostov V, Gregori A, Joy MK (2017) Alien freshwater fish species in the Balkans–Vectors and pathways of introduction. Fish Fish 19(1): 138–169 https://doi.org/10.1111/faf.12242

Pujin V, Sotirov S (1966) A study of the nutrition of the dwarf catfish (*Ictalurus nebulosus* Le Sueur) (In Serbian). Ann Sci Work Fac Agric Novi Sad (Serbia) 10:147–156

Radenković M (2019) Feeding and importance of predatory fish species in maintenance of ecosystem stability in reservoirs. Ph.D. thesis, University of Kragujevac, Kragujevac, Serbia

Radenković M, Stojković Piperac M, Miloŝković A, Kojadinović N, Đuretanović S, Veličković T, Jakovljević M, Nikolić M, Simić V (2022) Diet seasonality and food overlap of *Perca fluviatilis* (Actinopterygii: Perciformes: Percidae) and *Rutilus rutilus* (Actinopterygii: Cypriniformes: Cyprinidae) juveniles: A case study on Bovan Reservoir, Serbia. Acta Ichthyologica et Piscatoria 52(1):77–90. https://doi.org/10.3897/aiep.52.78215

Raković M (2015) Diversity of Molluscs in the Danube River (1260–863.5 rkm) and taxonomic analysis of the genera Planorbarius, Radix, Physella and Ferrissia (Pulmonata: Basommatophora). PhD thesis, Faculty of Biology, University of Belgrade, Belgrade, Serbia

Raković M, Popović N, Kalafatić V, Martinović-Vitanović V (2013) Spreading of *Dreissena rostriformis bugensis* (Andrusov 1897) in the Danube River (Serbia). Acta Zoologica Bulgarica 65:349–357

Simić V, Simić S, Stojković Piperac M, Petrović A, Milošević D (2014) Commercial fish species of inland waters: A model for sustainability assessment and management. Sci Total Environ 497–498:642–650. https://doi.org/10.1016/j.scitotenv.2014.07.092

Simić V, Bănăduc D, Curtean-Bănăduc A, Petrović A, Veličković T, Stojković-Piperac M, Simić S (2022) Assessment of the ecological sustainability of river basins based on the modified the ESHIPPO fish model on the example of the Velika Morava basin (Serbia, Central Balkans). Front Environ Sci 10:952692. https://doi.org/10.3389/fenvs.2022.952692

Simonović P (2001) Ribe Srbije. NNK International, Beograd. Zavod za zaštitu prirode Srbije i Biološki fakultet Univerziteta u Beogradu, 247 pp

Smirnov AK, Pavlov DD, Kodukhova YV, Karabanov DP (2020) Impact of Zebra mussel *Dreissena polymorpha* Pallas, 1771 (Bivalvia) appearance on fish populations in Lake Pleshcheevo, European Russia. Zoologicheskiĭ Zhurnal 99:1363–1374. https://doi.org/10.31857/S0044513420110070

Sotirov SK (1968) Ecology of the brown bullhead (*Ictalurus nebulosus nebulosus* Le Sueur, 1819) (In Serbian). PhD thesis, University of Belgrade

Stamenković O, Stojković Piperac M, Milošević D, Buzhdygan OY, Petrović A, Jenačković D, Đurđević A, Čerba D, Vlaičević B, Nikolić D, Simić V (2019) Anthropogenic pressure explains variations in the biodiversity of pond communities along environmental gradients: a case study in south-eastern Serbia. Hydrobiologia 838:65–83. https://doi.org/10.1007/s10750-019-03978-4

Stockner JG, Rydin E, Hyenstrand P (2000) Cultural oligotrophication: Causes and consequences for fisheries resources. Fish Mag 25(5):7–14. https://doi.org/10.1577/1548-8446(2000)025<0007:CO>2.0.CO;2

Taylor NG, Dunn AM (2017) Size matters: predation of fish eggs and larvae by native and invasive amphipods. Biol Invasions 19:89–107. https://doi.org/10.1007/s10530-016-1265-4

Thomaz SM, Kovalenko KE, Havel JE, Kats LB (2015) Aquatic invasive species: general trends in the literature and introduction to the special issue. Hydrobiologia 746:1–12. https://doi.org/10.1007/s10750-014-2150-8

Tomović J, Zorić K, Kračun M, Marković V, Vasiljević B, Simić V, Paunović M (2012) Freshwater mussels of the Velika Morava River. Water Res Manag 2(4):51–55

Tomović J, Simić V, Tubić B, Zorić K, Kračun M, Marković V, Paunović M (2013) Freshwater mussels of the Serbian stretch of the Tisa River. Water Res Manag 3(1):35–40

Tubić B (2016) Testing different methods for sampling macroinvertebrates in aquatic ecosystems and examination of the possibility for their standardization. Ph.D. thesis, University of Kragujevac, Kragujevac, Serbia

Van der Schalie WH, Shedd TR, Knechtges PL, Widder MW (2001) Using higher organisms in biologically early warning systems for real time toxicity detection. Biosens Bioelectron 16(7–8): 457–465. https://doi.org/10.1016/s0956-5663(01)00160-9

Vander Zanden MJ, Casselman JM, Rasmussen JB (1999) Stable isotope evidence for the food web consequences of species invasions in lakes. Nature 401:464–467. https://doi.org/10.1038/46762

Zhao D, Borkhanuddin MH, Wang W, Liu Y, Cech G, Zhai Y, Székely C (2016) The life cycle of *Thelohanellus kitauei* (Myxozoa: Myxosporea) infecting common carp (*Cyprinus carpio*) involves aurantiactinomyxon in *Branchiura sowerbyi*. Parasitol Res 115:4317–4325. https://doi.org/10.1007/s00436-016-5215-y

Zorić K, Jakovčev-Todorović D, Đikanović V, Vasiljević B, Tomović J, At-anacković A, Simić V, Paunović M (2011) Distribution of the Ponto-Caspian polychaeta Hypania invalida (Grube, 1860) in inland waters of Serbia. Aquat Invasions 6(1):33–38. https://doi.org/10.3391/ai.2011.6.1.04

Zorić K, Szekeres J, Csányi B, Kolarević S, Marković V, Paunović M (2015) Distribution of alien Bryozoan species, *Pectinatella magnifica* (Leidy 1851), in the river Danube. Acta Zoologica Bulgarica 67(2):241–247

Zorić K, Atanacković A, Tomović J, Vasiljević B, Tubić B, Paunović M (2020a) Diversity of alien macroinvertebrate species in Serbian waters. Water 12:3521. https://doi.org/10.3390/w12123521

Zorić K, Atanacković A, Ilić M, Tubić B, Csányi B, Paunović M (2020b) The spiny-cheek crayfish *Faxonius limosus* (Rafinesque, 1817) (Decapoda: Cambaridae) invades new areas in Serbian inland waters. Acta Zoologica Bulgarica 72(4):623–627

Part IV
Fish Stock Assessment Methods

Summary

Stock assessments of fishery resources are much better developed for marine than inland fisheries. The success and efficiency of the assessment largely depend on statistical data, which is a very weak link in the freshwater fisheries assessment process chain as it requires the participation of anglers, managers, and scientists. This chapter presents a model that targets managers of fisheries areas, in addition to scientists, because it uses basic fisheries parameters for fish stock assessment obtained from realistically collected catch data.

A Brief Overview of the Development of the ES–HIPPO Model for Assessing the Sustainability and Conservation Priorities of Fish, Fish Resources, and Inland Water Habitats

Vladica Simić, Snežana Simić, Ana Petrović, Tijana Veličković, Predrag Simović, Milica Stojković-Piperac, and Đurađ Milošević

Abstract This chapter contains a brief overview of the development of models for assessing the sustainability and conservation priorities for fish, fish resources, and inland water habitats in Serbia and the region. The basic version of the ES–HIPPO model consists of two elements. The first, Ecological Specialization (ES), reflects the level of taxon specialization in relation to habitat, diet, reproduction strategy, life cycle, body size, degree of endemism, and degree of isolation. The second factor, "HIPPO," refers to the impact of factors such as H-habitat alterations, I-invasive species, P-pollution, P-human population growth, and O-overexploitation on populations in the spatial and temporal dimensions. From the general model, advanced models were developed: *ES–HIPPOfish* intended for assessing the viability of fish species that are important for fishing, ES–HIPPOcrayfish to assess the sustainability of decapod crustacean populations, and the latest version of *ESE–HIPPOriverbasin* to assess the degree of habitat conservation for a sustainable fish community in a river basin.

Keywords Serbia; ES · HIPPO model; *ES · HIPPOfishing*; ES · HIPPOcrayfish; *ESE · HIPPOriverbasin*

V. Simić (✉) · S. Simić · A. Petrović · T. Veličković · P. Simović
Department of Biology and Ecology, Faculty of Science, University of Kragujevac, Kragujevac, Serbia
e-mail: vladica.simic@pmf.kg.ac.rs

M. Stojković-Piperac · Đ. Milošević
Department of Biology and Ecology, Faculty of Sciences and Mathematics, University of Niš, Niš, Serbia

V. Simić et al. (eds.), *Ecological Sustainability of Fish Resources of Inland Waters of the Western Balkans*, Fish & Fisheries Series 43,
https://doi.org/10.1007/978-3-031-36926-1_14

1 Introduction

Successful protection and conservation of ecosystems can only be achieved if there is a good diagnosis of ecosystem health (Pitcher 2015). Apart from the assessment of environmental conditions (biota), the assessment of the stability of the biotic component is necessary. In current conditions, when ecosystems are exposed to numerous stressors, mostly of anthropogenic origin, inland waters are susceptible and quickly lose ecological sustainability and loss of ecological services (Bănăduc et al. 2022). Fish, as the last or top members of the food chain, are good indicators of the health of aquatic ecosystems. The assessment is two-way, which means that the assessment of the condition of fish communities and populations reflects the health and sustainability of the ecosystem, and on the contrary, poor habitat conditions are often the cause of low sustainability of populations and fish stocks (Karr 1981, 1987; Jepsen and Pont 2007; Pont et al. 2007; Stojković et al. 2011, 2013; Radinger et al. 2019). However, the unsustainable fish community in good ecological habitat conditions is often a consequence of the overexploitation of the fish stocks by fishing.

Furthermore, while there are well-developed methods for assessing the state of fish stocks in marine ecosystems due to very intensive commercial fishing and the increasing depletion of fish and other biological resources, for inland waters, there are significantly fewer developed methods (Cooke et al. 2016). Often, methodology for marine is also used for inland fish resources with certain modifications (e.g., methods in the software packages FISAT 1 and 2). The European Water Directive (WFD) developed the European Ichthyological Index based on the Ichthyological Biotic Indices (IBI) in order to assess the ecological status of inland waters, leaving the possibility for each member state to develop modifications under the local and regional specificities of the fish fauna (Jepsen and Pont 2007). These methods have been significantly improved using eDNA methods (Pont et al. 2019).

Our long-term experience in assessing the state of the fish stocks and the water ecological status, using generally accepted WFD methods and assessment models (primarily FISAT 1 and 2) (Gayanilo et al. 2005), indicated that the hydrogeographic heterogeneity of the inland water habitats of Serbia requires certain modifications or the development of new assessment model. The next reason for the development of new models is the chronic lack or irregularity of data regarding the catches of commercial and recreational fishermen. That significantly complicates the development of sustainable management programs because the mentioned data obtained from the fishermen can hardly be used in the above-mentioned methods and models, especially for the assessment of fishing pressure. Finally, an important reason for developing new models is the more efficient implementation of conservation efforts and measures for the sustainable use of the fish stocks in accordance with the ecosystem approach to fisheries in inland waters (Kolding et al. 2016). Considering the stated reasons, the ES–HIPPO model system is designed not only for scientists but also for experts in managing fishing areas and fish stocks. It enables the management of fish stocks by all social actors and not only by the "top–down" system (Welcomme 2016).

1.1 *The ES–HIPPO Basic Model and its Importance in the Conservation of Fish Stocks*

The ES–HIPPO model (Simić et al. 2007) was created to supplement the assessment results of the degree of endangerment of aquatic organisms at the national level, which are obtained at the expense of the global IUCN criteria (IUCN 2022). The primary goal was expanded by assessing the degree of sustainability and priority for conservation at the national and population levels. The model is based on the ecological specialization of taxa, which is the result of adaptation through evolution and the ability to resist specific modern multiple stressors (Fisher and Owens 2004). The model has two elements; the first is the Ecological Specialization of taxa (ES) in relation to habitat, diet, breeding strategy, life cycle, body size, degree of endemism, and degree of isolation (Fisher and Owens 2004). The second factor, "HIPPO" refers to the impact of factors such as H-Habitat alterations, I-Invasive species, P-Pollution, P-human Population growth, and O-Overexploitation on population in spatial and temporal dimensions (Brennan and Withgott 2005). The degree of ecological specialization is evaluated on a three-level scale: 1—high ecological specialization of the taxon, 3—moderate, and 5—low specialization (cosmopolitan species). The sum of the ES + HIPPO factors for each population ($\sum$ES + HIPPO.np) of the taxon gives information regarding the degree of sustainability and the level of conservation priority. The higher sum indicates lower ecological sustainability and higher conservation priority (PP). The model was designed to assess the degree of sustainability of aquatic habitat taxa in general but has been most used for fish. The results obtained for the assessment of the degree of sustainability of fish species and the level of conservation priorities significantly helped in the implementation of the NATURA 2000 program for Serbia and the preparation of red books. Apart from this, the results were also used to propose taxa of macroalgae, macroinvertebrates, and fish for the national list of strictly protected species (Anonymous 2016) and the list of fish that need a permanent ban on fishing (Regulations on measures for the conservation and protection of fish stocks). The current list of autochthonous fish species with a certain degree of sustainability and conservation priority level based on the ES–HIPPO model is shown in Table 1. Particular conservation efforts should be focused on species characterized as High priority based on the ES–HIPPO model on the territory of Serbia (Table 1). Within those species, some have economic (e.g., Acipenseridae) or fisheries significance (e.g., huchen), but the model also implies that species with small-sized bodies and rare in Serbia but without economic value should be prioritized in conservation (e.g., vairone (Veličković et al. 2020) and Romanian golden loach (Marić et al. 2022)) (Table 1).

Table 1 Level of sustainability and priorities for fish conservation based on the ES–HIPPO model and in relation to global and official national status

Family/Fish species valid name according to FishBase	Global IUCN*	Appendix I and II of the Rulebook**	ES–HIPPO
Petromyzontidae			
Lamprey *Eudontomyzon danfordi* (Regan, 1911)	LC	I	No data
Drin brook lamprey *Eudontomyzon stankokaramani* (Karaman, 1974)	LC	I	No data
Ukrainian brook lamprey *Eudontomyzon mariae* (Berg, 1931)	LC	I	No data
Danubian brook lamprey *Eudontomyzon vladykovi* (Oliva & Zenandrea 1959)	LC	I	No data
River lamprey *Lampetra fluviatilis* (before *Petromyzon fluviatilis*) (Linnaeus, 1758)	LC		No data
European brook lamprey *Lampetra planeri* (Bloch, 1784)	LC		No data
Sea lamprey *Petromyzon marinus* (Linnaeus, 1758)	LC		No data
Acipenseridae			
Danube sturgeon *Acipenser gueldenstaedti* (Brandt and Ratzeburg, 1833)	CR	I	High
Fringebarbel sturgeon *Acipenser nudiventris* (Lovetsky, 1828)	CR	I	High
Sterlet sturgeon *Acipenser ruthenus* (Linnaeus, 1758)	EN	II	High
Starry sturgeon *Acipenser stellatus* (Pallas, 1771)	CR	I	High
Sturgeon *Acipenser sturio* (Linnaeus, 1758)	CR	I	High
Beluga sturgeon *Huso huso* (Linnaeus, 1758)	CR	I	High
Salmonidae			
Peled *Coregonus peled* (Gmelin, 1789)	LC	/	No data
Huchen *Hucho hucho* (Linnaeus, 1758)	EN	II	High
Marble trout *Salmo marmoratus* (Cuvier, 1829)	LC	I	No data
Brown trout *Salmo trutta* (Linnaeus, 1758)	LC	II	High–moderate
Salmo farioides (Karaman, 1938)	/	/	No data
Macedonian trout *Salmo macedonicus* (Karaman, 1924)	DD	/	Moderate
Arctic char *Salvelinus alpinus* (Linnaeus, 1758)	LC	/	
Brook trout *Salvelinus fontinalis* (Mitchill, 1814)	/	/	
Rainbow trout *Oncorhynchus mykiss* (Walbaum, 1792)	NE	/	
Thymalidae			
Grayling *Thymallus thymallus* (Linnaeus, 1758)	LC	II	Moderate
Angulidae			
European eel *Anguilla anguilla* (Linnaeus, 1758)	CR	I	High
Clupeidae			
Caspian shad *Alosa caspia* (Eichwald, 1838)	LC	/	High

(continued)

Table 1 (continued)

Family/Fish species valid name according to FishBase	Global IUCN*	Appendix I and II of the Rulebook**	ES–HIPPO
Pontic shad *Alosa immaculata* (Bennet, 1835)	VU	I	High
Esocidae			
Northern pike *Esox lucius* (Linnaeus, 1758)	LC	II	Moderate
Umbridae			
European mudminnow *Umbra krameri* (Walbaum, 1792)	VU	I	High–moderate
Cyprinidae			
Zope *Ballerus ballerus* (before *Abramis ballerus*) (Linnaeus, 1758)	LC	II	Low
Bream *Abramis brama* (Linnaeus, 1758)	LC	II	Low
White-eye bream *Ballerus sapa* (before *Abramis sapa*) (Pallas, 1814)	LC	II	Moderate
Barbel *Barbus barbus* (Linnaeus, 1758)	LC	II	
Round-scaled barbel *Barbus cyclolepis* (Heckel, 1837)	LC	I	High
Danube barbel *Barbus balcanicus* (Kotlík, Tsigenopoulos, Ráb and Berrebi, 2002)	LC	II	Low
Crucian carp *Carassius carassius* (Linnaeus, 1758)	LC	I	High
Carp *Cyprinus carpio* (Linnaeus, 1758)	VU	II	Low?
Goldfish *Carassius auratus* (before *Carassius auratus auratus*) (Linnaeus, 1758)	LC	/	
Prussian carp *Carassius gibelio* (before *Carassius aurata gibelio*) (Bloch, 1782)	/	/	
Grass carp *Ctenopharyngodon idella* (Valenciennes, 1844)	LC	/	
Xenocyprididae			
Silver carp *Hypophthalmichthys molitrix* (Valenciennes, 1844)	NT	/	
Bighead carp *Hypophthalmichthys nobilis* (Richardson, 1844)	DD	/	
Gobionidae			
White-finned gudgeon *Romanogobio albipinnatus* (before *Gobio albipinnatus*) (Lukasch, 1933)	LC	II	High
Gudgeon *Gobio gobio* (Linnaeus, 1758) *Gobio obtusirostris*	LC	II	Low
Kessler's gudgeon *Romanogobio kesslerii* (Dybowski, 1862)	LC	II	Moderate
Danube longbarbel gudgeon *Romanogobio uranoscopus* (Agassiz, 1828)	LC	II	High
Topmouth gudgeon *Pseudorasbora parva* (Temminck and Schlegel, 1842)	LC	/	
Leuciscidae			

(continued)

Table 1 (continued)

Family/Fish species valid name according to FishBase	Global IUCN*	Appendix I and II of the Rulebook**	ES–HIPPO
Spirlin, schneider *Alburnoides bipunctatus* (Bloch, 1782)	/	II	Low
Italian bleak *Alburnus albidus* (Costa, 1838)	VU	/	/
Bleak *Alburnus alburnus* (Linnaeus, 1758)	LC	/	Low
Alburnus scoranza (Bonaparte, 1845)	LC	/	/
Danube bleak *Alburnus chalcoides* (Güldenstadt, 1772)	/	I	No data
White bream *Blicca bjoerkna* (Linnaeus, 1758)	LC	/	Low
Nase *Chondrostoma nasus* (Linnaeus, 1758)	LC	II	Low
Aral asp *Leuciscus aspius* (before *Aspius aspius*) (Linnaeus, 1758)	LC	II	Low
Belica *Leucaspius delineatus* (Heckel, 1843)	/	I	No data
Chub *Squalius cephalus* (before *Leuciscus cephalus*) (Linnaeus, 1758)	LC	II	Low–moderate
Ide, orfe *Leuciscus idus* (Linnaeus, 1758)	/	II	Low–moderate
Common dace *Leuciscus leuciscus* (Linnaeus, 1758)	LC	/	No data
Vairone *Telestes souffia* (Risso, 1826)	LC	I	High
Pachychilon macedonicus (Steindachner, 1892)	DD	/	No data
Pachychilon pictum (Heckel & Kner, 1858)	LC	I	No data
Sichel *Pelecus cultratus* (Linnaeus, 1758)	LC	II	Nigh
Common minnow *Phoxinus phoxinus* (Linnaeus, 1758)	LC	/	High–moderate
Leucos basak (before *Rutilus basak)* (Heckel, 1843)	LC	/	No data
Pigo *Rutilus pigus* (Lacepede, 1804)	LC	II	Moderate
Rutilus virgo (Heckel, 1852)	LC	/	No data
Rutilus or common roach *Rutilus rutilus* (Linnaeus, 1758)	LC	/	Low
Scardinius erythrophthalmus (Linnaeus, 1758)	LC	/	Moderate
Scardinius graecus (Stephanidis, 1937)	CR	/	No data
Scardinius knezevici (Bianco and Kottelat, 2005)	LC	/	No data
Vimba *Vimba vimba* (Linnaeus, 1758)	LC	II	Low
Macedonian vimba *Vimba melanops* (Heckel, 1837)	DD	/	High?
Acheilognathidae			
Ammur bitterling *Rhodeus sericeus* (Pallas, 1776)	LC	I	Low
Tincidae			
Tench *Tinca tinca* (Linnaeus, 1758)	LC	I	High–moderate
Cobitidae			
Balkan loach *Cobitis elongata* (Heckel & Kner, 1858)	LC	I	Moderate
Spined loach *Cobitis taenia* (Linnaeus, 1758)	LC	II	Moderate
True loach *Misgurnus fossilis* (Linnaeus, 1758)	LC	I	Moderate

(continued)

Table 1 (continued)

Family/Fish species valid name according to FishBase	Global IUCN*	Appendix I and II of the Rulebook**	ES–HIPPO
Balkan spined loach *Sabanejewia balcanica* (Filippi, 1865)	LC	I	Moderate
Sabanejewia bulgarica	LC	I	No data
Romanian golden loach *Sabanejewia romanica* (Băcescu, 1943)	NT	/	High?
Nemacheilidae (before Balitoridae)			
Stone loach *Barbatula barbatula* (Linnaeus, 1758)	LC	/	Moderate
Struma stone loach *Oxynoemacheilus bureschi* (before *Barbatula bureschi*) (Drensky, 1928)	LC	I	High
Siluridae			
Wels catfish *Silurus glanis* (Linnaeus, 1758)	LC	II	Low
Ictaluridae			
Brown bullhead *Ameiurus nebulosus* (Le Sueur, 1819)	LC	/	
Black bullhead *Ameiurus melas* (Rafinesque, 1820)	LC	/	
Lotidae (before Gadidae)			
Burbot *Lota lota* (Linnaeus, 1758)	LC	II	Moderate
Syngnathidae			
Black-striped pipefish *Syngnathus abaster* (Risso, 1826)	LC	/	
Gasterosteidae			
Three-spined stickleback *Gasterosteus aculeatus* (Linnaeus, 1758)	LC	/	No data
Southern nine-spined stickleback *Pungitius platygaster* (Kessler, 1859)	LC	/	No data
Percidae			
Balon's ruffe *Gymnocephalus baloni* (Holčik & Hensel, 1974)	LC	I	High
Ruffe *Gymnocephalus cernua* (before *Gymnocephalus cernuus*) (Linnaeus, 1758)	LC	/	Low
Schraetzer *Gymnocephalus schraetser* (Linnaeus, 1758)	LC	II	Moderate
European perch *Perca fluviatilis* (Linnaeus, 1758)	LC	II	Low
Pikeperch *Sander lucioperca* (Linnaeus, 1758)	LC	II	Low
Volga pikeperch *Sander volgensis* (Gmelin, 1788)	LC	II	Moderate
Danube streber *Zingel streber* (Siebold, 1863)	DD	I	High
Zingel *Zingel zingel* (Linnaeus, 1766)	LC	I	Moderate
Centrarchidae			
Pumpkinseed *Lepomis gibbosus* (Linnaeus, 1758)	LC	/	
Largermouth black bass *Micropterus salmoides* (Lacepede, 1802)	LC	/	
Gobiidae			
Monkey goby *Neogobius fluviatilis* (Pallas, 1811)	LC	/	

(continued)

Table 1 (continued)

Family/Fish species valid name according to FishBase	Global IUCN*	Appendix I and II of the Rulebook**	ES–HIPPO
Racer goby *Babka gymnotrachelus* (Kessler, 1857)	LC	/	
Bighead goby *Ponticola kessleri* (Gunther, 1861)	LC	/	
Round goby *Neogobius melanostomus* (Pallas, 1811)	LC	/	
Tubenose goby *Proterorhinus marmoratus* (Pallas, 1811)	LC	/	
Blenniidae			
Freshwater blenny *Salaria fluviatilis* (Asso, 1801)	LC	II	No data
Cottidae			
European bullhead *Cottus gobio* (Linnaeus, 1758)	LC	I	Moderate
Odontobutidae			
Chinese sleeper *Perccottus glenii* (Dybowski, 1877)	LC	/	

*IUCN (2001) Red List categories and criteria, version 3.1 Critically Endangered (CR), Endangered (EN), Vulnerable (VU), Near Threatened (NT), Least Concern (LC), Data Deficient (DD). **Rulebook on declaration and protection of protected and strictly protected wild species of plants, animals, and fungi (Official Gazette of RS No. 5/2010, 47/2011, 32/2016, and 98/2016). Appendix (I—strictly protected, II—protected species)

2 A Model Adapted to Assess the Sustainability of Commercially Important Fish Species *ES–HIPPOfishing*

The *ES–HIPPOfishing* model (Simić et al. 2014) is a modification of the basic model (ES–HIPPO) designed to assess the sustainability of fish species exploited through commercial and recreational fishing in inland waters. A significant adjustment to this goal was achieved by introducing a new sub-element in the model, Index of Local Sustainability of Fish Populations (ILSFP). Spatial and temporal indicators of populations of commercially important fish species were used to estimate ILSFP, and they included dominance, frequency, biomass, maturity of females (%), number of age classes, the average length of fish stocks, and percentage of population located in protected areas. The ILSFP is estimated based on the trend of the mentioned indicators during ten years and spatially along the river courses of a catchment area. Stream sections with the highest ILSFP values for a particular fish species are designated as critical habitats for the sustainability of that species throughout the basin. The model has certain similarities in some elements with the work of Giam et al. (2011), Chantepie et al. (2011), Linke et al. (2012), and Chester and Robson (2013).

The results obtained for the sustainability of commercially important fish species based on the *ES–HIPPOfishing* model from the creation period in 2013/14 and the present (2021/22) are shown in Table 2.

After just under ten years, the sustainability of commercially important fish species obtained from the *ES–HIPPOfishing* model indicates some significant

Table 2 Degree of sustainability of commercial fish species in Serbian waters (Low—L, Moderate—M, High—H) based on calculated values of the *ES–HIPPOfishing* model during 2013/14 and estimates during 2021/22

Fish species	*ES–HIPPOfishing* 2014	*ES–HIPPOfishing* 2022
Huchen *Hucho hucho*	54 – VL	56 – VL
Sterlet sturgeon *Acipenser ruthenus*	54 – VL	60 – VL
Grayling *Thymallus thymallus*	50 – L	54 – VL
Brown trout *Salmo trutta*	50 – L	54 – VL
Northern pike *Esox lucius*	50 – L	52 – L
Barbel *Barbus barbus*	49 – L	52 – L
Pikeperch *Sander lucioperca*	46 – L	40 – ML
Carp *Cyprinus carpio*	46 – L	46 – L
Bream *Abramis brama*	42 – L	32 – M
Wels catfish *Silurus glanis*	42 – L	34 – M
Nase *Chondrostoma nasus*	38 – M	38 – M
Chub *Squalius cephalus*	32 – M	40 – ML
Aral Asp *Leuciscus aspius*	–	28 – H
Prussian carp *Carassius gibelio*	30 – M	28 – H

Abbreviations: VL (Very Low), ML (Moderate Low)

changes. The presented results show that the level of sustainability is decreasing for all salmonid species. In particular, this applies to brown trout, which has moved from a low sustainability level to a very low sustainability level. The reasons are multiple and highlighted in several chapters of this monograph. They include the habitat fragmentation due to the intensive construction of small hydro-power plants (SHPP), genetic contamination of the autochthonous Danubian genetic lineage with allochthonous Atlantic and Adriatic genes, climate changes affecting the hydrological and temperature regime of the salmonid area, pollution, and still present poaching. The model can be used to obtain key habitats for preserved autochthonous populations, which primarily include populations with unique haplotypes. Among the habitats where the viable and autochthonous trout populations are preserved, the Đetinja River and the upper reaches of the Temska River Basin in the Stara Planina National Park can be singled out. There is also observed the high sustainability of some populations contaminated with Atlantic lineage specimens, such as the Sokobanjska Moravica (South Morava Basin) and Gradac river (Kolubara basin) (Fig. 1).

In the case of cyprinid species, the model indicates reduced viability of chub populations (Fig. 2). Our research indicates that chub populations have a decreasing trend in the Morava basin, especially in the middle and lower reaches of larger tributaries (Simić et al. 2022).

On the other hand, there is a certain increase in the level of sustainability of fishing species such as bream, pikeperch, and wels catfish. Populations of these species with a higher level of sustainability occur in reservoirs in the Morava and Drina basins and the Sava River (Figs. 3 and 4). In addition to these species, a higher level of sustainability of populations of Aral asp, a predatory species from the

Fig. 1 Trout from the river Gradac (Kolubara Basin) (Photo by V. Simić)

Fig. 2 Chub from the recreational catch from Studenica River (Biosphere Reserve Golija-Studenica) from August 2020 (Photo by V. Simić)

Cyprinidae family, which has been a frequent catch in recent years, was also measured by commercial and recreational fishermen in the Danube, Sava, and Morava.

3 A Model Adapted to Assess the Viability of the Decapod Crustaceans ES–HIPPOcrayfish

Unlike some European and Scandinavian countries, Serbia does not have a tradition of harvesting or consuming freshwater decapod crustaceans. The situation is similar in other Western Balkans countries, but there are also cases where crayfish from these areas are caught by foreign concessionaires and exported to Western Europe (Rajković 2004; Rajković et al. 2006, 2012; see Chap. 8, Đuretanović et al. 2023). The crayfish plague decimated crayfish populations in Europe and only, to a slightly lesser extent, affected the Balkan Peninsula. Invasions of North American crayfish (*Pacifastacus leniusculus* and *Faxonius limosus*), pollution, and habitat destruction

Fig. 3 Catch of a commercial fisherman on the Sava River near Šabac with a dominant bream (Photo by M. Vlajković)

have further adversely affected crayfish populations. Research on the distribution of decapod crustaceans on the territory of Serbia by Simić et al. (2008), the first after 50 years, was a warning sign because it showed a critically endangered status for the noble crayfish *Astacus astacus*. In order to more effectively preserve decapod crustaceans in the area of the Balkan Peninsula, a modification of the ES–HIPPO model to *ES–HIPPOcrayfish* was formulated (Simić et al. 2015). The main modification was based on the introduction of the Index of the Local Adaptive Population of Crayfish (ILAP) element that assesses the level of the adaptive value of individual crayfish populations to local habitat conditions. ILAP indicators include genetic (phylogeny and nucleotide diversity), morphometric, and population data. The aim of the model was the determination of the level of sustainability of individual populations and the selection of the basic conservation unit and elemental conservation unit (ECU). The ECU represents the population that has priority in the conservation program. The model indicated very low sustainability of *A. astacus* populations in Serbia (Fig. 5), and the higher sustainability of *Austropotamobius torrentium* populations compared to *Austropotamobius pallipes/italicus* populations that inhabit the rivers of the Adriatic basin. The model also indicated the unsustainability of *A. astacus* populations in the reservoirs of the headwaters of

Fig. 4 Experimental fishing with standing nets for scientific research purposes in June 2020. The dominance of bream is evident in the Zavoj reservoir in the Stara Planina National Park (Photo by V. Simić)

the Zeta River in Montenegro if the rate of exploitation by foreign concessionaires continues (Rajković 2004; Rajković et al. 2006; Petrović et al. 2013). Similar considerations on crayfish can be read in Souty-Grosset et al. (2003), Bonin et al. (2007), Klobučar et al. (2013), Schrimpf et al. (2014), Parvulescu and Zaharia (2014), Lovrenčić and Maguire (2021), and Lovrenčić et al. (2022).

4 A Model for Assessing the Sustainability of River Basins *ESE–HIPPOriverbasin*

The latest version of the ES–HIPPO–*ESE–HIPPOriverbasin* model (Simić et al. 2022) has significant modifications. The primary goal and purpose of the model have been changed. Instead of assessing the sustainability of populations of aquatic organisms, especially decapod crustaceans and fish, the model deals with assessing the stability and sustainability of individual aquatic habitats up to the overall sustainability of river basins. As in the previous variants, the model has two elements: Ecological Stability of the Ecosystem (ESE) and "HIPPO." At the core of the assessment of the first element of the model, now called ESE, is the structure of the fish community. The indicators for ESE that may differ in the time dimension are presented in Table 3.

The first group of indicators (A set in the table) has a time dimension that implies that the value for each indicator of the structure of the fish community from a certain

Fig. 5 Drying up of small rivers in Central Serbia due to climatic changes, but also excessive exploitation of water due to the capture of springs for water supply, and irrigation, threatens the survival of the remaining populations of *Astacus astacus*. The picture presents the upper course of the Lepenica River and the remaining ponds during the summer of 2021 (**a**). Crayfish probably survive in these remaining ponds and later probably burrow into the wet mud (**b**) (Photo by V. Simić)

habitat can only be calculated if there is data on it for at least ten years (the longer the period, the more precise the data will be). In this case, a problem for model operation may be a lack of data. However, apart from scientists, the model is primarily designed for professional managers of fishing areas. If they use it, it requires constant monitoring of the state of the fish stock through real data collection on the catch of commercial and recreational fishermen. In this way, the model (if the manager uses it) can indirectly affect the chronic problem of lack of data on real catch and fishing pressure in freshwater fisheries. Another possibility enabled by the temporal version of indicator monitoring is alignment with ecosystem approaches to fisheries in inland waters. Primarily, this refers to the possibility of planning a balanced harvest in parts

Table 3 Structure of indicators of the *ESE–HIPPOriverbasin* model (taken from Simić et al. 2022)

ESE indicators (code/abbreviation)	Threshold and scoring system for each ESE indicator		
	1 (low)	3 (medium)	5 (high)
Fish community composition (species diversity)			
(% Au/his): Total number of autochthonous taxa (deviation between present and historical condition)[1]	Deviation >50%	Deviation <50%	No or minor change ±10%
The Shannon–Weaver Index (d)* (value d cannot be calculated when there is only one species)	d < 1	d (1–2)	d > 2
(Al. sp.%): Total number of allochthonous species[2]	Allochthonous species >50%	Allochthonous species (30–50%)	Allochthonous species <30%
The characteristics of target fish population which zonation concept is based on			
A (MF): Trend in abundance of target fish population in the past 10–20 years[3]	Decrease in abundance	Variation in abundance up to ±10%	Increase in abundance
B (MF): Trend in biomass target of fish population in the past 10–20 years	Decrease in biomass	Variation in biomass up to ±10%	Increase in biomass
Implementation of ABC (abundance biomass comparison) curves (Warwick and Clarke 1994)*	The pattern in the abundance and biomass shows heavily disturbed assemblages	The pattern in the abundance and biomass shows moderately disturbed assemblages	The pattern in the abundance and biomass shows undisturbed assemblages
TL (MF): Trend in total length of target fish population in the past 10–20 years[4]	Decrease in total length	Variation in total length up to ±10%	Increase in total length
Comparison of analyzed specimen's length with a common length of the species (FishBase)*	More than 50% of specimens TL < common length (FishBase)	% of specimens containing the common length: 50% ± 5% (FishBase)	More than 50% of specimens TL > common length (FishBase)
nAge (MF): Trend in number age classes in the past 10–20 years	Decrease in number age classes (1–2)	Stagnation in the number of age classes (3–4)	Increase in number age classes (>4)
The characteristics of accompanying fish population in particular zone			
A, B, TL, nAge (SMF)[5]	The same scoring system as used for target fish species		
Predator fish species			
(nPF): Total number of predator taxa (deviation between present and historical condition)[6] (%)	Decrease <50%	Stagnation or ≥ 1 species	Increase >30%

(continued)

Table 3 (continued)

ESE indicators (code/ abbreviation)	Threshold and scoring system for each ESE indicator		
	1 (low)	3 (medium)	5 (high)
(BPT): Trend in biomass predator fish population in the past 10–20 years	Decrease in biomass	Variation in biomass up to ±10%	Increase in biomass
Total number of predator taxa*	< 50% predatory fish characteristic for the fish zone	50% predatory fish characteristic for the fish zone	All / > 50% predatory fish characteristic for the fish zone
Biomass predator fish (BP) population in comparison to other fish populations (BO)*	BP/BO >10%	BP/BO (10–30%)	BP/BO >30%
Sensitive fish species			
(% SFT): Total number of sensitive fish species given in % and/or trend in the past 10–20 years [7–8]	ST ≤ 30% and/or decrease	ST 31–60% or ST ≤ 30%	ST > 60% and/or increase
	M ≤ 50% [8]	M > 50%	

In cases that there are lack of data for the evaluation of the indicators, alternative information for the evaluation are provided and marked with an asterisk
[1]Pančić (1860); Ristić (1972, 1977), Simonovic (2001). [2] Modification of Site-specific Biological Contamination (SBC) (Panov et al. 2008).[3] Fish river zone (Thienemann 1928; Huet 1949, 1959), characteristics of typical zone fish species: autochthonous species, fishing species, frequency > 90%, dominant >75%. [4] TL—Total body length. [5] Characteristics subtypical fish species: autochthonous species, fishing species, frequency > 75%, dominant >50%. [6] Primary and secondary piscivore fish species (excluding *Anguilla anguilla*): *Silurus glanis, Esox lucius, Sander lucioperca, Aspius aspius, Hucho hucho, Perca fluviatilis,* and *Salmo trutta* TL > 0,35 m. [7–8] FishBase data of Resilience (van Treeck et al., 2020); moderate fish species (%)

of the basin or the entire basin for certain fish species, several of them, or the total harvest. For example, by monitoring the trend of age classes and fish production in the waters exploited by fishing it is possible to assess the increased fishing pressure on the demographic structure (young fish, large older fish) in a certain time and accordingly make corrections in the management plans in terms of directing the pressure on the age classes with the largest current production (for further reading, see: Kolding et al. 2003; Kolding and van Zwieten 2011; Garcia et al. 2012; Law et al. 2012, 2013; Welcomme 2016; Kolding et al. 2016). Suppose there is no continuous data for the selected indicators, the model can be used for the current assessment of the ecological sustainability of fishing waters in the basin using data set marked in bold (Table 3). The model's significant role in preserving the fish stock is based on the possibility of determining parts of the watershed that can be declared as freshwater ecological reserves. Freshwater protected areas (FPAs) are also part of the ecosystem approach to fisheries in inland waters. Based on the indicators of the structure of the fish community and the intensity of the influence of the "HIPPO" factor, the model estimates the degree of ecological sustainability of the habitat in the watershed. Habitats with a higher degree of sustainability also support a sustainable fish community and, indirectly, the condition and sustainability

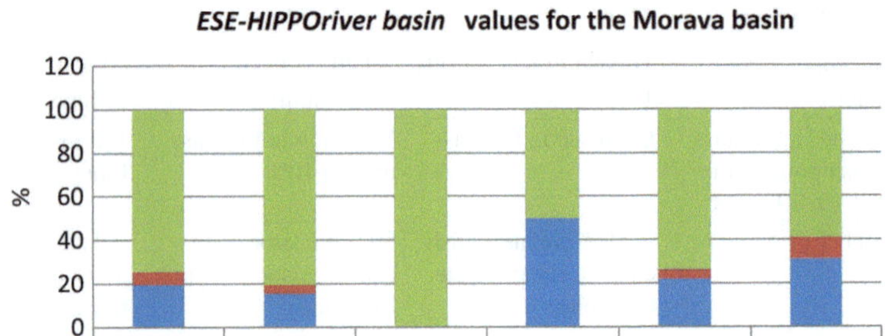

	HS General	HS Clch	HS BZ	HS THZ	HS DBZ	HS TZ
M	74.6	80.7	100	50	73.6	59.1
L	5.8	4.2	0	0	4.5	9.8
H	19.6	15.4	0	50	21.8	31.1

Fig. 6 Ecological sustainability of habitats in the Morava basin (Serbia) based on the *ESE–HIPPOriverbasin* model. Abbreviations: H—high level of sustainability, M—medium level of sustainability, L—low level of sustainability. HS—% of the length of the catchment stream which has high sustainability in relation to the total length. HS Clch—% HS under the influence of climate change. HS BZ—HS barbel zone, HS THZ—HS grayling zone, HS DBZ—HS Danube barbel zone, HS TZ—HS trout zone (taken from Simić et al. 2022)

of fish species important to fisheries and the fish stock as a whole. By monitoring and analyzing the level of habitat sustainability in the watershed, the schedule of FPAs can be planned. FPAs can be declared on parts of the reserves, watersheds with naturally preserved fish communities, or individual fish species to recover the fish stock from overfishing and/or some other stressor. In both cases, the reserves are spared from fishing and other anthropogenic influences and, over time, become places from which the "spillover" effect maintains the stability and sustainability of fish communities and fish stocks beyond their borders (Hannah et al. 2019). In the example of the Morava basin, which covers the largest part of Serbia, one can see the state of ecological sustainability of the basin and its units, which was obtained based on the *ESE–HIPPOriverbasin* model.

The results presented in Fig. 6 indicate an unexpectedly high percentage of habitats with a moderate and low level of ecological sustainability (about 50%) in the trout zone of this watershed. A careful analysis of the trend of the ESE and HIPPO factors shows major changes in brown trout populations in the last ten years, primarily due to massive habitat fragmentation, changes in the flow regime, and climatic factors that cause changes in temperature (greater warming of the water during the year) and hydrological regime (droughts or strong torrents).

Except for the basic ES–HIPPO model, all other modifications are supported by Kohonen artificial neural network (i.e., self-organizing maps—SOM) (Kohonen 1982).

For further reading on models that have been developed and used to assess fish stock status in freshwater ecosystems around the world, see Pitcher (2015). The model has certain similarities with the models and application of the Biotic Integrity

Index for some river basins of China (Yang et al. 2021; Huang et al. 2022). In the Balkans area, proposals and applications of the model for assessing the state of the fish stock in the Danube can be found in the works of Jarić et al. (2014a) and Smederevac-Lalić et al. (2017), as well as for the assessment of the state of sturgeon species (Jarić et al. 2009, 2010, 2014b, 2016) and the assessment of floodplain and marsh habitats (Stamenković et al. 2021).

5 The Application of Fish in the Assessment of the Ecological Status/Potential of Water Bodies

In addition to the previously mentioned publications that focus on the use of fish in assessing the ecological sustainability of fish populations and inland water ecosystems, publications that utilize fish and/or fish communities as indicators of the ecological status and potential of water bodies are also important, in compliance with the Water Framework Directive (WFD 2000).

For the waters of Serbia, notable findings on this topic are presented in the work of Stojković et al. (2013), where the fish communities in the South Morava River basin were analyzed. Based on the SOM analysis (Stojković et al. 2013), three clusters with indicator fish species that are indicative of specific ecological characteristics of the habitat were established (Table 4). The methodology and results presented in this manuscript can be applied to assess the ecological status/potential as well as to select priority areas for the conservation of biodiversity and fish stocks of rivers.

The use of fish communities in the operational monitoring of water bodies in Serbia is discussed in the report by Simonović et al. (2018). The report uses both fish and macrophytes to evaluate the ecological status and potential of water bodies in Serbia. The ecological status assessment was conducted using the Fish Index Slovakia (National method for evaluating the ecological status of streams based on fish populations) (Kováč 2015).

The results indicate the significant possibilities of applying this index in assessing the ecological status/potential of water bodies in Serbia and the need for its adaptation and modification in relation to the local geographical, hydrological, and hydroecological conditions of Serbia.

One of the possible approaches in the use of fish and fish communities in the assessment of the ecological status/potential of water bodies was obtained based on our research aimed at monitoring the relationship of high, moderate, and low susceptibility fish species according to van Treeck et al. (2020) in the rivers and reservoirs of Serbia (Fig. 7).

The results obtained indicate a significant correlation between the presence of sensitive fish species and the final assessment of the ecological status/potential of water bodies in Serbia, as determined by the communities of aquatic

Table 4 Species that differ by IndVal index, for cluster (I) and sub-cluster (Ks1, Ks2)

Y		X1		X2	Y
Rutilus rutilus	75.0***	*Barbus balcanicus*	52.6***	*Salmo trutta*	86***
Alburnus alburnus	61.8***	*Alburnoides bipunctatus*	40.8***	*Cottus gobio*	12.0*
Chondrostoma nasus	59.8***	*Barbatula barbatula*	34.4***	*Oncorhynchus mykiss*	12.0*
Barbus barbus	50.0***	*Cobitis elongata*	21.3**		
Leuciscus cephalus	45.0**				
Carassius auratus	39.6***				
Vimba vimba	38***				
Gobio gobio	36.0***				
Silurus glanis	36.0***				
Rhodeus sericeus	33.0***				
Perca fluviatilis	23.0***				
Aspius aspius	11.1**				
Scardinius erythrophthalmus	9*				
Esox lucius	8.9*				
Cobitis taenia					
Cyprinus carpio					
Lepomis gibbosus					
Leuciscus leuciscus					
Pseudorasbora parva					
Rutilus pigus					
Zingel zingel					

Bold letters indicate the species representative for SOM clusters that have IndVal values greater than 25%. Species without significant IndVal values are included at the end of the columns (taken from Stojković et al. 2013)

*Indicates the significance level: < 0.05** Indicates the significance level: <0.01***Indicates the significance level: < 0.001

macroinvertebrates and phytobenthos (diatoms) in accordance with the Water Framework Directive (WFD 2000).

In the large rivers Danube, Sava, and Morava, the ecological status ranges from good to moderate, and moderately sensitive fish species dominate, with some tolerant species present as well. In larger tributaries of these rivers, there is a stronger correlation between ecological status and the presence of sensitive fish species. Tributaries with poor ecological status, like the Despotovica (TP_3) are dominated by tolerant fish, while those with moderate status are medium sensitivity fish. Tributaries with good ecological status, such as (TP_3 Veliki Rzav and Vlasina rivers) and (TP_4 Djetinja, and Studenica rivers), have a significant presence of sensitive fish species. Sensitive fish species clearly dominate in salmonid rivers with excellent and good ecological status, such as the Gobeljska River, Samokovska River, and Barska River (TP_6). In the investigated lowland reservoirs (TP_3), where the ecological potential ranges from moderate to poor, moderately sensitive

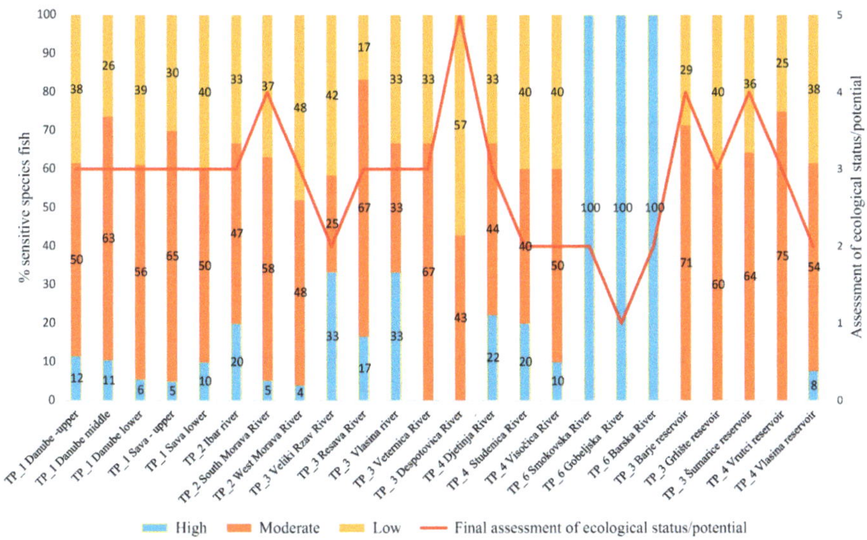

Fig. 7 Percentage of sensitive fish species in selected rivers and reservoirs of Serbia: **H**—High sensitivity species: *Salmo trutta, Oncorhynchus mykiss, Thymallus thymallus, Barbus barbus*; *Leuciscus idus*. **M**—Moderate sensitivity species: *Cottus gobio, Abramis brama, Barbus balcanicus, Ballerus sapa, Cyprinus carpio, Chondrostoma nasus, Leuciscus aspius, Phoxinus phoxinus, Squalius cephalus, Rutilus rutilus, Vimba vimba, Hypophthalmichthys nobilis, Hypophthalmichthys molitrix, Vimba vimba, Ctenopharyngodon idella, Scardinius erythrophthalmus, Pelecus cultratus, Perca fluviatilis, Zingel zingel, Sander lucioperca, Sander volgensis, Gymnocephalus schraetser, Acipenser ruthenus, Silurus glanis, Esox lucius*. **L**—Low sensitivity species: *Carassius gibelio, Alburnus alburnus, Alburnoides bipunctatus, Gobio obtusirostris, Pseudorasbora parva, Rhodeus amarus, Blicca bjoerkna, Rutilus pigus, Zingel streber, Cobitis taenia, Sabanejewia balcanica, Barbatula barbatula, Lepomis gibbosus, Ameiurus nebulosus*, (according to van Treeck et al. (2020). Abbreviations: TP_1—Type 1—large lowland rivers; TP_2—Type 2—large river; TP_3—Type 3–small and medium watercourses (altitude up to 500 m); TP_4—Type 4—small and medium watercourses; (altitude over 500 m); TP_6—Type 6—small watercourses outside the area of the Pannonian Plain that are not covered by Type 3 and 4 (Anonymus 2010). Final assessment of ecological status/potential: 1—high, 2—good, 3—moderate, 4—poor, 5—bad (Anonymus 2011)

fish species dominate, with a smaller or larger share of tolerant ones depending on the degree of ecological potential. The presence of sensitive fish species is greater only in mountain reservoirs, such as the Vlasina reservoir (TP_4) (Fig. 7).

6 Conclusion

All previous research and experience in utilizing fish to evaluate the ecological sustainability of fish stocks, the ecological status, and the potential of water bodies demonstrate the need for continued research and the creation of more effective models and methods that are adapted to the regional ecological conditions of Serbia, the Western Balkans, and the Balkan Peninsula in general.

Acknowledgments The investigation was supported by Grant (Agreement No. 451–03–68/2022–14/200122) funded by the Serbian Ministry of Education, Science and Technological Development.

References

Anonymous (2010) Regulation on the determination of water bodies of surface and groundwater waters (Official Gazette of RS, No. 96/10) (In Serbian)

Anonymous (2011) Regulation on the parameters of the ecological and chemical status of surface water and the parameters of the chemical and quantitative status of groundwater („Official Gazette of RS", No. 74/2011) (In Serbian)

Anonymous (2016) Rulebook on declaration and protection of strictly protected and protected wild species of plants, animals, and fungi (Official Gazette of RS No. 5/2010, 47/2011, 32/2016, and 98/2016) (In Serbian)

Bănăduc D, Simić V, Cianfaglione K, Barinova S, Afanasyev S, Öktener A, McCall G, Simić S, Bănăduc Curtean A (2022) Freshwater as a sustainable resource and generator of secondary resources in the 21st century: stressors, threats, risks, management and protection strategies, and conservation approaches. Int J Environ Res Public Health 19(24):16570. https://doi.org/10.3390/ijerph192416570

Bonin A, Nicole F, Pompanon C, Miaud C, Taberlet P (2007) Population adaptive index: a new method to help measure intraspecific, genetic diversity and prioritize populations for conservation. Conserv Biol 21:697–708. https://doi.org/10.1111/j.1523-1739.2007.00685.x

Brennan S, Withgott J (2005) Biodiversity and conservation biology. In: Environment; The Science Behind the Stories. Pearson, Benjamin Cummings, San Francisco

Chantepie S, Lasne E, Laffaille P (2011) Assessing the conservation value of waterbodies the example of the Loire floodplain (France). Biodivers Conserv 20:2427–2444

Chester E, Robson BJ (2013) Anthropogenic refuges for freshwater biodiversity: their ecological characteristics and management. Biol Conserv 166:64–75

Cooke SJ, Allison EH, Beard TDJ, Arlinghaus R, Arthington AH, Bartley DM, Cowx IG, Fuentevilla C, Leonard NJ, Lorenzen K, Lynch AJ, Nguyen VM, Youn SJ, Taylor WW, Robin L (2016) On the sustainability of inland fisheries: finding a future for the forgotten. Ambio 45:753–764. https://doi.org/10.1007/s13280-016-0787-4

Đuretanović S, Rajković M, Maguire (2023) Chapter 8: Freshwater crayfish of Western Balkan: is it possible to use them sustainably or do they need prompt conservation actions? In: Simić V, Simic S, Pešić V (eds) Ecological sustainability of fish resources of inland waters of the Western Balkans in the Anthropocene. Springer, Cham

Fisher OD, Owens PFI (2004) The comparative method in conservation biology. Trends Ecol Evol 9:391–339

Garcia SM, Kolding J, Rice J, Rochet MJ, Zhou S, Arimoto T, Beyer JE, Borges L, Bundy A, Dunn D, Fulton EA, Hall M, Heino M, Law R, Makino M, Rijnsdorp AD, Simard F, Smith ADM (2012) Reconsidering the consequence of selective fisheries. Science 335:1045–1047

Gayanilo FC, Sparre P, Pauly D (2005) FAO-ICLARM stock assessment tools II (FiSAT II). Revised version. User's guide. FAO computerized information series (fisheries). No. 8, revised version. Rome: FAO, p 168

Giam X, Ng TH, Lok FSLA, Ng HH (2011) Local geographic range predicted freshwater fish extinctions in Singapore. J Appl Ecol 48:356–363

Hannah L, Costello C, Elliot V, Owashi B, Nam S, Oyanedel R, Chea R, Vibol O, Phen C, McDonald G (2019) Designing freshwater protected areas (FPAs) for indiscriminate fisheries. Ecol Model 393:127–134
Huang X, Xu J, Liu B, Guan X, Li J (2022) Assessment of aquatic ecosystem health with indices of biotic integrity (IBIs) in the Ganjiang river system. China Water 14:278. https://doi.org/10.3390/w14030278
Huet M (1949) Aperçu des relations entre la pente et les populations piscicoles des eaux courantes. Schweiz Z Hydrol 11:333–351
Huet M (1959) Profiles and biology of Western European streams as related to fish management. Trans Am Fish Soc 88:155–163
IUCN (2022) The IUCN red list of threatened species. Version 2022-2. https://www.iucnredlist.org
IUCN (2001) IUCN red list categories and criteria: version 3.1. IUCN species survival commission. IUCN, Gland
Jarić I, Lenhardt M, Cvijanović G, Ebenhard T (2009) Population viability analysis and potential of its application to Danube sturgeons. Arch Biol Sci 61(1):123–128
Jarić I, Ebenhard T, Lenhardt M (2010) Population viability analysis of the Danube sturgeon populations in a vortex simulation model. Rev Fish Biol Fish 20:219–237. https://doi.org/10.1007/s11160-009-9151-0
Jarić I, Smederevac-Lalić M, Jovičić K, Jaćimović M, Cvijanović G, Lenhardt M, Kalauzi A (2014a) Indicators of unsustainable fishery in the middle Danube. Ecol Freshw Fish 25:86. https://doi.org/10.1111/eff.12193
Jarić I, Gessner J, Acolas ML, Lambert P, Rochard E (2014b) Modelling attempts utilized in sturgeon research: a review of the state-of-the-art. J Appl Ichthyol 30:1379–1386
Jarić I, Gessner J, Solow AR (2016) Inferring functional extinction based on sighting records. Biol Conserv 199:84–87
Jepsen N, Pont D (2007) Intercalibration of fish-based methods to evaluate river ecological quality report from an EU intercalibration pilot exercise. Report from an EU intercalibration pilot exercise. JRC, European Commission. https://doi.org/10.13140/RG.2.1.5148.5608
Karr J (1981) Assessment of biotic integrity using fish communities. Fisheries 6:21–27
Karr J (1987) Biological monitoring and environmental assessment: a conceptual framework. Environ Manag 11:249–256
Kohonen T (1982) Self-organized formation of topologically correct feature maps. Biol Cybern 43(1):59–69
Klobučar G, Podnar M, Jelić M, Franjević D, Faller M, Stambuk A, Gottstein S, Simić V, Maguire I (2013) Role of Dinaric Karst (western Balkans) in shipping the phylogeographic structure of the threatened crayfish *Austropotamobius torrentium*. Freshw Biol 58:1089–1105
Kolding J, Ticheler H, Chanda B (2003) The Bangweulu swamps – a balanced small-scale multi-species fishery. In: Jul-Larsen E, Kolding J, Nielsen JR, Overa R, van Zwieten PAM (eds) Management, co-management or no management? Major dilemmas in southern African freshwater fisheries, part 2: case studies. FAO, Rome, pp 34–66
Kolding J, van Zwieten PAM (2011) The tragedy of our legacy: how do global management discourses affect small-scale fisheries in the south? Forum Dev Stud 38:267–297
Kolding J, Law R, Plank M, Zwieten AMP (2016) The optimal fishing pattern. In: Graig FJ (ed) Freshwater fisheries ecology. John Wiley & Sons Ltd, Chichester, pp 468–482
Kováč V (2015) Metóda stanovenia ekologického stavu vôd podľa rýb –slovenský ichtyologický index FIS. In: Makovinská J, Mišíková Elexová E, Rajczyková E, Baláži P, Plachá M, Kováč V, Fidlerová D, Ščerbáková S, Lešťáková M, Očadlík M, Velická Z, Horváthová (eds)
Law R, Plank MJ, Kolding J (2012) On balanced exploitation of marine ecosystem: result from dynamic size spectra. ICES J Mar Sci 69:602–614
Law R, Kolding J, Plank MJ (2013) Squaring the circle: reconciling fishing and conservation of aquatic ecosystems. Fish Fish 16:160. https://doi.org/10.1111/faf.12056
Linke S, Kennard MJ, Hermoso V, Olden DJ, Stein J, Pusey JB (2012) Merging connectivity rules and large-scale condition assessment improves conservation adequacy in river systems. J Appl Ecol 49:1036–1045

Lovrenčić L, Maguire I (2021) Gap analysis revealed a moderate efficiency of protected areas for the conservation of the endangered noble crayfish in Croatia. Natura Croatica 30(2):529–536. https://doi.org/10.20302/NC.2021.30

Lovrenčić L, Temunović M, Gross R, Grgurev M, Maguire I (2022) Integrating population genetics and species distribution modeling to guide conservation of the noble crayfish, *Astacus astacus*, in Croatia. Sci Rep 12:2040. https://doi.org/10.1038/s41598-022-06027-8

Marić S, Bănăduc D, Gajić Đ, Šanda R, Veličković T (2022) *Sabanejewia romanica* (Băcescu, 1943) (Actinopterygii: Cobitidae), a new species for the ichthyofauna of Serbia. Acta Zoologica Bulgarica 74(3):369–377

Pančić J (1860) Ribe u Srbiji (Pisces Serbiae). Glasnik društva Srbske Slovesnosti, Beograd (In Serbian)

Panov VE, Alexandrov B, Arbaciauskas K, Binimelis R, Gordon H, Copp GH, Darius Daunys D, Rodolphe E, Gozlan RE, Grabowski M, Lucy FE, Leuven RSEW, Mastitsky S, Minchin D, Monterrosol I, Nehring S, Olenin S, Paunović M, Labajos BR, Semenchenko V, Son M (2008) Interim protocols for risk assessment of aquatic invasive species introductions via European inland waterways. Invasions Module (Aquatic Group)

Pậrvulescu L, Zaharia C (2014) Distribution and ecological preferences of noble crayfish in the Carpathian Danube basin: biogeographical insights into the species history. Hydrobiologia 726: 53–63

Petrović A, Rajković M, Simić S, Maguire I, Simić V (2013) Importance of genetic characteristics in the conservation and management of crayfish in Serbia and Montenegro. Bulgarian J Agr Sci 19:1095–1106

Pitcher T (2015) Assessment and modeling in freshwater fisheries. In: Craig JF (ed) Freshwater fisheries ecology. Wiley Blackwell, Oxford, pp 483–499

Pont D, Hugueny B, Rogers C (2007) Development of a fish-based index for the assessment of river health in Europe: the European fish index. Fish Manag 14:427–439. https://doi.org/10.1111/j.1365-2400.2007.00577

Pont D, Valentini A, Rocle M, Maire A, Delaigue A, Jean P, Dejean T (2019) The future of fish-based ecological assessment of European rivers: from traditional EU water framework directive compliant methods to eDNA metabarcoding-based approaches. J Fish Biol 98(2):354–366. https://doi.org/10.1111/jfb.14176

Radinger J, Britton JR, Carlson SM, Magurran AE, Alcaraz-Hernández JD, Almodóvar A, Benejam L, Fernandez-Delgado C, Nicola GG, Oliva Paterna F, Torralva M, Garcia-Berthou E (2019) Effective monitoring of freshwater fish. Fish 20. https://doi.org/10.1111/faf.12373

Rajković M (2004) Optimalni ekološki uslovi za razvoj riječno graka (*Astacus astacus* L.) u vodenim ekosistemima na području gornjeg toka rijeke Zete. Specijalistički rad. Univerzitet u Kragujevcu (In Serbian)

Rajković M, Simić V, Petrović A (2006) Length-weight gain of European crayfish *Astacus astacus* (L.) in the area of the upper course of the Zeta River, Montenegro. Arch Biol Sci 58:233–238

Rajković M, Petrović A, Maguire I, Simić V, Simić S, Paunović M (2012) Discovery of a new population of the species complex of the white-clawed crayfish, *Austropotamobius pallipes/italicus* (Decapoda, Astacidae) in Montenegro, range extension, endangerment and conservation. Crustaceana 85:333–347

Ristić M (1972) Rasprostranjenost riba u vodama Srbije, U Djuričić B (ed). Lov i ribolov u Srbiji (Borba, Beograd), pp 201–210 (In Serbian)

Ristić M (1977) Ribe i ribolov u slatkim vodama. Nolit, Beograd. (In Serbian)

Schrimpf A, Theissinger K, Dahlem J, Maguire I, Pa^rvulescu L, Schulz HK, Schulz R (2014) Phylogeography of noble crayfish (*Astacus astacus*) reveals multiple refugia. Freshw Biol 59: 761–776

Simić V, Simić S, Paunović M, Cakić P (2007) Model of assessment of critical risk of extinction and the priorities of protection of endangered aquatic species at the national level. Biodivers Conserv 16:2471–2493

Simić V, Petrović A, Rajković M, Paunović M (2008) Crayfish of Serbia and Montenegro - the population status and the level of endangerment. Crustaceana 81(10):1153–1176

Simić VM, Simić SB, Stojković Piperac M, Petrović A, Dj M (2014) Commercial fish species of inland waters: a model for sustainability assessment and management. Sci Total Environ 497: 642–650

Simić V, Maguire I, Rajković M, Petrović A (2015) Conservation strategy for the endangered crayfish species of the family Astacidae: the ESHIPPOcrayfish model. Hydrobiologia 760:1–13. https://doi.org/10.1007/s10750-015-2295-0

Simić V, Bănăduc D, Bănăduc Curtean A, Petrović A, Veličković T, Stojković Piperac M, Simić S (2022) Assessment of the ecological sustainability of river basins based on the modified the ESHIPPOfish model on the example of the Velika Morava basin (Serbia, Central Balkans). Front Environ Sci 10:952692

Simonović P, Hegediš A, Miljanović B, Vukov D et al (2018) Operativni monitoring površinskih i podzemnih voda Republike Srbije – partija 1 – operativni monitoring površinskih voda. Republika Srbija, Ministarstvo zaštite životne sredine, Biološki fakultet i Institut za multidisplinarna istraživanja Univerzitet u Beogradu; Prirodno-matematički fakultet, Univerzitet u Novom Sadu, 404-02-59/2018-02 (in Serbian)

Simonovic P (2001) Ribe Srbije. Zavod za zastitu prirode Srbije i Bioloski fakultet Univerziteta u Beogradu, Belgrade

Smederevac-Lalić M, Kalauzi A, Regner S, Lenghardt M, Naunović Z, Hegediš A (2017) Prediction of fish catch in the Danube River based on long-term variability in environmental parameters and catch statistics. Sci Total Environ 609(31):664–671

Souty-Grosset C, Grandjean F, Gouin N (2003) Keynote: involvement of genetics in knowledge, stock management and conservation of *Austropotamobius pallipes* in Europe. Bulletin Français de la Pêche et de la Pisciculture 370–371:165–179

Stamenković O, Simić V, Stojković Piperac M, Dj M, Simić S, Ostojić A, Đorđević N, Čerba D, Petrović A, Jenačković Gocić D, Đurđević A, Koh M, Buzhdygan OY (2021) Direct, water-chemistry mediated, and cascading effects of human-impact intensification on multitrophic biodiversity in ponds. Aquat Ecol 55:187–214. https://doi.org/10.1007/s10452-020-09822-5

Stojković M, Dj M, Simić V (2011) Ichthyological integral indices, the history of development and possible application on rivers in Serbia. Biologica Nyssana 2:59–66

Stojković M, Simić V, Milošević Đ, Mančev D, Penczak T (2013) Visualization of fish community distribution patterns using the self-organizing map: a case study of the great Morava River system (Serbia). Ecol Model 248:20–29. https://doi.org/10.1016/j.ecolmodel.2012.09.014

Thienemann A (1928) The oxygen in eutrophic and oligotrophic lakes. A contribution to the theory of lake types. The Inland Waters

van Treeck R, Van Wichelen J, Wolter C (2020) Fish species sensitivity classification for environmental impact assessment, conservation and restoration planning. *Sci Total Environ* 708: 135173. https://doi.org/10.1016/j.scitotenv.2019.135173

Veličković T, Simić V, Šanda R, Radenković M, Milošković A, Radojković N, Marić S (2020) New record of a population of *Telestes souffia* (Risso, 1827) (Actinopterygii: Cyprinidae) in Serbia. Acta Zoologica Bulgarica 72(1):13–20

Warwick RM, Clarke KR (1994) Relearning the ABC: taxonomic changes and abundance/biomass relationships in disturbed benthic communities. Mar Biol 118:739–744. https://doi.org/10.1007/BF00347523

Welcomme LR (2016) Fisheries governance and management. In: Graig FJ (ed). Freshwater Fisheries Ecology, pp 468–482

WFD (2000) Water framework directive - directive of European Parliament and of the council 2000/60/EC – establishing a framework for community action in the field of water policy. European Union, the European Parliament and Council, Luxembourg

Yang J, Yan D, Yang Q, Gong S, Shi Z, Qiu Q, Huang S, Zhou S, Hu M (2021) Fish species composition, distribution and community structure in the Fuhe river basin, Jiangxi province. China Glob Ecol Conserv 27:e01559. https://doi.org/10.1016/j.gecco.2021.e01559

Part V
Conservation and Aquaculture

Summary

The part includes chapters related to aquaculture and its importance for the sustainability of fish populations in wild waters. Other chapters present current national and international measures for *in situ* and *ex situ* conservation of fish and other aquatic animals. The importance of fresh water aquariums and specialized reprocenters in maintaining species diversity, autochthonous genes, and overall genetic diversity of wild fish populations (threatened by over-exploitation or other factors) is emphasized. Moreover, the entire *in situ* conservation process of huchen *Hucho hucho*, an endemic species important to fisheries in the Danube Basin, is presented in a separate chapter.

The Role and Importance of Aquaculture for the Ecological Sustainability of Fish Resources in the Inland Water of Serbia

Zoran Z. Marković and Vesna D. Poleksić

Abstract Aquaculture as a production of aquatic plants and animals in objects intended and custom-made for this purpose is furthermore one of the ways how to alleviate, rehabilitate, even entirely solve the problem of endangered freshwater fish species. Since the natural fish resources are limited globally, there is a raising demand for fish as food. This is the reason why today, from the total quantity of fish consumed globally, 50% originates from aquaculture. Besides this primary role of aquaculture, it became an important way to renew endangered fish and other aquatic organisms' populations in their natural aquatic habitats. This new task of aquaculture is of extreme importance for the preservation of ichthyofauna as well as other aquatic organisms' diversity, primarily endangered and protected species. In addition to positive effects on hydrobionts' biodiversity, preservation of autochthonous, primarily protected fish species in their natural environment opens possibilities for improvement of recreational and game fishery, and consequently socio-economic reinforcement of local communities placed near the natural aquatic ecosystems.

Keywords Aquaculture · Sustainability · Fish resources · Serbia

1 Introduction

While land surfaces have been almost entirely devoted to different purposes the needs of the raising human population cause increased use of aquatic surfaces of the Planet. Since 71% of the earth is covered with water, aquatic ecosystems will remain in focus of the humanity. In addition, from the total quantity of water, only 2.5% is occupied by inland freshwater (Babkin 2003). Land surfaces being occupied and inhabited by human population, it is clear that freshwater is under the greatest anthropogenic pressure.

Z. Z. Marković (✉) · V. D. Poleksić
Faculty of Agriculture, University of Belgrade, Belgrade, Serbia
e-mail: zoranmm@agrif.bg.ac.rs

V. Simić et al. (eds.), *Ecological Sustainability of Fish Resources of Inland Waters of the Western Balkans*, Fish & Fisheries Series 43,
https://doi.org/10.1007/978-3-031-36926-1_15

Exploitation of fish and other aquatic organisms from natural aquatic ecosystems has peaked; therefore, further increase of natural resources exploitation will lead to disturbances in biocenoses, and consequently in the entire ecosystems.

In addition to fish being exploited for food purposes from natural aquatic ecosystems, there is a huge indirect pressure on the ichthyofauna as well as on the other hydrobionts. This pressure is increased due to intensification of agriculture, urbanization, industrialization, traffic infrastructure improvement, and a whole range of human activities. Such activities could be called negligence and/or abandonment of aquatic ecosystems.

It is necessary to apply a number of existing and new regulations with the aim to establishing a regulatory framework within which a negative anthropogenic influence would be prevented and protection and renewal of aquatic biotopes, including reintroduction of endangered fish and other aquatic species.

Aquaculture plays an important role in this process, as an activity covering reproduction and culture of aquatic organisms with the aim of obtaining fish as food, as well as for the purpose of reintroduction in their natural habitats.

The significance of aquaculture is constantly increasing. As a result, each year aquaculture production is 7 to 9 percent greater than in the previous year, making this activity furthermore important in food production. In 2012, for the first time aquaculture production overtook the world beef production (Larsen and Roney 2013a, b). According to FAO (FAO FIPS 2014) in 2030, 62% of the fish consumed will be provided from aquaculture.

In addition to the fact that nowadays more than a half of fish for food consumption is originating from aquaculture (Naylor et al. 2009), the importance of aquaculture in renewal of endangered fish and other aquatic organisms' species in their natural habitat is becoming increasingly important.

2 Fish Resources in the Inland Water in Serbia with an Emphasis on Economically Important and Protected Fish Species

There are about 100 fish species living in Serbian waters (Simonović 2001; Šipoš et al. 2004; Cvijanović et al. 2005; Lenhardt et al. 2006a; Radulović et al. 2012;).

Economically important species in the Republic of Serbia are: sterlet (*Acipenser ruthenus*), carp (*Cyprinus carpio*), wels catfish (*Silurus glanis*), pikeperch/sander (*Sander lucioperca*), Northern pike (*Esox lucius*), bream (*Abramis brama*), Prussian carp (*Carassius gibelio*), gray bighead/bighead carp (*Aristichthys nobilis*), silver carp/white bighead (*Hypophthalmichthys molitrix*), barbel (*Barbus barbus*), Volga pikeperch (*Stizostedion volgensis*), grass carp (*Ctenopharyngodon idella*), asp (*Aspius aspius*), orfe/ide (*Leuciscus idus*), nase (*Chondrostoma nasus*), burbot (*Lota lota*), brown trout (*Salmo trutta*); Danubian salmon (*Hucho hucho*) and rainbow trout (*Oncorhynchus mykiss*) (Marković 2020).

From the 100 fish species in Serbia, over 60% are under a certain level of protection, i.e. 34 species are strictly protected and 29 species have a status of protected wild fish species (Marković et al. 2018), and according to the Law on nature protection 2010; Law on environmental protection, 2004, 2019 and the Rulebook on declaration and protection of severely protected and protected wild species of plants, animals and fungi, 2010 (Tables 1 and 2).

3 The most Important Factors that Jeopardize Ecological Sustainability of Fish Resources in the Inland Water in Serbia

3.1 Changes in Fish Habitat

The need for electrical energy from hydropower plants: building of the dams for water accumulation for different purposes (to decrease torrential flows, provide water supply, for agriculture, recreational needs, industry); building of the irrigation systems; traffic infrastructure; building of embankments to protect urban and industrial zones; management and flow regulation through urban zones; sand and gravel exploitation; etc., result primarily in the change of appearance of the fishery water. After physical changes of the biotope, soon its physical appearance changes such as: decrease or increase of the flow, depth, and volume of the aquatic ecosystem; change in bottom structure and temperature traits. These changes are followed by alteration in chemical water characteristics such as: aquatic gases regime; content of suspended and diluted matters in water. Changes in physical and/or chemical environmental factors affect aquatic communities, including ichthyocenoses of the ecosystem.

Various activities performed on Serbian water currents have negative effects on ichthyocenoses, primarily on economically important, as well as on protected and endangered fish species. The biggest such operation was building of the hydroelectrical plant Djerdap 1 and Djerdap 2 dams on the Danube, when fish migratory routes were interrupted. Effects on anadromous species, such as sturgeons and shads, are especially pronounced, and represent the biggest threat for these species' survival (Lenhardt et al. 2006b).

Besides Djerdap, a very important hydroaccumulation of the Serbian hydro energetic system, there are numerous examples of small hydroelectrical power plants built on small water currents, often with insufficiently functional fish tracks. Building small hydroelectrical power plants not only destroyed beautiful rural environment, but it also fragmented fish habitat, damaged physical integrity of the water current, resulting in changes in the aquatic communities, including fish. Among the number of examples of negative effects on fish communities, river courses arrangement, embankment building to protect urban, industrial, and agricultural zones from the main river courses, has fragmented/separated natural spawning sites (flood

Table 1 Strictly protected fish species in the Serbia, according to the national legislation (Official Gazette of RS, No. 5/2010, 47/2011 and 32/2016)

Order	Family	Latin name	Common name
ACIPENSERIFORMES	Acipenseridae	*Acipenser gueldenstaedtii*	Russian sturgeon
		Acipenser nudiventris	Ship sturgeon
		Acipenser stellatus	Stellate sturgeon
		Acipenser sturio	European sturgeon
		Huso huso	Huso huso
ANGUILLIFORMES	Anguillidae	*Anguilla anguilla*	European eel
CLUPEIFORMES	Clupeidae	*Alosa immaculata*	Pontic shad
		Alosa tanaica	Azov shad
CYPRINIFORMES	Balitoridae	*Oxynoemacheilus bureschi*	Barbata bureschi
	Cobitidae	*Cobitis elongata*	Balkan loach
		Cobitis taenia	Spined loach
		Misgurnus fossilis	Weatherfish
		Sabanejewia balcanica	Balkan loach
		Sabanejewia bulgarica	Bulgarian loach
	Cyprinidae	*Alburnus chalcoides*	Danube bleak
		Barbus cyclolepis	Eastern brook barbel
		Carassius carassius	Crucian carp
		Leucaspius delineatus	Sunbleak
		Pachychilon pictum	Moranec
		Pelecus cultratus	Checon
		Rhodeus amarus	Bitterling
		Romanogobio albipinnatus	Whitefin gudgeon
		Romanogobio banaticus	Banaticus gudgeon
		Romanogobio kessleri kessleri	Kessler's gudgeon
		Romanogobio uranoscopus	Danubian gudgeon
		Rutilus karamani	Roach
		Telestes souffia	Vairone
		Tinca tinca	Tench
ESOCIFORMES	Umbridae	*Umbra krameri*	European mudminnow
PERCIFORMES	Percidae	*Gymnocephalus baloni*	Danube ruffe
		Zingel streber	Danube streber
		Zingel zingel	Zingel
SALMONIFORMES	Salmonidae	*Salmo marmoratus*	Marble trout
SCORPAENIFORMES	Cottidae	*Cottus gobio*	European bullhead

zones) primarily of cyprinid species. This is one of the main reasons for the decrease of fish quantity in lowland rivers (primarily the Danube) in Serbia.

Table 2 Protected wild fish species in the Serbia, according to the national legislation (Official Gazette of RS, No. 5/2010, 47/2011 and 32/2016)

Order	Family	Latin name	Common name
ACIPENSERIFORMES	Acipenseridae	*Acipenser ruthenus*	Sterlet
CYPRINIFORMES	Cyprinidae	*Abramis brama*	Bream
		Alburnoides bipunctatus	Schneider
		Aspius aspius	Asp
		Ballerus ballerus	Blue bream
		Ballerus sapa	White-eye bream
		Barbus balcanicus	Danube barbel
		Barbus meridionalis	Mediterranean barbel
		Barbus peloponnesius	Brook barbel
		Barbus barbus	Barbel
		Chondrostoma nasus	Common nase
		Cyprinus carpio	Common carp
		Gobio obtusirostris	Gudgeon
		Leuciscus idus	Orfe
		Rutilus pigus	Danubian roach
		Squalius cephalus	European chub
		Vimba vimba	Vimba
ESOCIFORMES	Esocidae	*Esox lucius*	Northern pike
GADIFORMES	Gadidae	*Lota lota*	Burbot
PERCIFORMES	Blenniidae	*Salaria fluviatilis*	Freshwater blenny
	Percidae	*Gymnocephalus schraetser*	Striped ruffe
		Perca fluviatilis	Eurasian perch
		Sander lucioperca	Pikeperch
		Sander volgensis	Volga pikeperch
	Salmonidae	*Hucho hucho*	Danubian salmon
SALMONIFORMES		*Salmo dentex*	Toothtrout
		Salmo trutta	Brown trout
		Thymallus thymallus	Grayling
SILURIFORMES	Siluridae	*Silurus glanis*	Wels catfish

3.2 Pollution of Fishery Waters

Human population growth, migrations from rural to urban areas, enlargement of urban areas, industrial development, and intensification of agriculture are only part of the reason leading to the growing anthropogenic impact on aquatic organisms and their habitats. Since fish are on the top of the trophic pyramid of the aquatic ecosystems and being of greatest interest for men, effects of pollution enter in focus when fish kills appear.

Various human activities change abiotic and consequently biotic environmental factors, thus altering entire aquatic communities. Clean water communities are replaced by hydrobiocenoses from waters with deteriorated quality, i.e. diverse aquatic communities become replaced by the uniform biocenoses where fish are missing (Dulić et al. 2014).

The problem of pollution of fishery waters is particularly visible in poor and developing countries, such as the Republic of Serbia. The reason is the lack of financial possibilities for the installation of effective filters for depuration of communal, agricultural, and industrial wastewater, as well as insufficient consciousness of ecosystem preservation.

When municipal and industrial wastewater enters recipient waters in Serbia, examples of water purification and treatment before their discharge in fishery waters are extremely rare. From 173 municipalities in Serbia, including cities, only about 50 have facilities for water purification. Among them some are not functional and/or complete. Only 10% of wastewater is purified in Serbia. Serbian aquatic "aorta"—the Danube, along its flow across Serbia, 588 km long (which is 21% of the total river length) is the main final recipient of the majority of wastewater in our country. Although water quantity of the Danube increases about 2.5 times while flowing through Serbia (Martinović-Vitanović and Kalafatić 1995) and considering that Djerdap accumulation acts as precipitator and purifier, contributing to the river partial self-purification, still the Danube discharges annually approx. 35,000 tons of phosphorus and 400,000 tons of inorganic nitrogen, an important part coming from Serbia (data for the period 1988–2005) (Lenhardt et al. 2020). In addition to P and N, other pollutants such as heavy metals and persistent organic pollutants, directly or indirectly, affect the ecological sustainability of fish resources, especially since fish belong to the most sensitive organisms to toxic matters (Alibabić et al. 2007), and being on the top of the trophic pyramid in aquatic ecosystems (Yilmaz et al. 2007). Despite the lack of financing the remediation/restoration of ecosystems, Serbian scientists are detecting, recording, monitoring, and reporting inland water pollution and effects on fish to the scientific community (Rašković et al. 2015; Kostić et al. 2017; Rašković et al. 2018; Nikolić et al. 2021; Kostić-Vuković et al. 2021).

As we consider pollution as discharge of pollutants into the recipient, negative effects on fish community and the whole biocenosis may be pronounced through the discharge of warm water into the river, like the one from the Thermoelectrical plant "Nikola Tesla B" into the Sava River near Obrenovac. Taken out for the purposes of cooling the turbines, this process heats the water to between 8 and 12 °C. Heated water is then released back into the river without any cooling treatment (Marković et al. 2018).

3.3 Poaching and Overfishing

Poaching and overfishing are among the most important factors that endanger ecological sustainability of fish resources in the inland water in Serbia. These activities are particularly common among economically important fish species, such as Danubian salmon and brown trout in salmonid and sterlet and pikeperch in cyprinid waters.

Particular negative effect on ichthyocenoses is poaching by electrofishing, dynamite, or some narcotic substance. By the use of such pouching methods not only a population of economically important fish species is reduced but also the other fish species and other aquatic organisms.

Overfishing results in a long-lasting disturbances of the fish community. Since overfishing is particularly common among economically important species, rational and sustainable use of fishery waters is jeopardized in an extended period.

Besides human pressure on fish communities in fishery waters, quite important damage is made by ichthyophagous birds and mammals. Among the most important ichthyophagous bird species is the big cormorant—*Phalacrocorax carbo*, and among the ichthyophagous mammals is the Eurasian otter—*Lutra lutra* (Marković et al. 2018).

3.4 Introduction of Allochthonous Fish Species in Fishery Waters

Introduction of allochthonous fish species in fishery waters can be intentional or unintentional. From the total 22 fish species introduced in Serbian waters only 4 (rainbow trout—*Oncorhynchus mykiss*, grass carp—*Ctenopharyngodon idella*, bighead carp—*Aristichthys nobilis*, and silver carp—*Hypophthalmichthys molitrix*) are economically important. In addition, from the total of 22, 4 species (pumpkinseed—*Lepomis gibbosus*, brown bullhead—*Ameiurus nebulosus* and black bullhead—*Ameiurus melas*, Prussian carp—*Carassius gibelio*) occupy 51% of Serbian territory and 5 species (Grass carp—*Ctenopharyngodon idella*, bighead carp—*Aristichthys nobilis* and silver carp—*Hypophthalmichthys molitrix*; rainbow trout—*Oncorhynchus mykiss*, and topmouth gudgeon—*Pseudorasbora parva*) occupy between 21 and 50% of Serbian space (Lenhardt et al. 2011).

One of the most common intentional way of introducing fish species is for aquacultural purposes. Such species are: the rainbow trout—*Oncorhynchus mykiss*, the grass carp—*Ctenopharyngodon idella*, the bighead carp—*Aristichthys nobilis*, and the silver carp—*Hypophthalmichthys molitrix*. Allochthonous species enter natural waters by escaping from aquaculture objects or for the purpose of stocking the water currents. Some allochthonous species enter natural fishing waters released from aquaria when they overgrow aquarium size (such as the pumpkinseed—*Lepomis gibbosus*). Often anglers stock some species such as the brook trout *Salvelinus fontinalis* and the arctic char—*Salvelinus alpinus* because they want to catch these species (Simonović 2001). There is also stocking with allochthonous species such as the grass carp *Ctenopharyngodon idella*, the bighead carp—*Hypophthalmichthys nobilis*, and the silver carp—*Hypophthalmichthys molitrix*, for the purposes of protection against the macrophytes or phyto- and zooplankton overgrowth.

Deliberate introduction of new species is most often with good intentions but without sufficient knowledge about adverse effects on the existing ichthyocenoses and the whole autochthonous aquatic community. For some species introduced in Serbian waters (*Carassius gibelio*; *Pseudorasbora parva*; *Ameiurus nebulosus*; *Ameiurus melas*; *Lepomis gibbosus*) a negative effect on native species is known. For some introduced species such as *Neogobius fluviatilis*, there is competition for food and distribution area between members of the same family Gobiidae. Some of the species introduced (such as the *Hypophthalmichthys nobilis*) affect water quality. However, besides dominant negative effects, some species like *Neogobius fluviatilis* have positive effects on the ichthyocenoses as they increase food resources for the economically important species.

Introduction of non-native species affect autochthonous directly or indirectly. Most often direct influence is through nutrition, i.e., when the allochthonous species feeds on roe, larvae, fry, or adults of the autochthonous one, such as the brown bullhead (*A. nebulosus*) (Pujin and Sotirov 1966) or the pumpkinseed (*Lepomis gibbosus*) (Markovic and Simovic 1997). Indirect effect of allochthonous on autochthonous species is most often through reproduction or competition for food: grass carp fed on macrophytes reduces macrophytes abundance required for carp, a species which roe is phytophilic, in spawning zones (Janković 1998); Prussian carp affects crucian carp, tench affects common carp (Maletin et al. 1997). There is also indirect effect of the Prussian carp on growth and even survival of the tench as autochthonous species (Demény et al. 2009).

3.5 Introduction of Diseases

The health status of ichthyocenoses has a key impact on ecological sustainability of fish resources. Introduction of disease-causing agents in the new environment is a serious threat for fish resources sustainability, even for the survival of the domestic fish population (Bauer 1991). In Serbian fishery waters, causes of diseases are introduced naturally or by human action. Under "natural" we consider primarily disease introduction through infected fish or through parasites' transitional hosts, along large river courses (the Danube, the Sava, the Tisza). Although it is impossible to entirely prevent such infestation, care should be taken to avoid import of infected roe, fry, adult fish, or broodstock for aquaculture or aquaristics by strict enforcement of the law.

Diseases are transmitted among fish from aquaculture to fish from inland waters and represent a risk for sensitive species from natural populations (Radosavljević et al. 2014). Some authors (Peeler et al. 2011) consider that most of the newly emerged diseases of the autochthonous European fish species are the consequence of allochthonous species introduced for aquaculture. It is most probably the reason of the increased number of diseases in Serbia (Radosavljević et al. 2014, 2016, 2018).

With allochthonous species, new causative agents of the diseases are introduced, and since their immunological defense is not developed, this is a particular danger for the autochthonous species (Radosavljević 2020).

Investigation of parasite fauna in the 54 fish species inhabiting part of the flow of the Danube across Serbia during the past century, revealed 170 parasite species (Djikanović et al. 2012). Research of fish parasites is valuable for the assessment of the health status of ichthyocenoses (Lenhardt et al. 2020), population structure, even aquatic ecosystem functioning (Marcogliese 2004). Parasite fauna investigation: parasitic organisms' distribution, intensity of infestation, and parasites biology, gives information on the whole ecosystem health status (Britton and Pegg 2011).

4 Ecological Sustainability of Fish Resources in the Inland Water in Serbia

The main requirement for the ecological sustainability of fish resources, i.e. ichthyocenoses as integral part of biocenoses, is preservation of habitat, i.e. aquatic biotopes as fish environment.

Human activities mainly directly affect abiotic traits of the ecosystem (biotope or physical and chemical aquatic ecosystem characteristics), while communities including the ichthyocenoses change as a consequence of variation of abiotic factors (Živić et al. 2018). Before any changes in the biotopes are made (dam construction, river bottom, or cost management) it is important to first investigate all the effects on communities, then negative effects should be predicted and alleviated. However, although the laws of the Republic of Serbia (Law on the environmental impact assessment RS, Official Gazette RS No. 135/04 and 36/09) envisage carrying out assessment of impact on the environment, this assessment is often done by experts and/or institutions that are not fully competent, thus the assessment do not include the entire ecosystem. Therefore, the benefit of the intervention on the aquatic part of the ecosystem is not achieved. Often, the designers and later contractors do not fully respect prescribed measures of protection, due to lack of control and penalties, negligence, and possible corruption present. A particular problem concerning such construction works (such as dam for hydroenergetic power plant construction) is alteration of the biotope, and consequently abiotic living conditions. Therefore, there are consequences on the ecological stability of the fish community, primarily on reproduction and migrations, i.e. preservation of fish stocks.

When it comes to water quality as a precondition of fish community stability, despite existing laws in Serbia (Law on Protection and sustainable use of fish fund, Official gazette RS, No. 128/2014) that strictly forbid discharge of harmful and dangerous matters that could violate/disturb its physical, chemical, biological, and environmental traits, i.e. endanger fish health status, their lives and survival; practically there is no water current flowing through urban areas in Serbia where water quality is not deteriorated.

Considering aforementioned problems related to biotope preservation and maintenance of abiotic environmental factors, in the aim of preserving fishery funds, a question how to overcome problems and assure ichthyocenoses stability has to be asked. The answer remains that the laws should be consistently applied by competent institutions including: increase of ecological taxes; severe penal measures envisaged, and, if needed, new measures introduced. Raise of ecological awareness of every individual in the society would also contribute to the protection and sustainability of aquatic ecosystems including the fish community.

In addition to abiotic environmental factors the stability of the fish community is affected by other members of the biocenosis. The list is large: from plants, important as shelter, food supply or substrate for spawning; to animals: from invertebrates, bottom fauna, and plankton organisms, as food, till predators (fish, birds, mammals). Maintenance of the naturally established balance—stable biocenosis is, for sure, a precondition for the ecological sustainability of fish stocks.

When some of the environmental factors endanger ecological sustainability of fish resources, primarily economically important and threatened species are menaced.

Ways to overcome changes in ecological sustainability of fish resources are numerous and include various tactics from the removal or reduction of pressure from altered ecological factors to the enforcement of laws, and application of revitalization programs for endangered and degraded natural aquatic habitat and communities.

One of the ways to contribute to the ecological sustainability of fish resources in the inland water in Serbia is aquaculture. However, there are different opinions concerning the role of aquaculture for the ecological sustainability of fish resources. Some researchers think that production of fry from aquaculture in the aim of restocking inland waters is not effective for long-term maintenance of ecological sustainability of fish resources (Nehlsen et al. 1991). Others think that aquaculture is the problem solution, particularly for endemic species, when all the other methods are exhausted (Philippart 1995). Finally, some researchers consider aquaculture indispensable for maintenance of fish populations, and therefore maintenance of the entire fish communities (Holčik et al. 1988).

5 Interactive Relationship between Aquaculture and Ecological Sustainability of Fish Resources in the Inland Water in Serbia

Interaction between aquaculture and ecological sustainability of fish resources in the inland water is very complex, with various positive and negative effects on both aquaculture and fish resources in natural aquatic ecosystems (Marković et al. 2018).

Broodstock culture and spawning in aquaculture lakes and facilities with the aim of obtaining fry for restocking fishery waters have a positive effect on fish resources in natural aquatic ecosystems. Broodstock culture of the species with reduced abundance in natural waters (such as: sterlet, beluga, Danubian salmon), their reproduction, and fry culture for restocking natural aquatic ecosystems, help to achieve stability of ichthyocenoses. Of special importance is restocking of fishery waters where there is a lack of some species (primarily rare and endangered) by reintroduction using fry from aquaculture.

Increase of fish production on fish farms generates greater supply of freshwater fish. This has a positive effect on sustainability of fish resources since there is less pressure on natural resources. In addition, advance in economic efficiency on fish farms makes fish from aquaculture more affordable. This further decreases the activity from fishermen and poachers on natural populations, including their pressure on fish habitat and spawning sites.

By building new fish farms, primarily for warmwater species, not only unused and barren land surfaces are utilized (Marković 2010), but also new aquatic surfaces suitable for ichthyophagous birds (herons, cormorants, pygmy cormorants) and mammals (otters) are created, thus reducing pressure on fish natural populations. Actually, since stocking density in such fish farms is much higher than in natural waters, it attracts birds and mammals away from natural waters that remain relaxed from pressure coming from these animals.

Exploitation of quarry lakes established by mining: ore, clay, sand, and gravel for the purpose of fish farming has positive effect on reducing pressure on natural fish resources. Most often in such artificially formed aquatic ecosystems a cage system for fish farming is placed, as one of the ways for revitalization of degraded space. Cage culture increases fish offer on the market, and the space around the cages is used for angling, additionally decreasing pressure on natural fish resources. It is important to stress that very often, after the end of exploitation of ore, clay, sand, or gravel, the site becomes landfill for agricultural or urban waste deposition. Cleaning and using these surfaces/quarry lakes as fish farms contribute to environmental protection and reduction of risk from diseases coming from dead animals that are often thrown in such quarry lakes. Diseases transmission is particularly dangerous near river currents, therefore diseases could be transmitted on natural fish populations due to bottom permeability.

Positive effects of aquaculture on fish resources could be considered through exploitation of uneaten fish feed that originates from the farms (flow-through trout farms and cage systems) by fish living in the recipient water current or lake where the cage system is placed. Thus, natural fish communities are strengthened thanks to larger offer of feed and less competition for food.

Aquaculture is important for the socio-economic reinforcement of local communities, particularly in deprived rural areas (Marković et al. 2011). Employment of local population in aquaculture companies improves individual family's economic potential, with positive effects on reduction of poaching.

In addition to the aforementioned positive effects of aquaculture, there are numerous negative effects on fish resources in the inland water.

Some of the consequences: introduction of allochthonous species and introduction of diseases are already mentioned in previous chapters. Here, we would like to stress negative influence of aquaculture through introduction of allochthonous genotypes in inland water, effects of fish farms on recipient water, as well as introduction of drugs, hormones, and disinfectant in the inland water.

Introduction of allochthonous genotypes in the inland waters not only jeopardizes autochthonous fish resources but has other negative consequences. The so-called globalization of genotypes of certain fish species occur, thus the comparative advantage of autochthonous species is lost. Namely, in aquaculture fish are selected for fast growth, diseases resistance, increased tolerance for lower water quality, and some morphological characteristics (as example: taller height and smaller head in carp with the aim of obtaining more meat). Accordingly, alleles frequency in the genotype is changed compared to autochthonous individuals from the inland waters (Spasić et al. 2010). Arrival of such individuals from aquaculture in inland water through: discharged water; escape from aquaculture objects; as consequence of flood; broken meshes in cage aquaculture, etc., alters the genetic pool of the existing population inhabiting the aquatic ecosystem (Marković et al. 2018). An example of such negative influence of aquaculture on fish resources in Serbia is a particularly interesting case of the brown trout (*Salmo trutta*), cultured and reproduced in aquaculture objects for the purpose of restocking hilly mountain water currents. By such procedures, most often, the origin of the broodstock is not taken care of since spawning is done with the broodstock available on the trout farm, leading to genotypes mixture, i.e. loss of autochthonous genotypes, and the loss of autochthonous genotypes for the particular area.

When it comes to the negative effect of aquaculture on water quality in recipient waters in Serbia different systems should be considered: systems for intensive aquaculture; traditional flow-through for trout; and tank-systems for trout, carp, and sturgeon; as well as stagnant and slow-flowing aquatic ecosystems in which cages are placed (Dulić et al. 2009; Živić and Marković 2004; Stojanović et al. 2019; Božanić et al. 2019a). Contrary to mentioned intensive systems, the dominant in Serbia is a semi-intensive system for carp production. Since farm objects, the rearing ponds, are actually water precipitators, they do not dramatically affect recipient water. The unique period when one can be concerned by the outflowing water from fish farms is during the harvest and pond emptying at the end of the growing season, when a slight negative effect on water quality could be found (Dulić et al. 2010; Ćirić et al. 2015; Marković et al. 2016; Dulić et al. 2018; Božanić et al. 2019b).

As a consequence of production intensification, aquaculture is increasingly facing the need to use drugs, hormones, and disinfectants. There is not enough competent experts for fish diseases in Serbia, and therefore use of pharmaceutical treatment without previous detailed bacteriological and virological analyses and lack of antibiograms determination. As a consequence there is an increase of residues of pharmaceuticals in natural aquatic ecosystems. For the controlled reproduction of cyprinid species, recently of other families as well, almost exclusively dried hypophysis extract from sexually mature carps is used. While hypophysis is applied, part of the solution may enter natural waters with the discharged water from the hatchery. It can negatively affect the reproductive cycle of natural populations. Disinfectants should be mentioned as well: used for disinfection of tanks during roe incubation and maintenance of the object and working spaces, since these toxic and endocrine disrupting products are discharged as well in inland waters.

6 The Role of Aquaculture for the Ecological Sustainability of Fish Resources in the Inland Waters in Serbia in the Previous Period

Since over 60% of fish species inhabiting inland waters in Serbia are under a certain level of protection in the previous period, a limited number of species was used for repopulation and reintroduction in their natural habitat.

6.1 Family Acipenseridae: Sturgeon Species

The main reasons for the decrease of abundance and continuous increased vulnerability of populations of sturgeon species in the Danube are: building the dams (Djerdap 1 and Djerdap 2) to produce electricity; increased demand for caviar; political events in the 90s of the twentieth century in the countries of Lower Danube; interventions in the very riverbed; and pollution (Lenhardt et al. 2018.). Since fish ladders were not established as a way of compensation of the damage made by dam construction, a fish farm in Mala Vrbica near Kladovo was built, with the aim of producing fry, primarily sturgeons, as their reproduction was prevented by Danube partition. The farm faced a number of problems, such as: variation of the water column caused by water mirror level in the ponds, because of permeability of ponds' soil; unclear work program of the farm; staff not qualified for reproduction and weaning of sturgeons; unstable financing of the program; political influence on farm management; and many others. Shortly after opening, the first sturgeon farm was closed. Almost 2 decades later, at the beginning of the twenty-first century, the idea of reproducing species from the family Acipenseridae, primarily beluga was revived. Therefore, one single object that once belonged to the farm was reconstructed and adapted into a hatchery. Unfortunately, the project failed before serious trial for reproduction was attempted.

Today, there are officially five farms for sturgeon culture registered at the Veterinary Directorate of the Ministry of Agriculture, Forestry and Water Management of the Republic of Serbia (Marković 2020). However, these farms grow sterlets and Russian sturgeon until they reach consumable size from imported fry. There is no fry production of species from the Acipenseridae family for restocking fishery waters in Serbia.

6.2 Family Cyprinidae: Cyprinid Species

In the last three decades, there is a significant decrease of tench (*Tinca tinca*) population in aquatic ecosystems in Serbia due to intensive hydro-construction and water management works that have drastically decreased flood zones (Ćirković et al. 2009), degradation and disappearance of aquatic habitat rich in macrovegetation, and pollution (Simić et al. 2009). Although earlier, it was a second important species of cyprinid aquaculture, together with the decrease of a number of individuals of this species in slow flowing rivers, canal network, ponds and lakes in Serbia, in carp farms tench is rarely cultured. With the aim of repopulating inland waters by this species, efforts were made in the past decades to increase tench abundance in hydroaccumulation systems, such as Vlasina and Ćelije. Specimens used for restocking originated from our northern province of Vojvodina, obtained in one hatchery. In order to improve the hatching technology and research possibilities for this species, successful reproduction and fry culture, laboratory spawning of broodstock from wild populations collected from Ćelije hydroaccumulations was carried out in the Experimental hatchery of the Faculty of Natural Sciences in Kragujevac. Larvae and fry were cultured in the experimental pond "Ribarska zadruga Mošorin" during 2009 and 2010 (Maletin et al. 2011), as well as in 2010 in the laboratory conditions in aquaria of the Kragujevac Aquarium of the mentioned Faculty of the University of Kragujevac (Milošković et al. 2011). Unfortunately, after these successful attempts, activities to improve spawning technology and fry culture to repopulate and reintroduce the tench in the inland waters in Serbia were stopped.

In addition to the activities performed for tench reproduction and use of fry for repopulation, a commercial hatchery of the farm "Ribnjačarstvo" from Sremska Mitrovica should be mentioned. This hatchery owns a broodstock of tench and some other protected fish species (wild carp, barbel, crucian carp, pikeperch, wels, and northern pike), spawn and sell fry of these species for restocking.

Also, a Centre for Fishery and Applied Hydrobiology "Little Danube" of the Experimental school estate "Radmilovac" of the Faculty of Agriculture University of Belgrade holds a broodstock of wild carp, with tagged broodstock of known origin, from the Danube and the Tisza, that may be used for the increase of carp population in slow flowing and stagnant waters.

6.3 Family Salmonidae: Salmonid Species

As a consequence of anthropogenic pressure on salmonid species, primarily Danubian salmon and brown trout, by overfishing, deteriorated water quality, changes of salmonid habitat, and by building a number of dams for creating hydroaccumulation, population abundance of these noble species has dropped drastically.

Fig. 1 Hydroaccumulation “Bajina Bašta” on the Drina in Perućac

While the dams were constructed, mostly for production of electricity, in the earlier decades, fish ladders were not built. In the last few decades, when a number of small hydroelectric power plants on hilly mountain water currents of Serbia were made, fish ladders were built, but in some cases they were not functional enough. Lack of fish ladder and functional fish ladder represents an obstacle impossible to overcome for salmonid species while they migrate in the upper parts of salmonid flows. As an example, after building the “Bajina Bašta” biggest hydroaccumulation on the Drina in Perućac (Fig. 1) for electricity production, in 1966, as compensation for lack of fish ladder, a trout farm in Perućac was established in 1968 in order to spawn salmonids for the Drina repopulation. However, after the farm was built they started commercial production of rainbow trout within the company “Ribarsko gazdinstvo Beograd” (Fig. 2). After several decades of rainbow trout production, by enthusiasm of an ex-sportsman, Vanja Grbić, and with the professional help of the Department of Biology and Ecology of the Faculty of Natural Sciences, University of Novi Sad, and with collaboration of the management of the fishing area “Ekoribarstvo” from Valjevo, and finally with the support of the Ministry, after planning in 2010, in March 2011, Danubian salmon was spawned and fry cultured for restocking the Drina. In November 2011, 300 individuals of an average length of 10 cm, and about 1500 individuals in the next 2012, were used. Unfortunately, after project realization on the Perućac farm only rainbow trout culture continued.

Fig. 2 Trout farm in Perućac

In Serbia, for the first time artificial spawning and fry culture of the Danubian salmon and brown trout up until the age appropriate for restocking were successfully carried out in laboratories of the Centre for Multidisciplinary Research of the University of Belgrade, at the end of the twentieth century (Mićković et al. 1998). Then, several hundreds of individuals of both species were cultured in recirculating laboratory systems and the river Trešnjica, Drina's tributary, was restocked. Fish growth and their nutrition were monitored by the researchers of the Centre (Nikčević et al. 1998). After that, research in improvement of the spawning technology and fry culture of the Danubian salmon caught in the Trešnjica river, was successfully resumed on the trout farm "Riboteks" in Gornja Trešnjica (Mijović-Magdić and Hegediš 2000; Hegediš 2003). The same technology of experimental spawning, egg incubation, and fry culture was realized on the brown trout, using broodstock from the river Rača, Drina's tributary. As a result, about 1500 individuals of fry, average length 50 mm were obtained and used for the Rača river restocking (Hegediš 2003).

A positive practice of reproduction and culture of fry of the autochthonous cold water fish species on the farm hatchery "Braduljica" in Ivanjica, operating under the municipal organization of sport fisherman should be mentioned. They carry out successfully the program of spawning of the grayling, Danubian salmon, and brown trout, as well as fry growing for restocking.

In addition to the examples mentioned, in the last few decades, several smaller trout farms, periodically, carried out spawning and fry rearing of the Danubian salmon and brown trout.

Few years ago, in 2018, a public company "Srbijašume" at the source "Belo vrelo" of the Tolišnica built a reproduction center for production of fry of autochthonous salmonid species. Its capacity is 1,000,000 individuals per year.

6.4 *Family Umbridae*

Umbra krameri (Walbaum 1792)—European mudminnow, endemic species of the Danube and Dniester basin with population abundance decreasing rapidly in the last decades of the twentieth century, strictly protected species inhabiting only few localities in Serbia, has raised the interest of researchers in Serbia (Branković et al. 1996; Kostić et al. 2001; Sekulic et al. 1998; Simić et al. 2007; Sekulić 2013; Sekulić et al. 2013; Krpo-Ćetković et al. 2018). The last positive example of research on the European mudminnow was the collaboration of the research teams of the Faculty of Biology and the Faculty of Agriculture University of Belgrade, its Centre for Fishery and Applied Hydrobiology "Mali Dunav" (Fig. 3). Fry obtained was used for restocking the natural habitat of the species—Special reserve "Kraljevac."

7 The Role of Aquaculture for the Ecological Sustainability of Fish Resources in the Inland Water in Serbia

In the conditions of disturbed ecological balance of fish resources in inland water it is necessary to undertake measures for their recovery. The recovery may be done through the natural process with more or less human help or by the support of restocking fry obtained from aquaculture. The process of natural recovery of ichthyocenosis, as well as of the whole biocenosis takes place in a long period of time. It is most often based on elimination of causes of the disturbed ecological balance, in cases where it is possible. There is also establishment of new biotopes, such as new fish spawning sites, similar to previous ones (when physical changes of the biotope have occurred). The disadvantage is that this process requires a long time, and the ichthyocenosis will go through great instabilities. Under "instabilities" we consider the following: variation of abundance of some fish species; different reproductive potentials of different species; competitive interactions between different species; unequal predisposition of survival in new ecological conditions; different interactive relationships between members of the ichthyocenosis and other members of the aquatic community. All these contribute to fish community vulnerability. In such situations, the assistance provided by restocking with fry originated

Fig. 3 European mudminnow with eggs on an artificial substrate in the aquarium of the Center for Fishery and Applied Hydrobiology "Mali Dunav" at the Faculty of Agriculture University of Belgrade

from aquaculture based on scientific knowledge is the easiest and fastest, often the unique, way for ichthyocenosis recovery.

However, the use of the term "stocking" poses a number of questions, dilemmas, and problems. First, how to find broodstock with the genotype appropriate for the endangered aquatic ecosystem? When it comes to genotype as a result of a long evolution of biocenosis, if there are specimens preserved, then restocking is relatively easy: the broodstock may be caught. Care should be taken to use genetically "clean" breeding material from the autochthonous genotype of this very ecosystem. A much bigger problem arises when reintroduction has to be done, i.e. when species that need to be restocked in inland water do not live in this aquatic ecosystem any longer, and broodstock must be obtained from another water current. In such cases, the water current of choice has to be ecologically similar as much as possible to the one to be repopulated, to increase the chances for this genotype to be adapted to the new ecosystem. A long-term solution in such situation is establishment of the gene bank in a national reproductive center. Such a center should have a basis of tagged fish with the database on their origin, caught from different water currents, kept in aquaculture premises for the purpose of repopulation and reintroduction of protected and strictly protected fish species. In such a center, a bank of gametes may also be established where gametes are kept in liquid nitrogen instead of keeping live broodstock. This is particularly important for species difficult to keep in captivity or if their culture is particularly demanding. With such a gene bank inbreeding is avoided, although it happens when both females and males are caught in the same locations. Unfortunately, in Serbia, such a national center is not established and repopulation/reintroduction is carried out from existing commercial hatcheries or reproductive centers. In practice, restocking is done with fry that may be obtained, without taking care of their genetical traits and origin. After the decision is made where the broodstock will be obtained, it is important to choose the way of catch and transportation in order to reduce stress. Appropriate storage and accommodation of broodstock that should be in full reproductive condition is important. The choice of the spawning technology has to be scientifically based and appropriate for the species and/or family. Questions to be solved concern the choice of anaesthetics; use of hypophysis or synthetic hormones; ways of gametes extraction; when, how long, and where eggs and milt should be kept; how to carry out fertilization; what are the criteria for gametes fecundation. When roe and milt are obtained and eggs fertilized, new dilemma appears concerning incubation, use of the best logistics/equipment for incubation; decisions about water quantity and water flow, way how to overspread the fertilized eggs, light–dark periods; motion or not of the fertilized roe; and how to optimally hatch the larvae. When the larvae are obtained, the choice of feed is between the natural and artificial, depending on the phases of growth, feeding strategies; how to reduce cannibalism in ichthyophagous fish; reduce diseases; and finally when fry is raised, how to adapt to natural conditions, i.e. prepare for restocking.

Regardless whether the breeding technology for fish species identified for repopulation or reintroduction is known, completely or partly, or not developed at all, a multidisciplinary approach is necessary for successful breeding for repopulation/reintroduction, with the participation of experts from hydroconstructions, hydrochemistry, hydrobiology, hydroecology, ichthyology, and aquaculture. Yet, when the total number of strongly protected and protected fish species is compared with the number of attempts of reproduction and fry rearing, and the number of dilemmas that remained, it is clear that the role of aquaculture in maintaining stability of fish resources in inland water in Serbia is still small and would need much more attention in the future. In addition, when it comes to economic profitability of this venture, it is evident why there are no sufficient experts interested nor facilities (fish farms and reproductive centers). So far, the interest of both experts and farm and reprocenter owners is focussed on economically important species, while non-important species economically attract only scientists and enthusiasts.

8 The Need for Aquaculture Development from the Aspect of Preservation of Ecological Sustainability of Fish Resources

In Serbia, there are over 200 aquaculture facilities (Marković et al. 2011), among them officially registered at the Veterinary Directorate of the Ministry of Agriculture, Forestry and Water Management of the Republic of Serbia are 149 (77 carp, 68 trout, and 4 for sturgeon) (Marković 2020). From the total number of farms reproduction and fry rearing for restocking are carried out only in few. As mentioned earlier, it is evident that the approach to management of fishery/inland water has to be changed in the direction of aquaculture, as the very last defense for preservation of ecological sustainability of fish resources in the future.

In the process of protection of sustainability of fish resources in inland water it is necessary to establish sustainable management of fish resources based on the interest of all stakeholders, with the respect of all the laws concerning protection and preservation of the waters and the fishery fund, and with permanent monitoring and control from authorized professional and inspection services. In order to protect and improve fish resources, i.e. to provide sustainability in inland Serbian waters, funding is necessary for the support of such programs, together with investments in facilities and equipment. Funding for protection and improvement of the species that are not economically important should be assured from the budget of the Republic of Serbia through taxes from polluters and other sources that negatively affect strictly protected and protected fish species. Financing programs and investments for reproduction and fry growing should also be provided from interested business, from the users of fishery areas. Experiences from other countries such as Norway, Sweden, Finland, United Kingdom, France, Switzerland, Czech Republic, Austria, and Slovenia, countries in which such programs work well, should be analyzed, used, and followed. Namely, good functioning of such programs is based on interests of all stakeholders: the state, users of fishery areas, local population, and fishermen.

Aquaculture facilities and reproduction centers are very important links in the chain that aims to successfully connect the functionality of these programs. Their primary role is reproduction of autochthonous species and production of fry for restocking, development of fishery tourism, recreational and sport fishery. In addition to their primary role in reproduction and fry culture, all the mentioned premises could serve as educational centers for spreading knowledge on biology and ecology of protected and strictly protected species, thus raising awareness about the importance of these species, further contributing to their preservation as valuable members of the fish fauna in our inland waters.

A network of reproductive centers, fish farms, users of fishery waters, fishermen, and local population is the basis of functioning of the whole society/community of the local area. In such areas, there is benefit not only for the mentioned stakeholders, but also for the local population dealing as indirect support, such as transportation, repairs, hospitality industry, accommodation, etc. The above mentioned reflects on the economy of the local population, enabling growth and support to all activities, such as production and processing of fruit, vegetables, fish, met, souvenirs, etc. Besides, together with the fishery tourism there is an increase of other commercial offers attracting village, eco or adventure tourism. Thus, economic and ecological interests are connected, contributing to decrease of migrations toward big cities. Such a development, based on fishery tourism, improves the economy of the whole area, but also sustainability of aquatic ecosystems, since aquatic ecosystems are the backbone of their development.

Fish species that are the center of interest of potential investors in aquaculture, as well as users of fishery waters in hilly mountain areas are the following: Danubian salmon, brown trout, marble trout, and grayling. Water currents of particular interest are the: Drina, Moravica, Uvac, Đetinja, Vlasina, Rasina, Lopatnica, Veliki Rzav, Radovanska reka, Gradac, Grza, Temska, Rasina, as well as plenty of smaller currents.

Among the warm water fish species particularly interesting for the development of aquaculture are all protected and strongly protected from the family Acipenseridae and Clupeidae, such as the eel, tench, crucian carp, bream, asp, wild carp, chub, northern pike, and pikeperch. Concerning rivers, there are many extremely interesting destinations on the largest rivers in Serbia: Danube, Tisza, Sava, Velika Morava.

9 Concluding Remarks

From the times of migration of *Homo sapiens* on the European continent, over 40,000 years ago, man has inhabited areas near water currents. Besides the need for water, from these streams, swamps, and lakes, man caught fish for food. As human population grew, influence of men on aquatic ecosystems increased. In the last 100 years, the impact of men on aquatic ecosystems is bigger than previous 40,000 years. Intensification of agriculture, industrialization and urbanization, construction works for the management of water currents, building of dams on the rivers for establishing hydroaccumulations, building of traffic infrastructure, but also harvesting aquatic organisms, primarily fish, caused a serious degradation of the fishery funds.

Today, when Serbian waters are inhabited by 100 fish species, over 60% species are under certain level of protection (34 with the status of strongly protected and 29 protected wild fish species). This clearly shows the necessity of urgent measures to be taken in order to prevent further degradation of the fishery funds and ichthyocenoses renewal in lots of aquatic ecosystems in Serbia.

To prevent further degradation of the fishery fund, it is necessary to preserve biotopes and prevent pollution, as indirect ways of protection. A direct protection, includes protection from overfishing and different other negative effects (introduction of diseases, parasites, allochthonous species, allochthonous genotypes).

After prevention from further degradation of the aquatic ecosystems, and consequently the fishery fund, next step is their recovery. Concerning the natural recovery of aquatic ecosystems, a question to be asked is whether the system, upon protection from further degradation, should be left to recover through natural processes of recovery, or help from the main causative agent—the man, is necessary? Some experts think that it is sufficient to remove the reasons of ecosystem degradation and leave the aquatic community, including the ichthyocenosis to restore the original state. However, the process of natural recovery of the ecosystem with the biocenoses, with the focus on the fish community requires a long time. During this long period, recovery is not simple, many negative effects appear. This is the reason why some experts think that human action is necessary. Together with abiotic factors correction and maintenance, restocking is the general solution. When species have completely disappeared, stocking is the unique way to recover. Particular problem in such cases is to provide fry with the same genotype or similar to the autochthonous, so the species is adapted to the biotope quickly. In the case when the number of individuals of a certain species decreased and repopulation is needed, if there is a possibility to catch individuals with the autochthonous genotype from the same water current, then the problem is smaller. However, in practice, in Serbia, care is not taken about autochthonous genotypes. Restocking is done with the same species, originated from other water currents and/or from fish farms or reproductive centers, depending on where it is available. The question is whether any farm or reproductive center retains fry for restocking. Namely, although in Serbia there are over 200 fish farms, a small number is directed toward reproduction and rearing of the species that are not commercially important, and are not cultured for consumption. The reason for this is because there is weak and non-continuous demand for fry of autochthonous species from users and managers of fishery waters from the competent ministry (Ministry for Environmental Protection of the Republic of Serbia). The long list of strongly protected and protected fish species in Serbia on one side, and modest demand for fry of these species, including small number of fish farms and reproductive centers on the other, speaks in favor of the urgent necessity to change the approach to the management of fishery waters in Serbia.

In order to create a sustainable system, it is necessary to coordinate interests of all stakeholders in preserving and rationalizing the use of fish resources, together with strict respect of laws concerning environmental protection and water management, including the fishery funds and with continuous monitoring and control from competent experts and inspection services. Since financing is key to establishing sustainable development, it is of utmost importance to create a clear strategy of management of incomes and outcomes in the area. Since some species are attractive, there is economic interest for them (such as Danubian salmon), and not for some others (such as European mudminnow), logically strategies for financing should be different. For economically non-important species financing from the budget of the Republic of Serbia, obtained through budget incomes (such as ecological taxes, taxes obtained from regulation of fishery/angling); from the polluters; and other sources of negative effects on inland waters such as dam construction and traffic network establishment. For economically important species financing of the programs of reproduction and fry culture and investments would be provided from the budget of the Republic, and later from the interested business, i.e. incomes from fishery water users realized by sustainable exploitation of these species. Precisely,

Fig. 4 Reconstruction of houses from the culture period "Lepenski vir" (according to Pavlović 2013) in the Centre for Fishery and Applied Hydrobiology "Mali Dunav" at the Faculty of Agriculture University of Belgrade

good functioning of the mentioned programs has to be based on establishment of linkage between the state and the users of fishery waters, aquaculture and reproduction centers, local population, and fishermen. Aquaculture facilities and reproduction centers are the key link of the functional system of sustainable management of fishery waters. Besides this, the centers for reproduction and fry rearing may be very attractive education and touristic complexes that would give an important contribution to the knowledge of biology and ecology of strongly protected and protected species. Thus, the awareness about the importance of these species and their preservation would be significantly raised. Interesting stories concerning fish species and their importance in all aspects of life, from prehistoric times and further in history, trade, transportation, art, etc., would contribute to attractiveness and sustainability of such places. As an example, the story around beluga, large fish species from the Acipenseridae family and its importance in the establishment of first cultures on European ground such as the culture of Lepenski vir. The fact that people from the Mesolithic period, after the end of the last ice age, on the right bank of the Danube on the locality Lepenski vir had enough food thanks to the big beluga fish. This enabled more time to build first settlements (Fig. 4), as well as first works of art—fishlike human sculptures carved in sandstone (Fig. 5).

Fig. 5 Fishlike human sculptures carved in sandstone from the culture period "Lepenski vir"

Acknowledgment This research was financially supported by the Serbian Ministry of Education, Science and Technological Development (contract No 451-03-68/2022-14/200116).

References

Alibabić V, Valčić N, Bajramović M (2007) Bioaccumulation of metals in fish of Salmonidae family and the impact on fish meat quality. Environ Monit Assess 131:349–364

Babkin VI (2003) The earth and its physical features. Physical geography. In: Shiklomanov IA, Rodda JC (eds) World water resources at the beginning of the twenty–first century. University Press, Cambridge, pp 1–10

Bauer ON (1991) Spread of parasites and diseases of aquatic organisms by acclimatization: a short review. J Fish Biol 39(5):679–686

Božanić M, Dojčinović B, Živić M, Marković Z, Manojlović D, Živić I (2019a) Bioaccumulation of heavy metals in *Ephemera danica* larvae under influence of a trout farm outlet waters. Knowl Manag Aquat Ecosyst 420:50. https://doi.org/10.1051/kmae/2019040

Božanić M, Marković Z, Živić M, Dojčinović B, Perić A, Stanković M, Živić I (2019b) Mouthpart deformities of *Chironomus plumosus* larvae caused by increased concentrations of copper in sediment from carp fish pond. Turk J Fish Aquat Sci 19(3):251–259. https://doi.org/10.4194/1303-2712-v19_03_08

Branković D, Budakov Lj, Kovačev N, Mijović D, Mikeš B, Pavkov G, Puzović S, Sekulić N, Stojšić V, Habijan-Mikeš V, Mučenski V, Garovnikov B, & Stanković M (1996) Study – proposal for safeguard of the protected natural area "Zasavica" as special natural reserve. Institute for Natural Conservation of Serbia, Novi Sad

Britton JR, Pegg J (2011) Ecology of European barbel *Barus barbus*: implications for river, fishery, and conservation management. Rev Fish Sci 19(4):321–330

Ćirić M, Subakov-Simić G, Dulić Z, Bjelanović K, Čičovački S, Marković Z (2015) Effect of supplement feed type on water quality, plankton and benthos availability and carp (*Cyprinus carpio* L.) grown in semi-intensive monoculture ponds. Aquac Res 46(4):777–788

Ćirković M, Marković G, Simić V, Maletin S, Milošević N, Momirov D (2009) Reintroduction and repopulation of tench (*Tinca tinca* L.) in fish ponds and natural waters. Conference proceedings, pp.132–137. 4th international conference "fishery" Faculty of Agriculture, institute "Nofima-marine" Norway and institute "HAKI" – Hungary, Belgrade–Zemun, Serbia, 27–29 May 2009

Cvijanović G, Lenhardt M, Hegediš A (2005) The first record of black bullhead *Ameiurus melas* (Pisces, Ictaluridae) in Serbian waters. Arch Biol Sci, Belgrade 57(4):21–22

Demény F, Sipos S, Ittzés I, Szabó Z, Lévai P, Bodó I, Urbányi B, Müller T (2009) Observations of the crucian carp (*Carassius carassius*) pond culture. 4th international conference "fishery" Faculty of Agriculture, institute "Nofima–marine" Norway and institute "HAKI" – Hungary, Belgrade–Zemun, Serbia, 27–29 May 2009. Conference proceedings, pp 138–144

Djikanović V, Paunovic M, Nikolic V, Simonovic P, Cakic P (2012) Parasitofauna of freshwater fishes in the Serbian open waters: a checklist of parasites of freshwater fishes in Serbian open waters. Rev Fish Biol Fish 22(1):297–324

Dulić Z, Poleksić V, Rašković B, Lakić N, Marković Z, Živić I, Stanković M (2009) Assessment of the water quality of aquatic resources using biological methods. Desalin Water Treat 11:264–274

Dulić Z, Subakov-Simić G, Ćirić M, Relić R, Lakić N, Stanković M, Marković Z (2010) Water quality in semi-intensive carp production system using three different feeds. Bulg J Agric Sci Agric Acad 16(3):266–274

Dulić Z, Marković Z, Živić M, Ćirić M, Stanković M, Subakov-Simić G, Živić I (2014) The response of phytoplankton, zooplankton and macrozoobenthos communities to change in the water supply from surface to groundwater in aquaculture ponds. Ann Limnol Int J Lim 50:131–141

Dulić Z, Živić I, Pergal M, Živić M, Stanković M, Manojlović D, Marković Z (2018) Accumulation and seasonal variation of toxic and trace elements in tissues of *Cyprinus carpio* from semi–intensive aquaculture ponds. Ann Limnol Int J Limnol 54(4):1–10

FAO (2014) The State of World Fisheries and Aquaculture 2014. Rome. pp. 223

Hegediš A (2003) Artificial spawning and fry growth of native fish species in Serbia (huchen, brown trout). Seminar "trout and carp fishery" Faculty of Agriculture University of Belgrade and Aquaforsk Institute of Aquaculture Research, As, Norway, 23–24. September. Conference proceedings, pp. 78–84

Holčik J, Hensel K, Nieslanik J, Skacel L (1988) The Eurasian Huchen, Hucho hucho. Joint edition published by D.W. Junk Publishers, Netherland and VEDA, Bratislava. p. 239

Janković D (1998) Natural reproduction by Asiatic herbivorous fishes in the Yugoslav section of the river Danube. Ital J Zool 65(2):227–228

Kostić D, Pujin V, Popović E, Jurišić A (2001) Nutrition and growth of European mudminnow *Umbra krameri* (Walbaum, 1792) (Osteichtyes – Umbridae) Zasavica. In: Igić, R. & Gajin, S. (eds.), Zasavica 2001. Sremska Mitrovica, Department of Biology, Faculty of Sciences, University of Novi Sad. Proceedings, pp 76–79. (in Serbian)

Kostić J, Kolarević S, Kračun-Kolarević M, Aborgiba M, Gačić Z, Paunović M, Višnjić-Jeftić Ž, Rašković B, Poleksić V, Lenhardt M, Vuković-Gačić B (2017) The impact of multiple stressors on the biomarkers response in gills and liver of freshwater breams during different seasons. Sci Total Environ 601–602:1670–1168

Kostić-Vuković J, Kolarević S, Kračun-Kolarević M, Višnjić-Jeftić Ž, Rašković B, Poleksić V, Gačić Z, Lenhardt M, Vuković-Gačić B (2021) Temporal variation of biomarkers in common bream *Abramis brama* (L., 1758) exposed to untreated municipal wastewater in the Danube River in Belgrade, Serbia. Environ Monit Assess 193:1–8. https://doi.org/10.1007/s10661-021-09232-6

Krpo-Ćetković J, Sekulić N, Stojanović K, Galamboš D, Radošević D, Marić S (2018) Diet of European mudminnow (*Umbra krameri*) in Bačka, Mačva and Semberija. 8th international conference "Water & Fish" Faculty of Agriculture, University of Belgrade, Serbia and international aquatic veterinary biosecurity consortium Ludwig–Maximilians – University Munich, Germany, 13–15 June 2018. Conference Proceedings, pp. 255–257

Larsen J, Roney J M (2013a) Farmed Fish Production Overtakes Beef. http://www.earth-policy.org/plan_b_updates/2013/update114

Larsen J, Roney J M (2013b) Farmed Fish Production Overtakes Beef. http://grist.org/article/farmed-fish-production-overtakes-beef/

Law on Protection and sustainable use of fish fund, Official Gazette RS, No. 128/2014

Lenhardt M, Hegediš A, Mićković B, Višnjić–Jeftić Z, Smederevac M, Jarić I, Cvijanović G, Gačić Z (2006a) First record of the north American paddlefish (*Polyodon spathula* Walbaum, 1792) in the Serbian part of the Danube river. Arch Biol Sci 58(3): 27P–28P

Lenhardt M, Jaric I, Kalauzi A, Cvijanovic G (2006b) Assessment of extinction risk and reasons for decline in sturgeon. Biodivers Conserv 15:1967–1976

Lenhardt M, Markovic G, Hegediš A, Maletin S, Cirkovic M, Markovic Z (2011) Non–native and translocated fish species in Serbia and their impact on the native ichthyofauna. Rev Fish Biol Fish 21:407–421

Lenhardt M, Djikanović V, Hegediš A, Višnjić-Jeftić Ž, Skorić S, Smederevac-Lalić V (2018) Qualitative and quantitative changes in the ichthyofauna of the Danubian reservoirs after the construction of the iron gates hydropower plant dams. In Radmila Petanović (Ed.) Ecological and economic significance of fauna of Serbia. Proceeding of the scientific meeting, held on November 17, 2016. Serbian Academy of Sciences and Arts, Book 12, 173–197

Lenhardt M, Smederevac-Lalić M, Hegediš A, Skorić S, Cvijanović G, Višnjić-Jeftić Ž, Djikanović V, Jovičić K, Jaćimović M, Jarić I (2020) Human impacts on fish Fauna in the Danube River in Serbia: current status and ecological implications. In: Bănăduc D, Curtean-Bănăduc A, Pedrotti F, Cianfaglione K, Akeroyd JR (eds) Human impact on Danube watershed biodiversity in the XXI century. Springer, Cham, pp 257–280

Maletin S, Djukic N, Miljanović B, Ivanc B (1997) Status of allochthonous ichthyofauna of Pannonian basin in Yugoslavia. Ekologija 32:87–98

Maletin S, Lečić B, Ćirković M, Milošević N, Ljubojević D (2011) Reconstitution of tench natural populations tench (*Tinca tinca*) by application of laboratory spawning, 5th international conference "Aquaculture & Fishery". Faculty of Agriculture, Belgrade–Zemun, Serbia, 1–3 June 2011, Conference proceedings, pp 200–204

Marcogliese DJ (2004) Parasites: small players with crucial roles in the ecological theater. EcoHealth 1:151–164

Marković Z (2010) Carp, rearing in ponds and cage systems. Prof. Dr. Zoran Marković, p 152 (in Serbian)

Marković Z (2020) Fishery sector study for the IPARD programme. SEEDEV. Ministry of Agriculture, Forestry and Water Management Republic of Serbia, p 75

Markovic G, Simovic S (1997) Pumpkinseed (*Lepomis gibbosus* L. 1758) – unwelcome member to the ichthyofauna of the Yugoslav reservoirs. The proceedings of symposium "fisheries of Yugoslavia", pp 123–128 (in Serbian)

Marković Z, Dulić Z, Stanković M, Rašković B, Spasić M, Živić I, Poleksić V (2011) Freshwater aquaculture in function of development of (Agri)culture and rural areas. University of Banja Luka, Faculty of Agriculture. Book of abstracts XVI international scientific and professional conference of the agronomists of the Republika Srpska "natural resources in function of agriculture and rural development" Trebinje, p. 32 (in Serbian)

Marković Z, Stanković M, Rašković B, Dulić Z, Živić I, Poleksić V (2016) Comparative analysis of using cereal grains and compound feed in semi-intensive common carp pond production. Aquacult Int 24:1699–1723. https://doi.org/10.1007/s10499-016-0076-z

Marković Z, Stanković M, Rašković B, Sekulić N, Poleksić V (2018) Aquaculture in service of endangered fish species protection in Serbia. In Radmila Petanović (Ed.) ecological and economic significance of fauna of Serbia. Proceeding of the scientific meeting, held on November 17, 2016. Serbian Academy of Sciences and Arts, Book 12, pp 173–197

Martinović-Vitanović V, Kalafatić V (1995) Basic hydrological characteristics of the inland waters of Yugoslavia. In: Stevanović V, Vasić V (eds) Biodiversity of Yugoslavia with species of international importance overview. Faculty of Biology and Ecolibri, Belgrade, pp 97–117. (in Serbian)
Mićković B, Nikčević M, Hegediš A, Andjus RK (1998) Huchen fry growth in aquaculture and their natural habitat. Arch Biol Sci 50(4):35P–36P
Mijović-Magdić J, Hegediš A (2000) Contemporary status of the Danubian salmon (*Hucho hucho*) in Serbian waters and possibilities of controlled spawning and fry rearing in conditions of intensive fish production. "Contemporary Fishery of Yugoslavia" – Vršac September 20–22 2000. Proceedings of the IV Yugoslav Symposium "Fishery of Yugoslavia", 75–80 (in Serbian)
Milošković A, Pavlović M, Simić S, Simić V, Kovačević S, Radojković N (2011) Breeding of Tench fish (*Tinca tinca*) in laboratory. V international conference "Aquaculture & Fishery" Faculty of Agriculture University of Belgrade, Serbia, 1–3 June 2011. Conference proceedings, pp. 450–456
Naylor RL, Hardy RW, Bureau DP, Chiu A, Elliott M, Farrell AP, Forster I, Gatlin DM, Goldburg RJ, Hua K, Nichols PD (2009) Feeding aquaculture in an era of finite resources. PNAS 106: 15103–15110. https://doi.org/10.1073/pnas.0905235106
Nehlsen W, Williams JE, Lichatovitch JA (1991) Pacific salmon at the crossroad: stocks at risk from California, Oregon. Idaho and Washington Fish 16:21
Nikčević M, Mićković B, Hegediš A, Andjus RK (1998) Huchens (*Hucho hucho*) in river Trešnjica, a tributary of river Drina in Serbia; feeding habits of the fry. The Italian Journal of Zoology 65 (Suppl):231 233
Nikolić D, Skorić S, Poleksić V et al (2021) Sex–specific elemental accumulation and histopathology of pikeperch (*Sander lucioperca*) from Garaši reservoir (Serbia) with human health risk assessment. Environ Sci Pollut Res 28:53700. https://doi.org/10.1007/s11356-021-14526-w
Pavlović H (2013) House Lepenski Vir. IP Jovan, p 184. (in Serbian)
Peeler EJ, Oidtmann BC, Midtlyng PJ, Miossec L, Gozlan RE (2011) Nonnative aquatic animals introductions have driven disease emergence in Europe. Biol Invasions 13(6):1291–1303
Philippart JC (1995) Is captive breeding an effective solution for the preservation of endemic species? Biol Conserv 72:281–295
Pujin V, Sotirov S (1966) A study of the nutrition of the dwarf catfish (*Ictalurus nebulosus* Le Sueur). Ann Sci Work Fac Agric Novi Sad (Serbia) 10:147–156. (in Serbian)
Radosavljević V (2020) Health protection of fish and other aquatic animals in aquaculture. Research Institute for Veterinary Sciences of Serbia, Belgrade, p 250. (in Serbian)
Radosavljević V, Pavlović I, Žutić J, Nešić K, Marković Z, Ćirković M, Ljubojević D (2014) Exchange of pathogens between fish from open waters and reared in aquaculture. Ecologica, 76, XXI, str. 733–736
Radosavljević V, Milićević V, Labus T, Maksimović-Zorić J, Ljubojević D, Ćirković M, Marković Z (2016) Aquaculture and fish health in Serbia. In Second International Symposium of Veterinary Medicine (ISVM2016) Belgrade, 22–24 June, 2016. Proceedings, pp. 321–326
Radosavljević V, Milićević V, Maksimović-Zorić J, Nešić K, Novakov N, Marković Z (2018) Emerging fish diseases in Serbian aquaculture. 8th International Conference "Water & Fish" Faculty of Agriculture, University of Belgrade, Serbia and International Aquatic Veterinary Biosecurity Consortium Ludwig-Maximilians – University Munich, Germany, 13–15 June 2018. Conference proceedings, pp 109–114
Radulović S, Boon PJ, Laketić D, Simonović P, Puzović S, Živković M, Jurca T, Ovuka M, Malaguti S, Teodorović I (2012) Preliminary checklists for applying SERCON (System for evaluating rivers for conservation) to rivers in Serbia. Arch Biol Sci 64(3):1037–1056
Rašković B, Poleksić V, Višnjić-Jeftić Ž, Skorić S, Gačić Z, Djikanović V, Jarić I, Lenhardt M (2015) Use of histopathology and elemental accumulation in different organs of two benthophagous fish species as indicators of river pollution. Environ Toxicol 30(10): 1153–1161. https://doi.org/10.1002/tox.21988

Rašković B, Poleksić V, Skorić S, Jovičić K, Spasić S, Hegediš A, Vasić N, Lenhardt M (2018, May 30) Effects of mine tailing and mixed contamination on metals, trace elements accumulation and histopathology of the chub (*Squalius cephalus*) tissues: Evidence from three differently contaminated sites in Serbia. Ecotoxicol Environ Safety, 153, 238–247, issn 0147-6513, https://doi.org/10.1016/j.ecoenv.2018.01.058

Sekulić N (2013) Ecological characteristic and morphological-genetic differentiation in populations of European mudminnow (*Umbra krameri* Walbaum, 1792) from Bačka, Mačva and Semberija. Doctoral Dissertation, University of Belgrade, Faculty of Biology. p 177

Sekulic N, Lj B, Brankovic D (1998) Distribution of the European mudminnow *Umbra krameri* (Umbridae) in Serbia. Ital J Zool 65:381–382

Sekulić N, Marić S, Galambos L, Radošević D, Krpo-Ćetković J (2013) New distribution data and population structure of the European mudminnow *Umbra krameri* in Serbia and Bosnia and Herzegovina. J Fish Biol 83:659–666

Simić V, Simić S, Paunović M, Petrović A, Stanković M. (2007) Some endangered species in the special natural reserve "Zasavica" (*Umbra krameri*, Pisces and Batrachospermum gelatinosum, Rhodophyta). Proceedings. Scientific-professional meeting Zasavica 2007 with international participation. Movement for afforestation Sremska Mitrovica. Sremska Mitrovica, pp 99–107

Simić V, Simić S, Ćirković M, Pantović N (2009) Preliminary results of the research of the population of tench (*Tinca tinca*) in the water ecosystems of Serbia. 4th International Conference "Fishery" Faculty of Agriculture, Institute "Nofima–Marine" Norway and Institute "HAKI" – Hungary, Belgrade–Zemun, Serbia, 27–29 May 2009. Conference Proceeding, pp 219–223

Simonović P (2001) Fishes of Serbia. NNK international, institute for nature conservation of Serbia. Faculty of Biology, Belgrade, p 247. (In Serbian)

Šipoš Š, Miljanović B, Pejčić LJ (2004) The first record of Amur sleeper (*Perccottus glenii* Dybowsky 1877, fam. Odontobutidae) in the Danube River. 35rti. Conference IAD, Limnological reports, Vol. 35. Proceedings, Novi Sad. pp. 509–510

Spasić M, Poleksić V, Stanković M, Dulić Z, Rašković B, Živić I, Ćirić M, Relić R, Vukojević D, Bošković D, Marković Z (2010) Selective breeding programme of common carp (*Cyprinus carpio* L.) in Serbia – preliminary results. J Agric Sci 55:243–251

Stojanović K, Živić M, Marković Z, Đorđević J, Jovanović J, Živić I (2019) How changes in water quality under the influence of land-based trout farms shape chemism of the recipient streams – case study from Serbia. Aquac Int 27(6):1625–1641

Yilmaz F, Ozdemir N, Demirek A, Lement TA (2007) Heavy metal levels in two species *Leuciscus cephalus* and *Lepomis gibbosus*. Food Chem 100:830–835

Živić I, Marković Z (2004) The effect of trout pond on water quality in the recipient. Limnological reports, volume 35, Novi Sad, Serbia and Montenegro. International association for Danube research of the international association for theoretical and applied limnology, Limnological reports. Proceedings 35th IAD Conference, 389–395

Živić I, Ostojić A, Miljanović B, Marković Z (2018) Macroinvertebrates of Serbian streams and their significance as bioindicators in estimation of water quality. In Radmila Petanović (Ed.) Ecological and economic significance of fauna of Serbia. Proceeding of the scientific meeting, held on November 17, 2016. Serbian Academy of sciences and arts, Book 12, pp 199–229

Huchen *Hucho hucho* (Linnaeus, 1758) in Croatia: Distribution, Genetic Diversity, Threats and Conservation

Goran Jakšić, Neven Bočić, and Marina Piria

Abstract Huchen *Hucho hucho* (L.) is one of the largest salmonids in the world and one of the most endangered fish species inhabiting the Danube River Basin in Central Europe. The natural distribution range of the huchen included drainages in Austria, Bosnia and Herzegovina, Croatia, Czechia, Germany, Hungary, Montenegro, Poland, Romania, Serbia, Slovakia, Slovenia and Ukraine. In Croatia, huchen populations inhabit the Kupa and Una Rivers in a very narrow stretch of those rivers, threatened by a variety of human-caused effects which include hydropower development and river engineering, global climate change and deforestation, pollution of watercourses, genetic contamination, stocking and overfishing. Therefore, we discuss protective measures that could help to conserve huchen in Croatia, and our findings suggest removal of dams and obstacles with no function, determination of genetic structure and diversity of huchen, assessment of the number of spawning redds and individuals that spawn, establishment of ichthyological reserves and catch and release angling areas, aquarium activities oriented towards education on biological diversity and endangered species such as huchen.

Supplementary Information The online version contains supplementary material available at https://doi.org/10.1007/978-3-031-36926-1_16.

G. Jakšić (✉)
AQUATIKA – KARLOVAC FRESHWATER AQUARIUM Public Institute, Karlovac, Croatia
e-mail: gjaksic@aquariumkarlovac.com

N. Bočić
Department of Geography, Faculty of Science, University of Zagreb, Zagreb, Croatia

M. Piria
University of Zagreb Faculty of Agriculture, Department of Fisheries, Apiculture, Wildlife Management and Special Zoology, Zagreb, Croatia

Department of Ecology and Vertebrate Zoology, Faculty of Biology and Environmental Protection, University of Lodz, Lodz, Poland

V. Simić et al. (eds.), *Ecological Sustainability of Fish Resources of Inland Waters of the Western Balkans*, Fish & Fisheries Series 43,
https://doi.org/10.1007/978-3-031-36926-1_16

Keywords Huchen · Biology · Ecology · Distribution · Genetic diversity · Conservation

1 Introduction

Huchen *Hucho hucho* (Linnaeus, 1758) is the biggest salmonid in Europe and one of the most endangered fish species inhabiting the Danube Basin of Central Europe (Ihuţ et al. 2014). Huchen is threatened by a variety of human-caused effects, including river regulation, hydropower development, industrial and agricultural pollution, eutrophication, deforestation (Holčík 1990; Mrakovčić et al. 2006; Weiss and Schenekar 2016), global climate change (Döll and Zhang 2010), genetic contamination, stocking (Hindar et al. 1991) and overfishing (Witkowski et al. 2013). All these threats cause its very low abundance and continual decline (Holčík et al. 1988). In this respect, huchen has been listed by the IUCN *Red List of Threatened Species* as endangered (EN) (IUCN 2008).

Despite positive benefits from hydropower energy in terms of human well-being (domestic and industrial water supply, agricultural irrigation and electricity generation), it represents a major threat for global freshwater species diversity in general and the main threat for huchen decline due to dam construction (Witkowski et al. 2013). Hydropower plant dams and river engineering often divide river into segments that limit huchen areas of occurrence and block access to historical spawning grounds (Holčík 1990). The loss of connectivity also may reduce the genetic variability in the remaining populations of huchen (Schmutz et al. 2001).

Pollution of watercourses and increased eutrophication adversely affect huchen (Mrakovčić et al. 2006), which live in clear and cold, oxygen-rich waters (Holčík 1990), and directly threaten the survival of individuals through suffocation or physiological toxicity (e.g., toxic elements such as aluminium) and indirectly by degrading the food webs they depend on (Reid et al. 2013). Many small streams which were previously huchen spawning grounds are now abandoned because of pollution from fish farms and surrounding agricultural lands (Witkowski et al. 2013).

Climate change and global warming can lead to hydrological changes that affect thermal regimes in rivers, which is an environmental stressor for thermally sensitive huchen in mountain streams (Döll and Zhang 2010). The substantial extent of periods of high temperatures is most critical, as this might exhaust individuals' energy storage and capacity for growth, reproduction and survival (Marković et al. 2013).

Furthermore, huchen is a commercially important and prized species (Mrakovčić et al. 2006), and one of the greatest trophies of anglers (Klašterka 1976). The angler activities consequently caused a decline in almost all huchen populations (Holčík 1990; Mrakovčić et al. 2006) until recently, when the catch and release fishing (C&R) technique became popular in protecting populations from overfishing (Thorstad et al. 2004). However, that may have a cascading negative impact on the entire food chain and species diversity because huchen is an apex predator that controls abundance of small sized fishes (Reid et al. 2013).

Molecular analysis of wild populations and hatchery stocks of huchen across most of its range have shown low values of genetic diversity and intraspecific divergence (Geist et al. 2009; Weiss et al. 2011; Marić et al. 2014; Kuciński et al. 2015). At present, many huchen populations are isolated by recently constructed dams and weirs (Witkowski et al. 2013; Snoj et al. 2022), which leads to smaller groups of individuals with potentially small population size, below which they experience inbreeding and loss of genetic diversity, in turn increasing extinction risk (Frankel and Soulé 1981; Frankham 1999). Indeed, Croatian populations are also disconnected, isolated and threatened (Freyhof et al. 2015; Witkowski et al. 2013) and to date, no comprehensive research on huchen has been performed. Hence, an overview of the previous knowledge on huchen is needed in order to summarize the current information available in Croatia. The overall status on its biology, ecology, reproduction cycle, distribution and conservation genetics application will be described. Moreover, protective measures that could help to conserve this endangered species are discussed.

2 Biology and Ecology

The huchen *H. hucho* has an elongated and cylindrical body, the head long and laterally flattened, with a large mouth. The back is greyish green, the flanks lighter, the belly white to light silver grey. The back and flanks are spotted with sparse and irregular dark spots. The fins are relatively small, while the adipose fin is large (Fig. 1) (Mrakovčić et al. 2006; Ćaleta et al. 2015).

The typical habitat of the huchen is the mountainous sections of large rivers, where the average summer water temperature is about 15 °C. It inhabits deep eddies and pits under cascades or waterfalls (Movie V1). Smaller individuals live in the upper sections of rivers (Figs. 2 and 3), migrating downstream as they grow (Figs. 4 and 5) (Jakšić 2019). The species is sensitive to the physico–chemical properties of water and seasonal or other changes, which reflect on its physiology, reproduction, ecology and geographical distribution (Reid et al. 2013). Huchen prefers rapid and strong streams alternating with pools in places, with water rich in oxygen (O_2 content above 8.0–8.5 mgL^{-1}) (Holčík 1990), river width from 5 to 10 m, and mean discharge from 0.5 to 2 m^3s^{-1} (Ratschan 2014).

Huchen is the world's second largest salmonid, reaching up to 1.8 m standard length (L_S) where the largest documented size was 1.65 m L_S and 60 kg (Kottelat and Freyhof 2007; Sofradžija 2009). Based on the von Bertalanffy equation,[1]

[1] The equation is represented as $L_t = L_\infty[1 - e^{-k(t - t_0)}]$, where L_t is the length at age t, L_∞ is the asymptotic length, k is a growth coefficient and t_0 is a time coefficient at which length would theoretically be 0.

Fig. 1 Huchen *Hucho hucho* (total length ≈ 70 cm) swims in the upper flow of the Kupa River in Croatia (Photo by Marin Jarnjak)

Fig. 2 Huchen *Hucho hucho* in the Kupica River (Photo by Marin Jarnjak)

calculated from the mean values of linear growth, one may estimate that the nominate form may live as long as 60 years. A higher longevity value was obtained by entering the actual maximum body length into the von Bertalanffy equation, instead of the calculated average value of asymptotic length (Holčík et al. 1988). Also, it has been stated that huchen growth is extremely fast, rarely taking 20 years

Fig. 3 The upper course of the Kupa River is a typical habitat of the huchen *Hucho hucho* in Croatia (Photo by Goran Jakšić)

Fig. 4 A large specimen of huchen *Hucho hucho* (total length ≈ 115 cm) in the Kupica River in Croatia (Photo by Marin Jarnjak)

Fig. 5 The middle course of the Kupa River where large individuals huchen *Hucho hucho* migrate from the upper course (Photo by Krešimir Kuri)

of age (Hauer 2003). In Slovenia, for example, the measured mean value of 1.29 m in length and 22.5 kg in weight corresponded to the age of 15 years (Svetina 1968).

Huchen is a spring spawning fish (March–April), belongs to the ecological group of rock and gravel hiders (Balon 1990) and can migrate in order to search for suitable spawning habitat (0.3–1.5 m deep) in the main river (Holčík 1995). Huchen becomes sexually mature at an age of 3–5 years with associated mass 1–3 kg (Prawocheński and Kolder 1968) and lengths of 65–70 cm (Holčík 1995). In general, huchen may be characterized as a species which attains sexual maturity later than other salmonids. Males usually attain sexual maturity and spawn for the first time one year earlier than females and at smaller body dimensions (Prawocheński and Kolder 1968). Females and males use their tail fins to dig a spawning redd into the sandy or gravelly substrate, into which the female lays the eggs (Ćaleta et al. 2015) producing 1200–1500 eggs per kilogram body mass annually (Jungwirth 1978). In excavating the redd, the fish remove not only the surface impurities, sand and coarse gravel, but even bigger stones up to the size of a human head (Vokač 1959). Interestingly, during spawning the intensity of the mutual attacks may be so great as to cause serious injuries and even the death of the adversaries. This happened in a farm of the Slovak Angler's Union at Dubná Skala where a male 15 kg and a female 18 kg of two different pairs attacked each other and shortly thereafter succumbed to wounds caused by their bites (Holčík et al. 1988). Very little is known about the length of migration during spawning. Ivaška (1946) assumed long migrations of about 190 km from the Danube River into the Váh River to spawn. However, the

Fig. 6 Artistic expression of the huchen *Hucho hucho* diet (Author Mladen Merkaš)

spawning migration in the Drava River from one riverbank to another was observed as very short (Vokač 1959).

Juveniles feed on benthic invertebrates, but at 5–9 cm become piscivorous and, like adults, feed on fish, amphibians, reptiles, waterfowl and small mammals (Holčík 1990). Since it is a predator, the huchen will feed on practically any fish species. The most suitable fish species are those occurring in the environment typical of the huchen, i.e., in submontane streams: Eurasian minnow (*Phoxinus phoxinus*), grayling (*Thymallus thymallus*), common nase (*Chondrostoma nasus*), barbel (*Barbus barbus*), chub (*Squalius cephalus*) and schneider (*Alburnoides bipunctatus*) (Fig. 6; Movie V2) (Holčík et al. 1988).

Juveniles and adults are territorial (Kottelat and Freyhof 2007). An adult huchen defends its territory against incursions by individuals of the same species (Ivaška 1951) in the hunting area of 1–5 km, and a large pool can be occupied by several or even ten individuals (Svetina 1962). The number of individuals in a group is probably limited by the size of the territory, and indirectly also by angling intensity (Rudolf 1964). Recent observations in the Kupa and Kupica Rivers show that larger specimens of huchen inhabit a small space; usually at the depth of the water column between 2 and 4 meters, 2–7 individuals from 70 to 115 cm occupy the pit or hide under the waterfall (Mladen Merkaš, Marin Jarnjak, Goran Jakšić, personal communication) (Movie V3).

3 Distribution

The rivers inhabited by huchen belong to the Black Sea Basin (marginal Mediterranean Sea of the Atlantic Ocean) and include tributaries of the Danube River. In the Danube River Basin, huchen only inhabits submontane rivers draining the Alps, the Dinaric Alps, the Carpathians and partly also the Bohemian–Moravian Highlands, the Šumava (Bohemian Forest) Mountains and the Weinsberger Wald Hills (Holčík et al. 1988). Huchen is endemic to the Danube River Basin and historically was common over large areas of the drainage (Holčik 1995); however, the present range covers only some 33% of the original distribution range (Holčik 1990) and the distribution has become highly fragmented (Freyhof and Kottelat 2008). The range of huchen included drainages in Austria, Bosnia and Herzegovina, Croatia, Czechia, Germany, Hungary, Montenegro, Poland, Romania, Serbia, Slovakia, Slovenia and Ukraine (Fig. 7).

In Croatia, huchen inhabits the Kupa and Una Rivers, and in the past, it was also found in the Dobra, Drava and Sava Rivers (Fig. 8) (Ćaleta et al. 2015; Freyhof et al. 2015). The remaining huchen populations in the Kupa and Una Rivers still seem large but only in very narrow stretch of those rivers: in the Kupa from Turki to Blaževci and in the Una from Dvor to Hrvatska Dubica (Goran Jakšić, personal communication). The upper part of the Sava River in Slovenia still represents a good habitat for huchen and populations are still self-sustaining, but the estimated effective population size is below the accepted minimum, and additional fragmentation arising from dam construction would cause the genetic breakdown of the remaining

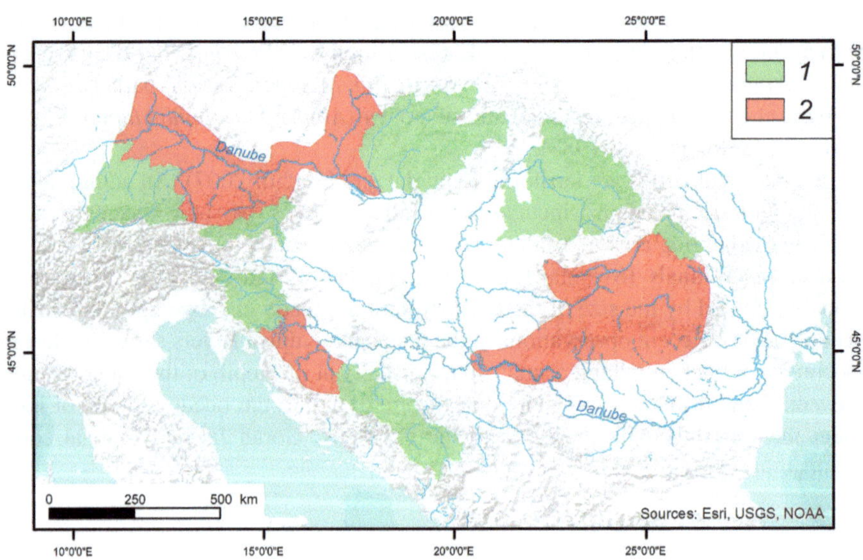

Fig. 7 Present and historical distribution of huchen in Europe (1) Present occurrence (Freyhof and Kottelat 2008); (2) historically documented occurrence (Holčik et al. 1988)

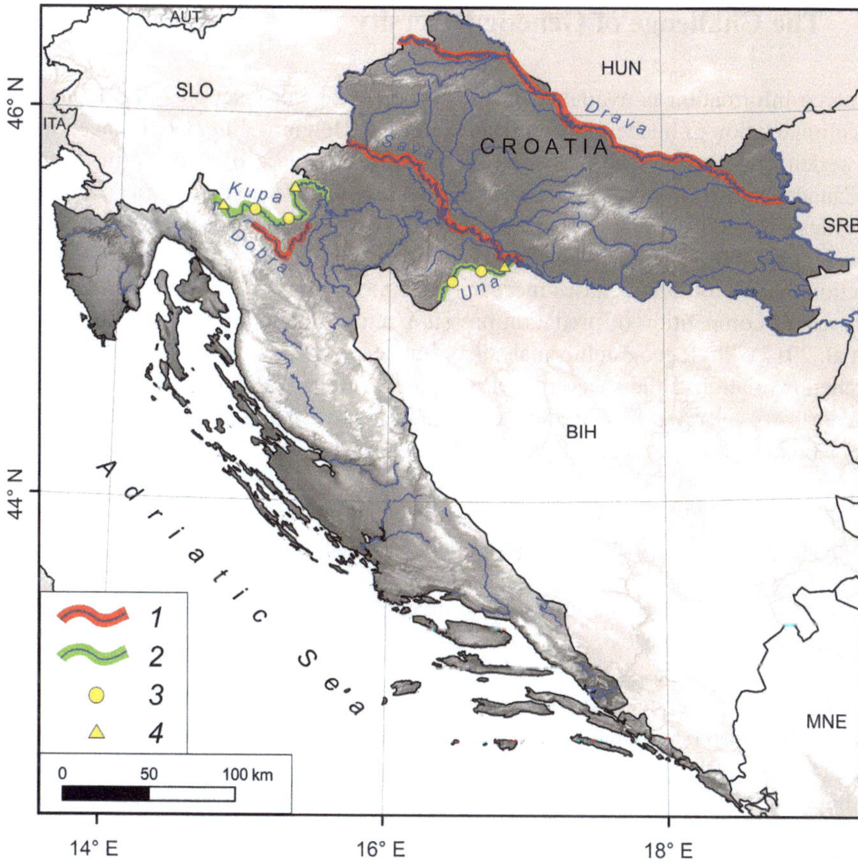

Fig. 8 Present and historical distribution of huchen in Croatia. River stretch in green represents the present distribution and in red represents the historical distribution; ○ = abundant populations; Δ = less abundant populations

huchen population (Snoj et al. 2022). However, in the stretch of the Sava River in Croatia, huchen is absent and cannot migrate because of hydropower plants upstream in the Sava River and because of inadequate water quality (Piria et al. 2019). Nevertheless, huchen is sporadically reported by anglers in Croatian part of the Sava River, whereby it is considered that their occasional occurrence may be a consequence of illegal stocking (Marina Piria, personal communication) or migration attempts from the Una River, which still represents adequate huchen habitat. The Drava River is considerably divided by several hydropower plants and dams (Witkowski et al. 2013) and in the last decade there have been no reports of huchen here. Furthermore, huchen has been lost from the Dobra River due to hydropower development since 1959 (Ćaleta et al. 2020).

4 The Challenge of Genetic Diversity

Scarce information is available on the outcomes of introduction by stocking and acclimatization of huchen in European countries (Belgium, England, France, Spain, Sweden and Switzerland), as well as in Africa (Morocco) and North America (Canada), much of which is scattered in local, poorly accessible journals, and thus evades attention (Holčík et al. 1988). However, stocking is considered not only an inadequate tool to manage or conserve natural populations of huchen, but also an action that all too often causes more harm than good in terms of genetic alterations, increased competition or predation pressure, and introduction of diseases (Freyhof et al. 2015). Phylogeographic analysis of huchen across its range has shown limited genetic variation at the species level compared to other salmonids. The populations in north-western Austria and Slovenia were found to differ significantly compared to the southern populations in Bosnia and Herzegovina and Montenegro and eastern populations in Slovakia and Ukraine (Weiss et al. 2011). Very large populations, or highly inbred populations, exhibit a higher probability of identical genotypes appearing in different individuals compared to smaller and/or outbred populations (Weiss and Schenekar 2016).

It may be possible that low genetic diversity, manifested in a small number of haplotypes (or in the extreme scenario of zero-value genetic diversity, where the number of haplotypes is equal to one) with continuous restocking leads ultimately to huchen extinction once restocking stops. Consequently, natural huchen populations may be managed without stocking whenever possible and the best conditions for demonstrating that a specific population is self-sustaining when there is no stocking (Ihuţ et al. 2014). However, while the genetic differentiation of individuals captured within current national boundaries appears low to non-existent with the set of genetic markers applied, a conservative approach to rearing and stocking the species is still advocated, to limit cross-basin transport to an absolute minimum (Weiss et al. 2011).

The potential for evolution and adaptation to new environments is limited by the genetic diversity of the population. Natural selection favours individuals that are better adapted to live in natural environment than in captivity. Thus, research programmes aimed at retaining the evolutionary genetic adaptive potential of huchen stocks are very important and the optimum degree of genetic differentiation and a high genetic diversity should be permanently controlled (Kuciński et al. 2015). A population genetic study on huchen in Bavaria (Germany) showed that both maintenance of hatchery stocks and conservation of wild populations are crucial for retaining a maximum of the species' genetic and evolutionary potential (Geist et al. 2009). However, it is important to note that the long-term assessment of results of historic stocking of huchen is very poorly known. Stocking activities have not been well documented and no official information is available regarding the purpose of stocking and the origin or numbers of stocked fish. Thus, the effects of hatchery fish on wild populations remain an open question and a topic of major concern (Kuciński et al. 2015).

Also, the impact of exogenous individuals on wild populations can be summarized in three general categories: interbreeding, competition and contamination (i.e., introduction of pathogens). A direct genetic effect on recipient populations is obvious in the first category, but there may be a genetic basis for the effects of the other two categories as well. Competition from exogenous fish may weaken a native population through severe reductions in population size that could result in inbreeding and loss of genetic variability or, in the extreme case, to extinction of its gene pool. Likewise, a genetically resistant stock may carry pathogens and introduce them following translocation to an area where resistance has not developed (Hindar et al. 1991). Local stocks should be used for restocking managed populations, with prior obligatory genotyping broodstock individuals to prevent further spreading of detected non-native genetic material (Marić et al. 2014). Strong negative effects on the gene by the common practice of mixed-milt fertilization (i.e., eggs from several females are combined with milt of several males in a single container) can be expected, due to sperm competition, a very powerful kind of sexual selection (Snook 2005; Wedekind et al. 2007). It has been suggested that hatchery-introduced sperm competition not only increases the loss of genetic variation due to an increased variation in individual reproductive success but may also induce artificial selection for traits that are linked to hatchery-specific breeding success that may not be adaptive in the wild (Campton 2004). Moreover, a loss of any reproductive skew may be positively linked to heritable viability in the wild, for example, due to 'good-genes' sexual selection (Neff and Pitcher 2005).

In around the city of Graz, Austria, Weiss and Scheneker (2016) concluded that the urban population of huchen is self-sustaining based on the little to no current stocking, presence of juvenile huchen in multiple year classes, sighting of spawning events, numerous redds, no deviation from Hardy–Weinberg equilibrium and limited number of recaptures. Similarly, in the middle Sava River in Slovenia, the huchen population is reproductively active and self-sustaining, apparently because of interbreeding between local and stocked huchen, and the relatively large population size of huchen in this stretch of river (Snoj et al. 2022).

Furthermore, Snoj et al. (2022) analysed specimens from the Kupa River stretch between Croatia and Slovenia and found very little genetic diversity and genetic structure clearly separated from other populations in Slovenia due to isolation by large distances and by unsuitable habitats. Clearly, the genetic structure of the remaining huchen populations in Croatia should be determined in order to establish adequate protection measures.

5 Conservation

Considering that huchen suitable habitat is decreasing and remaining populations are heavily threatened, it is important to define appropriate protective measures to prevent further decline and the worst scenario. However, in order to choose the best conservation options, it is very important to collect knowledge of specific genetic signatures of remaining huchen populations (Snoj et al. 2022) which in Croatia still not has been done.

Until then, protection may be directed to several protective solutions: (i) C&R fishing in order to protect huchen from overfishing (Witkowski et al. 2013); (ii) establishment of ichthyological reserves as a key strategy of balance between fishery and protecting biodiversity (Belharet et al. 2020); (iii) effective nature-like fish passages for huchen migration (Chanseau and Larinier 1999); (iv) effective responses to climate change (e.g., removal of dams and obstacles that reduce the velocity and oxygen concentration of the water) (Marković et al. 2013); and (v) education such as specific workshops in aquaria to transfer knowledge to the public (Maitland 1995).

5.1 Recreational Fishing

Huchen has always been attractive to recreational fishers mostly because of its size. However, recently the increasing popularity of catch and release (C&R) fishing (Fig. 9) has reduced the impact of anglers on declining huchen abundance that nowadays is smaller than it has been in the past (Cooke and Schramm 2007; Witkowski et al. 2013). C&R can be a sustainable economic strategy that protects fish stocks overfished due to recreational fishing, without significantly weakening

Fig. 9 Catch and release fishing for huchen *Hucho hucho* in the Kupa River in Croatia (Photo by Krešimir Kuri)

Fig. 10 Release of huchen *Hucho hucho* after the catch (Photo by Krešimir Kuri)

the reproductive success of fish (Fig. 10) (Thorstad et al. 2004). Numerous studies confirm that anglers come to relax in nature unencumbered by the quantity or quality of catches, which justifies C&R (Arlinghaus et al. 2002, 2007).

Besides C&R fishing as a measure of restriction of angling, the establishment of ichthyological reserves would also enable protection of huchen (Mrakovčić et al. 2006). The establishment of freshwater reserves primarily benefits the fishery, but also may provide benefits to biological conservation. Finding balance between fishing, socioeconomic activities and conservation is fundamental in protected area management to ensure buy-in among stakeholders (Belharet et al. 2020). An example of establishment of such a reserve can be found in Bosanska Krupa in Bosnia and Herzegovina the Una River. Angling is prohibited in the area of 600 m managed by the Sports Fishermen's Association. That practice was adopted from the former Yugoslavia, where reserves were present throughout Bosnia and Herzegovina (Anes Halkić, personal communication).

Fig. 11 Brod na Kupi (Kupa River) is a well-known destination in Croatia for huchen *Hucho hucho* fishing (Photo by Davorka Jakšić)

Nowadays huchen in Croatia is a high-value species in recreational fisheries and its fishing is regulated by Freshwater Fishing Act as follows: only artificial bait is allowed, fishing at night is prohibited, closed season is from 16 February to 30 September and minimal catch size allowed is 80 cm. However, it would be necessary in well-known huchen fishing areas such as Brod na Kupi in the Gorski Kotar region (Kupa River) (Fig. 11) and Hrvatska Dubica (Una River) to establish a C&R fishing regime, while in smaller tributaries, such as the Kupica and Curak Rivers, where huchen is concentrated in a relatively small area with low estimated population size, fishing should be prohibited.

5.2 Nature–Like Fish Passages

Dams, obstacles and regulation of watercourses prevent the migration of huchen. One solution is construction of fish passages (Mrakovčić et al. 2006) for migratory fish species (Parasiewicz et al. 1998; Schmutz et al. 2001).

Natural bypass channels and pool passes have been found to be the most effective fish passages (Chanseau and Larinier 1999). The location, design and water discharge of the fish passage entrance, the water discharge and velocity in the fish passage, as well as the fish passage design are important factors that determine how well the fish passages function (Lundqvist et al. 2008). Increasing the water discharge through the passage or locating the entrance closer to the waterfall could enhance its attraction capabilities and functionality (Karppinen et al. 2002). It was found that Atlantic salmon (*Salmo salar*) positioned in the confluence area between the power station outlet and bypass channel responded strongly to increased water discharge in the bypass channel in combination with low flows from the power station outlet (Lundqvist et al. 2008). Moreover, in regulated rivers that have possibilities of controlling the water discharge, artificial freshets may be used to stimulate the upstream migration (Thorstad et al. 2008).

Several studies have used radio telemetry to investigate the behaviour of Atlantic salmon migrating through fish passages (e.g. Karppinen et al. 2002; Gowans et al. 2003) and can be applied to behavioural research of migratory huchen in the fishway. Detailed monitoring of huchen's migratory prey such as grayling, common nase and barbel is also required. Certainly, controlled experimental studies in a laboratory setting are needed, for example on the cruising, sustained and burst swimming performances of huchen at different stages (Thorstad et al. 2008).

5.3 Climate Change Response

Salmonids as cold-water species are likely to be strongly affected by climate change, since their distribution area is additionally reduced (Mrakovčić et al. 2006). Climate change impacts salmonids not only by changing temperatures but also by changing water flow regimes (Döll and Zhang 2010). Many characteristics of the flow regime, in particular seasonality, interannual variability and timing of flow events, affect life-history patterns like spawning and recruiting (Dudgeon et al. 2006). Higher spring flows can be expected to lead to increased sediment transport and disruption of huchen's spawning (Scheurer et al. 2009). Also, lower summer flows reduce the oxygen concentrations and alteration of vegetation, leading to a decrease in young fish (Fig. 12) (Poff and Zimmerman 2010). Overall, warming conditions in temperate regions across the globe will probably narrow the distribution of wild salmonid stocks (Cochrane et al. 2009; Marić et al. 2022). That effect is already observed in the region (Knez et al. 2022).

Consequently, it is important to consider increase of the longitudinal and lateral ecological connectivity and dam removal enabling huchen to compensate for unfavourable conditions through migrations to more suitable habitats, as well as rehabilitation of riparian buffer stripes and riparian tree cover to increase shading especially at the smaller tributaries (Marković et al. 2013), such as Kupica and Curak in Croatia. Moreover, linear bands of permanent vegetation adjacent to an aquatic ecosystem are intended to maintain or improve water quality by trapping and removing various nonpoint source pollutants from both overland and shallow subsurface flow (Simonich and Hites 1994).

Fig. 12 Alteration of vegetation in the upper reaches of the Kupa River in Croatia (Photo by Marin Jarnjak)

If more fresh water is consumed through human activities than is restored by nature, the result is that the quantity of freshwater available in rivers and underground waters is reduced. This can cause serious depletion of or damage of the huchen's habitat and surrounding environment (Reid et al. 2013).

5.4 The Role of Aquaria-Education in Conservation

Aquarium educators have enormous opportunities to engage in conservation awareness-raising exercises for the aquatic environment and fishes. Educational panels and displays of endangered species can help to interpret threats and needs, raise awareness and provide resources for conservation (Reid et al. 2013). The education may be able to make a positive difference in children's conservation-related learning outcomes, although such learning outcomes may not fundamentally change conservation-related behaviour. Conservation biology learning may establish the basis for further engagement targeted at fostering proconservation social change (Jensen 2014). However, the assimilation of new ideas into a pupil's existing mental schema for understanding animals and habitats can be significantly enhanced through assistance from a more knowledgeable educator who plays a vital role in drawing children's attention in useful directions and providing conceptual tools that allow children to develop their conservation biology learning (Fig. 13) (Vygotsky 1987).

Croatia's only freshwater aquarium is situated in the city of Karlovac (Aquatika-Karlovac Freshwater Aquarium) and also serves as a public institute that, in

Fig. 13 Theatre performance at the Aquatika-Karlovac Freshwater Aquarium shows the anthropogenic effects on huchen (Photo by Denis Stošić)

cooperation with Croatian scientists, teaches school children and students about the huchen (Fig. 14). Furthermore, it enables scientists and artists (Fig. 15) to communicate the concept that freshwater is a living, dynamic system inhabited by beautiful, important and often unseen organisms such as huchen (Reid et al. 2013).

6 Conclusions

Water flow is extremely important for huchen, and its decline is the result of the construction of dams and due to climate change, and all possible barriers that do not have a function should be removed. Likewise, consideration is needed to reduce future construction of dams on rivers and their tributaries.

It is important to determine the genetic structure and diversity of huchen in Croatia to monitor the state of populations and to define adequate conservation measures.

In the Kupa and Una Rivers in Croatia, it is important to identify the locations of huchen spawning, assess the number of male and female individuals spawning and juveniles. Moreover, if necessary, restore huchen spawning areas.

Fig. 14 For molecular analyses, fin clips were obtained from free-living huchen to determine genetic diversity for scientific and educational purposes (Photo by Krešimir Kuri)

As a measure of protection, angling associations should define the areas in Croatia in the Kupa and Una Rivers that relate to catch and release fishing. Also, an ichthyological reserve should be established in the small tributary Kupica River, considering that a very suitable habitat for huchen has been identified there and a large number of individuals have been recorded.

Public institutions (e.g. Aquatika-Freshwater Aquarium Karlovac, Croatia) conduct educational activities for children, pupils and students about the importance of biological diversity and endangered endemic species of fish such as huchen.

Acknowledgements Involvement of MP in this chapter was supported by the EIFAAC/FAO. The authors would like to express their gratitude to the Mladen Merkaš, Marin Jarnjak, Roland Stanković, Krešimir Kuri, Anes Halkić, Vinko Krnić, Emil Pintar, Metka Povž, Admir Jeginović

Fig. 15 Artistic expression of huchen in its habitat in the Kupa River (Author Mladen Merkaš)

and Margarita Maruškić Kulaš for sharing their pearls of curiosity, knowledge, experience and hard work with us during this research.

References

Arlinghaus R, Mehner T, Cowx IG (2002) Reconciling traditional inland fisheries management and sustainability in industrialized countries, with emphasis on Europe. Fish Fish 3:261–316

Arlinghaus R, Cooke SJ, Lyman J, Policansky D, Schwab A, Suski C, Sutton SG, Thorstad EB (2007) Understanding the complexity of catch and release in recreational fishing: an integrative synthesis of global knowledge from historical, ethical, social, and biological perspectives. Rev Fish Sci 15:75–167

Balon EK (1990) Epigenesis of an epigeneticist: the development of some alternative concepts on the early ontogeny and evolution of fishes. Guelph Ichthyol Rev 1:1–42

Belharet M, Di Franco A, Calo A, Mari L, Claudet J, Casagrandi R, Gatto M, Lloret J, Seve C, Guidetti P, Melia P (2020) Extending full protection inside existing marine protected areas, or reducing fishing effort outside, can reconcile conservation and fisheries goals. J Appl Ecol 57: 1948–1957. https://doi.org/10.1111/1365-2664.12104

Ćaleta M, Buj I, Mrakovčić M, Mustafić P, Zanella D, Marčić Z, Duplić A, Mihinjač T, Katavić I (2015) Endemic fishes of Croatia. Croatian Environmental Agency, Zagreb, p 116

Ćaleta M, Mustafić P, Zanella D, Buj I, Marčić Z, Mrakovčić M (2020) Human impacts on the Dobra River (Croatia). In: Human impact on Danube watershed biodiversity in the XXI century. Springer, Cham, pp 151–168

Campton DE (2004) Sperm competition in salmon hatcheries: the need to institutionalize genetically benign spawning protocols. Trans Am Fish Soc 133:1277–1289

Chanseau M, Larinier M (1999) Étude du comportement du saumon atlantique (*Salmo salar* L.) au niveau de l'aménagement hydroélectrique de Baigts (gave de Pau) lors de sa migration anadrome. Bulletin Français de la Pêche et de la Pisciculture 353/354:239–262. (in France)

Cochrane K, De Young C, Soto D, Bahri T (2009) Climate change implications for fisheries and aquaculture. FAO Fisheries and Aquaculture Technical Paper 530:1–212. http://www.lis.edu.es/uploads/07483fb7_72a2_45ca_b8e7_48bf74072fd3.pdf

Cooke SJ, Schramm HL (2007) Catch and release science and its application to conservation and management of recreational fisheries. Fish Manag Ecol 14:73–79

Döll P, Zhang J (2010) Impact of climate change on freshwater ecosystems: a global-scale analysis of ecologically relevant river flow alterations. Hydrol Earth Syst Sci Discuss 7:1305–1342

Dudgeon D, Arthington AH, Gessner MO, Kawabata ZI, Knowler DJ, Lévêque C, Naiman RJ, Prieur-Richard AH, Soto D, Stiassny MLJ, Sullivan CA (2006) Freshwater biodiversity: importance, threats, status and conservation challenges. Biol Rev 81:163–182. https://doi.org/10.1017/S1464793105006950

Frankel OH, Soulé ME (1981) Conservation and evolution. Cambridge University Press, Cambridge, p 358

Frankham R (1999) Quantitative genetics in conservation biology. Genet Res 74(3):237–244. https://doi.org/10.1017/S001667239900405X

Freyhof J, Kottelat M (2008) *Hucho hucho* – in: IUCN 2022. IUCN Red List of Threatened Species. Version 2021.3. <www.iucnredlist.org>. Downloaded on 18 May 2022

Freyhof J, Weiss S, Adrović A, Ćaleta M, Duplić A, Hrašovec B, Kalamujić B, Marčić Z, Milošević D, Mrakovčić M, Mrdak D, Piria M, Simonović P, Šljuka S, Tomljanović T, Zabric D (2015) The Huchen *Hucho hucho* in the Balkan region: distribution and future impacts by hydropower development. RiverWatch & EuroNatur, 30 p.

Geist J, Kolahsa M, Gum B, Kuehn R (2009) The importance of genetic cluster recognition for the conservation of migratory fish species: the example of the endangered European huchen *Hucho hucho* (L.). J Fish Biol 75:1063–1078

Gowans ARD, Armstrong JD, Priede IG, Mckelvey S (2003) Movements of Atlantic salmon migrating upstream through a fish-pass complex in Scotland. Ecol Freshw Fish 12:177–189. https://doi.org/10.1034/j.1600-0633.2003.00018.x

Hauer W (2003) Fazination Huchen. Vorkommen, Fang, Anekdoten. Stocker Verlag, Graz. (in German)

Hindar K, Ryman N, Utter FM (1991) Genetic effects of cultured fish on natural fish populations. Can J Fish Aquat Sci 48:945–957

Holčík J (1990) Conservation of the huchen, *Hucho hucho* (L.), (Salmonidae) with special reference to Slovakian rivers. J Fish Biol 37(Suppl. A):113–121

Holčík J (1995) Threatened fishes of the world: *Hucho hucho* (Linnaeus 1758) (Salmonidae). Environ Biol Fish 43:105–106

Holčík J, Hensel K, Nieslanik J, Skacel L (1988) The Eurasian huchen, *Hucho hucho*: largest salmon of the world. Perspect Vertebrate Sci 5:239

Ihuț A, Zitek A, Weiss S, Ratschan C, Holzer G, Kaufmann T, Cocan D, Constantinescu R, Mireşan V (2014) Danube salmon (*Hucho hucho*) in central and South Eastern Europe: a review for the development of an international program for the rehabilitation and conservation of Danube salmon populations. Bull UASVM Animal Sci Biotechnologies 71(2):86–101

IUCN (2008) IUCN red list of threatened species. Available at: https://www.iucnredlist.org/species/10264/3186143. Accessed 11 July 2022

Ivaška S (1946) Naše ťažkosti s hlavátkami Čs Rybář 1:163–164. (in Slovak)

Ivaška S (1951) Hlavátka, jej lov a umelý chov. Tatran, Bratislava (in Slovak)

Jakšić G (2019) The freshwater fish of Aquatika. AQUATIKA – KARLOVAC FRESHWATER AQUARIUM Public Institute, Karlovac, p 182

Jensen E (2014) Evaluating children's conservation biology learning at the zoo. Conserv Biol 28(4): 1004–1011. https://doi.org/10.1111/cobi.12263

Jungwirth M (1978) The problem of farming and conservation of the Danube salmon (*Hucho hucho* L.). Environ Biol Fish 3:231–234

Karppinen P, Mäkinen TS, Erkinaro J, Kostin VV, Sadkovskij RV, Lupandin AI, Kaukoranta K (2002) Migratory and route-seeking behaviour of ascending Atlantic salmon in the regulated river Tuloma. Hydrobiologia 483:23–30. https://doi.org/10.1023/A:1021386319633

Klašterka V (1976) Sportski ribolov na rijekama i jezerima. Nakladni zavod Znanje, Zagreb, 310 (in Croatian)

Knez S, Štrbac S, Podbregar I (2022) Climate change in the Western Balkans and EU green Deal: status, mitigation and challenges. Energ Sustain Soc 12:1. https://doi.org/10.1186/s13705-021-00328-y

Kottelat M, Freyhof J (2007) Handbook of European freshwater fishes. Kottelat, Cornol, Switzerland and Freyhof, Berlin, Germany, p 646

Kuciński M, Fopp-Bayat D, Liszewski T, Svinger VW, Lebeda I, Kolman R (2015) Genetic analysis of four European huchen (*Hucho hucho* Linnaeus, 1758) broodstocks from Poland, Germany, Slovakia and Ukraine: implication for conservation. J Appl Genet 56(4):469–480

Lundqvist H, Rivinoja P, Leonardsson K, McKinnell S (2008) Upstream passage problems for wild Atlantic salmon (*Salmo salar* L.) in a regulated river and its effect on the population. Hydrobiologia 602(1):111–127. https://doi.org/10.1007/s10750-008-9282-7

Maitland PS (1995) The role of zoos and public aquariums in fish conservation. Int Zoo Yearbook 34:6–14

Marić S, Razpet A, Nikolić V, Snoj A, Simonović P (2014) Analysis of genetic structure of huchen (*Hucho hucho*) in Serbia inferred from mitochondrial and nuclear DNA. Acta Veterinaria–Beograd 64(2):236–244

Marić A, Špelić I, Radočaj T, Kanjuh T, Vidović Z, Vilizzi L, Piria M, Nikolić V, Škraba Jurlina D, Mrdak D, Simonović P (2022) Changing climate may affect the invasiveness risk of non-native salmonids in the Danube and Adriatic basins of the Balkan Peninsula (South-Eastern Europe). NeoBiota., in press 76:135

Marković D, Scharfenberger U, Schmutz S, Pletterbauer F, Wolter C (2013) Variability and alterations of water temperatures across the Elbe and Danube River basins. Clim Chang 119: 375–389

Mrakovčić M, Brigić A, Buj I, Ćaleta M, Mustafić P, Zanella D (2006) Red book of freshwater fish of Croatia. State Institute for Nature Protection, Zagreb, p 253

Neff BD, Pitcher TE (2005) Genetic quality and sexual selection: an integrated framework for good genes and compatible gens. Mol Ecol 14:19–38

Parasiewicz P, Eberstaller J, Weiss S, Schmutz S (1998) Conceptual guidelines for natural-like bypass channels. In: Jungwirth M, Schmutz S, Weiss S (eds) Fish migration and fish bypasses. Fishing News Books, Oxford, pp 348–362

Piria M, Simonović P, Zanella D, Ćaleta M, Šprem N, Paunović M, Tomljanović T, Gavrilović A, Pecina M, Špelić I, Matulić D, Rezić A, Aničić I, Safner R, Treer T (2019) Long–term analysis of fish assemblage structure in the middle section of the Sava River – the impact of pollution, flood protection and dam construction. Sci Total Environ 651:143–153. https://doi.org/10.1016/j.scitotenv.2018.09.149

Poff NL, Zimmerman JKH (2010) Ecological responses to altered flow regimes: a literature review to inform the science and management of environmental flows. Freshw Biol 55:194–205. https://doi.org/10.1111/j.1365-2427.2009.02272.x

Prawocheński R, Kolder W (1968) Synopsis of biological data on *Hucho hucho* (Linnaeus, 1758). FAO Fish Synopses 22(Suppl. 1):1–14

Ratschan C (2014) Aspekte zur Gefährdung und zum Schutz des Huchens in Österreich. Den 33 (in German)

Reid GMG, Contreras MacBeath T, Csatádi K (2013) Global challenges in freshwater–fish conservation related to public aquariums and the aquarium industry. Int Zoo Yearbook 47(1):6–45. https://doi.org/10.1111/izy.12020

Rudolf J (1964) Návštevou u ryb v slovenských vodách. Čs Rybářství 19:140–141. (In Slovak)

Scheurer K, Alewell C, Bänninger D, Burkhardt-Holm P (2009) Climate and land-use changes affecting river sediment and brown trout in alpine countries – a review. Environ Sci Pollut Res 16:232–242. https://doi.org/10.1007/s11356-008-0075-3

Schmutz S, Zitek A, Zobel S, Jungwirth M, Knopf N, Kraus E, Bauer T, Kaufmann T (2001) Integrated approach for the conservation and restoration of Danube salmon (*Hucho hucho* L.) populations in Austria. In: Collares Pereira MJ, Cowx IG, Coelho MM (eds) Freshwater fish conservation - options for the Future. Fishing News Books, Oxford, pp 157–173

Simonich SL, Hites RA (1994) Importance of vegetation in removing polycyclic aromatic hydrocarbons from the atmosphere. Nature 370:49–51

Snoj A, Bravničar J, Zabric D, Sušnik Bajec S (2022) Conservation genetics study of huchen in Slovenia recommends river system-based management and indicates self–sustainability of the middle Sava population. Aquat Conserv Mar Freshwat Ecosyst 32(7):1171–1183. https://doi.org/10.1002/aqc.3834

Snook RR (2005) Sperm in competition: not playing by the numbers. Trends Ecol Evol 20:46–53

Sofradžija A (2009) Slatkovodne ribe Bosne i Hercegovine. Vijeće Kongresa bošnjačkih intelektualaca, Sarajevo, p 353 (in Bosnian)

Svetina M (1962) Synopsis on biological data on *Hucho hucho* (Linnaeus, 1758). FAO Fish Synop 22(1):1–30

Svetina M (1968) Dinamika rasti sulca *Hucho hucho* (Linnaeus, 1758). Biološki vestnik 16(1): 103–114. (in Slovenian)

Thorstad EB, Hay CJ, Naesje TF, Chanda B, Økland F (2004) Effects of catch and release angling on large cichlids in the subtropical Zambezi River. Fish Res 69:141–144

Thorstad EB, Økland F, Aarestrup K, Heggberget TG (2008) Factors affecting the within-river spawning migration of Atlantic salmon, with emphasis on human impacts. Rev Fish Biol Fish 18:345–371. https://doi.org/10.1007/s11160-007-9076-4

Vokač O (1959) Sulci v Dravi. Ribič 18(10–13):34–35. (in Slovenian)

Vygotsky L (1987) The collected works of L.S. Vygotsky. Volume 4: the history of the development of higher mental functions, Plenum Press, New York, p 294. https://doi.org/10.1007/978-1-4615-5939-9

Wedekind C, Rudolfsen G, Jacob A, Urbach D, Müller R (2007) The genetic consequences of hatchery-induced sperm competition in a salmonid. Biol Conserv 137:180–188

Weiss S, Schenekar T (2016) Genetic evaluation of the self-sustaining status of a population of the endangered Danube salmon, *Hucho hucho*. Hydrobiologia 775(1):153–165

Weiss S, Maria S, Snoj A (2011) Regional structure despite limited mtDNA sequence diversity found in the endangered Huchen, *Hucho hucho* (Linnaeus, 1758). Hydrobiologia 658:103–110

Witkowski A, Bajić A, Treer T, Hegediš A, Marić S, Šprem N, Piria M, Kapusta A (2013) Past and present of and perspectives for the Danube huchen, *Hucho hucho* (L.), in the Danube basin. Arch Pol Fish 21:129–142

Future Perspective of Sustainable Development of Freshwater Fisheries and Conservation of Threatened Fish Species, Crustaceans, and Molluscs

Laszlo Galambos and Nenad Sekulić

Abstract Starting from the basic postulates of the Convention on Biological Diversity, through national legislation and continuous implementation of the Convention on the Conservation of European Wildlife and Natural Habitats and other ratified international documents, as well as the growing effect of European regulations, among which the most important are the Directive on the conservation of natural habitats and of wild fauna and flora and the Common Fisheries Policy of the EU, the countries of the Western Balkans will focus on achieving the goals of the EU Biodiversity Strategy for 2030.

Keywords Freshwater fisheries · Diversity · Threats · Management · Protection measures · Sustainability

1 Introduction

Resulting from the geographical and geomorphological features, as well as the climatic differences between parts of the region, the Balkan Peninsula is characterized with large heterogeneity of freshwater habitats. This has enabled the development of rich freshwater ichthyofauna, molluscs, and crustaceans, with a high proportion of endemic and threatened species. The high degree of endemism, compared to the rest of Europe, is perhaps the most remarkable feature of the Balkan Peninsula (Skoulikidis et al. 2009; Freyhof and Brooks 2011). As part of the Mediterranean Basin Biodiversity Hotspot, the area of the Balkans has been identified as one of the world's 36 biodiversity hotspots (Mittermeier et al. 2011; Bianco and Ketmaier 2016; Hrdina and Romportl 2017). The freshwater ecosystems are

L. Galambos (✉)
Institute for Nature Conservation of Vojvodina Province, Novi Sad, Serbia
e-mail: laszlo.galambos@pzzp.rs

N. Sekulić
Institute for Nature Conservation of Serbia, Belgrade, Serbia
e-mail: nenad.sekulic@zzps.rs

V. Simić et al. (eds.), *Ecological Sustainability of Fish Resources of Inland Waters of the Western Balkans*, Fish & Fisheries Series 43,
https://doi.org/10.1007/978-3-031-36926-1_17

under significant anthropogenic pressures and alterations, mainly caused by the construction of dikes, dams, barrages, hydropower plants, weirs, degradation and loss of habitats due to river management (river training works for navigation and transport), intensive agriculture and forestry, water management (construction of drainage channels, use of water for irrigation), pollution, and invasive alien species. During the last four decades, freshwater fish have become one of the most endangered groups of vertebrates in the world (Freyhof and Brooks 2011) and they require planning and implementation of conservation measures targeting species, as well as their habitats, in order to prevent the further decrease in their numbers.

2 Freshwater Fisheries

Inland fisheries have been a significant source of food for humankind, and many early cultures developed around major rivers and lakes (Welcomme 2016). Through the history, fisheries and fishing have gained huge economic, scientific, cultural, and historical importance. They provide food source and contribute to economic well-being via trade, tourism, and recreation and have an economic and recreational significance, but overall, their importance is usually underestimated. Fish are very often considered only as a resource that the humankind has benefits from. But as part of the global biodiversity, and a very important component of the aquatic biocenosis and ecosystems, they are a very valuable natural asset.

Inland waters are used for a number of fishery-related purposes. Traditionally, the large lowland rivers (e.g., Danube, Tisza, Sava) were used for commercial fishing of species with economic values (such as carp, zander, barbel, catfish, sturgeon species), and fish farming (mainly carp—*Cyprinus carpio*) in submerged cages on floodplain oxbow lakes. Nowadays, a smaller number of commercial fishermen still work on Danube, Tisza, and Sava rivers (mainly in Serbia), while the recreational fishing (angling) predominates. Aquaculture is represented with fish farming in artificial ponds. Two main types of aquaculture have been traditionally applied in Serbia, carp farming and trout farming. Carp farming is mainly typical for lowland regions, for example Vojvodina Province (northern part of Serbia), where fishponds were constructed on arable land or on soils with "not sufficient quality for agricultural production" (such as saline lakes, salt meadows). Also, former meanders and oxbows have been converted to support fish farming. In these aquaculture facilities, mainly cyprinid species are farmed, such as carp (*Cyprinus carpio*) and Asian carps (*Hypophthalmichthys molitrix*, *H. nobilis*, *Ctenopharyngodon idella*), along with native predatory fish such as catfish (*Silurus glanis*) and pike (*Esox lucius*). Trout farming is localized mainly in higher terrains (lower mountain regions and hilly terrain) in central parts of Serbia, south from the Danube and Sava rivers (Marković et al. 2018).

In developing countries, low priority is given to inland fisheries and their management as compared to other sectors (Welcomme 2016), because of low environmental awareness, different political and economic interests, as well as shortage of funds and personnel. In the Western Balkans countries, many fisheries have deteriorated as a consequence of mismanagement as a result of their turbulent and rough history during the past thirty years, as well as the due to the heritage from the former communist regimes (where nature protection was not considered important). Inland fisheries were not often a national or regional governance priority. As a result, inland capture fisheries are undervalued and largely overlooked (Cooke et al. 2016), and therefore they are threatened in both developing and developed countries. Management of fisheries was (and in some cases still is) based on the income from the sale of fishing permits, in order to provide financial support to the fishery managers, disregarding also the goals of nature protection, that is the needs of endangered species. The estimation of stock sizes is mostly done ad hoc, not fully complying the relevant methodology. Despite its crucial importance for sustainable fisheries, no relevant widespread data on the status of the stocks and the associated trends are currently available. The proposal and implementation of management measures is not sufficient and lacks financial backing from the governments. Serious management failures are present primarily because the decision-makers have chosen not to follow expert advice.

The major challenge for the future is to manage fisheries sustainably, particularly since many different objectives are involved. The interests of various stakeholders make it hard for decision-makers to protect fish stocks along with crustaceans and molluscs (organisms that are often out of the focus of management activities, but are inhabiting the same aquatic habitats), and to ensure long-term sustainable use.

3 Need for Sustainable Management

In developing countries, the achievement of economic development (through industrialization and market expansion) represents one of the main drivers of overexploitation and unsustainable use of fish stocks (Asche et al. 2018). On a European level, in the past fisheries also became threatened due to industrialization and the development of the economy. Maintaining or improving the biological health of fish populations is usually related to the limitation of fishing activities, other economic obligations (such as habitat protection, wastewater treatment) that require financial support or can cause profitability reduction.

The ecological sustainability of ecosystems cannot be dissociated from socio-economic challenges (Gozlan and Britton 2015) and it should be incorporated into legislation and environmental planning. At the same time, ecological focus alone does not guarantee long-term sustainability of any form, and fishery sustainability must consistently take on a socio-ecological perspective, if it is to be effective in the future. The sustainability of fisheries depends not on the abundance of a fish stock, but on the ability of the fishery management system to adjust fishing pressure and environmental pressures to appropriate levels.

The issues of sustainability globally in both terrestrial and marine ecosystems have long been tackled through the creation of protected areas (Gozlan and Britton

2015) to conserve threatened species, aiming to enable the renewal of exploited stocks and to protect biodiversity. After signing the Danube River Protection Convention (Convention on Cooperation for the Protection and Sustainable Use of the Danube river. Sofia, 1994), it came into force in 1998, and led to the foundation of an international commission for the integral transboundary management of the Danube River (International Commission for the Protection of the Danube River—ICPDR) by 14 states located in the Danube Basin, that were joined by the European Union (EU). The aim was to deliver a single sustainable and equitable water management framework that includes a conservation perspective through the implementation of the EU Water Framework Directive (EC 2000). In 2009, a network of 15 protected areas on the Danube River was established, which covers eight countries along the Danube. The goals of this network is to promote cooperation and the exchange of experiences, as well to enhance nature conservation, reduce pollution, and promote sustainable development.

4 Threats to Freshwater Fisheries

Rivers are under a wide variety of pressures, mainly driven by direct human activities such as land use changes for urban and industrial development, flood protection, water supply, hydropower generation, waste disposal (both solid waste and wastewater), agriculture, forestry, and river navigation, but also indirectly as a consequence of climate changes. The impact of main pressures is visible on the alteration of the hydromorphological conditions, e.g. hydrological regime, sediment balance, river morphology, migratory routes, trophic structures leading to the disruption of ecosystem stability, fish communities, and also fisheries productivity (Fig. 1).

During the period from the beginning of the 1970s in Serbia (former Yugoslavia), many fish habitats and aquatic ecosystems have been destroyed and permanently lost by the construction of dams, weirs, river regulations, release of municipal and industrial wastewater, deforestation, expansion of arable land, development of drainage channel system, and extraction of water for irrigation (Figs. 2 and 3). Large-scale changes in the rivers have influenced fish diversity, distribution, and abundance. Many fish species (such as the migratory sturgeons, fam. Acipenseridae) lost access to spawning grounds and their normal riverine habitats after the construction of hydropower dams (Iron Gate 1, Iron Gate 2, Gabčikovo). Connections between basins (e.g., Rhine–Main–Danube Canal) but also hydropower dams, allowed the invasion of alien species of fishes and other hydrobionts.

One of the most severe negative effects on natural habitats of fishes in Western Balkans are hydromorphological alterations of river. A serious ecological concern results from the construction of hydropower dams (Iron Gate 1 and Iron Gate 2), as a consequence, both abundance and species diversity have decreased (Fig. 4).

Nowadays, the quest for renewable energy has caused the destruction of many small mountainous streams and rivers with the construction of mini hydropower

Fig. 1 River training structures for navigation—groyne on the Danube River (Photo by L. Galambos)

Fig. 2 Regulated river—Channelized watercourse of Moravica River in Vojvodina Province, Serbia (Photo by L. Galambos)

Fig. 3 Weir on Karas River in Vojvodina Province, Serbia—Obstacle for longitudinal migration of fish species (Photo by L. Galambos)

Fig. 4 Barrier for longitudinal migration—Iron Gate 2 dam (Photo by L. Galambos)

plants in the region. The potential of mountainous watercourses of Western Balkans region for renewable energy is highly prioritized by the respective countries, besides the already constructed dams, hundreds and hundreds more are planned to be

Fig. 5 Forest plantation along the Danube River near Novi Sad—alteration of natural riparian vegetation (Photo by L. Galambos)

constructed in the following years. Many projects on mini hydropower plants developed some 40 years ago reemerge as part of the so-called national green efforts to obtain "clean" energy. Unfortunately, due to national subsidies, these power plants are built without appropriate procedures of assessment on the feasibility, cost–benefit analysis, and for what is worse, without (proper) environmental impact assessment. Many projects from the late 70s of the twentieth century are being implemented without revision, disregarding the changes in flow conditions and amount of water available in the rivers, as well as the needs of animal and plant communities in the rivers. In many cases, not even the needs of the local communities inhabiting the settlements and villages near the rivers are considered when constructing the mini hydropower plants, causing high level of discontent and protests. However, in recent years, in some countries (such as Serbia), the construction of mini hydropower plants in protected areas has been completely prohibited (The Law on nature protection ("Official Gazette of RS," no. 36/2009, 88/2010, 91/2010—amended, 14/2016 and 95/2018—other law, 71/2021)).

In the region of lowland rivers, intensive forestry represents another significant threat to rivers and fish habitats. Loss of floodplain habitats is a result of intensive forestry, mainly by conversion of natural forest habitats into forest plantations (Fig. 5). Coupled with human induced hydromorphological alterations and the effects of global warming, activities within the forestry sector have led to silting up of side arms, oxbows and marshes, vegetation succession on wet meadows. All the mentioned habitats have an important role as fish spawning and nursery areas for fry and juveniles. In the mountain regions, deforestation and clear cuttings are

Fig. 6 Strictly protected species: tench—*Tinca tinca* (Photo by L. Galambos)

causing flashfloods, landslides resulting with the destruction of in-stream habitats for aquatic species.

The river sediment, mainly sand and gravel, is a very important material for the industry and urban development. It is one of the globally most exploited natural resources. The effects of sediment extraction on rivers and fisheries have been unjustifiably neglected and underestimated for a long period of time, since they were considered as a "renewable resource," as rivers have constantly provided them. The resulting deepening of the riverbed and changes in its structure drastically alter the ecological conditions and thus endangering the existing animal and plant communities. This primarily refers to the benthic organisms, but also to other organisms, mainly fish species and molluscs. These effects are threatening not only the aquatic organisms, but also other organisms (plants, birds, mammals, reptiles) whose life cycle is linked to those habitats. Changes in the riverbed and loss of riparian vegetation are also critical, since they both provide refuge for fish, have an important role in the process of autopurification and contribute to the stability of the river banks and the ecological status of watercourses.

Human impacts have severely altered the original distribution of many native species and also have compromised their survival, thus erasing millions of years of evolutionary processes (Bianco and Ketmaier 2016). Many threatened species (among them endemic) have a high scientific value but are taken in little consideration in fisheries management, because of their low (or non-existing) economic value (e.g., European mudminnow—*Umbra krameri*, tench—*Tinca tinca*, crucian carp—*Carassius carassius*, eastern barbel—*Barbus cyclolepis*, Neretva spined loach—*Cobitis narentana*, marble trout—*Salmo marmoratus*) (Fig. 6).

In addition to uncontrolled fishing and overfishing, poaching (illegal fishing) is one of the most important actions that is causing the decrease (primarily) of the populations of economically important fish species (Fig. 7). Poaching not only has a

Fig. 7 Illegal fishing net on the Danube River collected by fish warden (Photo by L Galambos)

negative impact on the number of target species, but also on numerous other species (especially when using prohibited means and tools for fishing, such as electricity, explosives, or non-selective fishing gear). Illegal fishing is also contributing to overexploitation (Agnew et al. 2009) of freshwater fisheries. In low-income countries, it is a lucrative criminal activity, and one of the main reasons leading to the depletion of fish stocks along with habitat loss due to river regulations and water pollution.

5 The Legal Framework Promoting and Supporting Sustainable Use of Fisheries

Maintaining fisheries sustainability has a background in numerous legal documents adopted on national and international level, but also in obligatory international agreements, strategies, and programs, as well as policies.

Despite significant efforts by the EU Member States, biodiversity and the delivery of ecosystem services are declining. Recent data show that only 15% of habitat and 27% of species assessments at EU level have a good conservation status, while the majority continues to have poor or bad status. This is largely due to habitat loss and fragmentation, pressures from land use change, diffuse pollution, resource overexploitation, and increasing impacts from invasive alien species and climate change (State of nature report 2020). The European Union (EU) Biodiversity Strategy for 2030 (https://environment.ec.europa.eu/strategy/biodiversity-strategy-2030_

en) is a response to an alarming loss of nature that undermines our well-being and prosperity. It is a comprehensive, ambitious, and long-term plan to protect nature and reverse the degradation of ecosystems. The Strategy aims to put Europe's biodiversity on a path to recovery by 2030, and contains over 100 specific actions and commitments ranging across several policy areas. One of the aims is to support the more effective implementation of the "Nature Directives," the EU Birds and Habitats Directives.

The Council Directive 92/43/EEC on the Conservation of natural habitats and of wild fauna and flora (The Habitats Directive) is a directive adopted by the European Community in 1992 as a response to the Bern Convention. It has a key role in the protection of freshwater fish species. For non-EU states in the process of joining the European Union, from the aspect of nature protection and the establishment of the European ecological network Natura 2000, this is one of the most significant legal documents that regulates the preservation of natural habitats and certain wild plant and animal species in the territories of the member states. Along with the transposition of the Directive into national legislation and the application of additional technical documentation of the European Commission and the European Topic Center for Biodiversity (ETC/BD) in the process of establishing the Natura 2000 network, its implementation is also necessary in terms of implementing measures and activities for the protection and restoration of natural habitats and populations of wild species with the aim of preserving their favorable condition.

For fish species listed in Annex II of the Habitats Directive, core areas of their habitat—designated as Sites of Community Importance—must be protected under the Natura 2000 Network, and the sites managed in accordance with the ecological requirements of the species. The EU candidate country is preparing a national list of Proposed Sites of Community Importance, pSCIs within the Natura 2000 network. For species and sub-species listed in Annex IV (including many that are also listed in Annex II) a strict protection regime must be applied across their entire natural range within the EU, both within and outside Natura 2000 sites. Annex V species could require measures to ensure that their exploitation and taking in the wild is compatible with maintaining them in a favorable conservation status.

Other international documents related to the protection of fish fauna, crustaceans, and molluscs and their habitats are as follows:

- Bern Convention (Convention on the Conservation of European Wildlife and Natural Habitats (ETS No. 104)), a binding international legal instrument aiming to ensure conservation of wild flora and fauna species and their habitats.
- Convention on the Conservation of Migratory Species of Wild Animals (Official Journal L 210, 19/07/1982), also known as the Convention on Migratory Species or the Bonn Convention, one of the fundamental regulations in the field of nature protection that aims to conserve migratory species throughout their ranges, by the preservation and maintenance of suitable, and restoration of damaged or destroyed habitats. The Convention serves as a legal framework that enables separate international instruments for the conservation of migratory species or, more often, groups of species throughout their range.

- CITES Convention (Convention on International Trade in Endangered Species of Wild Fauna and Flora, March 3rd, 1973, 993 U.N.T.S. 243) aims to prevent the loss of biodiversity while implementing measures that would reduce the degree of threat for the survival of species, which is caused by the excessive exploitation and trade of the living world by humans. All species listed on CITES require control of international trade of endangered and protected wild species, as well as their parts and products.
- The Convention on Biological Diversity (The Convention on Biological Diversity, United Nations, Rio de Janeiro 1992) with its objectives supports the conservation of biological diversity and the sustainable use of its components. The Convention also gives guidelines on the prevention, introduction, and mitigation of the impact of non-native species. According to Article 8 of the Convention, each state party shall, to the extent possible and appropriate, prevent the introduction, control or eradicate non-native species that threaten ecosystems, habitats, or species.
- The EU Water Framework Directive (2000) represents another driver for achieving sustainable water management in Europe, and therefore the protection of the habitats of freshwater fish species in Europe. The Directive requires member states to achieve a good ecological status of surface (fresh) waters in Europe, with the allocation of protected areas that are functionally dependent on water status.

On the beginning of the twenty-first century, new documents (resolutions, declarations, conventions, agreements, programs, etc.) were adopted concerning the protection, preservation, and improvement of nature and natural values, sustainable use of natural resources, and sustainable development of the entire human community. Along with the immediate application of generally accepted rules of international law and confirmed international treaties as an integral part of the legal system at the national level, goals are also defined as obligations to which each state and society as a whole should strive. For example, within the process of incorporation of the Millennium Development Goals (UN General Assembly 2022) in the policy of Republic of Serbia, national goals and objectives were determined (Vlada Republike Srbije 2006). In relation to MDG 7: Ensure environmental sustainability, in the context of conservation and sustainable use of biodiversity, task 1 was highlighted along with specific tasks 1 and 3:

- Task 1: Incorporate the principles of sustainable development into national policies, stop the loss of natural resources, and encourage their revitalization
  - Specific task 1: By 2015, adopt and implement national programs, strategies, and laws that regulate the areas of sustainable development and environmental protection in the Republic of Serbia.
  - Specific task 3: By 2010, increase the area of protected natural resources to 10% of the total territory of the Republic of Serbia.

One of the ways of sustainable management of fish populations is the regulation of the use of economically important fish species, which ensures the preservation of stable populations and their survival. If there were no restrictions that regulate fishing (defined by national legislation), many species would have become extinct a long time ago. In addition to the preservation of these species, which are equally important from the point of view of commercial and recreational fishing, for food or material gain, certain laws also protect species that are rare or endangered and also that have of great importance in preserving the entire biodiversity as a natural value and parts of national identity.

The national legislation in the Republic of Serbia regulating the area of fish stock protection, as well as nature protection, is based on generally accepted rules of international law and confirmed international treaties and ratified conventions (Galamboš 2018).

Management of the "fish resources" (a legal definition in Serbia covering fish species, molluscs, crustaceans, and other aquatic organisms that are protected and sustainably used) in their habitats on the territory of the Republic of Serbia, is regulated by the Law on the Protection and Sustainable Use of Fish Resources ("Official Gazette of RS," No. 128/2014 and 95/2018—other law) and set of by-laws, the most significant of which is the Order on measures for the preservation and protection of fish stock ("Official Gazette of RS," no. 56/2015 and 94/2018). The Law enables the protection and use of wild fish species as natural wealth and as a goods of general interest, with the aim of ensuring the same structure and rate of use over a longer period of time. The Law on the Protection and Sustainable Use of the Fish Resources defines measures for the protection of fish and fry, also their habitats, as well as the obligation for designation of areas that are important for the life cycles and biological needs of fish (e.g., spawning, wintering, growth, feeding and movement, migrations), are called "specific habitats for fish." In order to preserve and protect the fish stock, the Law defines measures of restrictions, prohibitions, as well as measures whose implementation contributes to the improvement of the state of the fish stock. The use of poisonous, stunning, and explosive agents (e.g., electricity or explosives) for catching fish, as well as the use of tools (e.g., harpoons) and methods which can cause serious damage to fish (e.g., raking) are prohibited.

The main prerequisite for the approval of the implementation of works and construction on aquatic habitats, is to ensure unhindered reproduction, migration, and preservation of fish stocks. The law establishes the obligation to maintain the physical, chemical, biological, and ecological characteristics of aquatic habitats, to preserve water quality, and to prevent threats to the health status, life, and survival of fish and changes in the composition of fish fauna. It is forbidden to alter the hydrological regime of fish habitats that may have a negative effect on the life cycle, and especially on spawning of fish species. The introduction of non-native fish species into the fisheries is prohibited, as well as the introduction of fish species from regionally distant geographical areas, as well as species originating from geographically closer but mutually isolated watersheds. For the purposes of strengthening natural fish populations in fisheries, planned and controlled stocking is permitted, using exclusively indigenous fish species with certified origin. Important species protection measures are fish rescue and translocation, as well as selective fishing of invasive species (Fig. 8). If there are changes in the physical, chemical, or biological characteristics of water on the fisheries that threat or may threaten the fish species, it

Fig. 8 Selective fishing of non-native and invasive species—gibel carp—*Carassius gibelio* (Photo by L. Galambos)

is necessary to implement measures for preservation of the fish stocks. In these cases, it is possible to catch endangered fish for the purpose of saving them, as well as moving or translocating them to another parts of the fishery where there are suitable conditions for the survival of the fish. Selective fishing for the removal of fish species in order to prevent the spread of non-native species or the overpopulation of certain native fish species in the fishery can be carried out on the basis of a special permit.

The manner of recreational and commercial fishing and the catch of economically significant species are regulated by the Order on measures for the conservation and protection of fish stocks, which defines:

- species for which catch is permanently prohibited, i.e. species for which permanent ban on catch and landing have been established (Table 1), then
- species for which catch and landing is prohibited in a certain period during the year (Table 2) and catch and landing below a certain size (Table 3)
- the daily allowed catch amount of native fish species in recreational fishing, as well as
- species for which unlimited catch is allowed during recreational and commercial fishing

Apart from the preservation of the most endangered species of fish, which is carried out by a ban on their catch and landing during any type of fishing (with the

Table 1 Overview of fish species with established permanent catch ban

Species name
Acipenser gueldenstaedtii
Acipenser nudiventris
Acipenser ruthenus
Acipenser stellatus
Acipenser sturio
Alburnus sava (listed as *Alburnus chalcoides*)
Alosa immaculata
Alosa tanaica (listed as *Alosa caspia*)
Anguilla anguilla
Barbus cyclolepis
Carassius carassius
Cobitis elongata
Eudontomyzon danfordi
Eudontomyzon mariae
Eudontomyzon stankokaramani
Eudontomyzon vladykovi
Gymnocephalus baloni
Huso huso
Leucaspius delineatus
Leucos basak (listed as *Rutilus karamani*)
Misgurnus fossilis
Oxynoemacheilus bureschi
Pachychilon pictum
Sabanejewia balcanica
Sabanejewia bulgarica
Salmo marmoratus
Telestes souffia
Tinca tinca
Umbra krameri
Zingel streber
Zingel zingel

exception of scientific research purposes), the most important measures are aimed at preserving the fish habitats in relation to the space they inhabit and the quality of the habitat itself. On the other hand, the established catch and landing ban in a certain period of the year represents a ban on taking a given species at a specific time period (most often it includes the spawning period of the species), in order to ensure the possibility of reproduction and leaving offspring.

The minimum catch size of certain fish species allowed for catching and landing is determined in order to ensure the capture only of sexually mature, adult specimens that have already had the opportunity to spawn and leave offspring. The lengths of the fish per species are determined so that it is possible to catch only sexually mature fish.

Table 2 Overview of fish species for which a period of fishing ban has been established

Species name	Period of fishing ban
Abramis brama	Between 15th April and 31st May
Barbus barbus	Between 15th April and 31st May
Chondrostoma nasus	Between 15th April and 31st May
Cyprinus carpio	Between 1st April and 31st May
Esox lucius	Between 1st February and 31st March
Hucho hucho	Between 1st. March and 31st August
Leuciscus aspius (listed as *Aspius aspius*)	Between 15th April and 31st May
Leuciscus idus	Between 15th April and 31st May
Rutilus virgo (listed as *Rutilus pigus*)	Between 15th April and 31st May
Salmo farioides	Between 1st October and 1st March
Salmo macedonicus	Between 1st October and 1st March
Salmo trutta	Between 1st October and 1st March
Sander lucioperca	Between 1st March and 30th April
Sander volgensis	Between 1st March and 30th April
Silurus glanis	Between 1st May and 15th June
Squalius cephalus	Between 15th April and 31st May
Thymallus thymallus	Between 1st March and 31st May

In order to protect fish species subject to catch and landing, limits have been determined for the amount of fish that is allowed to be caught. Each angler is allowed to take up to 5 kg of native species in total. In case one caught specimen of one of the native species exceeds a weight of 5 kg, the limit of daily catch in number of specimens does not apply. In such cases, it is considered that the daily catch of an angler has been fulfilled. The aforementioned Order also allows unlimited fishing of all non-native species of fish present in Serbian fisheries during recreational and commercial fishing (Fig. 9).

The Rulebook on the methods, tools, and means used for commercial fishing, as well as on the method, tools, equipment, and means used for recreational fishing ("Official Gazette of RS," no. 9/2017 and 34/2018) regulates the use of certain types of fishing gear, as well as the temporal and spatial application of permitted fishing gear, which contributes to the prevention of excessive and indiscriminate fishing.

The Law on nature protection ("Official Gazette of RS," no. 36/2009, 88/2010, 91/2010—amended, 14/2016 and 95/2018—other law, 71/2021), along with the resulting by-laws, regulates the integral protection and preservation of native and authentic diversity of fish and their gene pool, as well as the protection of water as a resource and also as habitat of fish species and communities. The status of fish species as natural values is defined by the Rulebook on the proclamation and protection of strictly protected and protected wild species of plants, animals and mushrooms ("Official Gazette of the RS," No. 5/2010, 47/2011, 32/2016 and 98/2016). The protection of fish species with the status of "strictly protected wild species" is carried out by banning their use (such as the permanent ban that is defined for certain fish species in accordance with the regulations in the sector of protection

Table 3 Overview of fish species for which the minimum catch size (SL—standard body length) has been determined

Species name	Minimum catch size (SL/cm)
Abramis brama	20
Barbus balcanicus (navedena kao *B. peloponnesius*)	15
Barbus barbus	25
Chondrostoma nasus	20
Cyprinus carpio	30
Esox lucius	40
Gobio bulgaricus (listed as *Gobio* spp.)	10
Gobio obtusirostris (listed as *Gobio* spp.)	10
Hucho hucho	100
Leuciscus aspius (listed as *Aspius aspius*)	30
Leuciscus idus	20
Lota lota	25
Perca fluviatilis	10
Romanogobio kesslerii (listed as *Gobio* spp.)	10
Romanogobio uranoscopus (listed as *Gobio* spp.)	10
Romanogobio vladykovi (listed as *Gobio* spp.)	10
Rutilus virgo (listed as *Rutilus pigus*)	20
Salmo farioides	25
Salmo macedonicus	25
Salmo trutta	25
Sander lucioperca	40
Sander volgensis	25
Silurus glanis	60
Squalius cephalus	20
Thymallus thymallus	30
Vimba vimba	15

and use of fish resources) and also by prohibiting the implementation of all activities that may endanger these species and their habitats as well as activities that can lead to the destruction of habitats, and by implementing measures and activities to manage their populations. On the other hand, the protection of certain wild species of fish, for which the status of "protected wild species" has been established, is carried out by restricting their use under conditions and in a manner that are harmonized with the regulations in the sector of fish stocks protection, and also by prohibiting the destruction of habitats and implementation of other activities that cause damage to species and their habitats, and implementation of measures and activities to manage populations.

Measures for the protection for wild species of fish and their habitats (along with prohibitions or restrictions on the implementation of certain projects, works, and activities), as well as special conservation measures, are aimed at protected areas, areas within the ecological network of the Republic of Serbia (ecologically

Fig. 9 Non-native and invasive species—pumpkinseed (*Lepomis gibbosus*) (Photo by L. Galambos)

significant areas and ecological corridors of international importance, as well as ecological corridors of local importance), and also to identified habitats of certain wild fish species and specific habitat types.

The protection measures and conditions for nature protection to be followed are:

- prohibition of works and activities such as exploitation of natural resources and regulation of watercourses, in case it represent a threat to wild species of fish or other aquatic organisms that have a given status of protection at the national and international level
- limiting the implementation of works in a given period of the year in order to ensure a favorable protection status of some wild species of fish (that is, to enable the unhindered reproductive cycle)
- prohibition or restriction of works near the riverbanks and riparian areas (when one or more of rare and endangered wild species are present in those areas), in order to preserve the habitat for those species and
- time periods of bans or restrictions depend on the area and the wild species present and cannot be generalized

Conservation measures for the protection of priority types of aquatic and wetland habitats (terrestrial surface water areas, swamps, bogs, heaths, and salt marsh habitats) according to this Rulebook on the proclamation and protection of strictly protected and protected wild species of plants, animals, and mushrooms include the following obligations:

- preserve the aquatic and wetland habitats in natural or near natural state, and if necessary, carry out their revitalization
- preserve biological species important for habitat types
- prevent the introduction of foreign (non-native) species and genetically modified organisms
- preserve and improve the connectivity of watercourses
- ensure the necessary amount of water in aquatic and wetland habitats and to maintain a favorable water regime, which is necessary for their long-term existence and preservation
- preserve the favorable physic–chemical and biological properties of water for the given type of habitat, or take measures to improve them if they are unfavorable for the conservation of the habitat
- preserve the favorable composition and concentrations of nutrients (mineral and organic substances) in the water and soil of aquatic and wetland habitats
- prevent the processes of conversion of natural and semi-natural aquatic and wetland habitats for more intensive land uses practices
- ensure the maintenance of habitats diversity on watercourses by preserving natural banks (without embankment), shoals, rapids, waterfalls, etc., and favorable hydrodynamics (meandering, sediment transport, occasional natural flooding of backwaters, etc.)
- during the works on extraction of sand and gravel, river regulation, building of embankments, construction of canals and ponds, etc., the preservation of low, slightly sloping riverbank areas should be a priority. Also sediment extraction should be performed on elevated river terraces or in inactive floodplains, not in active riverbeds and floodplains
- in the preparation for the protection against the harmful effects of water (floods), priority should be given for the creation and use of natural retentions and watercourses as spaces for retaining floodwaters, and also to enable their runoff
- sediments should not be exploited from sandbars and shoals

In situ measures that imply the application of active protection measures (such as protection of populations in natural habitats, preservation of natural ecosystems, preservation and recovery of populations in their natural habitats) do not always have the fast enough effects required for the preservation of certain endangered species. In such cases, it is necessary to apply *ex situ* measures, i.e. conservation of genetic resources outside the natural habitats of the given species. These measures usually include the formation of a gene bank, the collection and cultivation of adult fish specimens (used for artificial reproduction in scientific laboratories for the purposes of reintroduction) and juveniles, artificial reproduction, as well as the development of plans and programs for the return of juveniles and also for the repopulation to protected or restored habitats. For certain wild fish species, such as sturgeon species, *ex situ* protection has no alternative. The rapid implementation of *ex situ* measures is especially important for economically significant species that are regularly exploited, as well as species that have become endangered due to traditional use and acquired the status of protected species, and which are still subject to illegal fishing.

Habitat conservation and rigorous fishing control represent one of the most important ways of protecting fish species and fisheries. However, controlled breeding and stocking are regularly considered as a necessary measure to preserve the populations of endangered fish species due to the large and intensive effects of threatening factors. This implies that it is necessary to "help" the degraded aquatic system artificially, using ecological engineering techniques (Galamboš 2018). If continuous scientific research on qualitative and quantitative composition of fish fauna, as well as monitoring of fish species, shows that the population of fish species is declining in a given area, then planned introductions of the given species are implemented. In this way it is possible to prevent the further decline in the number of the given population, but it requires large financial investments and the engagement of numerous experts.

From the around 100 species of freshwater fish species living in the aquatic and wetland habitats in Republic of Serbia (Simonović 2001), in accordance with the Law on nature protection and the Rulebook on the proclamation and protection of strictly protected and protected wild species of plants, animals and mushrooms (Official Gazette of the RS, No. 5/2010, 47/2011, 32/2016 and 98/2016) 67 species (63 species of fish and 4 species of cyclostomes) have a special protection status. Among them 38 have the status of strictly protected (Annex I), and 29 have the status of protected wild species (Annex II). Besides the fish species, two species of crayfish (European crayfish—*Astacus astacus*, and the stone crayfish—*Austropotamobius torrentium*), the thick shelled river mussel (*Unio crassus*), freshwater snails (shining ram's-horn snail—*Segmentina nitida*, narrow-mouthed whorl snail—*Vertigo angustior*) as well as several other Gastropoda species have the status of strictly protected species in Serbia.

One of the approaches used to preserve endangered species of aquatic organisms is aquaculture, the reproduction and rearing of juvenile aquatic organisms (primarily fish, but also other aquatic organisms such as molluscs, crabs, etc.) for the purpose of stocking of aquatic ecosystems (Marković et al. 2018). Although the expert public has a divided opinion on the effectiveness and justification of aquaculture in the long-term preservation of fish species (Epifanio and Waples 2015), this method of protection has recently been used more often worldwide in order to preserve the original genetic pool, when most environmentally favorable methods in the natural environment have been exhausted. In some cases, aquaculture may represent the last option for the preservation of endangered species of fish and other aquatic organisms. According to Marković et al. (2018), there is a growing need to give much more attention to the segment of aquaculture that deals with the preservation of broodstock (the parent fish from which fry and fingerlings are produced).

In aquaculture facilities, there is a lack of breeding and production of fry of native and endangered species of fish for stocking purposes, that could support the under-pressure populations in natural fisheries (Fig. 10). Based on market demands, breeding of these species is economically unjustified and is a long-term process. Their artificial spawning and propagation in fish farms is unprofitable and

Fig. 10 Specialized facility—reproduction center for the production of native species in Serbia (Photo by N. Sekulić)

unsustainable. This is caused by the absence of binding strategically defined programs and plans (e.g., action plans for species) and their continuous implementation, which would support the reduction of pressure on natural populations by fishing, but also to preserve and improve the condition of these species in the rivers (Galamboš 2018).

6 Protection Measures for Fish Species, Crustaceans, and Molluscs

In order to improve the protection status of fish species and prevent further loss of their populations, it is necessary to implement the already defined protection measures, as well as to establish new, more specific measures to reduce or even completely eliminate (where possible) the factors threatening fish. Measures should equally cover species and their habitats. It is very important to determine the appropriate means of protection of areas that are identified as the most important for the survival of fish species. For the protection of the most endangered species, it is necessary to develop and adopt strategic documents or action plans, which will define the goals, instruments, and measures of management and protection in more detail.

Similar action plans should be developed also for fish species (e.g., zander, grayling, pike, catfish, barbel, asp, carp, etc.) that are under a great pressure from anglers. These plans, along with the implementation of species monitoring, would enable a better understanding of the state of species populations, their distribution and vulnerability, which will serve as an expert basis for planning and developing programs for *ex situ* protection. The existing ban on the introduction of non-native species needs to be strictly enforced and controlled further on, while increasing the awareness of anglers first of all, but also aquarists and owners of commercial fishponds for fish farming about the importance of the prevention of the further spread and introduction of non-native species (which often prove to be very invasive). The basis for all of the measures is the legislative framework at the national level, that is, at the level of the European Union. In close cooperation with scientific and professional institutions, it is necessary to regularly update the laws and by-laws regarding the lists of endangered species, as well as the necessary protection measures based on the identified threatening factors.

Not all protection measures target directly the protection of habitats and/or species. Their contribution is significant for species protection indirectly, as a support to solve systemic problems within certain sectors, which are caused by insufficient financial resources, extensive administration and bureaucracy, low environmental awareness, non-compliance with regulations, and inconsistent and vague regulations:

- development of action plans for endangered species
- preparation of the Red List and the Red Book of Fishes
- improving research on the status and condition of fish species/populations
- preparing fishery management plans on a state level (e.g., ministry) as strategic documents
- assessing the condition of fish populations, especially endangered and protected species
- identification and definition of special habitats for the protection of certain endangered fish species
- preventing the reduction or the loss of habitat, or the blocking of migratory routes of numerous protected, but also economically important species due to the construction of dams and the formation of reservoirs on river courses
- regulation and management of fishing in transboundary watercourses
- establishing a unique methodology for the assessment of populations, biomass, and annual production of fish and determination of the permitted quantities for fish capture and landing through angling and commercial fishing, to avoid overfishing
- increasing the amount of compensation for damage caused to strictly protected and protected wild species of fish in order to discourage non-compliance with fishing bans and prevention of poaching
- subsidizing and starting artificial spawning and breeding of endangered and protected species of fish in aquaculture with the support and coordination of competent state authorities, for the purposes of stocking and improvement of species populations in open waters

- simplification of legal provisions related to artificial spawning and breeding of endangered and protected species of fish for the purposes of reintroduction and strengthening of native fish species populations
- defining the obligation and methods for marking fish (especially endangered and protected species) originating from fish farms, in order to prevent abuses in the sale of fish originating from poaching
- organized purchase of fish from commercial fishermen with the aim of monitoring the origin of landed fish and fish trade
- improving the recording of data on catches in angling, sport, and commercial fishing in full accordance with the actual catch, which enables the monitoring of trends in fish populations
- improvement of judicial practice regarding the appropriate processing of misdemeanors and criminal offenses in relation to wild species of fish, especially strictly protected and protected species
- providing education and strengthening the awareness of employees in the police, customs administration, courts and prosecutor's offices, with the aim of giving more importance to misdemeanors and criminal acts and imposing appropriate measures for violation of regulations
- conducting education and providing additional training for employees in the police, customs administration, prosecutor's office and courts, in order to work more efficiently in cases of poaching, illegal trade, and damage to fish species
- strengthening the capacity of inspections, primarily by increasing the number of inspectors and improving their field equipment
- improvement of criminal policy by stricter and more efficient enforcement of regulations, as well as by defining and applying more rigorous penalties
- strengthening the capacity of the fisheries sector in the competent administrative body (personnel and technical)
- strengthening the professional capacity of fishery managers
- strengthening the capacity of the fish warden service (number of staff, equipment, vehicles, and vessels)

The efforts for sustainability in freshwater ecosystems have been supported through the creation of protected areas for a long period of time. Certain mountain streams and rivers, lowland watercourses, lakes, reservoirs, canals, ponds, wetlands, and floodplains represent aquatic ecosystems of importance for the preservation of wild fish species populations, and are under a specific protection regime within protected areas. These areas were designated also with the aim to provide an area without human disturbance for the recovery of exploited stocks or declining species. The management goals provide the support for maintenance of biodiversity, and the recovery of threatened species, and should enable populations and ecosystems to return to their original, less disturbed states.

In modern times, protected areas covering aquatic habitats and wetlands are becoming more and more important for nature and biodiversity conservation, bearing in mind the negative impacts of human activities on natural habitats. However, fisheries' managers not always support the protected areas, since the limitations arising from the protected status often limit the fishing activities, therefore also the potential incomes. On a long term, though, protected areas support the fisheries by serving as hatcheries and nurseries for many fish species attractive for angling, and they also support fishery sustainability through the spillover of these fishes to areas outside of the protected areas. Both theory and empirical studies have shown (Hannah et al. 2019) that when fish stocks are heavily fished, protected areas can act as a substitute for active fishery management in delivering conservation benefits.

For the creation of sustainable freshwater fisheries, the designation of protected areas covering aquatic and wetland habitats is essential, in order to implement measures such as prohibition and/or limitation of fishing, the regulation of certain types and amounts of gear, restricted and controlled access to certain zones of the fishery, targeting increased fish yield available from existing, natural fish stocks.

Within most protected areas, areas in the process of protection designation, ecologically significant areas, and ecological corridors, there are water bodies inhabited by populations of different fish species. These areas in Serbia are home to most of the 67 protected species (63 species of fish and 4 species of cyclostomes).

In order to efficiently protect fish species (within the protected areas, as well as outside of them), it is necessary to designate special habitats for fish species that have a significant importance for the biological needs of fish, such as: spawning, wintering, growth, feeding, and movement/migration of fish. In order to preserve these habitats, it is necessary to determine their spatial layout, boundaries, and conservation measures. These areas are of special importance, because all types of fishing are permanently prohibited (except for scientific and research purposes), as well as all activities that interfere with the spawning, growth, and movement of fish.

For the protection of strictly protected species, it is necessary to define sufficient number of localities where their habitats would be protected. This could contribute to the improvement of the species' conservation status. In this regard, within the framework of the establishment of the national ecological network and area protection, emphasis should be placed on the preservation of habitats with the aim of improving their status.

With the exception of ecological corridors of international importance, Serbian authorities so far have not identified ecologically significant areas of the ecological network which are habitats of threatened migratory sturgeon fish species—*Acipenser gueldenstaedtii*, *A. nudiventris*, *A. stellatus*, *A. sturio*, and *Huso huso*, as well as the semi-migratory species *A. ruthenus* (Danube population). These species are listed on Appendix II of the Convention on the Conservation of Migratory Species of Wild Animals (Official Journal L 210, 19.07.1982) that have an unfavorable conservation status and for which management and conservation are the subject of international agreements.

7 Future Perspectives

The relations between sustainable development of freshwater fisheries and conservation of threatened fish species, crustaceans, and molluscs are closely intertwined. In an environment heavily influenced and modified by human activities, fisheries require planning and implementation of conservation action. It is also important to improve multi-sectoral communication and decision-making, integral catchment management and to limit degradation of fish habitats.

Unfortunately, people are not aware of the natural values living in the aquatic habitats in their surroundings. The environmental awareness is very low in developing countries (and countries with low incomes), so are the social standards, which have a direct negative impact on aquatic and wetland habitats, that are deteriorating in an alarming rate. The quest for green energy also contributes to habitats and fish population loss via the construction of mini hydropower plants, even though the regulations have become more strict. Special attention should be focused on ecosystem services and the benefits they provide to the society. This concept has a very important role in the modern approach to management of protected species and protected areas. The sustainability of ecosystems is inseparable with the socio-economic conditions. In the processes of environmental planning and policy development it is necessary to consider them both, as equal priorities. Ecosystem services enable stakeholders and decision-makers to have a better understanding of the goals of the protection and planned conservation measures. Focusing on the improvement of ecosystem services by the application of conservation measures (e.g., revitalization of aquatic habitats) would lead to a direct positive effect on the natural values, contributing to the long-term sustainability.

There is a need for the preparation of effective conservation strategy for the threatened species and their habitats. It is essential to have data on the taxonomic status of the species, as well as their biological needs. In some cases, there are scientific gaps regarding data on certain species, and further research is needed to gather necessary data about the fish species in order to plan and implement appropriate conservation measures. This is a long-time process, which requires scientific research and the involvement of scientists and experts, but also have significant financial requirements.

In addition to the basic role in providing–producing food, aquaculture also can contribute to the propagation and rearing of fry of endangered fish species, for reintroduction to fisheries. Given that almost 2/3 of the about 100 species of fish that living in aquatic habitats in Serbia are under a protection regime, and that they are under threat and the numbers of these species are decreasing, there is an increasing need for support to the threatened fish populations. The current practice in fisheries management has resulted in a deterioration of fish stocks, therefore there is a need for a more different approach to the conservation and protection of fish diversity, but also for significant financial support for long-term measures.

Aquaculture could help to minimize any future declines in freshwater fisheries, especially regarding the status of endangered species. Next to food production (the main role of aquaculture), aquaculture can have a very important role in producing also endangered fish species that do not have a commercial importance.

In the process of the harmonization of the legislative and institutional framework with the legal acquis of the EU, and with the Common Fisheries Policy (CFP) (Council Regulation (EC) N° 2371/2002), preservation of aquatic ecosystems and fish fauna is covered also by the negotiating chapter "Chap. 13," which includes issues of management of fish resources and their exploitation (https://europeanwesternbalkans.rs/poglavlje–13/). It also deals with issues of market regulation of fish products, structural policy, supervision and control, as well as international cooperation in fisheries. The CFP sets rules for the preservation of living aquatic resources, the limitation of the ecological impact of fisheries, and the conditions for access to waters and resources. It includes both structural policy and specific rules for fleet capacity management, control of fishing activities, and enforcement of its rules. It is based on sound scientific advice and provides a framework for the collection, management, and use of fisheries data.

In order to regulate this area, it is necessary to control illegal, unreported, and unregulated fishing, which creates the conditions for placing fishery products on the EU market, while preserving national water resources (https://eupregovori.bos.rs/progovori-o-pregovorima/poglavlje-13/1551/2016/02/11/poglavlje-13%2D%2D-ribarstvo-.html).

In accordance with the common policy of the EU, it is necessary to make efforts for consistent application of CFP on the national level, by establishing a legal framework, building administrative capacities for the use of European funds in the field of fisheries, and supporting local fishing action groups according to the EU model.

References

Agnew D, Pearce J, Ganapathiraju P, Peatman T, Watson R, Beddington J, Pitcher T (2009) Estimating the worldwide extent of illegal fishing. PLoS One 4:e4570. https://doi.org/10.1371/journal.pone.0004570

Asche F, Garlock TM, Anderson JL, Bush SR, Smith MD, Anderson CM, Chu J, Garrett KA, Lem A, Lorenzen K, Oglend A, Tveteras S, Vannuccini S (2018) Three pillars of sustainability in fisheries. PNAS 115(44):11221–11225. https://doi.org/10.1073/pnas.1807677115

Bianco PG, Ketmaier V (2016) Chapter 3.19: Nature and status of freshwater and estuarine fisheries in Italy and Western Balkans. In: Craig JF (ed) Freshwater fisheries ecology, 1st edn. Wiley

Common Fisheries Policy (CFP) (Council Regulation (EC) N° 2371/2002

Convention on Cooperation for the Protection and Sustainable use of the Danube river. Sofia, 1994

Convention on International Trade in Endangered Species of Wild Fauna and Flora, March 3rd, 1973, 993 U.N.T.S. 243

Convention on the Conservation of European Wildlife and Natural Habitats (ETS No. 104) (n.d.)

Cooke SJ, Allison EH, Beard TD Jr, Arlinghaus R, Arthington AH, Bartley DM, Cowx IG, Fuentevilla C, Leonard NJ, Lorenzen K, Lynch AJ, Nguyen VM, Youn SJ, Taylor WW, Welcomme RL (2016) On the sustainability of inland fisheries: finding a future for the forgotten. Ambio 45(7):753–764. https://doi.org/10.1007/s13280-016-0787-4

Council Directive 92/43/EEC on the conservation of natural habitats and of wild fauna and flora (Official Journal L 206, 22/07/1992)

EC (2000) Directive 2000/60/EC of the European Parliament and of the Council establishing a framework for community action in the field of water policy. (Official Journal L 327, 22.12.2000)

EEA Report No 10/2020: State of nature in the EU - Results from reporting under the nature directives 2013–2018

Epifanio J, Waples R (2015) Artificial propagation of freshwater fishes: benefits and risks to recipient ecosystems from stocking, translocation and re-introduction. In: Closs G, Krkosek M, Olden J (eds) Conservation of freshwater fishes (conservation biology). Cambridge University Press, Cambridge, pp 399–436. https://doi.org/10.1017/CBO9781139627085.014

European Union (EU) Biodiversity Strategy for 2030, https://environment.ec.europa.eu/strategy/biodiversity-strategy-2030_en

Freyhof J, Brooks E (2011) European red list of freshwater fishes. Publications Office of the European Union, Luxembourg

Galamboš L (2018) Studija - Analiza procene normi i procedura organa i institucija koje primenjuju i sprovode propise. Finalni izveštaj. WWF. LIFE15 GIE/AT/001004 projekat „Zaštita jesetri u donjem toku reke Dunav, kroz sprečavanje krivolova i ilegalne trgovine“

Gozlan RE, Britton JR (2015) Sustainable freshwater fisheries: the search for workable solutions. In: Craig J (ed) Freshwater fisheries ecology. Wiley-Blackwell, p 920. ISBN: 978-1-118-39442-7

Hannah L, Costello C, Elliot V, Owashi B, Nam S, Oyanedel R, Chea R, Vibol O, Phen C, McDonald G (2019) Designing freshwater protected areas (FPAs) for indiscriminate fisheries. Ecol Model 393:127–134, ISSN 0304-3800,. https://doi.org/10.1016/j.ecolmodel.2018.12.006

Hrdina A, Romportl D (2017) Evaluating global biodiversity hotspots – very rich and even more endangered. J Landsc Ecol 10(1):108–115. https://doi.org/10.1515/jlecol-2017-0013

Law on nature protection ("Official Gazette of RS", no. 36/2009, 88/2010, 91/2010 - amended, 14/2016 and 95/2018 - other law, 71/2021)

Law on the protection and sustainable use of fish resources ("Official Gazette of RS", No. 128/2014 and 95/2018 - other law)

Marković Z, Stanković M, Rašković B, Sekulić N, Poleksić V (2018) Akvakultura u službi zaštite ugroženih vrsta riba u Srbiji. Ekološki i ekonomski značaj faune Srbije. Naučni skupovi Srpske akademije nauka i umetnosti, knj. CLXXI, Odeljenje hemijskih i bioloških nauka, knj 12:173–197

Mittermeier RA, Turner WR, Larsen FW et al (2011) Global biodiversity conservation: the critical role of hotspots. In: Zachos FE, Habel JC (eds) Biodiversity hotspots. Springer, London, pp 3–22

Order on measures for the preservation and protection of fish stock ("Official Gazette of RS", no. 56/2015 and 94/2018)

Rulebook on the method, tools and means used for commercial fishing, as well as on the method, tools, equipment and means used for recreational fishing ("Official Gazette of RS", no. 9/2017 and 34/2018)

Rulebook on the proclamation and protection of strictly protected and protected wild species of plants, animals and mushrooms ("Official Gazette of the RS", No. 5/2010, 47/2011, 32/2016 and 98/2016)

Simonović P (2001) Ribe Srbije. Biološki fakultet, NNK International, Zavod za zaštitu prirode, p 248

Skoulikidis NTH, Alcibiades NE, Gritzalis KC, Stamatis Z (2009) Chapter 11. Rivers of the Balkans. In: Tockner K, Uehlinger U, Robinson CT (eds) Rivers of Europe, 1e. Academic Press, Cambridge, pp 421–466

The Convention on Biological Diversity, United Nations, Rio de Janeiro, 1992

The Convention on the Conservation of Migratory Species of Wild Animals (Official Journal L 210, 19.07.1982)

UN General Assembly (2022). United Nations Millennium Declaration, Resolution Adopted by the General Assembly, 18 September 2000, A/RES/55/2. Available at https://www.refworld.org/docid/3b00f4ea3.html. Accessed 8 Oct 2022

Vlada Republike Srbije (2006). Nacionalni milenijumski ciljevi razvoja u Republici Srbiji

Welcomme LR (2016) Chapter 5.1: Fisheries governance and management. In: Craig JF (ed) Freshwater fisheries ecology, 1st edn. Wiley

https://eupregovori.bos.rs/progovori-o-pregovorima/poglavlje-13/1551/2016/02/11/poglavlje-13-ribarstvo.html

https://europeanwesternbalkans.rs/poglavlje-13/

Part VI
Social Aspect

Summary

Fishing has accompanied people since their origins, first as a food source and later as an important economic activity. The chapter in this part critically examines the development of fishing and the relationship between the social communities throughout the Western Balkans.

Political and Socio-Economic Aspects of Fisheries in Inland and Coastal Waters of the Western Balkan

Marija Smederevac-Lalić, Ivan Špelić, Samir Đug, Polona Pengal, Mimoza Çobani, Danilo Mrdak, and Marina Piria

Abstract In the last decades in Europe, especially in the Balkan region, modernization of agriculture gave priority to economic growth without consideration of the environmental impact and sustainable development. Fisheries have traditional and essential importance of food and income in the Slovenia, Croatia, Serbia, Bosnia and Herzegovina, Montenegro, North Macedonia, and Albania (Western Balkans). The fisheries sector in Western Balkans is still not organized according to requirements of ecosystem integrity and sustainable utility. Practical application of the legislation indicates that the fishery has been developed without adequate harmonization with

Supplementary Information The online version contains supplementary material available at https://doi.org/10.1007/978-3-031-36926-1_18.

M. Smederevac-Lalić (✉)
Institute for Multidisciplinary Research, University of Belgrade, Belgrade, Serbia
e-mail: marijasmederevac@imsi.bg.ac.rs

I. Špelić
Faculty of Agriculture, Department of Fisheries, Apiculture, Wildlife Management and Special Zoology, University of Zagreb, Zagreb, Croatia

S. Đug
Faculty of Science, University of Sarajevo, Sarajevo, Bosnia and Herzegovina

P. Pengal
REVIVO, Institute for Ichthyological and Ecological Research, PE Ljubljana, Dob, Slovenia

M. Çobani
Ministry of Agriculture, Rural Development & Water Administration Albania, Tirana, Albania

D. Mrdak
Faculty of Science and Mathematics, Department of Biology, University of Montenegro, Podgorica, Montenegro

M. Piria
University of Zagreb Faculty of Agriculture, Department of Fisheries, Apiculture, Wildlife Management and Special Zoology, Zagreb, Croatia

Faculty of Biology and Environmental Protection, Department of Ecology and Vertebrate Zoology, University of Lodz, Lodz, Poland

V. Simić et al. (eds.), *Ecological Sustainability of Fish Resources of Inland Waters of the Western Balkans*, Fish & Fisheries Series 43,
https://doi.org/10.1007/978-3-031-36926-1_18

socio-economic characteristics, enforcement resources, and adequate control measures. Policymakers continue to be reluctant to give importance to the fishery sector, when allocating fishery resources among multifunctional users of water resources. In the following text, we gave review of the fisheries policy and socio-economic aspects of fishery sector in the Western Balkan countries. Critical fisheries policy and management measures are discussed and the propositions and recommendations of environmental sustainability measures under modern anthropogenic pressures are presented.

Keywords Artisanal fishery · Commercial fishery · Recreational fishery · Western Balkan countries

1 Introduction

The industrial revolution has been given priority to economic growth without special attention to sustainable development. Fisheries, although with a long tradition did not receive any attention, which has resulted in policy and decisions concerning watershed management, hydroelectric dams, dredging and irrigation, giving the lowest priority to fisheries (Visser et al. 2021; Cowx and van Zill de Jong 2004).

Fishing has great socio-economic importance where other occupations (agriculture and industry) are not sufficiently represented. The direct value of fish, from the anthropogenic aspect, is economical, because fish represents food, while another, no less important value is that fish represents an important component of biodiversity (Smederevac-Lalić 2013).

Fisheries globally are in the process of changes due to the effects of modernization and technical innovation but are stressed by regulation and resource depletion caused by failure of fisheries governance (Bavinck et al. 2013). Globalization of resources, trade and weak control of fishing activities have dramatically changed the fisheries circumstances in the last decades (Symes et al. 2015). Combined effects of overfishing, illegal fishing, poor control, poor governance, climate change, water abstraction, pollution, agriculture, and public water supply are putting the whole fisheries sector at risk and eroding the existence of those who are relying on this activity (OECD 2020). Despite the fact that the Balkan region has most pristine rivers, the region is the target of one of the most ambitious hydropower expansion plans in the world (Weiss et al. 2018) which have significant environmental and social consequences.

Improving policy is crucial to ensure that future generations will continue to benefit from these ecosystem services (OECD 2020). Fisheries are complex and their future sustainability is at risk, given the environmental, economic, political, and social instabilities around the world. Particular attention must be paid to the social aspects of fishing communities as this helps to better understand the changes and processes that shape vulnerability, resilience, adaptation, and issues in certain aspects of fisheries policy (Symes et al. 2015). Use of environmental and socio-economic information is among the major issues concerning management of waters (Berge and Dahl-Hansen 2005).

In the region of the Balkans, there are the last rivers in a pristine state at the European continent (Cormack 2020; EC 2020). The area is characterized by natural beauty and great biodiversity. Between 30 and 50% of the population live in rural areas (EC 2020). Fisheries are a source of food, employment, and livelihood, especially for coastal and rural communities (EC 2020). The close connection between local communities and the river as a natural resource indicates the need to regulate this relationship, in order to achieve the sustainability of the use of natural resources (Smederevac-Lalić et al. 2011b; Smederevac-Lalić 2013).

Most of the Balkan countries' population is living close to the water bodies and consequently is traditionally linked to fisheries. Therefore, this activity can be viewed through several social aspects: as a traditional activity, as a usual and expected activity but mainly as the basic economic activity of the local population on which the whole family mostly depends. Just for a small number it is just additional source of income (Smederevac-Lalić 2013). Fisheries are under increasing stress due to environmental conservation and multiple user demands. The resilience and the stability of the fishermen's livelihoods are at the risk. Droughts in recent years in the region have further emphasized the importance of ensuring water availability and resilience (EC 2020).

In the eighties of the last century, fishing in former Yugoslavia was identified as a developing sector (Spreizer and Rogelja Caf 2020). The post-socialist period brought many challenges for the fisheries sector and fishermen community in the Balkan countries. Market mechanisms and deregulation have replaced centralized government. During the transition, the fishermen also faced a new reality. The sector has been reduced and today it is only a small part of what it used to be (Teodorescu and van den Kommer 2020). Although the number of artisanal fishermen is decreasing, there has been an increase in recreational fisheries during the previous decades (Smederevac-Lalić 2013). Increasing number of recreational fishermen stresses the social importance of recreational fishery, highly appreciated throughout the world, as a socio-economic benefit for the society, a way to increase the income and development of tourism (Čaldarović 2005; Čaldarović 2006; Šprem et al. 2006). However, in the same period in Albania coastal commercial fishery increases due to migration of the population from inner to coastal areas with no visible recreational sector development (Çobani personal communication). Nonetheless, both types of fishing can contribute to economic benefits of the local and national economy (Arlinghaus et al. 2002). The disappearance of the fishing community is a result of the dilution of fishing activity to the point where it no longer dominates or represents the largest occupational group (Williams 2014). The term "fishing community" is more associated with the memory of the past than with a description of the present (Williams 2009). It is necessary to find a measure, so that fisheries policy can rebuild resilience from the perspective of tradition, economy, and sustainability. Proof of the importance of this sector and the robustness of social institutions is the survival of "small-scale" fisheries throughout Europe (Macfadyen et al. 2011, RPOA-SSF https://www.fao.org/gfcm/activities/fisheries/small-scale-fisheries/rpoa-ssf). The examples given in this chapter are an overview of how the absence of an effective state hindered fishing activities in the post-socialist years in the countries of the Western Balkans. It is important to clarify that in this chapter the particularities of certain types of fishing are mentioned under different definitions, but they are observed like a synonymous with the position and problems they face conducting of the activity (Rousseau et al. 2019). Therefore, the same conditions include small-scale fishing, commercial (professional) fishing, and artisanal fishing. The terminology is explained in the following paragraphs.

2 Terminology

In the literature and in legal acts, confused terms are used for fishing activities. Several terms were used, such as artisanal, traditional, commercial, professional, recreational, sport, catch-and-release, selective, and scientific fishing but usually is not clearly defined meaning of each one.

Hence, in the following text, the term artisanal fishing will be referred to small-scale freshwater fishery conducted by old traditional tools usually with low technology (Begossi 2014; Rousseau et al. 2019) for subsistence, survival, and livelihood purpose. Commercial fishing will be referred to commercial marine fishery for profit, whereas the term small-scale fishing will be used for individual household marine fishery that is used for livelihood and existence. Selective fishing is related to removal of particular species and sizes of fish from the ecosystem (Breen et al. 2016). This type of fishing is often called rehabilitation fishing. Scientific fishing is used only for research purposes, covered with particular permission from the fisheries management authority of a certain country. Selective and scientific fishing will not be covered by this review.

The term recreational fishing most often implies catch-and-kill but also including leisure, catch-and-release, and sport fishing. The sport, game fishing or angling, is fishing for leisure, but sometimes include competition (Aas et al. 2002). Hence, in following text, we will use the term recreational fishing for both recreational and sport fishing. In the developing world, recreational fisheries have been also referred as an alternative livelihood strategy (Bower et al. 2014).

3 Study Area

The study area that was covered in this review include the following countries (in this text, Western Balkans): Slovenia, Croatia, Serbia, Bosnia and Herzegovina, Montenegro, North Macedonia (known as former Yugoslavian countries), and Albania.

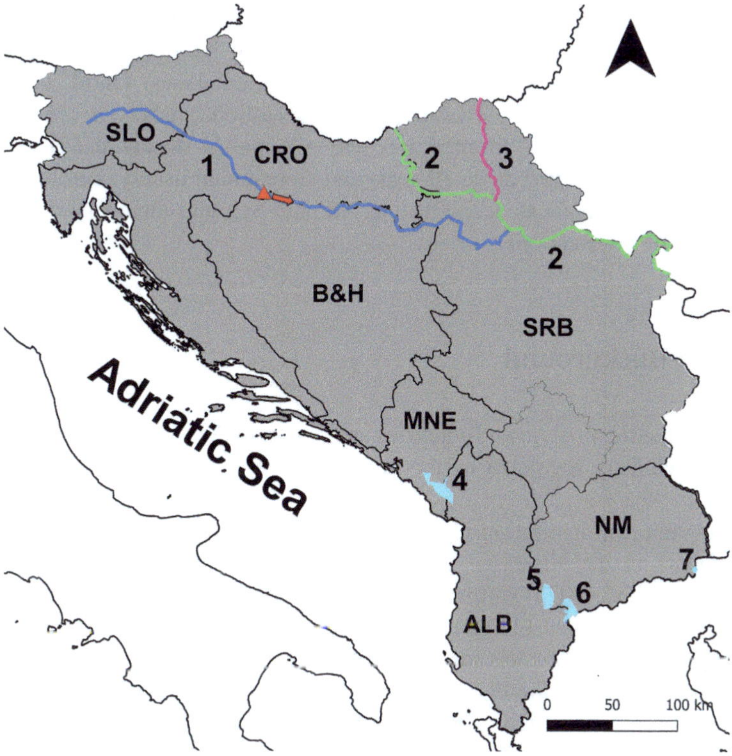

Fig. 1 Map of study area that includes eastern coast of the Adriatic Sea and freshwater bodies where artisanal fishery is allowed: (1) Sava River; (2) Danube River; (3) Tisza River; (4) Skadar Lake; (5) Ohrid Lake; (6) Prespa Lake; (7) Dojran Lake. Dark blue color represents the Sava River, green represents the Danube, and purple represents the Tisza River. Downstream of triangle in arrow direction begins artisanal fishing zone at the Sava River in Croatia

The only areas where artisanal fishing is allowed are the Sava, Danube, and Tisza Rivers and large transboundary lakes Skadar, Ohrid, Prespa, and Dojran (Fig. 1). Small-scale and commercial marine fisheries are activities related to the Adriatic Sea (Fig. 1). The recreational fishery has a wider scope and occurs almost on all water bodies (freshwater and at the Adriatic Sea) in the area that are not prohibited under specific protection measures and therefore is not shown separately on the map.

In Slovenia, artisanal freshwater fishery do not exist, but marine small-scale and commercial fishing is developed at the Adriatic Sea. The recreational fishery in Slovenia is developed at freshwater and marine waterbodies. Artisanal fishery in Croatia is conducted at the middle and lower zone of the Sava River and at the Danube River. In Bosnia and Herzegovina, practically there is no commercial fisheries activities in the traditional sense on inland waters, but exist considerably in illegal form at the Sava River (FAO 2015). In Serbia, artisanal fishery is conducted at the Tisza, Sava, and the Danube Rivers. The Skadar Lake is the only freshwater body for artisanal fishery in Montenegro which this country shares with

Albanian artisanal fishermen. Besides Skadar lake, in Albania at the Prespa and Ohrid Lakes also are conducted artisanal fishery. In North Macedonia, artisanal fishing is allowed on the three transboundary natural lakes: Ohrid, Prespa, and Dojran. Marine commercial, small-scale, and recreational fishery is conducted and well developed in Croatia, Montenegro, and Albania. In Bosnia and Herzegovina, marine fishing is restricted on small-scale and recreational fishery. Serbia and North Macedonia do not possess access to the Adriatic Sea and their fishery activity is restricted to freshwaters only.

4 Legal Background

All types of fishing in reviewed countries are subject to the issuance of permits by the authorized Ministry regulated by the related several Acts in force (Supplementary Table S1).

In Slovenia, the institution responsible for the fisheries policy (legislation and management) is the Fisheries Department within the Ministry of Agriculture, Forestry and Food (2022) (https://www.gov.si/en/policies/agriculture-forestry-and-food/fisheries/; Ministry of Agriculture, Forestry and Food Republic of Slovenia 2022). The professional basis for the preparation of the program and assessment of the situation is done by the Fisheries Research Institute of Slovenia with the help of the Fishermen's Association of Slovenia and the Institute of the Republic of Slovenia for Nature Protection. The Ministry of Environment and Spatial Planning is involved in the process of accepting the program, which gives consent to the aspect of environmental protection (Ministrstvo za kmetijstvo, gozdarstvo in prehrano) (Program upravljanja rib v celinskih vodah Slovenije za obdobje do leta 2021). In the field of freshwater fisheries in Slovenia, the policies aim to ensure sustainable management of fisheries resources in inland waterways, in particular the preservation and protection of fish stocks and the provisions of recreational fishing. Public services in the fields of freshwater and sea water fisheries expertise are provided by the Fisheries Research Institute of Slovenia. The Institute also carries out commercial activities related to sport and recreational fishing and fish farming and a range of research and expert tasks in fish biology. Fisheries supervision is provided by the Fisheries Inspection, which also supervises the implementation of fishing and farming, the operations of fishing clubs and the Fisheries Research Institute, and market in fisheries products. According to the Treaty on European Union (https://www.gov.si/en/policies/agriculture-forestry-and-food/fisheries), the EU has exclusive competence to preserve marine biological resources under the Common Fisheries Policy (CFP). The main aim of the CFP is to ensure that marine fisheries and aquaculture are environmentally sustainable and managed in accordance with economic and social objectives to provide a stable food supply (https://www.gov.si/en/policies/agriculture-forestry-and-food/fisheries/). As an EU member, Slovenia and Croatia are bound to honor the adopted commitments. The CFP measures differ according to individual EU marine regions. The aim is achieving maximum sustainable yield, i.e. the proportion of fish stock that can safely be removed from the stock while maintaining its potential for reproduction (https://www.gov.si/en/policies/agriculture-forestry-and-food/fisheries/). Another important element of CFP is fishing fleet management. Each member state is attributed a maximum level of fishing fleet capacity. New fishing vessels may only be added to the fleet after a vessel of the same capacity had been removed. The CFP also foresees the Common Market Organisation (CMO) in fisheries and aquaculture products, which aims at establishing a common EU market matching supply with demand. CMO interventions thus provide a balance between market needs and fishermen's interests and ensure a fair level playing field for all. At the international level, numerous regional organizations have been established for the management of fisheries. Slovenia and Croatia are members of the General Fisheries Commission for the Mediterranean (GFCM) (FAO 2022c; GFCM 2030).

In Croatia, fishing licenses for freshwater and marine recreational fishing are issued by the Ministry of Agriculture (2022) (https://ribarstvo.mps.hr/default.aspx?id=2802; Piria et al. 2022; Ministry of Agriculture Republic of Croatia 2022). The condition for freshwater recreational fishing is the payment of all costs of enforcing the fishing right, which each holder of the fishing right determines according to the provisions of the Freshwater Fisheries Act. The marine sport and recreational fishery is defined by the Marine Fisheries Act (Official gazette of Croatia 2017) as sport fishing for sport purposes and recreational fishing as fishing for recreation. Licenses for marine recreational and sport fishery can be purchased through web application. However, both fishing categories imply that license is required for performing these activities, but there are certain differences in legislation between these two categories in relation to fishing gear and equipment used, the fishing technique, and the targeted fish species (Soldo et al. 2018). Freshwater Fisheries Act defines the catch quotas for each artisanal important fish species, as well as their minimum catch sizes and the closed seasons during the spawning period for each species (Official Gazette of the Republic of Croatia 2019, No. 63). The number of licenses for artisanal fishery is fixed and they are issued by the Counties which have public authorization according to actual fisheries legislation. There are 49 licenses allowed for 7 fishing zones at the Sava River, and 50 licenses for 3 fishing zones at the Danube River (Piria et al. 2022). Artisanal fishermen are obliged to pay an annual fee for the utilization of the quota to the County on particular fishing ground and these financial funds could be used for improving the state of local freshwater fisheries. They are also obliged to collect and send to the authorized Ministry the catch data for certain fish species (Vehanen et al. 2020). To carry out freshwater recreational fishing in a specific fishing area or fishing zone, anglers purchase licenses from fishing right holders which are valid in the fishing area for which the fishing right holder has acquired management. The fishing right is given through public tenders and the contract lasts for 20 years. Holder of fishing right in Croatia is obliged to make Management plan, no later than a year after signing the contract with the Ministry and every 6 years a revision of each study must be conducted. Management plans are made by official scientific institutions. There are approximately 450 active sport-fishing associations, which are members of higher organizational units (county sport-fishing federations or communities) (Vehanen et al. 2020). The recreational inland fisheries are far more important than the artisanal fisheries, and the landings are about 4 times larger than in commercial fisheries (FAO 2022c). There is a certain interest in artisanal fishing with old, traditional fishing tools, traditional fishing gears, equipment, and methods, as a conservation of old gear-making crafts with educational and tourism purposes. There are two main electronic applications for the collection of the freshwater catch data. One for the angler catch data ("mRibič") and another for commercial fisherman catch data collection ("mAlas"), both in mobile and web form. This application is important to monitor catch quotas and reports fish trade and distribution of non-native fish (Piria et al. 2022). Competencies of the fisheries inspection, as well as other authorized persons for certain offenses were reviewed and harmonized with the current Law on Offenses (Official Gazette of the Republic of Croatia 2018, No 107/07, 39/13, 157/13, 110/15, 70/17, 118/18). Law emphasizes the responsible and sustainable use of natural resources—the freshwater fish stock, in ecologically, commercially, and socially justifiable ways through measures to protect, preserve, and restore resources and freshwater ecosystems, which will contribute to the sustainability and development of local rural communities through fishing tourism and supply of healthy fish to the market while ensuring its traceability. The coastal community and islands are highly dependent on the recreational, small-scale, and commercial fishing activities, while fish farming has been linked with the development of rural tourism. Croatia has a long tradition of small-scale coastal fishing, as indicated by the 55 different fishing gears officially listed as currently employed. The total amount of small-scale licenses was estimated at around 8500 in 1999 by national experts but at present it could be much higher (AdriaMed 2005). The total number of sport and recreational fishers both at sea and on rivers is estimated at some 40,000 (FAO 2022c). In addition, Croatian legislation allows for a special "subsistence fisheries at sea" category, allowing catch of marine organisms for personal consumption only, with a daily limit of 5 kg, and this activity is also subject to licensing and technical limits on gear (FAO 2022c). There is currently an estimated of 11–12,000 subsistence fisheries' licenses issued, but a far lesser number are actually active in this fishery (FAO 2022c).

There is no particular ministry responsible for the fisheries and aquaculture sector in Bosnia and Herzegovina. Fishery is managed by the Ministry of Foreign Trade and Economic Relations, which has the role of coordination, acting as an intermediary between entities and for international relations. Responsible water resources authorities at the entity level are Ministry of Agriculture, Water Management and Forestry Federation of Bosnia and Herzegovina (FBiH), Ministry of Agriculture, Forestry and Water Management of the Republic of Srpska. As for the Brčko District, the competent authority is the Department of the Spatial Planning. However, the fisheries laws applied in the entities still are not harmonized at state level. Environmental legislation has been prepared through the EU project, "Preparation of Environmental Law and Policy in Bosnia and Herzegovina" (Regional Environmental Center 2002). The environmental laws that include water and fishery issues are: Law on the Protection of the Environment; Law on the Protection of the Waters; Law on the Protection of the Nature; Law on Waste Management; Law on Sport

Fisheries. Regulation of aquaculture/fisheries sector is not fully elaborated and not harmonized between entities. Currently, there is no sufficient and reliable data on fishing activities, as well as on the ownership structure of inland waters. Traditional artisanal freshwater fishery is concentrated on the Sava River but is not permitted in rivers and lakes. However, illegal fishing is common. Many of the fish are caught in spawning sites. Gill nets, special fish traps, and long lines with hooks are used (FAO 2022b). Fish are sold through fish shops or directly to the consumer. Recreational fishing has a significant role in the life of communities close to rivers or lakes, and it is an important branch of sport tourism and generates income for the local community. Sport Fishing Societies help to improve the conditions that are necessary for development of tourism. The direct role of fisheries sector in rural development in Bosnia and Herzegovina cannot be determined easily, since the number of employees is low in comparison to other sectors and the fish is not a primary food item in the country (FAO 2022b). Fish consumption is low in the country, and most of the catch is exported. The contribution of fish and other marine products to livestock and food export was 6.8% (FAO 2022b). There are significant efforts on the state level to improve the conditions for utilization of fish resources through the establishment of effective legislation (FAO 2022b). Most of the achievement of the sector was in the development of aquaculture facilities.

Management of the freshwater fisheries sector in Serbia is divided between following ministries: Ministry of Environmental Protection, Ministry of Agriculture, Forestry and Water Management, and Ministry of Finance. The use and protection of fishing resources in Serbia is regulated by appropriate laws: Law on protection and sustainable use of the fish resources (Official Gazette of the Republic of Serbia 2018a, No 128/2014, 95/2018) and Law on Nature (Official Gazette of the Republic of Serbia 2018b, No, 36/2009, 88/2010, 91/2010, 14/2016, 95/2018) and set of by-laws. In accordance with the Law, 17 fishing areas have been established on the territory of the Republic of Serbia (except for fishing waters within the boundaries of protected areas) (Official Gazette of the Republic of Serbia 2018a, No 128/2014, 95/2018). Within the Ministry, there is a Directorate for Inspection, which monitors the work of assigned companies and has a direct supervision of fishing activities in the field. For certain occasions supervision might be carried out by the police too. Since 2016, fishing areas have been assigned to public or private companies, fisheries associations, and fishing unions for ten-years of sustainable use, management, and protection. Fishery right holders are obliged to make Management plan that is made by official scientific institutions engaged after bidding on the tender. The sector appears to be well established through numerous laws and regulations. However, the practical application of the legislation indicates that the sector has been developed without adequate adjustment to the socio economic characteristics of the community. Inspection and fishery ranger services have modest capacities and equipment, so the results in the protection of resources are insufficient (Smederevac-Lalić 2013). According to the Statistical Office of the Republic of Serbia (RZS 2022), the number of artisanal fishermen licenses sold in Serbia in the last 15 years varied between 1051 and 421, and recreational fishermen between 94,896 and 109,606 (RZS 2022). The amount of catches and landings can be considerably different, but only landings data are available. Freshwater fisheries catch data accessible in the Statistical Office are mostly based on the landing reports made by the public or private companies which were assigned to manage particular water resources (Smederevac-Lalić et al. 2011c).

In Montenegro, the umbrella laws governing fisheries and aquaculture are the Law on Marine Fisheries and Mariculture (Official Gazette of Montenegro 2015, No 56/09, 40/11, 47/15) and the Law on Freshwater Fisheries and Aquaculture (Official Gazette of Montenegro 2018, No 17/18), with accompanying regulations. Montenegro has valuable but underutilized resources (Fishery sector analysis in Montenegro for purposes of preparation of the Strategy of Agriculture and Rural Development 2021–2027 and IPARD III (2021–2027) programming). The sectoral policy is implemented by the Ministry of Agriculture, Forestry, and Water Management of the Republic of Montenegro, the Directorate for Fisheries. The government policy is based on the transposition of the EU legislative framework in domestic legislation, implementation of structural measures through the national budget, and projects. The main purpose is the increase in fishing intensity making sure that overfishing and fleet overcapacity are avoided, increase in production of fish and other aquatic organisms (aquaculture), use of the offshore areas for mariculture, increase in processing capacity, improve resource management, fight against the IUU fishing, protection of autochthonous species of fish and other aquatic organisms, product branding, achieving standards in food safety criteria, establishing a better climate investment, investigation of fish and shellfish resources, and introduction of new innovative environmental protection systems and new farming systems (Fishery sector analysis in Montenegro for purposes of preparation of the Strategy of Agriculture and Rural Development 2021–2027 and IPARD III (2021–2027) programming). The Agro-budget supports the development of the marine fisheries and mariculture sector, as well as the freshwater fisheries and aquaculture sector. Fisheries policy is implemented through the Directorate for Fisheries that comprises three administrative units: 1. Unit for Resource and Fleet Management; 2. Unit for Monitoring; and the 3. Unit for Structural Measures, State Aid, EU Funds, and the Market Organisation in Fisheries and Aquaculture. The Directorate for Fisheries defines and implements a policy that ensures a sustainable use of available fishery resources and manages the fishing fleet in a way that ensures long-term and economically cost-effective fisheries in Montenegro, monitors the national legislative activity, implements the measures and management systems based on the principles of the Common Fisheries Policy of the European Union, monitors the international cooperation. The Statistical Office of Montenegro (MONSTAT 2022, https://www.monstat.org/eng) is responsible for the general collection, processing, and delivery of statistical data, including the statistical data in the area of fisheries, collected by the Department of Statistics in Agriculture (MONSTAT 2022). The new Law on Marine Fisheries and Mariculture (Official Gazette of Montenegro 2015, No 56/09, 40/11, 47/15) is in the process of harmonization with the EU and is partly in compliance with the basic Regulations of the EU and Common Fisheries Policy. The water quality monitoring program and biomonitoring in farming sites are also governed by the same Law on Mariculture (Official Gazette of Montenegro 2015,

No 56/09, 40/11, 47/15). To a certain extent is harmonized with the Illegal, Unreported, and Unregulated (IUU) fishing regulations. In 2020, budget was introduced in order to continue fight against illegal, unreported, and unregulated fishing, which should improve monitoring and control system in fisheries. According to the Law on Freshwater Fisheries and Aquaculture (Official Gazette of Montenegro 2018, No 17/18), waters are assigned for the purpose of sports and recreational fishing, except waters in fish farms, accumulations, lakes, or flowing waters used for or intended to be used for public water supply by spatial planning documents. Fishing areas are designated for the purpose of carrying out artisanal fishing. Artisanal fishing may be carried out only in the fishing area with or without a vessel, based on the permit. The permit is issued by the authorized ministry on the basis of a public call, exception is artisanal fishing in protected natural zones when is issued by the protected resource area managing organization. A permit for conducting artisanal fishing can be issued to a business entity or an entrepreneur to person registered in the Central Register of Business Entities of Montenegro (CRPS) for conducting freshwater fishing activities, for a period of up to 6 years. Artisanal freshwater fishing is carried out on Lake Skadar, largest freshwater basin in Southeast Europe (Pešić et al. 2018). The fishing permit is issued by the National Parks of Montenegro—Skadar Lake National Park. Fishery can be conducted in compliance with the Law on Freshwater Fisheries and Aquaculture (Official Gazette of Montenegro 2018, No 17/18) and the Order on closed fishing areas, closed seasons, and minimum sizes of fish and other aquatic organisms (Official Gazette of Montenegro 2021, No 37/20, 9/21). Permits for freshwater recreational fishing are issued for the current year, for a specific area of the water body; time and/or for night carp fishing. The permit is issued by the user or manager of the water body. Montenegro established Fisheries Information System (FIS) for marine fishery, and until the end of 2020, there has been of 268 entities (Fishery sector analysis in Montenegro for purposes of preparation of the Strategy of Agriculture and Rural Development 2021–2027 and IPARD III (2021–2027) programming) which were in accordance with the Law registered as professional small-scale fishermen. Permit is issued for each vessel separately. It is allowed to have another workforce, called the crew that participates in the fishing activity. In this regard, FIS keeps records of persons engaged on the vessel, and the number of such persons engaged from 2011 to the end of 2020 was a total of 770 persons. By adding 770 persons engaged and 268 holders of permits, the number of small-scale fishermen was 1038 (Fishery sector analysis in Montenegro for purposes of preparation of the Strategy of Agriculture and Rural Development 2021–2027 and IPARD III (2021–2027) programming). There is no organized purchase of fish, but fishermen place their catch directly on the market, through markets, restaurants, because landing places and fishing ports are still not marked. Montenegro's fleet is composed mainly of old small lengths vessels, and the most fishery activities are carried out within the territorial waters and in the coastal zone. Public Enterprise Coastal Zone Management operates in the narrow coastal zone, working on the issuance of permits, monitoring of water quality, etc. In general, the fisheries sector is underdeveloped and of a coastal character. The total catch represents a small percentage of the total catch in the Adriatic and the Mediterranean (Fishery sector analysis in Montenegro for purposes of preparation of the Strategy of Agriculture and Rural Development 2021–2027 and IPARD III (2021–2027) programming). Permits for recreational fishing at sea are issued by sport fishing clubs that are members of the Federation. The Federation of Sea Angling of Montenegro has 23 clubs (Fishery sector analysis in Montenegro for purposes of preparation of the Strategy of Agriculture and Rural Development 2021–2027 and IPARD III (2021–2027) programming). During the six-month validity of the permit, the Federation submits to the administrative body a report on the number of licenses issued for sport and recreational fishing. Holders of licenses are obliged to keep the license with them all the time they are engaged in fishing. A license can only be used by the person to whom was issued. The form and method of distribution of permit forms is prescribed by the Ministry (MAFWM of Montenegro—Law on Marine Fisheries and Mariculture) (Official Gazette of the Republic of Montenegro 2015, No 56/09, 40/11, 47/15). The permit can be permanent and part-time. A part-time permit for sports and recreational fishing may be issued to a person who is not a citizen of Montenegro. Fishery plays a very important role in the economic development of Montenegro (FAO 2022d).

Ministry of Agriculture, Rural Development & Water Administration in Albania is the responsible state institution with the Directory of Agriculture Production and Trade Policies unit. Under this Directory is the sector of Fishery & Aquaculture Resources Policies. In Albania, the administration of the fishery and aquaculture sector is separated in two laws: the Law "On Fishing" (Official Gazette of Albania No 4/2019) as amended and the law "On Aquaculture" (Official Gazette of Albania No 103/2016). The Control Structures are organized in another Directorate within the Ministry, divided from Fishery Directorate. This structure is composed of 28 inspectors (Çobani personal communication). The Vessel Monitoring System (VMS) (http://gfcmsitestorage.blob.core.windows.net/documents/web/VMS/2011/ppt/Albania_VMS.pdf) is fully functional after the improving procedures, 2021 to control the fishing vessels larger than 12 meters. To support the VMS System, six system operators at the central level of the Directorate of Fisheries and Aquaculture Services were recruited, who monitor and report on the activity of fishing vessels in addition to the Fishery Monitoring Centre. Fishing is one of the main economic activities in the area of large transboundary lakes: Prespa, Ohrid, Skadar (Adhami 2015; Peveling et al. 2015; Talevski et al. 2009). Just 100 fishermen from the villages around the lake of Prespa on the Albanian side regularly fish (Grazhdani 2009). Unfortunately, their activities are disorganized and carried out on an individual basis, while half of them fish without licenses (Grazhdani 2009). Until the year 2000, artisanal traditional fisheries were disregarded; this was principally due to the small quantity of fish caught and the final destination of the production, which is almost always for consumption by the fisher and his immediate family. Furthermore, freshwater recreational fishing is included in the same category with artisanal fishery (Çobani 2016). In recent years due to the low employment that exists in Albania, commercial fishery sector at the sea started to develop (Çobani 2016). On the other hand, the measures taken by the Fishery Directorate to strengthen the application of fishing law along the coasts, especially concerning lagoons, created the necessity for other alternatives for the communities around those areas. Those fishermen who were refused a lagoon fishing license by the fishery administration (license board) found an alternative by switching to small-scale coastal fishing. The Fisheries and Aquaculture Development Plan (https://fliphtml5.com/cjmgl/jrso/basic) is a financial instrument with the purpose of distributing the available funds to fulfill the objectives of the Fisheries and Aquaculture Administration Plan (Sonila and Kosovrasti 2013). Fishery Management Organizations (FMOs) are being promoted in some water body categories, through the Pilot Project of Fishery Development (FAO 2022a) http://documents.worldbank.org/curated/en/240481468192881291/Albania-Pilot-Fishery-Development-Project, and their task is to manage fisheries in general. The presence of three transboundary lakes, made necessary establishment of the law "For the protection of trans-boundary lakes" (Law No. 9103 on the protection of transboundary lakes, http://extwprlegs1.fao.org/docs/pdf/alb60527E.pdf) which aims to protect the environment of transboundary lakes in their natural situation, to guarantee appropriate conditions for the maintenance of the ecosystem, and to prohibit activities that would threaten and encourage useful activities while respecting the principles of sustainable development (Law No. 9103 (2003)). In 2021, in the Ohrid Lake about 1.44 million Ohrid trout (*Salmo letnica*), fingerlings were stocked, and in Prespa Lake about 501,000 carp fingerlings were stocked financed by the state budget (Çobani personal communication). In order to provide reliable statistical support to the fisheries and aquaculture sector, preparatory work on improving the data collection system has been established ALBFISH Unit supported by the FAO AdriaMed Project (http://www.faoadriamed.org/html/adriamed_project.html). This will be implemented through the redesigning of the Fishery Research Institute projects' funding (Çobani 2016). Fisheries and aquaculture represent realistic and excellent development opportunity for Albania that can intertwine social, economic, biological, and environmental management (Prince 2021). Further development and need to study the economic and social impact of fisheries in Albania, including effects on poverty reduction and provision of high quality nutrition is necessary (FAO 2022a).

Fisheries in North Macedonia are regulated by the Law on Fishery and Aquaculture (Official Gazette of the Republic of North Macedonia 2011a, No 7/2008, 67/2010, 47/2011, 53/2011) which deals with artisanal fishing, sport fishing, and fish production (Official Gazette of the Republic of North Macedonia 2011a). The Ministry of Agriculture, Forestry and Water Management is responsible for issuing concession permits for commercial fishing companies and sport-fishing associations, as well as permits for the establishment and operation of fish ponds (Spirkovski and Bojadzieski 2013). The Ministry of Environment and Spatial Planning is also involved in matters of biodiversity conservation, because each lake has significant biological importance and commercial fish stocks, which require special management (FAO 2022e). Fishing company on each water body is awarded for the five-year concessions, and the company is expected to develop a master plan for the protection, enhancement, and exploitation of the relevant fish stock during the period of their license (Spirkovski and Bojadzieski 2013). Master plans must be submitted to the authorized Ministry before the license is issued, and the Ministry will request the opinion of the competent scientific institutions. According to the Law on Fisheries and Aquaculture, regulations govern the method of fishing, types of fishing gear, closed season, and fishing license costs (Official Gazette of the Republic of North Macedonia 2011b, No 7/2008, 67/2010, 47/2011, 53/2011). The only active government involvement in fisheries management is collecting taxes from the commercial concession companies and the Sport Fisheries Association (SFAs), although about half of the income is returned to the fishery fund. However, despite the decreasing catch from all the lakes, instead of using the fund for improving the fish stock situation, it has been used by fishing companies. There are no official limits on the total allowable catch. In North Macedonia, there is limited potential for fisheries (FAO 2022e), as the overall water resource are comparatively limited on the three large transboundary natural lakes: Ohrid, Prespa, and Dojran. The most valuable catch comes from Lake Ohrid. Lake Ohrid is shared with Albania, Prespa with Albania and Greece, and Dojran with Greece (FAO 2022e). The socio-political changes of the last decade in Balkan countries affected negatively fishery sector. Previously, it was organized through three subsectors: artisanal, fish farming, and recreation (sport) fishing, but from 1995 this arrangement collapsed. Artisanal fishery is limited to these lakes, as well as some of the reservoirs. There are six designated landing sites for the Lake Ohrid. Lake Prespa landings are dispersed. Lake Dojran has four designated landing sites for its artisanal fishery (FAO 2022e). There are designated landing sites for the artisanal fishery on the larger reservoirs, but there are no such sites for the recreational sector. Sport fishing is allowed on the lakes on the basis of an agreement between the Sport Fisheries Association (SFA) and the concessionaire. Recreational fishing sector was strong in the 1980s, but with the fundamental changes in the socio-political environment of the 1990s, membership has dropped. Sport-Fishing Associations operate with direct permission from the Ministry. Fishing on the lakes is strictly controlled, with a limited number of days available and a limit on the equipment allowed. Control is most stringent in the trout fishery. Recreational fishing on the rivers is also on a five-year basis concession, but directly with SFA. Both commercial companies and SFA must submit a five-year plan for the protection, improvement, and responsible utilization of fish resources. Within the Ministry, there is a Directorate for Inspection, which monitors the work of companies and SFA through reports and direct supervision in the field. Certain supervision of fishing activities is carried out by the police. There were difficulties in coordinating the fishing activities on the transboundary lakes. North Macedonia has enacted legislation to protect these lakes because of their biodiversity (Official Gazette of the Republic of Macedonia No. 62/1993; 51/2011) (Law for the protection of Ohrid, Prespa and Dojran Lake; Law for declaration of Lake Prespa as natural monument) (Official Gazette of the Republic of North Macedonia 1993). No recent stock estimates are available (FAO 2022e). Catch reports are unreliable (FAO 2022e). There are no fish markets. Fish from lake catches or ponds are sold at local markets, small shops in towns or directly to hotels and restaurants. There is a high demand for trout and eels from Lake Ohrid. Urgent action is needed to prevent overfishing of endangered species and valuable indigenous species, especially Lake Ohrid trout (FAO 2022e). There is an initiative within the Lake Ohrid Conservation

Project (Spirkovski et al. 2000) due to trout population decline to conduct a joint fish stock assessment between Macedonia and Albania. On Lake Prespa, the trilateral project (Project: Integrated ecosystem management (IEM) in the transboundary Prespa Lakes Basin of FYR-Macedonia, Albania, and Greece) for the protection of the area is in the stages of implementation (FAO 2022e). Aquaculture producers have difficulties in marketing their products due to low prices compared with cheap frozen imports in the market. There were some stocking activities with exotic species although they are prohibited (FAO 2022e). In July 2018, Macedonian Minister of Agriculture, Forestry and Water Economy signed agreement with FAO Country Programming Framework that should support progressive alignment of the country's agriculture and rural development with European Union accession requirements and the Sustainable Development Goals. The agreement matches the country's needs and priorities with the expertise and experience in particular to the sustainable and inclusive growth of smallholders and family farms, improved food safety and quality, sustainable and more resilient agriculture, and strong institutional capacity. Another key priority of the program is strengthening the country's ability to manage natural resources sustainably including water resources, forestry, and fisheries (FAO https://www.fao.org/europe/news/detail-news/en/c/1144995/).

5 Political Aspects of Fishery

Freshwater artisanal fishery is conducted in Serbia, Albania, Bosnia and Herzegovina, Croatia, Montenegro, and North Macedonia but has declining trend similarly as in whole Europe (Cowx 2015). In Slovenia, this type of fishery never had considerable impact and currently no longer exists. On the contrary, in all countries, freshwater recreational fishing is increasingly developing and the number of recreational fishermen is growing (Smederevac-Lalić et al. 2011b). The marine commercial, small-scale, and recreational fishery is important for Albania, Croatia, Montenegro, and Slovenia, i.e. countries that have access to coastal part on the eastern Adriatic Sea. Considering that Bosnia and Herzegovina possesses only short marine coastline, only little impact on economy has been given by small-scale and marine recreational fishery.

The region under research is covered by a considerable number of artisanal, commercial, and small-scale fishermen and intensive recreational fishing. It is important to emphasize that generally all of them are facing the same fate and mainly the difference stands in the water resource in which the activity is carried out.

Freshwater artisanal fishery in former Yugoslav countries is marginalized as an economic activity and occupation (Smederevac-Lalić et al. 2011b; Lenhardt et al. 2020). Martinović 1978 states that "the fisherman craft (the so-called alas) has almost disappeared from the suburban banks of the Sava and Danube, giving way to modern industrial farming and fish processing." The historical upheaval from a socially valued and socially necessary trade to a marginalized activity happened at the same time as the political disintegration of the former Yugoslav states and the war events during the nineties of the last century. Since 1990, there have been significant political and economic changes in former Yugoslav countries, leading to changes in fishing artisanal practices in this region (Spreizer and Rogelja Caf 2020). This period marked changes from the previous jurisdiction of the Republic of Yugoslavia in each country's authority. It was necessary to transfer leadership from the institutions of the former Yugoslavia and to establish new national fisheries institutions in each country. At the same time, the artisanal fishery in Albania has slightly developed since the 1990s as an alternative to unemployment and low income (Çobani 2016). However, the water quality and poor fishery policies and illegal fishing in Albania over the last fifty years have negatively affected the region's biodiversity in general and its fish species in particular (Grazhdani 2009). Poor policy and management have turned important lakes into reservoirs for a local, nationally economically important irrigation system. Hence, consequences on artisanal fishery sector may be expected.

According to Teodorescu and van den Kommer (2020), special attention has to be given to three emerging policy domains: economic restructuring, environmental planning, and EU policy implementation. These areas will provide a reason for an understanding of how post-socialist policies have failed to address and even ignored growing problems in fisheries sector. Inland fisheries have never been part of any high-profile global fisheries assessment (Teodorescu and van den Kommer 2020). The policy makers are largely unfamiliar with freshwater ecosystems, with fish resources, as well as the ecosystem services generated by inland fisheries (Cooke et al. 2016). This ignorance is particularly notable nowadays when emphasis is given to food–water–energy connection where important role of inland fish and fisheries is omitted despite important part in nutrition and livelihoods in low-income and food deficit countries (Cooke et al. 2016).

Artisanal fishermen in Balkan countries are not organized into cooperatives, associations, committees, or boards, through which they could have a much greater influence and participation in society and fisheries policy (Smederevac-Lalić 2013). Moreover, they are not a compact group that has common goals and it is difficult for them to take same positions among themselves and to represent common interests. The need to establish fishing cooperative exists among fishermen, but fishermen do not have the power to change decisions and represent interests in the management system (Pita et al. 2020).

Commercial and artisanal fishing is a significant source of income for many people in the Western Balkan countries (Mikuš et al. 2018; Matić-Skoko et al. 2017). This is a job that demands hard work and brings little income. Only a few of the commercial fishermen managed to form a market niche by way of which they can more successfully solve one of the key problems in their business—sale. They open their own fish shops and/or sell fish to restaurants. Also, many of them are aware of the need for association and initiative in business, by which they would replace the previous command concept of business that they remember, with a new one, based on higher input of individual entrepreneurship (Smederevac-Lalić 2013; Lenhardt et al. 2020). Measures defined by the state should give more incentive to such business concepts. Together with the definition and implementation of appropriate rules, this should lead to sustainable development of the industry.

Although the marine fishing sector in some countries (i.e., Slovenia) is small and has an insignificant influence on the national economy (Spreizer and Rogelja Caf 2020), it is still considered to have a strong social impact in terms of directly creating employment and give importance for maritime identity and tourism (Lopes et al. 2015). However, fisheries policies aim to ensure adequate working conditions and safety of fishermen, enhance the development of coastal fishing areas, and seek synergies between fisheries and tourism, cultural heritage and tradition (Miret-Pastor et al. 2021).

Countries of the Balkan that passed through the procedure of joining the EU had to find an adequate response to changes according to the EU policy. They had to make strategic decisions based on the analysis of benefits and losses and to implement the fisheries strategy. Firstly, strengthening of the administration capacity to propose and implement good decisions based on the data (Smederevac-Lalić et al. 2011b). The baseline conditions should be known by measuring the indicators, which should then be improved. Also, it is necessary to ensure their continuous monitoring.

In Slovenia, but also in Croatia, process of adaptation to the EU policy in the fishery sector resulted in negative perception of this policy that significantly and adversely influenced the behavior and emotional response of fishers, which, in turn, also influenced their resilience (Spreizer and Rogelja Caf 2020). In general, fishermen feel alienated from the decision-making process and their confidence in the policy-making is low. Example of that was the question of calling non-commercial fishing "small-scale for personal needs" or subsistence fishing, which according to current regulations should have been registered in the commercial category. Through this process, about 7500 fishermen in Slovenia have either joined the recreational fishing category or become inactive (Matić-Skoko et al. 2017). This happened because fishermen realized that EU policy makers do not recognize particular characteristics. The transition process lasted 2 years and had a very negative public perception (Spreizer and Rogelja Caf 2020). These transformations on water have implications for those who are allowed access to the resource (Boucquey 2017). Furthermore, Slovenian fishermen highlighted that social goals were almost completely neglected (Spreizer and Rogelja Caf 2020). Their opinion was that resource and habitat protection policies are introduced without consideration of socio-economic factors, particularly those that are in competition with small-scale fisheries (like the establishment of new Marine Protected Areas). Therefore, management authorities must find a way to meaningfully involve fishermen more in decision-making processes, reduce conflicts, so that political perception becomes positive and resilience of the fishermen community improved. The experience of the ministry in Slovenia during the EU accession process is that it was difficult to gain an understanding of the specifics of Slovenian fisheries within the EU. Recently, even the Slovenian state was brought before the European Court of Justice, because the sector did not develop an electronic log (fishing log system) (Spreizer and Rogelja Caf 2020).

Understanding resource user groups is crucial given their potential influence on policy choices and the real political and economic consequences of different access to resources (Adger et al. 2005). Artisanal fishermen use their own narratives to express importance of the fish in the social and economic context (Boucquey 2017). While artisanal fishermen cannot even unite through groups such as federations or associations, recreational fishing groups are increasingly visible in the political arena (Boucquey 2017). Understanding the roots and complexities of these moral arguments between two main categories of fishermen helps illuminate the conflicts that exist from local to international levels.

Real social and economic value of fisheries in terms of their contribution to livelihoods, recreation, and food security is still not well measured in Europe due to lack of appropriate data collection systems (Cowx and van Zill de Jong 2004). This is the reason why policy makers are still reluctant to give this sector the importance it deserves when making decisions affecting fish resources (Cowx and van Zill de Jong 2004). Since water resources are shared among competing users, the lack of reliable data, and thus the underestimation of fisheries, results in policy decisions that have historically favored the construction of dams to produce hydroelectric power, dredging to produce or transport sand, and flood control protection measures. The underestimation of inland fisheries due to the lack of accurate statistics contributes to the underestimation of these resources in political decisions regarding freshwater resources (Visser et al. 2021). Thus, there is a clear need for accurate statistics and sustainable data collection methodologies for fisheries managers which will serve the dual benefits of ensuring sustainable management of fisheries resources and equitable valuation of aquatic species when making decisions on the use and management of freshwater resources (Vehanen et al. 2020).

Although in most of European countries conflicts between commercial, artisanal, and recreational fishermen no longer exist (Arlinghaus et al. 2002), in the Balkans this problem reduces the potential for sustaining and developing fishery in the future. The recreational inland fisheries are far more important than the artisanal fisheries, and the landings are about 4 times larger than in artisanal fisheries. Besides, various forms of illegal fishing (poaching) are also present on a significant scale (Smederevac-Lalić et al. 2011a).

For economic fishery restructuring policies, we need to go back to the first period after 1989 and state what major changes took place. This period, also called "transition," marked a major metamorphosis of fisheries in the Balkan countries and included the privatization of state-owned companies, the closure of unprofitable fish processing industry, and the emergence of private sector fishing operations. Austerity measures implemented after 1989 led to a lack of control of fishing activities and rapid deterioration of vital infrastructure (Teodorescu and van den Kommer 2020). According to Teodorescu and van den Kommer (2020), also small-scale fisheries in Romania experienced failing of post-socialist economic and environmental policies and ignorance of the problems in the fisheries sector. As a result of misguided policies and environmental decline, resources were mismanaged. This period is characterized by a lack of interest in the development of new economic opportunities for fishermen and, more generally, an underestimation of the local importance of fisheries (Teodorescu and van den Kommer 2020).

With no EU common freshwater fisheries policy, every European country can arrange its fisheries policies according to its geomorphological, biological, ecological, ichthyological, but also socio-economic circumstances and conditions (EU Water Framework Directive 2000/60/EC). In addition, the success of any national strategy depends on the effective inclusion in other policies (agriculture, forestry, regional development, energy, transport, and fishing). The Balkan countries that are still not part of the EU, are encouraged to align their policies with the EU's 2030 Biodiversity Strategy and to support the EU's position in the international negotiations on the post-2020 global biodiversity framework (EC 2020).

The Balkan region is part of the EU Action Plan for the implementation of the Green Deal, defined as the Green Agenda for the Western Balkans (EU Commission Green Agenda). In October 2020, "Guidelines for the implementation of the Green Agenda for the Western Balkans" was published (EC 2020). The main focus related to agriculture where the fisheries sector belongs is the "greening" of existing programs, public administration reform, smart specialization, research and development, education, institutions' capacity development, and the market chain improvement. In most of the countries, agriculture, forestry, and fisheries contribute around 20% of the Gross Domestic Product (GDP), while 40% in the case of Albania (EC 2020). The EU is aware that the Balkan region is rich in habitats and species, including endemic species, and as such should make efforts to protect biodiversity and the ecosystem. The main challenges are the lack of political commitment to improve the implementation of biodiversity policy, the lack of financial resources, and the impact of economic activities, such as agriculture, forestry, and fishing (EC 2020). A comprehensive policy framework exists, but implementation lags. The European Green Deal proposed the establishment of a just transition mechanism and fund to support societies in the transition process (EC 2020). The mechanism could focus on the regions and cities most affected by the transition. Two EU macro-regional strategies with a strong innovation component are being implemented in the Western Balkans: European Union Strategy for the Danube Region (EUSDR) (EUSDR European Union Strategy for the Danube Region https://ec.europa.eu/regional_policy/en/policy/cooperation/macro-regional-strategies/danube/) and European Union Strategy for the Adriatic-Ionian Macro Regional Strategy (EUSAIR) (EUSAIR European Union Strategy for the Adriatic-Ionian Macro Regional Strategy EU Strategy for the Adriatic and Ionian Region—Regional Policy—European Commission) and should continue to support innovation and cross-border cooperation (EUSAIR n.d.; EUSDR n.d.).

6 Socio-Economic Aspects

6.1 Historical Background

The first signs of fishing activity on the territory of Western Balkans can be found since the Neolithic period (Ćurčić 1910; Smederevac-Lalić 2013). Also, since

Fig. 2 Fishing activity on the Sava River before Second World War (Petrović 1998)

ancient times, the Adriatic Sea has been used as food provider for people who have settled in these regions. In Dalmatian municipalities, particularly in Zadar in the tenth century, with a donation of fishing-grounds around the island of Molat, and in Telašćica of the island of Dugi Otok to the Benedictine monastery of St. Krševan in Zadar dated circa 995 in which, for the first time, Croatians are mentioned as involved in fishing along the coast (FAO AdriaMed Project (http://www.faoadriamed.org/html/adriamed_project.html) (FAO-AdriaMed Project 2022). As evidenced by numerous historical records, fishing has been a traditional activity for centuries (Petrović 1998). Archeological findings show that fishing was one of the main activities in the area of the Sava and Danube Rivers (Fig. 2). Fish was a symbol of life and one of the conditions of existence at the beginning of human civilization (Živaljević 2017). Inland fishery in Serbia represents "living-by-the-rivers" lifestyle as common and traditional activity of local inhabitants (Lopes et al. 2015). The main part of the area is comprised of a largely rural population and most households in the villages in the vicinity of the large rivers such as Danube, Sava, and Tisza depend on fishing because fishing is an integral part of their livelihood (Smederevac-Lalić et al. 2011a). Importantly, the former state of freshwater ecosystems differed hydrographically and hydrologically from present situation, primarily due to the draining of wetlands, the regulation of flows, and the construction and formation of artificial water systems such as dams. However, this did not diminish fishing importance, but it did have an impact on the characteristics of the fishing resource and fish stock (Smederevac-Lalić 2013).

The first regulation on fishing activity in the area was adopted in 1859, and the first Law on fisheries in 1898 (Petrović 1998; Smederevac-Lalić 2013). The fishermen's guild existed and at the beginning of the twentieth century. Within the Law on Freshwater Fisheries (Petrović 1998; Smederevac-Lalić 2013), there was a special article related to the protection and improvement of fisheries. There were fishing associations whose tasks were rational utilization, and improvement of fishing, aquaculture, processing, organization of sale and transport of fish, and general improvement of economic interests. The government supported these associations. Local administrations were obliged to collect statistical data and send them to the competent ministry. The law prescribed a fishery fund intended to be used for the improvement, organization, regulation, and fisheries research. After the Second World War, fishing on the large rivers Danube, Sava, and Tisza became an increasingly important economic branch (Smederevac-Lalić 2013). The abundance of fish was reduced by the regulation of the Danube, and later by the increasing development of industry along the banks of the Sava and Danube (Smederevac-Lalić 2013). Fish catch records are maintained since 1948, while the data on exact species composition was introduced in 1951 (Smederevac-Lalić 2013).

However, fishing has remained until present time the professional occupation of a part of the population, the basis of their diet, livelihood and the subject of trade (Smederevac-Lalić 2013).

6.2 *Present Socio-Economic Status*

Artisanal fishery in the Western Balkans has a long tradition and today it represents ethnological, traditional, sociological, cultural, and economical heritage (Lynch et al. 2016; FAO 2019; Pascual-Fernandez et al. 2020; RPOA-SSF (n.d.); RPOA-SSF https://www.fao.org/gfcm/activities/fisheries/small-scale-fisheries/rpoa-ssf). Inland fisheries have low economic value and a small impact on the economy (Smederevac-Lalić et al. 2011b; Lynch et al. 2016; Matić-Skoko and Stagličić 2020). Although there is demand for fresh fish, competition for space and conflicts with recreational fishermen reduces the potential for sustaining and developing artisanal fishery in the future (Gómez et al. 2006).

The most common indicators that illustrate the socio-economic importance of fishing are the catch and the number of fishermen (FAO 2019; GFCM 2030 https://www.fao.org/gfcm/2seas1vision/GFCM2030Strategy; GFCM n.d.). The decreasing trend of artisanal and small-scale fishermen can be linked with the legislation and corresponding regulations and obligations arising from it.

Artisanal fishermen are territorially unevenly distributed, and thus are not subject to the uniform rules of different users of the area, which makes it difficult for them to unite for their rights. Fishing is marginalized as an economic activity and occupation (Smederevac-Lalić et al. 2011b; FAO 2019). In this sense, aquaculture is far more represented as competition on the market.

Fishing is mostly done by people who have lost their jobs or are not qualified to do other occupations or have inherited and learned from family members (Chuenpagdee et al. 2013; Smederevac-Lalić 2013). Those who do not have another source of income or do not have the possibility to live in a different way remain in this business. Fishermen provide food for all levels of society. However, there are more poor citizens who are unable to buy other types of meat, so second quality fish is an option. Nowadays artisanal fishing in many countries is not considered by the government as an economically significant activity (Smederevac-Lalić 2013; Matić-Skoko and Stagličić 2020; Rakowski et al. 2020; FAO 2022b). The fish market has been disordered and unorganized for a couple of decades. Since there is no organized purchase of fish from fishermen, the sale of freshwater fish is most often carried out by fish farms (aquaculture facilities), private fish shops, restaurants, and resellers who buy fish from fishermen. More and more fish and fish products, and the low prices of imported fish compete with domestic freshwater fish. The market is small, and there are just a few companies involved in the processing of fish and its products (Smederevac-Lalić et al. 2015).

There is practically no official data on fishing resources, their richness, accurate fish stock assessments, condition, economic and recreational use, although the legal regulations manage the conditions of use and regular monitoring of resources (Visser et al. 2021; Vehanen et al. 2020). The absence of an organized fish purchase system and insufficient management of the fishery sector have resulted in poor data reliability. There is a lack of data on the historic catch per unit effort (CPUE) (Jarić et al. 2015). Knowledge about the present and historic fishery pressures based on the number of fishermen, their fishing effort, and catch quantities, is a prerequisite for the efficient management of fishery resources, which is commonly realized by keeping regular management statistics (Coates 2002; Smederevac-Lalić et al. 2011c; Jarić et al. 2015).

Based on the statistics (Visser et al. 2021), the number of recreational fishermen has been growing for years. While recreational fishing in European countries has only the character of a free time activity, directly reflecting standard of living, in Balkan countries it has a different character. For the majority it is an additional source of income or a supplement to the family's diet (Smederevac et al. 2006). Often exposed to political and governance changes, fisheries management institutions still lack adequate mechanisms to cope against corruption and poaching. Estimation of actual fishing pressure of both artisanal, small-scale and recreational fisheries is difficult. Recreational fishermen, unlike artisanal ones, with such a large number of fishermen, are extremely difficult to control. However, recreational fishing has a very significant impact on fishing resources and should not be neglected (Smederevac-Lalić 2013).

The socio-economic status of fishermen is an important factor affecting the sustainability of fishery (Cowx 2015; Boucquey 2017). It is important to note the constant and increasingly pronounced conflict between recreational fishermen and artisanal fishermen (Cooke and Cowx 2006; Boucquey 2017). Underlying that conflict lays a complicated set of reasons that can best be defined as the different economic and social positions of the two categories of resource users. The position of artisanal fishermen is clearer, because they catch fish on a professional basis and this is their primary economic activity (Fig. 3). However, their position does not depend only on the degree of success in catching (Supplementary Video S1), but

Fig. 3 Artisanal fisherman on the Danube River (photo: Smederevac-Lalić)

also on a series of legal measures that they must comply with, which is not the case with recreational fishermen to such an extent (Čaldarović 2005). Better conditions for the resource use (Rockland 1983) are given to recreational fishing, in the sense of a lower license price, less limitation of the territory where they can fish, minor restrictions on catches, tools, etc. Recreational fishers are also seen as "stewardship leaders" with a strong political voice, and they can have a strong influence over decision-makers (Bower et al. 2014). There are different reasons and motivation for recreational fishing: some return the catch to the water, others consume it, and there are also those who sell the catch. The motives for engaging in recreational fishing, and therefore the final result of the activity, are not always in accordance with the idea of recreation and relaxation. According to Cowx (1996), Eastern Europe recreational fishing is much more than simply a leisure activity and there is an increasing dependence on this activity as a source of food, as the transition from the communist regime proceeds. The motivation for those engaged in fishing, in general, depends on the social and economic position (Čaldarović 2005). Therefore, artisanal and recreational fishermen reflect two structurally and functionally different social positions in the society.

Results of the survey (Smederevac-Lalić et al. 2011b; Smederevac-Lalić 2013) conducted with the artisanal fishermen in Serbia, indicated that fishermen consider fishing has no future. The most significant problems according to the opinion of fishermen were lack of an organized system of fish purchase, poaching, high fishing licenses costs, poor regulations, and pollution (Smederevac-Lalić 2013). A number of artisanal fishermen experienced a trend of a constant decline since 1994, as the sector failed to find its proper place in the transition process and nowadays represents a marginalized branch of the industry. The freshwater fishery is an insufficiently regulated area that depends on personal commitment and individual responsibility (Smederevac-Lalić et al. 2011b, Smederevac-Lalić 2013).

The marine fishing sector has rarely been influenced by political issues. In the early 1990s, during the transitional period, socio-economic reasons led to a large number of people engaged in fishing. In the last decade, there has been no increase in the number of new small-scale fishers in Croatia (Matić-Skoko and Stagličić 2020). Despite small-scale fisheries not being very economically significant, they do have great social importance. Besides food supply, supplement to the household budget, this activity make a substantial contribution to employment among the rural population on islands and along the coast. Fisheries also provide an added value to tourism services, have significant cultural value and as such are part of the identity of the island community (Matić-Skoko and Stagličić 2020). Existence of the small-scale fisheries in the future is questionable due to potential conflicts with oil and gas industry, with the maritime transport sector and with the wind energy sector, with Marine Protected Areas (MPAs), which are expected to increase in size in the coming years (Matić-Skoko and Stagličić 2020).

Fishing communities exist in coastal villages all along the Adriatic coast (FAO 2022c). Coastal areas and particularly the islands are characterized by numerous fishing communities, where the majority of the population is employed in fisheries. In some areas, fishers are organized in cooperatives, trying to secure the best possible outcome for their activities (Matić-Skoko and Stagličić 2020). With the development of tourism, traditional fishing communities are difficult to identify, and generally there is still a rather low level of larger fisher organizations, such as cooperatives or other types of joint management. In order to fish, it is necessary to have a fishing license or fishing permit obtained from local offices or travel agents. Licenses are issued for recreational angling, underwater spear-gun fishing, hand trolling, long lining, and tuna fishing. Fishing in Croatia is strictly forbidden in the area of the National Parks (FAO 2022c). Also, Montenegro recently formed Marine Protected Area (MPA) where in particular time of the year fishing is banned (Fishery sector analysis in Montenegro for purposes of preparation of the Strategy of Agriculture and Rural Development 2021–2027 and IPARD III (2021–2027) programming). About 300 registered fishermen are engaged in freshwater artisanal fishing in Montenegro (Fishery sector analysis in Montenegro for purposes of preparation of the Strategy of Agriculture and Rural Development 2021–2027 and IPARD III (2021–2027) programming). They fish with quite outdated fishing vessels and tools. Freshwater artisanal fishermen use traditional fishing boats at the Skadar Lake, the so-called čun, maximum power engine of 10 hp, in accordance with the

Law on Navigation Safety (Official Gazette of the Republic of Montenegro 2014, No 62/2013 and 06/2014).

They argue for these restrictions, because they believe that 10 hp engines do not guarantee safety in bad weather conditions and request the change in the context of using higher power engines. There are no special moorings for fishing boats, docks, or other supporting infrastructure (warehouses, cold stores, ice machines, etc.) on the Skadar Lake in Montenegro. Fishermen sell their catch directly on the market (markets, restaurants), and to fish processing factories. Since 1987, statistics on fisheries have not been kept although there is strict law obligation for each fisherman to have log book with data regarding their catch (FAO 2022d).

For the majority of artisanal fishermen on the Skadar Lake in Montenegro, fishing is not their primary occupation, but rather an additional or supplementary occupation through which they earn a significant income (Fishery sector analysis in Montenegro for purposes of preparation of the Strategy of Agriculture and Rural Development 2021–2027 and IPARD III (2021–2027) programming). During the survey among commercial fishermen from Skadar Lake on the share of individual costs, no precise data were obtained to estimate costs and possible earnings. For most commercial fishermen on Skadar Lake, fishing is not the main occupation, but an additional or supplementary occupation with which they make a certain profit, and a certain number of respondents answered that they do not make a profit, i.e. do not sell the catch but use it exclusively for their own needs (Fishery sector analysis in Montenegro for purposes of preparation of the Strategy of Agriculture and Rural Development 2021–2027 and IPARD III (2021–2027) programming). When it comes to artisanal fishermen, the problems are: small and old fleet, poor equipment, as well as the lack and high price of raw materials and equipment for fishing, lack of specific places for moorings. An additional problem on the Skadar Lake is illegal fishing, so it is necessary to strengthen control and inspection because the current number of employees cannot ensure constant control. There is a need for the construction of facilities for the processing of traditional fishery products, among which the most important are smoked and marinated carp and smoked bleak, for which this region is famous (Fishery sector analysis in Montenegro for purposes of preparation of the Strategy of Agriculture and Rural Development 2021–2027 and IPARD III (2021–2027) programming).

Albania has the highest population growth rates in Europe and fishery and aquaculture can contribute in the poverty alleviation. Demographic movements led to an increase of population in coastal areas; consequently pressure on marine fishing grew rapidly (Bon et al. 2018). Fish farming is considered to be the most efficient measure to ensure that fish resources are not overexploited (FAO 2022a). Artisanal and commercial fishing in Albania has been a very important occupation since the time of the Illyrians (Çobani 2016). Coastal residents were mainly engaged in fishing and this activity has become one of the most important sources of food. The artisanal and commercial fishery has developed since the 1990s as an alternative to unemployment and low income in coastal areas (Çobani 2016). The data collection and reporting are obligatory for artisanal and commercial fishermen (Çobani personal communication). A recent field survey (Çobani 2016) of commercial, small-scale, and artisanal fisheries in Albania gave the insight into the importance of this sector in terms of species and quantities caught, socio-economic aspects, and relations with other categories of fisheries and management activities. About 85% of the fishing category is made up of illegal, unreported, and unregulated fishing (Çobani 2016). Recently, the fisheries inspection has begun to consider artisanal fishing more important. In order to get to know the situation and start the basic monitoring of this category of fishing, interviews were conducted with artisanal fishermen (Çobani 2016). It is not a primary economic activity in general in Albania, because fishermen are mostly employed in agriculture and other occasional activities (FAO 2022a). According to survey data (Çobani 2016) almost all fishermen have been engaged in this activity for more than 10 years, in some cases even 20 years, following the family tradition. Lagoon (commercial) fishermen exploit both the lagoon and the marine coastal zones by combining the lagoon's seasonal catch using traps with trammel net fishing along the coast. They comply with regulations regarding the stocking of their lagoons as it is considered fundamental to their activity. This category of fishermen proved to be appropriately licensed. Among the main reported constraints and problems affecting the small-scale fisheries are the lack of good quality fishing nets and the inability to directly access to the equipment (Çobani 2016). Fishing equipment imported from abroad does not offer a suitable range of quality and price (Çobani 2016). Illegal inshore trawling that causes damage or loss of small-scale fishing gear is also recognized as a problem. The sale of the catch to restaurants or fishing companies has been reported to be negatively affected due to small quantities sold. This category of fishermen is very poor. They are characterized by good expertise in maintenance of boats, making nets, and knowledge of the sea for forecasting and planning of fishing activities (Çobani 2016). Artisanal fishers know the fish behavior well and their biological seasonality, therefore this traditional local knowledge would need to be framed and valued by scientific programs and to be considered by policy makers in a way that this fishing category could be an integral part of management plans. Also, artisanal traditional fisheries may contribute to determine illegal fishing by fishing boats operating in shallow water near the coast (Çobani 2016).

Ecological and socio-economic changes in Balkan countries, increase of recreational fisheries, and the disappearance of traditional low-technology commercial fisheries led to a loss of the traditional ecological knowledge held by small-scale fishers (Matić-Skoko and Stagličić 2020). According to Cooke et al. (2016), inland fisheries have traditionally been forgotten compared to marine fisheries, but still there is a hope that in the future political will and public desire will properly assess and manage inland fisheries in an integrated water resources management. The active involvement of fishermen preserves and incorporates unavailable traditional and local knowledge into the decision-making process, and lends legitimacy to the rules governing fisheries and results in strategies that are respected and voluntarily followed (Dimech et al. 2009). Embedding different areas of social science in activities should rebuild resilience and adaptability within the fisheries sector and fishing community.

Experiences of other countries, opinions and knowledge of fishermen could be used as a good basis for sustainable planning, monitoring, and management of fish resources and fisheries sector (Smederevac-Lalić et al. 2011b; Smederevac-Lalić 2013).

7 Management

The main problems of fisheries worldwide are overfishing, unemployment and poverty, food insecurity and social injustice (Chuenpagdee et al. 2013). These problems can be felt in all countries and governments, so the approach must be holistic and to take into account principles, institutions, and conditions, analyzing the nature of fisheries management problems and challenges concerning the governance of fisheries, exploring where these problems lie and finding potential solutions (Chuenpagdee et al. 2013). Governance is an important concept in environmental policy and social science with increasing application to socio-ecological systems such as fisheries. Management analysis of fisheries has created innovative ways to implement the ideal of sustainability. However, despite progress, there are limitations in existing governance concepts (Song et al. 2018). This change is driven by the opinion that the state governance model is becoming increasingly inadequate in mobilizing effective and legitimate responses to the complexity of socio-ecological problems.

In general, there is still a lack of communication and trust among scientists, fishers, and fisheries managers. This should be the focus of any management process that could facilitate the desired outcome of the sustainable management of fisheries. The feeling of inequality leads to hopelessness and anger and a loss of confidence in future work. Perceptions of a lack of fairness in the distribution of costs and benefits of resources can have an impact on the way policies are perceived and implemented (Cochrane 2000). For example, recently Pešić et al. (2018) stressed that conflict over the sustainable use of limited aquatic resources has been recognized in the public as a consequence of the lack of an established set of equitable sharing principles among the contending parties (Pešić et al. 2020). This is particularly important for small-scale fisheries with non-transferable skills (Cochrane 2000). Management strategies based on fishermen's knowledge can provide invaluable practical information. The newly reformed Common Fisheries Policy (CFP) recognizes that securing the future small-scale fisheries is essential (CFP) (https://www.europarl.europa.eu/factsheets/en/sheet/114/the-common-fisheries-policy-origins-and-development) (The common fisheries policy (CFP) n.d.). Nevertheless, small-scale fishing has taken its rightful place in the wider picture of tourism, sustainability discourse, and neoliberal economy. They also had to deal with new challenges and opportunities at different levels, such as flow of the "projects." "Everything nowadays becomes a project," preferably sustainable, and it has become quite difficult to work outside of the project even if you are a fisherman. In this race for projects, fishermen come last, even though they are the ones on which the image of sustainable fisheries is based. They have to compete with aquaculture, with people who know how to apply for projects, or they have to get involved in the tourism industry to survive. Not everyone is ready or has the financial ability to make these changes (Spreizer and Rogelja Caf 2020).

Surveys of socio-economic characteristics of fisheries may help the national fisheries management authority by providing a better understanding of the situation in this sector. It can contribute to its management and the establishment of a more regular monitoring scheme (Çobani 2016).

There is a need for integrated water resources management where strategies should create incentives for resource-user communities to invest in the long-term conservation of habitats and fish stocks. Integrated management of water resources is necessary because it includes biophysical and socio-economic elements (Cooke et al. 2016). Limited availability of fish resources will require communication, proactive management strategies, and cooperation across sectors to ensure the sustainability and resilience of the socio-ecological system. In that context, management should be organized establishing an adaptive management strategy that includes avenues for stakeholder participation (Bower et al. 2014).

A considerable complication for the management and often a politically sensitive issue is when the water and fish stocks are shared by several countries. Sustainable management requires agreement among countries on unique management goals and measures, as well as a compatibility and regular exchange of data. This particular concern about fisheries management of shared water bodies has been expressed by several governments in the Balkan region (White et al. 2018). Governments are usually more motivated to monitor marine fisheries and aquaculture than freshwater because of the higher potential for tax revenue (Coates 2002).

8 Conclusions

By comparing the present situation in the Western Balkan countries, the situation in fishery sector seems similar. Fundamental changes in the socio-political environment on the territory covering most of the Balkan countries in the 1990s crucially affected regular life, economy, and consequently fisheries sector. The major common impression after the analysis of the data about fishery from Balkan countries is that fishing is in a prolonged process of transition from command economy under socialism to market economy. This period in fishing, as well as in the whole economy, is characterized by the absence of effective implementation of regulation, strengthening of illegal and informal activities and weak control. The only significant difference is pressure on resource, affected by the number of artisanal, commercial, small-scale, and recreational fishermen.

References

Aas Ø, Thailing CE, Ditton RB (2002) Controversy over catch-and-release recreational fishing in Europe. In: Pitcher TJ, Hollingworth CE (eds) Recreational fisheries: ecological, economic and social evaluation, vol 15. Blackwell Science, Oxford, pp 95–106

Adger WN, Brown K, Tompkins EL (2005) The political economy of cross-scale networks in resource co-management. Ecol Soc 10(2):14

Adhami E (2015) Initial characterisation of Lakes Prespa, Ohrid and Shkodra/Skadar

AdriaMed (2005) Adriatic Sea Small-scale Fisheries. Report of the AdriaMed Technical Consultation on Adriatic Sea Small-Scale Fisheries. Split, Croatia, 14th–15th October 2003. FAO-MiPAF Scientific Cooperation to Support Responsible Fisheries in the Adriatic Sea. GCP/RER/010/ITA/TD15. AdriaMed Technical Documents, 15:184 pp.

Arlinghaus R, Mehner T, Cowx IG (2002) Reconciling traditional inland fisheries management and sustainability in industrialized countries, with emphasis on Europe. Fish Fish 3:261–316

Bavinck M, Chuenpagdee R, Jentoft S et al (eds) (2013) Governability of fisheries and aquaculture: theory and applications. Springer, Dordrecht

Begossi A (2014) Ecological, cultural, and economic approaches to managing artisanal fisheries. Environ Dev Sustain 16:5–34. https://doi.org/10.1007/s10668-013-9471-z

Berge D, Dahl-Hansen G (2005) Management of fresh water fisheries on bordering rivers, EU-regulations in the field of water and fish management. Report, Norwegian Institute for Water Research, 27 p

Bon NG, Josipovič D, Kanjir U (2018) Linking geomorphological and demographic movements: the case of Southern Albania. Appl Geogr 100:55–67

Boucquey N (2017) 'That's my livelihood, it's your fun': the conflicting moral economies of commercial and recreational fishing. J Rural Stud 54:138–150. https://doi.org/10.1016/j.jrurstud.2017.06.018

Bower SD, Nguyen VM, Danylchuk AJ, Beard DT Jr, Cooke SJ (2014) Inter-sectoral conflict and recreational fisheries of the developing world: opportunities and challenges for co-operation. Rec. Fisheries Persp. http://www.fecpl.ca/wp-content/uploads/2014/08/Bower-et-al-2014_Rec-Fisheries-Persp.pdf

Breen M, Graham N, Pol M, He P, Reid D, Suuronen P (2016) Selective fishing and balanced harvesting. Fish Res 184:2–8. https://doi.org/10.1016/j.fishres.2016.03.014

Čaldarović O (2005) Sociological analysis of major socio-economic factors of freshwater fishing along bordering rivers Sava and Danube - as a part of the project "Management of freshwater fisheries on bordering Rivers – a pilot study with a holistic regional approach". University of Zagreb, Faculty of Philosophy, Department of Sociology, Zagreb

Čaldarović O (2006) The importance of key sociological factors for management of freshwater fishing. Paper presented at the 4th national scientific experts meeting with the international participation Ribarski dani "Osjek 2006", Conference proceedings, 59-66

Chuenpagdee R, Jentoft S, Bavinck M, Kooiman J (2013) Governability – new directions in fisheries governance. In: Bavinck M, Chuenpagdee R, Jentoft S, Kooiman J (eds) Governability of fisheries and aquaculture, MARE publication series, vol 7. Springer, Dordrecht. https://doi.org/10.1007/978-94-007-6107-0_1

Coates D (2002) Inland capture fishery statistics of Southeast Asia: current status and information needs. Asia-Pacific Fishery Commission, vol 11. RAP Publication, Bangkok, p 114

Çobani M (2016) Small-scale fisheries in Albania. AdriaMed technical documents, No.15. GCP/RER/010/ITA/TD-15 (9 pages) original. FAO. ICES CM 2016/G: small-scale fisheries in the Adriatic Sea: fishing strategies and eco-ethology of target species; small-scale fisheries in the Adriatic Sea: information gaps at biological, socio-economic and environmental level http://www.faoadriamed.org/pdf/publications/td15/NR_ALB.pdf

Cochrane KL (2000) Reconciling sustainability, economic efficiency and equity in fisheries: the one that got away? Fish Fish 1:3–21. https://doi.org/10.1046/j.1467-2979.2000.00003.x

Cooke SJ, Cowx IG (2006) Contrasting recreational and commercial fishing: searching for common issues to promote unified conservation of fisheries resources and aquatic environments. Biol Conserv 128:93–108

Cooke S, Allison E, Beard D, Arlinghaus R, Arthington A, Bartley D, Cowx I, Fuentevilla C, Loenard N, Lorenzen K, Lynch A, Nguyen V, Youn S, Taylor W, Welcomme R (2016) On the sustainability of inland fisheries: finding a future for the forgotten. Ambio 45:753–764. https://doi.org/10.1007/s13280-016-0787-4

Cormack M (2020) Rivers under siege in Western Balkans. European Data Journalism Network 43(3):32–32

Cowx IG (1996) Managing recreational fisheries in multiple aquatic resource user situations. EIFAC/OP32, 5–10

Cowx IG (2015) Characterisation of inland fisheries in Europe. Fish Manag Ecol 22:78–87. https://doi.org/10.1111/fme.12105

Cowx IG, van Zill de Jong M (2004) Rehabilitation of freshwater fisheries: tales of the unexpected? Fish Manag Ecol 11:243–249

Ćurčić V (1910) Narodno ribarstvo u Bosni i Hercegovini. Glasnik zemaljskog muzeja u Bosni i Hercegovini, Sarajevo. 109p

Dimech M, Darmanin M, Smith P, Kaiser M, Schembri P (2009) Fishers' perception of a 35-year old exclusive fisheries management zone. Biol Conserv 142:2691–2702. https://doi.org/10.1016/j.biocon.2009.06.019

European Commission (2020) (EC) Guidelines for the Implementation of the Green Agenda for the Western Balkans. https://neighbourhood-enlargement.ec.europa.eu/system/files/2020-10/green_agenda_for_the_western_balkans_en.pdf

EUSAIR (n.d.) European Union Strategy for the Adriatic-Ionian Macro Regional Strategy EU Strategy for the Adriatic and Ionian Region - Regional Policy - European Commission (europa.eu)

EUSDR (n.d.) European Union Strategy for the Danube Region https://ec.europa.eu/regional_policy/en/policy/cooperation/macro-regional-strategies/danube/

FAO (2015) Analiza sektora ribarstva i akvakulture u Bosni i Hercegovini. Projekat "Priprema analiza sektora šumarstva i ribarstva u Bosni i Hercegovini u svrhu IPARD-a", finansiran sredstvima Evropske unije https://www.fao.org/3/au016e/au016e.pdf

FAO (2019) Social protection for small-scale fisheries in the Mediterranean region https://www.fao.org/3/ca4712en/CA4712EN.pdf

FAO (2022a) Fishery and Aquaculture Country Profiles. Albania. Text by Cobani, M. Fisheries and Aquaculture Division [online]. Rome. https://www.fao.org/fishery/en/countrysector/al/en?lang=en

FAO (2022b) Fishery and Aquaculture Country Profiles. Bosnia and Herzegovina. Country Profile Fact Sheets. Fisheries and Aquaculture Division [online]. Rome. https://www.fao.org/fishery/en/facp/bih

FAO (2022c) Fishery and Aquaculture Country Profiles. Croatia. Country Profile Fact Sheets. Fisheries and Aquaculture Division [online]. Rome. https://www.fao.org/fishery/en/facp/hrv?lang=en

FAO (2022d) Montenegro. Text by Krasic, M., Mandic, M. Fisheries and Aquaculture Division [online]. Rome. https://www.fao.org/fishery/en/countrysector/me/en

FAO (2022e) Fishery and Aquaculture Country Profiles. North Macedonia. Country Profile Fact Sheets. Fisheries and Aquaculture Division [online]. Rome. https://www.fao.org/fishery/docs/DOCUMENT/fcp/en/FI_CP_MK.pdf

FAO-AdriaMed Project [online]. Rome. Updated. [Cited 28/10/2022]. http://www.faoadriamed.org/html/adriamed_project.html#

GFCM 2030 Strategy for sustainable fisheries and aquaculture in the Mediterranean and the Black Sea. https://www.fao.org/gfcm/2seas1vision/GFCM2030Strategy

Gómez S, DrJ L, Demestre M, Riera V (2006) The decline of the artisanal fisheries in Mediterranean coastal areas: the case of Cap de Creus (Cape Creus). Coast Manag 34(2):217–232. https://doi.org/10.1080/08920750500531389

Grazhdani D (2009) Social economic aspects of fishery and fishing activities in Albanian part of Prespa Lakes. J Int Environ Appl Sci 4(3):253–259

Jarić I, Smederevac-Lalić M, Jovičić K, Jaćimović M, Cvijanović G, Lenhardt M, Kalauzi A (2015) Indicators of unsustainable fishery in the middle Danube. Ecol Freshw Fish 25(1):86–98. https://doi.org/10.1111/eff.12193

Law No. 9103 on the protection of transboundary lakes, http://extwprlegs1.fao.org/docs/pdf/alb60527E.pdf, Law No. 35/2013 amending and supplementing Law No. 9103 on the protection of transboundary lakes (LIGJ Nr. 35/2013 për disa ndryshime dhe shtesa në ligjin nr. 9103, datë 10.7.2003 "për mbrojtjen e liqeneve ndërkufitare

Lenhardt M, Smederevac-Lalic M, Hegedis A, Skoric S, Cvijanovic G, Visnjic-Jeftic Z, Djikanovic V, Jovicic K, Jacimovic M, Jaric I (2020) Human impacts on fish fauna in the Danube River in Serbia: current status and ecological implications. In: Banaduc D et al (eds) Human impact on Danube watershed biodiversity in the XXI century. Springer Nature Switzerland AG, Cham, pp 257–279. https://doi.org/10.1007/978-3-030-37242-2_13

Lopes SPFM, Pacheco S, Clauzet M, Silvano RAM, Begossi A (2015) Fisheries, tourism, and marine protected areas: conflicting or synergistic interactions? Ecosyst Serv 16:333–340. https://doi.org/10.1016/j.ecoser.2014.12.003

Lynch A, Cooke S, Deines A, Bower S, Bunnell D, Cowx I, Nguyen V, Nohner J, Phouthavong K, Riley B, Rogers M, Taylor W, Woelmer W, Youn S, Douglas BT (2016) The social, economic, and environmental importance of inland fish and fisheries. Environ Rev 24:1–7. https://doi.org/10.1139/er-2015-0064

Macfadyen G, Salz P, Cappell R (2011) Characteristics of small-scale coastal fisheries in Europe (study). Policy department structural and cohesion policies. European Parliament, Brussels. https://www.europarl.europa.eu/document/activities/cont/201110/20111019ATT29772/20111019ATT29772EN.pdf

Martinović V (1978) Alati beogradskih alasa, Muzej grada Beograda, katalog izložbi – 19, Beograd

Matić-Skoko S, Stagličić N (2020) How is fisheries management perceived by Croatian small-scale fishers: should I stay or should I go? In: Pascual-Fernández J, Pita C, Bavinck M (eds) Small-scale fisheries in Europe: status, resilience and governance, MARE publication series, vol 23. Springer, Cham. https://doi.org/10.1007/978-3-030-37371-9_8

Matić-Skoko S, Ikica Z, Vrdoljak D, Peharda M, Tutman P, Dragičević B et al (2017) A comparative approach to the Croatian and Montenegrin small-scale fisheries (SSF) in the coastal eastern Adriatic Sea: fishing gears and target species. Acta Adriat 58(3):459–480

Mikuš O, Zrakić M, Kovačićek T, Jež Rogelj M (2018) Common fisheries policy and its impact on the fisheries sector in Croatia. Croat J Fish 76(1):41–50

Ministrstvo za kmetijstvo, gozdarstvo in prehrano Program upravljanja rib v celinskih vodah Slovenije za obdobje do leta 2021

Ministry of Agriculture, Forestry and Food Republic of Slovenia (2022) https://www.gov.si/en/policies/agriculture-forestry-and-food/fisheries/

Ministry of Agriculture Republic of Croatia (2022) Freshwater sports fisheries-angling in Croatia. https://ribarstvo.mps.hr/default.aspx?id=2802

Miret-Pastor L, Peiró-Signes A, Segarra-Oña M, Herrera-Racionero P (2021) The fisheries local action groups (flags) and the opportunity to generate synergies between tourism, fisheries and culture. In: Katsoni V, van Zyl C (eds) Culture and tourism in a smart, globalized, and sustainable world, Springer proceedings in business and economics. Springer, pp 687–694. https://doi.org/10.1007/978-3-030-72469-6_45

MONSTAT The statistical Office of Montenegro (2022) https://www.monstat.org/eng/

OECD (2020) Managing fisheries in OECD review of fisheries 2020. OECD Publishing, Paris. https://doi.org/10.1787/4e30b3e5-en

Official Gazette of Albania (2019) Law on Fisheries (Ligj për Peshkim) 64/2012, 80/2017, 4/2019; Law on Aquaculture (Ligj Për Akuakulturën) 103/2016
Official gazette of Croatia (2017) Marine Fisheries Act (Zakon o morskom ribarstvu) https://narodne-novine.nn.hr/clanci/sluzbeni/2017_06_62_1429.html
Official Gazette of Montenegro (2015) Law on marine fisheries and Mariculture Sl. list Crne Gore 56/09, 40/11, 47/15
Official Gazette of Montenegro (2018) Law on Freshwater Fisheries and Aquaculture Sl. list Crne Gore 17/18 (Zakon o slatkovodnom ribarstvu i akvakulturi)
Official Gazette of Montenegro (2021) Order on closed fishing areas, closed seasons, and minimum sizes of fish and other aquatic organisms below which catch is prohibited (Naredba o ribolovnim zabranima, lovostaju i minimalnim veličinama riba i drugih vodenih organizama ispod kojih je zabranjen ulov) Službeni list Crne Gore 37/20, 9/21
Official Gazette of the Republic of Croatia (2018) Law on Offenses (Prekršajni zakon) Narodne Novine 107/07, 39/13, 157/13, 110/15, 70/17, 118/18
Official Gazette of the Republic of Croatia (2019) Freshwater Fisheries Act (Zakon o slatkovodnom ribarstvu). Narodne Novine 2019 no. 63 (1232). https://narodne-novine.nn.hr/clanci/sluzbeni/2019_06_63_1232.html
Official Gazette of the Republic of Montenegro (2014) Law amending the Law on maritime navigation safety Sl. list Crne Gore 62/2013 and 06/2014
Official Gazette of the Republic of North Macedonia (1993) Law for the protection of Ohrid, Prespa and Dojran Lake, Služben vesnik na Republika Severna Makedonija 45/1977, 08/1980, 51/1988, 10/1990, 62/1993
Official Gazette of the Republic of North Macedonia (2011a) Law for declaration of Lake Prespa as natural monument, Služben vesnik na Republika Severna Makedonija 51/2011
Official Gazette of the Republic of North Macedonia (2011b) Law on Fishery and Aquaculture (LFA) Zakon o ribarstvu i akvakulturi Služben vesnik na Republika Severna Makedonija 7/2008, 67/2010, 47/2011, 53/2011
Official Gazette of the Republic of Serbia (2018a) Law on Protection and Sustainable Use of the Fish Resources (Zakon o zaštiti i održivom korišćenju ribljeg fonda). Službeni glasnik RS 128/2014, 95/2018. https://www.paragraf.rs/propisi/zakon_o_zastiti_i_odrzivom_koriscenju_ribljeg_fonda.html
Official Gazette of the Republic of Serbia (2018b) Law on Nature (Zakon o zaštiti prirode) Službeni glasnik RS, 36/2009, 88/2010, 91/2010, 14/2016, 95/2018. https://www.paragraf.rs/propisi/zakon_o_zastiti_prirode.html
Pascual-Fernandez J, Pita C, Bavinck M (2020) Small-scale fisheries in Europe: status, resilience and governance. Springer, Cham. https://doi.org/10.1007/978-3-030-37371-9
Pešić V, Karaman G, Kostianoy A (2018) Lake Skadar/Shkodra environment, The handbook of environmental chemistry, vol 80. Springer Nature, Cham, p 508
Pešić V, Paunović M, Kostianoy AG, Vukašinović-Pešić V (2020) The Rivers of Montenegro: from conflicts to science-based management. In: Pešić V, Paunović M, Kostianoy A (eds) The rivers of Montenegro. The handbook of environmental chemistry. Springer, Berlin, Heidelberg. https://doi.org/10.1007/698_2020_480
Petrović M (1998) Sabrana dela, Ribarstvo. Zavod za udžbenike i nastavna sredstva. Belgrade
Peveling R, Brämick U, Densky H, Parr B, Pietrock M (2015) Initial characterization of Lakes Prespa. Shkodra/Skadar, Ohrid and
Piria M, Suić J, Špelić I (2022) Inland fisheries in Croatia: historical aspects, fish resources, management and conservation, unpublished
Pita C, Pascual-Fernández JJ, Bavinck M (2020) Small-scale fisheries in Europe: challenges and opportunities. In: Pascual-Fernández J, Pita C, Bavinck M (eds) Small-scale fisheries in Europe: status, resilience and governance, MARE publication series, vol 23. Springer, Cham. https://doi.org/10.1007/978-3-030-37371-9_28
Prince B (2021) Concept environmental and social review summary (ESRS) - Albania blue economy development project - P176163 (English), World Bank Group. United States of

America. Retrieved from https://policycommons.net/artifacts/1817699/concept-environmental-and-social-review-summary-esrs/2555149/ on 28 Oct 2022. CID: 20.500.12592/h7n4sg

Rakowski M, Mytlewski A, Psuty I (2020) Small-scale fisheries in Poland. In: Pascual-Fernández J, Pita C, Bavinck M (eds) Small-scale fisheries in Europe: status, resilience and governance, MARE publication series, vol 23. Springer, Cham. https://doi.org/10.1007/978-3-030-37371-9_24

Rockland DB (1983) An economic analysis of Delaware's recreational/commercial fisheries conflict. University of Delaware

Rousseau Y, Watson RA, Blanchard JL, Fulton EA (2019) Defining global artisanal fisheries. Mar Policy 108:103634. https://doi.org/10.1016/j.marpol.2019.103634

RPOA-SSF (n.d.) Regional Plan of Action for Small-Scale Fisheries in the Mediterranean and the Black Sea. https://www.fao.org/gfcm/activities/fisheries/small-scale-fisheries/rpoa-ssf

RZS (2022) Pretraga diseminacione baze (stat.gov.rs) Statistical Office of the Republic of Serbia https://data.stat.gov.rs/Home/Result/13030202?languageCode=sr-Latn

Smederevac M, Višnjić-Jeftić Ž, Cvijanović G, Lenhardt M, Mićković B, Hegediš A (2006) A survey of fish resources management in rivers Danube, Sava and Tisa in Serbia during the period 1955–2005 Paper presented at the 4th national scientific experts meeting with the international participation Ribarski dani "Osjek 2006", Conference proceedings, 14–23

Smederevac-Lalić M (2013) Socio-ekonomske i biološke karakteristika privrednog ribolova na Dunavu [Doktorska disertacija, Univerzitet u Beogradu]. NaRDuS – National Repository of Dissertations in Serbia. https://nardus.mpn.gov.rs/handle/123456789/2707

Smederevac-Lalić M, Jarić I, Višnjić-Jeftić Ž, Skorić S, Cvijanović G, Gačić Z, Lenhardt M (2011a) Management approaches and aquaculture of sturgeons in the lower Danube region countries. J Appl Ichthyol 27(3):94–100. https://doi.org/10.1111/j.1439 0426.2011.01859.x

Smederevac-Lalić M, Pešić R, Cvejić S, Simonović P (2011b) Socio-economic features of commercial fishery in the bordering upper Danube River area of Serbia. Environ Monit Assess 184(5):2633–2646. https://doi.org/10.1007/s10661-011-2140-5

Smederevac-Lalić M, Višnjić-Jeftić Ž, Pucar M, Mićković B, Skorić S, Nikčević M, Hegediš A (2011c) Fishing circumstances on the Danube in Serbia. Water Research and Management, 4(1): 44-48. Serbian Water Pollution Control Society SWPCS, The Jaroslav Černi Institute for the Development of water resources-JCI, Siniša Stanković Institute for Biological Research-IBISS, The International Association of Water Supply Companies in the Danube River Catchment Area-IAWD

Smederevac-Lalić M, Rajković B, Zarić V (2015) Analysis of the distribution of freshwater fish in Serbia. Paper presented at the 7th International Conference "Water & fish", Faculty of Agriculture, Belgrade-Zemun, Serbia, Jun, 10–12, 2015, pp: 112–113

Soldo A, Fredotović M, Šaran A, Slišković M, Mihanović V, Jelić Mrčelić G (2018) Economic and social impact of marine sport and recreational fisheries in Croatia. Croat J Fish 76:154–163. https://doi.org/10.2478/cjf-2018-0019

Song AM, Johnsen JP, Morrison TH (2018) Reconstructing governability: how fisheries are made governable. Fish Fish 19:377–389. https://doi.org/10.1111/faf.12262

Sonila X, Kosovrasti A (2013) Development of fisheries and aquaculture in Albania. Proceedings international conference on economic and social studies (ICESoS'13), 10–11 May, 2013, Sarajevo

Spirkovski Z, Bojadzieski I (2013) Conservation and sustainable use of biodiversity of lakes Prespa, Ohrid and Skadar fisheries in Lakes Prespa and Ohrid – Macedonian part conference: conser vation and sustainable use of biodiversity of Lakes Prespa, Ohrid and Skadar CSBL Fishery MAFWE, Tirana. https://doi.org/10.13140/2.1.1507.0086

Spirkovski Z, Krstanovski Z, Selfo L, Sanxhaku M, Puka V (2000) The monitoring Programme of the Lake Ohrid conservation project. In: Ganoulis J, Murphy IL, Brilly M (eds) Transboundary water resources in the Balkans, NATO science series, vol 74. Springer, Dordrecht. https://doi.org/10.1007/978-94-011-4367-7_6

Spreizer AJ, Rogelja Caf N (2020) Small-scale fisheries in Slovenia (Northeastern Adriatic): From borders to Projects. In: Pascual-Fernández, J., Pita, C., Bavinck, M. (eds) Small-Scale Fisheries in Europe: Status, Resilience and Governance. MARE Publication Series, vol 23. Springer, Cham. 10.1007/978-3-030-37371-9_9

Šprem N, Piria M, Safner R, Treer T, Aničić I, Tomljanović T (2006) Recreational fisheries on bordering part of Sava and Danube River. Paper presented at the 4th national scientific experts meeting with the international participation Ribarski dani "Osjek 2006", Conference proceedings, 55-58.

Symes D, Phillipson J, Salmi P (2015) Sociologia Ruralis. Europe's coastal fisheries: instability and the impacts of fisheries policy. Special Issue: Resilience and Adaptation of Fishing Communities 55(3):245–257. https://doi.org/10.1111/soru.12096

Talevski T, Milošević D, Maric D, Petrovic D, Talevska M, Talevska L A (2009) Biodiversity of ichthyofauna from Lake Prespa, Lake Ohrid and Lake Skadar. Biotechnol. Biotec. Eq.. 23

Teodorescu D, van den Kommer M (2020) Economic decline, fishing bans, and obstructive politics: Is there a future for small-scale fisheries in Romania's Danube delta? In: Pascual-Fernández J, Pita C, Bavinck M (eds) Small-scale fisheries in Europe: status, resilience and governance, MARE Publication Series, vol 23. Springer, Cham. https://doi.org/10.1007/978-3-030-37371-9_3

The common fisheries policy (CFP) (n.d.) The common fisheries policy: origins and development Fact Sheets on the European Union | European Parliament (europa.eu) https://www.europarl.europa.eu/factsheets/en/sheet/114/the-common-fisheries-policy-origins-and-development

Vehanen T, Piria M, Kubečka J, Skov C, Kelly F, Pokki H, Eskelinen P, Rahikainen M, Keskinen T, Artell J, Romakkaniemi A, Suić J, Adámek Z, Heimlich R, Chalupa P, Ženíšková H, Lyach R, Berg S, Birnie-Gauvin K, Jepsen N, Koed A, Pedersen MI, Rasmussen G, Gargan P, Roche W, Arlinghaus R (2020) Data collection systems and methodologies for the inland fisheries of Europe. FAO Fisheries and Aquaculture Technical Paper No. 649. FAO, Budapest. https://doi.org/10.4060/ca7993en

Visser TAM, Valbo-Jorgensen J, Chomo V (2021) Good practices guidelines for data collection systems to support sustainable inland and recreational fisheries in the Western Balkans region. Fisheries and aquaculture circular No. 1218. Budapest. https://doi.org/10.4060/cb3261en

Water Framework Directive 2000/60/EC. https://www.eea.europa.eu/policy-documents/water-framework-directive-wfd-2000

Weiss S, Apostolou A, Đug S, Marčić Z, Mušović M, Oikonomou A, Shumka S, Škrijelj R, Simonović P, Vesnić A, Zabric D (2018) Endangered fish species in Balkan Rivers: their distributions and threats from hydropower development. Riverwatch & EuroNatur, 162 pp.

White O, Sadauskis R, Sheate W, Papadopoulou L (2018) Water use in the Western Balkans: regional outlooks and global megatrends. In: Künitzer A (ed) ETC/ICM technical report 2/2018. European Topic Centre on Inland, Coastal and Marine Waters, Magdeburg. 130 pp

Williams R (2009) Changing constructions of identity: fisher households and industry restructuring. Unpublished PhD thesis, Newcastle University, UK

Williams R (2014) The socio-cultural impact of industry restructuring: fishing identities in north East Scotland. In: Urquhart J, Acott TG, Symes D et al (eds) Social issues in sustainable fisheries management, MARE publication series, vol 9. Springer, New York, London, pp 301–317

Živaljević I (2017) Ribolov na Đerdapu u ranom holocenu (10. - 6. milenijum pre n. e.). Univerzitet u Beogradu

Fish Resources of the Western Balkans: Sustainable Use and/or Conservation?

Vladica Simić, Snežana Simić, and Vladimir Pešić

Abstract This concluding chapter contains a comparative overview of the state of fish resources, fisheries, and management models in the inland waters of the Western Balkans. The aim was to find a possible sustainable solution for freshwater fisheries under increasingly diverse anthropogenic stressors on the environment and ecosystems, as well as ongoing climate changes. Likely, a single model approach is not possible because of the heterogeneity of ecological and social conditions, even in such a relatively small area. However, the ecosystem approach could be a common basis for the integral management of water and water resources. This approach must be strictly scientifically and socially acceptable to balance sustainable use and conservation under different conditions.

Keywords Western Balkan · Fish stock · Sustainable use · Conservation

1 Contradictions: Significant Fish Resources: Relations between Social Community, Sustainable Use, and/or Conservation

The Western Balkans, as geographically and politically described in this book, comprises most of the Balkan Peninsula. This part of the European continent is ecologically, socially, culturally, and historically heterogeneous. Because of its great landscape and ecosystem diversity and, as important for the content of this book, the largest number of preserved (intact) aquatic ecosystems and specific biodiversity, it exceeds the importance for the European continent (Pešić et al. 2022) and has

V. Simić (✉) · S. Simić
Department of Biology and Ecology, Faculty of Science, University of Kragujevac, Kragujevac, Serbia
e-mail: vladica.simic@pmf.kg.ac.rs

V. Pešić
Department of Biology, Faculty of Sciences, University of Montenegro, Podgorica, Montenegro

V. Simić et al. (eds.), *Ecological Sustainability of Fish Resources of Inland Waters of the Western Balkans*, Fish & Fisheries Series 43,
https://doi.org/10.1007/978-3-031-36926-1_19

importance for the global biodiversity of the planet Earth. From the perspective of global biodiversity, the presence of a large number of endemic species in the aquatic habitats of the Adriatic and Aegean basins has contributed to the inclusion of the area in the Mediterranean "hot spot" area, one of the 25 significant points (as many as were originally recognized) for the global biodiversity of planet Earth (Myers 1990, 2003; Myers et al. 2000). Most endemic fish species are uninteresting for fishing due to their size, especially in the Adriatic karst area of Croatia (Dalmatia) and in a part of Bosnia and Herzegovina (Mrakovčić et al. 2006). However, the aquatic ecosystems from the Neretva River and further south through the large lakes Skadar, Ohrid, and Prespa to the marshy lagoons on the coast of Albania are habitats for endemic but significant fisheries species such as trout (soft mouth trout, trouts of Lake Ohrid and Skadar, marble trout), species of the genus *Chondrostoma*, *Scardinius*, *Alburnus*, and others. Overfishing in this part of the western Balkan, along with all other unfavorable factors, was one of the main reasons for the significant decline in endemic trout populations, the disappearance of endemic species of the genus *Chondrostoma* from Lake Skadar, and the significant reduction in European eel populations (Marić 2018).

1.1 The Danube Basin, State of the Fish Stock, Problems, Perspective

Most Western Balkans waters belong to the Black Sea basin, of which the Danube, one of the major European and world rivers, drains almost 70% into the Black Sea. The Danube River Basin is quantitatively and qualitatively the most important fishing area in the Western Balkans (Simić et al. 2014a; b). However, the basin has suffered and continues to suffer from significant changes in the fish communities and populations (Jarić et al. 2014; Simić et al. 2014a; b).

The huchen (*Hucho hucho*) (the "king fish" as it is called in Chap. 1, Pliberšek and Tavčar (2023) and described in detail in Chap. 16, Jakšić et al. 2023), an endemic fish of the Danube basin, has a vital population in the rivers belonging to the Sava basin as the largest tributaries of the Danube in the area of the Western Balkans. In contrast to its great ecological importance as a top predator, the pressure on this fish species has been great in the last 15 years in almost all of its habitats (Freyhof et al. 2015; see Chap. 16, Jakšić et al. 2023). Currently, there are no agreements, treaties, plans, or other measures between the Western Balkans countries, which would prevent the further decline of the stocks of these important fish species (see Chap. 16, Jakšić et al. 2023). Apart from huchen as the largest salmonid species, the Danube basin has been subjected to significant pressure on brown trout (*Salmo trutta*) populations (Danubian mitochondrial lineage) in the last 15 years, not so much due to fisheries but primarily due to fragmentation (partitions for small hydro-power plants), habitat loss, and hybridization (see Chap. 1, Pliberšek and Tavčar 2023; Chap. 3, Dekić et al. 2023; Chap. 4, Simić et al. 2023; Chap. 9,

Fig. 1 Part of the course of the Danube through Belgrade. The fish stock suffers from heavy pressure from recreational and commercial fishermen (Photo by V. Simić)

Schoffmann and Marić 2023). Additionally, the Danube Basin suffered the greatest change in ecology and fisheries due to the disappearance of populations of migratory species from the Acipenseridae family upstream of the Djerdapa area (see Chap. 2, Piria et al. 2023; Chap. 4, Simić et al. 2023; Chap. 10, Cvijanović et al. 2023). The Sterlet sturgeon, as the only stationary species, also suffered a sharp decline in populations, so much so that a moratorium was introduced on the territory of Serbia, and the species was placed on the list of strictly protected species (Chap. 4, Simić et al. 2023; Chap. 10, Cvijanović et al. 2023).

Fisheries in the Danube basin are primarily commercial and are now limited in quantity to catching less valuable species such as bream and non-native species such as Prussian carp, grass carp, silver carp, and bighead carp. Stocks of fishery and economically valuable species such as carp, wels catfish, pikeperch, and northern pike have generally declined compared to the period before the 1980s (see Chap. 2, Piria et al. 2023; Chap. 4, Simić et al. 2023) (Figs. 1 and 2). The extent of the decline, as well as the presence of sustainable populations, varies from country to country. For example, Croatia pays considerable attention to relatively well-preserved carp populations (see Chap. 2, Piria et al. 2023), while in Serbia, there is considerable sustainability of wels catfish and pikeperch populations in both the Morava and Drina basin reservoirs and the lower Danube (see Chap. 4, Simić et al. 2023).

Fig. 2 The catch of commercial fishermen on the part of the Danube near Belgrade, dominated by "white fish" such as *Abramis brama*, *Blicca bjoerkna*, and *Vimba vimba*, which are lower quality and economically less valuable (Photo by V. Simić)

2 Historical Effects on the Fisheries of the Western Balkans in the Past and the Impact on the Current State of the Fish Stock

The present state of fish stocks in the Western Balkans region is largely the result of past and present human activities. The destruction of fish stocks is also a consequence of various historical and socio-political circumstances in the area. The consequences of various past foreign influences are still visible today. Namely, the western parts of the area, represented today by the countries of Slovenia and Croatia, including a significant part of Vojvodina (today an autonomous province of the state of Serbia), were under the influence of the Austro-Hungarian monarchy and today, except for Vojvodina, are members of the EU. Today, compared to other areas of the Western Balkans, these areas do not possess more preserved fish stocks, however, the centuries-old greater social care, organization, and control of fish stocks are much more present and visible in these regions, starting with legislation, the presence of science and the implementation of control measures, and the current state of fish stocks (see Chap. 1, Pliberšek and Tavčar 2023; Chap. 2, Piria et al. 2023). Other Western Balkans countries were under the rule of the Ottoman Empire, and the exploitation of fish stocks was mainly done without control and legal regulation management of this biological resource. Fish, along with the great rivers and lakes, was an important food for the population (even though, in general, the

Fig. 3 Lake Skadar in Montenegro was an important fishing water for the local population in the past. The importance of fishing is still great today, but the fish stocks have significantly decreased (Photo by V. Simić)

total consumption was low) and available to all (see Chap. 4, Simić et al. 2023). However, the most valuable resources belonged to the Ottomans.

However, in the absence of quantitative data on the amount of fish caught over this long period of time, the impact of ubiquitous unlimited fishing on the current state of the fish stock cannot be fully understood. From the tools used, it can be concluded that large and high-quality fish were caught in a targeted and selective manner and that the population caught a large number of fish mainly for family or community consumption (e.g., the municipalities on the Lake Skadar, Ceklinska, Crmnica, and Krajina) (see Chap. 5, Mrdak and Milošević 2023) (Fig. 3).

The post-war period, characterized by the communist regime throughout the Western Balkans, is characterized by a managed planned economy in which fishing is a part of agriculture, with the main objective of exploitation and maximum yield both in the fishing waters and a strong incentive for fish production in aquaculture (see Chap. 4, Simić et al. 2023; Chap. 15, Marković and Poleksić 2023). This period is characterized by a clear favoring of commercial fishing, especially in the traditional fishing waters such as the major rivers Sava, Danube, Tisza, and Drina, as well as the Skadar, Ohrid, Prespa, and Dojran Lakes (Fig. 4a, b), but also by the production of significant quantities of fish for stocking and aquaculture.

Fishing had a clear legal and statutory basis throughout the area, but the catch quotas of a large number of commercial fishermen were high and based on the

Fig. 4 (**a**) Lake Ohrid. The fish stocks of the lake were significantly depleted by overfishing in the mid–twentieth century (Photo by V. Simić). (**b**). Lake Ohrid and fishing boats on the shore (Photo by V. Simić)

Fig. 4 (continued)

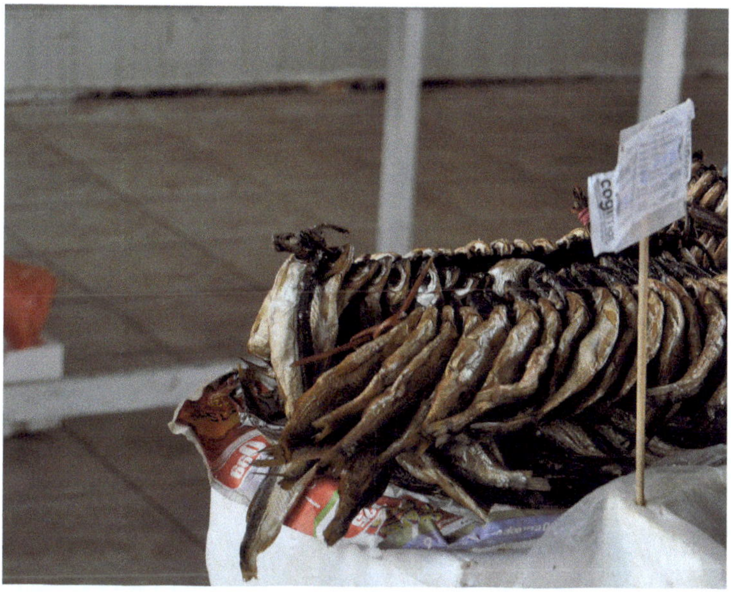

Fig. 5 Fish market in Bar, Montenegro. *Alburnus scoranza* prepared in the traditional way, salted and dried in the sun (the popular name for this product is "Saraga"), and brought out for sale (Photo by V. Simić)

conventional principle of maximum sustainable catch and minimum catch length (see Chap. 4, Simić et al. 2023; Chap. 6 Talevski et al. 2023). This period also saw the development of industries related to freshwater fishing, such as the opening of the fish processing factory "Ribarstvo" in Rijeka Crnojevića on Lake Skadar, which was in operation until 2013 (Figs. 5 and 6). Also, the fish farm "Đerdap" in Kladovo (Serbia) aimed to produce sturgeon for stocking the upper Danube, but also for producing highly prized caviar (see Chap. 4, Simić et al. 2023; Chap. 5, Mrdak and Milošević 2023). In addition to fish processing, companies and handicraft enterprises were also opened, which obtain silver coatings for pearls (Lake Skadar/Shkodra) or the production of whole pearls (Lake Ohrid) by processing the fish scales. Fish production in carp ponds in Serbia, Bosnia, and Croatia was at its peak in this period, and together with rainbow trout ponds, they made the former country of Yugoslavia an exporter of freshwater fish (see Chap. 2, Piria et al. 2023; Chap. 5, Mrdak and Milošević 2023; Chap. 6, Talevski et al. 2023). The collapse of this state and the war and economic crisis in the 1990s further led to the collapse of controlled fisheries in most Western Balkans (see Chap. 3, Dekić et al. 2023; Chap. 4, Simić et al. 2023; Chap. 6, Talevski et al. 2023). Fish from wild waters became an important food for the population, illegal fishing increased rapidly, and a large number of fish farms were closed or privatized. The most serious consequence of this crisis, in addition to the impoverishment of fish stocks, is the social marginalization of fish resources and fishing among the least important state resources, especially in the countries that now form the political framework of the Western

Fig. 6 A canned smoked *Alburnus scoranza* from Lake Skadar. One of the products of the no longer functioning fish processing factory "Ribarstvo" (Rijeka Crnojevića, Montenegro) (Photo by V. Simić).

Balkans (Bosnia and Herzegovina, Serbia, Montenegro, Macedonia, and Albania). In other words, fisheries have become a social category, and the effects of this treatment are still felt today.

Currently, fish resources and fisheries in the Western Balkans follow more or less the same global trends as in the world and Europe. In addition to the decline in the quantity and quality of fish stocks, there is also a significant decline in commercial fishing, which is completely absent in Slovenia, insignificant in Croatia, and has decreased by at least 50% in Serbia, Montenegro, and Northern Macedonia (see Chap. 1, Pliberšek and Tavčar 2023; Chap. 2, Piria et al. 2023; Chap. , Dekić et al. 2023; Chap. , Simić et al. 2023; Chap. 5, Mrdak and Milošević 2023; Chap. 6, Talevski et al. 2023; Chap. 7, Shumka et al. 2023). Only in Albania, commercial fishing is still almost the predominant form of fishing (see Chap. 7, Shumka et al. 2023). Sport and recreational fishing, on the other hand, is developing more and more in the opposite direction compared to commercial fishing. It is present only in Slovenia, growing in Croatia, Serbia, and partly in Bosnia, moderate in Montenegro and Northern Macedonia, and hardly developed in Albania (see Chap. 1, Pliberšek and Tavčar 2023; Chap. 2, Piria et al. 2023; Chap. 4, Simić et al. 2023; Chap. 5, Mrdak and Milošević 2023; Chap. 6, Talevski et al. 2023; Chap. 7, Shumka et al. 2023). Although poaching is illegal, it is a form of fishing pressure and is most prevalent in Albania, Montenegro, North Macedonia, and Bosnia; it is moderately present with a decreasing tendency in Serbia, and it is weakly present in Croatia and especially in Slovenia (see Chap. 1, Pliberšek and Tavčar 2023; Chap. 2, Piria et al. 2023; Chap. 3, Dekić et al. 2023; Chap. 4, Simić et al. 2023; Chap. 5, Mrdak and Milošević 2023; Chap. 6, Talevski et al. 2023; Chap. , Shumka et al. 2023).

2.1 *Legislation and Education Are Key to Choosing the Optimal Fish Resource Management Model*

The EU does not have a common strategy for the management of fish stocks in inland waters, so the management of fish stocks in the Western Balkan countries also differs from country to country (see Chap. 4, Simić et al. 2023; Chap. 18, Smederevac-Lalić et al. 2023). Nevertheless, a common feature in management is state ownership of inland waters, and the state delegates manage various services or organizations. In most countries, fisheries are the responsibility of the Ministry of Agriculture (see Chap. 1, Pliberšek and Tavčar 2023; Chap. 2, Piria et al. 2023; Chap. 5, Mrdak and Milošević 2023; Chap. 6, Talevski et al. 2023). In Serbia, the responsibilities are divided; The Ministry of Environmental Protection is responsible for fishing in wild waters while the. Ministry of Agriculture, Forestry and Water Management is responsible for aquaculture (see Chap. 4, Simić et al. 2023; Chap. 18, Smederevac-Lalić et al. 2023). There are also special state organizations responsible for fisheries, as in Albania (see Chap. 7, Shumka et al. 2023). A very complicated and largely dysfunctional system exists in Bosnia and Herzegovina, where each entity has its own state agency for fisheries. However, there is almost no mutual coordination between them, which at the end of the chain is transferred to the overall conditions of the fish stock (see Chap. 3, Dekić et al. 2023).

The structure of direct managers of fishing waters and fish stocks also varies. The predominant model includes the administrators of fisheries associations (which are non-governmental organizations in the legal sense) who are members of the Fishermen's Association as an umbrella organization. Almost all countries except Serbia have this model with more or fewer modifications (see Chap.4, Simić et al. 2023; Chap. 18, Smederevac-Lalić et al. 2023). At the beginning of this century, Serbia left such a system and moved to a centralized system, where the Ministry of Environmental Protection is at the head and managers can only be legal entities (i.e., state companies, and private companies or associations) but only if they are registered as legal entities, i.e. companies. In this way, the state receives revenue from taxes and fees on the gross income of the manager in the safest way. As legal entities, managers in Serbia organize commercial and recreational fishing in their territory. In contrast to this centralized system of management, in Slovenia and Croatia, decentralization takes place, and management passes to subordinate authorities (municipalities, civil society (see Chap. 2, Piria et al. 2023)). In the other states, Albania, Northern Macedonia, and Montenegro, the state selects concessionaires to operate commercial fishing in large natural lakes through a tender process (Skadar, Ohrid, Prespa, and Dojran lakes) and in the coastal lagoons of Albania (see Chap. 5, Mrdak and Milošević 2023; Chap. 6, Talevski et al. 2023; Chap. 7, Shumka et al. 2023). However, illegal fishing is still a major problem in these countries, and almost 50% of the concessionaires do not have a license to practice fishing (e.g., in Prespa Lake) (see Chap. 6, Talevski et al. 2023).

From the contents of this and other books (Crivelli and Catsadorakis 1997; Griffiths et al. 2004; Mrakovčić et al. 2006; Jelić et al. 2008; Simonović et al. 2010, 2015; Craig 2015; Bianco and Ketmaier 2015; Milačić et al. 2015; Simić et al. 2015; Marić 2018; Pešić et al. 2018, 2020, 2022; Visser et al. 2021, Stojković Piperac et al. 2022) it is evident that when dealing with the problem of fishing in inland waters, one cannot accept the prevailing opinion that this activity is less important than marine fishing. Especially since the demand for freshwater fish is increasing, considering its effect on human health, especially as a source of protein and healthy fatty acids (omega-3 polyunsaturated fatty acids), but also due to the relatively lower prices compared to marine fish (Khalili Tilami et al. 2018). In addition to its value for nutrition and recreation, which is becoming increasingly important, fishing is a particularly socially and economically important activity. It has a relaxing effect and thus increases the nation's general health, bringing people back to nature and creating a sense of importance for water, fish, and nature conservation. However, in the Western Balkans region, there is a problem with the profile of recreational anglers, their level of education, attitude, and knowledge about fish, fish resources, and fishing in general. In this educational process, Slovenia and Croatia, which are part of the EU, have the most success, where, among other things, the angler must have passed the fishing exam (see Chap. 3, Piria et al. 2023). On the other hand, the punishment and the legal enforcement of the pronounced penalty are more effective than in other Western Balkans countries. Therefore, in the coming period, other countries will have to take significant legal and practical measures to get a trained fisherman who confidently respects the Code of Measures for the protection and sustainable use of fish stocks (see Chap. 14, Simić et al. 2023).

In the entire process of improving the status of freshwater fisheries in the Western Balkans, the role of state governments is a first but crucial step. It is necessary for freshwater fishing to increase its social prestige and importance for the overall development and improvement of the social community. In the modern quest for rapid economic development, especially in the countries located in the political borders of the Western Balkans (Bosnia and Herzegovina, Serbia, Montenegro, Northern Macedonia, and Albania), the efforts of these countries to increase the importance of freshwater fishing must be greater and freshwater fishing must not be neglected at the expense of other economic sectors (industry, agriculture, etc.). Experiences from other countries can be used, and one of the models is the economic valuation of biological resources of inland waters (biodiversity, fish stocks, etc.). Fish resources have their own value, which is an important component of the overall ecological services of inland waters. The ecological services of the inland waters of the Western Balkans are of great importance worldwide (Bianco and Ketmaier 2015) and they depend only on the governments of the countries whether they will be recognized and valued in the right way or whether they will prevail against the further process of marginalization and the general trend of reduction of fish stocks.

3 Ecosystem Approach: A Common Basis for the Integral Management of Water and Water Resources

The involvement of science in the process of general conservation and sustainable use of fish stocks is crucial in the age of the Anthropocene, with a whole new complex of factors that it brings (Bănăduc et al. 2022). Two directions of science actions are important, the first being the development of methods for the best possible evaluation of fish stocks and the establishment of regular monitoring. The second is improving the data collection system for actual catches and evaluating fishing pressure by commercial and especially recreational fishermen (Simić et al. 2014a, b, 2023; Jarić et al. 2009, 2010; see Chap. 14, Simić et al. 2023). In addition to these current trends, in the coming period, scientists must propose strategic models for the conservation and management of fish stocks in inland waters that are consistent with all modern influences on aquatic ecosystems and fish stocks, as well as with modern trends in the social and economic development of the communities (Simić et al. 2023; see Chap. 4, Simić et al. 2023).

The ecosystem approach is the direction in which fisheries science in these areas should adapt to regional and local conditions (Craig 2015; Simonović et al. 2018; Jarić et al. 2014). It is evident that commercial fishing can reduce fish stocks to ecologically unsustainable levels sooner or later (examples from this monograph include the Danube, Sava, and the large natural lakes Skadar, Ohrid, Prespa, and Dojran (see Chap. 2 Piria et al. 2023; Chap. 4, Simić et al. 2023; Chap. 5, Mrdak and Milošević 2023; Chap. 6, Talevski et al. 2023). Moreover, the impact of increasing recreational fishing should not be ignored, which, if not based on scientific assessments, can also reduce stocks, especially for fishery-attractive fish species that often have great ecological importance for the fish communities. An ecological approach based on a balanced harvest according to the demographic structure of the fish community or attractive fishery species, with the introduction of an optimal sustainable harvest instead of a maximum sustainable harvest and a maximum catch length instead of a minimum length, can be a strategic direction for commercial fisheries in the traditional fishing waters of the Western Balkans (Danube, Sava, large natural lakes Skadar, Ohrid, Prespa, Dojran) (Garcia et al. 2012; see Chap. 4, Simić et al. 2023). Another and more likely option, given the state of fish stocks, is to reduce the scale of commercial fishing further and gradually adjust to minimum quotas that can only meet the needs of angling tourism, while strongly encouraging the use of traditional fishing methods and gear (see Chap. 2, Piria et al. 2023; Chap. 4, Simić et al. 2023). The wide variety of traditional fishing methods and gear in the area offers great opportunities for the development of fishing tourism and ecotourism (from special gear for small waters to large gear on the Danube and fishing with birds at Lake Dojran (see Chap. 2, Piria et al. 2023; Chap. 4, Simić et al. 2023; Chap. 6, Talevski et al. 2023).

Recreational fishing is becoming increasingly important in most Western Balkan countries (Slovenia, Croatia, Bosnia and Herzegovina, and Serbia) and represents the future of fishing in inland waters (see Chap. 1, Pliberšek and Tavčar 2023;

Chap. 2, Piria et al. 2023; Chap. 3, Dekić et al. 2023; Chap. 18, Smederevac–Lalić et al. 2023). Here, education of recreational fishermen (emphasized throughout this monograph) is the basis for confident acceptance and practical implementation of appropriate fishing models adopted based on scientific assessment of the state of fish stocks (see Chap. 14, Simić et al. 2023). Self-discipline in compliance with prescribed fishing rules and quotas by recreational anglers is the basis of sustainable inland fisheries. On the other hand, the work of managers must also be reorganized, primarily in two directions, a significant increase and improvement in the work of the fisheries control service, with full acceptance of the fact that organized and dedicated protection of fishing waters is the best way to preserve the fisheries than any other measures. Improving the work of fisheries managers requires a change in thinking about stocking as a traditionally necessary measure to improve the condition of the fish stock. Scientific research shows that stocking can have a number of adverse effects on native fish populations and fish communities (Simić et al. 2023, Marić et al. 2010; see Chap. 1, Pliberšek and Tavčar 2023; Chap. 2, Piria et al. 2023; Chap. 4, Simić et al. 2023; Chap. 9, Schöffmann and Marić 2023). Therefore, the prevailing opinion is that stocking is a last measure, implemented only when all previous measures have failed to produce positive results. Any necessary stocking measures must be carried out with scientific support and according to scientific principles that aim to preserve native gene pools and prevent the decline of genetic diversity and adaptability of native populations (see Chap. 1, Pliberšek and Tavčar 2023; Chap. 2, Piria et al. 2023; Chap. 4, Simić et al. 2023; Chap. 9, Schöffmann and Marić 2023).

4 The Main Threats to the Fish Stock in the Western Balkans Area

Inland water ecosystems and fish resources, as a particularly important biological component for human society, are under the same pressures in the Western Balkans as in Europe and worldwide. In their intensity, they are characterized by habitat fragmentation and other changes (construction of dams for small and large hydropower plants, regulation of river courses, melioration and loss of floodplain, exploitation of gravel and sand), which is a current threat to the whole area (Freyhof et al. 2015; see Chap. 1, Pliberšek and Tavčar 2023; Chap. 2, Piria et al. 2023; Chap. 3, Dekić et al. 2023; Chap. 4, Simić et al. 2023; Chap. 5, Mrdak and Milošević 2023; Chap. 6, Talevski et al. 2023; Chap. 7, Shumka et al. 2023). Non-native and invasive species are ubiquitous and undesirable, but their spread and presence in the rivers and lakes of the Adriatic and Aegean basins, rich in endemic fish species and other aquatic organisms, is ecologically particularly dangerous and unacceptable (see Chap. 1, Pliberšek and Tavčar 2023; Chap. 5, Mrdak and Milošević 2023; Chap. 6, Talevski et al. 2023; Chap. 9, Schöffmann and Marić 2023; Chap. 13, Radenković et al. 2023). Pollution associated with eutrophication is particularly pronounced in large cities, industrial zones, and zones of intensive agriculture (see Chap. 11, Simić et al. 2023; Chap. 12, Milošković and Kojadinović 2023). Overfishing occurs in almost all waters where commercial fishing occurs (Danube, Lakes Skadar, Ohrid, Prespa, and Dojran, coastal lagoons in Albania). Additionally, poaching is more intense in Bosnia and Herzegovina, Montenegro, northern Macedonia, and Albania (see Chap. 3, Dekić et al. 2023; Chap. 5, Mrdak and Milošević 2023; Chap. 6, Talevski et al. 2023; Chap. 7, Shumka et al. 2023). The intensity of all the above pressures is increasing significantly due to climate change, which significantly affects the qualitative and quantitative characteristics of the fish stock.

Such intense and diverse pressures on the inland water ecosystems of the Western Balkans require vigorous and science-based sustainable use and conservation measures. The sustainable use and conservation measures must be complementary and focused on maintaining aquatic habitats, biodiversity, and fish stocks, or more generally, on maintaining the structure and functionality of inland aquatic ecosystems. All countries of the Western Balkans, both EU members and non-EU members, more or less follow the guidelines and objectives of the EU Water Framework Directive (WFD) (WFD (2000)). However, the actual results of the basic guideline as "at least good ecological status of waters" are different. In relation to the area of the national territory, Serbia has the highest number of water bodies that do not meet the criteria of good ecological status. In other countries, it is a matter of individual catchment areas, such as the catchment area of the Bosna River in Bosnia and Herzegovina and the catchment areas of some rivers in Albania (see Chap. 3, Dekić et al. 2023; Chap. 7, Shumka et al. 2023; Chap. 11, Simić et al. 2023; Chap. 12, Milošković and Kojadinović 2023; Chap. 14, Simić et al. 2023).

One of the perceived deficiencies in the protection of inland waters throughout the Western Balkans are freshwater ecological reserves (see Chap. 1, Pliberšek and Tavčar 2023; Chap. 2, Piria et al. 2023; Chap. 4, Simić et al. 2023). Although, at first glance, there are various forms of protected areas throughout the region, in most cases, the essential function for the actual conservation of biodiversity and fish stocks has not been realized. Much progress has been made with the NATURA 2000 and EMERALD network programs, but these programs are implemented relatively slowly and they vary from country to country. The essence of freshwater ecological reserves is to recover fish stocks from overfishing, and they would be positioned primarily on exposed fishing waters. The expanded role of these reserves refers to portions or entire habitats of rivers or lakes where a healthy (sustainable level of stability and vitality) native fish community or populations of specific species are maintained (Simić et al. 2023; see Chap. 4, Simić et al. 2023).

The observed problem in protected areas is precisely the absence of this essential role of freshwater ecological reserves because the protection refers to the general area and more to terrestrial ecosystems, while the community in the aquatic habitat itself is left out or not appreciated in its proper importance. There are some similar forms of protection throughout the Western Balkans (Slovenia—fisheries protected areas, Croatia—two ichthyological reserves, protected areas Lonjsko polje and Kopački rit as natural fish spawning grounds, Serbia—special fish habitats, Special Nature Reserve "Zasavica" habitat of *Umbra krameri*, and protected peatlands in

Fig. 7 The special nature reserve "Zasavica" (Vojvodina, Srem) is the habitat of the protected species *Umbra krameri*, but also of preserved autochthonous populations of carp, pike, and true loach (Photo by V. Simić)

Vojvodina as natural fish spawning grounds, Bosnia and Herzegovina—Nacional Park "Una" and Nature park "Hutovo Blato") (see Chap. 2, Piria et al. 2023; Chap. 3, Dekić et al. 2023; Chap. 16, Marković and Poleksić 2023; Chap. 17, Galambos and Sekulić 2023) (Fig. 7). However, in general, there are almost no reserves specifically and functionally designed to protect fish habitats, fish stock, and the entire aquatic community, similar to such habitats for birds (The Important Bird Areas (IBAs)) or plants (Important Plant Areas (IPAs)) (see Chap. 4 Simić et al. 2023).

From the perspective of fisheries, freshwater ecological reserves would significantly impact fish stock recovery because they would represent protected habitats from which the "spillover" effect would restore fish populations beyond their boundaries (Simić et al. 2023). Moreover, there is no justification that such reserves were not formed for habitats with fish communities of international importance, such as the centers of endemic fish species (rivers and lakes in Adriatic and Aegean basins from Slovenia to Albania), huchen habitats in the Sava basin, or European eel in the Adriatic and Aegean basins. There are also a few centers that promote fishes and other aquatic organisms of inland waters, which can have important support and function in *ex situ* conservation programs. The freshwater aquariums in Kragujevac, Banja Luka, and the largest and best-equipped aquarium in Karlovac (Croatia) are the most important in the whole region (see Chap. 3, Dekić et al. 2023; Chap. 4, Simić et al. 2023; Chap. 16, Jakšić et al. 2023). Apart from the freshwater aquariums, there are only a few specialized centers that popularize the diversity of inland waters, such as in Serbia the Center for Fisheries and Applied Hydrobiology "Mali Dunav" in Radmilovac near Belgrade (see Chap. 15, Marković and Poleksić 2023) and the

Fig. 8 The Ecological Information Center "Jezero Gruža" (EIC) promotes the natural values of inland waters under the slogan "Life in the water" with the aim of more efficient conservation (Photo by V. Simić)

private ecological information center "Life in the Water" on the shore of the "Gruža" reservoir (Knić) (Fig. 8). In addition to specialized centers, inland waters biodiversity is also presented in natural museums such as the National Museum in Sarajevo, Natural History Museum in Belgrade, the Museum of Herzegovina in Trebinje (Bosnia and Herzegovina), the Natural History Museum in Struga (North Macedonia) and smaller exhibitions in the visitor centers of National Parks (NP Djerdap, Serbia) and others. On the other hand, it is incomprehensible that there are no large and significant centers (or there are only a few of them, e.g. Natural History Museum in Struga) for education and presentation of the natural values of these aquatic habitats. Moreover, there should be stricter regulations for in situ conservation on internationally important inland water bodies such as the Danube and the endemism centers of Lakes Skadar, Prespa, Ohrid, and Dojran.

Aquaculture is gaining importance throughout the region, especially in Bosnia and Herzegovina (see Chap. 3, Dekić et al. 2023), as its relation with fisheries can strengthen conservation if it does not use too many resources from wild waters. Aquaculture centers or reproduction centers dealing with the conservation of native genes of fish species are desirable, and their work must be based on strictly scientific principles (such as centers in Slovenia, ones in Bosnia and Herzegovina on the rivers Una and Neretva, reproduction centers for salmonid species in Serbia such as "Braduljica" and "Tolišnica", "Breznica" near Pljevlja in Montenegro, and North Macedonia hatchery within the Hydrobiological Institute on the shore of Lake Ohrid) (see Chap. 1, Pliberšek and Tavčar 2023; Chap. 3, Dekić et al. 2023;

Chap. 4, Simić et al. 2023; Chap. 5 Mrdak and Milošević 2023; Chap. 6, Talevski et al. 2023; Chap. 15, Marković and Poleksić 2023).

5 Future Perspectives

The future of inland fisheries in the Western Balkans lies mainly in the development of recreational fishing and the gradual reduction of commercial fishing to traditional methods and tools for angling tourism. The balance between the sustainable use of fisheries waters and conservation must be determined by science depending on each part of the Western Balkans' natural and social characteristics and peculiarities. Conservation priority is necessary for waters with endemic species, especially those with endemic but commercially important species, and for waters with endangered and potentially endangered commercially important fish species (such as Sterlet sturgeon and others from the family Acipenseridae, huchen, European eel, and brown trout). The extent and nature of conservation measures must be scientifically and separately determined for each aquatic habitat. Also, integration and harmonization of conservation measures with the NATURA 2000 and the EMERALD network are essential. Science should also play a dominant role in setting fishing activities, from the management model, education program, stock assessment of fish resources, and catch statistics to setting fishing regulations and quotas under the ecological potential of the fishing waters (ecosystem approach).

The inland waters of the Western Balkans still have the potential for fishing activities, with recreational fishing, even though a significant part of the waters needs conservation, so in the present time, a generally acceptable approach may be "sustainable use with an emphasis on conservation, especially for endemic trouts, huchen, sturgeons, and European eel."

References

Bănăduc D, Simić V, Cianfaglione K, Barinova S, Afanasyev S, Öktener A, McCall G, Simić S, Bănăduc Curtean A (2022) Freshwater as a sustainable resource and generator of secondary resources in the 21st century: stressors, threats, risks, management and protection strategies, and conservation approaches. Int J Environ Res Public Health 19(24):16570. https://doi.org/10.3390/ijerph192416570

Bianco GP, Ketmaier V (2015) Nature and status of freshwater and estuarine fisheries in Italy and Western Balkans. In: Craig J (ed) Freshwater fisheries ecology. Wiley Blackwell, pp 282–292. https://doi.org/10.1002/9781118394380

Craig J (2015) Freshwater fisheries ecology. Introduction. In: Graig J (ed) Freshwater fisheries ecology. Wiley Blackwell. 887 p

Crivelli A, Catsadorakis G (1997) (eds) Lake Prespa. A Unique Balkan Wetland. Springer Science + Business Media, Northwestern Greece, p 351

Cvijanović G, Đikanović V, Galambos L, Pengal P, Smederevac-Lalić M (2023) Chap. 10: Past and future of Sturgeon species (Acipenseridae) in Western Balkans – case for permanent conservation or sustainable management. In: Simić V, Simic S, Pešić V (eds) Ecological sustainability of fish resources of inland waters of the Western Balkans in the Anthropocene. Springer, Cham

Dekić R, Golub D, Kalamujić Stroil B, Bakrač A, Avdić S, Kanlić V, Škrijelj R, Adrović A (2023) Chap. 3: Fish resources and fisheries in Bosnia and Herzegovina – overview, sustainability, and conservation. In: Simić V, Simic S, Pešić V (eds) Ecological sustainability of fish resources of inland waters of the Western Balkans in the Anthropocene. Springer, Cham

Freyhof J, Weiss S, Adrović A, Ćaleta M, Duplić A, Hrašovec B, Kalamujić B, Marčić Z, Milošević D, Mrakovčić M, Mrdak D (2015) The huchen *Hucho hucho* in the Balkan region: distribution and future impacts by hydropower development. RiverWatch & EuroNatur 30

Galambos L, Sekulić N (2023) Chap. 17: Future perspective of sustainable development of freshwater fisheries and conservation of threatened fish species, crustaceans and molluscs. In: Simić V, Simic S, Pešić V (eds) Ecological sustainability of fish resources of inland waters of the Western Balkans in the Anthropocene. Springer, Cham

Garcia S, Kolding J, Rice J, Rochet M, Zhou S, Arimoto T, Beyer J, Borges L, Bundy A, Dunn D, Fulton E, Hall M, Heino M, Law R, Makino M, Rijnsdorp A, Simard F, Smith A (2012) Reconsidering the consequences of selective fisheries. Science 335(6072):1045–1047. https://doi.org/10.1126/science.1214594

Griffiths H, Kryštutek B, Reed J (eds) (2004) Balkan biodiversity pattern and process in the European hotspot. Springer Science + Business Media, 358 pp. https://doi.org/10.1007/978-1-4020-2854-0#toc

Jakšić G, Bočić N, Piria M (2023) Chap. 16: Huchen *Hucho hucho* (Linnaeus, 1758) In Croatia: distribution, genetic diversity, threats and conservation. In: Simić V, Simic S, Pešić V (eds) Ecological sustainability of fish resources of inland waters of the Western Balkans in the Anthropocene. Springer, Cham

Jarić I, Ebenhard T, Lenhardt M (2010) Population viability analysis of the Danube sturgeon populations in a Vortex simulation model. Rev Fish Biol Fish 20:219–237. https://doi.org/10.1007/s11160-009-9151-0

Jarić I, Lenhardt M, Cvijanović G, Ebenhard T (2009) Population viability analysis and potential of its application to Danube sturgeons. Arch Biol Sci Belgrade 61(1):123–128

Jarić I, Smederevac-Lalić M, Jovičić K, Jaćimović M, Cvijanović G, Lenhardt M, Kalauzi A (2014) Indicators of unsustainable fishery in the middle Danube. Ecol Freshw Fish 25:86. https://doi.org/10.1111/eff.12193

Jelić D, Duplić A, Ćaleta M, Žutinić P (2008) Endemske vrste riba Jadranskog sliva. Agencija za zaštitu okoliša, Zagreb, p 78

Khalili Tilami S, Sampels S, Zajíc T, Krejsa J, Másílko J, Mráz J (2018) Nutritional value of several commercially important river fish species from The Czech Republic. PeerJ 12(6):e5729. https://doi.org/10.7717/peerj.5729. PMID: 30345172; PMCID: PMC6187990

Marić D (2018) The Ichthyofauna of Lake Skadar/Shkodra: diversity, economic significance, condition, and conservation status. In: Pešić V, Karaman G, Kostianoy A (eds) The Skadar/Shkodra Lake environment. The handbook of environmental chemistry, vol 80. Springer, Cham. https://doi.org/10.1007/698_2018_238

Marić S, Simonović P, Razpet A (2010) Genetic characterization of broodstock brown trout from Bled fish-farm, Slovenia. Period Biol 112:145–148

Marković Z, Poleksić V (2023) Chap. 15: The role and importance of aquaculture for the ecological sustainability of fish resources in the inland water of Serbia. In: Simić V, Simic S, Pešić V (eds) Ecological sustainability of fish resources of inland waters of the Western Balkans in the Anthropocene. Springer, Cham

Milačić R, Ščančar J, Paunović M (eds) (2015) The Sava River. The Handbook of Environmental Sciences. Springer. https://doi.org/10.1007/978-3-662-44034-6

Milošković A, Kojadinović N (2023) Chap. 12: Pollution of fishing waters of the Western Balkan – potentially toxic elements and their impact on the ecological sustainability of fish resources. In:

Simić V, Simic S, Pešić V (eds) Ecological sustainability of fish resources of inland waters of the Western Balkans in the Anthropocene. Springer, Cham

Mrakovčić M, Brigić A, Buj I, Ćaleta M, Mustafić P, Zanella D (2006) Crvena knjiga slatkovodnih riba Hrvatske. Ministarstvo kulture i Državni zavod za zaštitu prirode, Zagreb

Mrdak D, Milošević D (2023) Chap. 5: The fish's resources and fisheries of Montenegro, management, sustainability and conservations. In: Simić V, Simic S, Pešić V (eds) Ecological sustainability of fish resources of inland waters of the Western Balkans in the Anthropocene. Springer, Cham

Myers N (1990) The biodiversity challenge: expanded hot-spots analysis. Environmentalist 10: 243–256

Myers N (2003) Biodiversity hotspots revisited. BioScience 53:916–917

Myers N, Mittermeier RA, Mittermeier CG, da Fonseca GA, Kent J (2000) Biodiversity hotspots for conservation priorities. Nature 403:853–858

Pešić V, Karaman G, Kostianoy GA (eds) (2018) The Skadar/Shkodra Lake environment, The handbook of environmental chemistry, vol 80. Springer, p 508

Pešić V, Milošević Đ, Miliša M (eds) (2022) Small water bodies of the Western Balkans. Springer Nature. 449 pp

Pešić V, Paunović M, Kostianoy AG, Vukašinović-Pešić V (2020) The rivers of Montenegro. The handbook of environmental chemistry. Springer, Berlin, Heidelberg. https://doi.org/10.1007/978-3-030-55712-6

Piria M, Suić J, Špelić I (2023) Chap. 2: Inland fisheries in Croatia: historical aspects, fish resources, management and conservation. In: Simić V, Simic S, Pešić V (eds) Ecological sustainability of fish resources of inland waters of the Western Balkans in the Anthropocene. Springer, Cham

Pliberšek K, Tavčar T (2023) Chap. 1: Fish resources of inland waters and fisheries in Slovenia, management, sustainability and conservation. In: Simić V, Simic S, Pešić V (eds) Ecological sustainability of fish resources of inland waters of the Western Balkans in the Anthropocene. Springer, Cham

Radenković M, Zorić K, Petrović A (2023) Chap. 13: Invasive macroinvertebrate and fish species and their impact on fish resources: a case study on fishing waters of Serbia. In: Simić V, Simic S, Pešić V (eds) Ecological sustainability of fish resources of inland waters of the Western Balkans in the Anthropocene. Springer, Cham

Schöffmann J, Marić M (2023) Chap. 9: Salmonid fish species – opportunities for sustainable use under multiple pressures and current climatic change. In: Simić V, Simic S, Pešić V (eds) Ecological sustainability of fish resources of inland waters of the Western Balkans in the Anthropocene. Springer, Cham

Shumka S, Šanda R, Dhora D, Grapci-Kotori L (2023) Chap. 7: Fish resources of inland waters and fisheries in Albania, management, sustainability and conservation. In: Simić V, Simic S, Pešić V (eds) Ecological sustainability of fish resources of inland waters of the Western Balkans in the Anthropocene. Springer, Cham

Simić V, Miljanović B, Petrović A, Radenković M, Stojković Piperac M, Veličković T, Jakovljević M, Simić S (2023) Chap. 4: Inland fisheries in Serbia: historical aspects, fish resources, management and conservation. In: Simić V, Simic S, Pešić V (eds) Ecological sustainability of fish resources of inland waters of the Western Balkans in the Anthropocene. Springer, Cham

Simić V, Petrović A, Erg B, Dimović D, Makovinska J, Karadžić B, Paunović M (2014a) Indicative status assessment, biodiversity conservation and protected areas within the Sava River Basin, in The Sava River. In: Milačič M, Ščančar J, Paunović M (eds) The handbook of environmental chemistry. Springer, p 490. https://doi.org/10.1007/978-3-662-44034-6

Simić VM, Simić SB, Stojković Piperac M, Petrović A, Milošević D (2014b) Commercial fish species of inland waters: a model for sustainability assessment and management. Sci Total Environ 497:642–650

Simonović P, Hegediš A, Miljanović B, Vukov D et al. (2018) Operativni monitoring površinskih i podzemnih voda Republike Srbije – partija 1 – operativni monitoring površinskih voda.

Republika Srbija, Ministarstvo zaštite životne sredine, Biološki fakultet i Institut za multidisplinarna istraživanja Univerzitet u Beogradu; Prirodno- matematički fakultet, Univerzitet u Novom Sadu, 404-02-59/2018-02. (in Serbian).

Simonović P, Povž M, Piria M, Treer T, Adrović A, Škrijelj R, Nikolić V, Simić V (2015) Ichthyofauna of the River Sava System, in The Sava River. In: Milačič M, Ščančar J, Paunović M (eds) The handbook of environmental chemistry. Springer, p 490. https://doi.org/10.1007/978-3-662-44034-6

Simonović P, Simić V, Simić S, Paunović M (2010) The Danube in Serbia -The Results of National Program of the Second Joint Danube Survey 2007. The Republic of Serbia, Ministry of Agriculture Forestry and Water Management – Republic Directorate for Water. DIS Public doo, Belgrade

Smederevac-Lalić M, Špelić I, Đug S, Pengal P, Çobani M, Mrdak D, Piria M (2023) Chap. 18: Political and socio-economic aspects of fisheries in inland and coastal waters of the Western Balkan. In: Simić V, Simic S, Pešić V (eds) Ecological sustainability of fish resources of inland waters of the Western Balkans in the Anthropocene. Springer, Cham

Stojković Piperac M, Milošević D, Čerba D, Simić V (2022) Fish communities over the Danube Wetlands in Serbia and Croatia. In: Pešić V, Milošević D, Miliša M (eds) Small water bodies of the Western Balkans. Springer Water. Springer, Cham. https://doi.org/10.1007/978-3-030-86478-1_16

Talevski T, Tasevska O, Talevska M, Trajčeski B (2023) Chap. 6: Inland fisheries in Northern Macedonia: historical aspects, fish resources, management and conservation. In: Simić V, Simic S, Pešić V (eds) Ecological sustainability of fish resources of inland waters of the Western Balkans in the Anthropocen. Springer, Cham

Visser TAM, Valbi-Jorgensen J, Chomo V (2021) Good practices guidelines for data collection systems to support sustainable inland and recreational fisheries in the Western Balkans region. Fisheries and Aquaculture Circular No. 1218. Fisheries and Aquaculture, Budapest

WFD (2000) Water framework directive - Directive of European Parliament and of the Council 2000/60/EC – Establishing a framework for community action in the field of water policy. European Union, the European Parliament and Council, Luxembourg

MIX
Papier aus verantwortungsvollen Quellen
Paper from responsible sources
FSC® C105338

If you have any concerns about our products,
you can contact us on
ProductSafety@springernature.com

In case Publisher is established outside the EU,
the EU authorized representative is:
Springer Nature Customer Service Center GmbH
Europaplatz 3, 69115 Heidelberg, Germany

Printed by Libri Plureos GmbH
in Hamburg, Germany